21st Century Astronomy

The Solar System

SEVENTH EDITION

21st Century Astronomy

The Solar System

SEVENTH EDITION

Stacy Palen

WEBER STATE UNIVERSITY

George Blumenthal

UNIVERSITY OF CALIFORNIA, SANTA CRUZ

W. W. NORTON & COMPANY

Independent Publishers Since 1923

W. W. Norton & Company has been independent since its founding in 1923, when William Warder Norton and Mary D. Herter Norton first published lectures delivered at the People's Institute, the adult education division of New York City's Cooper Union. The firm soon expanded its program beyond the Institute, publishing books by celebrated academics from America and abroad. By midcentury, the two major pillars of Norton's publishing program—trade books and college texts—were firmly established. In the 1950s, the Norton family transferred control of the company to its employees, and today—with a staff of five hundred and hundreds of trade, college, and professional titles published each year—W. W. Norton & Company stands as the largest and oldest publishing house owned wholly by its employees.

EDITOR: Rob Bellinger
PROJECT EDITOR: Laura Dragonette
DEVELOPMENTAL EDITOR: John Murdzek
ASSISTANT EDITOR: Selin Tekgurler
MANAGING EDITOR, COLLEGE: Marian Johnson
MANAGING EDITOR, COLLEGE DIGITAL MEDIA: Kim Yi
PRODUCTION MANAGERS: Eric Pier-Hocking, Sean Mintus, and Richard Bretan
MEDIA EDITORS: Miryam Chandler and Meg Leary
ASSOCIATE MEDIA EDITOR: Arielle Holstein
MEDIA PROJECT EDITOR: Danielle Belfiore
MEDIA EDITORIAL ASSISTANTS: Aly Grindall and Lily Edgerton
MARKETING MANAGER: Ruth Bolster
DESIGN DIRECTOR: Rubina Yeh
DESIGNER: Anne DeMarinis
PHOTO EDITOR: Amla Sanghvi
PHOTO RESEARCHER: Jane Sanders Miller
DIRECTOR OF COLLEGE PERMISSIONS: Megan Schindel
COLLEGE PERMISSIONS MANAGER: Bethany Salminen
PHOTO DEPARTMENT MANAGER: Stacey Stambaugh
PERMISSIONS ASSOCIATE: Elizabeth Trammell
COMPOSITION: Graphic World
MANUFACTURING: Transcontinental

Library of Congress Cataloging-in-Publication Data
Names: Palen, Stacy, author. | Blumenthal, George (George Ray), author.
Title: 21st century astronomy / Stacy Palen, Weber State University, George
    Blumenthal, University of California—Santa Cruz.
Other titles: Twenty-first century astronomy
Description: Seventh edition. | New York : W. W. Norton & Company, [2022] |
    Includes index.
Identifiers: LCCN 2021032451 | ISBN 9780393877021 (paperback) |
    ISBN 9780393877113 (epub)
Subjects: LCSH: Astronomy—Textbooks. | LCGFT: Textbooks.
Classification: LCC QB45.2 .P35 2022 | DDC 520—dc23
LC record available at https://lccn.loc.gov/2021032451

ISBN 978-0-393-53914-1 (pbk)

W. W. Norton & Company, Inc., 500 Fifth Avenue, New York, NY 10110
wwnorton.com
W. W. Norton & Company Ltd., 15 Carlisle Street, London W1D 3BS
1 2 3 4 5 6 7 8 9 0

Stacy Palen thanks everyone at Bellwether Farm for their support during this project that spanned the challenges of the COVID-19 pandemic.

George Blumenthal gratefully thanks his wife, Kelly Weisberg, and his children, Aaron and Sarah Blumenthal, for their support during this project. He also wants to thank Professor Robert Greenler for stimulating his interest in all things related to physics.

BRIEF CONTENTS

Chapters 15–23 are not included in this edition.

CONTENTS

PART I   INTRODUCTION TO ASTRONOMY

6 The Tools of the Astronomer 142

PART II THE SOLAR SYSTEM

7 The Formation of Planetary Systems 174

PART III   STARS AND STELLAR EVOLUTION

16 Evolution of Low-Mass Stars   456

17 Evolution of High-Mass Stars   486

PART IV   GALAXIES, THE UNIVERSE, AND COSMOLOGY

WORKING IT OUT

Chapters 15–23 are not included in this edition.

ASTROTOURS

INTERACTIVE SIMULATIONS

ASTRONOMY IN ACTION VIDEOS

Chapters 15–23 are not included in this edition.

AstroTour animations, Interactive Simulations, and Astronomy in Action videos are all available from the free Student Site at the Digital Resources Site, and they are also integrated into assignable Smartwork exercises. **digital.wwnorton.com/astro7.**

Dear Student,

Why is it a good idea to take a science course, and in particular, why is astronomy a course worth taking? Many people choose to learn about astronomy because they are curious about the universe. Your instructor likely has two basic goals in mind for you as you take this course. The first is to understand some basic physical concepts and how they apply to the universe around us. The second is to think like a scientist and learn to use the scientific method not only to answer questions in this course but also to be able to evaluate and understand new information you encounter in your life. We have written the Seventh Edition of *21st Century Astronomy* with these two goals in mind.

Throughout this book, we emphasize not only the content of astronomy (for example, the differences among the planets, the formation of chemical elements) but also *how* we know what we know. The scientific method is a valuable tool that you can carry with you and use for the rest of your life. One way we highlight how science works is in the **Process of Science Figures**. In each chapter, we have chosen one discovery and provided a visual representation illustrating the discovery or a principle of the process of science. In these figures, we develop the idea that science is not a tidy process that proceeds in a straight line from one idea to the next. Discoveries are made because people are trying to answer a question and show why or how we think something is the way it is. This is often surprising.

The most effective way to learn something is to "do" it. Whether playing an instrument or a sport or becoming a good cook, reading or listening can only take you so far. The same is true of learning astronomy. We have written this book to help you "do" as you learn. Sometimes this means doing a hands-on activity to apply a concept and sometimes we provide a tool to make reading a more active process.

The following tools in each chapter help you "do" as you learn:

- **Active Learning Figures** open each chapter and ask you to "do" science by setting up an experiment and then making either a prediction or an observation and recording the results. Keep in mind that the "answer" isn't the most important part of the activity. Rather, we want the experience of thinking about a physical phenomenon and predicting what will happen next to become a natural way for you to apply your knowledge and understand new concepts.

- **What if** questions throughout the chapter prompt you to apply what you have learned to situations both real and imagined. Many of these are based on questions that students have asked, when they wondered, "What if the universe were different than it is?" These questions are open-ended, so think of them as guides for how to wonder about scientific ideas, rather than tests of your preexisting knowledge.

- Each chapter's **What an Astronomer Sees** feature helps you understand how astronomers interpret astronomical images, obtaining enormous amounts of information from a single picture. This feature is accompanied by an end-of-chapter question to teach you to interpret astronomical images yourself. Similarly, we have identified all of the figure-based questions at the end of the

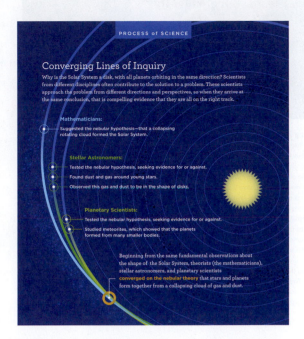

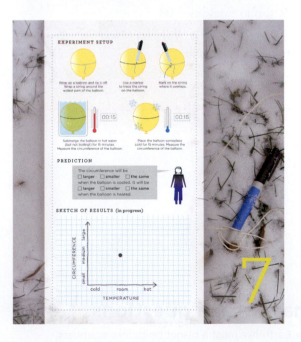

what if . . .

What if you observe a close pair of stars forming a binary system? How would you expect the disk around such a pair of stars to differ from the disk around a single star?

Figure 7.8 ★ WHAT AN ASTRONOMER SEES This gorgeous image from the Hubble Space Telescope shows a portion of the Carina Nebula. An astronomer looking at this image will immediately notice the colors, because color often indicates where different atoms or molecules are present. She will not necessarily know which atoms or molecules are represented by these colors because different astronomers will use a different palette. But even without reading any background on the image, she will know that there is something different about the hazy blue areas and the hazy pink or green ones. An astronomer will recognize and then mostly ignore the diffraction spikes (mentioned in Chapter 6) that form an X around each bright star. She will notice the brown clumps of material that are too dense to see through. These high-density regions indicate that star formation might be happening in this nebula, and this will be confirmed by the small oval protostar in the upper right corner and by the dense "fingers" that stick out in various places. These fingers point the way toward a source of interstellar wind outside the image to the upper right. That wind has eroded away the less dense material around these denser regions and may have triggered star formation by compressing the material at the top of each finger. Each finger is a dense blob of material that creates a wind "shadow" behind it. A new star has just formed at the top of the finger near the middle of the image. An astronomer will identify this new star because of the thin jets of material that are being ejected in opposite directions; new stars sometimes create such jets.

chapter, which should also help you develop your ability to interpret images and graphs.

- There are **Check Your Understanding** questions at the end of each chapter section. These questions are designed to be answered quickly if you have understood the previous section. The answers are provided in the back of the book so you can check your answer and decide whether further review is necessary.

CHECK YOUR UNDERSTANDING 7.1

Which of the following pieces of evidence support the nebular hypothesis? (Choose all that apply.) (a) Planets orbit the Sun in the same direction. (b) The Solar System is relatively flat. (c) Earth has a large Moon. (d) We observe disks of gas and dust around other stars.

Answers to Check Your Understanding questions are in the back of the book.

- **Reading Astronomy News** sections in each chapter include a news article or press release with questions to focus your attention on how the science is presented. As a citizen of the world, recognizing what is credible and questioning what is not are important skills. You make judgments about science, distinguishing between good science and pseudoscience, in order to make decisions in the grocery store, pharmacy, car dealership, and voting booth. You base these decisions on the presentation of information you receive through the media, which is very different from the presentation of information in class. The goal of Reading Astronomy News is to help you build your scientific literacy and your ability to challenge what you hear elsewhere.

- While we know a lot about the universe, science is an ongoing process, and we continue to search for new answers. To give you a glimpse of what we don't know, we provide **Unanswered Questions** features in each chapter. Most of these questions represent topics that scientists are currently studying.

unanswered questions

How Earth-like must a planet be before scientists declare it to be "another Earth"? An editorial in the science journal *Nature* cautioned that scientists should define "Earth-like" in advance—before multiple discoveries of planets "similar" to Earth are announced and a media frenzy ensues. Must a planet be of similar size and mass, be located in the habitable zone, and have spectroscopic evidence of liquid water before we call it "Earth 2.0"?

Reading Astronomy News

NASA's TESS Mission Uncovers Its 1st World With Two Stars
NASA

Discovering more planets means that we discover more unusual ones; like this planet that orbits two stars at once

The TOI 1338 system lies 1,300 light-years away in the constellation Pictor. The two stars orbit each other every 15 days. One is about 10% more massive than our Sun, while the other is cooler, dimmer and only one-third the Sun's mass.

Newly discovered TOI 1338 b is the only known planet in this binary star system. It's around 6.9 times larger than Earth, or between the sizes of Neptune and Saturn. The planet orbits in almost exactly the same plane as the stars, so it experiences regular stellar eclipses.

Scientists use the observations from TESS to generate graphs of how the brightness of stars change over time. When a planet transits in front of its star from our perspective, its passage causes a dip in the star's brightness.

Planets orbiting two stars are more difficult to detect than those orbiting one. TOI 1338 b's transits are irregular, between every 93 and 95 days, and vary in depth and duration thanks to the orbital motion of its stars. TESS only sees the transits crossing the larger star; the transits of the smaller star are too faint to detect.

"These are the types of signals that algorithms really struggle with," said lead author Veselin Kostov, a research scientist at the SETI Institute and Goddard. "The human eye is extremely good at finding patterns in data, especially non-periodic patterns like those we see in transits from these systems."

After identifying TOI 1338 b, the research team used a software package called eleanor, named after Eleanor Arroway, the central character in Carl Sagan's novel "Contact," to confirm the transits were real.

TOI 1338 had already been studied from the ground by radial velocity surveys. Kostov's team used this archival data to analyze the system and confirm the planet.

QUESTIONS

1. The period of the two stars in the binary pair is 15 days. Are these two stars very close to each other, or very far apart?

2. The planet's period is 95 days. Is the planet's orbit larger or smaller than the orbit of the two stars around one another?

3. Make a sketch of this system, including the two stars and the planet. Draw circles to show how the orbits are related, and whether or not they cross.

4. The article states that the discovery was confirmed using archival data from ground-based telescopes. Speculate: Why was the planet not discovered prior to this, using that ground-based data?

5. This system provides a window into the naming conventions for stars and planets. Study the names of the stars and planet. How are the two stars of a binary star system designated? How is the first planet in a planetary system designated?

Source: https://www.nasa.gov/feature/goddard /2020/nasa-s-tess-mission-uncovers-its-1st-world -with-two-stars.

working it out 7.1

Angular Momentum

In its simplest form, the angular momentum (L) of a system is given by

$$L = m \times v \times r$$

where m is the mass, v is the speed at which the mass is moving, and r represents how spread out the mass is.

Let's apply this relationship to the angular momentum of Jupiter in its orbit about the Sun. The angular momentum from one body orbiting another is called *orbital* angular momentum, $L_{orbital}$. The mass (m) of Jupiter is 1.90×10^{27} kilograms (kg), the speed of Jupiter in orbit (v) is 1.31×10^{4} meters per second (m/s), and the radius of Jupiter's orbit (r) is 7.79×10^{11} meters. Putting all that together gives

$$L_{orbital} = (1.90 \times 10^{27} \text{ kg}) \times (1.31 \times 10^{4} \text{ m/s}) \times (7.79 \times 10^{11} \text{ m})$$

$$L_{orbital} = 1.94 \times 10^{43} \text{ kg m}^2/\text{s}$$

Calculating the *spin* angular momentum of a spinning object, such as a skater, a planet, a star, or an interstellar cloud, is more complicated. Here, we must add up the individual angular momenta of *every tiny mass element* within the object. For a uniform sphere, the spin angular momentum is

$$L_{spin} = \frac{4\pi m R^2}{5P}$$

where R is the radius of the sphere and P is the rotation period of its spin.

Let's compare Jupiter's orbital angular momentum with the Sun's spin angular momentum to investigate the distribution of angular momentum in the Solar System. The Sun's radius is 6.96×10^{8} meters, its mass is 1.99×10^{30} kg, and its rotation period is 24.5 days $= 2.12 \times 10^{6}$ seconds. If we assume that the Sun is a uniform sphere, the spin angular momentum of the Sun is

$$L_{spin} = \frac{4 \times \pi \times (1.99 \times 10^{30} \text{ kg}) \times (6.96 \times 10^{8} \text{ m})^2}{5 \times (2.12 \times 10^{6} \text{ s})}$$

$$L_{spin} = 1.14 \times 10^{42} \text{ kg m}^2/\text{s}$$

$L_{orbital}$ of Jupiter is about 17 times greater than L_{spin} of the Sun. Thus, most of the angular momentum of the Solar System now resides in the orbits of its major planets.

For a collapsing sphere to conserve L_{spin}, its rotation period P must be proportional to R^2. As with the skater, when a sphere decreases in radius, its rotation period decreases; that is, it spins faster.

- The language of science is mathematics, and it can be as challenging to learn as any other language. The choice to use mathematics as the language of science is not arbitrary; nature "speaks" math. To learn about nature, you will need to speak its language. We don't want the language of math to obscure the concepts, however, so we have placed this book's mathematics in **Working It Out** boxes to make it clear when we are beginning and ending a mathematical argument, so that you can spend time with the concepts in the chapter text and then revisit the mathematics of the concept to study the formal language of the argument. You will also learn to work with data and identify when data aren't quite right. We want you to be comfortable reading, hearing, and speaking the language of science, and we provide you with tools to make it easier. At the end of each chapter, there are several mathematical problems. For each Working It Out box, there is a "scaffolded" end-of-chapter problem that guides you to predict the answers, calculate them, and then check your own work. Predicting the answer helps you develop intuition about mathematics, and checking your own work is a skill you will find useful later in life every time you need to calculate costs or estimate budgets.

- Each chapter concludes with an **Origins** section, which relates material or subjects found in the chapter to the origin of the universe and the origin of life. In recent years, astrobiologists have made much progress understanding these issues.

- At the end of each chapter, we have provided several types of questions, problems, and activities for you to practice your skills. The **Test Your Understanding**

questions focus on more detailed facts and concepts from the chapter. **Thinking about the Concepts** questions ask you to synthesize information and explain the "how" or "why" of a situation. **Applying the Concepts** problems give you a chance to practice the quantitative skills you learned in the chapter and to work through a situation mathematically.

- At the very end of each chapter, an **Exploration** activity shows you how to use the concepts and skills you learned in an interactive way. About half of the book's Explorations ask you to use animations and simulations found on the Student Site, while the others are hands-on, paper-and-pencil activities that use everyday objects such as ice cubes or balloons.

Digital resources linked to the book can help you understand and visualize many of the physical concepts described in the book. **AstroTours** and **Interactive Simulations** are represented by icons in the margins of the book (which are links in the ebook) and in Smartwork online homework. There is also a series of short **Astronomy in Action** videos embedded in the ebook (and represented by icons in the margins of the print book). The videos are available at the Student Site and in Smartwork online homework questions that your instructor might assign. These videos feature one of the authors (and several students) demonstrating physical concepts at work. Your instructor might assign these videos to you, or you might choose to watch them on your own to create a better picture of each concept in your mind.

Astronomy gives you a sense of perspective that no other field of study offers. The universe is vast, fascinating, and beautiful—filled with a wealth of objects that, surprisingly, can be understood using only a handful of principles. By the end of this book, you will have gained a sense of your place in the universe—both how incredibly small and insignificant you are and how incredibly unique and important you are.

Sincerely,
Stacy Palen
George Blumenthal

▶▶ **Interactive Simulation:** Habitable Zone

Astronomy in Action: Angular Momentum

Dear Instructor,

We wrote this book with a few overarching goals: to inspire students, to use active learning to build scientific literacy, and to create a useful and flexible tool that offers diverse approaches to the content.

As scientists and as teachers, we are passionate about the work we do. We hope to share that passion with students and inspire them to engage in science on their own. Through our own experience, familiarity with education research, and surveys of instructors, we have come to know a great deal about how students learn and what goals teachers have for their students. We have explicitly addressed many of these goals and learning styles in this book, sometimes in large, immediately visible ways, such as the inclusion of features, but also through less obvious efforts, such as in questions and problems that relate astronomical concepts to everyday situations or a fresh approach to organizing material.

One way we do this is through the *new* **Active Learning Figures** at the beginning of each chapter. These figures model student engagement and provide an opportunity for students to do an experiment or make an observation on their own using only everyday objects they can find around their house or dorm room. These are particularly useful in online classes, to encourage students to engage with the universe from their own home.

Many instructors state that they would like their students to become "educated scientific consumers" and "critical thinkers" or that their students should "be able to read a news story about science and understand its significance." We have specifically addressed these goals in our **Reading Astronomy News** feature, which presents a news article and a series of questions that guide a student's critical thinking about the article, the data presented, and the sources. Questions based on the Reading Astronomy News feature are available in Smartwork online homework.

Education research shows that the most effective way to learn is by doing. **Exploration** activities at the end of each chapter are hands-on, asking students to take the concepts they've learned in the chapter and apply them as they interact with animations and simulations on the Student Site or work through pencil-and-paper activities. Many of these Explorations incorporate everyday objects and can be used either in your classroom or as activities at home, and nearly all are assignable through Smartwork online homework.

We also believe students should be exposed to the more formal language of science—mathematics. We have placed the math in **Working It Out** boxes, so it does not interrupt the flow of the text or get in the way of students' understanding of conceptual material. But we've gone further by beginning with fundamental ideas in early Working It Out boxes and slowly building in complexity through the book. We've also worked to remove some of the stumbling blocks that affect student confidence by providing calculator hints, references to earlier Working It Out boxes, and detailed, fully worked examples. Many chapters include problems on reading and interpreting graphs. Appendix 1, "Mathematical Tools," summarizes some math concepts for students. Each Working It Out box is accompanied by a "scaffolded" problem at the end of the chapter that guides students to predict, solve, and check their work. Even students who are mathematically adept often forget to think about the problem before they attack it, or forget to check that their answer is sensible. Both of these skills are necessary for developing a mathematical intuition about the universe.

Discussion of basic physics is contained in Part I to accommodate courses that use the *Solar System* or *Stars and Galaxies* volumes. A "just-in-time" approach to introducing the physics is still possible by bringing in material from Chapters 2–6 as needed. For example, the sections on tidal forces in Chapter 4 can be taught along with the moons of the Solar System in Part II, or with mass transfer in binary stars in Part III, or with galaxy interactions in Part IV. Spectral lines in Chapter 5 can be taught with planetary atmospheres in Part II or with stellar spectral types in Part III, and so on.

In our overall organization, we have made several efforts to encourage students to engage with the material and build confidence in their scientific skills as they proceed through the book. For planets, stars, and galaxies, we have organized the material to cover the general case first and then delve into more details with specific examples. Thus, you will find "planetary systems" before our own Solar System, "stars" before the Sun, and "galaxies" before the Milky Way. This allows us to avoid frustrating students by making assumptions about what they know about stars or galaxies or forward-referencing to basic definitions and overarching concepts. This organization also implicitly helps students understand their place in the universe: our galaxy and our star are each one of many. They are specific examples of a physical universe in which the same laws apply everywhere. Planets have been organized comparatively to emphasize that science is a process of studying individual examples that lead to collective conclusions. All of these organizational choices were made with the student perspective in mind and a clear sense of the logical hierarchy of the material.

For example, we begin in Chapter 19 by introducing galaxies as a whole and our measurements of them, including recession velocities. Then we address the Milky Way in Chapter 20—a specific example of a galaxy that we can discuss in detail. This follows the repeating motif of moving from the general to the specific that exists throughout the text and gives students a basic grounding in the concepts of spiral galaxies, supermassive black holes, and dark matter before they need to apply those concepts to the specific example of our own galaxy. Chapter 21, "The Expanding Universe," covers the cosmological principle, the Hubble expansion, and the observational evidence for the Big Bang.

The **Origins** sections illustrate how astrobiologists and other scientists approach the study of a scientific question from the chapter related to the origin of the universe and of life. Material about exoplanets is introduced in Chapter 7, incorporated into other chapters when appropriate, and continued in Chapter 24. We revised each chapter, streamlining some topics and updating the science to reflect the progress in the field. This includes

- updates on new telescopes and space missions;
- new images and results from Solar System exploration missions to the Moon, Mars, Jupiter, and Venus;
- updated graphical data on climate change;
- results from Kepler and other missions on populations of exoplanets;
- results from the *Gaia* mission on stellar distances, the H-R diagram, and the Milky Way;
- results from LIGO/VIRGO on merging compact objects;
- many new articles for Reading Astronomy News; and
- revised simulations for some Exploration exercises, as well as new simulations.

Other items new to the Seventh Edition include the following:

- *New* **Active Learning Figures** at the start of each chapter, described previously.

- Each chapter contains a **What an Astronomer Sees** figure that demonstrates for students how an astronomer can derive a wealth of information from a single image. Each figure is accompanied by an end-of-chapter question and questions in Smartwork that further guide students in developing the skill of interpreting astronomical imagery.

- Multiple times in each chapter, we ask students "**What if**?" These "What if" questions often ask students to imagine the universe other than it is, or to extrapolate from the information they have learned. These advanced comprehension questions give students the opportunity to experiment with developing and testing hypotheses by asking, "If the universe were not as it is, what else would have to be true?" This feature is accompanied by a "guiding" question in Smartwork that helps students organize their thinking. The in-text questions can be particularly useful for sparking discussion, either in class or in an LMS.

- Teaching Astronomy by Doing Astronomy (**tada101.com**) is a blog for introductory astronomy instructors. Stacy Palen writes regular posts and provides suggestions for adding active learning to any size of class. Stacy also posts suggested Reading Astronomy News articles, discusses ways to integrate math into a course, and hosts discussions about recent developments in practical applications of astronomy education research. Join the community at tada101.com.

Many professors find themselves under pressure from accrediting bodies or internal assessment offices to assess their courses in terms of learning goals. Each chapter has Learning Goals and an end-of-chapter Summary to correspond to the chapter's Learning Goals. In Smartwork, questions and problems are tagged by type and can be sorted by Learning Goal. Smartwork contains more than 2200 questions and problems that are tied directly to this text, including the Check Your Understanding questions, versions of the Reading Astronomy News and Exploration questions, questions based on the Chapter-opening experiments, and many ranking, sorting, and labeling tasks. Smartwork now contains pre- and post-activity questions based on *Learning Astronomy By Doing Astronomy*, Second Edition, so you can help students prepare for each activity and assess learning afterwards.

Every question in Smartwork has hints and answer-specific feedback so that students are coached to work toward the correct answer. Smartwork is flexible: premade assignments based on each chapter make it easy to assign, and instructors can easily modify any of the provided questions, answers, and feedback or can create their own questions.

We've also created a series of 23 Astronomy in Action videos explaining and demonstrating concepts from the text, accompanied by questions in Smartwork. You might assign these videos prior to lecture—either as part of a flipped modality or as a "reading quiz." In either case, you can use the diagnostic feedback from the questions in Smartwork to tailor your in-class discussions. Or you might show them in class, to stimulate discussion. Or you might simply use them as a jumping-off point—to get ideas for activities to do with your own students.

We continue to look for better ways to engage students, so please let us know how these features work for your students!

Supporting Resources for Students

digital.wwnorton.com/astro7

Smartwork Online Homework

smartwork

Steven Desch, Guilford Technical Community College
Shimonee Kadakia, El Camino College
Ana M. Larson, Emerita, University of Washington
Violet Mager, Penn State Wilkes-Barre
Doug Stuffle, Tidewater Community College
Dave Wood, San Antonio College
Todd Young, Wayne State College
William Younger, Tidewater Community College

More than 2200 questions support *21st Century Astronomy*, Seventh Edition—all with answer-specific feedback, hints, and ebook links. Questions include Summary Self-Tests, versions of the Explorations (based on AstroTours and new Interactive Simulations), and questions based on the Reading Astronomy News features. Astronomy in Action video questions focus on getting students to come to class prepared and on overcoming common misconceptions. Process of Science Guided Inquiry Assignments help students apply the scientific method to important questions in astronomy, challenging them to think like scientists. The course also contains ranking, sorting, and labeling exercises based on art from the text. Questions span Bloom's Taxonomy, engaging and challenging students. Smartwork can be set up to work right in your LMS, with student scores flowing directly to your LMS gradebook. Your local Norton representative can help you set up LMS integration.

Norton Ebook

The *21st Century Astronomy*, Seventh Edition ebook provides students an enhanced reading experience at a fraction of the cost of a print textbook. Students are able to have an active reading experience and can take notes, bookmark, search, highlight, and even read offline. Instructors can add notes for students to see as they read the text. Norton Ebooks can be viewed on—and synced among—all computers and mobile devices. Every question in Smartwork provides a reference link to the ebook so that students have easy access to their textbook when completing homework assignments.

Twice per year, the Seventh Edition ebook will be updated with exciting new research from the field of astronomy, ensuring that the book remains timely and relevant. New Smartwork questions will support these features.

Student Site

W. W. Norton's student website features the following:

- Thirty AstroTour animations. These animations, some of which are interactive, use art from the text to help students visualize important physical and astronomical concepts. All are now tablet compatible.

- Eight Interactive Simulations, authored by Stacy Palen and paired with in-text Exploration activities, allow students to explore topics such as Moon phases, Kepler's laws, and the Hertzsprung-Russell diagram.
- Twenty-three Astronomy in Action videos feature author Stacy Palen demonstrating the most important concepts in a visual, easy-to-understand, and memorable way.

Learning Astronomy by Doing Astronomy, Second Edition: Collaborative Lecture Activities

Stacy Palen, Weber State University

Ana M. Larson, Emerita, University of Washington

Students learn best by doing. Devising, writing, testing, and revising suitable in-class activities that use real astronomical data, illuminate astronomical concepts, and pose probing questions that ask students to confront misconceptions can be challenging and time consuming. In this workbook, the authors draw on their experience teaching thousands of students in many different types of courses (large in-class, small in-class, hybrid, online, flipped, and so forth) to bring 36 field-tested activities that can be used in any classroom today. The activities have been designed to require no special software, materials, or equipment and to take no more than 50 minutes to do. Pre- and post-activity questions are now assignable within Smartwork.

If you are interested in packaging the workbook with this text, please contact your local Norton representative.

Starry Night Planetarium Software (College Version) and Workbook

Steven Desch, Guilford Technical Community College

Michael Marks, Bristol Community College

Starry Night is a realistic, user-friendly planetarium simulation program designed to allow students in urban areas to perform observational activities on a computer screen. Norton's unique accompanying workbook offers observation assignments that guide students' virtual explorations and help them apply what they've learned from the text's reading assignments. If you are interested in packaging the Starry Night software and workbook with this text, please contact your local Norton representative.

For Instructors

Instructor's Manual

Ana M. Larson, Emerita, University of Washington

This resource includes brief chapter overviews; suggested discussion points; notes on the Process of Science figures, AstroTour animations, Interactive Simulations, Astronomy in Action videos, Reading Astronomy News, and Explorations; and worked solutions to Check Your Understanding and end-of-chapter Questions and Problems. Also included are notes on teaching with *Learning Astronomy by Doing Astronomy: Collaborative Lecture Activities* and answers to the *Starry Night Workbook* exercises.

Teaching Astronomy by Doing Astronomy Blog, tada101.com

Teaching Astronomy by Doing Astronomy is a community-focused blog that features posts by Stacy Palen and contributing astronomy instructors on teaching ideas, classroom activities, and current events. Instructors looking for additional ways to incorporate active learning will find this blog to be an invaluable resource.

PowerPoint Lecture Slides

Allison Kirkpatrick, University of Kansas

These ready-made lecture slides integrate selected textbook art, all Check Your Understanding and Working It Out questions from the text, classroom response questions, and links to the AstroTour animations, Interactive Simulations, and Astronomy in Action videos. These lecture slides are fully editable and are available in Microsoft PowerPoint format.

Test Bank

Matthew Newby, Temple University

The Test Bank has been revised using Bloom's Taxonomy and provides over 2400 multiple-choice and short-answer problems. Each chapter of the Test Bank consists of five question levels classified according to Bloom's Taxonomy:

  Remembering

  Understanding

  Applying

  Analyzing

  Evaluating

Problems are further classified by section and difficulty level, making it easy to construct tests and quizzes that are meaningful and diagnostic. The Test Bank assesses a common set of Learning Objectives consistent with the textbook and Smartwork online homework.

Resources for Your LMS

These files, available for use in various Learning Management Systems (LMSs), add high-quality Norton digital resources to your course, including direct links to the ebook, Smartwork, AstroTours, Interactive Simulations, and Astronomy in Action videos. They also contain Flashcards for students' self-study. Resources for Your LMS are available in Blackboard, Canvas, Desire2Learn, and Moodle formats.

Acknowledgments

The authors would like to acknowledge the extraordinary efforts of the staff at W. W. Norton: Selin Tekgurler, Assistant Editor, who kept everything flowing smoothly while never losing sight of a single detail; Amla Sanghvi for managing the complex photo program; Elizabeth Trammell for managing text permissions; and the copyeditor, Carla Barnwell, who made sure that all the grammar and punctuation survived the multiple rounds of the editing process. We would especially like to thank John Murdzek for the developmental editing process, and Laura Dragonette, who shepherded the manuscript through production.

Rob Bellinger was the editor. Eric Pier-Hocking, Sean Mintus, and Richard Bretan managed the production; Anne DeMarinis designed the book; and Rubina Yeh was the design director. Miryam Chandler, Meg Leary, Arielle Holstein, Aly Grindall, and Lily Edgerton worked on the assessment and visualization resources, and Ruth Bolster will help get this book into the hands of people who can use it.

We gratefully acknowledge the contributions of the authors who worked on previous editions of *21st Century Astronomy*: Jeff Hester, Gary Wegner, the late Dave

Burstein, Ron Greeley, Brad Smith, and Howard Voss, with special thanks to Dave for starting the project, to Jeff for leading the original authors through the First Edition, and to Brad for leading the Second and Third Editions. Laura Kay led the team through the Fourth, Fifth, and Sixth Editions, and we miss her insightful and critical eye.

Stacy Palen
George Blumenthal

We would also like to thank the reviewers, whose input at every stage improved the book:

Seventh Edition Reviewers

Amy Abe, Lake Forest College
Paulo Afonso, American River College
Marilyn Akins, Bluegrass Community Technical College–Newtown
Heather Appleby, Richland College
Scott Armel, Trident Technical College
Brandon Bear, Virginia Polytechnic Institute
Raymond Benge, Tarrant County College
Philip Bennett, Dalhousie University
Ryan Bennett, University of North Texas
Katie Berryhill, Los Medanos College
Brett Bochner, Hofstra University
Julie Bray-Ali, Mt. San Antonio College–Walnut
Robert Brandenberger, McGill University
Michael Briley, Appalachian State University
Meade Brooks, College of Mainland
Shea Brown, University of Iowa
Spencer Buckner, Austin Peay State University
Philip Choi, Claremont College–Pomona
Micol Christopher, Mount San Antonio College
Asantha R Cooray, University of California–Irvine
Jonathan Craig, Horry Georgetown Technical College–Conway
James Dickinson, Clackamas Community College–Oregon City
Edmund Douglass, Farmingdale State College
Donald D. Driscoll, Kent State University
Jay Dunn, Georgia State University–Dunwoody
Robert Egler, North Carolina State University
Emmanuel Fonseca, West Virginia University
Douglas Gobeille, University of Rhode Island
Jose J. Gomez, The Citadel
Dirk Grupe, Morehead State University
George Hassel, Sienna College
David John Helfand, Columbia University
Jerry Horne, College of Southern Nevada–West Charleston
Manish Jadhav, Dutchess Community College–Poughkeepsie
Jennifer Jones, Arapahoe Community College
Charles Kerton, Iowa State University
John LaBrasca, Clark College
Peter Lanagan, College of Southern Nevada–Henderson
Denis Leahy, University of Calgary
Jane H. MacGibbon, University of North Florida
James Maxin, Louisiana State University–Shreveport
John McClain, Temple College

Marcio B. Melendez, Anne Arundel Community College
Sharon Morsink, University of Alberta
Irina Mullins, Houston Community College
Alexis Nduwimana, Georgia State University–Decatur
Jisun Park, Kingsborough Community College
Victoria Plaveti, Manhattan College
Valerie A. Rapson, Schenectady County Community College
Elliot Richmond, Austin Community College
Edward Rhoads, Indiana University–Purdue
Allen Rogel, Bowling Green State University
Ajani Luister Walden Ross, Wright State University–Dayton
Takashi Sato, Kwantlen Polytechnic University
Eric Schlegel, University of Texas at San Antonio
Arthur Schneider, Amarillo College
Parandis Tajbakhsh, University of Toronto–Mississauga
William K. Teets, Vanderbilt University
Jose Vazquez, Washington State University–Vancouver
Jordan Watkins, Brookhaven College
Casey Watson, Millikin University
Melinda Weil, City College of San Francisco
Lee Widmer, Xavier University
Nelka Wijesinghe, Long Star College–Montgomery
Jeffrey Wilson, Texas A&M University–Commerce
David Wood, San Antonio College
William Younger, Tidewater Community College–Portsmouth/
    Virginia Beach

Reviewers of Previous Editions

Manuel Alvarado, El Paso Community College
Gagandeep Anand, Boston University
Scott Atkins, University of South Dakota
Simon Balm, Santa Monica College
Timothy Barker, Wheaton College
Peter A. Becker, George Mason University
Timothy C. Beers, Michigan State University
Raymond Benge, Tarrant College
Ryan Bennett, University of North Texas
David Bennum, University of Nevada–Reno
Edwin Bergin, University of Pittsburgh
William Blass, University of Tennessee
Steve Bloom, Hampden Sydney College
Brett Bochner, Hofstra University
Daniel Boice, University of Texas at San Antonio

Bram Boroson, Clayton State University
David Branning, Trinity College
Julie Bray-Ali, Mt. San Antonio College
Jack Brockway, Radford University
Shea Brown, University of Iowa
Spencer Buckner, Austin Peay State University
Suzanne Bushnell, McNeese State University
Paul Butterworth, George Washington University
Juan E. Cabanela, Minnesota State University–Moorhead
Amy Campbell, Louisiana State University
C. Austin Campbell, El Paso Community College
Michael Carini, West Kentucky University
Christina Cavalli, Austin Community College
Gerald Cecil, University of North Carolina–Chapel Hill
Supriya Chakrabarti, Boston University
Robert Cicerone, Bridgewater State College
David Cinabro, Wayne State University
Scott Cochran, Central Michigan University
Judith Cohen, California Institute of Technology
Eric M. Collins, California State University–Northridge
Tara Cotton, University of Georgia
John Cowan, University of Oklahoma–Norman
Robert Coyne, Texas Tech University
Debashis Dasgupta, University of Wisconsin–Milwaukee
Steve Desch, Arizona State University
Robert Dick, Carleton University
Gregory Dolise, Harrisburg Area Community College
Michael Endl, Austin Community College
Tom English, Guilford Technical Community College
David Ennis, The Ohio State University
Duncan Farrah, University of Hawaii
Sunil Fernandes, University of Texas at San Antonio
John Finley, Purdue University
Matthew Francis, Lambuth University
Keigo Fukumura, James Madison University
Kevin Gannon, College of Saint Rose
Todd Gary, O'More College of Design
Christopher Gay, Santa Fe College
Ken Gayley, University of Iowa
Christopher Gerardy, University of North Carolina–Charlotte
Parviz Ghavamian, Towson University
Martha Gilmore, Wesleyan University
Guillermo Gonzalez, Ball State University
Bill Gutsch, St. Peter's College
Karl Haisch, Utah Valley University
Ioannis Haranas, Wilfrid Laurier College
Javier Hasbun, University of West Georgia
Charles Hawkins, Northern Kentucky University
Sebastian Heinz, University of Wisconsin–Madison
Jim Higdon, Georgia Southern University
Scott Hildreth, Chabot College
Barry Hillard, Baldwin Wallace College
Paul Hintzen, California State University–Long Beach
Paul Hodge, University of Washington
Christopher Hodges, Sacramento State University

William A. Hollerman, University of Louisiana at Lafayette
Hal Hollingsworth, Florida International University
Mike Hood, Mt. San Antonio College
Olencka Hubickyj-Cabot, San Jose State University
Kevin M. Huffenberger, University of Miami
James Imamura, University of Oregon
Adam Johnston, Weber State University
Jennifer Jones, Arapahoe Community College
Bruno Jungweirt, North Carolina State University
Steven Kawaler, Iowa State University
Patrick Kelly, Dalhousie University
Bethuel Khamala, El Paso Community College
Monika Kress, San Jose State University
Jessica Lair, Eastern Kentucky University
Rafael Lang, Purdue University
Kenneth Lanzetta, Stony Brook University
Lee LaRue, Paris Junior College
Alex Lazarian, University of Wisconsin–Madison
Denis Leahy, University of Calgary
Hector Leal, University of Texas–Rio Grande Valley
Hyun-chul Lee, University of Texas–Rio Grande Valley
Kevin Lee, University of Nebraska–Lincoln
Ludwik Lembryk, University of Toledo
Matthew Lister, Purdue University
M. A. K. Lodhi, Texas Tech University
Leslie Looney, University of Illinois at Urbana–Champaign
Isaac Lopez, Boston University
Jack MacConnell, Case Western Reserve University
Kevin Mackay, University of South Florida
Dale Mais, Indiana University–South Bend
Michael Marks, Bristol Community College
Norm Markworth, Stephen F. Austin State University
Kevin Marshall, Bucknell University
Petrus Martens, Georgia State University
Stephan Martin, Bristol Community College
Justin Mason, Old Dominion University
Amanda Maxham, University of Nevada–Las Vegas
Chris McCarthy, San Francisco State University
Douglas McElroy, Chaffey College
Ben McGimsey, Georgia State University
Charles McGruder, West Kentucky University
Janet E. McLarty-Schroeder, Cerritos College
Stanimir Metchev, Stony Brook University
Chris Mihos, Case Western Reserve University
Milan Mijic, California State University–Los Angeles
J. Scott Miller, University of Louisville
Scott Miller, Sam Houston State University
Kent Montgomery, Texas A&M University–Commerce
Robert Morehead, Texas Tech University
Andrew Morrison, Illinois Wesleyan University
Edward M. Murphy, University of Virginia
Kentaro Nagamine, University of Nevada–Las Vegas
Hon-kie Ng, Florida State University
Merav Opher, Boston University
Robert Parks, Louisiana State University

Jon Pedicino, College of the Redwoods

Nicolas Pereyra, University of Texas–Rio Grande Valley

Ylva Pihlström, University of New Mexico

Jascha Polet, California State Polytechnic University

Bob Powell, University of West Georgia

Dora Preminger, California State University–Northridge

Daniel Proga, University of Nevada–Las Vegas

Laurie Reed, Saginaw Valley State University

Barry Rice, Sierra College

Judit Györgyey Ries, University of Texas at Austin

Allen Rogel, Bowling Green State University

Kenneth Rumstay, Valdosta State University

Masao Sako, University of Pennsylvania

Samir Salim, Indiana University–Bloomington

Ruben Sandapen, Acadia University

Ata Sarajedini, University of Florida

Eric Schlegel, University of Texas at San Antonio

Paul Schmidtke, Arizona State University

Ann Schmiedekamp, Pennsylvania State University

Michael Schwartz, Santa Monica College

Jonathan Secaur, Kent State University

Ohad Shemmer, University of North Texas

Caroline Simpson, Florida International University

Parampreet Singh, Louisiana State University

Paul P. Sipiera, William Rainey Harper College

Ian Skilling, University of Pittsburgh

Tammy Smecker-Hane, University of California–Irvine

Allyn Smith, Austin Peay State University

Jason Smolinski, State University of New York at Oneonta

Roger Stanley, San Antonio College

Ben Sugerman, Goucher College

Neal Sumerlin, Lynchburg College

James Sy, El Paso Community College

Catherine Tabor, El Paso Community College

Angelle Tanner, Mississippi State University

Christopher Taylor, California State University–Sacramento

Benjamin Team, University of West Georgia

Fiorella Terenzi, Florida International University

Donald Terndrup, The Ohio State University

Todd Thompson, The Ohio State University

Glenn Tiede, Bowling Green State University

Frances Timmes, Arizona State University

Trina Van Ausdal, Salt Lake Community College

Lennart Van Haaften, Texas Tech University

Walter Van Hamme, Florida International University

Karen Vanlandingham, West Chester University

Nilakshi Veerabathina, University of Texas at Arlington

Paul Voytas, Wittenberg University

Ezekiel Walker, University of North Texas

Jasper Wall, University of British Columbia

Colin Wallace, University of North Carolina–Chapel Hill

G. Scott Watson, Syracuse University

James Webb, Florida International University

Julia Wickett, Collin College

Paul Wiita, Georgia State University

Richard Williamon, Emory University

Kurt Williams, Texas A&M University

Fred Wilson, Collin College

David Wittman, University of California–Davis

David Wood, San Antonio College

ABOUT THE AUTHORS

Stacy Palen is an award-winning professor in the Department of Physics and Astronomy at Weber State University. She received her BS in physics from Rutgers University and her PhD in physics from the University of Iowa. As a lecturer and postdoc at the University of Washington, she taught Introductory Astronomy more than 20 times over four years. Since joining Weber State, she has been very active in science outreach activities ranging from star parties to running the state Science Olympiad. Stacy does research in formal and informal astronomy education and the death of Sun-like stars. She spends much of her time thinking, teaching, and writing about the applications of science in everyday life. She then puts that science to use on her small farm in Ogden, Utah.

George Blumenthal has been a professor of astronomy and astrophysics at the University of California, Santa Cruz, since 1972, and served as the university's Chancellor from 2006 to 2019. He received his BS degree from the University of Wisconsin–Milwaukee and his PhD in physics from the University of California, San Diego. As a theoretical astrophysicist, George's research encompasses several broad areas, including the nature of the dark matter that constitutes most of the mass in the universe, the origin of galaxies and other large structures in the universe, the earliest moments in the universe, astrophysical radiation processes, and the structure of active galactic nuclei such as quasars. Besides teaching and conducting research, he has served as chair of the UC Santa Cruz Astronomy and Astrophysics Department, has chaired the Academic Senate for both the UC Santa Cruz campus and the entire University of California system, and has served as the faculty representative to the UC Board of Regents. He is currently the director of UC Berkeley's Center for Studies in Higher Education and serves on several nonprofit boards, including the California Institute for Regenerative Medicine and the California Association for Research in Astronomy, which he chairs.

21st Century Astronomy

The Solar System

SEVENTH EDITION

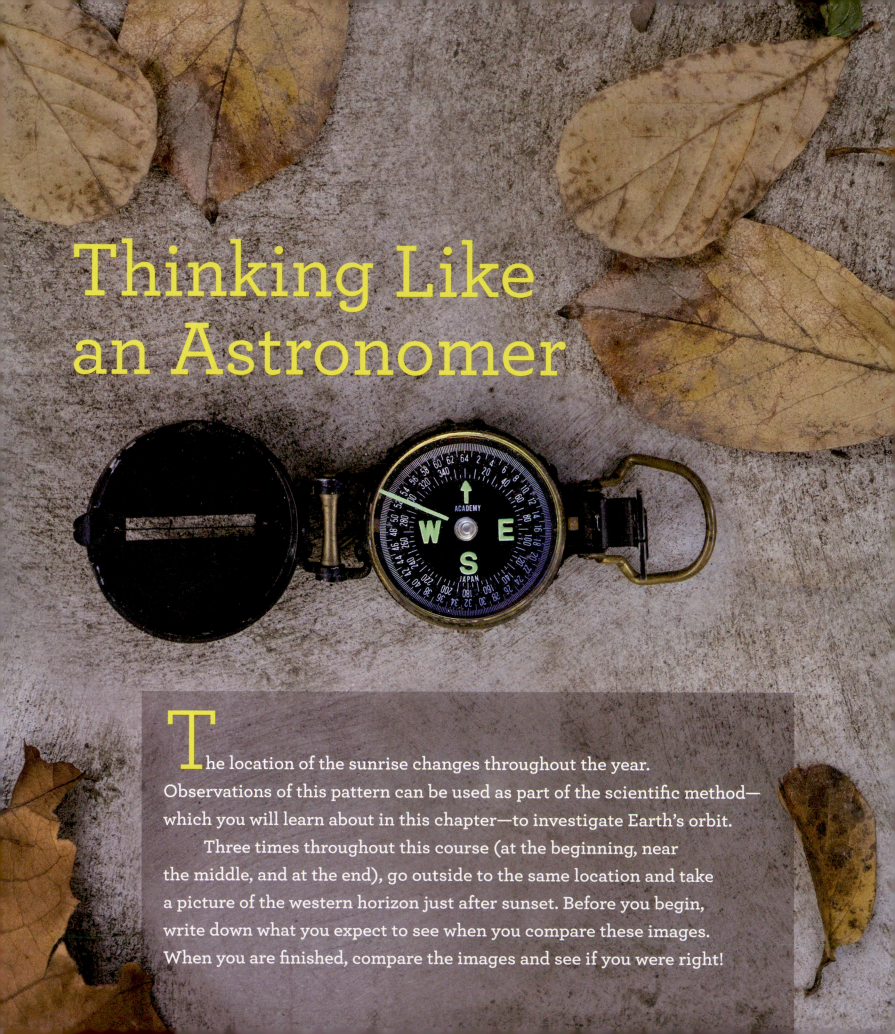

Thinking Like an Astronomer

The location of the sunrise changes throughout the year. Observations of this pattern can be used as part of the scientific method—which you will learn about in this chapter—to investigate Earth's orbit.

Three times throughout this course (at the beginning, near the middle, and at the end), go outside to the same location and take a picture of the western horizon just after sunset. Before you begin, write down what you expect to see when you compare these images. When you are finished, compare the images and see if you were right!

EXPERIMENT SETUP

PHOTO 1

PHOTO 2

PHOTO 3

Go to the same location and take three photos of the horizon at a time of day when the Sun is just below the horizon. Be sure that you have a stationary object (like a tree) in each photo. Take a photo at:

1 beginning of semester
2 middle of semester
3 end of semester

PREDICTION

1

2

3

When I compare the three images, I expect to see:

_____

_____

SKETCH OF RESULTS

1

L
oosely translated, the word **astronomy** means "patterns among the stars." But modern astronomy—the astronomy we talk about in this book—is about far more than looking at the sky and cataloging the visible stars. The contents of the universe, its origin and fate, and the nature of space and time have become the subjects of rigorous scientific investigation. All of these subjects are related to questions humans have long had about our *origins*. How and when did the Sun, Earth, and Moon form? Are other galaxies, stars, planets, and moons similar to our own? The answers that scientists are finding to these questions are changing our view of both the cosmos and ourselves.

LEARNING GOALS

In this chapter, we will begin studying astronomy by exploring our place in the universe and the methods of science. By the end of Chapter 1, you should be able to:

(1) Describe the size and age of today's universe and Earth's place in it.

(2) Explain how astronomers use the scientific method to study the universe.

(3) Show how scientists use mathematics, including graphs, to find patterns in nature.

(4) Summarize our astronomical origins.

what if . . .

What if our Solar System were located in the Andromeda Galaxy? What would our cosmic address be then?

1.1 Earth Occupies a Small Place in the Universe

Locating Earth in the larger universe is the first step in learning the science of astronomy. In this section, you will get a feel for the neighborhood in which Earth is located, and begin to develop a framework to organize your growing knowledge of the universe by both size and distance from Earth. You will also begin to explore the scale of the universe in both space and time.

Our Place in the Universe

Many people receive their postal mail at an address—house or building number, street, city, state, and country. If we expand our view to include the enormously vast universe, our "cosmic address" might include our planet, star, galaxy, galaxy group or cluster, and galaxy supercluster. Follow along in **Figure 1.1** as we describe your cosmic address.

We reside on a planet called Earth, which is orbiting under the influence of gravity around a star called the **Sun**. The Sun is an ordinary middle-aged star. It is more massive and luminous than some stars but less massive and luminous than others. The Sun is extraordinary only because of its importance to us within our own **Solar System**. Our Solar System consists of eight planets (listed in order of distance from the Sun): Mercury, Venus, Earth, Mars, Jupiter, Saturn, Uranus, and Neptune. It also contains many smaller bodies, such as dwarf planets (for example, Pluto, Ceres, and Eris), asteroids (for example, Ida and Eros), and comets (for example, Halley). All those objects are gravitationally bound to the Sun.

The Sun is one of several hundred billion stars in the **Milky Way Galaxy**, a pancake-shaped disk of stars, gas, and dust. The Sun is located about halfway

Figure 1.1 Our place in the universe is given by our cosmic address: Earth, Solar System, Milky Way Galaxy, Local Group, Virgo Supercluster, and Laniakea Supercluster.

★ **WHAT AN ASTRONOMER SEES** In this type of figure, an astronomer will be especially sensitive to the arrows, which show that the figure "zooms out" from panel to panel. While each panel is the same size on the page, they represent dramatically different sizes in space. This figure is representative, without precision, but an astronomer will know that the Laniakea structure in the last panel is much larger—more than 100,000,000,000,000,000 times larger—than Earth, in the first panel. Learning to work with large numbers and ranges of size is one of the challenges of thinking like an astronomer.

out from the center of this disk. Like the Sun, most stars in the Milky Way Galaxy have planets in orbit around them.

The Milky Way is a member of a collection of a few dozen galaxies, including the Andromeda and Triangulum Galaxies, similar to the Milky Way, and about 20 smaller galaxies, called the **Local Group**. Most galaxies in that group are much smaller than the Milky Way. The Local Group, in turn, is part of a vastly larger collection of thousands of galaxies—a **supercluster**—called the Virgo Supercluster, which astronomers have more recently learned is part of an even larger grouping called the Laniakea Supercluster. (Unusually for astronomy, there are not special names that distinguish various sizes of superclusters.) The observable universe has millions of superclusters.

We can now define our cosmic address—Earth, Solar System, Milky Way Galaxy, Local Group, Virgo Supercluster, Laniakea Supercluster. Yet even that address is incomplete because it encompasses only the *local universe*. The part of the universe that we can see—the *observable universe*—extends to many times the size of Laniakea in every direction. This observable part of the universe contains roughly 2 trillion (two thousand billion) galaxies. The entire universe is much larger than the local universe and contains much more than the observed planets, stars, and galaxies. Astronomers estimate that about 95 percent of the universe is made up of matter that does not interact with light, known as *dark matter*, and a form of energy that permeates all space, known as *dark energy*. Dark matter and dark energy aren't well understood, and they are among the many exciting areas of research in astronomy.

The Scale of the Universe

As shown in Figure 1.1, the size of the universe completely dwarfs our human experience. We can start trying to understand astronomical size scales by comparing astronomical sizes and distances to something more familiar. For example, the diameter of our Moon (3474 kilometers [km]) is slightly greater than the distance between New York, New York, and Ogden, Utah (**Figure 1.2a**), about 80 percent of the way across the United States. The distance from Earth to the Moon is about 100 times that distance, which is comparable

unanswered questions

What makes up the universe? We have listed planets, stars, and galaxies as components of the universe, but astronomers now have evidence that 95 percent of the universe is in the form of dark matter and dark energy, which we do not yet understand. Scientists are using the largest telescopes and particle colliders on Earth, as well as telescopes and experiments in space, to explore what makes up dark matter and what constitutes dark energy.

a.

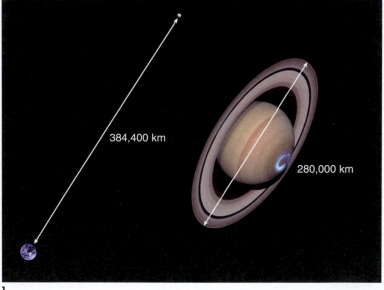

b.

Figure 1.2 a. The diameter of the Moon is about the same as the distance between New York, New York, and Ogden, Utah. **b.** The size of Saturn, including the rings, is about 70 percent of the distance between Earth and the Moon.

Credit (part b.): NASA, ESA, J. Clarke (Boston University, USA), and Z. Levay (STScI). https://esahubble.org/products/calendars/cal200604/. https://creativecommons.org/licenses/by/4.0/.

to the diameter of the planet Saturn with its majestic rings (**Figure 1.2b**). The distance from Earth to the Sun is about 400 times the Earth–Moon distance, and the distance to the planet Neptune is about 30 times the Earth–Sun distance. We could multiply all those relationships together, to find that the distance from the Sun to Neptune is 1.2 million times greater than the distance from New York, New York to Ogden, Utah.

The enormous distances beyond the edge of the Solar System are even more difficult to comprehend, but astronomers have a useful "trick" to help with this problem. In common language, we often use distance and time interchangeably; for example, if someone asks you how far it is to the nearest city, you might say 100 km or you might say 1 hour. In either case, you will have given that person an idea of how far away the city is. It would be unusual to say that the city is one "car-hour" away, but inventing that term would be one way to be more precise in your language. The city is as far away as a car can travel in one hour; it is one car-hour away.

In astronomy, the speed of a car on the highway is far too slow to be useful. Instead, astronomers use the fastest speed in the universe—the speed of light—to express the vast distances. Light travels at 300,000 kilometers per second (km/s). Traveling at the speed of light, you could travel all the way around Earth—a distance of 40,000 km—in just under $\frac{1}{7}$ of a second. Astronomers would then say that the circumference of Earth is $\frac{1}{7}$ of a "light-second." Most distances in astronomy are so vast that they are measured in units of **light-years (ly)**: the distance light travels in 1 year—about 9.5 trillion km or 6 trillion miles. Pause for a moment to think about how much longer a year is than a second; this is how much bigger a light-year is than a light-second.

Because light takes time to reach us, we see astronomical objects not as they are now, but as they used to be. We see them as they were in the past, when the light left them to begin its travel toward Earth. Because light from the Moon takes $1\frac{1}{4}$ seconds to reach us, we see the Moon as it was $1\frac{1}{4}$ seconds ago. Because light from the Sun takes $8\frac{1}{3}$ minutes to reach us, we see the Sun as it was $8\frac{1}{3}$ minutes ago. We see the nearest star as it was more than 4 years ago and we see the Andromeda Galaxy as it was 2.5 million years ago. The light from the most distant observable objects has been traveling for almost the age of the universe—nearly 13.8 billion years. **Figure 1.3** shows distances ranging from the size of Earth to the size of the observable universe, expressed as the time it takes light to travel that far. This is often called the "light travel time."

These vast distances show that we occupy a very small part of the space in the universe. We also occupy only a very small part of time. Imagine that the entire history of the universe took place within a single day. The universe begins the cosmic day at midnight, and hydrogen and helium cool enough to combine with electrons within the first 2 seconds. The first stars and galaxies appear within the first 10 minutes. Generations of stars are born and die before our Solar System forms from recycled gas and dust at about 4 P.M. The first bacterial

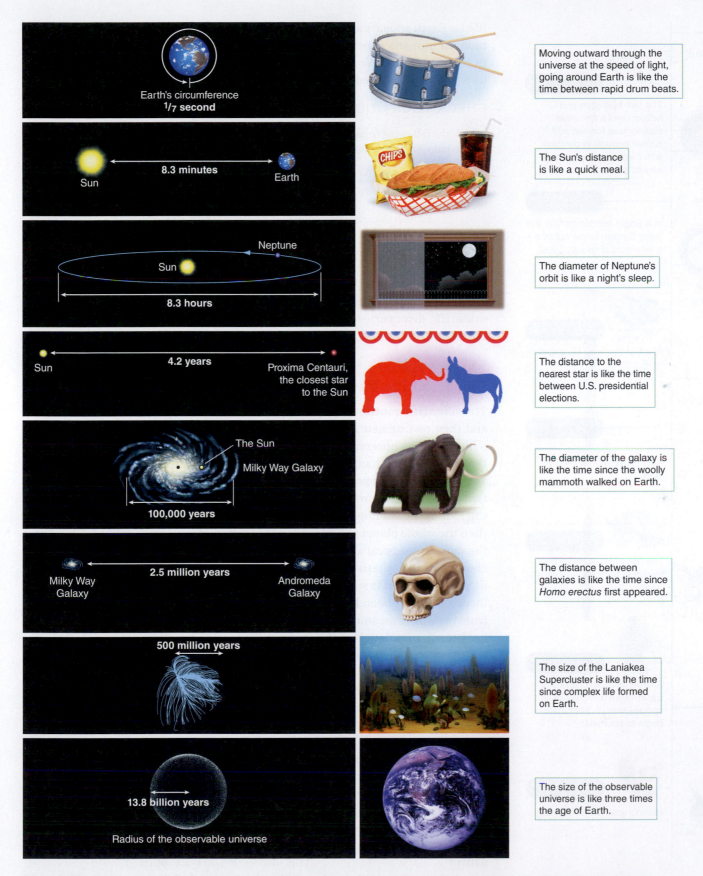

Earth's circumference
1/7 second

Moving outward through the universe at the speed of light, going around Earth is like the time between rapid drum beats.

8.3 minutes
Sun   Earth

The Sun's distance is like a quick meal.

Neptune
Sun
8.3 hours

The diameter of Neptune's orbit is like a night's sleep.

4.2 years
Sun   Proxima Centauri, the closest star to the Sun

The distance to the nearest star is like the time between U.S. presidential elections.

The Sun
Milky Way Galaxy
100,000 years

The diameter of the galaxy is like the time since the woolly mammoth walked on Earth.

2.5 million years
Milky Way Galaxy   Andromeda Galaxy

The distance between galaxies is like the time since *Homo erectus* first appeared.

500 million years

The size of the Laniakea Supercluster is like the time since complex life formed on Earth.

13.8 billion years
Radius of the observable universe

The size of the observable universe is like three times the age of Earth.

Figure 1.3 Thinking about the time light takes to travel between objects helps us comprehend the vast distances in the observable universe.

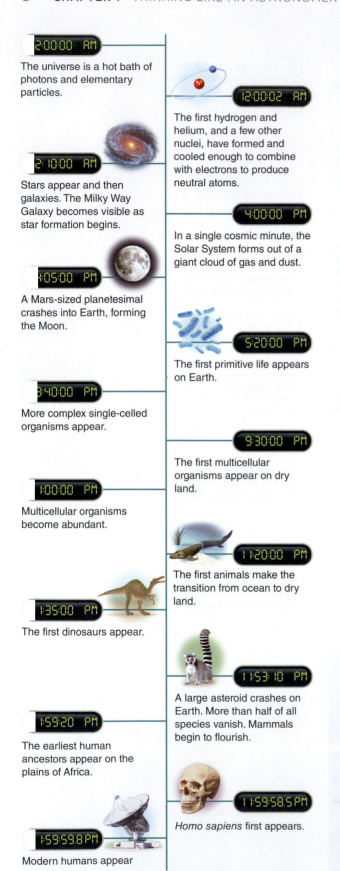

2:00:00 AM

The universe is a hot bath of photons and elementary particles.

2:10:00 AM

Stars appear and then galaxies. The Milky Way Galaxy becomes visible as star formation begins.

1:05:00 PM

A Mars-sized planetesimal crashes into Earth, forming the Moon.

3:40:00 PM

More complex single-celled organisms appear.

1:00:00 PM

Multicellular organisms become abundant.

1:35:00 PM

The first dinosaurs appear.

1:59:20 PM

The earliest human ancestors appear on the plains of Africa.

1:59:59.8 PM

Modern humans appear

12:00:02 AM

The first hydrogen and helium, and a few other nuclei, have formed and cooled enough to combine with electrons to produce neutral atoms.

4:00:00 PM

In a single cosmic minute, the Solar System forms out of a giant cloud of gas and dust.

5:20:00 PM

The first primitive life appears on Earth.

9:30:00 PM

The first multicellular organisms appear on dry land.

11:20:00 PM

The first animals make the transition from ocean to dry land.

11:53:10 PM

A large asteroid crashes on Earth. More than half of all species vanish. Mammals begin to flourish.

11:59:58.5 PM

Homo sapiens first appears.

Figure 1.4 This cosmic timeline presents the history of the universe as a 24-hour day.

life on Earth appears at 5:20 P.M., the first land animals appear at 11:20 P.M., and modern humans appear at 11:59:59.8 P.M.—that is, with just $\frac{1}{5}$ of a second left in the cosmic day. We humans occupy only a sliver of time in the history of the universe, as illustrated in **Figure 1.4**.

CHECK YOUR UNDERSTANDING 1.1

Rank the following from smallest to largest: (a) a light-minute, (b) a light-year, (c) a light-hour, (d) the radius of Earth, (e) the distance from Earth to the Sun, (f) the radius of the Solar System.

Answers to Check Your Understanding questions are in the back of the book.

1.2 Science Is a Way of Viewing the Universe

To view the universe through the eyes of an astronomer, you need to understand how science itself works. Throughout this book, we emphasize not only scientific discoveries but also the *process* of science. This section outlines the scientific method.

The Scientific Method

The **scientific method** is a systematic way of exploring the world by developing and then testing new ideas or explanations. You might begin a scientific study with a fact—an observation or a measurement. For example, you might observe that the weather changes predictably each year and wonder why that happens. You then create a **hypothesis**, a testable explanation of the observation: "I think that it is cold in the winter and warm in the summer because Earth is closer to the Sun in the summer." You come up with a test: if your hypothesis is correct, then the whole planet will be cold in the winter—Australia should have winter at the same time of year as the United States. This is a prediction that you can use to check your hypothesis! In January, you travel from the United States to Australia and find that it is summer in Australia. Your hypothesis has just been proved incorrect; it has been **falsified**. (Notice that this usage of "falsified" is different from the word's meaning in common usage. In common usage, "falsified" evidence has been manipulated to misrepresent the truth. Here, it just means the hypothesis has been shown to be incorrect.) Your test has two important elements that all scientific tests share. Your observation is reproducible: anyone who goes to Australia will find the same result. And your result is repeatable: if you conducted a similar test next year or the year after, you would get the same result. Because you have falsified your hypothesis, you must revise or replace it to be consistent with the new data.

Any idea that is not testable—that is, not falsifiable—must be accepted or rejected based on intuition alone, so it is not a scientific idea. A scientific idea does not have to be testable using current technology, but we must be able to imagine an experiment or observation that could prove the idea wrong. These tests must be repeatable over time and reproducible by everyone. As continuing tests support a hypothesis by failing to disprove it, scientists come to accept the hypothesis as a theory.

A **theory** is a well-developed idea that agrees with known physical laws and makes testable predictions. As with "falsified," the scientific meaning of "theory" is different from the meaning in common usage. In everyday language, theory may mean a guess: "Do you have a theory about who did it?" In everyday language, a theory can be something we don't take very seriously. "After all," people say, "it's only a theory." In stark contrast, scientists use the word *theory* to mean a carefully constructed proposition that accounts for every piece of relevant data as well as our entire understanding of how the world works. A theory has been used to make testable predictions, and all those predictions have come true. Sometimes, competing theories exist to explain a phenomenon. The success or failure of the predictions is the deciding factor between competing theories.

Einstein's theory of general relativity, which underlies the modern understanding of gravity, is an example of a scientific theory. For more than a century, scientists have tested the predictions of Einstein's theory of general relativity and have not been able to falsify it. As Einstein himself noted, a theory that fails only one test is proved false. Even after 100 years of verification, if a prediction of the theory of general relativity failed tomorrow, the theory would require revision or replacement. In that sense, all scientific knowledge is subject to challenge. This openness to challenge is one of the greatest strengths of science.

Let's pause to summarize this science-specific vocabulary. An *idea* is a notion about how something might be. A *fact* is an observation or measurement—for example, the measured value of Earth's radius is a fact. A *hypothesis* is an idea that leads to testable predictions. A hypothesis may lead to a scientific theory, may be based on an existing theory, or both. A *theory* is an idea that has been examined carefully, is consistent with all existing theoretical and observational knowledge, and makes testable predictions. Scientists also often use the term "law." A scientific *law* is a series of observations that can be used to predict a phenomenon but has no underlying explanation of why the phenomenon occurs. A *law* of daytime might say that the Sun rises and sets once each day, whereas a *theory* of daytime might say that the Sun rises and sets once each day because Earth spins on its axis. A **model** describes the properties of a particular object or system in terms of known physical laws or theories. These models are often computational, and use computers to predict the behavior of a complex system, like a system of multiple stars or planets. Scientists themselves can be sloppy about how they use these words, so you may sometimes see them used differently from how we have defined them here.

As the **Process of Science Figure** shows, the steps of the scientific method are interrelated. Scientists often begin with an observation or idea, followed by analysis, followed by a hypothesis, followed by a prediction, followed by further observations or experiments to test the prediction. Typically, the process of testing a theory is never completely finished; there is always another test to be performed under another set of conditions.

Scientific Principles

Scientific **principles** are general rules about the universe that provide guidelines for the formulation of scientific theories. Two important principles used in astronomy are the *cosmological principle* and *Occam's razor*.

The **cosmological principle** assumes that matter and energy behave throughout space and time as they do today on Earth. This means that the same physical laws and theories that we observe and apply in laboratories on Earth can be used to understand what goes on in the centers of stars or in distant galaxies. The principle also implies that the universe has no special locations or directions. In a

what if . . .

Would a theory still be a scientific theory if it were not likely to be falsifiable in many human lifetimes?

The Scientific Method

The scientific method is a formal procedure used to test the validity of scientific hypotheses and theories.

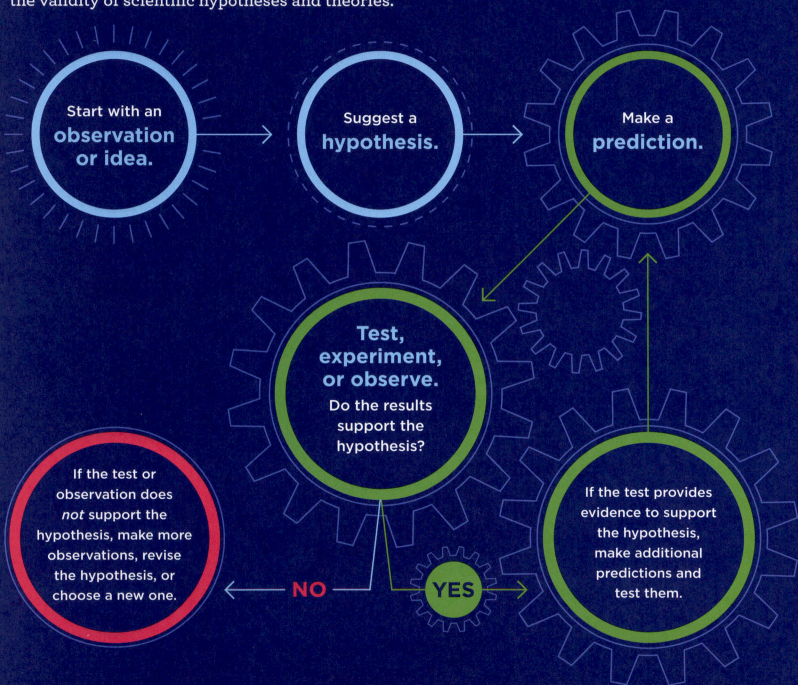

Start with an observation or idea.

Suggest a hypothesis.

Make a prediction.

Test, experiment, or observe. Do the results support the hypothesis?

If the test or observation does *not* support the hypothesis, make more observations, revise the hypothesis, or choose a new one.

NO

YES

If the test provides evidence to support the hypothesis, make additional predictions and test them.

An idea or observation leads to a falsifiable hypothesis. This hypothesis is tested by observation or experiment, and is either accepted as a tested theory or rejected based on the results. The **green loop** repeats indefinitely as scientists continue to test the theory.

sense, the cosmological principle is tested each time we apply our theories to new observations of astronomical objects. Thus far, observations of the universe around us have only added to scientists' confidence in the validity of the cosmological principle. If reasonable experimental evidence ever challenges the validity of the cosmological principle, scientists will construct a new description of the universe that takes those new data into account.

How should you choose between two hypotheses if both explain all of the observations equally well? According to **Occam's razor**, you should choose to adopt the hypothesis that requires the fewest assumptions until there is evidence to the contrary. For example, the nucleus of an atom is known to be positively charged, here in the Milky Way. Suppose you have two competing hypotheses about the charge of the nucleus of an atom in the Andromeda Galaxy: one hypothesis is that the nucleus is positively charged, like nuclei in the Milky Way; the second hypothesis is that the nucleus is negatively charged, opposite to nuclei in the Milky Way. The negative-nucleus hypothesis would require you to make additional assumptions about the location of the boundary between Andromeda-like matter and Milky Way–like matter and about why atoms on the boundary between the two regions did not destroy each other. You also would need an assumption about how the atoms in the two regions came to be constructed so differently. You also would need an assumption about why the Andromeda Galaxy is "special," which violates the cosmological principle. It is far simpler, and requires fewer assumptions, to hypothesize that atoms in the Andromeda Galaxy are the same as atoms in the Milky Way Galaxy.

In many sciences, researchers can conduct controlled experiments to test hypotheses. That experimental method is typically unavailable to astronomers. Astronomers cannot change the tilt of Earth or the temperature of a star to see what happens. Instead, astronomy is an observational science: astronomers apply physical theories based on Earth-bound experiments to observations of astronomical objects. Astronomers typically make multiple observations using as many methods as possible, and then create mathematical and physical models based on established science to explain the observations.

One example of observation leading to new theories comes from the study of planets orbiting stars other than the Sun. Planets orbiting other stars are called *exoplanets*. The first exoplanets to be discovered were giant planets (similar to Jupiter) orbiting very close to their star. However, planets like these are not found in our own Solar System, where the giant planets are all far from the Sun. Those discoveries challenged existing ideas about how our Solar System formed. As different observers using multiple telescopes found more and more of these unexpected planets, astronomers realized they needed new ideas of planet formation to explain how such large planets could wind up so close to their star. Astronomers could not actually build different systems of stars and planets to run a controlled experiment, but they could use the known laws of physics to create computer simulations of planetary systems. When researchers did that, they discovered that the orbits of planets can migrate, becoming closer or farther from the central star. Planetary scientists are searching for evidence that planets migrated early in the history of our own Solar System. The new theory of star and planet formation, which includes planetary migration, has successfully explained thousands of planetary systems, and is at present the best theory available to explain all of the known facts.

Sometimes, science proceeds from theory to observation, instead of from observation to theory. One example is the discovery of black holes. In the late 18th century, John Mitchell (1724–1793) and Pierre-Simon Laplace (1749–1827) independently hypothesized the existence of "dark stars": massive objects having

what if . . .

What if we discovered a region in the universe where the laws of physics are different from the ones on Earth (that is, the cosmological principle is wrong): How should we then think about the laws of physics in yet other parts of the universe?

Reading Astronomy News

Astronomers Find Massive Black Hole in the Early Universe

Brooks Hays

Even though you may not yet know much about black holes, you can begin to make sense of some of the numbers astronomers use to talk about them.

June 25 (UPI)—With the help of a trio of Hawai'ian telescopes, astronomers have imaged the 13-billion-year-old light of a distant quasar—the second-most distant quasar ever found.

Scientists gave the new quasar an indigenous Hawaiian name, Pōniuā'ena, which means "unseen spinning source of creation, surrounded with brilliance." Researchers described the brilliant object in a new paper, which is available in preprint format online and will soon be published in the *Astrophysical Journal Letters*.

Quasars are like lighthouses, their beams hailing from far away in the ancient universe. Powered by supermassive black holes at the center of galaxies, quasars are some of the brightest objects in the universe.

As astronomers peer deeper into the cosmos, they're able to see what the universe was like during its earliest days. In this instance, the Pōniuā'ena's lighthouse-like beacon hails from a period when the universe was still in its infancy—just 700 million years after the Big Bang.

The light of J1342+0928, a quasar spotted in 2018, is older and more distant, but the power and size of Pōniuā'ena is unmatched in the early universe. Spectroscopic observations of Pōniuā'ena, recorded by the Keck and Gemini observatories, revealed a supermassive black hole with a mass 1.5 billion times that of the sun.

"Pōniuā'ena is the most distant object known in the universe hosting a black hole exceeding one billion solar masses," lead study author Jinyi Yang, postdoctoral research associate at the University of Arizona's Steward Observatory, said in a news release.

According to Yang and colleagues, for a black hole to grow to such a tremendous size so early in the history of the universe, it would have needed to start out as a 10,000-solar-mass "seed" black hole, born no later than 100 million years after the Big Bang.

"How can the universe produce such a massive black hole so early in its history?" said Xiaohui Fan, associate head of the astronomy department at the University of Arizona. "This discovery presents the biggest challenge yet for the theory of black hole formation and growth in the early universe."

The light of distant objects, including quasars and massive galaxies in the early universe, can help scientists pinpoint the reionization of the universe. Astrophysicists estimate reionization occurred between 300 million years and one billion years after the Big Bang, but astronomers haven't been able to determine exactly when and how quickly it happened.

The phenomenon describes the ionization of hydrogen gas as the first stars, quasars, galaxies, and black holes came into existence. Prior to the reionization, the universe was without distinct light sources. Diffuse light dominated, and most radiation was absorbed by neutral hydrogen gas.

"Pōniuā'ena acts like a cosmic lighthouse," said study coauthor Joseph Hennawi, a cosmologist and an associate professor in the department of physics at the University of California, Santa Barbara. "As its light travels the long journey towards Earth, its spectrum is altered by diffuse gas in the intergalactic medium which allowed us to pinpoint when the Epoch of Reionization occurred."

Pōniuā'ena was initially spotted by a deep universe survey using the observations of the University of Hawai'i Institute for Astronomy's Pan-STARRS1 telescope on the Island of Maui. Later, scientists used the Gemini Observatory's GNIRS instrument, as well as the Keck Observatory's Near Infrared Echellette Spectrograph, to confirm the identify of Pōniuā'ena.

"The preliminary data from Gemini suggested this was likely to be an important discovery," said study coauthor Aaron Barth, a professor in the physics and astronomy department at the University of California, Irvine. "Our team had observing time scheduled at Keck just a few weeks later, perfectly timed to observe the new quasar using Keck's NIRES spectrograph in order to confirm its extremely high redshift and measure the mass of its black hole."

QUESTIONS

1. The article describes the light from this object as being 13 billion years old. Is this object inside or outside of the Milky Way Galaxy?

2. Which panel of Figure 1.3 would be most useful for showing someone else how far away this object is?

3. This supermassive black hole has a mass of 1.5 billion times that of the Sun. Write this number in scientific notation.

4. This black hole must have been born from a "seed" that formed no later than 100 million years after the Big Bang. In an astronomical context, is that "soon" after the Big Bang, or a "long time" after the Big Bang? How do you know?

5. What is the new question that Xiaohui Fan is asking, now that this observation has been made? Where do questions like this fit into the scientific method shown in the Process of Science Figure and discussed in Section 1.2?

Source: https://www.upi.com/Science_News/2020/06/25/Astronomers-find-massive-black-hole-in-the-early-universe/7781593108062/.

such strong gravity that light could not escape. At that time, there was no way to test that hypothesis. More than 100 years later, in the early 20th century, Karl Schwarzschild (1873–1916) studied Einstein's relativity equations and calculated that, despite their mass, those dark stars would be very small, with a radius of only a few kilometers. Fifty years after that, those objects were named *black holes*. No evidence of their existence was available until the 1970s and 1980s, when the new technology of space-based X-ray telescopes made possible the observations needed to test the hypothesis. In this case, the theory preceded the observation by nearly 200 years.

The scientific method provides a mechanism for testing new scientific ideas, but it offers no insight into where the idea came from in the first place or how an experiment was designed. Scientists discussing the creation of new ideas and experiments use words such as *insight*, *intuition*, and *creativity*. Scientists speak of a beautiful theory in the same way that an artist speaks of a beautiful painting or a musician speaks of a beautiful song. Science has an aesthetic that is as human and as profound as any found in the arts.

Scientific Revolutions

Scientific inquiry is necessarily dynamic. Scientists must constantly refine their ideas in response to new data and new insights. This vulnerability of scientific knowledge may seem like a weakness. "Gee, you really don't know anything," the cynical person might say. But that apparent vulnerability is actually a great strength, because it means that science self-corrects. New information eventually overturns incorrect ideas. In science, even our most cherished ideas about the nature of the physical world remain subject to challenge by new evidence. Many of history's best scientists earned their status by falsifying a universally accepted idea. That is a powerful motivation for scientists to challenge old ideas constantly—to formulate and test new explanations for their observations.

For example, the classical physics that Sir Isaac Newton developed in the 17th century to explain motion, forces, and gravity withstood the scrutiny of scientists for more than 200 years. During the late 19th and early 20th centuries, however, a series of scientific revolutions completely changed our understanding of the nature of reality. The work of Albert Einstein (**Figure 1.5**) is representative of those scientific revolutions. Einstein's special and general theories of relativity replaced Newton's mechanics. Einstein showed that Newton's theories were a special case of a far more general and powerful set of physical theories. Einstein's revolutionary new ideas unified the concepts of mass and energy and destroyed the conventional notion of space and time.

Throughout this text, you will encounter many other discoveries that forced scientists to abandon accepted theories. Einstein himself never embraced the view of the world offered by *quantum mechanics*—a second revolution he helped start. Yet quantum mechanics, a statistical description of the behavior of particles smaller than atoms, has held up for more than 100 years. In science, all authorities are subject to challenge, even Einstein.

Science is a way of thinking about the world. It is a search for the relationships that make our world what it is. A scientist assumes that order exists in the universe and that the human mind can grasp the essence of the rules underlying that order. Scientists build on those assumptions to make and then test predictions, finding the underlying rules that allow humanity to solve problems, invent new technologies, or find a new appreciation for the natural world. Scientific knowledge

Figure 1.5 Albert Einstein is perhaps the most famous scientist of the 20th century, and he was *Time* magazine's selection for Person of the Century. Einstein helped usher in two scientific revolutions.

working it out 1.1

Mathematical Tools

Scientists use mathematics to understand the patterns they observe and to communicate that understanding to others. The following mathematical tools will be useful in our study of astronomy:

Units. Scientists use the metric system of units because most metric units are related to each other by multiplying or dividing by 10. Thus, converting from one unit to another means simply moving the decimal point to the right or to the left. Metric measurements also have prefixes that identify the relationship of the units. There are 100 *centi*meters (cm) in a meter, 1000 meters in a *kilo*meter (km), and 1000 grams in a *kilo*gram (kg). (A more complete listing of metric prefixes is located inside the front cover of this book.) One way to check your own work is to think about whether the answer should be larger or smaller than the original number. The number of centimeters, for example, should always be larger than the number of meters because there is more than 1 cm in a meter. Often, a quick estimate can also help. For example, 1 ft is about an "order of magnitude" (a power of 10) larger than a centimeter; in other words, there are a few tens—not hundreds—of centimeters in a foot. If you have converted 1.2 ft to centimeters and calculated an answer of 3200 cm, you should try again. You can also include the units in every step of the calculation, and at each step, strike out the ones that cancel. For example, if you are multiplying 10 minutes by 60 seconds per minute, the unit *minutes* appears in the numerator and the denominator and divides out.

$$10 \ \text{minutes} \ \times \ \frac{60 \ \text{seconds}}{1 \ \text{minute}} = 600 \ \text{seconds}$$

This can help you remember if you should multiply or divide by the conversion factor; you want the answer in the correct units.

Scientific notation. Scientific notation is how we handle numbers of vastly different sizes. Writing out 7,540,000,000,000,000,000,000 in standard notation is inefficient. Scientific notation uses the first few digits (the *significant* ones) and counts the number of decimal places to create the condensed form 7.54×10^{21}. Similarly, instead of writing out 0.000000000005, we write 5×10^{-12}. The exponent on the 10 is positive or negative depending on whether the decimal point moves left or right, respectively. For example, the average distance to the Sun is 149,600,000 km, but astronomers usually express that value as 1.496×10^{8} km.

Ratios. Ratios are a useful way to compare objects. A star may be "10 times as massive as the Sun" or "10,000 times as luminous as the Sun." Those expressions are ratios.

Proportionality. Often, understanding a concept amounts to understanding the *sense* of the relationships that it predicts or describes. "If you have twice as far to go, getting there will take you twice as long." "If you have half as much money, you will be able to buy only half as much gas." Those are examples of proportionalities.

Appendix 1 further explains the mathematical tools used in this book.

---

is an accumulated collection of ideas about how the universe works, yet scientists are always aware that what is known today may be superseded tomorrow. Science has found such a central place in our civilization because science makes the most accurate predictions about how natural systems will behave.

CHECK YOUR UNDERSTANDING 1.2

The scientific method is a process by which scientists: (a) prove theories to be known facts; (b) gain confidence in theories by failing to prove them wrong; (c) turn theories into laws; (d) survey what most scientists think about a theory.

1.3 Astronomers Use Mathematics to Find Patterns

Scientific thinking allows scientists to make predictions. Once a pattern has been observed, such as the daily rising and setting of the Sun, scientists can predict what will happen next. Making an accurate prediction typically means making a numerical prediction. To do that, scientists need math.

The rhythms of nature produce patterns in our lives, and those patterns give us clues about the nature of the physical world. The Sun rises, sets, and then rises again at predictable times and in predictable locations. Spring turns into summer, summer turns into autumn, autumn turns into winter, and winter turns into spring again. Astronomers identify and characterize those patterns and use them to understand the world around us. As shown in **Figure 1.6**, the visible star patterns in the sky change predictably with the seasons. You will learn many other examples of patterns in the sky in Chapter 2.

Astronomers use mathematics to analyze patterns of all kinds and to communicate complex material compactly and accurately. Many people find mathematics to be a major obstacle that prevents them from appreciating the beauty and elegance of the world as seen through the eyes of a scientist. In this book, we strive to explain any necessary math in everyday language. We describe what equations mean and help you use them in a way that allows you to connect scientific concepts to the world. You will learn to appreciate the beauty of mathematics if you accept the challenge and make an honest effort to understand the mathematical material. **Working It Out 1.1** introduces units and scientific notation. Units are not only necessary to understand relationships between numbers (how large is an object a meter in diameter compared to one that is a kilometer in diameter?), but they also provide a convenient way to check your own work! Numbers in astronomy are typically so large that we must express them in scientific notation. **Working It Out 1.2** reviews some basics of mathematical tools and graphs. In modern society, the ability to read and interpret graphs is indispensable, whether they are graphs of natural systems or graphs of systems like the stock market. You will need the tools in these two boxes as you learn astronomy, whether or not your course is quantitative. But more importantly, you will need these tools in your later life, whether or not you ever pick up another astronomy book. In addition to these tools, Appendix 1 at the back of the book gathers together some essential mathematics, and Appendix 2 collects the physical constants of nature. Other appendixes contain data tables with key information about planets, moons, and stars.

Figure 1.6 Since ancient times, people recognized that patterns in the sky change with the seasons. Those and other patterns shape our lives. These star maps show the sky in the Northern Hemisphere during each season. Find the constellation Ursa Major in each star map to see a pattern over the course of the year.

CHECK YOUR UNDERSTANDING 1.3

When you see a pattern in nature, it is usually evidence of: (a) a theory's being displayed; (b) a breakdown of random clustering; (c) an underlying physical law; (d) a coincidence.

working it out 1.2

Reading a Graph

Scientists often convey complex information and mathematical patterns in graphical form. Graphs typically have two axes: a horizontal axis (the *x*-axis) and a vertical axis (the *y*-axis). The *x*-axis usually shows an independent variable, the one a researcher might have control over in an experiment, whereas the *y*-axis shows the dependent variable, which is typically the variable a researcher is studying.

Graphs can take different shapes. Suppose we plot the distance a car travels over time, as shown in **Figure 1.7a**. In a linear graph, each interval on an axis represents the same-sized step. Each step on the horizontal axis of the graph in Figure 1.7a represents 5 minutes. Each step on the vertical axis represents a distance of 5 km traveled by the car. Data are plotted on the graph, with one dot for each observation; for example, after 20 minutes, the car has traveled 20 km.

Drawing a line through those data indicates the trend (or relationship) of the data. To understand what the trend means, scientists often find the slope of the line, which is the relationship of the line's rise along the *y*-axis to its movement along the *x*-axis. To find the slope, we look at the change between two points on the vertical axis divided by the change between the same two points on the horizontal axis. For example, finding the slope of the line gives

$$\text{Slope} = \frac{\text{Change in vertical axis}}{\text{Change in horizontal axis}}$$

$$\text{Slope} = \frac{(15 - 10)\ \text{km}}{(15 - 10)\ \text{min}}$$

$$= 1\ \text{km/min}$$

There, the trend tells us that the car is traveling at 1 kilometer per minute (km/min), or 60 kilometers per hour (km/h).

Many observations of natural processes do not yield a straight line on a graph. When you catch a cold, for example, you feel fine when you get up in the morning at 7 A.M. At 9 A.M. you feel a little tired. By 11 A.M. you have a bit of a sore throat or a sniffle and think, "I wonder whether I'm getting sick," and by 1 P.M., you have a runny nose, congestion, fever, and chills. The illness progresses faster with each hour as time goes by. That process is exponential because the virus that has infected you reproduces exponentially.

For the sake of this discussion, suppose that the virus produces one copy of itself each time it invades a cell. (Viruses actually produce 1000–10,000 copies each time they invade a cell, so the exponential curve is much steeper.) One virus infects a cell and multiplies, so now two viruses exist—the original and a copy. Those viruses invade two new cells, and each one produces a copy. Now four viruses are there. After the next cell invasion, we have eight. Then 16, 32, 64, 128, 256, 512, 1024, 2048, and so on. That behavior is plotted in **Figure 1.7b**.

Seeing what's happening in the early stages of an exponential curve can be difficult because the later numbers are so much larger than the earlier ones. That's why we sometimes plot that type of data *logarithmically*, by putting the logarithm (roughly the exponent of the 10 in scientific notation) of the data on the vertical axis, as shown in **Figure 1.7c**. Now each step on the axis represents 10 times as many viruses as the previous step. Even though we draw all the steps the same size on the page, they represent different-sized steps in the data (for example, the number of viruses). We often use that technique in astronomy because it has a second, related advantage: very large variations in the data can easily fit on the same graph.

Each time you see a graph, you should first understand the axes—what data are plotted on the graph? Then you should check whether the axes are linear or logarithmic. Finally, you can look at the actual data or lines in the graph to understand how the system behaves.

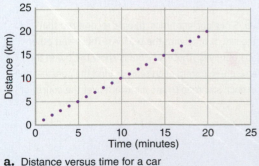

a. Distance versus time for a car

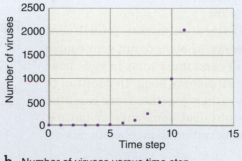

b. Number of viruses versus time step

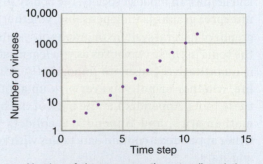

c. Number of viruses versus time step (log plot)

Figure 1.7 Graphs such as these show relationships between quantities. **a.** The relationship between time and distance traveled. **b.** The relationship between time and number of viruses. **c.** The data in part b. plotted logarithmically.

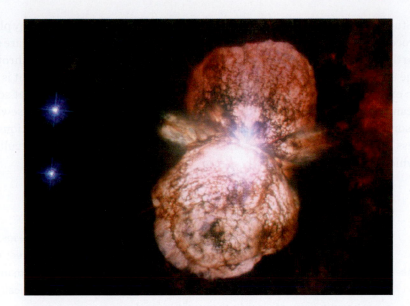

Figure 1.8 You and everything around you are composed of atoms forged in the interiors of stars that lived and died before the Sun and Earth formed. The supermassive star Eta Carinae, shown here, is ejecting a cloud of chemically enriched material just as earlier generations of stars once did to enrich the gas that would become our Solar System.
Credit: ESA/Hubble & NASA. https://esahubble.org/images/potw1208a/. https://creativecommons.org/licenses/by/4.0/.

Origins: An Introduction

How and when did the universe begin? What combination of events led to the existence of humans as sentient beings living on a small, rocky planet orbiting a typical middle-aged star? Are others like us scattered throughout the galaxy?

In these "Origins" sections, which conclude each chapter, we look into the origin of the universe and the origin of life on Earth. We also examine the possibilities of life elsewhere in the Solar System and beyond—a subject called **astrobiology**. This origins theme includes the discovery of planets around other stars and how they compare with the planets of our own Solar System.

Later in the book, we present observational evidence that supports the **Big Bang** theory, which states that the universe started expanding from an infinitesimal size about 13.8 billion years ago. In the early universe, only the lightest chemical elements were found in substantial amounts: hydrogen and helium, and tiny amounts of lithium and beryllium. However, we live on a planet with a central core consisting mostly of heavy elements such as iron and nickel, surrounded by outer layers made up of rocks containing large amounts of silicon and various other elements—all heavier than the original elements. The human body contains carbon, nitrogen, oxygen, calcium, phosphorus, and a host of other chemical elements—all of which are heavier than hydrogen and helium. If those heavier elements that make up Earth and our bodies were not present in the early universe, then where did they come from?

The answer to that question lies within the stars themselves (**Figure 1.8**). Earlier "generations" of stars supplied the building blocks for the chemical processes that we see in the universe, including life. In the core of a star, light elements, such as hydrogen, combine to form more massive atoms, which eventually leads to atoms such as carbon. When a star nears the end of its life, it often loses much of its material—including some of the new atoms formed in its interior—by ejecting it back into interstellar space. That material combines with material lost from other stars—some of which produced even more massive atoms as they exploded—to form large clouds of dust and gas. Those clouds go on to make new stars and planets, similar to our Sun and Solar System. The phrase "we are stardust" is not just poetry; we are literally made of recycled stardust.

what if . . .

What if Earth formed 4 billion years after the Big Bang, rather than 9.3 billion years after it: How might Earth be different?

unanswered questions

Does life as we know it exist elsewhere in the universe? At the time of this writing, no scientific evidence indicates that life exists on any other planet.

Studying origins also provides examples of the process of science. Many physical processes in chemistry, planetary science, physics, and astronomy that are seen on Earth or in the Solar System are observed across the Milky Way and throughout the universe. As of this writing, though, the only biology we know about is that which exists on Earth. At this point in human history, then, much of what scientists can say about the origin of life on Earth and the possibility of life elsewhere is reasoned extrapolation and educated speculation. In these "Origins" sections, we address some of those hypotheses and try to be clear about which are speculative and which have been tested.

SUMMARY

Astronomy seeks answers to many compelling questions about the universe. Astronomy uses all available tools to follow the scientific method. The process of science is based on objective reality, physical evidence, and testable hypotheses. Scientists continually strive to improve their understanding of the natural world and must be willing to challenge accepted truths as new information becomes available. Understanding numerical measurements is an important part of understanding science.

(1) Describe the size and age of today's universe and Earth's place in it. We reside on a planet orbiting a star at the center of a solar system in a vast galaxy that is one of many in an ancient universe that is roughly three times as old as the Sun. We occupy a very tiny part of the universe in space and time.

(2) Explain how astronomers use the scientific method to study the universe. The scientific method is an approach to learning about the physical universe. The method includes observing phenomena, forming hypotheses, making predictions to test and refine those hypotheses, and repeated testing of theories. All scientific knowledge is subject to challenge. Like art, literature, and music, science is a creative human activity; it is also a remarkably powerful, successful, and aesthetically beautiful way of viewing the world.

(3) Show how scientists use mathematics, including graphs, to find patterns in nature. Mathematics provides many of the tools that astronomers need to understand the patterns we see and to communicate that understanding to others.

(4) Summarize our astronomical origins. We are a product of the universe: Except for hydrogen, the very atoms we're made of formed in stars that died long before the Sun and Earth formed. The atoms that make up ordinary matter come from particles that have been around since the Big Bang, processed through stars, and collected into the Sun and the Solar System.

QUESTIONS AND PROBLEMS

TEST YOUR UNDERSTANDING

1. Rank the following in order from smallest to largest.
   a. Local Group
   b. Milky Way
   c. Solar System
   d. universe
   e. Sun
   f. Earth
   g. Laniakea Supercluster
   h. Virgo Supercluster

2. If an event were to take place on the Sun, approximately how long would the light it generates take to reach us?
   a. 8 minutes
   b. 11 hours
   c. 1 second
   d. 1 day
   e. It would reach us instantaneously.

3. *Understanding* in science means that
   a. we have accumulated lots of facts.
   b. we can connect facts through an underlying idea.
   c. we can predict events on the basis of accumulated facts.
   d. we can predict events on the basis of an underlying idea.

4. The cosmological principle states that
   a. on a large scale, the universe is the same everywhere at a given time.
   b. the universe is the same at all times.
   c. our location is special.
   d. every location in the universe has its own laws and theories.

5. The Sun is part of
   a. the Solar System.
   b. the Milky Way Galaxy.
   c. the universe.
   d. all of the above.

6. A light-year is a measure of
   a. distance.
   b. time.
   c. speed.
   d. mass.

7. Occam's razor states that
   a. the universe is expanding in all directions.
   b. the laws of nature are the same everywhere in the universe.
   c. if two hypotheses fit the facts equally well, the one with fewer assumptions is preferred.
   d. patterns in nature are really manifestations of random occurrences.

8. The circumference of Earth is $\frac{1}{7}$ of a light-second. Therefore, if you were traveling at the speed of light,
   a. you would travel around Earth's equator 7 times in 1 second.
   b. you would travel around Earth's equator once in 7 seconds.
   c. you would travel across Earth's diameter 7 times in 1 second.
   d. you would travel across Earth's radius once in 7 seconds.

9. According to the graphs in Figures 1.7b and c, by how much did the number of viruses increase in four time steps? 👁
   a. It doubled.
   b. It tripled.
   c. It quadrupled.
   d. It went up more than 10 times.

10. Any explanation of a phenomenon that includes a supernatural influence is not scientific because
    a. it is not based on a hypothesis.
    b. it is wrong.
    c. people who believe in the supernatural are not credible.
    d. science is the study of the natural world.

11. "All scientific knowledge is provisional." This means:
    a. Everything we know will be proved wrong, eventually.
    b. Scientists don't really know anything.
    c. Science is no better at predicting the natural world than any other way of knowing.
    d. Science is always improving through challenge and further experimentation.

12. When we observe a star that is 10 light-years away, we are seeing that star
    a. as it is today.
    b. as it was 10 days ago.
    c. as it was 10 years ago.
    d. as it was 20 years ago.

13. Which of the following elements formed in the Big Bang? (Choose all that apply.)
    a. hydrogen   c. beryllium
    b. lithium   d. carbon

14. "We are stardust" means that
    a. Earth exists because two stars collided.
    b. the atoms in our bodies have passed through (and often formed in) stars.
    c. Earth is formed primarily of material that used to be in the Sun.
    d. Earth and the other planets will eventually form a star.

15. According to our current understanding of the universe, the following astronomical events led to the formation of you. Place them in order of their occurrence over astronomical time.
    a. Stars die and distribute heavy elements into the space between the stars.
    b. Hydrogen and helium are made in the Big Bang.
    c. Enriched dust and gas gather into clouds in interstellar space.
    d. Stars are born and process light elements into heavier ones.
    e. The Sun and planets form from a cloud of interstellar dust and gas.

THINKING ABOUT THE CONCEPTS

16. ★ **WHAT AN ASTRONOMER SEES**   Figure 1.1 repeatedly "zooms out" to show larger and larger segments of space. Compare the steps shown in this figure with the size of the steps shown in Figure 1.3. Each step of "zoom" shown in Figure 1.1 does not show the same increase in size. Why was this figure drawn this way? 👁

17. Suppose you lived on the planet named "Tau Ceti e" that orbits Tau Ceti, a nearby star in our galaxy. How would you write your cosmic address?

18. Imagine living on a planet orbiting a star in a very distant galaxy. What does the cosmological principle tell you about the physical laws at that distant location?

19. Consider Figure 1.3. If the Sun suddenly exploded, how soon after the explosion would we know about it? 👁

20. Consider Figure 1.3. If a star exploded in the Andromeda Galaxy, how long would that information take to reach Earth? 👁

21. Give an example of a scientific theory that a newer theory has superseded. As scientists developed that new theory, where on the Process of Science Figure did a change occur so that the old theory became invalid and the new theory was accepted? 👁

22. Some people have proposed the theory that extraterrestrials (aliens) visited Earth in the remote past. Can you think of any tests that could support or refute that theory? Is it falsifiable?

23. Explain how the word *theory* is used differently in the context of science than it is in everyday language.

24. What is the difference between a *hypothesis* and a *theory* in science?

25. Suppose the tabloid newspaper at your local supermarket claimed that children born under a full Moon become better students than children born at other times.
    a. Is that theory falsifiable?
    b. If so, how could it be tested?

26. A textbook published in 1945 stated that light from the Andromeda Galaxy takes 800,000 years to reach Earth. In this book, we assert that it takes 2,500,000 years. What does that difference tell you about a scientific "fact" and how our knowledge evolves?

27. Astrology makes testable predictions. For example, it predicts that the horoscope for your star sign on any day should fit you better than horoscopes for other star signs. Read the daily horoscopes for all the astrological signs in a newspaper or online. How many of them might fit the day you had yesterday? Repeat the experiment every day for a week and record which horoscopes fit your day each day. Did your horoscope sign consistently best describe your experiences?

28. A scientist on television states that it is a known fact that life does not exist beyond Earth. Would you consider that scientist's statement reputable? Explain your answer.

29. Some astrologers use elaborate mathematical formulas and procedures to predict the future. Does doing so show that astrology is a science? Why or why not?

30. Why can it be said that we are made of stardust? Explain why current theory supports that statement.

APPLYING THE CONCEPTS

31. The distance to the Sun is 8.3 light-minutes.
   a. If you converted this to light-hours, should you get a larger or a smaller number?
   b. Convert the distance to the Sun from light-minutes to light-hours.
   c. Check your work by comparing the value in light-hours to the value in light-minutes.

32. Suppose that you want to travel 100 km at 30 km/h, and you want to know how much time it will take.
   a. Your answer should have units of "hours." Why?
   b. If you multiply the two numbers together, what units will the result have? Are these the units of time?
   c. If you divide the speed by the distance, what units will the result have? Are these the units of time?
   d. If you divide the distance by the speed, what units will the result have? Are these the units of time?
   e. Carry out the operation (multiplication or division) that will give you the units of time that you want for your answer. How long will it take you to drive 100 km at 30 km/h?

33. a. Write 86,400 (the number of seconds in a day) and 0.0123 (the Moon's mass compared to Earth's) in scientific notation.
   b. In each case, check your work by carefully thinking about the exponent of 10, which should be negative if the number is less than 1 and positive if the number is greater than 1.

34. The average distance from Earth to the Moon is 384,000 km. In the late 1960s, astronauts reached the Moon in about 3 days. How fast (on average) must they have been traveling (in kilometers per hour) to cover this distance in this time? Compare this speed to that of a jet aircraft (800 km/h). ●─●─●
   a. The problem asks for a speed in kilometers per hour, but you have been given a time in days. How many hours are in a day?
   b. Should the number of hours be larger or smaller than the number of days? Do you want to multiply by the number of hours per day, or divide by the number of hours per day to convert 3 days into hours?
   c. The problem is asking for a speed. What are the units of speed?
   d. How should you combine the units of these numbers to find a speed? Multiply or divide them? If divide, which way should you divide them?
   e. Combine the distance in kilometers and the time in hours to find a speed in kilometers per hour.
   f. Compare your answer to the speed of a jet aircraft.

35. Light takes about 8⅓ minutes to travel from the Sun to Earth, and, when it is closest, Neptune is 30 times farther from Earth than the Sun is. How long does light from Neptune take to reach Earth? ●─●─●
   a. Make a prediction: Should light take more or less time to reach Earth from Neptune than from the Sun?
   b. How should you combine the numbers 30 and 8⅓ to get a number consistent with your answer to a? That is, should you multiply or divide them? If you should divide them, which way should you divide them so that the units come out in minutes?
   c. Calculate the time it takes for light to travel from Neptune to Earth.
   d. Check your work by verifying that your units are units of time (minutes or hours), and your answer is consistent with your prediction.

36. Review Working It Out 1.1., and then convert the following numbers to scientific notation:
   a. 7,000,000,000   **c.** 1238
   b. 0.00346

37. Review Working It Out 1.1, and then convert the following numbers to standard notation:
   a. 5.34×10^8 **c.** 6.24×10^{-5}
   b. 4.1×10^3

38. If a car is traveling at 35 km/h, how far does it travel in
   a. 1 hour?   **c.** 1 minute?
   b. half an hour?

39. Review Appendix 1. The surface area of a sphere is proportional to the square of its radius. How many times larger is the surface area if the radius is
   a. doubled?   **c.** halved (divided by 2)?
   b. tripled?   **d.** divided by 3?

40. The average distance from Earth to the Moon is 384,400 km. How many days would it take you, traveling at 800 km/h—the typical speed of jet aircraft—to reach the Moon?

41. The distance from Earth to Mars varies from 56 million km to 400 million km. How long does a radio signal traveling at the speed of light take to reach a spacecraft on Mars when Mars is closest and when Mars is farthest away?

42. The surface area of a sphere is proportional to the square of its radius. The radius of the Moon is only about one-quarter that of Earth. How does the surface area of the Moon compare with that of Earth?

43. A remote Web page may sometimes reach your computer by going through a satellite orbiting approximately 3.6×10^4 km above Earth's surface. What is the minimum delay, in seconds, before the Web page shows up on your computer?

44. Imagine that you have become a biologist, studying rats in Indonesia. Usually, Indonesian rats maintain a constant population. Every half century, however, those rats suddenly begin to multiply exponentially. Then the population crashes back to the constant level. Sketch a graph that shows the rat population over two of those episodes.

45. The circumference of a circle is given by $C = 2\pi r$, where r is the radius of the circle.
   a. Calculate the approximate circumference of Earth's orbit around the Sun, assuming that the orbit is a circle with a radius of 1.5×10^8 km.
   b. A year has 8766 hours, so how fast, in kilometers per hour, does Earth move in its orbit?
   c. How far along in its orbit does Earth move in 1 day?

EXPLORATION The Scale of the Universe

Neptune is about 30 times farther from the Sun than Earth is. The nearest star is about 9000 times farther from the Sun than Neptune is. The diameter of our Milky Way Galaxy is 30,000 times the distance to that nearest star. The Andromeda Galaxy, the nearest large galaxy similar to the Milky Way, is about 600,000 times farther than that nearest star. The diameter of the Local Group of galaxies is about 4 times the distance to Andromeda, and the diameter of the Laniakea Supercluster, which includes the Local Group and many other galaxy groups and clusters, is 50 times larger than the Local Group. That supercluster is just one of millions in the observable universe. The radius of the observable universe is about 110 times larger than the Laniakea Supercluster.

These distances are incredibly difficult to visualize. In order to better picture the relative sizes and distances of things, astronomers often "scale" them. This means that they imagine them much smaller, in the same way that we imagine our town is much smaller when we make a map of it. To get a better sense of the problem in visualizing these distances, imagine a model in which the Sun is 1 centimeter away from Earth—about the size of a peanut M&M. In this model, Earth would be so small as to be barely visible. Calculate the distances to the other objects in the universe, on this scale. We've done the distance to Neptune and the distance to Andromeda for you, as an example. Practice changing units and switching to scientific notation when the numbers get large.

1. At this scale, how far is Neptune from the Sun?

    Neptune is 30 times farther, so 30 cm.

2. How far is the nearest star?

3. What is the diameter of the Milky Way?

4. How far away is the Andromeda Galaxy, on this scale?

    Andromeda is 600,000 times farther than the nearest star, so

    1,620,000,000 km, or 1.6×10^9 km, or 1.6 billion km.

5. What is the diameter of the Local Group, on this scale?

6. What is the diameter of the Laniakea Supercluster, on this scale?

7. What is the radius of the observable universe, on this scale?

As you can see, even when we imagine Earth so small as to be barely visible, the sizes of structures in the universe are unimaginably large.

Patterns in the Sky— Motions of Earth and the Moon

Each month, the Moon changes its appearance in a predictable way. For 1 month, go outside on 10 different dates to the same location and take a picture of the Moon. You will have to think about what time of day to go outside to take these photos. Label each photo with the date and time of the observation. For each photograph, make a sketch of the relative positions of the Sun, Earth, and the Moon when you took the photograph. Think about where you must have been standing on Earth to see that phase of the Moon at that location in your sky, and add a stick figure to your Earth sketch to show where you must have been standing.

EXPERIMENT SETUP

S	M	T	W	T	F	S
				1	2	3
4	5	6	7	8	9	10
11	12	13	14	15	16	17
18	19	20	21	22	23	24
25	26	27	28	29	30	

DAY 10
Moon Rise
1:55 PM

DAY 13
Moon Rise
4:14 PM

DAY 16
Moon Rise
6:07 PM

For 1 month, go outside on 10 different dates to the same location, and take a picture of the moon. You will have to think about what time of day to go outside to take these photos.

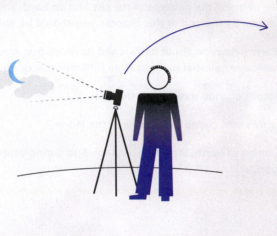

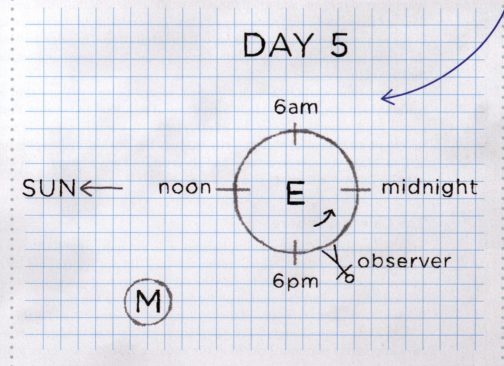

DAY 14

DAY 5

Label each photo with the date and time of the observation.

SKETCH OF RESULTS

DAY 5

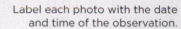

6am

SUN ← noon — E — midnight

6pm

observer

Ⓜ

2

Ancient peoples learned to use the patterns they observed in the sky to predict how the length of day, the seasons, and the appearance of the Moon would change. Some people understood those patterns well enough to create complicated calendars and tables of predictions of rare eclipses. Now, we see those patterns with the perspective of modern science, and we know that those changes are caused by the relative motions of Earth and the Moon. Discovering the causes of those patterns has shown humankind the way outward into the universe.

LEARNING GOALS

Here in Chapter 2, we examine the patterns in the sky and on Earth and the underlying motions that cause them. By the end of this chapter, you should be able to:

① Describe how Earth's rotation about its axis and its revolution around the Sun affect how we perceive celestial motions from different places on Earth.

② Explain why seasons change throughout the year.

③ Describe the factors that create the phases of the Moon.

④ Sketch the alignment of Earth, the Moon, and the Sun during eclipses of the Sun and the Moon.

2.1 Earth Spins on Its Axis

Ancient humans may not have known that they were "stardust," but they did sense a connection between their lives on Earth and the sky above. By watching the repeating patterns of the Sun, Moon, and stars in the sky, people could predict when the seasons would change and when the rains would come and the crops would grow. Some of those early observations and ideas live on today in the names of stars, in the apparent grouping of stars we call **constellations**, in calendars based on the Moon and Sun that many cultures still use, and in the astronomical names of the days of the week.

Across the world, the archaeological record holds evidence that ancient cultures built structures to study astronomical positions and events. **Figure 2.1** shows some examples. From the 8th through 17th centuries, people used pretelescopic astronomical observatories to study the sky for timekeeping and navigation. Many of those structures and observatories are now national historical or UNESCO World Heritage sites.

The Celestial Sphere

Aristotle and other Greek philosophers knew, in antiquity, that Earth is a sphere. However, because Earth seems stationary, they did not realize that its motion caused the changes they observed in the sky from day to day and year to year. As we explain here, Earth's rotation on its axis determines the rising and setting of the Sun, Moon, and stars—one of the rhythms that governs life on Earth.

As Earth rotates about its axis, the planet's surface is moving quite fast—about 1674 kilometers per hour (km/h) at the equator. Like the ancients, you do not

a.

b.

Figure 2.1 **a.** One suspected use of Stonehenge 4000 years ago was to keep track of celestial events. **b.** The Mayan El Caracol at Chichén Itzá in Mexico (906 CE) is believed to have been designed to align with Venus.

feel that motion. Because the motion is uniform, you do not feel it any more than you would feel the motion of a car with a perfectly smooth ride cruising down a straight highway. Nor do you feel the direction of Earth's spin, although the apparent daily motion of the Sun, Moon, and stars across the sky reveals that direction. Imagine that you are in space far above Earth's **North Pole**, which is located at the north end of Earth's rotation axis. From there you would see Earth complete a counterclockwise rotation, once each 24-hour period, as **Figure 2.2** shows. As the rotating Earth carries an observer on the surface from west to east, objects in the sky appear to move in the opposite direction, from east to west. As seen from Earth's surface, each object seems to move across the sky in a path called its **apparent daily motion**.

Several special terms are needed to describe the sky that you can see above you while standing on Earth. The **zenith** is the point in the sky directly above you (**Figure 2.3a**). Wherever you are on Earth, you can see only half the sky. The **horizon** is the boundary between the part of the sky you can see and the other half of the sky that is blocked by Earth (**Figure 2.3b**). You can find the horizon by standing up and pointing your right hand at the zenith and your left hand straight out from your side. Turn in a complete circle. Your left hand has traced out the entire horizon. In common language, the "horizon" is thought of as where the sky meets the ground. Objects like mountains or tall buildings can make that complicated. So astronomers ignore those complications, and consider the horizon to be perfectly straight, as if you were on a boat in the middle of a calm ocean. The **meridian** is an imaginary line that runs from north to south, through the zenith. It divides the sky into an eastern half and a western half. The meridian is shown as a dashed line in Figure 2.3a. The meridian line continues around the far side of Earth, through the **nadir** (the point directly below you), and back to the starting point due north.

To help visualize the apparent daily motions of the Sun and stars, think of the sky as a huge imaginary sphere, called the **celestial sphere**, with the stars painted on its surface and Earth at its center. From ancient Greek times to the Renaissance, many people believed that to be a true representation of the heavens. The celestial sphere is a useful concept because it is easy to visualize, but never forget that it is, in fact, imaginary.

For an Earth-bound observer, each point on the celestial sphere indicates a direction in space (**Figure 2.4**). An observer looking straight up from Earth's North Pole is looking in the direction of the **north celestial pole (NCP)**. Directly above Earth's **South Pole**, which is at the south end of Earth's rotation axis, is the **south celestial pole (SCP)**. Directly above Earth's **equator** is the **celestial equator**, an imaginary circle that divides the sky into a northern half and a southern half. Just as Earth's North Pole is 90° away from Earth's equator, the north celestial pole is 90° away from the celestial equator. If you are in the Northern Hemisphere and you point one arm toward the celestial equator and one arm toward the north celestial pole, your arms will always form a right angle. The same holds true for the Southern Hemisphere: the angle between the celestial equator and the south celestial pole is always 90° as well.

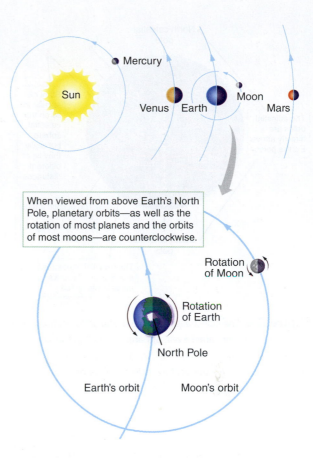

Figure 2.2 Motions in the Solar System, as viewed from above Earth's North Pole (not drawn to scale).

▶❙❙ **AstroTour:** The Celestial Sphere and the Ecliptic

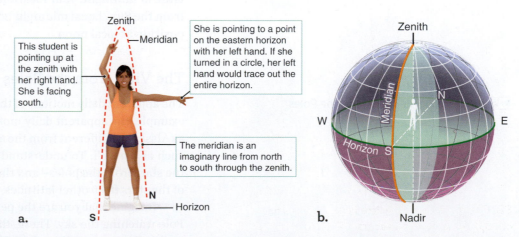

Figure 2.3 **a.** The meridian is a line on the celestial sphere that runs from north to south, dividing the sky into an east half and a west half. **b.** At any location on Earth, the part of the sky above the horizon is divided into an east half and a west half by an imaginary meridian projected onto the celestial sphere.

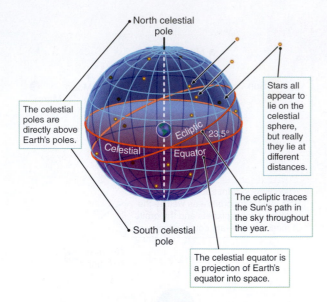

North celestial pole

The celestial poles are directly above Earth's poles.

Stars all appear to lie on the celestial sphere, but really they lie at different distances.

Ecliptic   23.5°

Celestial   Equator

South celestial pole

The ecliptic traces the Sun's path in the sky throughout the year.

The celestial equator is a projection of Earth's equator into space.

Figure 2.4 The celestial sphere is a useful fiction in thinking about the stars' appearance and apparent motion in the sky. Because Earth's axis is tilted 23.5°, there is an angle of 23.5° between the ecliptic and the celestial equator.

🎥 **Astronomy in Action:** Vocabulary of the Celestial Sphere

▶❚❚ **AstroTour:** The View from the Poles

Between the celestial poles and the equator, objects have positions on the celestial sphere with coordinates analogous to latitude and longitude on Earth. **Latitude** indicates distance north or south from Earth's equator. You find your latitude by imagining one line from the center of Earth to your location on the surface, and a second line from Earth's center to the point on the equator closest to you. The angle between those two lines is your latitude. At the North Pole, for example, those two imaginary lines form a 90° angle. At the equator, they form a 0° angle. So the latitude of the North Pole is 90° north, and the latitude of the equator is 0°. The South Pole is at latitude 90° south. Similarly, on the celestial sphere, **declination** indicates the distance of an object north or south of the celestial equator (from 0° to ±90°).

On Earth, **longitude** measures position east or west from the Royal Observatory in Greenwich, England. Similarly, on the celestial sphere, **right ascension** is an eastward and westward measure. The **ecliptic** is the Sun's path in the sky throughout the year. Notice in Figure 2.4 that the ecliptic is tilted relative to the celestial equator. The two circles in the sky intersect in two locations. Right ascension is equal to zero at one of these crossing points, and increases towards the east. Right ascension and declination are used to locate objects in the sky as precisely as anyone might require. In Appendix 7, we discuss detailed descriptions and illustrations of latitude and longitude as well as coordinates used with the celestial sphere.

Take a moment to visualize all those locations in space. To see how to use the celestial sphere, consider the Sun at noon and at midnight. Local noon occurs when the Sun crosses the meridian at your location. That is the highest point above the horizon that the Sun will reach on any given day. The highest point is almost never the zenith. You have to be in a specific place on a specific day for the Sun to be directly overhead at noon. One such place and time is at latitude 23.5° north of the equator on June 20.

From our perspective on Earth, the celestial sphere appears to rotate, carrying the Sun across the sky to its highest point at noon, over toward the west to set in the evening. In *reality*, the Sun remains in the same place in space through the entire 24-hour period, and Earth rotates so that any given location on Earth faces a different direction at every moment. When it is noon where you live, Earth has rotated so that you face most directly toward the Sun. Half a day later, at midnight, your location on Earth has rotated to face most directly away from the Sun. Local midnight occurs when the Sun is precisely opposite from its position at local noon.

The View from the Poles

The apparent daily motions of the stars and the Sun depend on where you live. For example, the apparent daily motions of celestial objects in a northern place such as Alaska are different from the apparent daily motions seen from a tropical island such as Hawai'i. To understand why location matters, let's examine the view of the stars from the poles—and then use them to guide our thinking about the view of the stars from other latitudes.

Imagine that you are the person in **Figure 2.5a** who is standing on the North Pole watching the sky. There, the north celestial pole is directly overhead at the zenith. You are standing where Earth's axis of rotation intersects its surface, which is like standing at the center of a rotating carousel. As Earth rotates, the spot directly above you remains fixed over your head while everything else in the sky appears

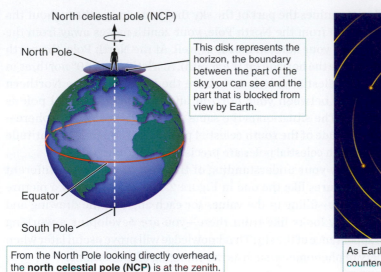

North celestial pole (NCP)

North Pole

This disk represents the horizon, the boundary between the part of the sky you can see and the part that is blocked from view by Earth.

Equator

South Pole

From the North Pole looking directly overhead, the **north celestial pole (NCP)** is at the zenith.

a.

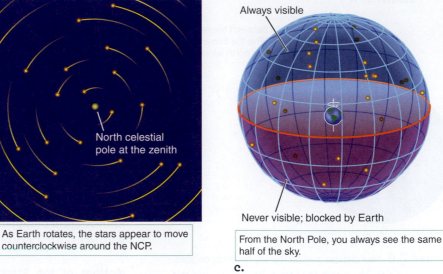

North celestial pole at the zenith

As Earth rotates, the stars appear to move counterclockwise around the NCP.

b.

Always visible

Never visible; blocked by Earth

From the North Pole, you always see the same half of the sky.

c.

Figure 2.5 As viewed from **a.** Earth's North Pole, **b.** stars move throughout the night in counterclockwise, circular paths about the zenith. **c.** The same half of the sky is always visible from the North Pole.

to revolve counterclockwise around that spot. **Figure 2.5b** depicts that view of the sky from the ground, when the Sun is below the horizon.

If you are standing at the North Pole, the zenith and the horizon are always indicating the same locations in space. Objects visible from the North Pole follow circular paths that always have the same **altitude**, or angle above the horizon. Objects close to the zenith appear to follow small circles, whereas objects near the horizon follow the largest circles (Figure 2.5b). The view from the North Pole is special: you will always see the same half of the celestial sphere from there because nothing rises or sets each day as Earth turns (**Figure 2.5c**).

The view from Earth's South Pole is much the same—with two differences. First, the South Pole is on the opposite side of Earth from the North Pole, so the visible half of the sky at the South Pole is precisely the half hidden from view at the North Pole. Second, stars appear to move clockwise around the south celestial pole rather than counterclockwise as they do at the north celestial pole. To visualize why those motions are different, stand up and spin around from right to left. As you look at the ceiling, things appear to move counterclockwise, but as you look at the floor, they appear to be moving clockwise.

CHECK YOUR UNDERSTANDING 2.1a

No matter where you are on Earth, stars appear to rotate about a point called the:
(a) zenith; (b) celestial pole; (c) nadir; (d) meridian.

Answers to Check Your Understanding questions are in the back of the book.

The View Away from the Poles

Away from the poles, the visible half of the sky changes constantly as Earth carries you around; your horizon cuts across different points in space, and your zenith moves across the sky as Earth rotates. Suppose that you leave the North Pole to travel south to lower latitudes. Your latitude is equal to the angle of your position from the equator, as shown for the person standing at 60° north latitude in **Figure 2.6**.

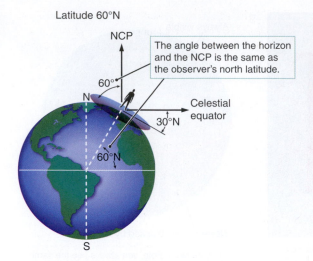

Latitude 60°N

NCP

The angle between the horizon and the NCP is the same as the observer's north latitude.

60°

N

30°N

Celestial equator

60°N

S

Figure 2.6 Your perspective on the sky depends on your location on Earth. The locations of the celestial poles and the celestial equator in an observer's sky depend on the observer's latitude. Here, an observer at latitude 60° north sees the north celestial pole at an altitude of 60° above the northern horizon and the celestial equator 30° above the southern horizon.

Your latitude determines the part of the sky that you can see throughout the year. As you move south from the North Pole, your zenith moves away from the north celestial pole, and your horizon moves as well. At the North Pole, the north celestial pole is 90° above the horizon (at the zenith). At a latitude of 60° north, as in Figure 2.6, the north celestial pole is 60° above the horizon. In the Northern Hemisphere, the angle between your horizon and the north celestial pole is equal to your latitude. The situation is the same in the Southern Hemisphere—your latitude is the altitude of the south celestial pole. At the equator, at a latitude of 0°, the north and south celestial poles are precisely on the horizon.

One way to solidify your understanding of the view of the sky at different latitudes is to draw pictures like the one in Figure 2.6. If you can draw a picture like that for any latitude—filling in the values for each angle in the drawing and imagining what the sky looks like from there—you are developing a working knowledge of the appearance of the sky. That knowledge will prove useful later when we discuss a variety of phenomena, such as the changing of the seasons.

Motions of the Stars and the Celestial Poles   The stars' apparent motions about the celestial poles also differ from latitude to latitude. At the poles, all the stars are **circumpolar**; they go "around the pole" and never rise or set. As an observer moves farther from the poles, they will see fewer circumpolar stars. Two time-lapse views of the sky from different latitudes are shown in **Figure 2.7**. More stars are circumpolar for the more northern latitude, in Canada. The image taken from Utah is farther south, so it has fewer circumpolar stars. The Utah image also demonstrates the added complication that objects on the horizon often block low altitudes in the sky. Even stars that ought to be visible all the time, because they are circumpolar, might drop behind nearby objects that block the horizon.

From the Canada and Utah latitudes, if we focus our attention on stars near the north celestial pole, we see much the same thing we saw from Earth's North Pole. All the stars appear to move throughout the night in counterclockwise, circular paths around the north celestial pole. The north celestial pole is no longer directly overhead as it was at the North Pole, however, so the apparent circular paths of the stars are now tipped with respect to the horizon. (More correctly, your horizon is now tipped with respect to the apparent circular paths of the stars.)

Recall from Figure 2.6 that your latitude is equal to the altitude of the north celestial pole. Stars closer to the north celestial pole than that angle are above the horizon all the time, so they are circumpolar. Another group of stars, near the south celestial pole, are below the horizon all the time; they are never visible at all from your location. Stars between those that never rise and those that are circumpolar can be seen for only part of each 24-hour day; they appear to rise and set as Earth rotates.

Figure 2.8 shows the orientation of the sky as seen by observers at the North Pole, the equator, 30° north, and 30° south. The view from the North Pole (Figure 2.8a) was previously shown in Figure 2.5, but it is reproduced here for ease of comparison. For an observer at the equator (Figure 2.8b), the celestial poles are both at the horizon, and all the stars are visible in a 24-hour period, rising straight up and setting straight

Figure 2.7 Time exposures of the sky showing the apparent motions of stars through the night. Note the difference in the circumpolar portion of the sky as seen from the two northern latitudes.

From a location in the Canadian woods, the north celestial pole appears high in the sky…

…but at a lower latitude in Utah, the north celestial pole appears closer to the horizon.

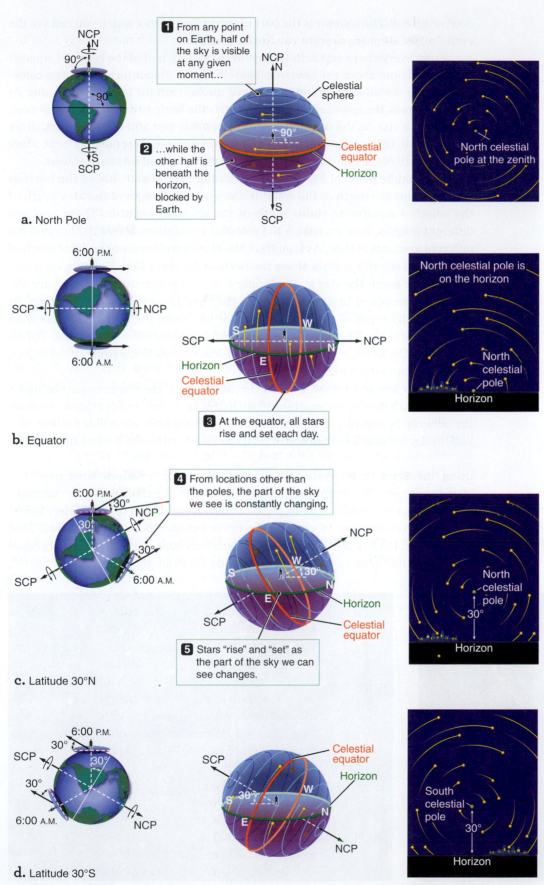

1 From any point on Earth, half of the sky is visible at any given moment...

2 ...while the other half is beneath the horizon, blocked by Earth.

NCP
N
90°
90°
S
SCP

NCP
N
Celestial sphere
90°
Celestial equator
Horizon
S
SCP

a. North Pole

North celestial pole at the zenith

6:00 P.M.
SCP ◄—||—► NCP
6:00 A.M.

SCP ◄———
W
S
E
N
Horizon
Celestial equator
———► NCP

3 At the equator, all stars rise and set each day.

b. Equator

North celestial pole is on the horizon

North celestial pole

Horizon

4 From locations other than the poles, the part of the sky we see is constantly changing.

6:00 P.M.
30°
NCP
30°
SCP ◄—
30°
6:00 A.M.

NCP
S
W
30°
E
N
Horizon
SCP
Celestial equator

5 Stars "rise" and "set" as the part of the sky we can see changes.

c. Latitude 30°N

North celestial pole
30°
Horizon

6:00 P.M.
30°
SCP ◄—
30°
30°
6:00 A.M.
NCP

Celestial equator
SCP ◄—
Horizon
S
30°
W
E
N
NCP

d. Latitude 30°S

South celestial pole
30°
Horizon

Figure 2.8 The celestial sphere is shown here as viewed by observers at four latitudes. At all locations other than the poles, stars rise and set as the part of the celestial sphere that we see changes during the day.

what if . . .

What if Earth did not rotate at all relative to the stars as it orbited the Sun? What effect would that have on the science of astronomy?

down each day. The equator is the only place on Earth from which you can see the entire celestial sphere over the course of 24 hours.

From everywhere on Earth (except at the poles), half of the celestial equator is always visible above the horizon. Therefore, any object located on the celestial equator is visible half the time—above the horizon for 12 hours each day. At other latitudes, the celestial equator intersects the horizon due east and due west. Therefore, a star on the celestial equator rises due east and sets due west. Stars located north of the celestial equator rise north of east and set north of west. Stars located south of the celestial equator rise south of east and set south of west.

The middle panel of Figure 2.8c shows the path of stars above the horizon for a location 30° north of the equator. Compare the length of the arcs north of the celestial equator to those south of the celestial equator. These arcs are different lengths, so stars not on the celestial equator are above the horizon for different amounts of time. As seen from the Northern Hemisphere, stars north of the celestial equator remain above the horizon for more than 12 hours each day. The farther north the star is, the longer it stays up. Circumpolar stars are the extreme example of that phenomenon; they are always above the horizon. In contrast, stars south of the celestial equator are above the horizon for less than 12 hours each day. The farther south a star is, the less time it is visible. For an observer in the Northern Hemisphere, stars located close to the south celestial pole never rise above the horizon.

Figures 2.8c and d show that stars in the observer's hemisphere are visible for more than half the day because more than half of each star's path in the sky is above the horizon. In contrast, stars in the opposite hemisphere are visible for less than half the day because less than half of each star's path in the sky is above the horizon.

Using the Stars to Navigate   Since ancient times, travelers have used the stars to navigate. They would find the north or south celestial poles by recognizing the stars that surround them. In the Northern Hemisphere, a moderately bright star happens to be located close to the north celestial pole (**Figure 2.9a**). That star is called Polaris, the "North Star." Polaris's altitude in the sky is nearly equal to your latitude. If you are in Phoenix, Arizona, for example (latitude 33.5° north),

Figure 2.9 Groups of stars near the pole stars in the sky can be used to locate a pole star. **a.** Two bright "pointer stars" in the cup of the Big Dipper point toward Polaris, the "North Star." **b.** In the Southern Hemisphere, the constellation Crux and two of its bright pointer stars can be used to locate the relatively faint southern pole star.

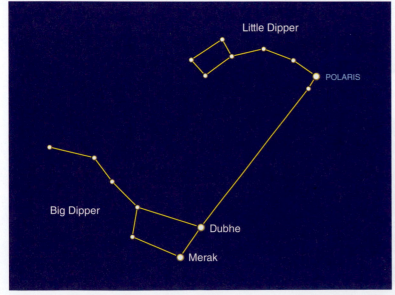

a.

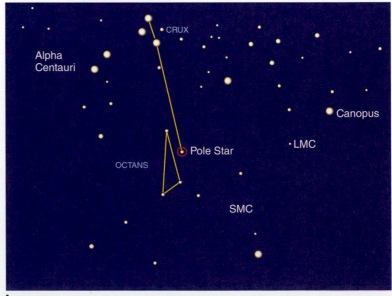

b.

working it out 2.1

How to Estimate Earth's Size

We can use the location of the north celestial pole in the sky to estimate Earth's size. Suppose we start out in Phoenix, Arizona, and we observe the north celestial pole to be 33.5° above the horizon. If we head north about 290 km to the Grand Canyon, the north celestial pole has risen to about 36° above the horizon. That difference between 33.5° and 36° (2.5°) is 1/144 of the way around a circle. (A circle is 360°, and 2.5°/360° = 1/144.)

That means we must have traveled 1/144 of the way around Earth's circumference, C, by traveling the 290 km between Phoenix and the Grand Canyon. In other words,

$$\frac{1}{144} \times C = 290 \text{ km}$$

When we rearrange the expression, Earth's circumference is given by

$$C = 144 \times 290 \text{ km} \approx 42,000 \text{ km}$$

Earth's actual circumference is just over 40,000 km, so our simple calculation was close. The circumference of a circle is equal to $2\pi r$, where r is its radius. Earth's radius, then, is given by

$$\text{Radius} = \frac{C}{2\pi} = \frac{40,000 \text{ km}}{2\pi} = 6400 \text{ km}$$

In about 230 BCE, using much the same method, the Greek astronomer Eratosthenes (276–194 BCE) was the first to accurately measure Earth's size. As **Figure 2.10** illustrates, Eratosthenes used the distance between his home city of Alexandria, Egypt, and the city of Syene (now Aswân, also in Egypt), which was 5000 "stadia." He noticed that on the first day of summer in Syene, the sunlight reflected directly off the water in a deep well, so the Sun must have been nearly at the zenith. By measuring the shadow of the Sun from an upright stick in Alexandria, he saw that the Sun was about 7.2° south of the zenith on the same date. Assuming that Earth was spherical and that Syene was directly south of Alexandria, he determined the distance between the two cities to be 7.2° divided by 360, or 1/50 of Earth's circumference.

Although historians know Eratosthenes concluded that the cities were 5000 stadia apart, they are still not sure of the value of his stadion unit. Other sources indicate that it may have been about 185 meters. We can use this conversion factor to compare Eratosthenes' value to modern values:

$$\frac{1}{50} \times C = 5000 \text{ stadia} \times 185 \text{ meters/stadion}$$

$$= 925,000 \text{ meters} = 925 \text{ km}$$

$$C = 50 \times 925 \text{ km} = 46,250 \text{ km}$$

That result is only about 16 percent higher than the modern value.

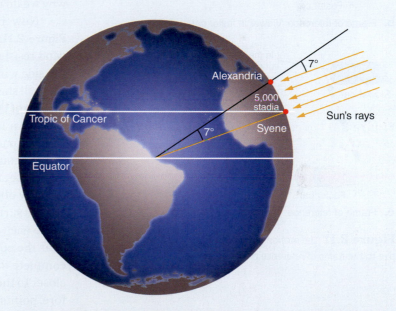

Figure 2.10 Eratosthenes used observations and basic calculations to estimate Earth's size.

Polaris has an altitude very close to 33.5°. In Fairbanks, Alaska (latitude 64.6° north), Polaris sits much higher, with an altitude very close to 64.6°. In the Southern Hemisphere, the constellation Crux (commonly called the Southern Cross) points to a faint star near the south celestial pole (**Figure 2.9b**). A navigator who has located a pole star can identify north and south and therefore east, west, and her latitude. Doing so enables the navigator to determine which direction to travel. The location of the north celestial pole in the sky was also used to measure Earth's

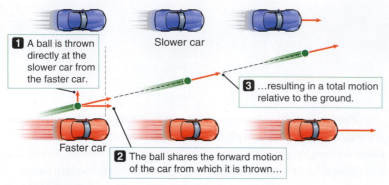

1 A ball is thrown directly at the slower car from the faster car.

Slower car

3 …resulting in a total motion relative to the ground.

Faster car

2 The ball shares the forward motion of the car from which it is thrown…

a. Frame of reference: Viewer on the street

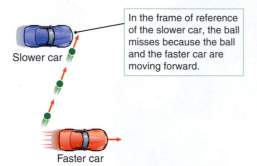

In the frame of reference of the faster car, the ball misses because the slower car is moving backward.

Slower car

Faster car

b. Frame of reference: Viewer in faster car

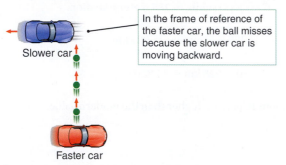

In the frame of reference of the slower car, the ball misses because the ball and the faster car are moving forward.

Slower car

Faster car

c. Frame of reference: Viewer in slower car

Figure 2.11 The motion of an object is always measured relative to the frame of reference of the observer.

size, as described in **Working It Out 2.1**. Because of Earth's rotation, determining longitude by astronomical methods is much more complicated. Longitude cannot easily be determined from astronomical observation alone.

Relative Motions and Frame of Reference

Why don't we feel motion as Earth spins on its axis and moves through space in its orbit around the Sun? The answer lies in the concept of a **frame of reference**, a coordinate system within which an observer measures positions and motions. The difference in motion between two individual frames of reference is called the **relative motion**. For example, imagine that you are riding in a car traveling down a straight section of highway at a constant speed. Under those conditions, without looking out the window or feeling road vibrations, you cannot easily tell the difference between being in the moving car and sitting in a parked car. Because everything inside the car is moving together, you can measure only the relative motions between objects in the car, and they are all zero—no motion is observed. Similarly, the resulting relative motions between objects that are near each other on Earth are zero because they have the same rotational motion. That is why we do not notice Earth's motion.

Now imagine two cars driving down the road at different speeds, as **Figure 2.11a** shows. For the moment, ignore any real-world complications, such as wind resistance. If you were to throw a ball from the faster-moving car directly out the side window at the slower-moving car as the two cars passed, you would miss. The ball shares the forward motion of the faster car, so the ball outruns the forward motion of the slower car. From your perspective in the faster car (**Figure 2.11b**), the slower car lagged behind the ball. From the slower car's perspective (**Figure 2.11c**), your car and the ball sped on ahead.

Although you cannot feel Earth's rotation, it influences motions around you as diverse as weather patterns and the flight of artillery shells toward a distant target. Any object that travels a significant distance northward or southward above Earth's surface will be affected. Here's why: All points on Earth's surface move in a circle each day around the planet's rotation axis. That circle is larger for points near the equator and smaller for points closer to one of the poles, but all points complete their circular motion in the same amount of time: exactly 1 day. A point closer to the equator has farther to go each day than one nearer a pole does. Therefore, points nearer the equator must be moving faster than ones at higher latitude. If an object starts out at one latitude and then moves to another, that difference in the speed of the surface influences the apparent direction of the object's motion.

Imagine a cannonball launched directly north from a point in the Northern Hemisphere, as in **Figure 2.12a**. Because the cannon is closer to the equator than the target, the cannon is moving toward the east faster than its target. Even though the cannonball is fired toward the north, it shares the eastward velocity of the cannon itself, just like the ball thrown from the faster car in Figure 2.11. Therefore, the cannonball is *also* moving toward the east faster than its target. As the cannonball flies north, it moves toward the east faster than the ground underneath it does. To an observer on the ground, the cannonball appears to curve toward the east as it outruns the eastward motion of the ground it is crossing. The farther north the cannonball flies, the greater the difference between its eastward velocity and the ground's. The cannonball follows a path that appears to curve more and more

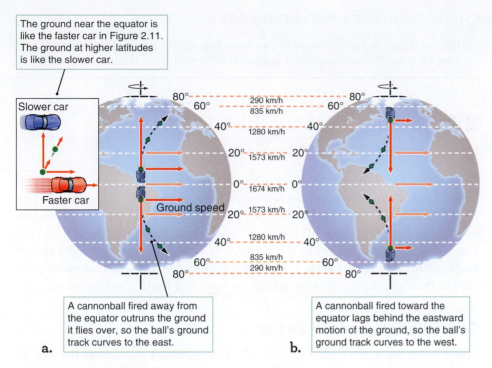

The ground near the equator is like the faster car in Figure 2.11. The ground at higher latitudes is like the slower car.

Slower car

Faster car

Ground speed

80°
60°   290 km/h
       835 km/h
40°   1280 km/h
20°   1573 km/h
0°   1674 km/h
20°   1573 km/h
40°   1280 km/h
       835 km/h
60°   290 km/h
80°

a. A cannonball fired away from the equator outruns the ground it flies over, so the ball's ground track curves to the east.

b. A cannonball fired toward the equator lags behind the eastward motion of the ground, so the ball's ground track curves to the west.

Figure 2.12 The Coriolis effect causes objects to be deflected as they move across Earth's surface. The green dashed curve shows the cannonball's path as measured by a local observer.

to the east the farther north it goes. The cannonball also curves east if shot toward the south in the Southern Hemisphere, as shown in Figure 2.12a.

If you are located in the Northern Hemisphere and fire a cannonball *south* toward the equator, as in **Figure 2.12b**, the opposite effect will occur. Now the cannon is moving toward the east more slowly than its target. As the cannonball flies toward the south, its eastward motion lags behind that of the ground underneath it, so the cannonball appears to curve toward the west.

That curving motion of objects as a result of the difference in Earth's rotation speeds at different latitudes is called the **Coriolis effect**. In the Northern Hemisphere, the Coriolis effect causes a cannonball fired north to drift to the east as seen from the ground. In other words, the cannonball appears to curve to the right, as viewed by the cannon operator. A cannonball fired south appears to curve to the west, which is also to the right as observed by the cannon operator. In the Northern Hemisphere, the Coriolis effect seems to deflect things to the *right*. Viewed from space, this deflection will produce a clockwise motion of weather patterns. In the Southern Hemisphere, it seems to deflect things to the *left*. Viewed from space, this will produce a counterclockwise motion of weather patterns. In between, at the equator itself, the Coriolis effect vanishes.

On Earth, the Coriolis effect is large enough to deflect a fly ball hit north or south deep into the outfield by about a half a centimeter. At some time or other, then, the Coriolis effect has probably affected the outcome of a baseball game. On the other hand, the story about water spiraling in different directions down a drain in different hemispheres is just a myth. On the scale of sinks or toilets, the Coriolis effect is far too small to overcome other effects, such as the shape of the basin itself. However, this story often helps people remember that the Coriolis effect works oppositely in each hemisphere, so it may be worth remembering, even if it isn't true.

2.2 Revolution about the Sun Leads to Changes during the Year

Earth orbits (or **revolves**) around the Sun in the same direction that Earth spins about its axis—counterclockwise as viewed from above Earth's North Pole (see Figure 2.2). A **year** is defined as the time Earth takes to complete one revolution around the Sun—it's a little bit less than 365.25 days. Earth's motion around the Sun is responsible for many of the patterns of change we see in the sky and on Earth, including changes in the stars we see overhead. Because of that motion, the stars in the night sky change throughout the year, and Earth experiences seasons.

Constellations and the Zodiac

About 9000 stars are visible to the naked eye—spread out across the sky. As Earth orbits the Sun, our view of these stars changes. To follow the patterns of the Sun and the stars, early humans grouped together stars that formed recognizable patterns, called constellations. People from different cultures saw different patterns and projected ideas from their own cultures onto what they saw in the sky. Constellations are creations of the human imagination.

If you noted the position of the Sun in relation to the stars each day for a year, you would find that the Sun traces out a great circle against the background of the stars (**Figure 2.13**). On September 1, the Sun appears to be in the direction of the constellation Leo. Six months later, on March 1, Earth is on the other side of the Sun, and the Sun appears from our perspective on Earth to be in the

Figure 2.13 As Earth orbits the Sun, different stars appear in the night sky, and the Sun's apparent position against the background of stars changes, blocking our view of stars in that direction. The imaginary circle that the Sun's annual path traces is called the ecliptic. Constellations along the ecliptic form the zodiac.

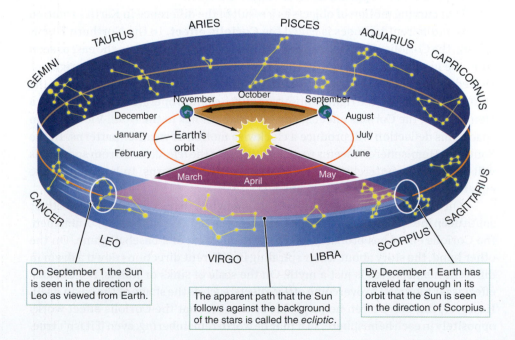

On September 1 the Sun is seen in the direction of Leo as viewed from Earth.

The apparent path that the Sun follows against the background of the stars is called the *ecliptic*.

By December 1 Earth has traveled far enough in its orbit that the Sun is seen in the direction of Scorpius.

direction of the constellation Aquarius. Recall that the Sun's apparent path against the background of the stars is called the ecliptic (the blue band in Figure 2.13). The constellations that lie along the ecliptic through which the Sun appears to move are called the constellations of the **zodiac**.

Today, astronomers use an officially sanctioned set of 88 constellations to serve as a road map of the sky. Constellations visible from the Northern Hemisphere draw their names primarily from the list of constellations that the Alexandrian astronomer Ptolemy compiled 2000 years ago. Constellation names in the Southern Hemisphere come from European explorers who visited the Southern Hemisphere during the 17th and 18th centuries. Every star in the sky lies within the borders of a single constellation. This makes it possible to catalog all the stars using the names of their parent constellation. Following the Greek alphabet, astronomers call the brightest star in a constellation *alpha*; the second brightest, *beta*; and so forth. For example, Sirius, the brightest star in the sky, lies within the constellation Canis Major (meaning "big dog"). So Sirius is also called "Alpha Canis Majoris," indicating that it is the brightest star in Canis Major. Appendix 7 includes sky maps showing the constellations.

Earth's Axial Tilt and the Seasons

To understand why the seasons change, we need to consider the combined effects of Earth's rotation on its axis and its revolution around the Sun. Many people believe that seasons change because Earth is closer to the Sun in summer and farther away in winter. How can we test that hypothesis? We might predict: if the distance between Earth and the Sun caused the seasons, then all of Earth should experience summer at the same time of year. But the United States has summer in June, whereas Australia has summer in December. In modern times, we can directly measure the distance between Earth and the Sun, and we find that Earth is actually closest to the Sun at the beginning of January, by about 5 million km. Yet, in the Northern Hemisphere, we have winter in January. We have just falsified the hypothesis tying seasonal change to distance in two ways, so we need to look for a different explanation that accounts for all the available facts.

Earthbound observers can see that as the year passes, the Sun is above the horizon longer in summer than in winter. We often say "the days are longer in the summer", meaning there are more daytime hours, and fewer nighttime ones. The Sun is also higher in the sky as it crosses the meridian in summer than in winter. Observing more closely, the Earthbound observer finds that the Sun appears to move along the ecliptic, which is tilted 23.5° with respect to the celestial equator.

Looking at Earth from space, **Figure 2.14** shows that as Earth moves around the Sun over the course of the year, its rotation axis always points in the same direction, toward distant Polaris. This axis is tilted 23.5° from the perpendicular to Earth's orbital plane, causing the ecliptic to likewise be tilted 23.5° from the celestial equator. As it orbits, Earth is sometimes on one side of the Sun and sometimes on the other. As a result, Earth's North Pole is sometimes tilted more toward the Sun, and other times the South Pole is tilted more toward the Sun. This accounts for the changing altitude of the Sun as it crosses the meridian, and the changing length of daytime and nighttime. When Earth's North Pole is tilted toward the Sun, an observer on Earth views the Sun north of the celestial equator. Therefore, for observers in the Northern Hemisphere, the Sun is above the horizon more than

▶❚❚ **AstroTour:** The Earth Spins and Revolves

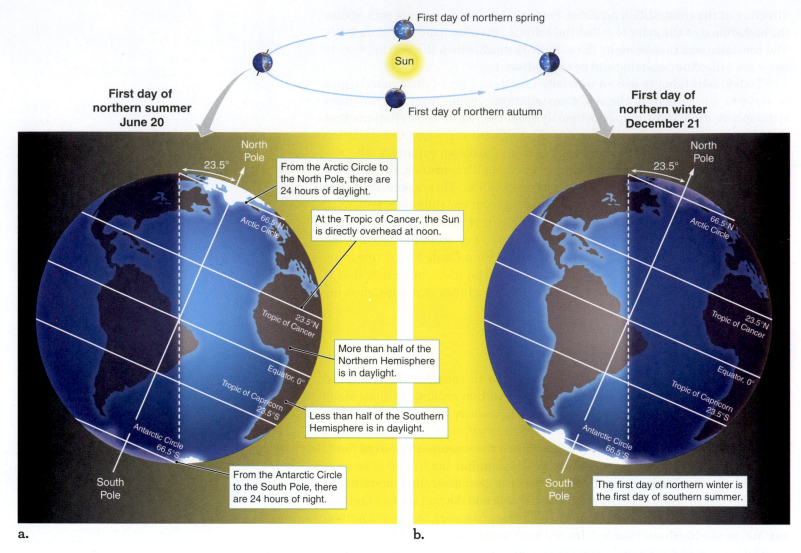

Figure 2.14 **a.** On the first day of the northern summer (around June 20, the summer solstice), the northern end of Earth's axis is tilted most nearly toward the Sun, whereas the Southern Hemisphere is tipped away. **b.** Six months later, on the first day of the northern winter (around December 21, the winter solstice), the situation is reversed. Seasons are opposite in the two hemispheres.

🎥 **Astronomy in Action:** The Cause of Earth's Seasons

12 hours each day, thus making the daytime longer than 12 hours. Six months later, when Earth's North Pole is tilted away from the Sun, an observer in the same place views the Sun south of the celestial equator and the daytime is less than 12 hours.

In the preceding paragraph, we specified the *Northern* Hemisphere because seasons are opposite in the Southern Hemisphere. Look again at Figure 2.14. Around June 20, while the Northern Hemisphere is enjoying the long days and short nights of summer, Earth's South Pole is tilted away from the Sun. It is winter in the Southern Hemisphere; less than half of the Southern Hemisphere is illuminated by the Sun, and the days are shorter than 12 hours. On December 21, Earth's South Pole is tilted toward the Sun. It is summer in the Southern Hemisphere, so its days are long and its nights are short.

To understand how the combination of Earth's axial tilt and its path around the Sun creates seasons, consider what would happen if Earth's spin axis were exactly perpendicular to the plane of Earth's orbit (the **ecliptic plane**). First, the Sun would always be on the celestial equator. Then, at every latitude (except the poles), the Sun would follow the same path through the sky every day, rising due east each morning and setting due west each evening. The Sun would be above the horizon exactly half the time, so days and nights would always be exactly 12 hours long everywhere on Earth. In short, Earth would have no seasons.

The changing length of days through the year only partly explains seasonal temperature changes. A more important effect relates to the angle at which the Sun's rays strike Earth. The Sun is higher in the sky during summer than during winter, and sunlight strikes the ground *more directly* during summer than during winter. To see why that is important, study **Figure 2.15**. During summer, Earth's surface is more nearly face-on to the incoming sunlight. The light is concentrated and bright so more energy falls on each square meter of ground each second. During winter, the surface is more tilted with respect to the sunlight, so the light is more diffuse. Less energy falls on each square meter of the ground each second. For example, at latitude 40° north, which stretches across the United States from northern California to New Jersey, more than twice as much solar energy falls on each square meter of ground per second at noon on June 20 as falls there at noon on December 21. That variation is the main reason why summer is hotter and winter is colder.

The **Process of Science Figure** demonstrates how determining the causes of seasonal change requires accounting for all the known facts. The two effects of Earth's tilted axis—the directness of sunlight and the different lengths of the night—mean that the Sun heats a hemisphere more at one time of year than another, causing summer and winter.

The Solstices and the Equinoxes

Four days during Earth's orbit (two solstices and two equinoxes) mark special moments in the year. The day when the Sun is highest in the sky as it crosses the meridian is called the **summer solstice**. On that day, the Sun rises farthest north of east and sets farthest north of west. In the Northern Hemisphere, that occurs each year near June 20, the first day of summer. Figure 2.14a shows that orientation of Earth and Sun.

Six months after the summer solstice, around December 21, the North Pole is tilted away from the Sun (Figure 2.14b). That day is the **winter solstice** in the Northern Hemisphere—the shortest day of the year and the first day of winter in the Northern Hemisphere. Almost all cultural traditions in the Northern Hemisphere include some sort of major celebration in late December. Those winter festivals celebrate the return of the source of Earth's light and warmth. The days have stopped growing shorter and are beginning to get longer, and spring will come again.

Between the two solstices, the ecliptic crosses the celestial equator on two days. On those days, the Sun lies directly above Earth's equator. We call those days *equinoxes*, which means "equal night," because the entire Earth experiences 12 hours of daylight and 12 hours of darkness. Halfway between summer solstice and winter solstice, the **autumnal equinox** marks the beginning of fall in the Northern Hemisphere; it occurs around September 22. Halfway between winter solstice and summer solstice, the **vernal equinox** marks the beginning of spring in the Northern Hemisphere; it occurs around March 20. In the Southern Hemisphere, the dates for the summer and winter solstices, and the autumnal and vernal equinoxes, are reversed from those in the Northern Hemisphere.

what if . . .

What if humans establish a colony on Mars, where the axial tilt is 25.19°? Considering only this single piece of information, would you expect seasonal variations to be larger, smaller, or about the same as seasonal variations on Earth? Explain.

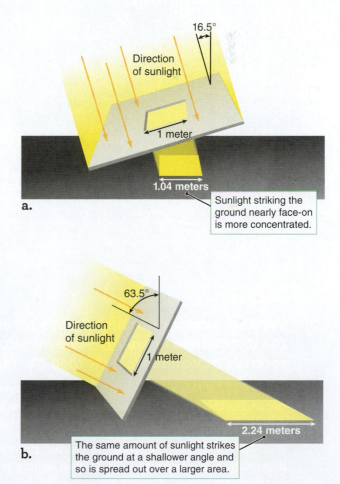

Figure 2.15 Local noon at latitude 40° north. **a.** On the first day of northern summer, sunlight strikes the ground almost face-on. **b.** On the first day of northern winter, sunlight strikes the ground more obliquely, and less than half as much sunlight falls on each square meter of ground each second.

Theories Must Fit All the Known Facts

Many people misunderstand the phenomenon of changing seasons because they do not account for all the relevant facts.

HYPOTHESIS 1

We have seasons because Earth is closer to the Sun in the summer and farther away in winter.

THE TEST If this were true, both the Northern and Southern Hemispheres would have summer in July. However, the Northern and Southern Hemispheres experience opposite seasons.

THE CONCLUSION The hypothesis is falsified.

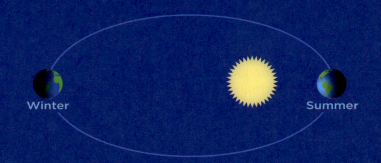

Winter   Summer

HYPOTHESIS 2

Distance 1

Distance 2

We have seasons because the tilt of Earth's axis causes one hemisphere to be significantly closer to the Sun than the other.

THE TEST   If this were true, the distances would have to be very different to cause such a large effect. Earth is tiny compared to its distance from the Sun: The distances between each hemisphere and the Sun differ by less than 0.004 percent.

THE CONCLUSION   The hypothesis is falsified.

HYPOTHESIS 3

We have seasons because Earth's tilt changes the distribution of energy—one hemisphere receives more intense light than the other.

THE TEST If this were true, the amount of sunlight striking the ground in the summer would have to be more than in the winter, and the days would have to be longer in summer.

THE CONCLUSION Seasons are caused primarily by a change in illumination due to Earth's tilt. During winter, less energy falls on each square meter of ground per second.

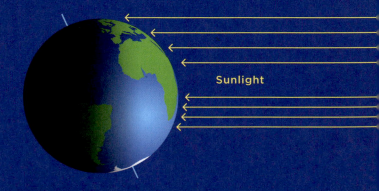

Sunlight

New information often challenges misconceptions. Incorporating this new information may alter current theories to improve the explanation of observed phenomena.

Figure 2.16 shows the solstices and equinoxes from two perspectives, oriented so that Earth's North Pole points straight up. Figure 2.16a is in the reference frame of the Sun. It shows Earth in orbit around a stationary Sun, whereas Figure 2.16b is in the reference frame of Earth. It shows the Sun's apparent motion along the celestial sphere, around a stationary Earth. Practice shifting between those perspectives. You will know that you understand them when you can look at a position in either panel and predict the corresponding positions of the Sun and Earth in the other panel.

Just as a pot of water on a stove takes time to heat up when the burner is turned on and to cool when the burner is turned off, Earth takes time to respond to changes in heating from the Sun. The hottest months of northern summer are usually July and August, which come after the summer solstice, when days are growing shorter. Similarly, the coldest months of northern winter are usually January and February, which occur after the winter solstice, when days are growing longer. Temperature changes on Earth lag behind changes in the amount of heating we receive from the Sun.

That picture of the seasons must be modified somewhat near Earth's poles. At latitudes north of 66.5° north and south of 66.5° south, the Sun is circumpolar for a part of the year surrounding the first day of summer. Those lines of latitude are the **Arctic Circle** and the **Antarctic Circle**, respectively (see Figure 2.14). When the Sun is circumpolar, it is above the horizon 24 hours per day, earning the polar regions the nickname "land of the midnight Sun" (**Figure 2.17**). An equally long period surrounds the first day of winter, when the Sun never rises and the nights are 24 hours long. The Sun never rises very high in the Arctic or Antarctic sky, so the sunlight is never very direct. Even with the long days at the height of summer, the Arctic and Antarctic regions remain relatively cool.

In contrast, on the equator, *all* stars, including the Sun, are above the horizon approximately 12 hours per day. Days and nights there are 12 hours long throughout the year. The Sun passes directly overhead on the first day of spring and the first day of autumn because on those days the Sun is on the celestial equator. Sunlight is most direct, perpendicular to the ground, at the equator on those days. On the summer solstice, the Sun is at its northernmost point along the ecliptic. On that day, and on the winter solstice, the Sun is farthest from the zenith at noon, and therefore sunlight is least direct at the equator.

Latitude 23.5° north is called the Tropic of Cancer, and latitude 23.5° south is called the Tropic of Capricorn (Figure 2.14). The band between those latitudes is called the **tropics**. If you live in the tropics—in Rio de Janeiro or Honolulu, for example—the Sun will be directly overhead at noon twice during the year.

Precession of the Equinoxes

Two thousand years ago, when Ptolemy and his associates were formalizing their knowledge of the positions and motions of objects in the sky, the Sun appeared in the constellation Cancer on the first day of northern summer and in the constellation

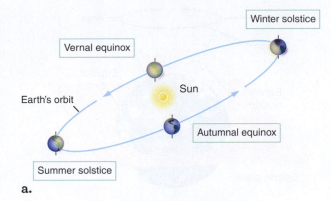

Vernal equinox

Winter solstice

Sun

Earth's orbit

Autumnal equinox

Summer solstice

a.

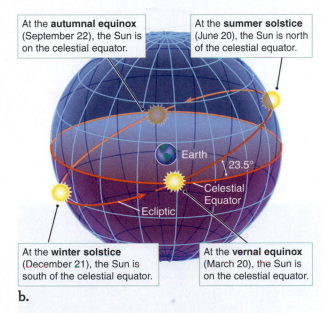

At the **autumnal equinox** (September 22), the Sun is on the celestial equator.

At the **summer solstice** (June 20), the Sun is north of the celestial equator.

Earth

23.5°

Celestial Equator

Ecliptic

At the **winter solstice** (December 21), the Sun is south of the celestial equator.

At the **vernal equinox** (March 20), the Sun is on the celestial equator.

b.

Figure 2.16 Earth's motion around the Sun from the frame of reference of **a.** the Sun and **b.** Earth.

Astronomy in Action: The Earth-Moon-Sun System

Figure 2.17 This composite photo shows the midnight Sun, which is visible in latitudes above 66.5° north (or south). In the 360-degree panoramic view, the Sun moves 15° each hour.

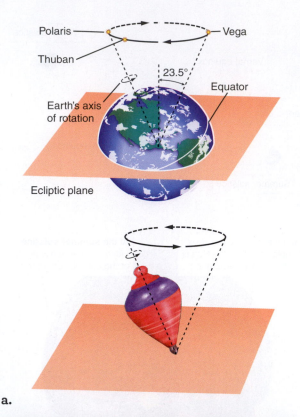

a.

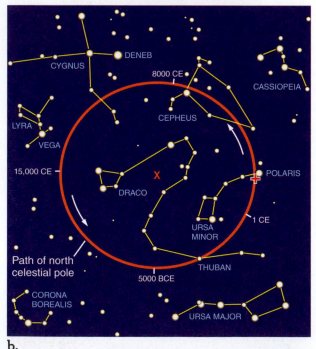

b.

Figure 2.18 a. Earth's axis of rotation changes orientation in the same way that the axis of a spinning top changes orientation. Polaris is the current North Pole star, Thuban was the North Pole star in 3000 BCE, and Vega will be the pole star in 14,000 CE. **b.** That precession causes the projection of Earth's rotation axis to move in a circle, centered on the north ecliptic pole (orange *X* in the center). The red cross on the circle marks the projection of Earth's axis on the sky in the early 21st century.

Capricornus on the first day of northern winter. Today, the Sun is in Taurus on the first day of northern summer and in Sagittarius on the first day of northern winter. Why have the constellations in which solstices appear changed? Two motions are actually associated with Earth and its axis: Earth spins on its axis, but its axis also wobbles like the axis of a spinning top slowing down (**Figure 2.18a**). The wobble is very slow: the north celestial pole takes about 26,000 years to complete one circle in the sky. Polaris is the star we now see near the north celestial pole. However, if you could travel several thousand years into the past or the future, the point about which the northern sky appears to rotate would no longer be near Polaris. Instead, the stars would rotate about another point on the path shown in **Figure 2.18b**. That figure shows the path of the north celestial pole through the sky during one cycle of that wobble.

The celestial equator is perpendicular to Earth's axis. Therefore, as Earth's axis wobbles, the celestial equator also must wobble. As it does so, the locations where the celestial equator crosses the ecliptic—the equinoxes—change as well. During each 26,000-year wobble of Earth's axis, the locations of the equinoxes make one complete circuit around the celestial equator. That change of the position of the equinox, due to the wobble of Earth's axis, is called the **precession of the equinoxes**.

CHECK YOUR UNDERSTANDING 2.2

If Earth's axis were tilted by 45°, instead of its actual tilt of 23.5°, how (if at all) would the seasons be different from what they are now? (a) The seasons would remain the same. (b) Summers would be colder. (c) Winters would be shorter. (d) Winters would be colder.

2.3 The Moon's Appearance Changes as It Orbits Earth

After the Sun, the most prominent object in our sky is the Moon. Just as Earth orbits around the Sun, the Moon orbits around Earth once every 27.32 days. In this section, we discuss the phases of the Moon as seen from Earth.

The Changing Phases of the Moon

The Moon and its changing aspects have been the frequent subject of mythology, art, literature, and music. In mythology, the Moon was the Greek goddess Artemis, the Roman goddess Diana, and the Inuit god Igaluk. We speak of the "man in the Moon," the "harvest Moon," and sometimes a "blue Moon."

Unlike the Sun, the Moon has no light source of its own; instead, the Moon shines by reflected sunlight. As the Moon orbits Earth, our view of the illuminated portion of the Moon constantly changes. Those different appearances of the Moon are called **phases**. During a new Moon, when the Moon is between Earth and the Sun, the side facing away from us is illuminated, and during a full Moon, when Earth is between the Sun and the Moon, the side facing toward us is illuminated. The rest of the time, we can see only part of the illuminated portion from Earth. Sometimes the Moon appears as a circular disk in the sky. Other times it is nothing more than a thin sliver, or its face appears dark.

To help you visualize the changing phases of the Moon, use a sphere (such as an orange or a softball), a lamp, and your head (**Figure 2.19**). Your head is Earth, the sphere is the Moon, and the lamp is the Sun. Turn off all the other lights in the room, and step back as far from the lamp as you can. Hold up the sphere slightly above your head so that the lamp illuminates it from one side. Move the sphere clockwise around your head and watch how the appearance of the sphere changes. When you are between the sphere and the lamp, the face of the sphere that is toward you is fully illuminated. The sphere appears to be a bright, circular disk. As the sphere moves around its circle, you will see a progression of lighted shapes, depending on how much of the bright side and how much of the dark side of the sphere you can see. That progression of shapes mimics the changing phases of the Moon.

Figure 2.20 shows the changing phases of the Moon, from two perspectives. The orbit is shown as viewed from a point above Earth's North Pole; the outer circle of Moon images then shows how the Moon appears from Earth at that phase. The **new Moon** occurs when the Moon is between Earth and the Sun. The far side of the

Figure 2.19 A sphere (such as an orange) and a lamp can help you visualize the changing phases of the Moon.

🎥 **Astronomy in Action:** Phases of the Moon

▶️❚❚ **AstroTour:** The Moon's Orbit: Eclipses and Phases

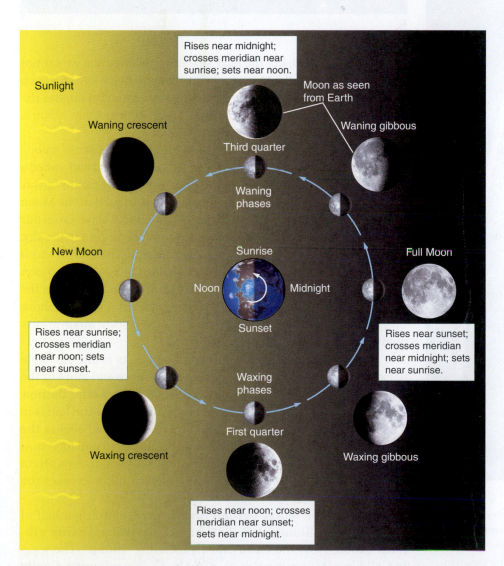

Figure 2.20 The inner circle of images (connected by blue arrows) shows the Moon as it orbits Earth, as seen by an observer far above Earth's North Pole. The Sun is on the left. The outer ring of images shows the corresponding phases of the Moon as seen from the Northern Hemisphere of Earth.

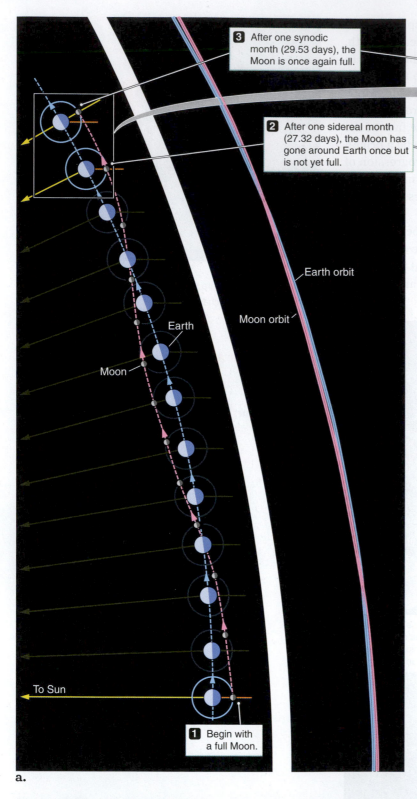

3 After one synodic month (29.53 days), the Moon is once again full.

2 After one sidereal month (27.32 days), the Moon has gone around Earth once but is not yet full.

Earth orbit

Moon orbit

Earth

Moon

To Sun

1 Begin with a full Moon.

a.

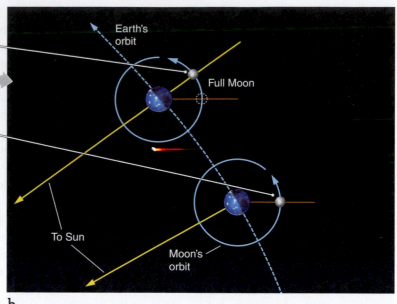

Earth's orbit

Full Moon

To Sun

Moon's orbit

b.

Figure 2.21 **a.** The Moon completes one sidereal orbit in 27.32 days, but the synodic period (the period between phases seen from Earth) from one full Moon to the next is 29.53 days. The horizontal orange line to the right of the Moon indicates a fixed direction in space. **b.** Here the orbits of Earth and the Moon are shown to scale, but the sizes of Earth and the Moon are not.

Moon is illuminated, but the near side is in darkness and we cannot see it. At new phase, the Moon is close to the Sun in the sky, so it is up in the daytime with the Sun: it rises in the east at sunrise, crosses the meridian near noon, and sets in the west near sunset. A new Moon is never above the horizon in the nighttime sky.

A few days after a new Moon, the Moon has moved farther along its orbit of Earth. At this point, a sliver of its illuminated half, called a **waxing crescent Moon**, becomes visible. *Waxing* here means "growing in size and brilliance"; the name refers to the fact that the Moon appears to be "filling out" from night to night at that time. From our perspective, the Moon has also moved away from the Sun in the sky. Because the Moon travels around Earth in the same direction in which Earth rotates, we now see the Moon following the Sun, so the Moon is east of the Sun in the sky. A waxing crescent Moon is visible in the western sky in the evening, near the setting Sun but remaining above the horizon after the Sun sets. The "horns" of a crescent Moon always point directly away from the Sun.

As the Moon moves even farther along in its orbit, the angle between the Sun and Moon grows larger, so more and more of the Moon's near side becomes illuminated. About a week after the new Moon, half of the near side of the Moon is illuminated and half is in darkness. This phase is called a **first quarter Moon** because the Moon has moved a quarter of the way around Earth and has completed the first quarter of its cycle from new Moon to new Moon (Figure 2.20). The first quarter Moon rises at noon, crosses the meridian at sunset, and sets at midnight.

As the Moon moves beyond first quarter, more than half of its near side is illuminated. This phase is called a **waxing gibbous Moon**, from the Latin *gibbus*, meaning "hump." The waxing gibbous Moon continues nightly to "grow" until finally we see the entire near side of the Moon illuminated—a **full Moon**. Earth is now between the Sun and the Moon, so they appear opposite each other in the sky when viewed from Earth. As a result, the full Moon rises as the Sun sets, crosses the meridian at midnight, and sets in the morning as the Sun rises.

The second half of the Moon's orbit is the reverse of the first half. The Moon continues in its orbit, again appearing gibbous but now becoming smaller each night. This phase is called a **waning gibbous Moon**, where *waning* means "becoming smaller." When the Moon is waning, the left side—as viewed from the Northern Hemisphere—appears illuminated. A **third quarter Moon** occurs when half of the near side is illuminated by sunlight and half is in darkness. A third quarter Moon rises at midnight, crosses the meridian near sunrise, and sets at noon. The cycle continues with a **waning crescent Moon** in the morning sky, until the new Moon once again rises and sets with the Sun, and the cycle begins again. Notice that when the Moon is farther from the Sun than Earth is, it is in gibbous (or full) phases, whereas when the Moon is closer to the Sun than Earth is, it is in crescent (or new) phases.

You can always tell a waxing Moon from a waning Moon because the illuminated side is always the side facing the Sun. When the Moon is waxing, it appears in the evening sky, so its western side is illuminated. That is the right side as viewed from the Northern Hemisphere. Conversely, when the Moon is waning, it appears in the morning sky, so the eastern side appears bright. That is the left side as viewed from the Northern Hemisphere.

Figure 2.21 illustrates two types of lunar periods. The first one is based on the Moon's orbit in space, and the second is based on the alignment of the moving Moon, Earth, and Sun. The Moon completes one orbit around Earth in 27.32 days. That **sidereal period** is how long the Moon takes to return to the same location in its orbit. However, the relationships among Earth, the Moon, and the Sun change as a result of Earth's orbital motion, so going from one full Moon to the next takes 29.53 days. That cycle is known as the Moon's **synodic period** and is the basis for our "month," because it is what we can easily observe from Earth.

Do not try to memorize all possible combinations of where the Moon is in the sky at what phase and at what time of day. Instead, work on understanding the motion and phases of the Moon, and then use your understanding to figure out the specifics of any given case. To study the phases of the Moon, draw a picture like Figure 2.20, and use it to follow the Moon around its orbit. From your drawing, figure out what phase you would see and where it would appear in the sky at a given time of day. You might also try the simulations described in "Exploration: Phases of the Moon" at the end of the chapter.

The Moon's Visible Face

Although the Moon's illumination varies as it orbits, one aspect of the Moon's appearance does *not* change: the face that we see. If we were to go outside next week, next month, 20 years from now, or 200 centuries from now, we would still see the same side of the Moon that we see tonight. That is because the Moon rotates on its axis exactly once for each revolution that it makes around Earth.

Imagine walking around a tree while always keeping your face toward the tree. By the time you complete one circle around the tree, your head has turned completely around once. When you were south of the tree, you were facing north; when you were east of the tree, you were facing west; and so on. But someone looking at you from the tree would always see your face. The Moon does the

what if . . .

What if the Moon were a cube rather than a sphere? How would this change the appearance of the Moon's phases?

▶▶ **Interactive Simulation:** Phases of the Moon

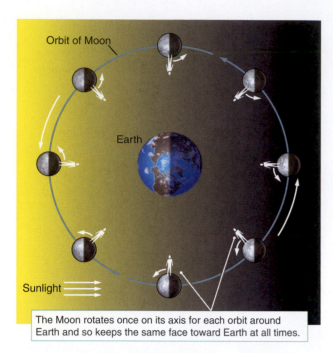

The Moon rotates once on its axis for each orbit around Earth and so keeps the same face toward Earth at all times.

Figure 2.22 The Moon rotates once on its axis for each orbit around Earth—an effect called synchronous rotation. Here, the Sun is far to the left of the Earth–Moon system.

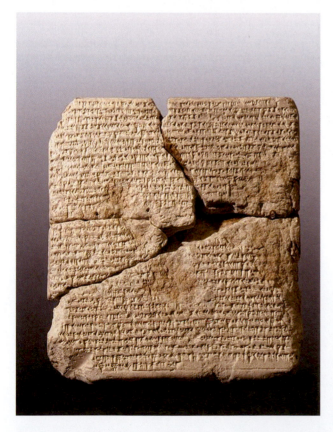

Figure 2.23 Hieroglyphic calendar at the Kom Ombo Temple. The calendar used a system of 12 months, plus festival days.

same thing, rotating on its axis once per revolution around Earth, always keeping the same face toward Earth (**Figure 2.22**). Objects whose revolution and rotation are synchronized (or "in sync") with each other are said to be in **synchronous rotation**. It occurs as a result of the gravitational and tidal forces between Earth and the Moon (Section 4.4). In later chapters we discuss other examples of synchronous rotation.

The Moon's *far side*, facing away from Earth, is often called the "dark side of the Moon." In fact, no side of the Moon is always dark. At any given time, half of the Moon is in sunlight and half is in darkness—just as for Earth. The side of the Moon that faces away from Earth, the "far side," spends just as much time in sunlight as the side of the Moon that faces Earth.

CHECK YOUR UNDERSTANDING 2.3

If you see the Moon rising just as the Sun is setting, what is the phase of the Moon? What is the phase if the Moon is setting at noon?

2.4 Calendars Are Based on the Day, Month, and Year

Archaeologists tell us that developing agriculture was crucial for the rise of human civilization, and keeping track of the seasons and best times of the year to plant and harvest was critical to successful farming. Records going back to the dawn of humanity suggest that people kept track of time by following the patterns in the sky, especially those of the Sun, Moon, and stars. Some anthropologists have speculated that notches on fragments of bone found in southern France represent a 33,000-year-old lunar calendar. In this section, we examine some different calendars.

Lunar and Lunisolar Calendars

As civilizations developed around the globe, different cultures tried to solve the problem of keeping track of time. The rates of rotation of Earth and the revolutions of the Moon around Earth and of Earth around the Sun are not integer multiples of one another. A lunar cycle (from full Moon to full Moon) is 29.5 days, and a solar cycle—the time the Sun takes to appear to move from and return to its highest possible point in the sky at noon on the summer solstice—is 365.24 days. One solar cycle, then, has 12.38 lunar cycles. Those fractions of days and months make calendars complicated.

Some of the oldest known calendars come from the Egyptians (**Figure 2.23**), the Babylonians, and the Chinese. The ancient Egyptians used a system of 12 months of 30 days each—which added up to 360 days—and then added five "festival days" to the end of the year. Without leap years, the seasons in that 365-day year drifted, so an extra month was added when necessary. When we consider how some Western countries celebrate the days between the modern December holidays and the new year, an end-of-year calendar break for festivals seems like a good solution!

The Babylonians started the 24-hour day and 7-day week—7 for the Sun, the Moon, and the 5 planets visible with the naked eye. They created the first **lunisolar calendar**, in which a month began with the first sighting of the lunar crescent, and a 13th month was added when needed to catch up to the solar year.

As the Babylonians developed mathematics, they discovered that 235 lunar months equals 19 solar years (and 6940 days). Then they created a calendar cycle that consisted of 19 years, 12 of which have 12 months and 7 of which have 13 months, and then the cycle repeats. The ancient Hebrew calendar adopted that lunisolar cycle, and the Jewish calendar still uses it today. That type of calendar keeps holidays in the same season from year to year, even though the dates change.

The ancient Chinese calendar, dating back several thousand years, is lunisolar, too. By about 500 BCE, the Chinese were using a year of 365.25 days and a system similar to that of the Babylonians of adding a 13th month to some years. To mark out a year, a few other cultures used stellar calendars, which followed the position of a bright star such as Sirius or certain prominent groups of stars in their sky, such as the Pleiades or the Big Dipper.

The Islamic calendar is purely lunar, with no added 13th lunar month. Their 12 months of 29 or 30 days each add up to 354 days—11.24 days short of the solar year. For that reason, the Islamic New Year and all other holidays drift earlier in each solar year. In the Islamic calendar, a holiday may fall in the winter in some years, and then a few years later it will have moved back to autumn.

The Modern Civil Calendar

The international civil calendar used today is a solar calendar known as the **Gregorian calendar**. It is based on the **tropical year**, which measures the 365.242 solar days between vernal (spring) equinoxes. A **solar day** is the 24-hour period of Earth's rotation that brings the Sun back to the same local meridian. That measure is in contrast to the **sidereal day**, the time Earth takes to make one rotation and face the same star on the meridian. The sidereal day is about 23 hours 56 minutes and is slightly shorter than the solar day because of Earth's motion around the Sun. This is similar to the two measures of the time it takes the Moon to orbit (Figure 2.21).

The Gregorian calendar includes a system of **leap years**—years in which a 29th day is added to February—decreed by Julius Caesar in 45 BCE to make up for the extra fraction of a day. Leap years prevent the seasons from slowly sliding through the year to become increasingly out of sync with the months, so that we don't end up experiencing winter in December one year and in August in later years.

The Gregorian calendar is named for Pope Gregory XIII. He was concerned that the Easter holiday—the first Sunday after the first full Moon after March 21—was drifting away from the vernal equinox. Julius Caesar's rule of one leap year every 4 years resulted in an average year of 365.25 days, but the actual year is 365.242 days. That difference of 0.008 day is about 11.5 minutes per year, or about 3 days every 400 years, and by Gregory's time it had caused the date of the vernal equinox to drift in the Julian calendar by about 10 days. In 1582, therefore, Pope Gregory decreed that 10 days would be deleted from the calendar to move the vernal equinox back to March 21. To make that system work out better in the future, he declared that only century years divisible by 400 are leap years, thereby deleting 3 leap years (and the 3 days) every 400 years. Catholic countries followed that system immediately, but Protestant countries did not adopt it until the 1700s. Eastern Orthodox countries, including Russia, did not switch from the Julian to the Gregorian calendar until the 1900s. One slight further revision—making years divisible by 4000 into common 365-day years—has been proposed so the modern Gregorian calendar will slip by about only 1 day in 20,000 years.

what if . . .

What if the astronomical cycles were even multiples of each other, so that a cycle of lunar phases was precisely 30 days and Earth's orbital period was precisely 12 months? In such a system, how would the dates of "wandering" holidays, such as Chinese New Year or Ramadan, change?

Despite international adoption of the Gregorian calendar, billions of people still celebrate holidays and festivals according to a lunisolar or lunar calendar. Chinese New Year, Passover, Easter, Ramadan, Rosh Hashanah, and Diwali, among others, have dates that change from one year to the next because they are based on lunar months from lunisolar or lunar calendars. The astronomy of people from long ago is still in use today.

CHECK YOUR UNDERSTANDING 2.4

Suppose Earth orbited the Sun twice as fast as it currently does. Would the solar day be longer or shorter than it currently is, in that case?

2.5 Eclipses Result from the Alignment of Earth, the Moon, and the Sun

Ancient peoples must have been terrified to see an eclipse, as they watched the Sun or the Moon being eaten away as if by a giant dragon. An **eclipse** is the total or partial obscuration of one celestial body, or the light from that body, by another celestial body. In this section, we describe the types of eclipses and their frequency.

Solar Eclipses

A **solar eclipse** occurs when the Moon passes between Earth and the Sun; observers on Earth in the shadow of the Moon will see the eclipse. Three types of solar eclipses exist: *total*, *partial*, and *annular*. Consider the structure of the shadow of the Sun cast by a round object such as the Moon, as shown in **Figure 2.24**. An observer at

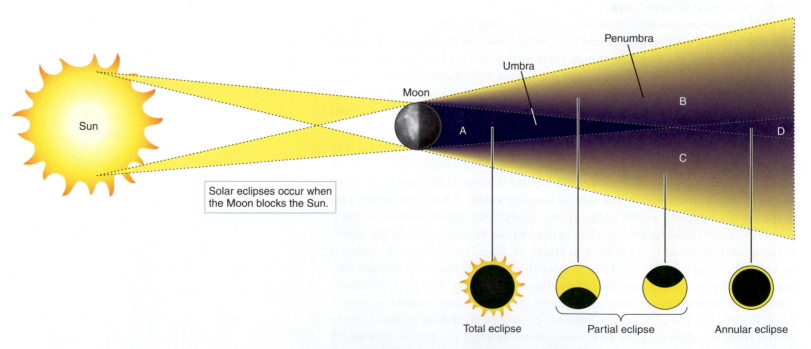

Figure 2.24 Different parts of the Sun are blocked at different places within the Moon's shadow. An observer on Earth in the umbra (point A) sees a total solar eclipse, observers in the penumbra (points B and C) see a partially eclipsed Sun, and observers at point D see an annular solar eclipse. Not to scale.

point A would see the Sun blocked. That darkest, inner part of the shadow is called the **umbra**. If a location on Earth passes through the Moon's umbra, the Moon blocks all the Sun's light, and a **total solar eclipse** will be observed (**Figures 2.25** and **2.26a**). At points B and C in Figure 2.24, an observer can see one side of the disk of the Sun but not the other. Only partially in shadow, that outer region is the **penumbra**. If a location on Earth passes through the Moon's penumbra, viewers there will observe a **partial solar eclipse**, in which the Moon's disk blocks the light from a portion of the Sun's disk.

In the third type of solar eclipse, called an **annular solar eclipse**, the Sun appears as a bright ring surrounding the dark disk of the Moon (**Figure 2.26b**). An observer at point D in Figure 2.24 is far enough from the Moon that it appears smaller than the Sun. An object's apparent size in the sky depends on the object's actual size and its distance from us. The Sun is about 400 times the diameter of the Moon, and the distance between the Sun and Earth is about 400 times more than the distance between the Moon and Earth. As a result, the Moon and Sun have almost the same apparent size in the sky. In addition, the Moon's orbit is not a perfect circle. When the Moon and Earth are a bit closer together than average, the Moon appears slightly larger in the sky than the Sun. An eclipse occurring then will be total for some observers. When the Moon and Earth are farther apart than average, the Moon appears smaller than the Sun, so eclipses occurring then will be annular for some observers. Among all solar eclipses, one-third are total at some location on Earth, one-third are annular, and one-third are only partial.

Figure 2.27 shows the geometry of a solar eclipse when the Moon's shadow falls on Earth's surface. Figures like this one usually show Earth and the Moon much closer together than they really are. The page is too small to draw them to scale and still show the critical details. The relative sizes and distances between Earth and the Moon are roughly equivalent to the difference between a basketball and a tennis ball, respectively, placed 7 meters apart. Figure 2.27b shows the geometry of a solar eclipse with Earth, the Moon, and the separation between them

Figure 2.25 The full spectacle of a total solar eclipse.

a.

b.

Figure 2.26 Time-lapse images of the Sun taken **a.** during a total solar eclipse and **b.** during the annular solar eclipse of May 20, 2012. The Sun set during the ending phases.

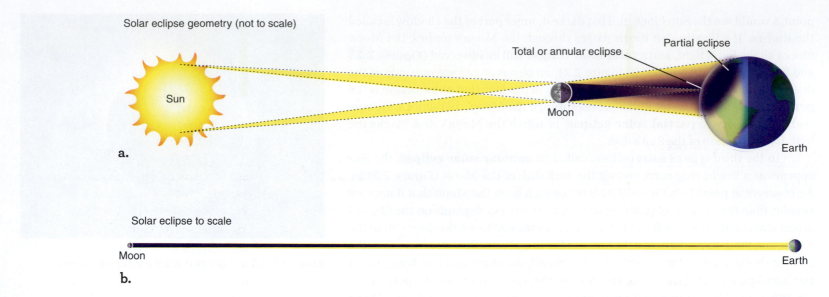

Figure 2.27 **a.** A solar eclipse occurs when the shadow of the Moon falls on the surface of Earth. **b.** The same as part a., but now the Moon and Earth are drawn to scale.

drawn to scale. Compare that drawing with Figure 2.27a to see why drawings of Earth and the Moon are rarely drawn to scale. If the Sun were drawn to scale in Figure 2.27a, it would be bigger than your head and located almost 64 meters off the left side of the page.

From any particular location, you are more likely to observe a partial solar eclipse than a total solar eclipse. Where the Moon's penumbra touches Earth, it has a diameter of almost 7000 km—large enough to cover a substantial fraction of Earth. Thus, a partial solar eclipse is often visible from many locations on Earth. In contrast, the path along which a total solar eclipse is visible, shown in **Figure 2.28**,

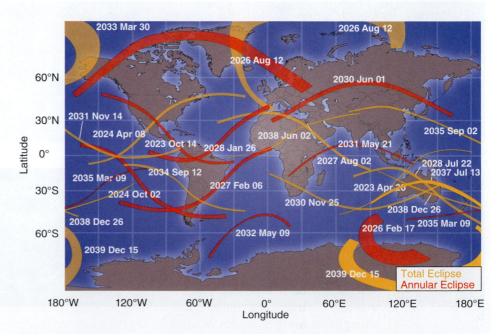

Figure 2.28 The paths of total solar eclipses through 2032. Solar eclipses occurring in Earth's polar regions cover more territory because the Moon's shadow hits the ground obliquely.

Reading Astronomy News

2500 Miles of Citizen Scientists

Laura Reed, sciencenode.org

Citizen scientists across the nation captured images of the Sun's corona to create the largest dataset of its kind.

On August 21 [2017] many of us stood outside and marveled at the solar eclipse.

But for some, the eclipse was more than just a show. It was an opportunity to participate in the National Solar Observatory (NSO) Citizen Continental-America Telescopic Eclipse (CATE) Experiment.

The Citizen CATE Experiment is funded by federal, corporate, and private sources, including several companies that donated equipment to the project. Volunteer observers and schools get to keep the equipment used in the study.

A grand experiment

Scientists, students, and volunteers assembled along the 2500-mile path of eclipse totality tracked the Sun using 68 identical telescopes.

Using specialized software and instrument packages, they produced more than 1000 images from the start of the partial solar eclipse until totality.

The goal was to capture 90 minutes of continuous, high-resolution, rapid-cadence images detailing a difficult-to-capture region of the solar atmosphere: the Sun's inner corona.

This is the first time scientists have collected research-quality observations of the corona during the eclipse's entire transit across the United States.

"This dataset is extraordinary," says Matt Penn, principal investigator for the project. "Normally during a solar eclipse, we get about 2 minutes of data in the region closest to the photosphere. But Citizen CATE allows us to get an hour and a half of data."

My corona

The corona is the Sun's outer atmosphere. It is difficult for scientists to study because the photosphere, or solar surface, is so bright that it overpowers the faint corona. We can only see the corona when something obscures the photosphere.

(Think of a light so bright that it makes it difficult to see an object close to you. Blocking the light with your hand allows a better view of the object.)

Scientists can create artificial eclipses using an instrument called a coronagraph that covers the Sun's bright disk. The proximity of the instrument to the observer distorts the Sun's edge, making precise observation and measurement difficult.

During a real eclipse, the Moon blocks the Sun. The Moon's great distance lets scientists measure and study the corona in greater detail.

The process

On the day of the event, skies were clear for 58 of the 68 observation sites.

Observers, spaced about 50 miles from each other, started observations when the Moon's shadow appeared on their horizon. They captured images every 10 seconds during totality.

The rapid cadence of imagery along with a 2-arcsecond pixel resolution should help scientists understand the intensity of the corona over an extended period as well as the motions of prominences, coronal inflows, coronal mass ejections, and other active regions.

The next total solar eclipse in the United States will be April 8, 2024. The path of totality will begin in Texas, moving north from Mexico, and will exit the U.S. via Maine.

Start making plans to join the next team of citizen scientists.

QUESTIONS

1. Why does a solar eclipse move across the surface of the Earth?

2. Why did scientists want data from along the whole track of the eclipse?

3. Why can the corona be seen only during an eclipse?

4. Would you have expected a lunar eclipse near the date of the solar eclipse? Explain.

5. See whether any results are available on the project website (http://citizencate.org/). What did participants learn from this experiment?

Source: https://sciencenode.org/feature/2,500-miles-of-citizen-scientists.php.

covers only a tiny fraction of Earth's surface. Even when the distance between Earth and the Moon is at a minimum, the umbra is only 270 km wide on Earth. As the Moon moves along in its orbit, that tiny shadow sweeps across Earth at a few thousand kilometers per hour. In addition, the Moon's shadow falls on Earth's curved surface, causing the region shaded by the Moon during a solar eclipse to

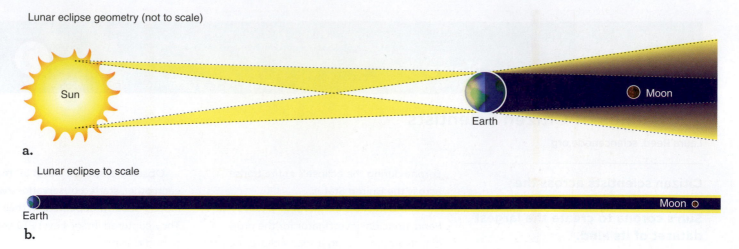

Lunar eclipse geometry (not to scale)

a.

Lunar eclipse to scale

b.

Figure 2.29 a. A lunar eclipse occurs when the Moon passes through Earth's shadow. **b.** The same as part a., but now Earth and the Moon are drawn to scale.

be elongated by various amounts. The curvature can even cause an eclipse that started out as annular to become total.

As a result of those factors, a total solar eclipse can never last longer than $7\frac{1}{2}$ minutes and is usually significantly shorter. Even so, it is one of the most amazing sights in nature. People all over the world flock to the most remote corners of Earth to witness the fleeting spectacle of the bright disk of the Sun blotted out of the daytime sky. Perhaps you saw the total eclipse visible from much of the United States in August 2017. The next total solar eclipse visible in the continental United States will take place in 2024. Annular eclipses will be visible from parts of the United States in 2021 and 2023. Viewing a solar eclipse should be on your lifetime to-do list!

Lunar Eclipses

Lunar eclipses occur when the Moon moves through Earth's shadow. **Figure 2.29a** shows the geometry of a lunar eclipse and is drawn to scale in **Figure 2.29b.** During a lunar eclipse, Earth is between the Sun and the Moon. Because Earth is much larger than the Moon, the dark umbra of Earth's shadow at the distance of the Moon is more than 2 times the Moon's diameter. A **total lunar eclipse**, in which the Moon is entirely within Earth's shadow, lasts as long as 1 hour 40 minutes. In a total lunar eclipse, the Moon often appears red (**Figure 2.30a**). That "blood-red Moon," as literature and poetry have called it, occurs because red light from the Sun is bent as it travels through Earth's atmosphere and then illuminates the Moon. Earth's atmosphere absorbs or scatters other colors of sunlight away from the Moon and therefore does not illuminate it.

A **penumbral lunar eclipse** occurs when the Moon passes through the penumbra of Earth's shadow; those are noticeable only from a very dark location or when the Moon passes within about 1000 km of the umbra. If Earth's shadow incompletely covers the Moon, some of the Moon's disk remains bright and some of it is in shadow. The result is a **partial lunar eclipse**. **Figure 2.30b** shows a composite of images taken at different times during a partial lunar eclipse. In the middle frame, Earth's shadow nearly completely eclipses the Moon.

unanswered questions

How long will Earth continue to have total solar eclipses? Those occur because the Moon and the Sun are coincidentally the same size in our sky, but will that always be the case? An object's observed size in the sky depends on its actual diameter and its distance from us. One or both of those can change. The Moon is slowly moving away from Earth by about 4 meters per century. Over time, the Moon will appear smaller in the sky, and it won't be able to cover the full disk of the Sun. Although we can measure how fast the Moon is moving away from Earth now, we are less certain of how that rate may change with time. A lesser and more uncertain effect comes from the Sun—which will continue to brighten slowly, as it has throughout its history. With that brightening, the Sun's actual diameter will slightly increase, and it will appear larger in our sky. A more distant Moon and a larger Sun will eventually (in hundreds of millions of years) end total eclipses on Earth.

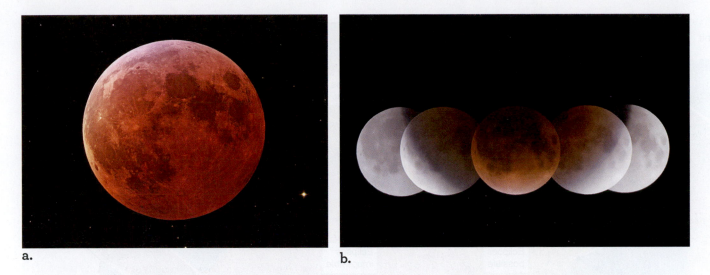

a.   b.

Figure 2.30 **a.** During a total lunar eclipse, the Moon often appears blood red. **b.** A time-lapse series of photographs of a partial lunar eclipse clearly shows Earth's shadow.

★ **WHAT AN ASTRONOMER SEES**   Astronomers learn to be sensitive to color variations. An astronomer looking at these two images for the first time would immediately conclude from the difference in color between the images that the Moon in a. is in full lunar eclipse. An astronomer would also notice in b. that the images of the Moon were carefully aligned, so that in each image Earth's shadow remained stationary. Combining the images in this way makes it possible to see (or measure) the size of Earth's shadow relative to the Moon.

Many more people have observed a total lunar eclipse than have observed a total solar eclipse. To see a total solar eclipse, you must be located within that very narrow band of the Moon's shadow as it moves across Earth's surface. In contrast, when the Moon is immersed in Earth's shadow, anyone located in the hemisphere facing the Moon can see it. As a result, total eclipses of the Moon are relatively common from any location, so you may have seen at least one.

Frequency of Eclipse Seasons

How could some people in ancient cultures predict eclipses? From their understanding of lunar and solar cycles for making calendars, they computed cycles of eclipses. Imagine Earth, the Moon, and the Sun all sitting on the same flat tabletop. If the Moon's orbit were in the same plane as Earth's orbit, the Moon would pass directly between Earth and the Sun at every new Moon. The Moon's shadow would pass across the face of Earth, and we would see a solar eclipse. Similarly, Earth would pass directly between the Sun and the Moon every synodic month, and a lunar eclipse would occur at each full Moon.

Solar and lunar eclipses do *not* happen every month, however, because the Moon's orbit does *not* lie in the same plane as Earth's orbit. As **Figure 2.31** shows, the plane of the Moon's orbit around Earth is inclined by about 5.2° with respect to the plane of Earth's orbit around the Sun. The line along which the orbital planes of the Sun and the Moon intersect is called the **line of nodes**. For part of the year, the line of nodes points generally toward the Sun. During those times, called **eclipse seasons**, a new Moon passes directly between the Sun and Earth, casting its shadow on Earth's surface and causing a solar eclipse. Similarly, a full Moon occurring during an eclipse season passes through Earth's shadow, causing a lunar eclipse. An eclipse season lasts only 38 days. That's how long the Sun is close enough to the line of nodes for eclipses to occur. Usually, the line of nodes

what if . . .

What if Earth had two moons, each having the same size and the same orbit, but located 120° apart in that orbit? Would we expect to see more solar eclipses in this case? Would the eclipses for both moons occur during the same eclipse seasons?

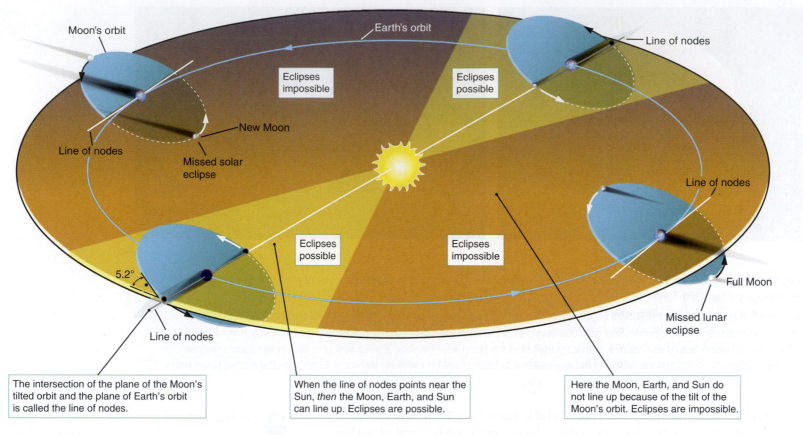

The intersection of the plane of the Moon's tilted orbit and the plane of Earth's orbit is called the line of nodes.

When the line of nodes points near the Sun, *then* the Moon, Earth, and Sun can line up. Eclipses are possible.

Here the Moon, Earth, and Sun do not line up because of the tilt of the Moon's orbit. Eclipses are impossible.

Figure 2.31 Eclipses are possible only when the Sun, the Moon, and Earth lie along (or very close to) an imaginary line known as the line of nodes. When the Sun does not lie along the line of nodes, Earth passes under or over the shadow of a new Moon, and a full Moon passes under or over the shadow of Earth.

points farther away from the Sun, so Earth, the Moon, and the Sun cannot line up closely enough for an eclipse to occur. At those times, a solar eclipse cannot take place because the shadow of a new Moon passes "above" or "below" Earth. Similarly, no lunar eclipse can occur because a full Moon passes "above" or "below" Earth's shadow.

If the plane of the Moon's orbit always had the same orientation, eclipse seasons would occur twice a year, as suggested in Figure 2.31. In actuality, eclipse seasons occur about every 5 months 20 days. The roughly 10-day difference occurs because the plane of the Moon's orbit slowly wobbles, much like the wobble of a spinning plate balanced on the end of a circus performer's stick. As the Moon's orbital plane wobbles, the line of nodes changes direction. That wobble rotates in the direction opposite the direction of the Moon's motion in its orbit. That is, the line of nodes moves clockwise as viewed from above Earth's orbital plane. One wobble of the Moon's orbit takes 18.6 years, so we say that the line of nodes regresses by 360° every 18.6 years, or 19.4° per year. That rate amounts to about a 20-day regression each year. If January 1 marks the middle of an eclipse season, the next eclipse season will be centered around June 20, and the one after that around December 10.

CHECK YOUR UNDERSTANDING 2.5

If the Moon were in its same orbital plane but twice as far from Earth, which of the following would happen? (a) The Moon would not go through phases. (b) Total eclipses of the Sun would not be possible. (c) Annular eclipses of the Sun would not be possible. (d) Total eclipses of the Moon would not be possible.

Origins: The Obliquity of Earth

Earth's various motions give rise to the most basic of patterns faced by life on Earth. Earth's rotation is responsible for the cycle of night and day. Earth's axial tilt and its passage around the Sun bring the change of the seasons. As life evolved on Earth, it had to adapt to those patterns.

Earth's range of climate, based on distance from the equator, probably has contributed to the broad diversity of life on our planet. Earth's biodiversity includes life that adapted to the long, cold polar nights and to the much higher temperatures of the tropics. Earth's life adapted to seasonal patterns in rain and drought, leading to acquired seasonal patterns of migration and reproduction. If Earth had no axial tilt, the poles would continuously be in winter and probably too cold for humans. Latitudes near the equator would be consistently even warmer than they are now.

Periodic changes in Earth's axial tilt might have affected life. If Earth's tilt were larger than 23.5°, the seasonal variation would be even stronger. If the tilt were smaller, the seasonal variation would be weaker. Chinese, Indian, Greek, and Arabic records going back 3000 years indicate that people estimated the tilt by measuring the length of the shadow from a vertical pole on the day of the solstice. Earth's axial tilt actually varies from 22.1° to 24.5° over a 41,000-year cycle. Currently, the tilt is about midway between the two extremes and getting smaller. It will reach its minimum value of 22.1° in about 10,000 years. Scientists are studying whether that variation in tilt correlates with periods of temperature change on Earth, especially the times of ice ages.

SUMMARY

The motions of Earth and the Moon are responsible for many of the repeating patterns that can be observed in the sky. Calendars keep track of time by using those patterns. Earth's rotation about its axis causes daily patterns of rising and setting, and Earth's revolution around the Sun causes yearly patterns of the stars in the sky and marks the passage of the seasons. Earth's tilt on its axis changes both the length of daytime and the intensity of sunlight, causing the seasons. Life on Earth adapted to those seasonal variations. Earth's axial tilt varies slightly over tens of thousands of years. The Moon's revolution around Earth causes the monthlong pattern of the phases of the Moon. Occasionally, alignments of Earth, the Moon, and the Sun cause eclipses.

(1) **Describe how Earth's rotation about its axis and its revolution around the Sun affect how we perceive celestial motions from different places on Earth.** Earth's daily rotation on its axis causes the apparent daily motion of the Sun, Moon, and stars. Our location on Earth and Earth's location in its orbit around the Sun determine which stars we see at night. You can determine your latitude from the altitude of the pole star. When observing objects in our sky, we need to consider how Earth moves in relation to other objects. The ecliptic is the path that the Sun appears to take through the stars.

(2) **Explain why seasons change throughout the year.** A year is the time Earth takes to complete one revolution around the Sun.

Constellations are patterns of stars that reappear in the same place in the sky at the same time of night each year. The tilt of Earth's axis determines the seasons by changing the angle at which sunlight strikes Earth's surface in different locations. The changing angle of sunlight and the differing length of the day cause the seasonal variations on Earth. Equinoxes and solstices mark the changing seasons.

(3) **Describe the factors that create the phases of the Moon.** The relative locations of the Sun, Earth, and the Moon determine the phases of the Moon. The Moon takes one sidereal month to complete one revolution around Earth and one synodic month to go through a cycle of phases. The Moon's motion around Earth causes it to be illuminated differently at different times. When farther from the Sun than Earth is, the Moon is in gibbous phases. When closer to the Sun than Earth is, the Moon is in crescent phases.

(4) **Sketch the alignment of Earth, the Moon, and the Sun during eclipses of the Sun and the Moon.** A solar eclipse occurs when the new Moon is in the plane of Earth and the Sun, and the shadow of the Moon falls on Earth. A lunar eclipse occurs when the full Moon is in the plane of Earth and the Sun, and the shadow of Earth falls on the Moon. Twice a year, at new or at full Moon, the Moon is exactly in line between Earth and the Sun. At those times, called eclipse seasons, eclipses can occur.

QUESTIONS AND PROBLEMS

TEST YOUR UNDERSTANDING

1. Constellations are groups of stars that
   a. are close to one another in space.
   b. are bound to each other by gravity.
   c. are relatively close to one another in Earth's sky.
   d. all have the same composition.

2. Where on Earth can you stand and, over the course of a year, see the entire sky?
   a. only at the North Pole
   b. at either pole
   c. at the equator
   d. anywhere

3. Day and night are caused by
   a. the tilt of Earth on its axis.
   b. the rotation of Earth on its axis.
   c. the revolution of Earth around the Sun.
   d. the revolution of the Sun around Earth.

4. Polaris, the North Star, is unique because
   a. it is the brightest star in the night sky.
   b. it is the only star in the sky that doesn't move throughout the night.
   c. it is always located at the zenith, for any observer.
   d. it has a longer path above the horizon than any other star.

5. An angle exists between the ecliptic and the celestial equator because
   a. Earth's axis is tilted with respect to its orbit.
   b. Earth's orbit is tilted with respect to the orbits of other planets.
   c. the Sun follows a rising and falling path through space.
   d. the Sun's orbit is tilted with respect to Earth's.

6. Earth's axial tilt causes the seasons because
   a. one hemisphere of Earth is closer to the Sun in summer.
   b. the days are longer in summer.
   c. the rays of light strike the ground more directly in summer.
   d. both a and b
   e. both b and c

7. Which of the following are true on the vernal and autumnal equinoxes? (Choose all that apply.)
   a. Every nonpolar place on Earth has 12 hours of daylight and 12 hours of darkness.
   b. The Sun rises due east and sets due west.
   c. The Sun is located on the celestial equator.
   d. The motion of the stars in the sky is different from that on other days.

8. We always see the same side of the Moon because
   a. the Moon does not rotate on its axis.
   b. the Moon rotates on its axis once for each revolution around Earth.
   c. the other side of the Moon is unlit when it is facing Earth.
   d. the other side of the Moon is on the opposite side of Earth when it is facing Earth.

9. If you see the Moon on the meridian at sunrise, the phase of the Moon is
   a. waxing gibbous.
   b. full.
   c. first quarter.
   d. third quarter.

10. A lunar eclipse occurs when _____ shadow falls on _____.
    a. Earth's; the Moon
    b. the Moon's; Earth
    c. the Sun's; the Moon
    d. the Sun's; Earth

11. Different cultures created different calendars because
    a. they had measured different lengths of the day, month, and year.
    b. they used different definitions of the day, month, and year.
    c. the number of days in a month and the number of days and months in a year are not integers.
    d. calendars are completely arbitrary.

12. Which stars we see at night depends on
    a. our location on Earth.
    b. Earth's location in its orbit.
    c. the time of the observation.
    d. all of the above.

13. On the summer solstice in June, the Sun will be directly above _____ and all locations north of _____ will experience daylight all day.
    a. the Tropic of Cancer; the Antarctic Circle
    b. the Tropic of Capricorn; the Arctic Circle
    c. the Tropic of Cancer; the Arctic Circle
    d. the Tropic of Capricorn; the Antarctic Circle

14. The Sun, Moon, and stars
    a. appear to move each day because the celestial sphere rotates about Earth.
    b. change their relative positions over time.
    c. rise north or south of west and set north or south of east, depending on their location on the celestial sphere.
    d. always remain in the same position with respect to one another.

15. If you see the first quarter Moon on the meridian, the Sun must be
    a. on the western horizon.
    b. on the eastern horizon.
    c. below the horizon.
    d. on the meridian.

THINKING ABOUT THE CONCEPTS

16. Seafarers such as Columbus used Polaris to navigate as they sailed from Europe to North America. When Magellan sailed the South Seas, he could not use Polaris. Explain why.

17. If you were standing at Earth's North Pole, where would you see the north celestial pole in relation to your zenith?

18. Figure 2.13 shows that observers in the Northern Hemisphere see the zodiacal constellation Gemini in the winter. Why don't they see it in the summer? ⓞ

19. Imagine that you are flying along in a jetliner.
    a. Describe ways to tell that you are moving.
    b. If you look down at a building, which way is it moving relative to you?

20. Astronomers are sometimes asked to serve as expert witnesses in court cases. Suppose you are called as an expert witness, and the defendant states that he could not see the pedestrian because the full Moon was casting long shadows across the street at midnight. Is that claim credible? Why or why not?

21. Imagine that one person was developing a theory of seasons as described in the three hypotheses in this chapter's Process of Science Figure. Compare that process with the flowchart of Chapter 1's Process of Science Figure. How would the development of that theory look on the diagram? 👁★

22. Why is the winter solstice *not* the coldest time of year?

23. Earth spins on its axis and wobbles like a top.
   a. How long does it take to complete one spin?
   b. How long does it take to complete one wobble?

24. At what approximate time of day do you see the full Moon near the meridian? At what time is the first quarter (waxing) Moon on the eastern horizon? Use a sketch to help explain your answers.

25. Assume that the Moon's orbit is circular, and imagine that you are standing on the side of the Moon that faces Earth.
   a. How would Earth appear to move in the sky as the Moon made one revolution around Earth?
   b. How would the "phases of Earth" appear to you compared with the phases of the Moon as seen from Earth?

26. If people on Earth were observing a lunar eclipse, what would you see from the Moon?

27. From any given location, why are you more likely to witness a partial eclipse of the Sun than a total eclipse?

28. Why don't we see a lunar eclipse each time the Moon is full or witness a solar eclipse each time the Moon is new?

29. ★ **WHAT AN ASTRONOMER SEES**   Figure 2.30b shows the shadow of Earth, projected onto the Moon over the course of a partial lunar eclipse. From this shadow, what can you determine about the shape of Earth? Many observations like this, taken from many different locations on Earth in different years, show the same shadow shape. From those observations, what can you determine about the shape of Earth? 👁★

30. How would Earth's temperature variation be different if its axis were tilted 90° to the plane of its orbit (like the planet Uranus)?

APPLYING THE CONCEPTS

31. If, as some historians believe, Eratosthenes used the Egyptian stadion, about 157.5 meters, what would he have computed for the size of Earth?
   a. In Working It Out 2.1, the circumference of Earth is calculated using 1 stadion = 185 m. Make a prediction: Do you expect your answer to be larger or smaller than the answer from Working It Out 2.1? Will it differ by a lot or a little?
   b. Follow Working It Out 2.1 to calculate the circumference of Earth.
   c. Check your work by comparing your answer to your prediction.

32. A point on the surface of Earth spins along at 1674 km/h at the equator. Use that fact, along with the length of the day, to calculate Earth's equatorial diameter. ●–●–●
   a. Study Working It Out 2.1 to make a prediction about how large Earth's diameter should be. Is it about 1000 km, about 10,000 km, or about 100,000 km?
   b. The units of speed given here are km/h. You have the hint about using the length of the day to calculate diameter. In what units should you write the length of the day (seconds, minutes, or hours)? Should you multiply or divide the speed by the length of the day to get a distance in km for an answer?
   c. Calculate Earth's diameter.
   d. Check your work by comparing the diameter you calculated with the radius from Working It Out 2.1.

33. The Moon's apparent diameter in the sky is approximately ½°. About how long does the Moon take to move a distance equal to its own diameter across the sky? (Hint: How long does it take the Moon to move 360°—that is, through one orbit?)
   a. Make a prediction: Compare ½° to 360°; do you expect it to take the Moon more or less time to move through ½° than 360°? Do you expect the difference in time to be large or small?
   b. Reason: The ratio of the angles should equal the ratio of the times, because the Moon is moving at constant speed through those angles.
   c. Calculate: Set up the ratio and calculate the time for the Moon to move through ½°.
   d. Check your work by comparing your answer to your prediction.

34. An object's apparent size in the sky is proportional to its actual diameter divided by its distance. The Moon has a radius of 1737 km, with an average distance of 3.780×10^5 km from Earth. The Sun has a radius of 696,000 km, with an average distance of 1.496×10^8 km from Earth. Show that the apparent sizes of the Moon and the Sun in our sky are approximately the same. ●–●–●
   a. Make a prediction: If the apparent sizes of the Moon and the Sun in the sky are approximately the same, what value would you expect to get if you divided one apparent size by the other?
   b. Calculate the apparent size of the Moon and the Sun, and divide the apparent size of the Moon by the apparent size of the Sun to find the ratio of sizes.
   c. Check your work by comparing your answer to your prediction.

35. Earth has an average radius of 6371 km, while the radius of the Moon is 1737 km. An object's apparent size in the sky is proportional to its actual diameter divided by its distance. If you were standing on the Moon, how much larger would Earth appear in the lunar sky than the Moon appears in our sky?
   a. Reason: How does the distance from Earth to the Moon compare to the distance from the Moon to Earth? In the calculation of the apparent size of Earth as seen from the Moon, then, what is the difference in the calculation of the apparent size of the Moon as seen from Earth? Set up an algebraic equation for the ratio of the apparent size of Earth (as seen from the Moon) to the apparent size of the Moon (as seen from Earth).
   b. Make a prediction: Compare the radius of Earth to the radius of the Moon. Do you expect the apparent size of Earth to be a little larger, a few times larger, or hundreds of times larger than the apparent size of the Moon?
   c. Calculate the ratio of the apparent size of Earth (as seen from the Moon) to the apparent size of the Moon (as seen from Earth).
   d. Check your work by comparing your result to your prediction.

36. Determine the latitude where you live. Draw and label a diagram showing that your latitude is the same as (a) the altitude of the north celestial pole and (b) the angle (along the meridian) between the celestial equator and your local zenith. What is the Sun's altitude at noon as seen from your home at the winter solstice and at the summer solstice?

37. If the angle between your zenith and Polaris is 40°, what is the altitude of Polaris? What is your latitude?

38. The southernmost star in a group of stars known as the Southern Cross lies approximately 65° south of the celestial equator. What is the farthest-north latitude from which the entire Southern Cross is visible? Can it be seen in any of the U.S. states? If so, which ones?

39. Imagine that you are standing on the South Pole at the time of the southern summer solstice.
    a. How far above the horizon will the Sun be at noon?
    b. How far above (or below) the horizon will the Sun be at midnight?

40. Suppose the tilt of Earth's equator with respect to its orbit were 10° instead of 23.5°. At what latitudes would the Arctic and Antarctic circles and the Tropics of Cancer and Capricorn be located?

41. The Moon's orbit is tilted by about 5° with respect to Earth's orbit around the Sun. What is the highest altitude in the sky that the Moon can reach, as seen in Philadelphia (latitude 40° north)?

42. Suppose you would like to witness the midnight Sun (when the Sun appears just above the northern horizon at midnight), but you don't want to travel any farther north than necessary.
    a. How far north (that is, to which latitude) would you have to go?
    b. At what time of year would you make that trip?

43. ⊙
    a. The vernal equinox is now in the zodiacal constellation Pisces. The precession of Earth's axis will eventually cause the vernal equinox to move into Aquarius. How long, on average, does the vernal equinox spend in each zodiacal constellation?
    b. Stonehenge was erected roughly 4000 years ago. Referring to the zodiacal constellations shown in Figure 2.13, identify the constellation in which those ancient builders saw the vernal equinox.

44. Use Figure 2.18 to estimate when Vega, the fifth-brightest star in our sky (excluding the Sun), will once again be the northern pole star. ⊙

45. How would the length of the eclipse season change if the plane of the Moon's orbit were inclined less than its current 5.2° to the plane of Earth's orbit? Explain your answer.

EXPLORATION Phases of the Moon

digital.wwnorton.com/astro7

In this Exploration, we will be examining the phases of the Moon. Visit the Student Site at the Digital Resources page and open the Phases of the Moon Interactive Simulation in Chapter 2. This simulator animates the orbit of the Moon around Earth, allowing you to control the simulation speed and a number of other parameters.

Begin by starting the animation to explore how it works. Examine all three image frames. The large frame shows the Earth-Moon system, as looking down from far above Earth's North Pole. The upper right frame shows what the Moon looks like to the person on the ground. The lower right frame shows where the Moon appears in the person's sky. Stop the animation, and press "Reset Animation."

1. What time of day is this for the person shown on Earth?

_____

_____

_____

_____

2. What phase is the Moon in?

_____

_____

_____

_____

3. Where is the Moon in this person's sky?

_____

_____

_____

_____

Run the animation until the Moon reaches waxing crescent phase.

4. As viewed from Earth, which side of the Moon is illuminated (the left or the right)?

_____

_____

_____

_____

5. The person shown on Earth will observe this waxing crescent Moon either after sunset or before sunrise. At which of these times can the person see the waxing crescent Moon?

_____

_____

_____

_____

Run the animation until the Moon reaches first quarter and the Sun is setting for the person on Earth. (Hint: You may want to slow the animation rate!)

6. How many full days have passed since new Moon?

_____

_____

_____

_____

7. At this instant, where is the first quarter Moon in the person's sky?

_____

_____

_____

_____

8. If an astronaut were standing on the near side of the Moon at this time, what phase of Earth would he see?

_____

_____

_____

_____

Three observations about the phases of the Moon are connected: the location of the Moon in the sky, the time for the observer, and the phase of the Moon. If you know two of these, you can figure out the third. Use the animation to fill in the missing pieces in the following situations:

9. An observer sees the Moon in _____ phase, overhead, at midnight.

10. An observer sees the Moon in third quarter phase, rising in the east, at _____.

11. An observer sees the Moon in full phase, _____, at 6:00 A.M.

Motion of Astronomical Bodies

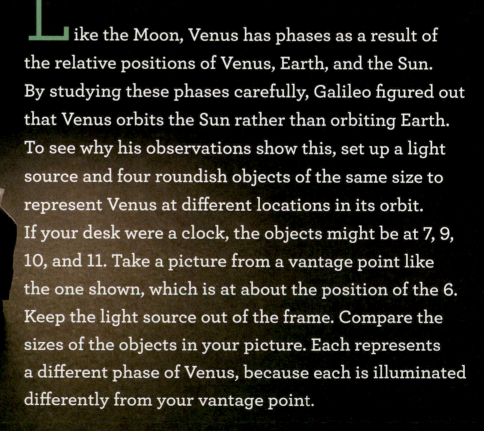

Like the Moon, Venus has phases as a result of the relative positions of Venus, Earth, and the Sun. By studying these phases carefully, Galileo figured out that Venus orbits the Sun rather than orbiting Earth. To see why his observations show this, set up a light source and four roundish objects of the same size to represent Venus at different locations in its orbit. If your desk were a clock, the objects might be at 7, 9, 10, and 11. Take a picture from a vantage point like the one shown, which is at about the position of the 6. Keep the light source out of the frame. Compare the sizes of the objects in your picture. Each represents a different phase of Venus, because each is illuminated differently from your vantage point.

EXPERIMENT SETUP

Use a desk lamp and a few roundish objects to set up a model of the orbit of Venus. Make a prediction: How will the size of the crescent phase "Venus" compare to the size of "Venus" showing nearly full phase? Take a picture, keeping the light source out of the frame.

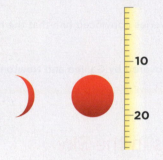

Using your picture, measure the size of "Venus" at crescent and full phase. What does this tell you about the relative distances of these objects from your camera? The real Venus also behaves this way, so what can you conclude about the orbit of Venus? Specifically, is it centered on the Sun or centered on Earth?

PREDICTION

I predict that when the object is in "full" phase, it will appear

☐ larger   ☐ smaller

than when it is at "crescent" phase.

SKETCH OF RESULTS (in progress)

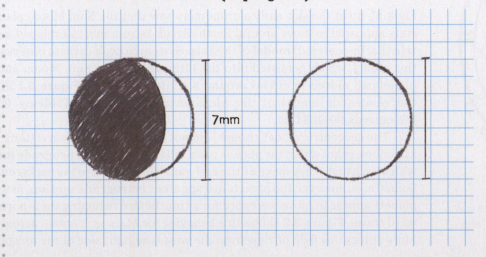

7mm

3

The birth of modern astronomy dates back to when astronomers and mathematicians discovered regular patterns in the motions of the planets. A successful theory of how Earth moved and how it fit in with its neighbors in the Solar System was the first step toward understanding our planet's place in the universe.

LEARNING GOALS

Here in Chapter 3, we examine how astronomers came to understand that Earth and other planets orbit the Sun. By the end of this chapter, you should be able to:

(1) Compare the geocentric and heliocentric models of the Solar System.

(2) Use Kepler's laws to describe how objects in the Solar System move.

(3) Explain how Galileo's astronomical discoveries convinced him that the heliocentric model was correct.

(4) Describe the physical laws of motion discovered by Galileo and Newton.

3.1 The Motions of Planets in the Sky

When people in ancient times looked up at the sky, they saw the Sun, Moon, and stars rise in the east and set in the west and *appear* to move around Earth. The ancient peoples were aware of five planets (*planet* means "wandering star") because from one night to the next they moved in a generally eastward direction among the stars, whose positions appeared fixed on the celestial sphere. One thing the ancients did *not* know, however, was that Earth was similar to those planets. Developing a successful theory of how Earth and the planets move and how Earth fits in with its neighbors in the Solar System was the first step to understanding Earth's place in the universe. The history of how those ideas evolved—from Earth at the center of all things to Earth as just an ordinary planet—is a good example of how science is self-correcting.

The Geocentric Model

Looking up at the sky, early astronomers saw that the Sun, Moon, planets, and stars appeared to move around Earth. Greek astronomers developed a **geocentric** (Earth-centered) **model** of the Solar System to explain these observations. In this model, which persisted into the 17th century, the Sun, Moon, and known planets all moved in circles around a stationary Earth. **Figure 3.1** illustrates the geocentric model of the Alexandrian astronomer Ptolemy (Claudius Ptolemaeus, 90–168 CE).

The geocentric model, though, did not account for all observations. Ancient astronomers knew that the planets usually have an eastward **prograde motion**, in which each night they move a little eastward with respect to the background stars. Sometimes, however, the planets had apparent **retrograde motion**, in which the planets appear to move westward for a time before resuming their normal eastward travel. **Figure 3.2** shows this behavior for Mars as it moves across the sky. Mars enters this field of view from the west (the right side of the image), makes

Figure 3.1 In the Ptolemaic view of the heavens, Earth is at the center, orbited by the Moon, Mercury, Venus, the Sun, Mars, Jupiter, and Saturn.

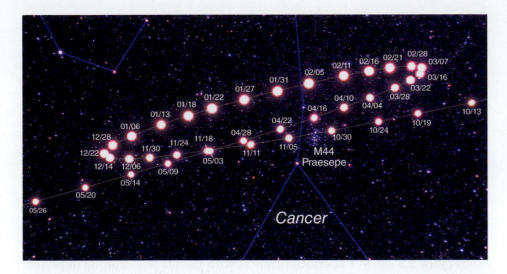

Figure 3.2 This time-lapse photographic series shows Mars as it moves in apparent retrograde motion.

★ **WHAT AN ASTRONOMER SEES** An astronomer will notice from the dates shown that the photographic series begins with Mars on the right side of the image. As time passes, Mars moves toward the left, then completes the loop, and leaves the image on the left, about 7.5 months later. An astronomer will also notice that Mars is larger and brighter on the image in the retrograde part of the loop, indicating that Earth and Mars are closer together at that time. This color image clearly shows that some stars are red and some stars (like the ones in the group of stars called M44) are blue.

a loop, and leaves the field of view to the east (the left side of the image) about seven months later.

The apparent retrograde motion of the five "naked eye" planets—Mercury, Venus, Mars, Jupiter, and Saturn—created a puzzling problem for Ptolemy's geocentric model, as it stood in 150 CE. Because the geocentric model in its simplest form failed to explain the apparent retrograde motion of the planets, Ptolemy added an embellishment called an *epicycle*—a small circle superimposed on each planet's larger circle (**Figure 3.3**). As the planet travels along its larger circle around Earth, it also moves along its epicycle. When moving along the smaller circle in the opposite direction of the forward motion of the larger circle, the planet's forward motion would be reversed and it would appear to move backward in the sky. This embellishment made the model reasonably successful at predicting the positions of planets in the sky. For nearly 1500 years, Ptolemy's model, in which the Sun, Moon, and planets all moved in perfect circles around a stationary Earth, with the "fixed stars" located far beyond the planets, was the accepted paradigm in the Western world.

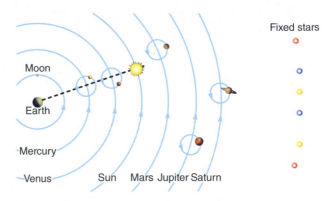

Figure 3.3 To reconcile retrograde motion with the geocentric model of the Solar System, Ptolemy added loops called epicycles to each planet's circular orbit around Earth.

CHECK YOUR UNDERSTANDING 3.1a

How did the ancients know the planets were different from the stars?

Answers to Check Your Understanding questions are in the back of the book.

Copernicus Proposes a Heliocentric Model

Nicolaus Copernicus (1473–1543—**Figure 3.4**) was not the first person to suggest that the Sun might be at the center of the Solar System. A few ancient Greek and medieval Arab astronomers, such as Aristarchus of Samos (310–230 BCE), had briefly considered the idea when they saw problems with the Earth-centered model. These astronomers lacked the observational or mathematical tools to test the Sun-centered hypothesis, and most astronomers found the fact that they could not feel Earth's motion around the Sun to be a powerful argument in favor of the Earth-centered model.

Copernicus, however, was the first to develop a comprehensive mathematical model with the Sun at the center and that later astronomers could test. That **heliocentric** (Sun-centered) **model** was the beginning of the Copernican

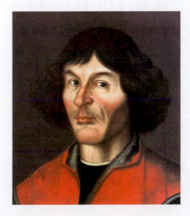

Figure 3.4 Nicolaus Copernicus rejected the ancient Greek model of an Earth-centered universe and replaced it with a model that centered on the Sun.

○
what if . . .

What if we launch an Earth satellite that orbits at half the distance to the Moon? Would astronauts aboard that satellite observe retrograde motion of the Moon?

Revolution. Through the work of 16th- and 17th-century scientists such as Tycho Brahe, Galileo Galilei, Johannes Kepler, and Isaac Newton, the heliocentric model of the Solar System became one of the best-corroborated theories in all of science.

Copernicus was multilingual and highly educated: he studied philosophy, canon (Catholic) law, medicine, economics, mathematics, and astronomy in his native Poland and in Italy. He conducted astronomical observations from a small tower, and sometime around 1514 he started writing about a heliocentric model. Eighteen years later, he completed his manuscript. He did not publish the book because he knew his ideas would be controversial: philosophical and religious views of the time held that humanity, and thus Earth, must be the center of the universe. Late in his life, Copernicus was finally persuaded to publish his ideas, and his great work *De revolutionibus orbium coelestium* ("On the Revolutions of the Heavenly Spheres") appeared in 1543, the year he died.

Figure 3.5 shows Copernicus's model with the planets orbiting in perfect circles around the Sun. That model explained the observed motions of Earth, the Moon, and the planets, including apparent retrograde motion, much more simply than the geocentric model did. In this model, relative motion is important; when you are in a car or train and you pass a slower-moving car or train, the other vehicle seems as though it is moving backward.

Similarly, in the Copernican model, planets farther from the Sun appear to have retrograde motion when Earth overtakes them in their orbits. **Figure 3.6** illustrates that effect for Mars; compare this diagram to the observations shown in Figure 3.2. Conversely, planets closer to the Sun than Earth is—Mercury and Venus—appear to have retrograde motion when overtaking Earth. Except for the Sun and the Moon, all Solar System objects exhibit apparent retrograde motion. The effect diminishes with increasing distance from Earth. Apparent retrograde motion is caused by the relative motion between Earth and the other planets.

Although Copernicus correctly placed the Sun at the center of the Solar System, he still conceived of the planets as moving in perfectly circular orbits at constant speeds, so he needed to use some epicycles to match the observations. His model made testable predictions of the location of each planet on a given night. Those predictions were at least as accurate as those of the geocentric model. Overall, however, the heliocentric model was simpler than the geocentric model and became the basis for further refinements in understanding how Earth moved. As copies of *De Revolutionibus* and Copernicus's ideas slowly spread across Europe, other scientists began to consider and then accept the heliocentric model, and a scientific revolution began.

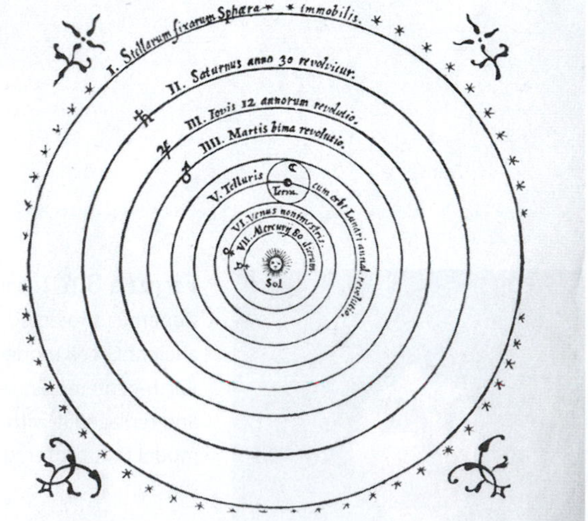

Figure 3.5 The Copernican heliocentric view of the Solar System (II–VII) and the fixed stars (I). The Sun is at the center, orbited by Mercury, Venus, Earth, Mars, Jupiter, and Saturn. The Moon orbits Earth.

Scaling the Solar System

From his observations, Copernicus deduced the correct order of the planets and concluded that planets closer to the Sun travel faster than planets farther out. He also realized that he needed to consider two categories of planets: **inferior planets**, which are closer to the Sun than Earth is, and **superior planets**, which are farther from the Sun than Earth is.

In Copernicus's model, Earth, another planet, and the Sun periodically align in space to form either a line or a right triangle. There are special words for each of these alignments, as shown in **Figure 3.7**. When a superior planet is in line with the Sun and Earth but on the other side of the Sun from Earth, the planet is in *conjunction* (Figure 3.7a) A superior planet in conjunction rises and sets in the sky with the Sun. When a superior planet is in conjunction, it is at its farthest from Earth, so it is at its faintest, too. Exactly at conjunction, however, you won't see the superior planet at all because it's behind the Sun in the sky.

In contrast, when a superior planet is in line with the Sun and Earth on the same side of the Sun as Earth, the configuration is an *opposition* (Figure 3.7a). At opposition, the superior planet is "opposite" the Sun in the sky. Like a full Moon, the planet rises when the Sun sets and sets when the Sun rises. When the superior planet is in opposition, it is at its closest to Earth during that orbit and thus at its brightest; therefore, opposition is the best time to observe the planet in the sky. Opposition occurs during the time when the planet exhibits retrograde motion because that is exactly when Earth is overtaking the planet in its orbit. *Quadrature* occurs when Earth, the Sun, and a superior planet form a right triangle in space (Figure 3.7a).

For an inferior planet, the configurations are slightly different (Figure 3.7b). When the inferior planet is between Earth and the Sun, it is closest to Earth and the configuration is called *inferior conjunction*. If the inferior planet is on the other side of the Sun from Earth, it is farthest from Earth, and the configuration is called a *superior conjunction*. *Greatest elongation* occurs when the inferior planet forms a right triangle with Earth and the Sun, and thus is the farthest it gets from the Sun in the sky. The inner planets are always close to the Sun in the sky, so Mercury and Venus are visible only within a few hours of sunrise or sunset. The best time to observe those planets is at greatest elongation because they will have the greatest separation from the Sun in the sky.

Copernicus realized that two types of orbital periods existed. We name these periods with terms similar to those used for lunar orbits. A planet's *sidereal period* is how long the planet takes to make one orbit around the Sun with respect to the stars and return to the same point in space. A planet's *synodic period* is how long the planet takes to return to the same configuration with the Sun and Earth, such as from inferior conjunction to inferior conjunction or from opposition to opposition.

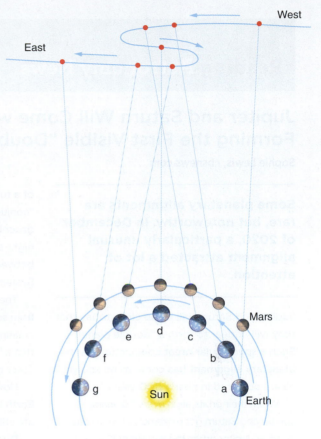

Figure 3.6 The Copernican model explains the apparent retrograde motion of Mars (see Figure 3.2) as seen in Earth's sky when Earth passes Mars in its orbit. (Not drawn to scale.)

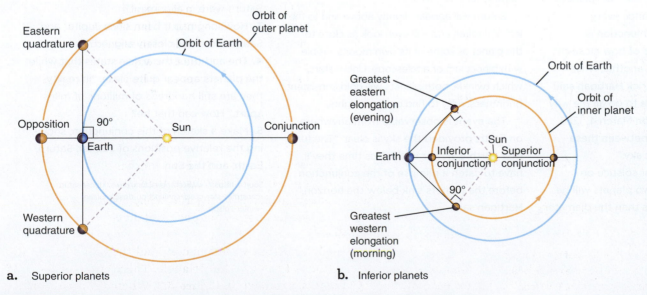

a. Superior planets

b. Inferior planets

Figure 3.7 Planetary configurations for **a.** superior (outer) planets and **b.** inferior (inner) planets.

Reading Astronomy News

Jupiter and Saturn Will Come within 0.1 Degrees of Each Other, Forming the First Visible "Double Planet" in 800 Years

Sophie Lewis, cbsnews.com

Some planetary alignments are rare, but noteworthy. In December of 2020, a particularly unusual alignment attracted a lot of attention.

Jupiter and Saturn will be so close today that they will appear to form a "double planet." Such a spectacular great conjunction, as the planetary alignment has come to be known, hasn't occurred in nearly 800 years.

When their orbits align every 20 years, Jupiter and Saturn get extremely close to one another. Jupiter orbits the sun every 12 years, while Saturn's orbit takes 30 years, so every few decades Jupiter laps Saturn, according to NASA.

The 2020 great conjunction is especially rare—the planets haven't been this close together in nearly 400 years, and haven't been observable this close together at night since medieval times, in 1226.

"Alignments between these two planets are rather rare, occurring once every 20 years or so, but this conjunction is exceptionally rare because of how close the planets will appear to one another," Rice University astronomer Patrick Hartigan said in a statement. "You'd have to go all the way back to just before dawn on March 4, 1226, to see a closer alignment between these objects visible in the night sky."

Aligning with the winter solstice on December 21, 2020, the two planets will be just 0.1 degrees apart—less than the diameter

of a full Moon, EarthSky said. The word "conjunction" is used by astronomers to describe the meeting of objects in our night sky, and the great conjunction occurs between the two largest planets in our Solar System: Jupiter and Saturn.

The planets will be so close, they will appear, from some perspectives, to overlap completely, creating a rare "double planet" effect. So close, that a "pinkie finger at arm's length will easily cover both planets in the sky," NASA said.

However, while they may appear from Earth to be very, very close, in reality, they are still hundreds of millions of miles apart.

During the last great conjunction in 2000, Jupiter and Saturn were so close to the Sun that the event was difficult to observe. But skywatchers should have a clearer view of the celestial event this time around. The great conjunction will be shining bright shortly after sunset, low in the southwestern sky, as viewed from the Northern Hemisphere, NASA said.

Saturn will appear slightly above and to the left of Jupiter, and will even look as close to the planet as some of its own moons, visible with binoculars or a telescope. Unlike stars, which twinkle, both planets will hold consistent brightness, easy to find on clear nights.

The event is observable from anywhere on Earth, provided the sky is clear. "The further north a viewer is, the less time they'll have to catch a glimpse of the conjunction before the planets sink below the horizon," Hartigan said.

The planets will appear extremely close for about a month, giving skywatchers plenty of time to witness the spectacular alignment throughout the holiday season. The event coincidentally aligns with the December solstice, marking the shortest day of the year in the Northern Hemisphere.

This will be the "greatest" great conjunction for the next 60 years, until 2080. Hartigan said that, following that conjunction, the duo won't make such a close approach until sometime after the year 2400.

QUESTIONS

1. The use of the word "conjunction" in this context is more general than the formal usage introduced in the text of this chapter. In the chapter, we use "conjunction" to mean "solar conjunction." Compare that usage with the usage in this article. What does the term mean more generally?

2. How long had it been since Jupiter and Saturn were in alignment?

3. How long had it been since Jupiter and Saturn were this closely aligned?

4. The author of the article states that while the planets appear quite close, "in reality, they are still hundreds of millions of miles apart." How can that be?

5. Make a sketch of this conjunction, showing the relative positions of Jupiter, Saturn, Earth, and the Sun.

Source: https://www.cbsnews.com/news/jupiter-saturn-christmas-star-great-conjunction-double-planet-winter-solstice/.

The synodic period is what can be observed directly from Earth. In **Figure 3.8a**, Earth and the superior planet are in opposition at point A. Superior planets move around the Sun more slowly than Earth does, so Earth orbits the Sun once and then catches up to the superior planet to form the next opposition at point B. In **Figure 3.8b**, Earth and the inferior planet are in inferior conjunction at point A. An inferior planet moves around the Sun faster than Earth does, so it completes one sidereal period and then must continue in its orbit to catch up to Earth for the next inferior conjunction at point B.

Copernicus used the geometry of these alignments along with his observations of the positions of the planets in the sky, including their altitudes and the times they rose and set, to estimate the planet–Sun distances. He could not figure this out in terms of meters or feet, but instead determined these distances in multiples of the Earth–Sun distance. The average distance from Earth to the Sun is so useful that it has its own unit, the **astronomical unit** (**AU**). **Table 3.1** shows that the relative distances Copernicus calculated are remarkably close to the distances modern methods yield. **Working It Out 3.1** describes the numerical details. Copernicus's model not only predicted planetary positions in the sky but also could be used to accurately compare the distances between the planets and the Sun.

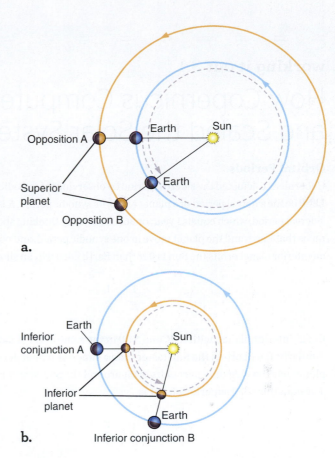

a.

b.

CHECK YOUR UNDERSTANDING **3.1b**

The planet Uranus will be observed in retrograde motion when: (a) Uranus is closest to the Sun; (b) Uranus is farthest from the Sun; (c) Earth overtakes Uranus in its orbit; (d) Uranus overtakes Earth in its orbit.

Figure 3.8 The synodic periods of planets indicate how long they take to return to the same configuration with Earth and the Sun. **a.** Earth completes one orbit around the Sun first and then catches up to the superior planet. **b.** Inferior planets complete a full orbit around the Sun first and then catch up to Earth.

3.2 Kepler's Laws Describe Planetary Motion

Copernicus did not understand *why* the planets move about the Sun, but he realized that his heliocentric model offered a way to compute the planets' relative distances. His theory is an example of **empirical science**, which seeks to describe patterns in nature as accurately as possible even if it's not yet possible to explain why those patterns exist. Copernicus's work was revolutionary because he challenged the accepted geocentric model and proposed that Earth is just one planet among many. His conclusions paved the way for other great empiricists, including Tycho Brahe and Johannes Kepler.

Tycho Brahe's Observations

Tycho Brahe (1546–1601—**Figure 3.9**) was a Danish astronomer of noble birth who entered university at age 13 to study philosophy and law. After seeing a partial solar eclipse in 1560, Tycho (conventionally referred to by his first name) became interested in astronomy. A few years later, he observed Jupiter and Saturn near each other in the sky, though not exactly where they were expected to be from the astronomical tables based on Ptolemy's model. Tycho gave up studying law and devoted himself to making better tables of the planets' positions in the sky.

The king of Denmark granted Tycho the island of Hven, between Sweden and Denmark, to build an observatory. Tycho designed and built new instruments,

Table 3.1	Copernicus's Scale of the Solar System	
Planet	**Copernicus's Value (AU)**	**Modern Value (AU)**
Mercury	0.38	0.39
Venus	0.72	0.72
Earth	1.00	1.00
Mars	1.52	1.52
Jupiter	5.22	5.20
Saturn	9.17	9.58

AU = Astronomical Unit.

working it out 3.1

How Copernicus Computed Orbital Periods and Scaled the Solar System

Orbital Periods

Copernicus calculated the sidereal period by observing the synodic period. Let P be the sidereal period of a planet and S, its synodic period. E is Earth's sidereal period, which equals 1 year, or 365.25 days. By thinking about the distance that Earth and the planet move in one synodic period, and noting that an inferior planet orbits the Sun faster than Earth does, we can show that

$$\frac{1}{P} = \frac{1}{E} + \frac{1}{S}$$

for an inferior planet, with P, E, and S all in the same units of days or years. Similarly, Earth orbits the Sun faster than a superior planet does, so the planet has traveled only part of its orbit around the Sun after 1 Earth year. The equation for a superior planet is

$$\frac{1}{P} = \frac{1}{E} - \frac{1}{S}$$

For Saturn, the time that passes between oppositions—the date of maximum brightness—shows that Saturn's synodic period (S) is 378 days, or $378 \div 365.25 = 1.035$ years. Then, to compute Saturn's sidereal period (P) in years, we use $S = 1.035$ yr and $E = 1$ yr in the equation for a superior planet:

$$\frac{1}{P} = \frac{1}{1\,\text{yr}} - \frac{1}{1.035\,\text{yr}} = 1 - 0.966\,\text{yr}^{-1} = 0.034\,\text{yr}^{-1}$$

Thus,

$$P = \frac{1}{0.034\,\text{yr}^{-1}} = 29.4\,\text{yr}$$

Saturn's sidereal period is 29.4 years, which means that Saturn takes 29.4 years to travel around the Sun and return to where it started in space.

Scaling the Solar System

Copernicus used the configurations of the planets shown in Figure 3.7 along with their sidereal periods to compute the relative distances of the planets. For the superior planets, he measured the fraction of the circular orbit that the planet completed in the time between opposition and quadrature, and then he used trigonometry to solve for the planet–Sun distance in astronomical units (see Figure 3.7a). For the inferior planets, he had a right triangle at the point of greatest elongation, and then he used right-triangle trigonometry to solve for the planet–Sun distance in astronomical units (see Figure 3.7b). Copernicus's values are impressively similar to modern values (see Table 3.1). Copernicus still did not know the actual value of the astronomical unit in miles or kilometers, but he was the first to accurately compute the relative distances of the planets from the Sun.

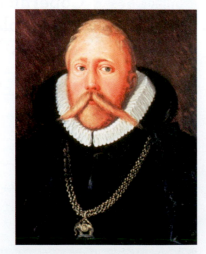

Figure 3.9 Tycho Brahe, known commonly as Tycho, was one of the greatest astronomical observers before the invention of the telescope.

operated a printing press, and taught students and others how to conduct observations. With the help of his sister Sophie, Tycho carefully measured the precise positions of planets in the sky over several decades, developing the most comprehensive set of planetary data then available. He created his own geo-heliocentric model, shown in **Figure 3.10**. In Tycho's model, the planets orbit the Sun, and the Sun and Moon orbit Earth. His model gained some acceptance among people who preferred to keep Earth at the center for philosophical or religious reasons. Tycho lost his financial support when the king died, and in 1600 he relocated to Prague.

Kepler's Laws

In 1600, Tycho hired a more mathematically inclined astronomer, Johannes Kepler (1571–1630—**Figure 3.11**), as his assistant. Kepler, who had studied Copernicus's ideas, was responsible for the next major step toward understanding the planets' motions. When Tycho died, Kepler inherited the records of his observations. Working first with Tycho's observations of Mars, Kepler deduced three rules that accurately describe how the planets move. These are now generally referred to as **Kepler's laws**. Kepler's laws are empirical: they use existing data to make predictions about future behavior but do not include an underlying theory of why the objects behave as they do.

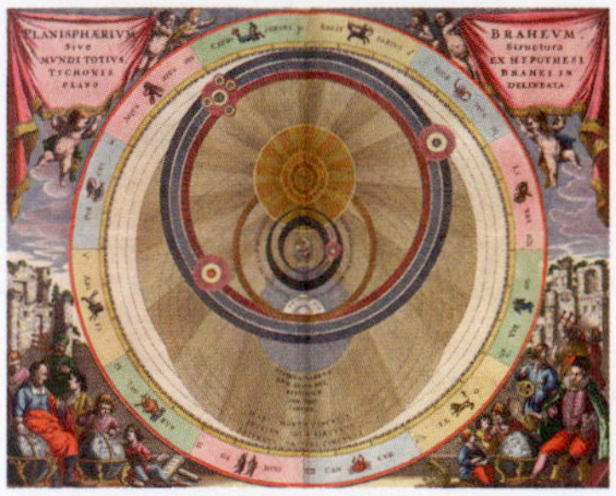

Kepler's First Law   Comparing Tycho's extensive planetary observations with predictions from Copernicus's heliocentric model, Kepler expected the data to confirm circular orbits for planets orbiting the Sun. Instead, he found disagreements between his predictions and Tycho's observations. He was not the first to notice such discrepancies. Rather than discard Copernicus's model, Kepler revised it.

By replacing Copernicus's circular orbits with *elliptical* orbits, Kepler could predict the positions of planets for any day, and his predictions fit Tycho's observations almost perfectly. An **ellipse** is an oval that is symmetric from right to left and from top to bottom. As shown in **Figure 3.12a**, you can draw an ellipse by attaching the two ends of a piece of string to a piece of paper, stretching the string tight with the tip of a pencil, and then drawing around those two points while keeping the string tight. Each point at which the string is attached to the paper is a **focus** (plural: **foci**) of the ellipse. If the two foci are close together, then the ellipse is more circular (**Figure 3.12b**). If the two foci are farther apart, the ellipse is more elongated. The long axis of the ellipse is called the **major axis**, and the short axis is called the **minor axis**. The **eccentricity** (*e*) of an ellipse measures that elongation; it is determined by the distance from the center of the ellipse to a focus, divided by half the length of the major axis. A circle has an eccentricity of 0 because the two foci coincide at the center. The more elongated the ellipse becomes, the closer its eccentricity gets to 1.

Kepler's first law of planetary motion states that the orbit of each planet is an ellipse with the Sun located at one focus. The other focus, as shown in **Figure 3.13**, is nothing but empty space. The ellipse in Figure 3.13 has very high eccentricity, compared to actual planetary orbits, in order to better distinguish its features. The dashed lines represent the major and minor axes of the ellipse. Half the length of the major axis of the ellipse is called the **semimajor axis**, *A*. The average distance between the Sun and a planet is equal to the semimajor axis of the planet's orbit.

Figure 3.10 Tycho's geo-heliocentric model, showing the Moon and Sun orbiting Earth, with the other planets orbiting the Sun.

Figure 3.11 Johannes Kepler explained the motions of the planets with three empirically determined laws.

▶❚❚ **AstroTour:** Kepler's Laws

▶▶ **Interactive Simulation:** Planetary Orbits

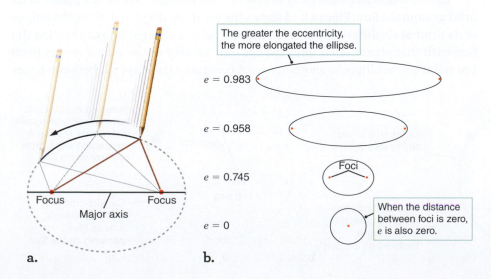

The greater the eccentricity, the more elongated the ellipse.

e = 0.983

e = 0.958

e = 0.745

e = 0

Foci

When the distance between foci is zero, *e* is also zero.

Focus   Focus

Major axis

a.   **b.**

Figure 3.12 a. We can draw an ellipse by attaching a length of string to a piece of paper at two points (called foci) and then pulling the string around as shown. The long axis is called the major axis. **b.** Ellipses range from circles (*e* = 0) to elongated eccentric shapes. *e* = eccentricity.

Figure 3.13 According to Kepler's first law, planets move in elliptical orbits with the Sun at one focus. (Nothing is at the other focus.) The orbit's eccentricity is the center-to-focus distance divided by the semimajor axis.

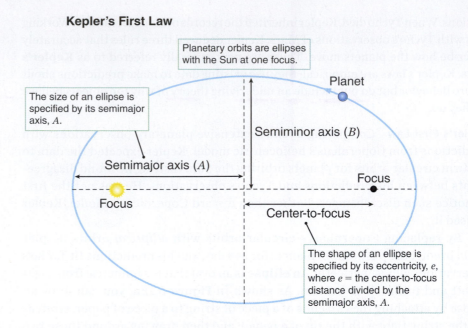

Kepler's First Law

Planetary orbits are ellipses with the Sun at one focus.

The size of an ellipse is specified by its semimajor axis, A.

Planet

Semiminor axis (B)

Semimajor axis (A)

Focus

Focus

Center-to-focus

The shape of an ellipse is specified by its eccentricity, e, where e = the center-to-focus distance divided by the semimajor axis, A.

what if . . .

What if Earth's orbit were highly elliptical, with Earth being closest to the Sun in northern winter: what factors might cause life to be different in the Northern and Southern hemispheres in that case?

Most planetary objects in our Solar System have nearly circular orbits (that is, they are ellipses with very low eccentricities). As **Figure 3.14a** shows, Earth's orbit, with an eccentricity of 0.017, is very nearly a circle centered on the Sun; therefore, the variation in distance between Earth and the Sun is small. In contrast, dwarf planet Pluto's orbit (**Figure 3.14b**) has an eccentricity of 0.249. The Sun is noticeably offset from center, and the orbit is elongated.

The **Process of Science Figure** traces the steps that led to the development of Kepler's first law. Along the way, several theories were falsified. Recall from Chapter 1 that in order for a theory to be scientific, it must be falsifiable.

Kepler's Second Law   By analyzing Tycho's observational data of changes in the planets' positions, Kepler found that a planet moves fastest when closest to the Sun and slowest when farthest from the Sun. We now know that Earth's average speed in its orbit around the Sun is 29.8 kilometers per second (km/s). When closest to the Sun, Earth travels at 30.3 km/s. When farthest from the Sun, Earth travels at 29.3 km/s.

Kepler found an elegant way to describe the changing speed of a planet in its orbit around the Sun. **Figure 3.15** shows the location of a planet along the ellipse of its orbit at six different times (t_1 to t_6). Imagine a straight line connecting the Sun with that planet. That line "sweeps out" an area as the planet moves from one point on the ellipse to another. Area A (in orange) is swept out between times

Figure 3.14 The orbits of **a.** Earth and **b.** Pluto in comparison with circles around the Sun. e = eccentricity.

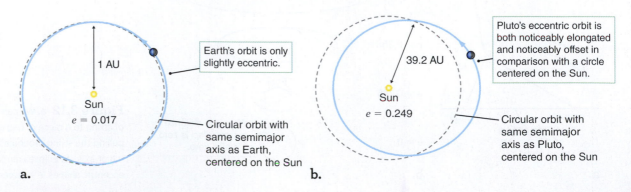

1 AU

Sun

e = 0.017

Earth's orbit is only slightly eccentric.

Circular orbit with same semimajor axis as Earth, centered on the Sun

39.2 AU

Sun

e = 0.249

Pluto's eccentric orbit is both noticeably elongated and noticeably offset in comparison with a circle centered on the Sun.

Circular orbit with same semimajor axis as Pluto, centered on the Sun

a.

b.

Theories Are Falsifiable

Early astronomers studied the motions of the planets,
but did not understand why the planets behaved as they do.

Copernicus
1473–1543

The Hypothesis
Copernicus proposed that planets moved in
circular orbits (with epicycles) around the Sun.

Tycho
1546–1601

The Observation
Tycho, born three years after Copernicus died, observed and
collected enormous amounts of data about planet positions.

Kepler
1571–1630

The Prediction
Kepler, 25 years younger than Tycho, used Copernicus's model
and Tycho's data to predict where the planets should be.

The Test
When Kepler compared the predictions with actual data,
they disagreed. Copernicus's idea was falsified!

The New Hypothesis
Planet orbits are not circular. They are elliptical.

Newton
1643–1726

This new hypothesis gained acceptance a century later
when Newton showed that planetary orbits are an
inevitable consequence of gravitational attraction.

In order for a theory to be scientific, it must be falsifiable, even if the test can't
be carried out until decades or centuries later. Disproving an old theory is part of
the self-correcting nature of science: It always leads to deeper understanding.

Kepler's Second Law

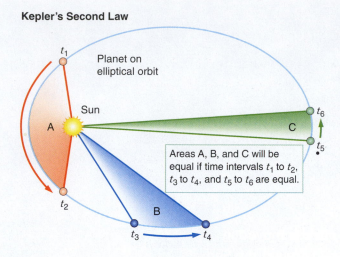

Figure 3.15 An imaginary line between a planet and the Sun sweeps out an area as the planet orbits. According to Kepler's second law, the three areas A, B, and C will be the same if the three time intervals shown are equal.

t_1 and t_2, area B (in blue) is swept out between times t_3 and t_4, and area C (in green) is swept out between times t_5 and t_6. Kepler realized that if the three time intervals in the figure are equal (that is, $t_1 \rightarrow t_2 = t_3 \rightarrow t_4 = t_5 \rightarrow t_6$), then the three areas A, B, and C must be equal as well. In order for this to be true, the planet must move fastest when it is closest to the Sun (area A), because it must cover the largest distance in its orbit. When the planet is farthest from the Sun (Area C), it must move more slowly because it covers a smaller distance in its orbit. This change in speed balances the change in distance so the area remains the same.

Kepler's second law, also called Kepler's **law of equal areas**, states that the imaginary line connecting a planet to the Sun sweeps out equal areas in equal times, regardless of where the planet is along the ellipse of its orbit. That law applies to only one planet at a time. The area swept out by Earth in a given time interval is always the same. Likewise, the area swept out by Mars in a given time is always the same. But the area swept out by Earth and the area swept out by Mars in a given time are *not* the same. Kepler's second law can be used to find the speed of a planet anywhere in its orbit.

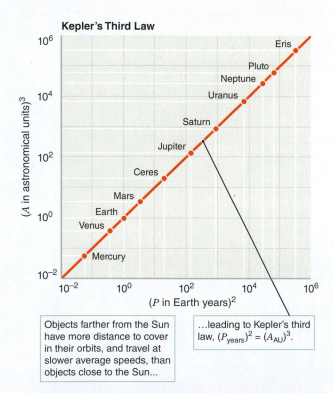

Figure 3.16 A plot of A^3 versus P^2 for objects in our Solar System shows that they obey Kepler's third law. (By plotting powers of 10 on each axis, we can fit both large and small values on the same plot. We will use that approach often.)

Table 3.2	Kepler's Third Law: $P^2 = A^3$		
Planet	**Period P (years)**	**Semimajor Axis A (AU)**	$\dfrac{P^2}{A^3}$
Mercury	0.241	0.387	$\dfrac{0.241^2}{0.387^3} = 1.00$
Venus	0.615	0.723	$\dfrac{0.615^2}{0.723^3} = 1.00$
Earth	1.000	1.000	$\dfrac{1.000^2}{1.000^3} = 1.00$
Mars	1.881	1.524	$\dfrac{1.881^2}{1.524^3} = 1.00$
Ceres	4.599	2.765	$\dfrac{4.599^2}{2.765^3} = 1.00$
Jupiter	11.86	5.204	$\dfrac{11.86^2}{5.204^3} = 1.00$
Saturn	29.46	9.582	$\dfrac{29.46^2}{9.582^3} = 0.99*$
Uranus	84.01	19.201	$\dfrac{84.01^2}{19.201^3} = 1.00$
Neptune	164.79	30.047	$\dfrac{164.79^2}{30.047^3} = 1.00$
Pluto	247.92	39.482	$\dfrac{247.92^2}{39.482^3} = 1.00$
Eris	557.00	67.696	$\dfrac{557.00^2}{67.696^3} = 1.00$

*Ratios are not exactly 1.00 because of slight perturbations from the gravity of other planets.

working it out 3.2

Kepler's Third Law

Kepler's third law states that the square of the period of a planet's orbit measured in years, P_{years}, is equal to the cube of the semimajor axis of the planet's orbit measured in astronomical units, A_{AU}. As an equation, the law says

$$(P_{years})^2 = (A_{AU})^3$$

For mathematical convenience, Kepler used units based on Earth's orbit—astronomical units and years. If other units were used, such as kilometers and days, P^2 would still be proportional to A^3, but the constant of proportionality would not be 1.

To calculate the average size of Neptune's orbit in astronomical units, you first need to find out how long Neptune's period is in Earth years,

which you can determine by observing the synodic period and from that computing its sidereal period (using Working It Out 3.1). Neptune's sidereal period is 165 years. Plugging that number into Kepler's third law gives the following result:

$$(P_{years})^2 = (165)^2 = 27{,}225 = (A_{AU})^3$$

To solve that equation, you must first square 165 to get 27,225 and then take its cube root (see Appendix 1 for calculator hints). Then

$$A_{AU} = \sqrt[3]{27{,}225} = 30.1$$

The semimajor axis of Neptune's orbit—that is, the average distance between Neptune and the Sun—is 30.1 AU.

Kepler's Third Law   Kepler looked for patterns in the planets' orbital periods. Compared with planets closer to the Sun, he found that planets farther from the Sun have longer orbits *and* move more slowly in those orbits. Kepler discovered a mathematical relationship between a planet's sidereal period—how many years the planet takes to go around the Sun and return to the same position in the Solar System—and its average distance from the Sun in astronomical units. **Kepler's third law** states that the square of the sidereal period (P) is equal to the cube of the semimajor axis (A). This is true in the Solar System, where the period is measured in Earth years, and the semimajor axis is measured in AU.

Table 3.2 lists the periods and semimajor axes of the orbits of the eight classical planets and three of the dwarf planets, along with the values of the ratio P^2 divided by A^3. Those data are also plotted in **Figure 3.16**. Kepler referred to that relationship as his **harmonic law** or, more poetically, as the "Harmony of the Worlds." Kepler's third law is explored further in **Working It Out 3.2**. Kepler's laws enhanced the heliocentric mathematical model of Copernicus and led to its greater acceptance.

what if . . .

What if you read online that "experts have discovered a new planet with a distance from the Sun of 2 AU and a period of 4 years"? Use Kepler's third law to argue that this is impossible.

CHECK YOUR UNDERSTANDING **3.2**

Order the following from largest to smallest semimajor axis: (a) a planet with a period of 84 Earth days; (b) a planet with a period of 1 Earth year; (c) a planet with a period of 2 Earth years; (d) a planet with a period of 0.5 Earth years.

3.3 Galileo's Observations Supported the Heliocentric Model

Galileo Galilei (1564–1642—**Figure 3.17**) was the first to use a telescope to make and report significant discoveries about astronomical objects. Galileo's telescopes were small, yet they were sufficient for him to observe spots on the Sun, the uneven

Figure 3.17
Galileo Galilei laid the physical framework for Newton's laws.

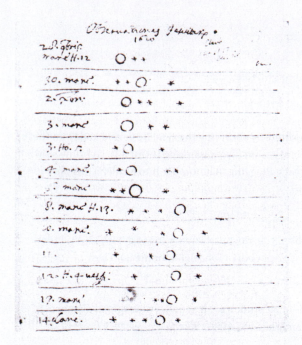

Figure 3.18 A page from Galileo's notebook shows his observations of Jupiter's four largest moons.

surface and craters of the Moon, and the many stars in the band of light in the sky called the Milky Way.

Galileo's Observations

Galileo provided the first observational evidence that some objects in the sky do not orbit Earth. When Galileo turned his telescope on Jupiter, he observed four "stars" in a line near the planet. Over time he observed that the objects remained near Jupiter, but changed position from night to night (**Figure 3.18**). Galileo hypothesized that those objects were moons orbiting Jupiter. Those four "Galilean moons" are the largest of Jupiter's many moons. Galileo also estimated the relative distance of each moon from Jupiter and the periods of their orbits, and he showed that the square of the period was proportional to the cube of the radius of the orbit for each moon, consistent with Kepler's third law.

Galileo also observed that Venus went through an entire set of phases like the Moon. He noticed that as the phases of Venus changed, so did the size of Venus in his telescope. In a geocentric model in which Venus orbits Earth like the Moon does, the apparent size of Venus would change only slightly, when it looped an epicycle. In the heliocentric model, however, the Earth–Venus distance varies by a lot, and Venus's size changes accordingly. When Venus is in its gibbous to full phases, it is farther away, on the other side of the Sun from Earth, and smaller in the sky. When Venus is in its crescent to new phases, it is closer, on the same side of the Sun as Earth, and larger in the sky (**Figure 3.19**). In the experiment that opens this chapter, you can build a model that demonstrates this relationship. Galileo's observations of Jupiter's moons and the phases of Venus in particular convinced Galileo that Copernicus was correct to place the Sun at the center of the Solar System.

In addition to his astronomical observations, Galileo did important work on the motion of objects. Unlike natural philosophers, who thought about objects in motion but did not actually experiment with them, Galileo conducted experiments with falling and rolling objects. As with his telescopes, Galileo improved or developed new technology to enable him to conduct those experiments. For example, by carefully rolling balls down an inclined plane and by dropping various objects from a height, he found that a falling object travels a distance proportional to the square of the time it has been falling. If he simultaneously dropped two objects of different masses, they reached the ground at the same time, showing that all objects falling to Earth accelerate at the same rate, independent of their mass.

Galileo's observations and experiments with many types of moving objects, such as carts and balls, led him to disagree with the Greek philosophers about when and why objects continue to move or come to rest. Before Galileo, it was thought that an object's natural state was to be at rest. But he found that an object naturally does what it was doing until a force acts on it. That is, an object in motion continues moving along a straight line with a constant speed until a force acts on it to change that state of motion. That idea of *inertia*, which Newton later adopted as his first law of motion, has implications for not only the motion of carts and balls but also the orbits of planets.

Figure 3.19 Modern photographs of the phases of Venus show that when we see Venus more illuminated, it also appears smaller, implying that Venus is farther away then.

Dialogue Concerning the Two Chief World Systems

Galileo faced considerable danger because of his work. His later life was consumed by conflict with the Catholic Church over his support of the Copernican system. In 1632, Galileo published his best-selling book, *Dialogo sopra i due massimi sistemi del mondo* ("Dialogue Concerning the Two Chief World Systems"). The *Dialogo* presents a brilliant philosopher named Salviati as the champion of the Copernican heliocentric view of the universe. The defender of an Earth-centered universe, Simplicio—who uses arguments made by the classical Greek philosophers and the pope—sounds silly and ignorant.

Galileo, a religious man with two daughters in a convent, thought he had the Catholic Church's tacit approval for his book. But when he placed several of the pope's geocentric arguments in the unflattering mouth of Simplicio, the perceived attack on the pope got the church's attention. Galileo was put on trial for heresy, sentenced to prison, and eventually placed under house arrest. To escape a harsher sentence, Galileo was forced to publicly recant his belief in the Copernican theory that Earth moves around the Sun. According to one story, as he left the courtroom, Galileo stamped his foot on the ground and muttered, "And yet it moves!"

The *Dialogo* was placed on the pope's Index of Prohibited Books, along with Copernicus's *De Revolutionibus*. Nevertheless, Galileo's work traveled across Europe, was translated into other languages, and was read by other scientists. Galileo spent his final years compiling his research on inertia and other ideas into the book *Discourses and Mathematical Demonstrations Relating to Two New Sciences*, which was published in 1638 in Holland, outside the Catholic Church's jurisdiction. (In 1992, Pope John Paul II apologized for the "Galileo Case.")

CHECK YOUR UNDERSTANDING 3.3

Which of Galileo's astronomical observations did Copernicus's model explain better than Ptolemy's? (a) sunspots; (b) craters on the Moon; (c) the moons of Jupiter; (d) the apparent size and phases of Venus

3.4 Newton's Three Laws Help Explain How Celestial Bodies Move

Empirical laws, such as Kepler's laws, describe *what* happens, but they do not explain *why*. Kepler described the orbits of planets as ellipses, but he did not explain why they should be so. To take that next step in the scientific process, scientists use basic physical principles and the tools of mathematics to derive the empirically determined laws. Alternatively, a scientist might start with physical laws and predict relationships, which are then verified or falsified through experiment and observation. If those predictions are verified, the scientist may have determined something fundamental about how the universe works.

Sir Isaac Newton (1642–1727—**Figure 3.20**) took that next step in explaining the nature of motion. Newton was a student of mathematics at Cambridge University when it closed because of the Great Plague and students were sent home to the safer countryside. Over the next 2 years, he studied on his own, and at the age of 23 he invented calculus, which would become crucial to his development of the physics of motion. (The German mathematician Gottfried Leibniz independently developed calculus around the same time.)

what if . . .

What if Galileo had found that instead of obeying Kepler's third law, the moons of Jupiter behave in such a way that the orbital period is proportional to the radius of the orbit? Does this observation falsify Kepler's third law for the planets, and what can you conclude from the difference between these results?

▶❚❚ **AstroTour:** Velocity, Acceleration, Inertia

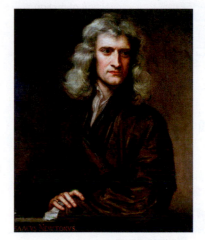

Figure 3.20 Sir Isaac Newton formulated three laws of motion.

unanswered questions

Would the history of scientific discoveries in physics and astronomy have been different if the Inquisition had not prosecuted Galileo? Galileo wrote the *Dialogo* after the Catholic Church in 1616 ordered him not to "hold or defend" the idea that Earth moves and the Sun is still. And he wrote his equally famous *Discorsi e Dimostrazioni Matematiche* (often shortened in English to "Two New Sciences") while under house arrest after his trial. However undeterred Galileo appeared to be, the effects of the decrees, prohibitions, and prosecutions might have dissuaded other scientists in Catholic countries from pursuing that type of work. Indeed, after Galileo's experiences, the center of the scientific revolution moved north to Protestant Europe.

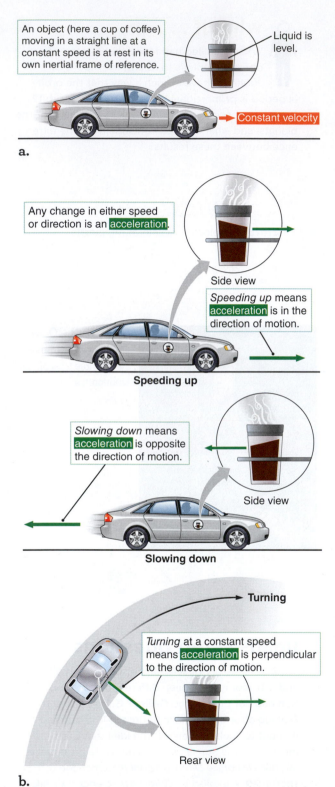

An object (here a cup of coffee) moving in a straight line at a constant speed is at rest in its own inertial frame of reference.

Liquid is level.

Constant velocity

a.

Any change in either speed or direction is an acceleration.

Side view

Speeding up means acceleration is in the direction of motion.

Speeding up

Slowing down means acceleration is opposite the direction of motion.

Side view

Slowing down

Turning

Turning at a constant speed means acceleration is perpendicular to the direction of motion.

Rear view

b.

Figure 3.21 **a.** An object moving in a straight line at a constant speed is at rest in its own inertial frame of reference. **b.** Any change in an object's velocity is an acceleration. When you are driving, for example, any time your speed changes or you follow a curve in the road, you are accelerating. (Throughout the text, velocity arrows are red and acceleration arrows are green.)

Building on the work of Kepler, Galileo, and others, Newton proposed three physical laws that govern the motions of all objects in the sky and on Earth. To understand how the planets and all other celestial bodies move, you must understand these three laws.

Newton's First Law: Objects at Rest Stay at Rest; Objects in Motion Stay in Motion

A **force** (**F**) is a push or a pull on an object. Two or more forces can oppose one another such that they are perfectly balanced and cancel out. For example, gravity pulls down on you as you sit in your chair, but the chair pushes up on you with an exactly equal and opposite force. As a result, you stay motionless. Forces that cancel out do not affect an object's motion. When forces combine to produce an effect, we often use the term *net force*, or sometimes just *force*.

Imagine that you are driving a car, and your book is on the seat next to you. A rabbit runs across the road in front of you, so you hit the brakes hard. You feel the seat belt tighten to restrain you. At the same time, your book flies off the seat and hits the dashboard. You have just experienced what Newton describes in his first law of motion. **Inertia** is an object's tendency to maintain its state—either of uniform motion or of rest—until a net force pushes or pulls it. In the stopping car, you did not hit the dashboard because the force of the seat belt slowed you down. The book, however, hit the dashboard because no such force acted upon it.

Newton's first law of motion describes inertia and states that an object in motion tends to stay in motion, in the same direction, until a net force acts upon it; and an object at rest tends to stay at rest until a net force acts upon it. Galileo's law of inertia became the cornerstone of physics as Newton's first law.

Recall from Section 2.1 the concept of a frame of reference. Within a frame of reference, only the relative motions between objects have any meaning. Without external clues, you cannot tell the difference between sitting still and traveling at constant speed in a straight line. For example, if you close your eyes while riding in the passenger seat of a quiet car on a smooth road, you feel as though you are sitting still. For you, the book on the seat beside you was "at rest," whereas a person standing by the side of the road would see the book moving past at the same speed as you and the car. People in a car approaching you would see the book moving even faster than the person by the side of the road—namely, at the speed they are traveling plus the speed you are traveling! All those perspectives are equally valid, and all those speeds of the book are correct when measured in the appropriate reference frame.

A reference frame moving in a straight line at a constant speed is called an **inertial frame of reference**. Any inertial frame of reference is as good as another. In the inertial frame of reference of a cup of coffee, for example, **Figure 3.21a** shows the cup is at rest in its own frame even if the car is moving quickly down the road.

Newton's Second Law: Motion Is Changed by Forces

What happens if a net force does act? In the earlier example, you were traveling in the car, and your motion slowed when the force of the seat belt acted upon you. Forces change an object's motion—by changing either the speed or the direction. That effect reflects **Newton's second law of motion**: if a net force acts on an object, the object's motion changes.

In the driver's seat of a car, you have several controls, including an accelerator and a brake pedal, which you use to make the car speed up or slow down. A *change in speed* is one way the car's motion can change. But you also have the steering wheel in your hands. When you are moving down the road and you turn the wheel, your speed does not necessarily change, but the direction of your motion does. A *change in direction* also is a kind of change in motion.

Together, an object's speed and direction are called **velocity (v)**. "Traveling at 50 kilometers per hour (km/h)" indicates speed, whereas "traveling north at 50 km/h" indicates velocity. **Acceleration (a)** describes the changes in an object's velocity. For example, if you go from 0 to 100 km/h in 4 seconds, you feel a strong push from the seat back as it shoves your body forward, causing you to accelerate along with the car. However, if you take 2 minutes to go from 0 to 100 km/h, the acceleration is so slight that you hardly notice it.

Partly because a car's gas pedal is often called the accelerator, some people think *acceleration* always means that an object is speeding up. In physics, however, *any change in speed or direction is an acceleration.* **Figure 3.21b** illustrates that point by showing what happens to the coffee in a cup as the car speeds up, slows down, or turns. Slamming on your brakes and going from 100 to 0 km/h in 4 seconds is just as much an acceleration as going from 0 to 100 km/h in 4 seconds. Similarly, the acceleration you experience as you go through a tight turn at a constant speed is every bit as real as the acceleration you feel when you slam your foot on the accelerator or the brake. Whether speeding up, slowing down, turning left, or turning right—if you are not moving in a straight line at a constant speed, you are accelerating.

Newton's second law of motion says that a net force causes acceleration. An object's acceleration depends on two things. First, as shown in **Figure 3.22**, the acceleration depends on the strength of the net force acting on the object to change its motion. If the forces acting on the object do *not* add up to zero, a net force is present and the object accelerates (Figure 3.22a). The stronger the net force, the greater the acceleration (Figure 3.22b). If you push on something twice as hard, it experiences twice as much acceleration. Push on something three times as hard and its acceleration will be three times as great. The acceleration occurs in the direction the net force points. Push an object away from you, and it will accelerate away from you.

An object's acceleration also depends on its inertia. You can push some objects easily, such as an empty box from a new refrigerator. But you can't easily shove an actual refrigerator around, even though it is about the same size as the box. Figure 3.22c shows that the greater the mass, the greater the inertia, and the *less* acceleration that will occur in response to the same net force. That relationship among acceleration (*a*), force (*F*), and mass (*m*) is expressed mathematically in **Working It Out 3.3**.

Newton's Third Law: Whatever Gets Pushed Pushes Back

Imagine that you are standing on a skateboard and pushing yourself along with your foot. Each shove of your foot against the ground sends you faster along your way. But why? You accelerate because as you push on the ground, the ground pushes back on you.

Part of Newton's genius was his ability to see patterns in such everyday events. Newton realized that *every* time one object exerts a force on another, the second object exerts a matching force on the first. That second force is as strong as the first force but is in the opposite direction. When you accelerate yourself on the skateboard,

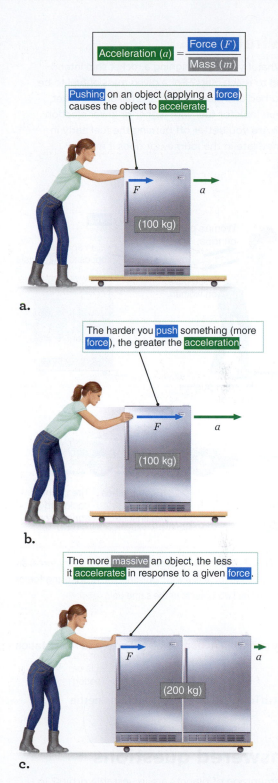

$$\text{Acceleration } (a) = \frac{\text{Force } (F)}{\text{Mass } (m)}$$

Pushing on an object (applying a force) causes the object to accelerate.

(100 kg)

a.

The harder you push something (more force), the greater the acceleration.

(100 kg)

b.

The more massive an object, the less it accelerates in response to a given force.

(200 kg)

c.

Figure 3.22 According to Newton's second law of motion, an object's acceleration is the force acting on the object divided by the object's mass. (Throughout the text, force arrows are blue.)

what if . . .

What if you are designing a rocket ship intending to reach Mars? For a given mass of fuel to be burned and ejected from the tail of the rocket, should you look for fuel with higher or lower ejection velocity? Are you better off burning the fuel early in the journey, late in the journey, or does it matter?

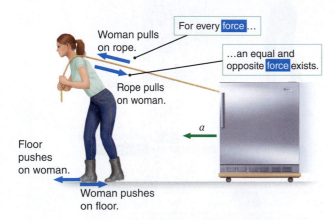

Figure 3.23 Newton's third law states that for every force, an equal and opposite force is always present. Those opposing forces always act on the two objects in the same pair.

🎥 **Astronomy in Action:** Velocity, Force, and Acceleration

▶⏸ **AstroTour:** Velocity, Acceleration, Inertia

unanswered questions

What percentage of planets are in unstable orbits? In younger planetary systems, planets might migrate in their orbits because of the presence of other massive planets nearby. We explain in Chapter 7 that Uranus and Neptune might have migrated in that way. Some planets have been discovered moving through the galaxy without any obvious orbit around a star. These planets did not originate in the stable orbits we see in our own Solar System.

you push backward on Earth, and Earth pushes you forward. As shown in **Figure 3.23**, when a woman moves a load on a cart by pulling a rope, the rope pulls back, and when a car tire pushes back on the road, the road pushes forward on the tire. Similarly, when Earth pulls on the Moon, the Moon pulls on Earth, and when a rocket engine pushes hot gases out of its nozzle, those hot gases push back on the rocket, propelling it into space.

All those force pairs exemplify **Newton's third law of motion**, which says forces always come in pairs, with those two forces always equal in strength but opposite in direction. The forces in those pairs always act on two objects. Your weight pushes down on the floor, and the floor pushes back up on you with the same amount of force. For every force, an equal force in the opposite direction is *always* there.

CHECK YOUR UNDERSTANDING 3.4

If a planet moves in a perfectly circular orbit around the Sun, is that planet accelerating? (a) Yes, because it is constantly changing its speed. (b) Yes, because it is constantly changing its direction. (c) No, because its speed is not constantly changing. (d) No, because planets do not accelerate.

Origins: Planets and Orbits

In addition to the planets in our own Solar System, researchers have detected thousands of planets orbiting stars other than our own Sun. Those planets' orbits can be calculated and understood by applying the same three Kepler's laws that we have discussed here. Astrobiologists think that the distance of a planet's orbit around its star affects the chances of life developing.

A planet close to its star receives more energy than a planet farther out. If Earth were closer to the Sun, it would be hotter throughout the year—probably so hot that water would evaporate and no longer exist as a liquid. If Earth were farther from the Sun, it would be colder and all surface water would probably freeze. Liquid water was crucial for life to form on Earth, so some astronomers look for planets at a distance from their star such that liquid water can exist (that distance varies depending on the star's temperature and size).

Next, recall from Figure 3.14a that Earth's elliptical orbit differs from a circle by less than 2 percent. Thus, Earth's distance from the Sun varies by only about 5 million km throughout the year; and, as we saw in Chapter 2, seasons change on Earth because of the tilt of Earth's axis, not because of slight changes in distance from the Sun. Mars has about the same axial tilt as Earth, but the orbit of Mars is more eccentric, so Mars has greater seasonal variation. The distance between Mars and the Sun varies from 1.38 AU (207 million km) to 1.67 AU (249 million km)—an eccentricity of 9 percent (or ~40 million km). As a result, the seasons on Mars are unequal. They are shorter when Mars is closer to the Sun and moving faster, and longer when Mars is farther from the Sun and moving slower. The inequality of the martian seasons affects the overall stability of the planet's temperature and climate. When we look at planets orbiting other stars, we see that many have orbital eccentricities even higher than that of Mars and therefore large variations in temperature.

Earth is at the right distance from the Sun to have temperatures that permit water to be liquid, and its orbital eccentricity is low enough that the average planetary temperature does not change much during its orbit. Those orbital characteristics have contributed to making the conditions on Earth suitable for life to develop.

working it out 3.3

Using Newton's Laws

Your acceleration is calculated by dividing how much your velocity changes by how long that change takes to happen:

$$\text{Acceleration} = \frac{\text{How much velocity changes}}{\text{How long the change takes to happen}}$$

For example, if an object's speed goes from 5 to 15 meters per second (m/s), the change in velocity is 10 m/s. If that change happens in 2 seconds, the acceleration is given by

$$a = \frac{15\,\text{m/s} - 5\,\text{m/s}}{2\,\text{s}} = 5\,\text{m/s}^2$$

To determine how an object's motion is changing, we need to know two things: what net force is acting on the object and the object's resistance to that force. We can put that idea into equation form as follows:

$$(\text{An object's acceleration}) = \frac{(\textit{The force acting to change the object's motion})}{(\textit{The object's resistance to that change})}$$

$$= \frac{\textit{Force}}{\textit{Mass}}$$

Newton's second law above is often written as Force = mass × acceleration, or $F = ma$. The units of force are called **newtons (N)**, so 1 N = 1 kg m/s².

Suppose you are holding two blocks of the same size, but the block in your right hand has twice the mass of the block in your left hand. When you drop the blocks, they both fall with the same acceleration, as Galileo showed, and they hit your two feet at the same time. Which will hit with more force: the block falling onto your right foot or the one falling onto your left foot? The block in your right hand, with twice the mass, will hit your right foot with twice the force that the other block hits your left foot.

To see how Newton's three laws of motion work together, study **Figure 3.24**. An astronaut is adrift in space, motionless with respect to the nearby space station. With no tether to pull on, how can the astronaut get back to the station? Suppose the 100-kg astronaut throws a 1-kg wrench directly away from the station at a speed of 10 m/s. Newton's second law says that to change the wrench's motion, the astronaut must apply a force to it in the direction away from the station. Newton's third law says that the wrench must therefore push back on the astronaut with as much force but in the opposite direction. The force of the wrench on the astronaut causes the astronaut to begin drifting toward the station. How fast will the astronaut move? Turn to Newton's second law again. Because the astronaut has more mass, he or she will accelerate less than the wrench will. A force that causes the 1-kg wrench to accelerate

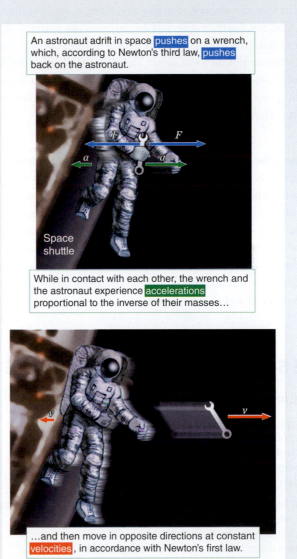

An astronaut adrift in space pushes on a wrench, which, according to Newton's third law, pushes back on the astronaut.

Space shuttle

While in contact with each other, the wrench and the astronaut experience accelerations proportional to the inverse of their masses…

…and then move in opposite directions at constant velocities, in accordance with Newton's first law.

Figure 3.24 According to Newton's laws, if an astronaut adrift in space throws a wrench, the two will move in opposite directions. Their speeds will depend on their masses: the same force will produce a smaller acceleration of a more massive object than of a less massive object. (Acceleration and velocity arrows not drawn to scale.)

to 10 m/s will have much less effect on the 100-kg astronaut. Because acceleration equals force divided by mass, the 100-kg astronaut will experience only 1/100 as much acceleration as the 1-kg wrench. The astronaut will drift toward the station, but only at the leisurely rate of 1/100 × 10 m/s, or 0.1 m/s.

SUMMARY

Early astronomers hypothesized that Earth was stationary at the center of the Solar System. Later astronomers realized that a Sun-centered Solar System was much simpler and could explain what people observed in the sky. Planets, such as Jupiter, orbit the Sun, not Earth. Kepler's laws describe the planets' elliptical orbits around the Sun, including details about how fast a planet travels at various points in its orbit. Those laws helped Newton develop his laws of motion, which govern how all objects move (not just orbiting ones). Orbital semimajor axis, eccentricity, and stability may affect a planet's suitability to foster life.

(1) **Compare the geocentric and heliocentric models of the Solar System.** Earth's motion is hard to detect, so before the Copernican Revolution, most people accepted a geocentric model of the Solar System, in which all objects orbit Earth. But in that model, the planets' apparent retrograde motion was hard to explain. Copernicus created the first comprehensive mathematical model of the Solar System with the Sun at the center, called a heliocentric model. His model explained apparent retrograde motion as a phenomenon caused when an inner planet passes an outer planet in their orbits.

(2) **Use Kepler's laws to describe how objects in the Solar System move.** Using Tycho's observational data, Kepler developed empirical rules to describe the motions of the planets. Kepler's three laws state that (1) planets move in elliptical orbits around the Sun; (2) planets move fastest when closest to the Sun and slowest when farthest from the Sun, so that the planets sweep out equal areas in equal times; and (3) the orbital period (P) of a planet squared equals the semimajor axis (A) of its orbit cubed: $P^2 = A^3$.

(3) **Explain how Galileo's astronomical discoveries convinced him that the heliocentric model was correct.** Galileo used telescopes to observe sunspots, craters on the Moon, and moons orbiting Jupiter. He also saw Venus going through phases, like the Moon, but changing its apparent size in each phase. The observations of Venus were hard to explain with Ptolemy's geocentric model but were consistent with a heliocentric model.

(4) **Describe the physical laws of motion discovered by Galileo and Newton.** Galileo studied the physics of falling objects and discovered the principle of inertia. Newton's laws state that (1) objects do not change their motion unless they experience a net force, (2) force = mass × acceleration, and (3) every force has an equal and opposite force. Net forces cause accelerations (changes in motion). Inertia resists changes in motion.

QUESTIONS AND PROBLEMS

TEST YOUR UNDERSTANDING

1. An *empirical science* is one based on
   a. hypothesis.
   b. calculus.
   c. computer models.
   d. observed data.

2. When Earth catches up to a slower-moving outer planet and passes it in its orbit, the outer planet
   a. exhibits retrograde motion.
   b. slows down because it feels Earth's gravitational pull.
   c. becomes dimmer as it passes through Earth's shadow.
   d. moves into a more elliptical orbit.

3. Copernicus's model of the Solar System was superior to Ptolemy's because
   a. it had a mathematical basis and computed the spacing of the planets.
   b. it was much more accurate.
   c. it did not require epicycles.
   d. it fit the telescopic data better.

4. An inferior planet is one that is
   a. smaller than Earth.
   b. larger than Earth.
   c. closer to the Sun than Earth.
   d. farther from the Sun than Earth.

5. The time a planet takes to come back to the same position in space in relation to the Sun is called its _____ period.
   a. synodic
   b. sidereal
   c. heliocentric
   d. geocentric

6. Suppose a planet is discovered orbiting a star in a highly elliptical orbit. While the planet is close to the star, it moves _____, but while it is far away, it moves _____.
   a. faster; slower
   b. slower; faster
   c. retrograde; prograde
   d. prograde; retrograde

7. If a superior planet is observed from Earth to have a synodic period of 1.2 years, what is its sidereal period?
   a. 0.54 years
   b. 1.8 years
   c. 4.0 years
   d. 6.0 years

8. A net force must be acting when an object
   a. accelerates.
   b. changes direction but not speed.
   c. changes speed but not direction.
   d. all of the above

9. For Earth, $P^2/A^3 = 1.0$ (in appropriate units). Suppose a new dwarf planet is discovered that is 14 times farther from the Sun than Earth is. For that planet,
   a. $P^2/A^3 = 1.0$.
   b. $P^2/A^3 > 1.0$.
   c. $P^2/A^3 < 1.0$.
   d. we can't know the value of P^2/A^3 without more information.

10. Galileo observed that Venus had phases that correlated with the size of its image in his telescope. From that information, you may conclude that Venus
    a. is the center of the Solar System.
    b. orbits the Sun.
    c. orbits Earth.
    d. orbits the Moon.

11. Kepler's second law says that
    a. planetary orbits are ellipses with the Sun at one focus.
    b. the square of a planet's orbital period equals the cube of its semi-major axis.
    c. net forces cause changes in motion.
    d. planets move fastest when closest to the Sun.

12. The average speed of a new planet in orbit has been reported to be 33 km/s. When it is closest to its star it moves at 31 km/s, and when it is farthest from its star it moves at 35 km/s. Those data must be in error because
    a. the average speed is far too fast.
    b. Kepler's third law says the planet has to sweep out equal areas in equal times, so the speed of the planet cannot change.
    c. Kepler's second law says the planet must move fastest when closest to its star, not when farthest away.
    d. with those numbers, the square of the orbital period will not be equal to the cube of the semimajor axis.

13. Galileo observed that Jupiter has moons. From that information, you may conclude that
    a. Jupiter is the center of the Solar System.
    b. Jupiter orbits the Sun.
    c. Jupiter orbits Earth.
    d. some things do not orbit Earth.

14. If you start from rest and accelerate at 10 mph/s and end up traveling at 60 mph, how long did it take?
    a. 1 second
    b. 6 seconds
    c. 60 seconds
    d. 0.6 seconds

15. Planets with high eccentricity may be unlikely candidates for life because
    a. the speed varies too much.
    b. the period varies too much.
    c. the temperature varies too much.
    d. the orbit varies too much.

THINKING ABOUT THE CONCEPTS

16. ★ **WHAT AN ASTRONOMER SEES** In Figure 3.2, Mars appears larger in the part of the loop that is retrograde, when Mars is moving from left to right. Why does Mars appear larger at this time? What conclusion can you make about the cause of the changing size of Venus in Figure 3.19? ◉

17. Study Figure 3.2. During normal motion, does Mars move toward the east or west? Which direction does it travel when in apparent retrograde motion? For how many days did Mars move in apparent retrograde motion? If one of the martian missions were photographing *Earth* in the sky during those days, what would it have observed? ◉

18. Copernicus and Kepler engaged in what is called empirical science. What do we mean by *empirical*?

19. Explain why Saturn's synodic period is very close to 1 Earth year (a sketch may help).

20. Experiment with falling objects as Galileo did. When you drop pairs of objects with different masses, do they reach the ground at the same time? Do they hit the ground with the same force? Does that approach work with a sheet of paper or a tissue—why or why not?

21. A planet's orbital speed around the Sun varies. When is the planet moving fastest in its orbit? When is it moving slowest?

22. The Moon's orbit around Earth is elliptical, too, with an eccentricity of 0.05. How does that value compare with the eccentricity of Earth's orbit? How do those elliptical orbits explain the types of solar eclipses discussed in Chapter 2?

23. Galileo came up with the concept of inertia. What do we mean by *inertia*?

24. If Kepler had lived on Mars, would he have deduced the same empirical laws for the motion of the planets? Explain.

25. What is the difference between speed and velocity? Between velocity and acceleration?

26. When involved in an automobile collision, a person not wearing a seat belt will move through the car and often strike the windshield directly. Which of Newton's laws explains why the person continues forward, even though the car stopped?

27. When riding in a car, we can sense changes in speed or direction through the forces that the car applies on us. Do we wear seat belts in cars and airplanes to protect us from speed or from acceleration? Explain your answer.

28. An astronaut standing on Earth can easily lift a wrench having a mass of 1 kg, but not a scientific instrument with a mass of 100 kg. On the International Space Station, an astronaut can manipulate both, although the scientific instrument responds much more slowly than the wrench. Explain why.

29. The Process of Science Figure illustrates that scientific ideas are always open to challenge. Construct an argument that the constant process of challenging and falsifying ideas is a strength of science, rather than a weakness. ◉

30. How might you expect conditions on Earth to be different if the eccentricity of its orbit were 0.17 instead of 0.017?

APPLYING THE CONCEPTS

31. Dwarf planet Ceres is 2.77 AU from the Sun. Its synodic period is 1.278 years. What is its sidereal period? ●–●–●
   a. Make a prediction: There is enough information in this question to calculate the sidereal (orbital) period in two different ways! How do you expect those answers to compare?
   b. Calculate: Use Working It Out 3.1 to find the sidereal period in years.
   c. Calculate: Use Kepler's law to find the sidereal period in years.
   d. Check your work: Compare your results for (a) and (b).

32. Suppose you are riding an electric scooter with a top speed of 32 km/hr. A typical electric scooter takes about ten seconds to accelerate to this speed. What is the acceleration of an electric scooter? ●–●–●
   a. Make a prediction: Do you expect this acceleration to be larger or smaller than the acceleration due to gravity (9.8 m/s²)? You may think you have no idea about this, but if you compare riding an electric scooter to falling off of a building, you might be able to guess.
   b. Calculate: Follow Working It Out 3.3 to calculate the acceleration of the scooter. Be careful about the units you are using!
   c. Check your work: Compare your answer to the acceleration due to gravity on Earth (9.8 m/s²).

33. Venus has a synodic period of 1.6 years. What is the sidereal period of Venus?
   a. Make a prediction: Is Venus closer or farther from the Sun than Earth? Therefore, do you expect its sidereal (orbital) period to be longer or shorter than Earth's? Will it be much longer or shorter?
   b. Follow Working It Out 3.1 to find the sidereal period, P, of Venus. Choose carefully which equation to use for Venus.
   c. Check your work by comparing your answer to your prediction.
   d. Repeat these steps and this calculation for Mars, which has a synodic period of 2.1 years. Again, think carefully before choosing which equation to use from Working It Out 3.1.

34. Suppose a dwarf planet is discovered orbiting the Sun with a semimajor axis of 50 AU. What would be the planet's orbital period?
   a. Make a prediction: Table 3.2 lists the period and semimajor axis of many objects in the Solar System. Examine this table to predict whether this dwarf planet will have an orbital period in the tens, hundreds, or thousands of years. ★
   b. Calculate: Use Kepler's third law to find the orbital period in years.
   c. Check your work: Compare your answer to the period of some of the objects in Table 3.2. ★

35. Suppose you are pushing a small refrigerator of mass 50 kg on wheels. You push with a force of 100 N. Assume the refrigerator starts at rest. How long will the refrigerator accelerate before it is moving faster than you can run (about 10 m/s)?
   a. Make a prediction: Do you expect this time to be small (a few seconds) or large (tens of minutes)? Study Working It Out 3.3. This is a two-step problem; you'll first have to find the acceleration, and then find the time. What units do you expect for each of these answers? Do you expect the acceleration due to your push to be larger or smaller than the acceleration due to Earth's gravity (9.8 m/s₂)?
   b. Calculate: What is the refrigerator's acceleration?
   c. Check your work: Does your acceleration have the correct units (m/s₂), and is this a reasonable acceleration?
   d. Calculate: Assume the refrigerator starts at rest. How long will the refrigerator accelerate before it is traveling at 10 m/s?
   e. Check your work: Does your answer have the correct units of seconds, and is it a reasonable length of time?

36. Is the graph in Figure 3.16 linear or logarithmic? From the data on the graph, find the approximate semimajor axis and period of Saturn. Show your work. ★

37. Study Figure 3.19, which shows that Venus's apparent size changes as it goes through phases. Approximately how many times larger is Venus in the sky at the thinnest crescent than at the gibbous phase shown? Approximately how many times closer is Venus to us at the phase of that thinnest crescent than at the gibbous phase? ★

38. The orbital period of Uranus is 84 years. Compute the semimajor axis of its orbit. How much time passes between oppositions of Uranus?

39. Suppose you read online that "experts have discovered a new planet with a distance from the Sun of 4 AU and a period of 3 years." Use Kepler's third law to argue that this is impossible.

40. Show, as Galileo did, that Kepler's third law applies to the four Galilean moons of Jupiter by calculating P^2 divided by A^3 for each moon. (Data on the moons can be found in Appendix 4.)

41. In a period of 3 months, a planet travels 30,000 km with an average speed of 3.8 m/s. Some time later, the same planet travels 65,000 km in 3 months. How fast is the planet traveling at that later time? During which period is the planet closer to the Sun?

42. If you were on Mars, how often would you see retrograde motion of *Earth* in the martian night sky? (You can view a picture at https://mars.jpl.nasa.gov/allaboutmars/nightsky /retrograde.)

43. The elliptical orbit of a comet that a spacecraft recently visited is 1.24 AU from the Sun at its closest approach and 5.68 AU from the Sun at its farthest.
   a. Sketch the comet's orbit. When is it moving fastest? Slowest?
   b. What is the semimajor axis of its orbit? How long does the comet take to go around the Sun?
   c. What is the distance from the Sun to the "center" of the ellipse? What is the eccentricity of the comet's orbit?

44. You are driving down a straight road at a speed of 90 km/h, and you see another car approaching you at a speed of 110 km/h along the road.
   a. With respect to your own frame of reference, how fast is the other car approaching you?
   b. With respect to the other driver's frame of reference, how fast are you approaching the other driver's car?

45. Sketch the orbit of Mars by using the information provided in the "Origins: Planets and Orbits" section of the chapter for the closest and farthest distances of Mars from the Sun.
   a. What is the orbit's major axis? Its semimajor axis?
   b. What is the distance from the "center" of the orbit to the Sun? Compute the orbit's eccentricity. Compare that value with the eccentricity of Earth's orbit.

EXPLORATION Kepler's Laws

digital.wwnorton.com/astro7

In this Exploration, we examine how Kepler's laws apply to the orbit of Mercury. Visit the Digital Resources Page and on the Student Site open the "Planetary Orbits" Interactive Simulation in Chapter 3. That simulator animates the orbits of the planets, enabling you to control the simulation speed, as well as several other parameters. Here we focus on exploring Mercury's orbit, but you may wish to spend some time examining the orbits of other planets as well.

Kepler's First Law

In the top left panel, use the drop-down menu to select "Mercury" and then click "OK." Click the "Kepler's First Law" tab at the bottom of the control panel. Use the radio buttons to select "Show Empty Focus" and "Show Center."

1. How would you describe the shape of Mercury's orbit?

   _____

   _____

Deselect "Show Empty Focus" and "Show Center," and select "Show Semiminor Axis" and "Show Semimajor Axis." Under "Visualization Options," select "Show Grid."

2. Use the grid markings to estimate the ratio of the semiminor axis to the semimajor axis.

   _____

   _____

3. Calculate the eccentricity of Mercury's orbit from that ratio by using $e = [1 - (\text{Ratio})^2]^{1/2}$.

   _____

   _____

Kepler's Second Law

Click on "Reset" in the bottom left panel, set parameters for Mercury, and click "OK." Then click on the "Kepler's Second Law" tab at the bottom of the control panel. Click "Play."

4. Observe the speed of the planet in its orbit. Where does the planet travel fastest? Where does the planet travel slowest?

   _____

   _____

Kepler's Third Law

Click on "Reset" in the bottom left panel, set parameters for Mercury, and then click on the "Kepler's Third Law" tab at the bottom of the control panel. Select "Show Solar System Orbits" in the "Visualization Options" panel. Study the graph. Use the eccentricity slider to change the simulated planet's eccentricity. Make the eccentricity first smaller and then larger.

5. Did anything in the graph change?

   _____

   _____

6. What do your observations of the graph tell you about the dependence of the period on the eccentricity?

   _____

   _____

Click on "Reset," and set the parameters back to those for Mercury. Now use the semimajor axis slider to change the semimajor axis of the simulated planet.

7. What happens to the period when you make the semimajor axis smaller?

   _____

   _____

8. What happens when you make it larger?

   _____

   _____

9. What do those results tell you about the dependence of the period on the semimajor axis?

   _____

   _____

Gravity

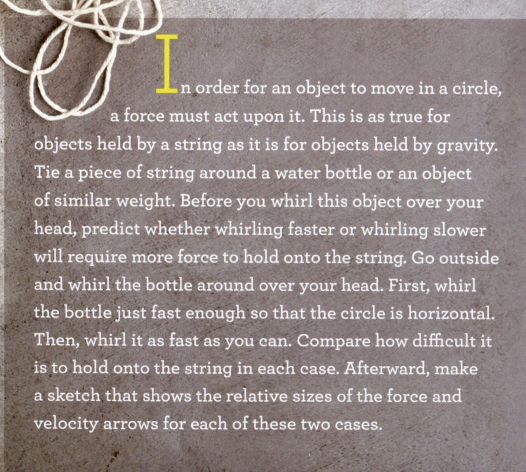

In order for an object to move in a circle, a force must act upon it. This is as true for objects held by a string as it is for objects held by gravity. Tie a piece of string around a water bottle or an object of similar weight. Before you whirl this object over your head, predict whether whirling faster or whirling slower will require more force to hold onto the string. Go outside and whirl the bottle around over your head. First, whirl the bottle just fast enough so that the circle is horizontal. Then, whirl it as fast as you can. Compare how difficult it is to hold onto the string in each case. Afterward, make a sketch that shows the relative sizes of the force and velocity arrows for each of these two cases.

EXPERIMENT SETUP

SLOWER

First, spin the bottle just fast enough so that the circle is horizontal.

FASTER

Then, spin it as fast as you can.

PREDICTION

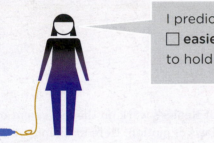

I predict that the string will be

☐ easier   ☐ harder

to hold when the water bottle is faster.

SKETCH OF RESULTS

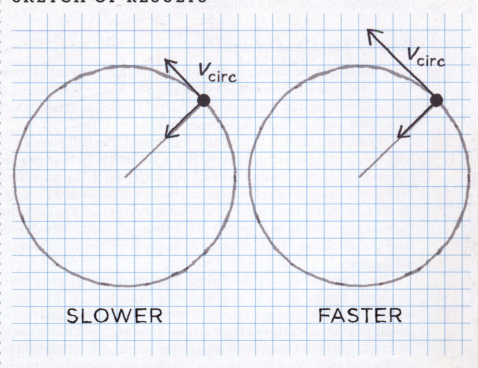

V_{circ}

V_{circ}

SLOWER   FASTER

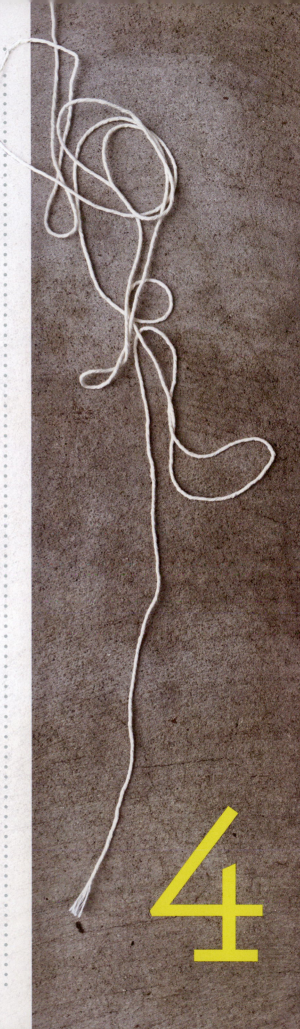

4

ere in Chapter 4, we explore the physical laws that explain the regular patterns in the motions of the planets. Because the Sun is far more massive than all the other parts of the Solar System combined, its gravity shapes the motions of every object in its vicinity, from the almost circular orbits of some planets to the extremely elongated orbits of comets.

LEARNING GOALS

By the end of this chapter, you should be able to:

① Identify the elements of Newton's universal law of gravitation.

② Use the laws of motion and gravitation to relate planetary orbits, trajectories of rocket flights, and the escape velocity from a system.

③ Explain how gravitational forces from the Sun and Moon create Earth's tides.

④ Describe how tidal forces affect solid bodies.

4.1 Gravity Is an Attractive Force between Objects

In Chapter 3, you learned about Kepler's work on the movement of the planets around the Sun and Newton's laws of motion. Here in Chapter 4, you will build on those concepts to understand Newton's universal law of gravitation. Although some properties of gravity were observed before Newton, his work connected the everyday phenomenon of falling objects to the motion of the planets around the Sun. Newton's law of gravity, when combined with Newton's own laws of motion, inevitably produce Kepler's empirical laws.

Gravity, Mass, and Weight

Many forces that we see in everyday life involve direct contact between objects. The cue ball on a pool table strikes the eight ball. When a person runs, their shoe presses directly against the pavement. When physical contact occurs between two objects, the source of the forces between them is easy to see.

If you drop a ball, it falls toward the ground. The ball picks up speed as it falls, accelerating downward toward Earth. According to Newton's second law, an accelerating object must have a force acting on it. But what force causes the ball to accelerate? The force causing the ball to fall toward Earth is an example of a force that acts at a distance without direct contact. The ball falling toward Earth is accelerating in response to the force of gravity. **Gravity**, one of the fundamental forces of nature, is the mutually attractive force between objects having mass.

Recall from Chapter 3 that Galileo discovered that all freely falling objects accelerate toward Earth at the same rate, regardless of their mass. Drop a marble and a book at the same time and from the same height, and they will hit the ground together. Note that air resistance becomes a factor at higher speeds, but the effect is negligible for dense, slow objects. The acceleration of falling objects due to gravity near the surface of Earth, which Galileo also measured experimentally, is usually written as g (lowercase) and has an average value on the surface of Earth of 9.8 meters per second squared (m/s²).

Newton's second law (Section 3.4) states that acceleration (a) equals force (F) divided by mass (m), or $a = F/m$. The acceleration due to gravity can be the same for all objects only if the value of the force divided by the mass is the same for all objects. In other words, an object twice as massive has double the gravitational force acting on it, an object three times as massive has triple the gravitational force acting on it, and so on. If all objects, regardless of mass, fall with the same acceleration, the gravitational *force* on an object must be determined by the object's *mass*.

The gravitational force acting on an object attracted by a planet is called the object's **weight**. In common language, we often use the words *weight* and *mass* interchangeably. To be more scientifically precise, scientists use *mass* to refer to the amount of matter in an object and *weight* to refer to the force that the planet's gravitational pull exerts on that object. On the surface of Earth, weight equals mass multiplied by the acceleration of gravity at Earth's surface, g:

$$F_{\text{weight}} = m \times g$$

where F_{weight} is an object's weight in newtons (N), the metric unit of force; m is the object's mass in kilograms (kg); and g is Earth's constant for acceleration due to gravity, 9.8 m/s². On Earth, then, an object with a *mass* of 1 kg has a *weight* of 9.8 N (**Figure 4.1**).

Your mass is the same no matter what planet or moon you are on, but your weight will depend on the local acceleration due to gravity. On the Moon, for example, the acceleration due to gravity is 1.6 m/s², which is about one-sixth of its value on Earth, so a 1-kg mass has a weight of 1.6 N on the Moon. Similarly, your weight on the Moon would be about one-sixth of your weight on Earth.

The value of g varies slightly across Earth's surface, ranging from 9.78 m/s² at the equator to 9.83 m/s² at the poles. That variation exists because Earth is not a perfect sphere: its rotation makes it flatter at the poles, so the radius of Earth is smaller there. A similar effect operates in locations at high altitude, causing g to be slightly smaller there.

Newton's Law of Gravity

As Newton told the story, he saw an apple fall from a tree to the ground. He reasoned that if gravity is a force that depends on mass, a gravitational force should exist between *any* two masses, including between a falling apple and Earth. That great insight came from applying his third law of motion to gravity. According to Newton's third law (Chapter 3), every force is paired with another force that is equal and opposite to it. Therefore, if Earth exerts a force of 9.8 N on a 1-kg mass sitting on its surface, that 1-kg mass must also exert a force of 9.8 N on Earth. Drop a 7-kg bowling ball and it falls toward Earth, but at the same time Earth falls toward the 7-kg bowling ball. We don't notice the planet's motion because Earth is very massive, so it has a lot of inertia. In the time a 7-kg bowling ball takes to fall to the ground from a height of 1 kilometer (km), Earth has "fallen" toward the bowling ball by only a tiny fraction of the size of an atom.

Newton reasoned that if doubling the mass of any object doubles the gravitational force between the object and Earth, doubling the mass of Earth ought to do the same thing. In short, the gravitational force between Earth and an object equals the product of the two masses multiplied by *something*:

Gravitational force = Something × Mass of Earth × Mass of object

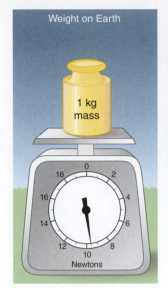
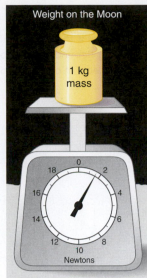

Figure 4.1 On the Moon, a mass of 1 kg has ⅙ the weight (displayed in newtons) that it has on Earth.

unanswered questions

What range of gravities will support human life? Humans have evolved to live on Earth's surface, in Earth's "surface gravity," but what happens when humans go elsewhere? What are the limits for our hearts, lungs, eyes, and bones? At the higher end of human tolerance, fighter pilots have been trained to experience about 10 times the normal surface gravity on Earth for very short periods (too long and they black out). Astronauts who spend several months in near-weightless conditions experience medical problems such as bone loss. On the Moon or Mars, humans will weigh much less than on Earth. Many science fiction tales have been written about what happens to children born on a space station or on another planet or moon with low surface gravity: would their hearts and bodies ever be able to adjust to the higher surface gravity of Earth, or must they stay in space forever?

what if . . .

What if the gravitational force on an object did not depend on the object's mass? How would objects in our world behave differently?

If the mass of the object is 2 times greater, the force of gravity will be 2 times greater. Likewise, if the mass of Earth happened to be 3 times what it is, the force of gravity also would have to be 3 times greater. If the mass of both Earth and the object were greater by those amounts, the gravitational force would increase by a factor of 2×3, or 6 times. Because objects fall toward the center of Earth, we know that gravity is an attractive force acting along a line between the two masses.

If gravity is a force that depends on mass, a gravitational force should exist between *any* two masses. Suppose we have two masses—call them mass 1 and mass 2 (or m_1 and m_2 for short). The gravitational force between them is *something* multiplied by the product of the masses:

$$\text{Gravitational force between two objects} = \text{Something} \times m_1 \times m_2$$

We have gotten this far just by combining Galileo's observations of falling objects with (1) Newton's laws of motion and (2) Newton's belief that Earth is a mass just like any other mass. But what's that "something" in the previous expression?

Kepler had already thought about that question. He reasoned that because the Sun is the focal point for planetary orbits, the Sun must exert an influence over the motions of the planets. Kepler speculated that the influence must grow weaker with distance from the Sun because the planets closer to the Sun moved much faster than the ones farther away. Kepler did not know about forces or inertia or gravity as the cause of celestial motion, but he thought that geometry alone suggested how that solar "influence" might change for planets progressively farther from the Sun.

To see why the influence must become weaker, imagine you have a certain amount of paint to spread in a 2-mm-thick layer over the surface of a small sphere. The surface area of a sphere depends on the square of the sphere's radius: double the radius of a sphere, and the sphere's surface area becomes 4 times what it was. Suppose you wish to paint a larger sphere, with twice the radius of the small sphere. The paint must cover 4 times as much area, and the thickness of the paint will be only a fourth of what it was on the smaller sphere. If you triple the radius of the sphere, the sphere's surface will be 9 times larger and the coat of paint will be only one-ninth as thick.

Kepler reasoned that as the influence of the Sun extended farther and farther into space, that influence would have to spread out to cover the surface of a larger and larger imaginary sphere centered on the Sun. The Sun's influence should diminish with the square of the distance from the Sun—this is an example of a relationship known as an **inverse square law**.

Kepler had an interesting idea, but not a scientific hypothesis with testable predictions. He couldn't explain how the Sun influences the planets, and he lacked the mathematical tools to calculate how an object would move under such an influence. Newton, however, had both. If gravity is a force between *any* two objects, a gravitational force should then exist between the Sun and each of the planets. If that gravitational force were the same as Kepler's "influence," gravity might behave according to an inverse square law.

Newton's expression for gravity came to look like this:

$$\text{Gravitational force between two objects} = \text{Something} \times \frac{m_1 \times m_2}{(\text{Distance between objects})^2}$$

The expression still has a "something," and that something is a constant of proportionality. That constant determines the strength of gravity between objects, and it is the same for all pairs of objects. Newton named it the **universal gravitational constant**. He estimated the gravitational constant, written as G (uppercase), by using Galileo's measurement of g, estimates of Earth's radius, and a guess at the mass of Earth by assuming it had about the same density as typical rocks. Not until about a century later was the actual value of G first measured. Today the value of G is accepted as 6.67×10^{-11} N m^2/kg^2 or its equivalents: 6.67×10^{-11} m^3/(kg s^2) or 6.67×10^{-20} km^3/(kg s^2).

A Universal Law of Gravitation

Newton's **universal law of gravitation**, illustrated in **Figure 4.2**, states that gravity is a force between any two objects having mass. The force due to gravity has the following properties:

1. It is an attractive force acting along a straight line between the two objects.

2. It is proportional to the mass of one object (m_1) multiplied by the mass of the other object (m_2). If we double m_1, the force (F) increases by a factor of 2. Likewise, if we double m_2, F increases by a factor of 2.

3. It is inversely proportional to the square of the distance r between the centers of the two objects. The graph in **Figure 4.3** shows that if we double r, F decreases by a factor of 4. If we triple r, F falls by a factor of 9.

Written as a mathematical formula, the universal law of gravitation is

$$F_{\text{grav}} = G \times \frac{m_1 \times m_2}{r^2}$$

where F_{grav} is the force of gravity between two objects; m_1 and m_2 are the masses of objects 1 and 2, respectively; r is the distance between the centers of mass of the two objects; and G is the universal gravitational constant. The relationship between the force of gravity and the masses of and distance between two objects is further explored in **Working It Out 4.1**.

Gravity pulls you toward the center of Earth. Gravity holds the planets and stars together and keeps the thin blanket of air we breathe close to Earth's surface. Gravity holds the planets, including Earth, in orbit around the Sun. Gravity caused a vast interstellar cloud of gas and dust to collapse 4.5 billion years ago to form the Sun, Earth, and the rest of the Solar System. Gravity binds colossal groups of stars into galaxies. Gravity shapes space and time, and it can affect the ultimate fate of the universe. We return often to the concept of gravity because it is central to an understanding of the universe.

CHECK YOUR UNDERSTANDING 4.1a

If the distance between Earth and the Sun were cut in half, the gravitational force between them would: (a) decrease by a factor of 4; (b) decrease by a factor of 2; (c) increase by a factor of 2; (d) increase by a factor of 4.

Answers to Check Your Understanding questions are in the back of the book.

Gravity Differs from Place to Place within an Object

As you sit reading this book, you are exerting a gravitational attraction on every other fragment of Earth, just as every other fragment of Earth is exerting a gravitational attraction on you. Your

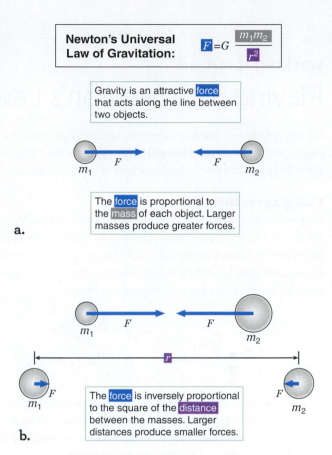

Newton's Universal Law of Gravitation: $F = G\dfrac{m_1 m_2}{r^2}$

Gravity is an attractive force that acts along the line between two objects.

The force is proportional to the mass of each object. Larger masses produce greater forces.

a.

The force is inversely proportional to the square of the distance between the masses. Larger distances produce smaller forces.

b.

Figure 4.2 Gravity is an attractive force between two objects. The force of gravity depends on the masses of the objects, m_1 and m_2, and the distance, r, between them.

Figure 4.3 As two objects move apart, the gravitational force between them decreases by the inverse square of the distance between them.

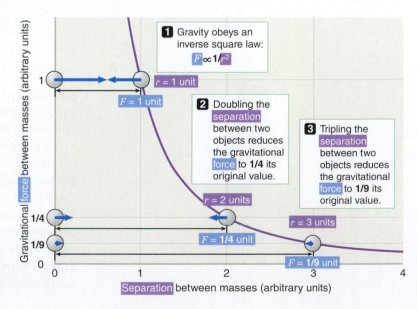

1 Gravity obeys an inverse square law: $F \propto 1/r^2$

$r = 1$ unit
$F = 1$ unit

2 Doubling the separation between two objects reduces the gravitational force to **1/4** its original value.

3 Tripling the separation between two objects reduces the gravitational force to **1/9** its original value.

$r = 2$ units
$F = 1/4$ unit

$r = 3$ units
$F = 1/9$ unit

Gravitational force between masses (arbitrary units)

Separation between masses (arbitrary units)

working it out 4.1

Playing with Newton's Laws of Motion and Gravitation

For any two objects, the force of gravity is directly proportional to the masses and inversely proportional to the *square* of the distance between them. Let's look at a few examples of how to use this equation:

Changing the Distance

How would the gravitational force between Earth and the Moon change if the distance between them were doubled? In this example, the masses of Earth and the Moon stay the same, but r becomes $2r$. We can calculate how the force changes by writing the equation for distance r and again for distance $2r$, and then taking a ratio to compare them:

$$\frac{F_{\text{grav at distance } 2r}}{F_{\text{grav at distance } r}} = \frac{G \times \dfrac{M_{\text{Earth}} M_{\text{Moon}}}{(2r)^2}}{G \times \dfrac{M_{\text{Earth}} M_{\text{Moon}}}{r^2}}$$

We can cancel out the constant G and the masses of Earth and the Moon, which do not change. Then we need to multiply both the numerator and denominator of the fraction by r^2, remembering that both the 2 and the r get squared in $(2r)^2 = 4r^2$. The equation becomes

$$\frac{F_{\text{grav at distance } 2r}}{F_{\text{grav at distance } r}} = \frac{\cancel{G} \times \cancel{M_{\text{Earth}}} \cancel{M_{\text{Moon}}}}{\cancel{G} \times \cancel{M_{\text{Earth}}} \cancel{M_{\text{Moon}}}} \times \frac{r^2}{(2r)^2}$$

$$\frac{F_{\text{grav at distance } 2r}}{F_{\text{grav at distance } r}} = \frac{r^2}{4r^2} = \frac{1}{4}$$

Doubling the distance reduced the force by a factor of 4; that is, the force is ¼ as strong.

Gravitational Acceleration

We have two ways to think about the gravitational force that Earth exerts on an object with mass m located on the surface of Earth. Recall Newton's second law of motion: $F = m \times a$. Here we are specifically considering the gravitational force and the acceleration due to gravity, so we might write $F_{\text{grav}} = m \times g$. The other way to think about the force is from the perspective of the universal law of gravitation:

$$F_{\text{grav}} = G \times \frac{M_{\text{Earth}} \times m}{R_{\text{Earth}}^2}$$

Here, M_{Earth} is the mass of Earth, and R_{Earth} is the radius of Earth. The two expressions describing that force must be equal. Therefore,

$$m \times g = G \times \frac{M_{\text{Earth}} \times m}{R_{\text{Earth}}^2}$$

The mass m is on both sides of the equation, so we can cancel it out. The equation then becomes

$$g = G \times \frac{M_{\text{Earth}}}{R_{\text{Earth}}^2}$$

This expression shows that Earth's mass and radius determine the gravitational acceleration (g) experienced by an object of mass m on the surface of Earth. The mass of the object itself (m) appears nowhere in the expression, so changing m does not affect the gravitational *acceleration* of an object on Earth.

gravitational interaction is strongest with the parts of Earth closest to you. The parts of Earth on the other side of our planet are much farther from you, so their pull on you is correspondingly less.

The net effect of all those forces is to pull you (or any other object) toward Earth's center. If you drop a hammer, it falls directly toward the ground. Earth is nearly spherical, so for every piece of Earth pulling you toward your right, a corresponding piece of Earth is pulling you toward your left with just as much force. For every piece of Earth pulling you forward, a corresponding piece of Earth is pulling you backward. Because Earth is almost spherically symmetric, all those "sideways" forces (see the blue arrows in **Figure 4.4a**) cancel out, leaving behind an overall force that points toward Earth's center (**Figure 4.4b**).

Some parts of Earth are closer to you and others are farther away, but an average distance exists between you and all the small fragments of Earth pulling on you. That average distance is the distance between you and the center of Earth. As illustrated in Figure 4.4b, the overall pull that you feel is the same as it would be if the entire mass of Earth were concentrated at a single point located at the very center of the planet.

That relationship is true for any spherically symmetric object. Outside the object, its gravity behaves as though all the mass of that object were concentrated at a point at its center. That relationship is important in many applications. For example, when you estimate your weight on another planet, you are calculating the force of gravity between you and the planet. The "distance" in the gravitational equation will be the distance between you and the center of the planet, which is just the radius of the planet plus your altitude.

You can think of Earth as a collection of small masses, each of which feels a gravitational attraction toward every other small part of Earth. The mutual gravitational attraction that occurs among all parts of the same object is called *self-gravity*. Self-gravity of a spherically symmetric object, like a planet or a roughly spherical cloud, is calculated much the same way as in the preceding discussion. For any bit of mass within the object, all the parts that are closer to the center pull that bit of mass toward the center, and the gravitational force acts as though they actually were at the center. All the parts that are farther away from the center than the bit of mass pull in different directions, and their forces cancel out.

CHECK YOUR UNDERSTANDING 4.1b

If Earth shrank to a smaller radius but kept the same mass, would the gravitational force between Earth and the Moon: (a) decrease; (b) increase; or (c) stay the same? Would everyone's weight at Earth's surface: (a) increase; (b) decrease; or (c) stay the same?

4.2 An Orbit Is One Body "Falling Around" Another

Newton used his laws of motion and his proposed law of gravity to predict the paths of planetary orbits. His calculations showed that those orbits should be ellipses with the Sun at one focus, that planets should move faster when closer to the Sun, and that the square of the period of a planet's orbit should vary as the cube of the semimajor axis of that elliptical orbit. Thus, Newton's universal law of gravitation indicated that planets should orbit the Sun in just the way that Kepler's empirical laws described. By explaining Kepler's laws, Newton found important corroboration for his law of gravitation.

Gravity and Orbits

Newton's laws tell us how forces change an object's motion and how objects interact with one another through gravity. To know where an object will be at any given time, we have to "add up" changes in the object's motion over time. Newton invented calculus to do just that, but here we aim for a conceptual understanding of how gravity causes orbits.

In Newton's time, the closest thing to making a heavy object fly was shooting cannonballs out of a cannon, so he used cannonballs in "thought experiments"

what if . . .

What if you jump from a very high tower toward a very deep body of water at an amusement park? As you fall, how would your experience compare to that of an astronaut in orbit?

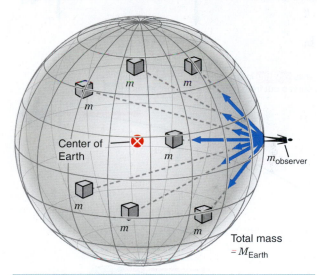

An object on the surface of a spherical mass (such as Earth) feels a gravitational attraction toward each small part of the sphere.

a.

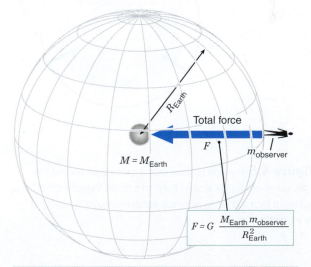

The net force is the same as if we scooped up the mass of the entire sphere and concentrated it at a point at the center.

b.

Figure 4.4 Outside a sphere, the net gravitational force due to a spherical mass is the same as the gravitational force from the same mass concentrated at a point at the center of the sphere.

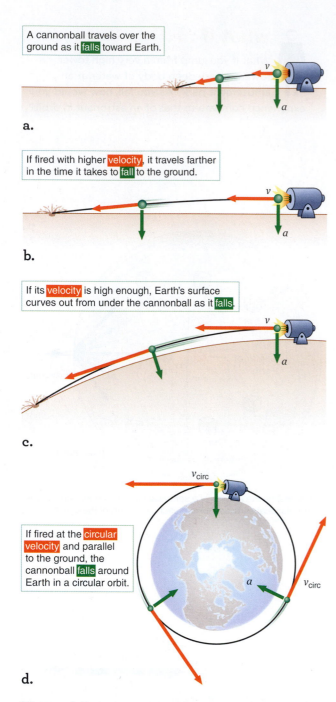

A cannonball travels over the ground as it falls toward Earth.

a.

If fired with higher velocity, it travels farther in the time it takes to fall to the ground.

b.

If its velocity is high enough, Earth's surface curves out from under the cannonball as it falls.

c.

If fired at the circular velocity and parallel to the ground, the cannonball falls around Earth in a circular orbit.

d.

Figure 4.5 Newton realized that a cannonball fired at the right speed would fall around Earth in a circle. Velocity (*v*) is indicated by a red arrow, whereas acceleration (*a*) is indicated by a green arrow.

about planetary motions. A dropped cannonball falls directly to the ground, like any other mass does. A cannonball fired out of a cannon that is level with the ground behaves differently, however, as shown in **Figure 4.5a**. The cannonball still falls to the ground in the same amount of time as it does when it is dropped, but while falling it also travels a horizontal distance over the ground, following a curved path. The faster the cannonball moves when it is fired from the cannon, the farther it will go before it hits the ground (**Figure 4.5b**).

To travel through air, the cannonball must push air out of its way—an effect normally referred to as *air resistance*—which slows it down. But we can ignore such real-world complications in this thought experiment. Instead imagine that, having inertia, the cannonball continues along its course until it runs into something. The faster the cannonball moves when it is fired, the farther it goes before hitting the ground. If the cannonball flies far enough, Earth's surface curves out from under it, as shown in **Figure 4.5c**. Eventually a speed is reached (**Figure 4.5d**) at which the cannonball is flying so fast that the surface of Earth curves away from the cannonball at the same rate at which the cannonball is falling toward Earth. When that occurs, the cannonball, which always falls toward the center of Earth, is orbiting. An **orbit** is the path of one object that freely falls around another.

Why do astronauts appear to float freely about the cabin of a spacecraft? It is not because they have escaped Earth's gravity; Earth's gravity is what holds them in their orbit. Instead, the answer lies in Galileo's early observation that all objects fall with the same acceleration, regardless of their mass. The astronauts and the spacecraft are both in orbit around Earth, moving in the same direction, at the same speed, and experiencing the same gravitational acceleration, so they fall around Earth together. **Figure 4.6** demonstrates that point. The astronaut is orbiting Earth just as the spacecraft is orbiting Earth. On the surface of Earth, your body tries to fall toward the center of Earth, but the ground gets in the way. You feel your weight when you are standing on Earth because the ground pushes on you to oppose the force of gravity, which pulls you downward. In the spacecraft, however, nothing interrupts the astronaut's fall because the spacecraft is falling around Earth in the same orbit as the astronaut. The astronaut is in **free fall**, falling freely in Earth's gravity. The **Process of Science Figure** illustrates the universality of Newton's law of gravitation.

When two objects are closer to having the same mass—such as dwarf planet Pluto and its moon Charon, or a large planet and a star, or two stars—both objects experience significant accelerations in response to their mutual gravitational attraction. The two objects are both orbiting about a common point located between them, called the **center of mass**, so we now must think of them as falling around each other. Each mass is moving on its own elliptical orbit around the two objects' mutual center of mass. From measuring the size and period of any orbit, we can calculate the *sum* of the masses of the orbiting objects. Almost all knowledge about the masses of astronomical objects comes directly from applying Newton's version of Kepler's third law.

What Velocity Is Needed to Reach Orbit?

How fast must Newton's cannonball be fired for it to fall around the world? The cannonball would be in **uniform circular motion**, meaning it would move along a circular path at constant speed. That type of motion is discussed in more depth in Appendix 8. Another example of uniform circular motion is a ball whirling around your head on a string. If you let go of the string, the ball will fly off in a

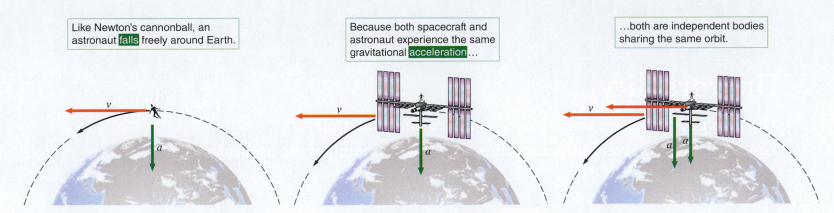

Like Newton's cannonball, an astronaut **falls** freely around Earth.

Because both spacecraft and astronaut experience the same gravitational **acceleration**…

…both are independent bodies sharing the same orbit.

Figure 4.6 A "weightless" astronaut has not escaped Earth's gravity. Rather, an astronaut and a spacecraft share the same orbit as they fall around Earth together.

straight line in whatever direction it is traveling at the time, just as Newton's first law predicts for an object in motion, as shown in **Figure 4.7a**. The string prevents the ball from flying off by constantly changing the direction the ball is traveling. The central force of the string on the ball is called a **centripetal force**: a force toward the center of the circle. Using a more massive ball, speeding up its motion, and making the string shorter so that the turn is tighter all are ways to increase the force needed to keep a ball moving in a circle.

For Newton's cannonball (or a satellite), no string holds the ball in its circular motion. Instead, gravity supplies the force, as illustrated in **Figure 4.7b**. The force of gravity must be just enough to keep the satellite moving on its circular path. Because that force has a specific strength, the satellite must therefore be moving at a particular speed around the circle, which we call its **circular velocity** (v_{circ}). If the satellite were moving at any other velocity, it would not be moving in a circular orbit. Remember the cannonball: if it moves too slowly, it will drop below the circular path and hit the ground. Similarly, if the cannonball moves too fast, its motion will carry it above the circular orbit. Only a cannonball moving at just the right velocity—the circular velocity—will fall around Earth on a circular path (see Figure 4.5d).

Newton's thought experiment became a reality in 1957, when the Soviet Union (the USSR) launched the first artificial object to orbit Earth. The USSR used a rocket to lift Sputnik 1, an object about the size of a basketball, high enough

▶ **AstroTour:** Newton's Laws and Universal Gravitation

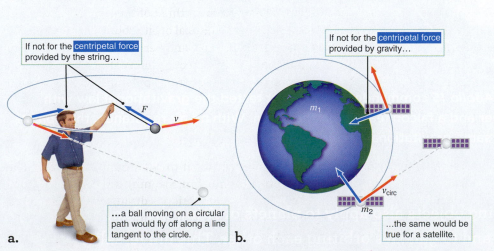

If not for the centripetal force provided by the string…

F

v

…a ball moving on a circular path would fly off along a line tangent to the circle.

a.

If not for the centripetal force provided by gravity…

m_1

v_{circ}

m_2

…the same would be true for a satellite.

b.

Figure 4.7 a. A string provides the centripetal force that keeps a ball moving in a circle. (We are ignoring the smaller force of gravity that also acts on the ball.) **b.** Gravity provides the centripetal force that holds a satellite in a circular orbit about Earth.

Universality

The laws of physics are the same everywhere and at all times.
The principle of universality underlies our understanding of the natural world.

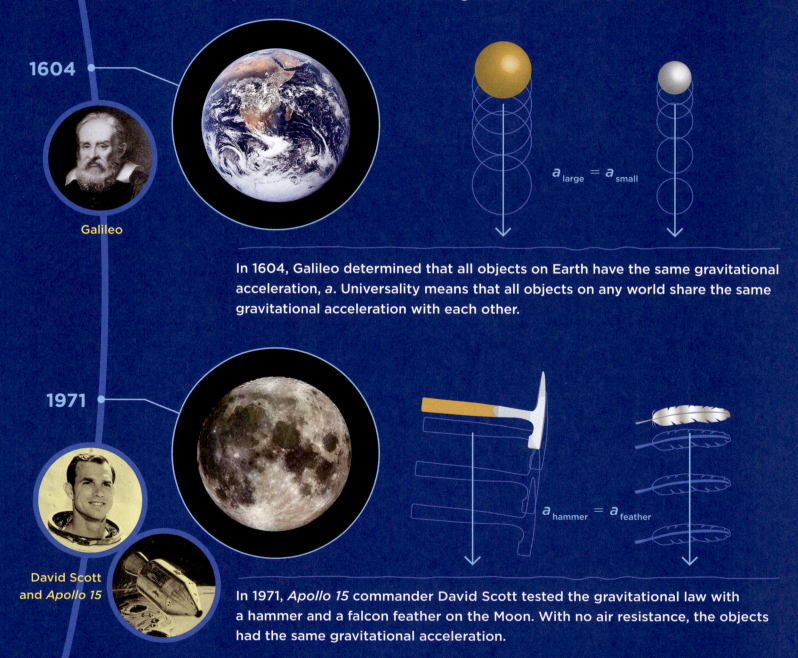

1604

Galileo

$$a_{large} = a_{small}$$

In 1604, Galileo determined that all objects on Earth have the same gravitational acceleration, *a*. Universality means that all objects on any world share the same gravitational acceleration with each other.

1971

David Scott and *Apollo 15*

$$a_{hammer} = a_{feather}$$

In 1971, *Apollo 15* commander David Scott tested the gravitational law with a hammer and a falcon feather on the Moon. With no air resistance, the objects had the same gravitational acceleration.

The physical laws that apply to falling objects also apply to planets orbiting the Sun, to stars orbiting within the galaxy, and to galaxies orbiting each other. This universality allows us to draw conclusions about very distant objects.

above Earth's upper atmosphere that air resistance wasn't an issue. Sputnik 1 achieved a high enough speed that it fell around Earth, just like Newton's imaginary cannonball.

When one object is falling around a much more massive object, we say that the less massive object is a **satellite** of the more massive object. Planets are satellites of the Sun, and moons are natural satellites of planets. Newton's imaginary cannonball and Sputnik 1 were satellites (*sputnik* means "satellite" in Russian). An Earth-orbiting spacecraft and the astronauts inside it are independent satellites of Earth that share the same orbit.

▶‖ AstroTour: Elliptical Orbit

The Shape of Orbits

Some Earth satellites travel a circular path at constant speed. Just like the ball on the string, satellites traveling at the circular velocity always stay the same distance from Earth, neither speeding up nor slowing down in orbit. But what would happen if the satellite then fired its rockets and started traveling *faster* than the circular velocity? Earth's pull is as strong as ever, but because the satellite has greater speed, Earth's gravity does not bend the satellite's path sharply enough to hold it in a circle. When that happens, the satellite begins to climb above a circular orbit.

As the distance between the satellite and Earth increases, the satellite slows down. Think about a ball thrown upward into the air, as shown in **Figure 4.8a**. As the ball climbs higher, the pull of Earth's gravity opposes its motion, slowing the ball down. The ball climbs more and more slowly until its vertical motion stops for an instant and then is reversed; the ball then begins to fall back toward Earth, speeding up along the way. A satellite does the same thing. As the satellite climbs above a circular orbit and begins to move away from Earth, Earth's gravity opposes the satellite's outward motion, slowing the satellite down. The farther the satellite is from Earth, the more slowly the satellite moves—just like the ball thrown into the air. Also just like the ball, the satellite reaches a maximum height on its curving path and then begins falling back toward Earth. When it does, Earth's gravity speeds the satellite up as it gets closer and closer to Earth. The satellite's orbit has changed from circular to elliptical.

Any object in an elliptical orbit, including a planet orbiting the Sun (**Figure 4.8b**), will therefore lose speed as it pulls away from what it is orbiting and then gain that speed back again as gravity causes it to fall inward toward what it is orbiting. Gravity, therefore, explains Kepler's second law from Section 3.2—namely, that a planet moves fastest when closest to the Sun and slowest when farthest from the Sun.

Newton's laws do more than explain Kepler's laws: they predict different types of orbits beyond Kepler's empirical experience. **Figure 4.9a** shows a series of satellite orbits, each with the same point of closest approach to Earth but with different velocities at that point, as indicated in **Figure 4.9b**. The more speed a satellite has at its closest approach to Earth, the farther the satellite can pull away from Earth, and the more eccentric its orbit becomes. As long as it remains elliptical, no matter how eccentric, the orbit is called a **bound orbit** because the satellite is gravitationally bound to the object it is orbiting.

In that sequence of faster and faster satellites, a point comes at which the satellite is moving so fast that gravity cannot reverse its outward motion, so the

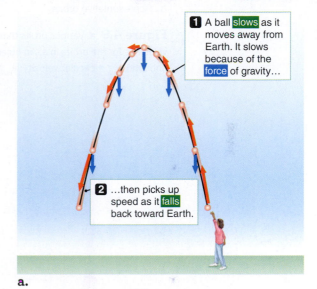

1 A ball slows as it moves away from Earth. It slows because of the force of gravity…

2 …then picks up speed as it falls back toward Earth.

a.

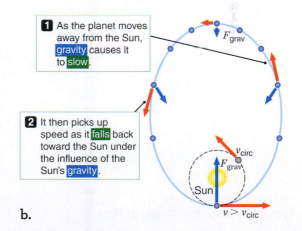

1 As the planet moves away from the Sun, gravity causes it to slow.

F_{grav}

2 It then picks up speed as it falls back toward the Sun under the influence of the Sun's gravity.

v_{circ}

F_{grav}

Sun

b.

$v > v_{circ}$

Figure 4.8 a. A ball thrown into the air slows as it climbs away from Earth and then speeds up as it heads back toward Earth. **b.** A planet in an elliptical orbit around the Sun does the same thing. (Although no planet has an orbit as eccentric as the one shown here, the orbits of comets can be far more eccentric.) A planet in an elliptical orbit will move fastest at its closest approach to the Sun. That speed is faster than the speed of a planet in a circular orbit of that radius.

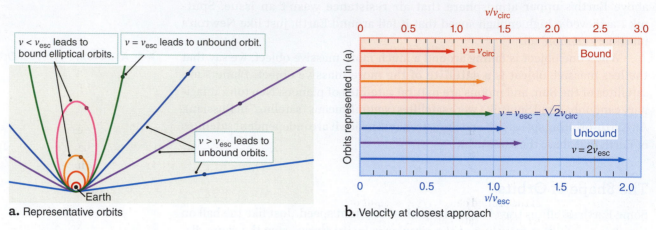

a. Representative orbits

b. Velocity at closest approach

Figure 4.9 a. Various orbits that share the same point of closest approach but differ in velocity at that point.   **b.** Closest-approach velocities for the orbits in a. An object's velocity at closest approach determines the orbit's shape and whether the orbit is bound or unbound. v_{circ} = circular velocity; v_{esc} = escape velocity.

object travels away from Earth, never to return. The lowest speed at which an object can permanently leave the gravitational grasp of another mass is called the **escape velocity**, v_{esc}. Once a satellite's velocity at closest approach equals or exceeds v_{esc}, and it is no longer gravitationally bound to the object it was orbiting, we say it travels on an **unbound orbit**.

As Figure 4.9a shows, an object with a velocity *less* than the escape velocity (v_{esc}) has an elliptically shaped orbit and follows the same path over and over again, whereas unbound orbits do not close like an ellipse. An object such as a comet on an unbound orbit makes only a single pass around the Sun and then continues away from the Sun into deep space, never to return. Circular velocity and escape velocity are further explored in **Working It Out 4.2**.

Estimating Mass by Using Newton's Version of Kepler's Laws

Newton's calculations opened up a new way to investigate the universe. He showed that the same physical laws that describe the flight of a cannonball on Earth—or the legendary apple falling from a tree—also describe the motions of the planets through the heavens. His laws of motion and gravitation predict all three of Kepler's empirical laws of planetary motion. Newton's version of Kepler's laws can be used to estimate the mass of the Sun from the orbit of Earth, as demonstrated in **Working It Out 4.3**.

When Earth, a much less massive object, is orbiting the Sun, a much more massive object, the Sun's gravity has a strong influence on Earth, but Earth's gravity has little effect on the Sun. Therefore, the Sun's motion due to Earth's orbit around it is tiny.

Astronomy in Action: Center of Mass

CHECK YOUR UNDERSTANDING 4.2

Rank the circular velocities of the following objects from smallest to largest: (a) a 5-kg object orbiting Earth halfway to the Moon; (b) a 10-kg object orbiting Earth just above Earth's surface; (c) a 15-kg object orbiting Earth at the same distance as the Moon.

working it out 4.2

Circular Velocity and Escape Velocity

Circular Velocity

In Appendix 8, we show that the circular velocity (v_{circ}) is given by

$$v_{\text{circ}} = \sqrt{\frac{GM}{r}}$$

where M is the mass of the orbited object, r is the radius of the circular orbit, and G is the universal gravitational constant. A cannonball moving at just the right velocity—the circular velocity—will fall around Earth on a circular path.

We can use that equation to show how fast Newton's cannonball would have to travel to stay in its circular orbit. The average radius of Earth is 6370 km, the mass of Earth is 5.97×10^{24} kg, and the gravitational constant is 6.67×10^{-20} km³/(kg s²). Inserting those values into the expression for v_{circ}, we get

$$v_{\text{circ}} = \sqrt{\frac{[6.67 \times 10^{-20}\,\text{km}^3/(\text{kg s}^2)] \times (5.97 \times 10^{24}\,\text{kg})}{6370\,\text{km}}} = 7.91\,\text{km/s}$$

To stay in its circular orbit, Newton's cannonball would have to be traveling about 8 kilometers per second (km/s)—more than 28,000 kilometers per hour (km/h). That's well beyond the range of a typical cannon, but rockets routinely reach those speeds.

Now let's compare that speed with the speed needed to launch a satellite from the Moon into orbit just above the lunar surface. The radius of the Moon is 1740 km, and its mass is 7.35×10^{22} kg. Those values give a considerably lower circular velocity than the one for Earth:

$$v_{\text{circ}} = \sqrt{\frac{(6.67 \times 10^{-20}\,\text{km}^3/(\text{kg s}^2)) \times (7.35 \times 10^{22}\,\text{kg})}{1740\,\text{km}}} = 1.68\,\text{km/s}$$

Escape Velocity

Sending a spacecraft to another planet requires launching it with a velocity greater than Earth's escape velocity. The escape velocity is a factor of $\sqrt{2}$, or approximately 1.41, times the circular velocity. That relation can be expressed as

$$v_{\text{esc}} = \sqrt{2} \times v_{\text{circ}} = \sqrt{\frac{2GM}{R}}$$

Using the numbers in the above example, we can calculate the escape velocity from the surface of Earth:

$$v_{\text{esc}} = \sqrt{2} \times v_{\text{circ}} = 1.41 \times 7.91\,\text{km/s} = 11.2\,\text{km/s}$$

To leave Earth, a rocket must have a speed of at least 11.2 km/s, or 40,100 km/h.

As with weight, the escape velocity from other astronomical objects is different from the escape velocity from Earth. Ida is a small asteroid orbiting the Sun between the orbits of Mars and Jupiter. Ida has an average radius of 15.7 km and a mass of 4.2×10^{16} kg. Therefore,

$$v_{\text{esc}} = \sqrt{\frac{2 \times [6.67 \times 10^{-20}\,\text{km}^3/(\text{kg s}^2)] \times (4.2 \times 10^{16}\,\text{kg})}{15.7\,\text{km}}}$$

$$v_{\text{esc}} = 0.019\,\text{km/s} = 68\,\text{km/h}$$

A baseball thrown at about 130 km/h (81 mph) would easily escape from Ida's surface and fly off into interplanetary space.

4.3 Tidal Forces Are Caused by Gravity

The rise and the fall of the oceans are called Earth's **tides**. Coastal dwellers long ago noted that the strength of the tides varies with the phase of the Moon. Tides are strongest during a new or a full Moon and are weakest during first quarter or third quarter Moon. Here in Section 4.3, we see how tides result from differences between the strength of the gravitational pull of the Moon and Sun on one part of Earth in comparison with their pull on other parts of Earth.

 **Astronomy in Action:** Tides

Tides and the Moon

Recall from Figure 4.4 that each small part of an object feels a gravitational attraction toward every other small part of the object, and that self-gravity differs from place to place. In addition, each small part of an object feels a gravitational attraction

working it out 4.3

Calculating Mass from Orbital Periods

Newton's Version of Kepler's Third Law

The time a planet takes to complete one orbit around the Sun (the orbital period, P) equals the distance traveled divided by the planet's speed. For simplicity, let's assume the orbit is circular. Thus, the time an object takes to make one trip around the Sun is the circumference of the circle ($2\pi r$) divided by the object's speed (v). The speed of the planet must be equal to the circular velocity discussed in Working It Out 4.2. Combining these two relationships, we have

$$\text{Orbital period }(P) = \frac{\text{Circumference of orbit}}{\text{Circular velocity}} = \frac{2\pi r}{\sqrt{\dfrac{GM_{\text{Sun}}}{r}}}$$

Squaring both sides of the equation gives

$$P^2 = \frac{4\pi^2 r^2}{\dfrac{GM_{\text{Sun}}}{r}} = \frac{4\pi^2}{GM_{\text{Sun}}} \times r^3$$

The square of the period of an orbit is equal to a constant ($4\pi^2/GM_{\text{Sun}}$) multiplied by the cube of the radius of the orbit. That is Kepler's third law ($P^2 = \text{constants} \times A^3$) applied to circular orbits. When Kepler used Earth units of years and astronomical units for the planets, he was taking a ratio of their periods and orbital radii with those for Earth, so the constants canceled out. Using calculus, Newton similarly could derive Kepler's third law for elliptical orbits with semimajor axis A instead of radius r:

$$P^2 = \frac{4\pi^2}{GM_{\text{Sun}}} \times A^3$$

Mass of the Sun

If we can measure the size and period of any orbit, we can use Newton's universal law of gravitation to calculate the mass of the object being orbited. To do so, we rearrange Newton's form of Kepler's third law above to read

$$M = \frac{4\pi^2}{G} \times \frac{A^3}{P^2}$$

Everything on the right side of that equation is either a constant (4, π, and G) or a measurable quantity (the semimajor axis A and period P of an orbit). The left side of the equation is the mass of the object at the focus of the ellipse. For example, we can find the mass of the Sun by noting the period and semimajor axis of the orbit of a planet around the Sun. Let's use the numbers for Earth. Whenever we have an equation with G, putting everything else into the same units as G (km, kg, s) tends to simplify the math. So first we must compute the number of seconds in 1 year: $P = 1$ yr $= 365.24$ day/yr $\times 24$ h/day $\times 60$ min/h $\times 60$ s/min $= 3.16 \times 10^7$ s. The semimajor axis $A = 1$ AU $= 1.5 \times 10^8$ km. Then the mass of the Sun can be computed:

$$M_{\text{Sun}} = \frac{4\pi^2}{G} \times \frac{A^3}{P^2} = \frac{4\pi^2}{6.67 \times 10^{-20}\ \text{km}^3/(\text{kg s}^2)} \times \frac{(1.5 \times 10^8\ \text{km})^3}{(3.16 \times 10^7\ \text{s})^2}$$

$$M_{\text{Sun}} = 2.0 \times 10^{30}\ \text{kg}$$

If we had used the period and semimajor axis of any other planet in the Solar System, we would have gotten the same result for the mass of the Sun.

▶▶ **Interactive Simulation:** Tides on Earth

▶‖ **AstroTour:** Tides and the Moon

toward every other mass in the universe, and those external forces also differ from place to place within the object.

The Moon pulls more strongly on the part of Earth that is closest (the near side) than it does on the part of Earth that's farthest (the far side). The pull of the Moon on the near side of Earth is about 7 percent greater than its pull on the far side of Earth. This is because the side of Earth that faces the Moon is closer to the Moon than the rest of Earth is, so it feels a stronger-than-average gravitational attraction toward the Moon. In contrast, the side of Earth facing away from the Moon is farther than average from the Moon, so it feels a weaker-than-average attraction toward the Moon.

To understand the consequence of that variation in the pull of the Moon, imagine three rocks being pulled by gravity toward the Moon. A rock closer to the Moon feels a stronger force than a rock farther from the Moon. Now suppose the three rocks are connected by springs (**Figure 4.10a**). As the

Reading Astronomy News

Earth Has Acquired a Brand-New Moon That's about the Size of a Car

Leah Crane

In the Solar System, whether an orbit is stable depends on interactions between objects that orbit the Sun. Sometimes even Earth gets a temporary extra moon!

Earth might have a tiny new moon. On 19 February, astronomers at the Catalina Sky Survey in Arizona spotted a dim object moving quickly across the sky. Over the next few days, researchers at six more observatories around the world watched the object, designated 2020 CD3, and calculated its orbit, confirming that it has been gravitationally bound to Earth for about three years.

An announcement posted by the Minor Planet Center, which monitors small bodies in space, states that "no link to a known artificial object has been found," implying that it is most likely an asteroid caught by Earth's gravity as it passed by.

This is just the second asteroid known to have been captured by our planet as a mini-moon—the first, 2006 RH120, hung around between September 2006 and June 2007 before escaping.

Our new moon is probably between 1.9 and 3.5 meters across, or roughly the size of a car, making it no match for Earth's primary moon. It circles our planet about once every 47 days on a wide, oval-shaped orbit that mostly swoops far outside the larger moon's path.

The orbit isn't stable, so eventually 2020 CD3 will be flung away from Earth. "It is heading away from the Earth-moon system as we speak," says Grigori Fedorets at Queen's University Belfast in the UK, and it looks likely it will escape in April.

However, there are several different simulations of its trajectory and they don't all agree—we will need more observations to accurately predict the fate of our mini-moon and even to confirm that it is definitely a temporary moon and not a piece of artificial space debris. "Our international team is continuously working to constrain a better solution," says Fedorets.

QUESTIONS

1. About how large (in meters) is the mini-moon? How large is this in feet?

2. How much time elapsed between when the mini-moon began orbiting Earth and its discovery in 2020? Is that disturbing? Why or why not?

3. What is the predicted ultimate fate of this mini-moon?

4. Look up this mini-moon on the Internet, by searching for it by name: "2020 CD3." What has happened to the mini-moon since this article was written?

Source: https://www.newscientist.com/article/2235427 -earth-has-acquired-a-brand-new-moon-thats-about-the -size-of-a-car/.

---

rocks are pulled toward the Moon, the purple rock pulls away from the red rock, and the red rock pulls away from the blue rock. Therefore, the differences in the gravitational forces they feel will stretch *both* springs (**Figure 4.10b**). Now, instead of springs, imagine that the rocks are at different places on Earth (**Figure 4.10c**). On the side of Earth away from the Moon, the force is smaller (as indicated by the shorter arrow), so that part gets left behind (**Figure 4.10d**). Those differences in the Moon's gravitational attraction on different parts of Earth are called **tidal forces**. Tidal forces cause an object to stretch.

Figure 4.11 shows how Earth is stretched, causing a tidal bulge. The Moon is not pulling on the far side of Earth as hard as it is pulling on the planet as a whole. The far side of Earth is "left behind" as the rest of the planet is pulled more strongly toward the Moon. Figure 4.11 shows that a net force also is squeezing inward on Earth in the direction perpendicular to the line between Earth and the

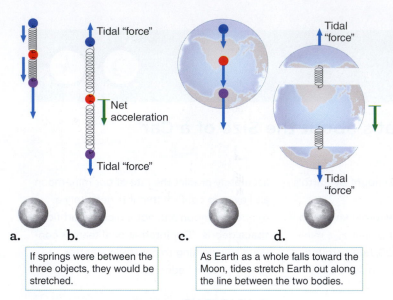

a.   b.   c.   d.

If springs were between the three objects, they would be stretched.

As Earth as a whole falls toward the Moon, tides stretch Earth out along the line between the two bodies.

Figure 4.10 **a.** Imagine three objects connected by springs. **b.** The springs are stretched as though forces were pulling outward on each end of the chain. **c.** Similarly, three locations on Earth experience different gravitational attractions toward the Moon. **d.** The difference in the Moon's gravitational attraction across Earth causes Earth's tides.

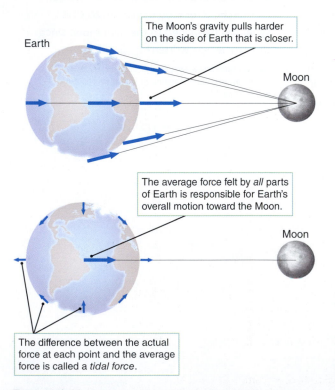

The Moon's gravity pulls harder on the side of Earth that is closer.

Earth

Moon

The average force felt by *all* parts of Earth is responsible for Earth's overall motion toward the Moon.

Moon

The difference between the actual force at each point and the average force is called a *tidal force*.

Figure 4.11 Tidal forces stretch Earth along the line between Earth and the Moon but compress Earth perpendicular to that line.

Moon. Together, the stretching by tidal forces along the line between Earth and the Moon and the squeezing by tidal forces perpendicular to that line distort Earth's shape, like a rubber ball caught in the middle of a tug-of-war.

If the surface of Earth were perfectly smooth and covered with an ocean of uniform depth and Earth did not rotate, the Moon would pull our oceans into an elongated **tidal bulge** like that shown in **Figure 4.12a**. The water would be at its deepest on the side toward the Moon and on the side away from the Moon, and the water would be at its shallowest midway between. But Earth is *not* covered with perfectly uniform oceans, and Earth *does* rotate. As any point on Earth rotates through the ocean's tidal bulges, that point experiences the ebb and flow of the tides. In addition, friction between the spinning Earth and its tidal bulge drags the oceanic tidal bulge around in the direction of Earth's rotation, as illustrated in **Figure 4.12b**.

Figure 4.12c shows the tides you would experience throughout a day. Recall that Earth's rotation period of one day is much shorter than the Moon's orbital period of about a month. Over one day, the Moon moves only a little way in its orbit, but you travel around Earth once. Each day, the Moon rises 50 minutes later than it did on the previous day, so the time between the Moon's return to the same position in the sky from the previous day is 24 h 50 m. Begin as the rotating Earth carries you through the tidal bulge on the Moonward side of the planet. Because Earth's rotation drags the tidal bulge, the Moon is not exactly overhead but is instead high in the western sky. When you are at the high point in the tidal bulge, the ocean around you has risen higher than average—called a *high tide*. The rotation of Earth then carries you around to a point where the ocean is lower than average—called a *low tide*. Later, you pass through the region where ocean water is "left behind" (with respect to Earth as a whole) in the tidal bulge on the side of Earth that is away from the Moon. At this high tide, the Moon at that time is hidden from view on the far side of Earth. After the Moon has risen above the eastern horizon, you pass through a second low tide. After 24 h 50 m—the amount of time the Moon takes to return to the same point in the sky from which it started—you again pass through the tidal bulge on the near side of the planet. Each of these tides occurs about $6\frac{1}{4}$ hours after the previous one, because there are four tides, and $\frac{1}{4}$ of 24 h 50 m is about $6\frac{1}{4}$ hours. That is the age-old pattern by which mariners have lived their lives for millennia: the twice-daily coming and going of high tide, shifting through the day in lockstep with the passing of the Moon.

Ocean currents and local geography, such as the shapes of Earth's shorelines and ocean basins, complicates the simple picture of tides. In addition, ocean-wide oscillations occur, similar to water sloshing around in a basin. As the oceans respond to the tidal forces from the Moon, they flow around the various landmasses that break up the water covering our planet. Some places, such as the Mediterranean Sea and the Baltic Sea, are protected from tides by their relatively small sizes and the narrow passages connecting those bodies of water to the larger ocean. In other places, the shape of the land funnels the tidal surge from a large region of ocean into a relatively small area, concentrating its effect, as at eastern Canada's Bay of Fundy (**Figure 4.13**).

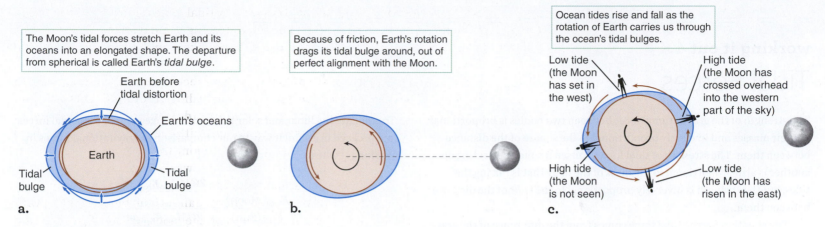

The Moon's tidal forces stretch Earth and its oceans into an elongated shape. The departure from spherical is called Earth's *tidal bulge*.

Earth before tidal distortion

Earth's oceans

Earth

Tidal bulge

Tidal bulge

a.

Because of friction, Earth's rotation drags its tidal bulge around, out of perfect alignment with the Moon.

b.

Ocean tides rise and fall as the rotation of Earth carries us through the ocean's tidal bulges.

Low tide (the Moon has set in the west)

High tide (the Moon has crossed overhead into the western part of the sky)

High tide (the Moon is not seen)

Low tide (the Moon has risen in the east)

c.

Figure 4.12 a. Tidal forces pull Earth and its oceans into a tidal bulge **b.** Earth's rotation pulls its tidal bulge slightly out of alignment with the Moon. **c.** As Earth's rotation carries us through those bulges, we experience the ocean tides. The magnitude of the tides has been exaggerated here for clarity. In these figures, the observer is looking down from above Earth's North Pole. Sizes and distances are not to scale.

Solar Tides

Tides resulting from the pull of the Moon are called **lunar tides**, but the Sun also influences Earth's tides. The gravitational pull of the Sun causes Earth to stretch along a line pointing approximately in the direction of the Sun. The side of Earth closer to the Sun is pulled toward the Sun more strongly than the side of Earth away from the Sun, just as the side of Earth closest to the Moon is pulled more strongly toward the Moon. Tides on Earth resulting from differences in the gravitational pull of the Sun are called **solar tides**. The absolute strength of the Sun's pull on Earth is nearly 200 times greater than the strength of the Moon's pull on Earth. However, the Sun's gravitational attraction does not change by much from one side of Earth to the other because the Sun is much farther away than the Moon. As a result, solar tides are only about half as strong as lunar tides (**Working It Out 4.4**).

a.

b.

Figure 4.13 The world's most extreme tides are found in the Bay of Fundy in eastern Canada. Water rocks back and forth in the bay with a period of about 13 hours, close to the 12.5-hour period of the tides. The shape of the basin amplifies the tides so that the difference in water depth between **a.** low tide and **b.** high tide is extreme—typically about 14.5 meters and as much as 16.6 meters.

working it out 4.4

Tidal Forces

The strength of the gravitational force between two bodies is proportional to their masses and inversely proportional to the square of the distance between them. The strength of tidal forces caused by one body acting on another is also proportional to the mass of the body that is raising the tides, but the strength is inversely proportional to the *cube* of the distance between them.

The equation for the tidal force comes from the difference of the gravitational force on one side of a body compared with the force on the other side. We can approximate the tidal force of the Moon acting on a mass m on the surface of Earth with

$$F_{tidal}(\text{Moon}) = \frac{2GM_{\text{Moon}}R_{\text{Earth}}m}{d_{\text{Earth-Moon}}^3}$$

where R_{Earth} is Earth's radius and $d_{\text{Earth-Moon}}$ is the distance between Earth and the Moon.

Let's compare the Moon's tidal force acting on Earth (given by the preceding equation) with the Sun's tidal force acting on Earth, which is given by the following equation:

$$F_{tidal}(\text{Sun}) = \frac{2GM_{\text{Sun}}R_{\text{Earth}}m}{d_{\text{Earth-Sun}}^3}$$

To compare the lunar and solar tides, we can take a ratio of the tidal forces and proceed in a similar way to our comparison of gravitational forces in Working It Out 4.1:

$$\frac{F_{tidal}(\text{Moon})}{F_{tidal}(\text{Sun})} = \frac{\dfrac{2GM_{\text{Moon}}R_{\text{Earth}}m}{d_{\text{Earth-Moon}}^3}}{\dfrac{2GM_{\text{Sun}}R_{\text{Earth}}m}{d_{\text{Earth-Sun}}^3}}$$

Canceling out the constant G and the terms common in both equations (m and R_{Earth}) gives

$$\frac{F_{tidal}(\text{Moon})}{F_{tidal}(\text{Sun})} = \frac{M_{\text{Moon}}}{M_{\text{Sun}}} \times \frac{d_{\text{Earth-Sun}}^3}{d_{\text{Earth-Moon}}^3} = \frac{M_{\text{Moon}}}{M_{\text{Sun}}} \times \left(\frac{d_{\text{Earth-Sun}}}{d_{\text{Earth-Moon}}}\right)^3$$

Using the values from Appendixes 2 and 4, $M_{\text{Moon}} = 7.35 \times 10^{22}$ kg, $M_{\text{Sun}} = 2 \times 10^{30}$ kg, $d_{\text{Earth-Moon}} = 384{,}400$ km, and $d_{\text{Earth-Sun}} = 1.5 \times 10^8$ km, gives

$$\frac{F_{tidal}(\text{Moon})}{F_{tidal}(\text{Sun})} = \frac{7.35 \times 10^{22}\text{ kg}}{2 \times 10^{30}\text{ kg}} \times \left(\frac{1.5 \times 10^8\text{ km}}{384{,}400\text{ km}}\right)^3 = 2.2$$

Thus, the tidal force between Earth and the Moon is 2.2 times stronger than the tidal force between Earth and the Sun. The Sun is still an important factor, though, and that's why the tides change depending on the alignment of the Moon and the Sun with Earth.

what if . . .

What if the Moon's orbit was gradually shrinking? As the Moon's orbital radius shrank to half, or even less, of its current size, how would the tides change, and what effects might that have on life on Earth?

Solar and lunar tides interact. When the Moon and the Sun are lined up with Earth (**Figure 4.14a**), at either new or full Moon, the lunar and solar tides on Earth overlap. That alignment creates more extreme tides; both higher high tides and lower low tides. The extreme tides near the new or full Moon are called **spring tides**—not because of the season but because the water appears to spring out of the sea. Conversely, when the Moon, Earth, and Sun make a right angle (**Figure 4.14b**), at the Moon's first and third quarters, the lunar and solar tidal forces stretch Earth in different directions, creating less extreme tides known as **neap tides**. The word *neap* is derived from the Saxon word *neafte*, which means "scarcity": at those times of the month, shellfish and other food gathered in the tidal region are less accessible because the low tide is higher than at other times. Neap tides are only about half as strong as average tides and only a third as strong as spring tides.

CHECK YOUR UNDERSTANDING **4.3**

Rank the following in order of weakest to strongest total tides: (a) new Moon in July; (b) first quarter Moon in July; (c) full Moon in January; (d) third quarter Moon in January.

4.4 Tidal Forces Affect Solid Bodies

In Section 4.3, you learned how the liquid of Earth's oceans moves in response to the tidal forces from the Moon and Sun. But those tidal forces also affect the solid body of Earth. As Earth rotates through its tidal bulge, the solid body of the planet is constantly being deformed by tidal forces. Earth is somewhat elastic (like a rubber ball), and tidal stresses cause a vertical displacement of about 30 centimeters (cm) between high tide and low tide, or roughly a third of the displacement of the oceans.

Deforming the shape of a solid object requires energy. For a practical demonstration of that fact, hold a rubber ball in your hand and squeeze and release it a few dozen times; it will gradually become warmer. The energy from the deformation is converted into thermal energy because of friction. Similarly, for Earth, friction from the tidal deformation opposes and takes energy from the rotation of Earth, causing Earth to gradually slow. Earth's internal friction adds to the slowing caused by friction between Earth and its oceans as the planet rotates through the tidal bulge of the oceans. As a result, Earth's days are currently getting about 1.7 milliseconds (ms) longer every century. That sounds insignificant, but it adds up: when dinosaurs ruled 200 million years ago, the day was closer to 23 hours long, and 200 million years into the future, the day will be close to 25 hours.

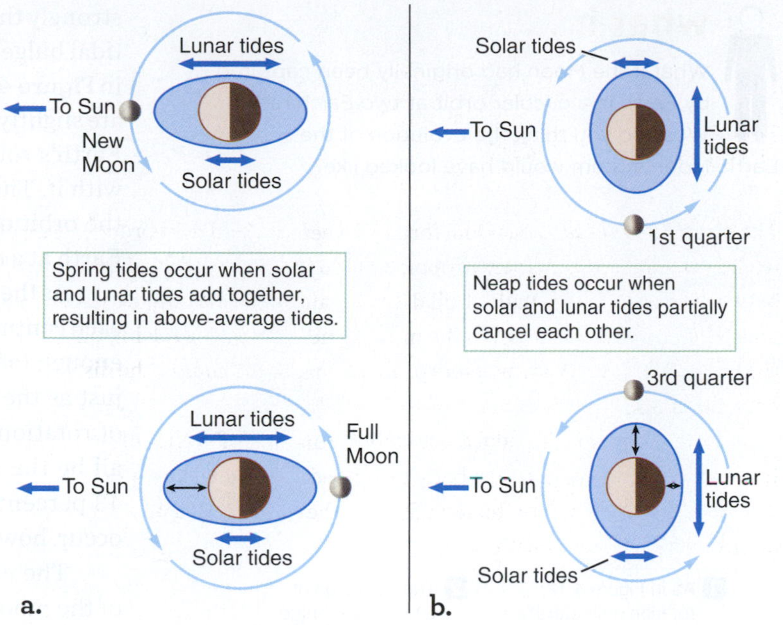

Figure 4.14 Solar tides are about half as strong as lunar tides. The interactions of solar and lunar tides result in either **a.** spring tides when they are added together or **b.** neap tides when they partially cancel each other.

Tidal Locking

Other solid bodies besides Earth experience tidal forces. For example, the Moon has no bodies of liquid to make tidal forces obvious, but its shape is distorted in the same manner as Earth's. Because of Earth's much greater mass, Earth's tidal effects on the Moon are stronger than the Moon's tidal effects on Earth. Given that the average tidal deformation of Earth is about 30 cm, the average tidal deformation of the Moon should be about 6 meters. However, what we actually observe on the Moon is a tidal bulge of about 20 meters. That unexpectedly large displacement exists because the Moon's tidal bulge was "frozen" into its relatively rigid crust at an earlier time, when the Moon was closer to Earth and tidal forces were much stronger than they are today. Planetary scientists sometimes call that deformation the Moon's *fossil tidal bulge*.

Recall that the Moon's rotation period equals its orbital period. That synchronous rotation of the Moon is a result of **tidal locking**. Early in the Moon's history, the period of its rotation was almost certainly different from its orbital period. As the Moon rotated through its extreme tidal bulge, however, friction within the Moon's crust was tremendous, rapidly slowing the Moon's rotation. After a fairly short time, the period of the Moon's rotation equaled the period of its orbit. When its orbital and rotation periods became equalized, the Moon no longer rotated with respect to its tidal bulge. Instead, the Moon and its tidal bulge rotated *together*, in lockstep with the Moon's orbit around Earth. As illustrated in **Figure 4.15**, that scenario continues today as the tidally distorted Moon orbits Earth, always keeping the same face and the long axis of its tidal bulge toward Earth.

Tidal forces affect not only the rotations of the Moon and Earth but also their orbits. Because of its tidal bulge, Earth is not perfectly spherical. Therefore, the material in Earth's tidal bulge on the side nearer the Moon pulls on the Moon more

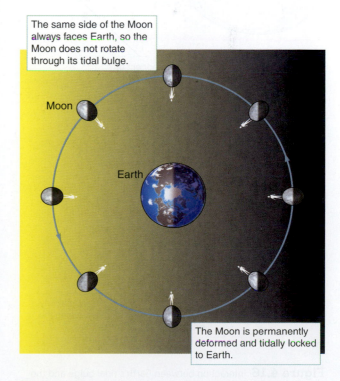

The same side of the Moon always faces Earth, so the Moon does not rotate through its tidal bulge.

The Moon is permanently deformed and tidally locked to Earth.

Figure 4.15 Tidal forces due to Earth's gravity lock the Moon's rotation to its orbital period.

what if . . .

What if the Moon had originally been captured by Earth in a circular orbit at two Earth radii? What do you think the evolution of the Earth–Moon system would have looked like?

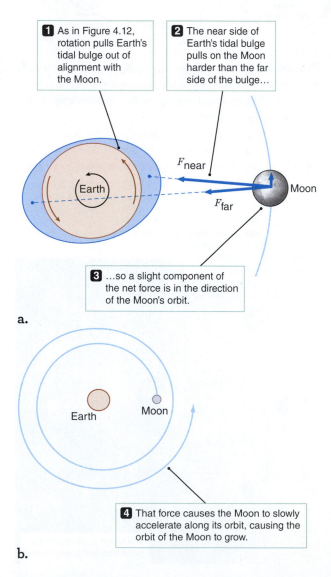

1 As in Figure 4.12, rotation pulls Earth's tidal bulge out of alignment with the Moon.

2 The near side of Earth's tidal bulge pulls on the Moon harder than the far side of the bulge...

F_{near}

Earth

Moon

F_{far}

3 ...so a slight component of the net force is in the direction of the Moon's orbit.

a.

Earth Moon

4 That force causes the Moon to slowly accelerate along its orbit, causing the orbit of the Moon to grow.

b.

Figure 4.16 Interaction between Earth's tidal bulge and the Moon causes the Moon to accelerate in its orbit and the Moon's orbit to grow.

strongly than does material in the tidal bulge on the back side of Earth. Because the tidal bulge on the Moonward side of Earth "leads" the Moon somewhat, as shown in **Figure 4.16**, the gravitational attraction of the bulge causes the Moon to accelerate slightly along the direction of its orbit around Earth and simultaneously causes Earth's rotation to slow down. The rotation of Earth is dragging the Moon along with it. The acceleration of the Moon in the direction of its orbital motion causes the orbit of the Moon to grow larger. At present, the Moon is drifting away from Earth at a rate of 3.83 cm per year.

As the Moon grows more distant, the lunar month gets about 0.014 second longer each century. If that increase in the radius of the Moon's orbit were to continue long enough (about 50 billion years), Earth would become tidally locked to the Moon, just as the Moon is now tidally locked to Earth. At that point, the Earth's period of rotation, the Moon's period of rotation, and the Moon's orbital period would all be the same—about 47 of our current days—and the Moon would be about 43 percent farther from Earth than it is today. That situation will never actually occur, however, because the Sun will have burned itself out long before then.

The effects of tidal forces are apparent throughout the Solar System. Most of the moons in the Solar System are tidally locked to their parent planets, and for dwarf planet Pluto and its largest moon, Charon, each is tidally locked to the other.

Tidal locking is only one way that orbits and rotations can be coupled. Tidal forces have coupled the planet Mercury's rotation to its very elliptical orbit around the Sun. Unlike the Moon's synchronous rotation, however, Mercury spins on its axis three times for every two trips around the Sun. The period of Mercury's orbit—87.97 Earth days—is exactly 1½ times the 58.64 days that Mercury takes to spin once on its axis. When Mercury comes to the point in its orbit that is closest to the Sun, one hemisphere faces the Sun, and then in the next orbit, the other hemisphere faces the Sun.

Tidal Forces on Many Scales

We normally think of the effects of tidal forces as small in comparison with the force of gravity holding an object together, yet tidal effects can be extremely destructive. Consider for a moment the fate of a small moon, asteroid, or comet that wanders too close to a massive planet such as Jupiter or Saturn. All objects in the Solar System larger than about a kilometer in diameter are held together by their self-gravity. However, the self-gravity of a small object such as an asteroid, a comet, or a small moon is feeble. In contrast, the tidal forces close to a massive object such as Jupiter can be very strong. If the tidal forces trying to tear an object apart become greater than the self-gravity trying to hold the object together, the object will break into pieces.

The **Roche limit** is the distance at which a planet's tidal forces become greater than the self-gravity of a smaller object—such as a moon, asteroid, or comet—causing the object to break apart. For a smaller object having the same density as the planet, the Roche limit is about 2.45 times the planet's radius. Such an object bound together solely by its own gravity can remain intact when it is outside a planet's Roche limit, but not when it is inside the limit. Objects such as the International Space Station and other Earth satellites are not torn apart, even though they orbit well within Earth's Roche limit, because chemical and physical bonds hold them together, not just self-gravity.

Tidal forces exist throughout the Solar System and the universe. Any time two objects of significant size or two collections of objects interact gravitationally, the gravitational forces will differ from one place to another within the objects, giving rise to tidal effects. Tidal disruption of small bodies is the source of the particles that make up the rings of the giant planets. Tidal interactions can cause material from one star in a binary pair to be pulled onto the other star. Tidal effects can strip stars from clusters consisting of thousands of stars. Galaxies can pass close enough together to strongly interact gravitationally. When that happens, as in **Figure 4.17**, tidal effects can grossly distort both galaxies. Tidal forces even play a role in shaping huge collections of galaxies—the largest known structures in the universe.

CHECK YOUR UNDERSTANDING 4.4

The Moon always keeps the same face toward Earth because of: (a) tidal locking; (b) tidal forces from the Sun; (c) tidal forces from Earth and the Sun.

Origins: Tidal Forces and Life

Earth's rotation is slowing down as the Moon slowly moves away into a larger orbit, so in the distant past the Moon was closer and Earth rotated faster. Tides at that time would have been stronger and the interval between high tides would have been shorter. Tides are also affected by the configuration of the continents and oceans on Earth, so precisely how much faster Earth rotated billions of years ago is unknown. What is known, though, is that the stronger and more frequent tides would have provided additional energy to the oceans of the young Earth.

Scientists debate whether life on Earth originated deep in the ocean, on the surface of the ocean, or in hot springs on land (see Chapter 24). The tides shaped the regions in the margins between land and ocean, such as tide pools and coastal flats. Some researchers think that those border regions, which alternate between wet and dry with the tides, could have been places where concentrations of biochemicals periodically became high enough for more complex reactions to take place. Those complex reactions were important to the earliest life. Later, those border regions may have been important as advanced life moved from the sea to the land.

Elsewhere in the Solar System, the giant planets Jupiter and Saturn are far from the Sun and thus very cold. Both Jupiter and Saturn have many moons, and the closest moons would experience strong tidal forces from their respective planets. Several of those moons are thought to have a liquid ocean underneath an icy surface. Tidal forces from Jupiter or Saturn supply the heat to keep the water in a liquid state. Astrobiologists think that those subsurface liquid oceans, such as the ones on Saturn's moon Enceladus, are perhaps the most probable location for life elsewhere in the Solar System.

On Earth, solar tides are about half as strong as lunar tides. A planet with a closer orbit would experience much stronger tidal forces from its star. Many planets detected outside the Solar System have orbits very close to their stars (see Chapter 7), so those exoplanets experience strong tidal forces. The planets might be tidally locked, too, so they have a synchronous rotation like the Moon does around Earth, with one side of the planet always facing the star and one side facing away. How might life on Earth have evolved differently if half the planet was in perpetual night and half in perpetual day?

Figure 4.17 ★ WHAT AN ASTRONOMER SEES
An astronomer looking at this image will instantly identify that there are two galaxies here, indicated by the bright white blobs. She will see that these galaxies are in motion, because the tidal "tails" that stream roughly diagonally across the image are characteristic of tidal interactions between galaxies. These are made of stars, gas, and dust from each galaxy that have been left behind as the galaxies come together. To an astronomer, this picture is dynamic, providing a snapshot of a time during which these galaxies are evolving from two distinct galaxies into one larger one.

Credit: NASA, Holland Ford (JHU), the ACS Science Team and ESA. https://esahubble.org/images/heic0206b/. https://creativecommons.org/licenses/by/4.0/.

SUMMARY

Objects stay in orbit because of gravity. Newton's laws of motion and his proposed law of gravity predict the paths of planetary orbits and explain Kepler's laws. Newton's calculations showed that those orbits should be ellipses with the Sun at one focus, that planets should move faster when closer to the Sun, and that the square of the period of a planet's orbit should vary as the cube of the semimajor axis of that elliptical orbit. Newton also mathematically confirmed Galileo's conclusion that falling objects have an accelerated motion independent of their mass. Tidal forces provide energy to Earth's oceans. Tide pools on Earth may have been a site of early biochemical reactions. Some moons in the outer Solar System might have liquid water because of tidal heating from their respective planets.

(1) **Identify the elements of Newton's universal law of gravitation.** Gravity is a force between any two objects with mass. Gravity is one of the fundamental forces of nature—it binds the universe together. The force of gravity is proportional to the mass of each object and inversely proportional to the square of the distance between them.

(2) **Use the laws of motion and gravitation to relate planetary orbits, trajectories of rocket flights, and the escape velocity from a system.**

An orbit is one body "falling around" another. Planets orbit the Sun in elliptical orbits. All objects affected by gravity move either in bound elliptical orbits or unbound paths. If an object acquires escape velocity, it is moving fast enough to be on an unbound path. Orbits and trajectories are ultimately given their shape by the gravitational attraction of the objects involved, which in turn reflects the masses of those objects.

(3) **Explain how gravitational forces from the Sun and Moon create Earth's tides.** Tides on Earth are the result of differences between how hard the Moon and Sun pull on one part of Earth in comparison with their pull on other parts of Earth. The primary cause of tides is the Moon, which stretches out Earth. The tides are the strongest when the Sun, Moon, and Earth are aligned. As Earth rotates, tides rise and fall twice each day.

(4) **Describe how tidal forces affect solid bodies.** Tidal forces lock the Moon's rotation to its orbit around Earth. Tidal forces can break up an object if it gets too close to a more massive object. Tidal forces are observed throughout the universe—in planets and moons, pairs of stars, and interacting galaxies.

QUESTIONS AND PROBLEMS

TEST YOUR UNDERSTANDING

1. In Newton's universal law of gravitation, the force is
   a. proportional to both masses.
   b. proportional to the radius.
   c. proportional to the radius squared.
   d. inversely proportional to the orbiting mass.

2. If we move a satellite to a new, higher altitude above Earth, the satellite would now
   a. have a faster speed.
   b. have a higher mass.
   c. have a lower mass.
   d. have a slower speed.

3. An object in a(n) _____ orbit in the Solar System will remain in its orbit forever. An object in a(n) _____ orbit will escape from the Solar System.
   a. unbound; bound
   b. circular; elliptical
   c. bound; unbound
   d. elliptical; circular

4. Compared with your mass on Earth, your mass on the Moon would be
   a. lower because the Moon is smaller than Earth.
   b. lower because the Moon has less mass than Earth.
   c. higher because of the combination of the Moon's mass and size.
   d. the same—mass doesn't change.

5. If you went to Mars, your weight would be
   a. higher because you are closer to the center of the planet.
   b. lower because Mars has two small moons instead of one big moon, so less tidal force is present.
   c. lower because the lower mass and smaller radius of Mars combine to a lower surface gravity.
   d. the same as on Earth.

6. Venus has about 80 percent of Earth's mass and about 95 percent of Earth's radius. Your weight on Venus would be
   a. 20 percent more than on Earth.
   b. 20 percent less than on Earth.
   c. 10 percent more than on Earth.
   d. 10 percent less than on Earth.

7. The connection between gravity and orbits enables astronomers to measure the _____ of stars and planets.
   a. distances
   b. sizes
   c. masses
   d. compositions

8. If the Moon were twice as massive, how would the strength of lunar tides change?
   a. The highs would be higher, and the lows would be lower.
   b. Both the highs and the lows would be higher.
   c. The highs would be lower, and the lows would be higher.
   d. Nothing would change.

9. If Earth had half its current radius, how would the strength of lunar tides change?
   a. The highs would be higher, and the lows would be lower.
   b. Both the highs and the lows would be higher.
   c. The highs would be lower, and the lows would be higher.
   d. Both the highs and the lows would be lower.

10. If the Moon were 2 times closer to Earth than it is now, the gravitational force between Earth and the Moon would be
    a. 2 times stronger.   c. 8 times stronger.
    b. 4 times stronger.   d. 16 times stronger.

11. If the Moon were 2 times closer to Earth than it is now, the tides would be
    a. 2 times stronger.   c. 8 times stronger.
    b. 4 times stronger.   d. 16 times stronger.

12. If two objects are tidally locked to each other,
    a. the tides always stay on the same place on each object.
    b. the objects always remain in the same place in each other's sky.
    c. the objects are falling together.
    d. both a and b would be true.

13. Spring tides occur only when
    a. the Sun is near the vernal equinox.
    b. the Moon's phase is new or full.
    c. the Moon's phase is first quarter or third quarter.
    d. it is either spring or fall.

14. If an object crosses from farther to closer than the Roche limit, it
    a. can no longer be seen.
    b. begins to accelerate very quickly.
    c. slows down.
    d. may be torn apart.

15. Self-gravity is
    a. the gravitational pull of a person.
    b. the force that holds objects such as people and lamps together.
    c. the gravitational interaction of all the parts of a body.
    d. the force that holds objects on Earth.

THINKING ABOUT THE CONCEPTS

16. Both Kepler's laws and Newton's laws tell us something about the motion of the planets, but they have a fundamental difference. What is that difference?

17. Explain the difference between circular velocity and escape velocity. Which of those must be larger? Why?

18. Explain the difference between weight and mass.

19. Weight on Earth is proportional to mass. Weight is proportional to mass on the Moon, too, but the multiplier is different on the Moon from what it is on Earth. Why? Explain why that difference does not violate the universality of physical law, as described in the Process of Science Figure. ★

20. Suppose a satellite has its orbit disrupted so that it begins traveling at a speed of twice the circular velocity. Study Figure 4.9 to determine if the orbit of the satellite remains bound, or if it is unbound. ★

21. What is the advantage of launching satellites from spaceports located near the equator? Would you expect satellites to be launched to the east or to the west? Why?

22. Explain how to use celestial orbits to estimate an object's mass. What observational quantities do you need to make that estimation?

23. What determines the strength of gravity at various radii between Earth's center and its surface?

24. The best time to dig for clams along the seashore is when the ocean tide is at its lowest. What phases of the Moon would be best for clam digging? What would be the best times of day during those phases?

25. The Moon is on the meridian at your seaside home, but your tide calendar does not show that it is high tide. Use Figure 4.12 to explain that apparent discrepancy. ★

26. We may have an intuitive feeling for why lunar tides raise sea level on the side of Earth facing the Moon, but why is sea level also raised on the side facing away from the Moon?

27. Tides raise and lower the level of Earth's oceans. Can they do the same for Earth's landmasses? Explain your answer.

28. Lunar tides raise the ocean surface by less than 1 meter. How can tides as large as 5–10 meters occur?

29. Most commercial satellites are well inside the Roche limit as they orbit Earth. Why are they not torn apart?

30. ★ WHAT AN ASTRONOMER SEES   In Figure 4.17, two galaxies are merging. The tidal tail of the galaxy on the right shows its trajectory and indicates that these galaxies are about to pass each other. Why, then, would an astronomer say these galaxies are going to merge? ★

APPLYING THE CONCEPTS

31. Mars has about one-tenth the mass of Earth and about half of Earth's radius. What is the value of gravitational acceleration on the surface of Mars in comparison with that on Earth? Compare your estimated mass and weight on Mars with your mass and weight on Earth. Do Hollywood movies showing people on Mars accurately portray that difference in weight? ●━●━●
    a. Make a prediction: Mars is less massive, but also smaller than Earth. Will these factors work together, or will they somewhat cancel out? Do you expect the gravitational acceleration on Mars to be very much bigger, very much smaller, or about the same as the gravitational acceleration on Earth?
    b. Set up a ratio. Because the first part of the question asks for a number "in comparison," the question is asking for a ratio. Use the formula for gravitational acceleration on Earth, rewrite it for Mars, and then set up to find the ratio of g_{Mars}/g_{Earth}.
    c. Solve the right-hand-side of the ratio.
    d. Check your work by comparing your answer to your prediction.
    e. Answer the other questions in the problem; be sure to think about the difference between mass and weight!

32. Earth speeds along at 29.8 km/s in its orbit. Neptune's nearly circular orbit has a radius of 4.5×10^9 km, and Neptune takes 164.8 Earth years to make one trip around the Sun. Calculate how fast Neptune moves along in its orbit. ●–●–●
    a. Make a prediction: Neptune is much farther from the Sun than Earth. Do you expect Neptune to move faster or slower in its orbit?
    b. Calculate: Follow Working It Out 4.2 to calculate Neptune's circular velocity. (Ponder: Why can you use the "circular velocity" formula here?)
    c. Check your work: Check the units of your final answer, and compare your answer to your prediction to check your work.

33. Venus's circular velocity is 35.03 km/s, and its orbital radius is 1.082×10^8 km. Use that information to calculate the mass of the Sun. ●–●–●
    a. Make a prediction: Do you expect the mass of the Sun to be very large or very small? What units do you expect to have for "mass"?
    b. Calculate: Follow Working It Out 4.3 to calculate the mass of the Sun.
    c. Check your work: Check your units and compare to your prediction to check your work. In this instance, the answer is well known, so you can also compare your answer to the actual mass of the Sun to see if your answer is approximately in agreement. (Ponder: You have made an assumption in this problem that the orbit of Venus is perfectly circular. Given that assumption, might it be reasonable for your answer to be slightly "off" from the accepted value?)

34. Some astrologers claim that your destiny is determined by the "influence" of the planets that are rising above the horizon at the moment of your birth. Compare the tidal force of Jupiter (mass = 1.9×10^{27} kg; distance = 7.8×10^8 km) with that of the doctor in attendance at your birth (mass = 80 kg; distance = 1 meter = 0.001 km). ●–●–●
    a. Make a prediction: Consider that in our discussion of the tides, we only worried about the Sun and the Moon. Given that, what might you conclude about the tidal force of Jupiter at the distance of Earth—is it likely to be larger or smaller?
    b. Calculate: Follow Working It Out 4.4 to calculate the ratio of the tidal force from Jupiter to the tidal force from the doctor.
    c. Check your work: Compare your answer to your prediction.

35. At the surface of Earth, the escape velocity is 11.2 km/s. What would be the escape velocity at the surface of a very small asteroid having a radius 10^{-4} that of Earth and a mass 10^{-12} that of Earth?
    a. Make a prediction: This asteroid is less massive, but also much smaller than Earth. Will these factors work together, or will they somewhat cancel out? Do you expect the escape velocity from an asteroid to be very much bigger, very much smaller, or about the same as the escape velocity from Earth?
    b. Set up a ratio: All the information in the problem is given in terms of those values for Earth. Follow Working It Out 4.2 to set up a ratio of the escape velocity for the asteroid divided by the escape velocity from Earth. On the right-hand side, the mass of Earth and the radius of Earth will divide out.
    c. Calculate: Solve for the escape velocity of the asteroid.
    d. Check your work: Compare your answer to your prediction to check your work.

36. When a spacecraft is sent to Mars, it is first launched into an Earth orbit with circular velocity.
    a. Describe the shape of that orbit.
    b. What minimum velocity must we give the spacecraft to send it on its way to Mars?

37. Earth's average radius is 6370 km and its mass is 5.97×10^{24} kg. Show that the acceleration of gravity at the surface of Earth is 9.81 m/s².

38. Using 6370 km for Earth's radius, compare the gravitational force acting on a NASA rocket when it is sitting on its launchpad with the gravitational force acting on it when it is orbiting 350 km above Earth's surface.

39. The International Space Station travels on a nearly circular orbit 350 km above Earth's surface. What is its orbital speed?

40. Using the last equation in Working It Out 4.1, calculate the value of g on Earth. Compare that result with the measured value.

41. How long does Newton's cannonball, moving at 7.9 km/s just above Earth's surface, take to complete one orbit around Earth?

42. The asteroid Ida (mass = 4.2×10^{16} kg) is attended by a tiny asteroidal moon, Dactyl, which orbits Ida at an average distance of 90 km. If you ignore the mass of the tiny moon, what is Dactyl's orbital period in hours?

43. Suppose you go skydiving.
    a. Just as you fall out of the airplane, what is your gravitational acceleration?
    b. Would that acceleration be bigger, smaller, or the same if you were strapped to a flight instructor and so had twice the mass?
    c. Just as you fall out of the airplane, what is the gravitational force on you? (Assume that your mass is 70 kg.)
    d. Would that force be bigger, smaller, or the same if you were strapped to a flight instructor and so had twice the mass?

44. Assume that a planet just like Earth is orbiting the bright star Vega at a distance of 1 astronomical unit (AU). Vega has twice the mass of the Sun.
    a. How long in Earth years will that planet take to complete one orbit around Vega?
    b. How fast is the Earth-like planet traveling in its orbit around Vega?

45. Suppose that in the past the Moon was 80 percent of the distance from Earth that it is now. Calculate how much stronger the lunar tides would have been. How would the neap and spring tides be different from now?

EXPLORATION Newton's Laws

digital.wwnorton.com/astro7

In the Exploration of Chapter 3, we used the "Planetary Orbits" Interactive Simulation to explore Kepler's laws for Mercury. Now that we know how Newton's laws explain why Kepler's laws describe orbits, we will revisit the simulator to explore the Newtonian features of Mercury's orbit. Visit the Digital Resources Page and on the Student Site open the "Planetary Orbits" Interactive Simulation in Chapter 3.

Acceleration

To begin exploring the simulation, set parameters for "Mercury" in the top left panel and then click "OK." Click the "Newtonian Features" tab at the bottom of the control panel. Select "Show Solar System Orbits" and "Show Grid" under "Visualization Options." Slow down the animation rate, and select the "Play" button.

Examine the graph at the bottom of the panel.

1. Where is Mercury in its orbit when the acceleration is smallest?

2. Where is Mercury in its orbit when the acceleration is largest?

3. What are the values of the largest and smallest accelerations?

In the "Newtonian Features" graph, mark the boxes for "Vector" and "Line" that correspond to the acceleration. Checking these boxes will insert an arrow that shows the direction of the acceleration and a line that extends the arrow.

4. To what Solar System object does the arrow point?

5. In what direction is the force on the planet?

Velocity

Examine the graph at the bottom of the panel again.

6. Where is Mercury in its orbit when the velocity is smallest?

7. Where is Mercury in its orbit when the velocity is largest?

8. What are the values of the largest and smallest velocities?

Add the velocity vector and line to the simulation by clicking on the boxes below the graph window. Study the resulting arrows carefully.

9. Are the velocity and the acceleration always perpendicular (is the angle between the arrows always 90°)?

10. If the orbit were a perfect circle, what would be the angle between the velocity and the acceleration?

Hypothetical Planet

In the top left panel, change the semimajor axis to 0.8 AU.

11. How does that imaginary planet's orbital period now compare with Mercury's? (Check on the Kepler's Third Law tab.)

Now change the semimajor axis to 0.1 AU.

12. How does the planet's orbital period now compare with Mercury's?

13. Summarize your observations of the relationship between the orbital period of an orbiting object and its semimajor axis.

Light

Light bends as it passes through a gap or past a sharp edge. To see this, wrap some tape around the top of a pencil, and then hold another pencil tightly against it to make a small slit. Hold the slit between the pencils up to your eye, and look through it at an LED light, such as the power light on your television or from some other electronic source. You will see a bright line perpendicular to the slit, caused by the light bending as it passes each side of the gap. (If it's difficult to see, you may need to darken the rest of the room.) Rotate the slit, and you will see that the line rotates, too. If you look carefully, you will see a series of bumps in the line. The slit is spreading the light out into a spectrum, and in some places the light overlaps to add together, whereas in other places it overlaps to cancel out. Because the spectrum of an LED has only one color in it, the bumps are very clear and of only one color. Now try this again with a small light source that has multiple colors (like the flashlight of your phone, which appears white). It's more difficult to see, but now the bumps are rainbow colored, with red at one end and blue at the other. You can try this out with all kinds of light sources, to see how they are the same or different.

EXPERIMENT SETUP

Wrap some tape around the top of a pencil...

SLIT →

...and then hold another pencil upside down and tightly against it to make a small slit.

LED power light on your television or other electronics.

Hold the slit between the pencils up to your eye, and look through it at the LED light. You will see a bright line perpendicular to the slit, caused by the light bending as it passes each side of the gap.

PREDICTION

I predict that the bumps on each side of the slit between the pencils will move

☐ **farther apart**   ☐ **closer together**

if I make the slit smaller by squeezing the pencils closer together.

SKETCH OF RESULTS

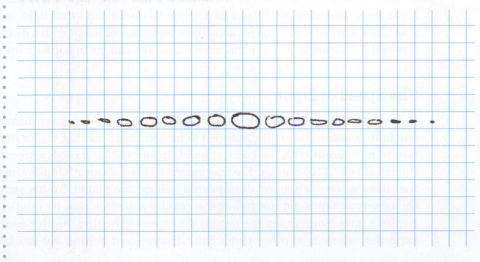

5

Our knowledge of the universe beyond Earth comes from light emitted, absorbed, or reflected by astronomical objects. Light carries information about the temperature, composition, and speed of the objects. Light also tells us about the nature of the material that the light passed through on its way to Earth. Light, however, plays a far larger role in astronomy than as a messenger. Light is one of the primary means by which energy is transported throughout the universe. Stars, planets, and vast clouds of gas and dust filling the space between the stars heat up as they absorb light and cool off as they emit light. Light carries energy generated in the heart of a star outward through the star and off into space. Light transports energy from the Sun outward through the Solar System, heating the planets; and light carries energy away from each planet, allowing each one to cool. The balance between those two processes establishes each planet's temperature and therefore a planet's possible suitability for life.

LEARNING GOALS

An astronomer must try to understand the universe by studying the light and other particles that reach Earth from distant objects. By the end of this chapter, you should be able to:

(1) Compare the wave and particle properties of light, and describe the electromagnetic spectrum.

(2) Describe how to measure the chemical composition of distant objects by using the unique spectral lines of different types of atoms.

(3) Apply the Doppler effect and use it to measure the motion of distant objects.

(4) Explain how the spectrum of light that an object emits depends on its temperature.

(5) Differentiate luminosity from brightness, and illustrate how distance affects each.

5.1 Light Brings Us the News of the Universe

Is light a **wave**—a disturbance that travels from one point to another—or is it made up of particles? Scientists have come to understand that light sometimes acts like a wave and sometimes acts like a particle. We begin with a discussion of how fast light travels. We then discuss its wavelike and particle-like properties.

The Speed of Light

In the early 1600s, Galileo tried to measure the speed of light as it traveled from one hilltop to another. This distance was far too small, and light is far too fast, for him to measure it with the tools at hand. In the 1670s, Danish astronomer Ole Rømer (1644–1710) studied the movement of the moons of Jupiter, measuring the times when each moon disappeared behind the planet. To his amazement, the observed times did not follow the regular schedule that he predicted from Kepler's laws. Sometimes the moons disappeared behind Jupiter sooner than expected, whereas at other times they disappeared behind Jupiter later than

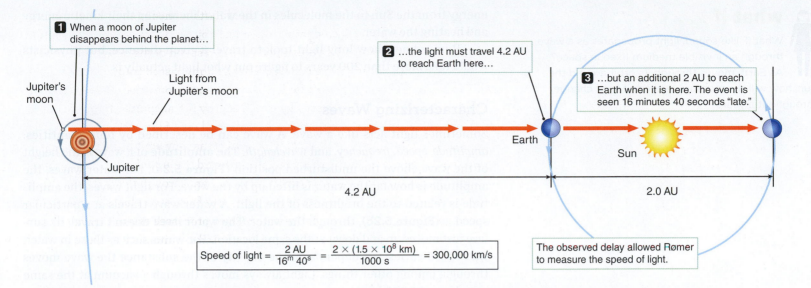

1 When a moon of Jupiter disappears behind the planet…

2 …the light must travel 4.2 AU to reach Earth here…

3 …but an additional 2 AU to reach Earth when it is here. The event is seen 16 minutes 40 seconds "late."

Jupiter's moon

Light from Jupiter's moon

Jupiter

Earth

Sun

4.2 AU

2.0 AU

$$\text{Speed of light} = \frac{2\ \text{AU}}{16^m\ 40^s} = \frac{2 \times (1.5 \times 10^8\ \text{km})}{1000\ \text{s}} = 300{,}000\ \text{km/s}$$

The observed delay allowed Rømer to measure the speed of light.

Figure 5.1 Danish astronomer Ole Rømer realized that apparent differences between the predicted and observed orbital motions of Jupiter's moons depend on the distance between Earth and Jupiter. He used those observations to measure the speed of light. (The superscript letters in "16ᵐ 40ˢ" stand for minutes and seconds of time, respectively.)

expected. Rømer realized that the difference depended on where Earth was in its orbit. If he began tracking the moons when Earth was closest to Jupiter, the moons were almost 17 minutes "late" by the time Earth was farthest from Jupiter. When Earth was once again closest to Jupiter, the moons again passed behind Jupiter at the predicted times.

Rømer correctly concluded that his observations did not represent a failure of Kepler's laws. Instead, he was seeing the first clear evidence that light travels at a finite speed. As shown in **Figure 5.1**, the moons appeared "late" when Earth was farther from Jupiter because of the time needed for light to travel the extra distance between the two planets. Over the course of Earth's yearly trip around the Sun, the distance between Earth and Jupiter changes by 2 astronomical units (AU). The speed of light equals that distance divided by Rømer's 16.7-minute delay, or about 3×10^5 kilometers per second (km/s). The value that Rømer actually announced in 1676 was a bit on the low side—2.25×10^5 km/s—because the size of Earth's orbit was not well known. Modern measurements of the speed of light give a value of 2.99792458×10^5 km/s in a **vacuum** (a region of space devoid of matter). The speed of light in a vacuum is one of nature's fundamental constants, usually written as c (lowercase). The speed of light through any medium, such as air or glass, is always less than c. The International Space Station moves around Earth at a speed of about 28,000 kilometers per hour (km/h), taking 91 minutes to complete one orbit. Light travels almost 40,000 times faster than that and can circle Earth in only $\frac{1}{7}$ of a second.

Because light is so fast, and its speed is constant, it is convenient to express cosmic distances in terms of the time it takes light to travel that far, in units such as light-seconds, light-hours, or light-years. Light takes $1\frac{1}{4}$ seconds to travel between Earth and the Moon, so the Moon is $1\frac{1}{4}$ *light-seconds* from Earth. The Sun is $8\frac{1}{3}$ *light-minutes* away, and the next-nearest star is $4\frac{1}{3}$ *light-years* distant. A *light-year*—how far light travels in 1 year—is about 9.5 trillion km.

While traveling at that high speed, light carries energy from place to place. **Energy** is the ability to do work, and it comes in many forms. **Kinetic energy (E_k)** is the energy of moving objects. **Thermal energy** is closely related to kinetic energy and is the sum of all the random motion of atoms, molecules, and particles in some region, which we characterize by its temperature. For example, when light from the Sun strikes a body of water, the water heats up. Light carried that

unanswered questions

Has the speed of light always been 300,000 km/s? Some theoretical physicists have questioned whether light traveled much faster earlier in the history of our universe. The observational evidence that may test that idea comes from studying the spectra of the most distant objects—whose light has been traveling for billions of years—and determining whether billions of years ago chemical elements absorbed light differently from how they do today. So far, no evidence exists that the speed of light has changed.

what if . . .

What if, like sound, light propagates as a wave through an invisible medium fixed in space? As Earth moves through its orbit around the Sun, how will the measured speed of light change throughout a year?

energy from the Sun to the molecules in the water, increasing their kinetic energy and heating the water.

Rømer knew how long light took to travel a given distance, but physicists would take more than 200 years to figure out what light actually is.

Characterizing Waves

Sometimes light acts like a wave. A wave can be described by four quantities: *amplitude, speed, frequency,* and *wavelength*. The **amplitude** of a wave is the height of the wave above the undisturbed position (**Figure 5.2a**). For water waves, the amplitude is how far the water is lifted up by the wave. For light waves, the amplitude is related to the brightness of the light. A water wave travels at a particular speed, v (**Figure 5.2b**), through the water. The water itself doesn't travel; its surface just moves up and down at the same location. For waves such as those in water, that speed varies and depends on the density of the substance the wave moves through, among other things. Light always moves through a vacuum at the same speed, $c \approx 300,000$ km/s.

The distance from one crest of a wave to the next is the **wavelength**, usually denoted by the Greek letter lambda, λ (**Figures 5.2c** and **5.2d**). The number of wave crests passing a point in space each second is the wave's **frequency**, f. The unit of frequency is cycles per second, which is called **hertz (Hz)** after the 19th-century physicist Heinrich Hertz (1857–1894), who was the first to experimentally confirm theoretical predictions about electromagnetic radiation.

Figures 5.2c and 5.2d show that waves with longer (larger) wavelengths have lower (smaller) frequencies, whereas waves with shorter (smaller) wavelengths have higher (larger) frequencies. Higher-frequency waves carry more energy. Imagine standing on an ocean beach with the waves lapping up against you: the amount of energy you feel from the waves increases as the frequency of the ocean waves increases.

Waves travel a distance of one wavelength each cycle, so the speed of a wave can be found by multiplying the wavelength and the frequency; $v = \lambda f$. The speed of light in a vacuum is always c, so once the wavelength or frequency is known, its frequency or wavelength can be found from this equation. Because light travels at constant speed, its wavelength and frequency are inversely proportional to each other. Thus, if the wavelength increases, the frequency decreases,

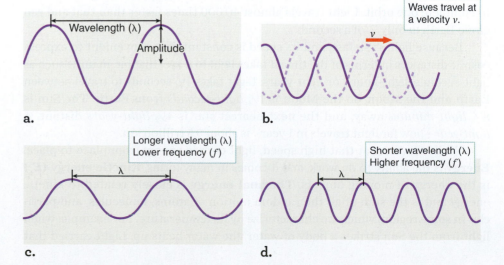

Figure 5.2 A wave is characterized by **a.** the distance from one peak to the next (wavelength, λ), the maximum deviations from the medium's undisturbed state (amplitude), and **b.** the velocity (v) at which the wave pattern travels from one place to another. The wavelength and the frequency (f) of the wave are inversely proportional: when one is large, the other is small, as shown in **c.** and **d.** In an electromagnetic wave, the amplitude is the maximum strength of the electric field, and the speed of light is written as c.

and vice versa. We revisit the idea of a wave in multiple contexts throughout this text—from seismic waves inside Earth to light to pressure waves in stars.

Light as an Electromagnetic Wave

In the late 19th century, the Scottish physicist James Clerk Maxwell (1831–1879) introduced the concept that electricity and magnetism are two components of the same physical phenomenon. An **electric force** is the push or pull between electrically charged particles that make up atoms, such as protons and electrons, arising from their electric charges. Particles with opposite charges attract, whereas those with like charges repel. A **magnetic force** is a force between electrically charged particles arising from their motion.

To describe those electric and magnetic forces, which act at a distance from the charges that create them, Maxwell used the concept of a "field." He thought of electric charges as creating a field around themselves that interacts with other charges, which then experience a force. The **electric field** is caused by a stationary charge and determines the electric force on a charge at any point in space. The **magnetic field** is caused by a moving charge and determines the magnetic force acting on a moving charge at any point in space.

Maxwell summarized the behavior of electric fields and magnetic fields in four elegant equations. Among other things, those equations indicate that a changing electric field causes a magnetic field and that a changing magnetic field causes an electric field. An acceleration (that is, a change) in the motion of a charged particle causes a changing electric field, which causes a changing magnetic field, which causes a changing electric field, and so on. This interaction is illustrated in **Figure 5.3**. Once the process starts, a self-sustaining procession of oscillating electric and magnetic fields moves out in all directions through space. In other words, an accelerating charged particle gives rise to an **electromagnetic wave**. Maxwell's equations also predict the speed at which an electromagnetic wave should travel, which agrees with the measured speed of light (c).

Maxwell's wave description of light also suggests how light originates and how it interacts with matter. When a drop of water falls from the faucet into a sink full of water, it causes a disturbance, or wave, like the one shown in **Figure 5.4a**. The wave moves outward as a ripple on the surface of the water. Similarly, an oscillating particle carrying a charge causes electromagnetic waves that move out through space away from their source in much the same way (**Figure 5.4b**). The ripples in the sink, however, are distortions of the water's surface, and they require a **medium**—a substance to travel through. Light waves do not require a medium. They move through empty space—what we call a vacuum.

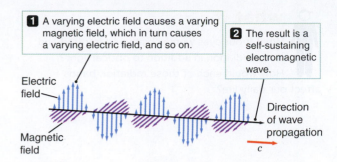

Figure 5.3 An electromagnetic wave consists of oscillating electric and magnetic fields that are perpendicular both to each other and to the direction in which the wave travels.

1 A varying electric field causes a varying magnetic field, which in turn causes a varying electric field, and so on.

2 The result is a self-sustaining electromagnetic wave.

Electric field

Magnetic field

Direction of wave propagation

c

Figure 5.4 **a.** A drop falling into water generates waves that move outward across the water's surface. **b.** Similarly, an oscillating (accelerated) electric charge generates electromagnetic waves that move away at the speed of light.

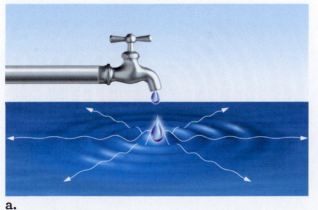

a.

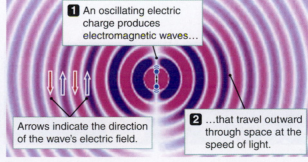

1 An oscillating electric charge produces electromagnetic waves…

Arrows indicate the direction of the wave's electric field.

2 …that travel outward through space at the speed of light.

b.

Now imagine that a cork is floating in the sink (**Figure 5.5a**). The cork remains stationary until the ripple from the dripping faucet reaches it. As the ripple passes by, the rising and falling water causes the cork to rise and fall. That can happen only if the wave is carrying energy—a conserved quantity that can give objects and particles the ability to change their state. Light waves similarly carry energy through space and cause other electrically charged particles to vibrate, as in **Figure 5.5b**.

The Electromagnetic Spectrum

For visible light, different wavelengths correspond to different colors. Most light signals are made up of many wavelengths. For example, a rainbow is created when white light interacts with water droplets and is spread out by wavelength into its component colors. Red light has a long wavelength, and therefore a low frequency (and a low energy), compared to violet light, which has a short wavelength and a high frequency (and a high energy). Light spread out by wavelength, as in a rainbow, is called a **spectrum**. A commonly used unit for the wavelength of light is the **nanometer (nm)**. A nanometer is one-billionth (10^{-9}) of a meter.

The light-sensitive cells in our eyes respond to visible light, but light can have wavelengths much shorter or much longer than what our eyes can perceive. The whole range of wavelengths of light—collectively called the **electromagnetic spectrum**—is shown in **Figure 5.6**. Most of the electromagnetic spectrum—and therefore most of the information in the universe—is invisible to the human eye. Specialized detectors of various kinds are required to detect light outside the visible range, as we explain in Chapter 6.

Refer to Figure 5.6 as we tour the electromagnetic spectrum, beginning with the shortest wavelengths and working our way to the longest ones. The very shortest wavelengths of light are called **gamma rays**, or sometimes gamma radiation. Because such radiation has the shortest wavelengths, it has the highest frequency and the highest energy, so it penetrates matter easily. Wavelengths between 0.1 and 40 nm are called **X-rays**. You have probably encountered X-rays at the dentist's office or in a hospital's emergency room—X-ray light has enough energy to penetrate skin and muscle but is stopped by denser bone. **Ultraviolet (UV) radiation** has wavelengths between 40 and about 380 nm—longer than X-rays but shorter than visible light. You are familiar with that type of light if you have ever tanned or been sunburned. Thus, UV light has enough energy to penetrate your skin, but not much deeper.

Figure 5.5 a. When waves moving across the surface of water reach a cork, they cause the cork to bob up and down. **b.** Similarly, a passing electromagnetic wave causes an electric charge to oscillate in response to the wave.

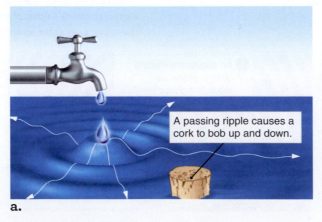

A passing ripple causes a cork to bob up and down.

a.

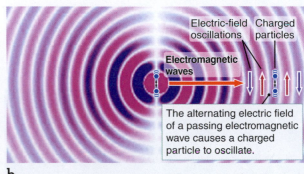

Electric-field oscillations   Charged particles

Electromagnetic waves

The alternating electric field of a passing electromagnetic wave causes a charged particle to oscillate.

b.

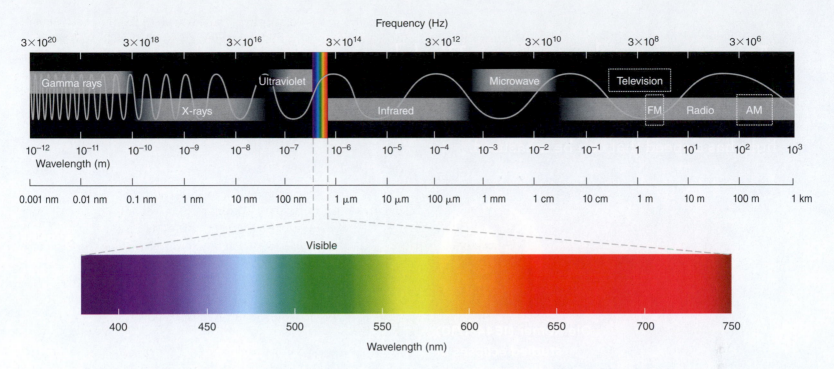

Figure 5.6 By convention, the electromagnetic spectrum is broken into loosely defined regions ranging from gamma rays to radio waves. Throughout the rest of this book, a labeled icon appears below individual astronomical images to identify what part of the spectrum was used to take the image: gamma rays (G), X-rays (X), ultraviolet (U), visible (V), infrared (I), or radio (R). If more than one region of the spectrum was used, multiple labels are highlighted in the icon.

Visible light, the type that the human eye is sensitive to, lies between violet (about 380 nm) and red (750 nm). Stretched out between violet and red are the other colors of the rainbow.

Infrared (IR) radiation (or infrared light) has longer wavelengths (and less energy) than the reddest wavelengths in the visible range. You often feel infrared radiation as heat. When you hold your hand next to a hot stove, for example, some of the heat you feel is carried to your hand by infrared radiation emitted from the stove. In that sense, you could think of your skin as being a giant infrared eyeball—it is sensitive to infrared wavelengths. Infrared radiation is also used in television remote controls, and night vision goggles detect infrared radiation from warm objects such as animals. A useful unit for infrared light is the **micron (μm**, where μ is the Greek letter mu). One micron is 1000 nm, or one-millionth (10^{-6}) of a meter. Infrared wavelengths are longer than about 0.75 microns (the red end of the visible range), and shorter than 500 microns.

Microwave radiation has even longer wavelengths (and less energy) than those of infrared radiation. The microwave in your kitchen heats the water in food by using the light of those wavelengths. The longest-wavelength light, which has wavelengths longer than a few centimeters, is called **radio waves**. The light of those wavelengths in the form of FM, AM, television, and cell phone signals is used to transmit information.

★ **WHAT AN ASTRONOMER SEES** An astronomer looking at this figure will notice that there are two axes shown: frequency is on top and wavelength is on the bottom. She will pause to notice that wavelength is measured in meters and frequency is measured in hertz (Hz). Both scales are logarithmic, so that one tick mark corresponds to a factor of 10 change in the frequency or wavelength. The visible portion of the spectrum is tiny, compared to the rest of the spectrum, and has been "zoomed" below the main graph. Yellow light marks the middle of this visible part of the spectrum, at about 550 nm, just a bit larger than 3×10^{14} Hz.

CHECK YOUR UNDERSTANDING **5.1a**

Rank the following in order from longest wavelength to shortest wavelength: (a) gamma rays; (b) visible light; (c) infrared light; (d) ultraviolet light; (e) radio waves

Answers to Check Your Understanding questions are in the back of the book.

Light as a Particle

Up to now, we've discussed light as a wave but sometimes it acts like a particle. A particle might have mass and sometimes has charge. An electron, for example, is a subatomic

Agreement between Fields

Scientists working on very different problems in different fields all found the same results: light has a speed that can be measured.

Ole Rømer (1644–1710) studied eclipses of Jupiter's moons.
Rømer calculated the speed of light from the eclipse delays of Jupiter's moons.

In 1727, James Bradley (1693–1762) studied the apparent motions of stars, which appeared to make small circles because of the relative motion of Earth.
A half century after Rømer's measurement, Bradley's motion studies led to a more accurate measurement of the speed of light.

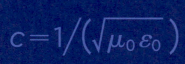

$$c = 1 / \left(\sqrt{\mu_0 \varepsilon_0} \right)$$

James Clerk Maxwell (1831–1879) studied electricity and magnetism.
Maxwell determined that the speed of electromagnetic waves must be a constant, and is equal to the speed of light.

$$t' = \frac{t}{\sqrt{1 - \dfrac{v^2}{c^2}}}$$

Einstein (1879–1955) explored the consequences of the fact that the speed of light is the same for all observers.
Predictions from this theory have been tested repeatedly since 1905.

Astronomers and physicists converged on an understanding that photons travel at the "speed of light"—a fundamental constant of the universe that is the same for all observers.

particle that carries a negative charge. If it is moving, a particle has velocity, which means that it also has momentum and kinetic energy. Particles are often visualized as tiny baseballs. This is not entirely correct, but it's useful for distinguishing particle-like behaviors from wavelike behaviors. Particles are *quantized*; like baseballs, they come in individual units. **Quantum mechanics** is a branch of physics that deals with particles and the quantization of energy and of other properties of matter.

The work of Albert Einstein and other scientists showed that light not only acts like a wave, but also acts like a particle. In 1905, Einstein explained the *photoelectric effect*. This effect occurs when light (photo-) strikes a surface and liberates electrons (-electric). The photoelectric effect has two important properties. First, electrons are only liberated when the light has a frequency (and therefore energy) above a certain threshold. The value of the threshold frequency depends on the type of material in the surface. The more the incoming light's frequency is above the threshold, the faster the liberated electrons move. Second, as long as the frequency of the light is above the threshold, the rate at which electrons are emitted depends only on the brightness of incoming light, not the frequency. In order to explain these observations, Einstein proposed that light is made up of massless particles called **photons** (*phot-* means "light," as in *photograph*; and *-on* signifies a particle). Photons always travel at the speed of light (**Process of Science Figure**), and they carry energy. This concept explains the photoelectric effect because it means that each electron interacts with only one photon. If a photon has enough energy to lift the electron out of the atom, any extra energy is carried away in the kinetic energy of the electron. At a fixed frequency above the threshold, brighter light has more photons, and liberates more electrons, but all those electrons have the same "extra" energy, so they have the same speed. This work earned Einstein the Nobel Prize in Physics in 1921.

Several times, we've noted that higher-frequency light has more energy. In fact, energy and frequency are directly proportional, so if a photon has twice as much energy, it has twice the frequency. The constant of proportionality between the energy (E) and the frequency (f) is called Planck's constant (h), which is equal to 6.63×10^{-34} joule-seconds (a joule is a unit of energy):

$$E = hf$$

The wavelength (λ) and frequency (f) of electromagnetic waves are inversely proportional (remember: $c = \lambda f$), so the photon energy is inversely proportional to the wavelength: $E = hc/\lambda$. High-energy visible light is blue, whereas low-energy visible light is red. **Working It Out 5.1** explores the relationships among the wavelength, frequency, and energy of light.

In the particle description of light, the electromagnetic spectrum is a spectrum of photon energies. The higher the frequency of the electromagnetic wave, the greater the energy carried by each photon. Photons of shorter wavelength (higher frequency) carry more energy than that carried by photons of longer wavelength (lower frequency). For example, photons of blue light carry more energy than photons of longer-wavelength red light. Ultraviolet photons carry more energy than photons of visible light, and X-ray photons carry more energy than ultraviolet photons. The lowest-energy photons are radio wave photons. For a beam of red light to carry just as much energy as a beam of blue light, the red beam must have more photons than the blue beam. This concept is illustrated in **Figure 5.7**, which also shows that the relationship between pennies and quarters is similar. The number of pennies required to pay for an item is larger than the number of quarters, because each quarter carries more value.

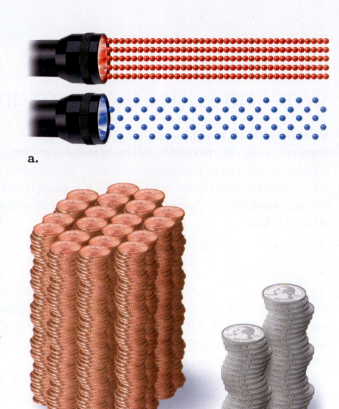

a.

b.

Figure 5.7 a. Photons of red light carry less energy than photons of blue light, so it takes more red photons than blue photons to make a beam of a particular intensity. **b.** Similarly, pennies are worth less than quarters, so it takes more pennies than quarters to add up to $10.

▶❚❚ **AstroTour:** Light as a Wave, Light as a Photon

working it out 5.1

Working with Electromagnetic Radiation

Wavelength and Frequency

When you tune a radio to, say, 770 AM, you are receiving an electromagnetic signal that travels at the speed of light and is broadcast at a frequency of 770 kilohertz (kHz), or 7.7×10^5 Hz. We can use the relationship between wavelength and frequency, $c = \lambda f$, to calculate the wavelength of the AM signal:

$$\lambda = \frac{c}{f} = \left(\frac{3 \times 10^8 \, \text{m/s}}{7.7 \times 10^5 \text{/s}}\right) = 390 \, \text{m}$$

FM frequencies are in megahertz (MHz), a factor of 1000 higher than AM stations. Therefore, FM wavelengths are much shorter than AM wavelengths. For example, a station at 89.5 FM broadcasts signals with a frequency of 89.5 MHz and a wavelength of 3.4 m.

The human eye is most sensitive to light in green and yellow wavelengths, about 500–590 nm. If we examine green light with a wavelength of 530 nm, we can compute its frequency:

$$f = \frac{c}{\lambda} = \left(\frac{3 \times 10^8 \, \text{m/s}}{530 \times 10^{-9} \, \text{m}}\right) = 5.66 \times 10^{14} \text{/s} = 5.66 \times 10^{14} \, \text{Hz}$$

That frequency corresponds to 566 *trillion* wave crests passing by each second.

Photon Energy

How does the energy of an X-ray photon with a wavelength of 1 nm compare with the energy of a visible-light photon with a wavelength of 530 nm (the same green light used in the previous calculation)? The equation for the energy of a photon is $E = hf$. Because $f = c/\lambda$, substituting c/λ for f yields the inverse relationship, $E = hc/\lambda$. Because we are making a *comparison*, we can take a ratio, and then the constants h and c cancel out:

$$\frac{E_{\text{X-ray photon}}}{E_{\text{visible photon}}} = \frac{hc/\lambda_{1\,\text{nm}}}{hc/\lambda_{530\,\text{nm}}} = \frac{hc}{hc} \times \frac{\lambda_{500\,\text{nm}}}{\lambda_{1\,\text{nm}}} = \frac{530 \, \text{nm}}{1 \, \text{nm}} = 530$$

The X-ray photon has 530 times the energy of the visible-light photon.

CHECK YOUR UNDERSTANDING 5.1b

As wavelength increases, the energy of a photon _____ and its frequency _____. (a) increases; decreases (b) increases; increases (c) decreases; decreases (d) decreases; increases

5.2 The Quantum View of Matter Explains Spectral Lines

Matter is anything that occupies space and has mass. Light and matter interact, and that interaction allows us to detect matter even at great distances in space. To understand that interaction, we must first understand the building blocks of matter. Here in Section 5.2, we review atomic structure and the process by which astronomers identify the chemical elements in astronomical objects.

Atomic Structure

The matter around you is composed of atoms. An atom is composed of a central massive **nucleus**, which contains **protons** with a positive charge and **neutrons**, which have no charge. A cloud of negatively charged **electrons** surrounds the nucleus. Atoms with the same number of protons are all of the same type of

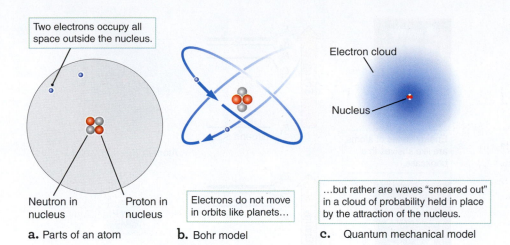

Two electrons occupy all space outside the nucleus.

Electron cloud

Nucleus

Neutron in nucleus

Proton in nucleus

Electrons do not move in orbits like planets…

…but rather are waves "smeared out" in a cloud of probability held in place by the attraction of the nucleus.

a. Parts of an atom

b. Bohr model

c. Quantum mechanical model

Figure 5.8 a. An atom (here, helium) is made up of a nucleus consisting of positively charged protons and electrically neutral neutrons and is surrounded by much less massive negatively charged electrons. **b.** Atoms are often drawn as miniature "solar systems," but this Bohr model is incorrect. **c.** Electrons are actually smeared out around the nucleus in quantum mechanical clouds of probability.

chemical **element**. For example, an atom with two protons, shown in **Figure 5.8a**, is the element helium. An atom with six protons is the element carbon, one with eight protons is the element oxygen, and so forth. An element may have many **isotopes**: atoms with the same number of protons but different numbers of neutrons. **Molecules** are groups of atoms bound together by shared electrons. A single teaspoon of water contains about 10^{23} molecules—about as many molecules as the number of stars in the observable universe.

Electrons have much less mass than protons or neutrons, so almost all the mass of an atom is found in its nucleus. The Danish physicist Niels Bohr (1885–1962) proposed a model of the atom in 1913 in which a massive nucleus sits in the center and the smaller electrons orbit around it, much as planets orbit around the Sun (**Figure 5.8b**). For an atom to be electrically neutral, it must have the same number of electrons as protons.

The **Bohr model**, however, is an incomplete description of the atom. Just as waves of light have particle-like properties, particles of matter also have wavelike properties. Once that was understood, the Bohr model was modified so that the positively charged nucleus is surrounded by electron "clouds" or "waves," as shown in **Figure 5.8c**. That model conveys the concept that it is not possible to know precisely where an electron is in its orbit. The wave characteristics of particles make it impossible to pin down simultaneously both their exact location and their exact velocity; some uncertainty will always exist. That is why a featureless cloud is used to represent electrons in orbit around an atomic nucleus.

▶❚❚ **AstroTour:** Atomic Energy Levels and the Bohr Model

Atomic Energy Levels

Each atom has a series of energy states, like the shelves of the bookcase depicted in **Figure 5.9a**. The energy of an atom might correspond to the energy of one state or to the energy of the next state, but the energy of the atom is never found between the two states, just as a book can be on only one shelf at a time and cannot be partly on one shelf and partly on another. A given atom may have many energy states available to it, but those states are *discrete*. When electrons in an atom gain or lose energy, the atom shifts from one energy state to another. Because the atom's energy states are discrete, the electrons must gain or lose energy in particular amounts, corresponding to the difference between energy levels in the atom. An electron cannot gain an amount of energy that puts the atom between states.

Figure 5.9 a. Energy states of an atom are analogous to shelves in a bookcase. You can move a book from one shelf to another, but books can never be placed between shelves. **b.** Atoms exist in one allowed energy state or another but never in between. The ground state is the lowest possible level.

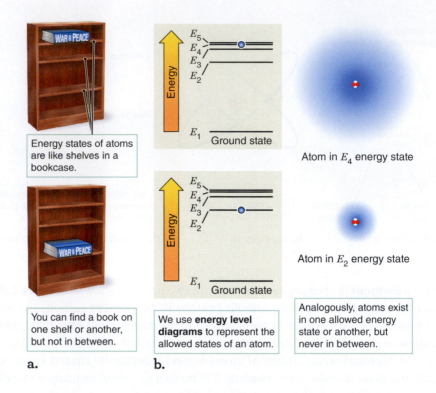

Energy states of atoms are like shelves in a bookcase.

Atom in E_4 energy state

You can find a book on one shelf or another, but not in between.

We use **energy level diagrams** to represent the allowed states of an atom.

Analogously, atoms exist in one allowed energy state or another, but never in between.

Atom in E_2 energy state

a.   b.

Astronomers keep track of the allowed states of an atom by using energy level diagrams, as shown in **Figure 5.9b.** Both the bookcase and the energy level diagram are simplifications of the possible energies of a three-dimensional system. The lowest possible energy state for an atom is called the **ground state**. When the atom is in the ground state, the electron has its minimum energy. It can't give up any more energy to move to a lower state because a lower state does not exist. An atom will remain in its ground state forever unless it gets energy from outside itself. In the bookcase analogy, a book sitting on the bottom shelf at the floor is in its ground state. It has nowhere left to fall, and it cannot jump to one of the higher shelves on its own.

Energy levels above the ground state are called **excited states**. Just as a book on an upper shelf might fall to a lower shelf, an atom in an excited state might **decay** to a lower state by getting rid of some of its extra energy. A common way for an atom to do that is to emit a photon. The photon carries away exactly the amount of energy that the atom loses as it goes from the higher energy state to the lower energy state. Similarly, atoms moving from a lower energy state to a higher energy state can absorb only certain specific energies. Unlike in the analogy of the bookcase, where you might happen to see the book between shelves as it falls, the transition of an atom between energy states is instantaneous: When the atom drops from a higher state to a lower one, the difference in energy between the two states is carried off all at once. We commonly refer to this change as a transition of the electron between energy states, and we may say that "an electron dropped from state 3 to state 2, emitting a photon."

To make an analogy with money, suppose you have a penny (1 cent), a nickel (5 cents), and a dime (10 cents), totaling 16 cents. Now imagine that you give away the nickel and are left with 11 cents. You never had exactly 13 cents or 13.6 cents. You had 16 cents and then 11 cents. Atoms don't accept and give away money to change energy states, but they do accept and give away photons with well-defined energies.

Emission Spectra   Imagine a hypothetical atom that has only two available energy states. The energy of the lower energy state (the ground state) is E_1, and the energy of the higher energy state (the excited state) is E_2. The energy levels of that atom

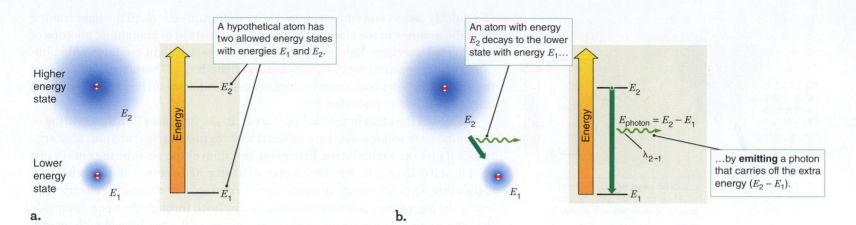

a.

b.

Figure 5.10 **a.** The energy levels of a hypothetical two-level atom. **b.** A photon with energy $E_{photon} = E_2 - E_1$ is emitted when an atom in the higher energy state (E_2) decays to the lower energy state (E_1).

can be represented in an energy level diagram such as the one in **Figure 5.10a**. An atom in the excited state moves to the ground state by getting rid of the "extra" energy all at once. In that **emission** process, a photon is released.

In **Figure 5.10b**, the green arrow pointing straight down indicates that the atom went from the higher state (E_2) to the lower state (E_1). The atom lost an amount of energy equal to $E_2 - E_1$, the difference between the two states. Because energy is never truly lost or created, the energy the atom lost has to show up somewhere. Here, the energy shows up in the form of a photon emitted by the atom. The energy of the emitted photon, E_{photon}, matches the energy lost by the atom; that is, $E_{photon} = E_2 - E_1$.

An atom can emit photons with energies corresponding *only* to the difference between two of its allowed energy states. Because the energy of a photon is related to the frequency or wavelength of electromagnetic radiation, a photon of energy $E_{photon} = E_2 - E_1$ has a specific frequency $f_{2\rightarrow1}$ and a specific wavelength $\lambda_{2\rightarrow1}$ for the transition from level 2 to level 1. Therefore, those emitted photons have a specific color, and every photon emitted in any transition from E_2 to E_1 will have that color. The energy level structure of an atom determines the wavelengths of the photons it emits—the color of the light that the atom gives off.

Figure 5.11 illustrates the light coming from a cloud of gas consisting of the hypothetical two-state atoms in Figure 5.10. Any atom in the higher energy state

Figure 5.11 A cloud of gas containing atoms with two energy states, E_1 and E_2, emits photons with an energy $E_{photon} = E_2 - E_1$, which appear in the spectrum (right) as a single bright *emission line*.

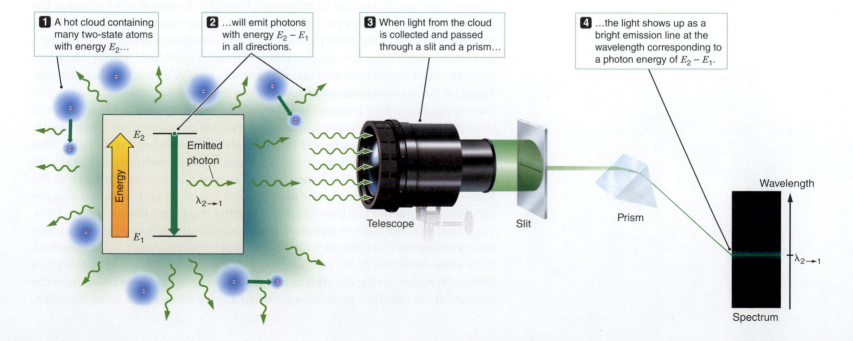

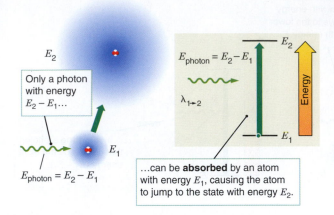

Figure 5.12 An atom in a lower energy state (E_1) may absorb a photon of energy $E_{photon} = E_2 - E_1$, leaving the atom in a higher energy state (E_2).

(E_2) quickly decays and emits a photon in a random direction. All the light coming from those atoms in the cloud is the same color. Instead of containing photons of all different energies—light of all different colors—that light contains only photons with the specific energy $E_2 - E_1$ and wavelength $\lambda_{2\to1}$. Therefore, if you spread the light out into its component colors, only one color would be present—a single bright line called an **emission line**.

Why was the atom in the excited state E_2 in the first place? An atom sitting in its ground state will remain there unless it absorbs just the right amount of energy to kick it up to an excited state. In general, the atom either absorbs the energy of a photon or it collides with another atom or an unattached electron and absorbs some of the other particle's energy. In a neon sign, for example, an alternating electric field inside the glass tube pushes electrons back and forth through the neon gas inside the tube. Some of those electrons crash into atoms of the gas, knocking them into excited states. The atoms then drop back down to their ground states by emitting photons, causing the gas inside the tube to glow a red-orange color characteristic of the element neon.

Absorption Spectra   An atom in a low energy state can absorb the energy of a passing photon and move to a higher energy state, as shown in **Figure 5.12**. Once again, the energy required to go from E_1 to E_2 is the difference in energy between the two states, $E_2 - E_1$. The only photons that can excite atoms from E_1 to E_2 are photons with exactly $E_{photon} = E_2 - E_1$. Those absorbed photons have precisely the same energy as the photons emitted by the atoms when they decay from E_2 to E_1. That is not a coincidence. The energy difference between the two levels is the same whether the atom is emitting a photon or absorbing one, so the energy of the photon involved is the same in either case. In both cases, the photons have a corresponding frequency ($f_{1\to2} = E_{photon}/h$) and wavelength ($\lambda_{1\to2} = hc/E_{photon}$), so they have the same color.

In **Figure 5.13a**, white light (which has all wavelengths of photons in it) passes directly through a glass prism and spreads out into a rainbow of colors, called a **continuous spectrum**. If the white light passes through a cool cloud composed of the hypothetical gas of two-state atoms (**Figure 5.13b**), however, some photons will be absorbed. Almost all the photons will pass through the cloud of gas unaffected because they do not have the right energy ($E_2 - E_1$) to be absorbed by atoms of the gas. But photons with just the right amount of energy can be absorbed, so those photons will be missing from the light passing through the prism. That means a particular color will be missing from the continuous spectrum. A sharp, dark line appears at the wavelength corresponding to the energy of the missing photons. This dark line is called an **absorption line**, and the process by which atoms capture the energy of passing photons is called **absorption**. **Figure 5.14a** shows the absorption lines in the spectrum of a star. Each color corresponds to a wavelength, and the spectrum has a particular brightness at each of those wavelengths. This relationship between brightness and wavelength can be redrawn as a graph, as shown in **Figure 5.14b**. The brightness in Figure 5.14b drops abruptly at the wavelengths corresponding to the dark lines in Figure 5.14a. Places between the dark lines are brighter and therefore higher on the graph than the absorption lines.

After an atom absorbs a photon, it may quickly decay to its previous lower energy state, emitting a photon with the same energy as the photon it just absorbed. If the atom reemits a photon just like the one it absorbed, why does the absorption matter? Why doesn't the absorption line get "filled in" by the reemission from the atoms? All the absorbed photons were originally traveling in the same direction,

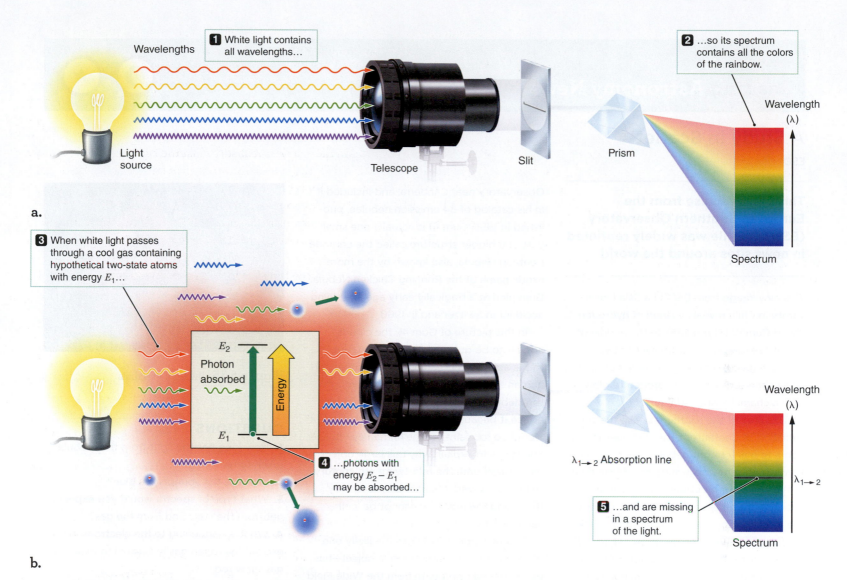

Figure 5.13 **a.** When passed through a prism, white light produces a spectrum containing all colors. **b.** When light of all colors passes through a cloud of hypothetical two-state atoms, photons with energy $E_{photon} = E_2 - E_1$ may be absorbed, leading to the dark absorption line in the spectrum.

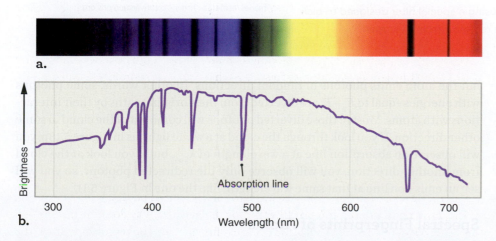

Figure 5.14 Absorption lines in the spectrum of a star as an image **a.** and a graph **b.**

Reading Astronomy News

A Study in Scarlet

ESO

This press release from the European Southern Observatory (ESO) in Chile was widely reprinted in news sites around the world.

This new image from ESO's La Silla Observatory in Chile reveals a cloud of hydrogen called Gum 41 (**Figure 5.15**). In the middle of this little-known nebula, brilliant hot young stars are giving off energetic radiation that causes the surrounding hydrogen to glow with a characteristic red hue.

This area of the southern sky, in the constellation of Centaurus (The Centaur), is home to many bright nebulae, each associated with hot newborn stars that formed out of the clouds of hydrogen gas. The intense radiation from the stellar newborns excites the remaining hydrogen around them, making the gas glow in the distinctive shade of red typical of star-forming regions. Another famous example of this phenomenon is the Lagoon Nebula, a vast cloud that glows in similar bright shades of scarlet.

The nebula in this picture is located some 7300 light-years from Earth. Australian astronomer Colin Gum discovered it on photographs taken at the Mount Stromlo

Observatory near Canberra, and included it in his catalog of 84 emission nebulae, published in 1955. Gum 41 is actually one small part of a bigger structure called the Lambda Centauri Nebula, also known by the more exotic name of the Running Chicken Nebula. Gum died at a tragically early age in a skiing accident in Switzerland in 1960.

In this picture of Gum 41, the clouds appear to be quite thick and bright, but this is actually misleading. If a hypothetical human space traveler could pass through this nebula, it is likely that they would not notice it because—even at close quarters—it would be too faint for the human eye to see. This helps to explain why this large object had to wait until the mid-twentieth century to be discovered—its light is spread very thinly and the red glow cannot be well seen visually.

This new portrait of Gum 41—likely one of the best so far of this elusive object—has been created using data from the Wide Field Imager (WFI) on the MPG/ESO 2.2-meter telescope at the La Silla Observatory in Chile. It is a combination of images taken through blue, green, and red filters, along with an image using a special filter designed to pick out the red glow from hydrogen.

Figure 5.15 The Gum 41 Nebula.

QUESTIONS

1. How long has the light from that nebula taken to reach us?

2. Why are the young stars blue?

3. What type of spectra would you expect to get from the stars and from the gas?

4. What is happening to the electrons in the excited hydrogen gas in Gum 41 to make the gas glow red?

5. Would you be able to see Gum 41 from your location? Why or why not?

Source: "Photo Release: A Study in Scarlet," ESO.org, April 16, 2014. CC by 4.0.
Photo credit: ESO, https://www.eso.org/public /images/eso1413a/. https://creativecommons.org /licenses/by/4.0/.

AstroTour: Atomic Energy Levels and Light Emission and Absorption

but the atom emits photons in random directions. In other words, some photons with energies equal to $E_2 - E_1$ are diverted from their original paths by their interactions with atoms. Most of those diverted photons will come out of the cloud in some other direction. If you look through the cloud at a white light, as in Figure 5.13b, you will observe an absorption line at a wavelength of $\lambda_{2\to1}$, but if you look at the cloud from another direction, you will observe only the redirected photons, so you will see an emission line at that same wavelength, like the one in Figure 5.11.

Spectral Fingerprints of Atoms

In the mid-19th century, Gustav Kirchoff (1824–1887) first observed emission, continuous, and absorption spectra from the three types of sources shown in

Figures 5.11 and 5.13. He did not know about energy levels in atoms, so he could not create a theory about emission. He summarized his findings as three empirical laws. Kirchoff, together with Robert Bunsen (1811–1899), concluded that the dark-line (absorption) spectrum of the Sun was the "reverse" of the bright-line (emission) spectrum that would be produced by the Sun's atmosphere alone and identified some of the elements on the Sun. (A few years later the element helium was discovered in the solar spectrum.) Others observed the stars. Noteworthy among them were William (1824–1910) and Margaret Lindsay (1848–1915) Huggins, who published an atlas of stellar spectra in 1899 and who showed that the types of atoms seen in the stars are the same as those found on Earth. Spectra are how we know what makes up the stars and planets, and that we on Earth are composed of the same elements.

An atom's allowed energy states are determined by the interactions among the electrons and the nucleus. Real atoms can occupy many more than just two possible energy states; therefore, any given type of atom can emit and absorb photons at many wavelengths. An atom with three energy states, for example, might jump from state 3 to state 2, or from state 3 to state 1, or from state 2 to state 1. The three distinct emission lines in the spectrum from a gas made up of those atoms would have wavelengths of $hc/(E_3 - E_2)$, $hc/(E_3 - E_1)$, and $hc/(E_2 - E_1)$, respectively.

For example, every neutral hydrogen atom consists of a nucleus containing one proton, plus a single electron in a cloud surrounding the nucleus. Therefore, every hydrogen atom has the same energy states available to it, and all hydrogen atoms have the same emission and absorption lines. **Figure 5.16a** shows the energy

Figure 5.16 a. The energy states of the hydrogen atom. Decays to level E_2 emit photons in the visible part of the spectrum. **b.** This visible emission spectrum is what you might see if you looked at the light from a hydrogen lamp projected through a prism onto a screen. **c.** The brightness at every wavelength can be measured to produce a graph of the brightness of spectral lines versus their wavelength. **d.** Emission spectra from gaseous sodium, helium, neon, and mercury.

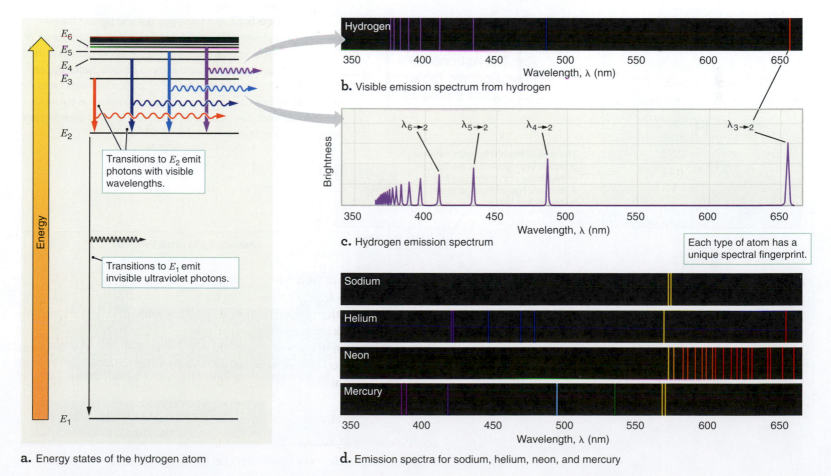

a. Energy states of the hydrogen atom

b. Visible emission spectrum from hydrogen

c. Hydrogen emission spectrum

d. Emission spectra for sodium, helium, neon, and mercury

what if . . .

What if the abundance of elements on Earth were similar to what we see in the Sun? How might this have affected the evolution of life on Earth?

level diagram of hydrogen. **Figure 5.16b** shows the visible emission spectrum from hydrogen, and **Figure 5.16c** displays that same information as a graph.

Each type of atom—that is, each chemical element—has a unique set of available energy states and therefore a unique set of wavelengths at which it can emit or absorb radiation. **Figure 5.16d** shows the emission spectra of the elements sodium, helium, neon, and mercury. Those unique sets of wavelengths serve as unmistakable spectral "fingerprints" for each chemical element.

Cecilia Payne-Gaposchkin applied this analysis to the stars in 1925. She figured out the types of atoms (or molecules) in distant objects by looking at the spectra of light from those objects. If the spectral lines of hydrogen, helium, carbon, oxygen, or any other element are visible in the light from a distant object, then that element is present in that object. Payne-Gaposchkin took the analysis further, by studying the strengths of spectral lines. The "strength" of a line describes how deep and wide it is, if it's an absorption line, or how high and wide it is if it's an emission line. The strength of a line is determined in part by how many atoms of that type are present in the source. The strengths of the lines from different types of atoms in the spectrum of a distant object can be used to infer the relative amounts of elements that make up the object. This analysis led Payne-Gaposchkin to the then-controversial conclusion that the universe is primarily made of hydrogen and helium. Astronomers use the relative abundance of the elements in the Sun (called the **solar abundance**) as a standard reference. Hydrogen (H) is the most abundant element in the Sun (**Figure 5.17**), followed by helium (He) and 13 others. Those 15 elements make up 99.99 percent of the mass of the Sun. The other elements on the regular periodic table (in the lower right) make up

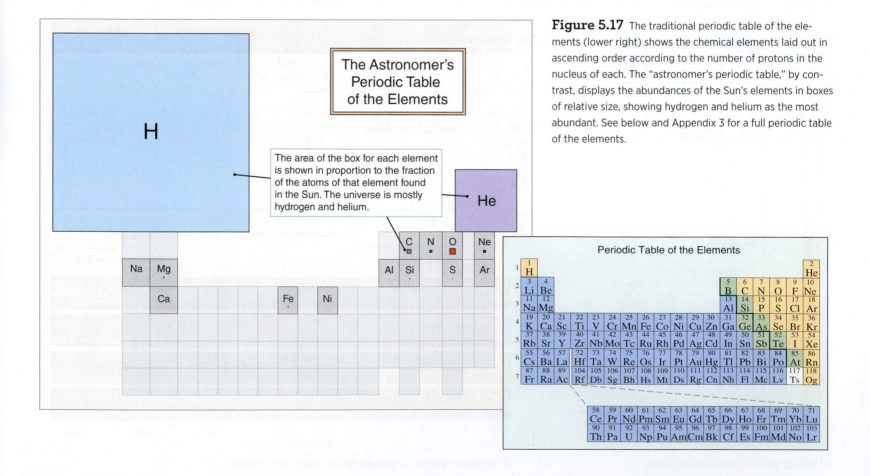

Figure 5.17 The traditional periodic table of the elements (lower right) shows the chemical elements laid out in ascending order according to the number of protons in the nucleus of each. The "astronomer's periodic table," by contrast, displays the abundances of the Sun's elements in boxes of relative size, showing hydrogen and helium as the most abundant. See below and Appendix 3 for a full periodic table of the elements.

less than 0.01 percent of the mass of the Sun. In addition, by looking at the relative strengths of lines from the same element, astronomers often can determine the temperature, density, and pressure of the material.

Excitement and Decay

If a book on a level shelf is not disturbed, it will sit there forever; something must *cause* the book to fall off the shelf. However, while sometimes an atom in a higher energy state can be "stimulated" into emitting a photon, usually nothing causes the atom to jump to the lower energy state. Instead, the atom decays *spontaneously*. Although scientists can determine on average how long a specific atom is likely to remain in the excited state, they cannot predict exactly when a particular atom will decay. We can observe the moment when an atom decays, but beforehand, we can only use the average lifetime to roughly estimate when the atom might decay. Rather than use the average lifetime, scientists typically use the "half-life" of an atomic state; this describes the time until half of all the atoms will have decayed.

Toys that glow in the dark are an example of spontaneous decay. Photons in sunlight or from a lightbulb are absorbed by phosphorescent atoms in the toy, exciting those atoms to upper energy states. The excited states of the atoms in the toy last for many seconds. If the half-life of an excited state is a minute, then on average half of the atoms will have decayed within one minute. It is impossible to say exactly which atoms will decay in which minute, but about half of the trillions and trillions of atoms in the toy will decay within 1 minute, and the brightness of the glow from the toy will have dropped to half of what it was. After each minute, half of the remaining excited atoms decay, and the glow from the toy drops to half of what it was 1 minute earlier. Thus, the glow from the toy slowly fades away.

In deep space, where atoms can remain undisturbed for long periods, certain excited states of atoms last, on average, for tens of millions of years or even longer. An atom may have been in such an excited energy state for a few seconds, a few hours, or 50 million years when, in an instant, it decays to the lower energy state without anything having caused it to do so. Space is extremely large, and has a lot of atoms in it, so even rare atomic decays are regularly observed. Physicists can calculate the *probability* that a decay will take place somewhere along a line of sight through space, but cannot predict when any individual atom will decay.

CHECK YOUR UNDERSTANDING 5.2

How can spectra tell us the chemical composition of a distant star?

5.3 The Doppler Shift Indicates Motion Toward or Away from Us

You have already seen that light can reveal a wealth of information about the physical state of material located tremendous distances away. Here in Section 5.3, we explain how light can be used to determine whether a distant astronomical object is moving away from us or toward us, and at what speed.

Have you ever listened to an ambulance speed by with its siren blaring? As the ambulance approaches, the siren has a certain high pitch, but as it passes by, the pitch of the siren drops noticeably. If you close your eyes and listen, you have no trouble knowing when the ambulance passed; the change in the pitch of

what if . . .

We've discussed the changing pitch you hear as an ambulance speeds by. What if the ambulance accelerated and reached, or even exceeded, the speed of sound? What might you hear?

Astronomy in Action: Doppler Shift

AstroTour: The Doppler Effect

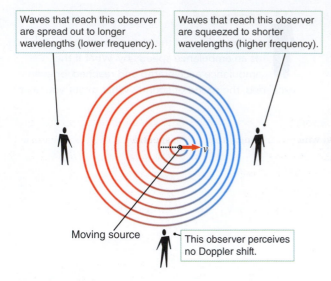

Waves that reach this observer are spread out to longer wavelengths (lower frequency).

Waves that reach this observer are squeezed to shorter wavelengths (higher frequency).

Moving source

This observer perceives no Doppler shift.

Figure 5.18 Motion of a light or sound source relative to an observer may cause waves to be spread out (redshifted, for light, or lower in pitch, for sound) or squeezed together (blueshifted, for light, or higher in pitch, for sound). A change in the wavelength of light or the frequency of sound is called a *Doppler shift*.

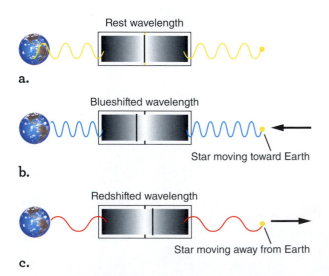

Rest wavelength

a.

Blueshifted wavelength

Star moving toward Earth

b.

Redshifted wavelength

Star moving away from Earth

c.

Figure 5.19 **a.** Spectral lines from a star at rest will be observed at their rest wavelength. **b.** Spectral lines from a star moving towards Earth will be blueshifted, while **c.** spectral lines from a star moving away from Earth will be redshifted.

its siren indicates that it has passed you by. You do not even need an ambulance to hear that effect. The sound of normal traffic behaves in the same way. As a car drives past, the pitch of the sound that it makes suddenly drops. (The same effect would happen if you were the one driving past a stationary siren.)

Like light, sound is a wave. The pitch of a sound is like the color of light: it is determined by the wavelength or, equivalently, the frequency of the sound wave. Sound waves with higher frequencies and shorter wavelengths have higher pitch. Sound waves with lower frequencies and longer wavelengths have lower pitch. When an object is moving toward you, the waves that it emits "crowd together" in front of the object. You can see how that process works in **Figure 5.18**, which shows the locations of successive wave crests emitted by a moving object. If you are standing in front of the object, the waves that reach you have a shorter wavelength and therefore a higher frequency than the waves given off by the object when it is not moving. Conversely, if an object is moving away from you, the waves reaching you from the object are spread out (longer λ, lower f). That change in frequency as a result of motion is known as the **Doppler effect** and is named after physicist Christian Doppler (1803–1853).

The Doppler effect applies to light waves as well as sound waves. If a star is at rest relative to you, then it emits light with the **rest wavelength** (λ_{rest}), as shown in **Figure 5.19a**. If a star is moving *toward* you, the light reaching you from the star has a shorter wavelength than its rest wavelength. The light is "bluer" than the rest wavelength, and the light is **blueshifted**, as shown by the blue waves in **Figure 5.19b**. In contrast, light from a star moving *away* from you is shifted to longer, redder wavelengths. The light is **redshifted**, as shown by the red waves in **Figure 5.19c**. The amount by which the wavelength of light is shifted by the Doppler effect is called the light's **Doppler shift**, which depends on the speed of the object emitting the light; faster objects have larger shifts.

The Doppler shift provides information only about the **radial velocity** (v_r) of the object, which is the part of the motion that is toward you or away from you. The radial velocity is the rate at which the distance between you and the object is changing: if v_r is positive, the object is getting farther away from you; if v_r is negative, the object is getting closer. At the moment an ambulance is passing you, it is getting neither closer nor farther away, so the pitch you hear is the same as the pitch heard by the crew riding on the truck (see the observer at the bottom of Figure 5.18). Similarly, a distant object moving slowly across the sky does not move toward or away from you, and so its light will not be Doppler shifted from your point of view.

Doppler shifts are easiest to measure when an object has prominent spectral lines. These lines are shifted, just like the rest of the light. Because the spectral lines from each element have a particular pattern, astronomers can find that pattern no matter how far it has been shifted from its rest position. The amount of the shift indicates how rapidly the object is moving toward or away from Earth. To determine that velocity, astronomers first identify a spectral line from a certain chemical element. The rest wavelengths (λ_{rest}) of most spectral lines have been measured in a lab on Earth. An astronomer measures the observed wavelength (λ_{obs}) in the spectrum of the distant object. The difference between the rest wavelength and the observed wavelength is used to find the object's radial velocity, as shown in detail in **Working It Out 5.2**.

CHECK YOUR UNDERSTANDING **5.3**

Which of the following Doppler shifts indicates the fastest-approaching object?
(a) 0.04 nm; (b) 0.06 nm; (c) −0.04 nm; (d) −0.06 nm

working it out 5.2

Making Use of the Doppler Effect

The Doppler formula for objects moving at a radial velocity (v_r) much less than the speed of light is given by

$$v_r = \frac{\lambda_{obs} - \lambda_{rest}}{\lambda_{rest}} \times c$$

A prominent spectral line of hydrogen atoms has a rest wavelength, λ_{rest}, of 656.3 nm (see Figure 5.16b). Suppose that you measure the wavelength of that line in the spectrum of a distant object and find that instead of seeing the line at 656.3 nm, you see the line at a wavelength, λ_{obs}, of 659.0 nm. What is its radial velocity? Using the above equation,

$$v_r = \frac{659.0\,\text{nm} - 656.3\,\text{nm}}{656.3\,\text{nm}} \times (3 \times 10^5\,\text{km/s})$$

$$v_r = 1200\,\text{km/s}$$

Thus, the object is moving away from you with a radial velocity of 1200 km/s.

Suppose instead that you know the velocity and want to compute the wavelength at which you would observe the spectral line. Earth's nearest stellar neighbor, Proxima Centauri, is moving toward us at a radial velocity of −21.6 km/s. What is the observed wavelength, λ_{obs}, of a magnesium line in Proxima Centauri's spectrum that has a rest wavelength, λ_{rest}, of 517.27 nm? We can rearrange the Doppler formula to solve for λ_{obs}:

$$\lambda_{obs} = \left(1 + \frac{v_r}{c}\right) \times \lambda_{rest}$$

$$\lambda_{obs} = \left(1 + \frac{-21.6\,\text{km/s}}{3 \times 10^5\,\text{km/s}}\right) \times 517.27\,\text{nm} = 517.23\,\text{nm}$$

Although the observed Doppler blueshift ($\lambda_{obs} - \lambda_{rest} = 517.23 - 517.27$) is only −0.04 nm, it is easily measured with modern instrumentation.

5.4 Temperature Affects the Spectrum of Light That an Object Emits

The balance between heating and cooling in an object determines its temperature. If an object's temperature is constant, the heating and cooling must be in balance. Here in Section 5.4, we examine that balance and see how we can use it to predict the temperatures of planets and stars.

Astronomy in Action: Changing Equilibrium

Equilibrium and Balance

In a tug-of-war contest between two perfectly matched teams, each team pulls on the rope, but the force of one team's pull is only enough to match, not overcome, the force exerted by the other team. A picture taken now and another taken 5 minutes from now would not differ in any significant way. In that static equilibrium, opposing forces balance each other exactly. Static equilibrium can be stable, unstable, or neutral. A marble in a bowl is in a stable equilibrium: if it moves, it will return to its original position at the bottom of the bowl. A book standing on its edge, unsupported on either side, is in an unstable equilibrium: if you nudge it, it will fall over, not settle back into its original position. When an unstable equilibrium is disturbed, it moves further away from equilibrium rather than back toward it.

Equilibrium can be dynamic, so that one source of change is exactly balanced by another source of change, and the configuration of the system remains the same. In **Figure 5.20**, a can with a hole near the bottom has been placed under an

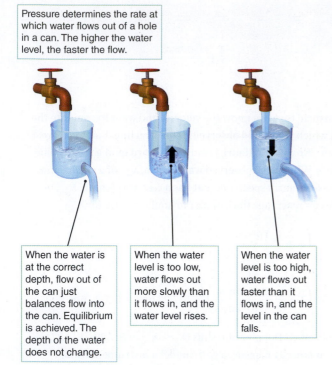

Pressure determines the rate at which water flows out of a hole in a can. The higher the water level, the faster the flow.

When the water is at the correct depth, flow out of the can just balances flow into the can. Equilibrium is achieved. The depth of the water does not change.

When the water level is too low, water flows out more slowly than it flows in, and the water level rises.

When the water level is too high, water flows out faster than it flows in, and the level in the can falls.

a. Equilibrium   **b.** Water level low   **c.** Water level high

Figure 5.20 The relative rates at which water flows into and out of a can determine the water level in the can. This is an example of dynamic equilibrium. The pressure in the can changes with the depth of the water, increasing or decreasing the flow through the hole at the bottom. **a.** When the water flows out at the same rate it flows in, the water level is in equilibrium. **b.** If the water flows out more slowly than it flows in, the water level will rise, increasing the pressure until the rates are equal. **c.** Conversely, if the water flows out more quickly than it flows in, the water level will fall, decreasing the pressure until the rates are equal.

open water faucet. This system provides an example of dynamic equilibrium, because the depth of the water in the can determines how fast water pours out through the hole near the bottom. When the water reaches just the right depth (Figure 5.20a), water pours out of the hole in the bottom of the can at the same rate it pours into the top of the can from the faucet. The water leaving the can balances the water entering, and the system is in equilibrium. If you took a picture now and another picture in a few minutes, little of the water in the can would be the same, but the pictures would be indistinguishable.

Suppose the level of the water in the can is too low (Figure 5.20b). Since the depth determines the rate at which water flows through the hole, water will not flow out of the bottom of the can fast enough to balance the water flowing in. The water level will begin to rise until the depth increases so much that the amount going out equals the amount coming in. Conversely, if the water level in the can is too high (Figure 5.20c), water will flow out of the can faster than it flows in. The water level will begin to fall until the amount going out equals the amount coming in. This is a dynamic equilibrium. One source of change is balanced by another source of change, so that the system finds a new state of equilibrium.

Your body is heated by the release of chemical energy inside it. For your body temperature to remain stable, the heating must be balanced by cooling. Typically, your body cools by radiating energy from your skin. If a given day is particularly hot, you may also perspire, so that your skin cools by evaporation of water on its surface.

A system is in **thermal equilibrium** if its heating (energy in) is balanced by its cooling (energy out). Planets have a dynamic but stable thermal equilibrium, and electromagnetic radiation plays a crucial role in maintaining that balance. Energy from sunlight heats the surface of a planet, driving its temperature up, whereas the planet emits thermal radiation into space, cooling it down. For a planet to remain at the same average temperature over time, the energy it radiates into space must exactly balance the energy it absorbs from the Sun. **Figure 5.21** shows how the equilibrium temperature of a planet is analogous to the water level in Figure 5.20. We return to planetary equilibrium later in the chapter. Many kinds of equilibrium exist besides thermal equilibrium, some of which we encounter later in the book.

Temperature

In everyday life, we define hot and cold subjectively: something is hot or cold when it feels hot or cold. When we measure *temperature*, we specify degrees on a thermometer, but the way we define a degree is arbitrary.

The air around you is composed of vast numbers of atoms and molecules. Those particles are moving about every which way. Some particles move slowly, and some move more rapidly, but all of them are constantly in motion. The kinetic energy (E_K) of a particle is given by $E_K = \frac{1}{2}mv^2$, where m is the mass of the particle and v is its velocity. **Temperature** is a measurement of the average kinetic energy of all the atoms and molecules in an object. The more energetically the atoms or molecules are bouncing about, the higher the object's temperature. In fact, the random motions of atoms and molecules are often called their **thermal motions** to emphasize the connection between those motions and temperature. **Figure 5.22** shows that when the temperature of a gas is increased, the kinetic energy is increased, in which case the atoms move faster.

The atoms and molecules in a solid body (like you) cannot move about freely like a gas, but they still move back and forth around their average location, and temperature measures the amount of that movement. If an object is hotter than you are, thermal energy flows from that object into you. At the atomic level, that means the object's atoms are bouncing more energetically than the atoms in your body are, so if you touch the object, its atoms collide with your atoms, causing the atoms in your body to move faster. Your body gets hotter as thermal energy flows from the object to you. At the same time, those collisions rob the particles in the object of some of their energy. Their motions slow down, and the hotter object cools. Heating processes increase the average thermal energy of an object's particles, whereas cooling processes decrease the average thermal energy of those particles.

There are three commonly used scales for measuring temperature. If you grew up in the United States, you probably think of temperatures in degrees Fahrenheit (°F), whereas if you grew up almost anywhere else in the world, you think of temperatures in degrees Celsius (°C). On the Fahrenheit scale, water at sea level has a freezing point of 32°F and a boiling point of 212°F. On the Celsius scale, water freezes at 0°C and boils at 100°C. Because the number of degrees between freezing and boiling on those two scales is different (180°F vs. 100°C), a 1-degree change measured in °F is not the same as a 1-degree change measured in °C.

As the motions of the particles in an object slow down, the temperature decreases more and more. The lowest possible temperature, at which all thermal motions would stop, is called **absolute zero**. Absolute zero corresponds to −273.15°C and −459.57°F. It also is the zero point of the **Kelvin temperature scale**. The size of one unit on the Kelvin scale, called a **kelvin (K)**, is the same as the Celsius degree. To convert between °C and K, just add 273.15 to the temperature in °C. Thus, water freezes at 273.15 K and water boils at 373.15 K.

Scientists use the Kelvin temperature scale because when temperatures are measured in kelvins, the average thermal energy of particles is proportional to the measured temperature. Thus, the average thermal energy of the atoms in an object with a temperature of 200 K is twice the average thermal energy of the atoms in an object with a temperature of 100 K. Just as the thermal energy cannot be negative, the Kelvin scale has no negative temperatures. Note that increments on the Kelvin temperature scale are called "kelvins" and not "degrees Kelvin."

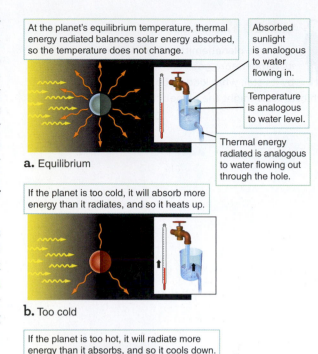

At the planet's equilibrium temperature, thermal energy radiated balances solar energy absorbed, so the temperature does not change.

Absorbed sunlight is analogous to water flowing in.

Temperature is analogous to water level.

Thermal energy radiated is analogous to water flowing out through the hole.

a. Equilibrium

If the planet is too cold, it will absorb more energy than it radiates, and so it heats up.

b. Too cold

If the planet is too hot, it will radiate more energy than it absorbs, and so it cools down.

c. Too hot

Figure 5.21 Planets are heated by absorbing sunlight (and sometimes by internal heat sources) and cooled by emitting thermal radiation into space. In the absence of other sources of heating or means of cooling, the equilibrium between those processes determines the temperature of the planet. Compare parts **a**, **b**, and **c** of this figure with those of Figure 5.20. Just as water level adjusts in a can until the rate at which water flows out is equal to the rate at which it flows in, the temperature of a planet will adjust until the rate at which energy flows out is equal to the rate at which it flows in.

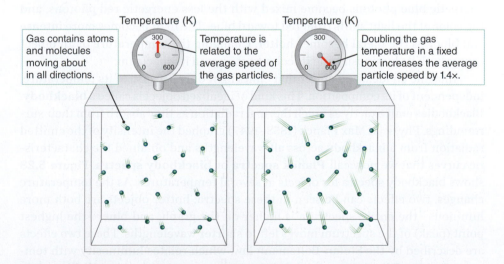

Gas contains atoms and molecules moving about in all directions.

Temperature (K)

Temperature is related to the average speed of the gas particles.

Temperature (K)

Doubling the gas temperature in a fixed box increases the average particle speed by 1.4×.

Figure 5.22 Higher gas temperatures correspond to faster-moving atoms.

what if . . .

What if a planet emitted *less* radiation flux as its temperature goes up? If the planet is heated by its star, what kind of equilibrium will it achieve? How might the planet change that result?

Temperature, Luminosity, and Color

We have seen the way discrete atoms emit and absorb radiation, which leads to a useful understanding of emission lines and absorption lines that tell us about the physical state and motion of distant objects. But not all objects have spectra dominated by discrete spectral lines. As you saw in Figure 5.13a, if you pass the light from a lightbulb through a prism, instead of discrete bright and dark bands you will see light spread out smoothly from the blue end of the spectrum to the red. Similarly, if you look closely at the spectrum of the Sun, you will see absorption lines, and you will see light smoothly spread out across all colors of the spectrum—the continuous spectrum noted earlier. How is this continuous spectrum produced?

A dense material is a collection of charged particles that are constantly jostled as their thermal motions cause them to run into their neighbors. The hotter the material is, the more violently its particles are jostled; they speed up, slow down, and change direction. In short, they are accelerated. Each particle will be accelerated differently, across an entire range of changes of speed and direction. Recall from Figure 5.4 that a charged particle radiates anytime it accelerates. The energy of the radiation (and therefore the frequency and the wavelength) depends on how much the particle is accelerated, so the jostling of particles that results from their thermal motions causes them to emit an entire range of electromagnetic radiation. Any material dense enough for its particles to be jostled by their neighbors emits light simply because of its temperature. That kind of continuous radiation is called **thermal radiation**.

Luminosity is the total amount of light emitted each second (energy per second, measured in watts, W) from a source. The hotter the object, the more energetically the charged particles within it move, and the more energy they emit in the form of electromagnetic radiation. In other words, *an object is more luminous when it is hotter*.

As an object gets hotter, the thermal motions of its particles become more energetic, producing not only more, but also more energetic photons. As the average energy of the photons that it emits increases, the average wavelength of the emitted photons becomes shorter, and the light from the object gets bluer. Thus, *hotter objects are bluer*. If you heat a piece of metal, the metal will glow—first a dull red, then orange, and then yellow. The hotter the metal becomes, the more the highly energetic blue photons become mixed with the less energetic red photons, and the color of the light shifts from red toward blue. The light becomes more intense and bluer as the metal becomes hotter. Changing the temperature of an object causes both its luminosity and its color to change in its spectrum.

Imagine an idealized object that emits light only because of its temperature, independent of its composition. This kind of idealized object is called a **blackbody**. Blackbodies emit exactly as much thermal radiation as they absorb from their surroundings. Physicist Max Planck (1858–1947) graphed the intensity of the emitted radiation from a blackbody across all wavelengths and obtained the characteristic curves that we now call **Planck spectra** or **blackbody spectra**. Figure 5.23 shows blackbody spectra for objects at several temperatures. As the temperature changes, two effects can be seen in these spectra: hotter objects are both more luminous—the entire spectrum is higher on the graph, and bluer—the highest point (peak) of the spectrum moves left to shorter wavelengths. These two effects are described by the Stefan-Boltzmann law, which relates luminosity with temperature, and Wien's law, which relates temperature with color. We'll explain both of these in the next subsection.

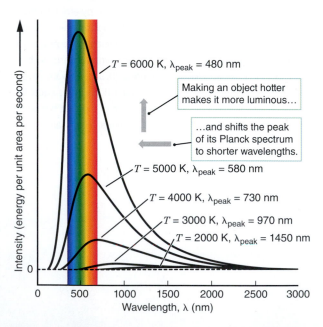

Figure 5.23 This graph shows blackbody spectra emitted by sources with temperatures of 2000, 3000, 4000, 5000, and 6000 K. At higher temperatures, the peak of the spectrum shifts toward shorter wavelengths, and the amount of energy radiated per second from each square meter of the source increases.

Blackbody Laws

Spectra from stars such as the Sun and the thermal radiation from planets often closely resemble blackbody spectra. The spectra of both types of objects can be described by the Stefan-Boltzmann law and Wien's law.

Stefan-Boltzmann Law   It would be difficult to directly measure all the photons emitted by Earth in all possible directions. But it is relatively easy to measure the number of photons leaving a small area. That number can then be multiplied by the area of the entire Earth to find Earth's luminosity. The **flux** ($\mathcal{F}$) is the amount of energy radiated by each square meter of the surface of an object each second. Multiplying the flux by the total surface area gives the luminosity. According to the **Stefan-Boltzmann law**, the flux is given by ($\mathcal{F} = \sigma T^4$, where σ (the Greek letter sigma) is the **Stefan-Boltzmann constant**. It equals 5.67×10^{-8} W/(m² K⁴), where 1 watt (W) = 1 joule per second (J/s).

The Stefan-Boltzmann law was discovered in the laboratory by physicist Josef Stefan (1835–1893) and derived mathematically by his student Ludwig Boltzmann (1844–1906). The Stefan-Boltzmann law says that an object rapidly becomes more luminous as its temperature increases. If the temperature of an object doubles, the amount of energy being radiated each second increases by a factor of $2^4 = 16$. If the temperature of an object increases by a factor of 3, the energy being radiated by the object each second increases by a factor of $3^4 = 81$. A lightbulb with a filament temperature of 3000 K radiates 16 times as much light as it would if the filament temperature were 1500 K. Even modest changes in temperature can result in large changes in the luminosity of an object.

Wien's Law   Look again at Figure 5.23. The wavelength where the blackbody spectrum is at its peak, λ_{peak}, indicates the wavelength where the electromagnetic radiation from an object is greatest. As the temperature, T, increases, the peak of the spectrum shifts toward shorter, bluer, wavelengths. For example, $\lambda_{peak} = 970$ nm for a 3000 K object, but only 480 nm for a 6000 K object. The physicist Wilhelm Wien (1864–1928) found that the peak wavelength in the spectrum is inversely proportional to the temperature of the object ($\lambda_{peak} \propto 1/T$). **Wien's law** says that if you double the temperature, the peak wavelength becomes half of what it was. If you increase the temperature by a factor of 3, the peak wavelength becomes a third of what it was. The Stefan-Boltzmann law and Wien's law are further explored in **Working It Out 5.3**. We use both laws later in the chapter to estimate the temperatures of the planets.

Astronomy in Action: Wien's Law

CHECK YOUR UNDERSTANDING 5.4

When you look at the sky on a dark night you see stars of different colors. Rank them from coolest to hottest. (a) orange; (b) red-orange; (c) yellow; (d) red; (e) blue

5.5   The Brightness of Light Depends on the Luminosity and Distance of the Source

Whereas luminosity is the amount of light *leaving* a source, the **brightness** is the amount of light *arriving* at a particular location. The brightness depends on the luminosity and the distance of the light source. For example, if you needed more light to read this book, you could replace the bulb in your lamp with a more

working it out 5.3

Working with the Stefan-Boltzmann Law and Wien's Law

The Stefan-Boltzmann law can be used to estimate the flux and luminosity of Earth. Earth's average temperature is 288 K, so the flux from its surface is

$$\mathcal{F} = \sigma T^4$$
$$\mathcal{F} = (5.67 \times 10^{-8} \text{ W/m}^2 \text{ K}^4) \times (288 \text{ K})^4$$
$$\mathcal{F} = 390 \text{ W/m}^2$$

The luminosity is the flux multiplied by the surface area (A) of Earth. The surface area of a spherical object is given by $4\pi R^2$, where R is the radius of the object. The radius of Earth is 6378 km, or 6.378×10^6 meters, so the luminosity is

$$L = \mathcal{F} \times A = \mathcal{F} \times 4\pi R^2$$
$$L = (390 \text{ W/m}^2) \times [4\pi (6.378 \times 10^6 \text{ m})^2]$$
$$L \approx 2 \times 10^{17} \text{ W}$$

Earth emits the equivalent of the energy used by 2,000,000,000,000,000 (2 million billion) hundred-watt lightbulbs. That value is still not anywhere close to the amount emitted by the Sun.

If astronomers measure the spectrum of an object emitting thermal radiation and find where the peak in the spectrum is, Wien's law can be used to calculate the temperature of the object. Wien's law can be written as

$$T = \frac{2{,}900{,}000 \text{ nm K}}{\lambda_{\text{peak}}}$$

What, for example, is the surface temperature of the Sun? The spectrum of the light coming from the Sun peaks at a wavelength of $\lambda_{\text{peak}} = 500$ nm, so

$$T = \frac{2{,}900{,}000 \text{ nm K}}{500 \text{ nm}} = 5800 \text{ K}$$

What is the peak wavelength at which Earth radiates? Using Earth's average temperature of 288 K in Wien's law gives

$$\lambda_{\text{peak}} = \frac{2{,}900{,}000 \text{ nm K}}{288 \text{ K}} = 10{,}100 \text{ nm} = 10.1 \text{ µm}$$

Thus, Earth's radiation peaks in the infrared region of the spectrum.

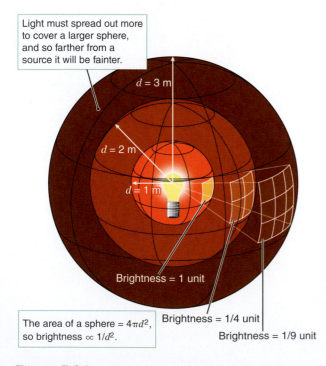

Light must spread out more to cover a larger sphere, and so farther from a source it will be fainter.

$d = 3$ m

$d = 2$ m

$d = 1$ m

Brightness = 1 unit

The area of a sphere = $4\pi d^2$, so brightness $\propto 1/d^2$.

Brightness = 1/4 unit

Brightness = 1/9 unit

Figure 5.24 Light obeys an inverse square law as it spreads away from a source. Twice as far means one-fourth as bright, three times as far means one-ninth as bright, and so on.

luminous bulb or you could move the book closer to the light. Conversely, if a light were too bright, you could move away from it, or replace the bulb with a less luminous one.

Suppose you had a piece of cardboard that measured 1 meter by 1 meter. To make the light falling on the cardboard twice as bright, you would need to double the number of photons that hit the cardboard each second. Tripling the brightness of the light would mean increasing the number of photons hitting the cardboard each second by a factor of 3, and so on. Brightness depends on the number of photons falling on each square meter of a surface each second.

Now imagine a lightbulb sitting at the center of a spherical shell (**Figure 5.24**). Photons from the bulb travel in all directions and strike the inside of the shell. To find the number of photons landing on each square meter of the shell during each second—that is, to determine the brightness of the light—take the *total* number of photons given off by the lightbulb each second and divide by the area over which those photons have to be spread. The surface area of a spherical shell is given by the formula $A = 4\pi d^2$, where d is the distance between the bulb and the surface of the sphere. The number of photons striking one square meter each second is equal to the total number of photons emitted each second divided by the surface area $4\pi d^2$.

Imagine that you increase the size of the spherical shell. As the shell becomes larger, the photons from the lightbulb must spread out to cover a larger surface area. Each square meter of the shell receives fewer photons each second, so the brightness of the light decreases. If the shell's surface is moved twice as far from the light, d is twice as large, so the area over which the light must spread increases by a factor of $2^2 = 2 \times 2 = 4$. The photons from the bulb spread out over 4 times as much area, so the number of photons falling on each square meter each second becomes $\frac{1}{4}$ of what it was. If the surface of the sphere is 3 times as far from the light, the area over which the light must spread increases by a factor of $3^2 = 3 \times 3 = 9$, and the number of photons per second falling on each square meter becomes $\frac{1}{9}$ of what it was originally. The brightness of the light from an object is inversely proportional to the square of the distance from the object. Twice as far means one-fourth as bright. (You have seen this "inverse square" relationship before, when you learned about gravity in Chapter 4.)

The idea of photons streaming and spreading onto a surface from a light explains why brightness follows an inverse square law. In practice, however, talking about the average *energy* falling on a surface each second, rather than the number of photons, is usually more useful to an astronomer.

The luminosity of an object is the total number of photons given off by the object multiplied by the energy of each photon. So, instead of thinking about how the number of photons must spread out to cover the surface of a sphere, we can think about how the energy carried by the photons must spread out to cover the surface of a sphere. The brightness of the light is the amount of energy falling on a square meter in a second, and it equals the luminosity L divided by the area of the sphere, which depends on the radius squared. That tells us, for example, that the brightness of the Sun on a given planet depends on the inverse square of the planet's distance from the Sun. That relationship factors in when we estimate the equilibrium temperatures of the planets in **Working It Out 5.4.**

CHECK YOUR UNDERSTANDING 5.5

The average distance of Mars from the Sun is 1.4 AU. How bright is the Sun on Mars compared with its brightness on Earth? (a) 1.4 times brighter; (b) about 2 times brighter; (c) about 2 times fainter; (d) 1.4 times fainter

Origins: Temperatures of Planets

In the previous chapters, we discussed how a planet's axial tilt and the shape of its orbit affect its temperature and thus its prospects for life. Now let's get more specific about the temperatures of planets, using what you learned in this chapter about thermal radiation. For a planet at an equilibrium temperature, the energy radiated by a planet exactly balances the energy absorbed by the planet. If the planet is hotter than that equilibrium temperature, it will radiate energy faster than it absorbs sunlight, and its temperature will decrease. If the planet is cooler than that temperature, it will radiate energy slower than it absorbs sunlight, and its temperature will increase.

Planets at different distances from the Sun will have different temperatures, and the temperature should be inversely proportional to the square root of the distance, as you saw in Working It Out 5.4. **Figure 5.25** plots the actual and predicted temperatures of nine solar system objects. Each vertical orange bar shows the range of temperatures found on the surface of the object or, for the giant planets, at the top of the planet's clouds. The black dots show the predictions made using the equation in Working It Out 5.4. For most objects, the predictions are not too far off, indicating that our basic understanding of *why* planets and dwarf planets have the

what if . . .

What if the Sun's luminosity suddenly increased by a factor of two? We know the new equilibrium temperature of Earth would be higher and would be independent of Earth's radius. Do you think the length of time to achieve that new equilibrium will depend on Earth's radius?

Astronomy in Action: Inverse Square Law

working it out 5.4

Using Radiation Laws to Calculate Equilibrium Temperatures of Planets

The temperature of a planet is determined by a balance between the amount of sunlight being absorbed and the amount of energy being radiated back into space. We begin with the amount of sunlight being absorbed. When viewed from the Sun, a planet looks like a circular disk with a radius equal to the radius of the planet, R_{planet}. The area of the planet that is lit by the Sun is

$$\text{Absorbing area of planet} = \pi R^2_{planet}$$

The amount of energy striking a planet also depends on the brightness of sunlight at the distance at which the planet orbits. The brightness of sunlight at a distance d from the Sun is equal to the luminosity of the Sun (L_{Sun}, in watts) divided by $4\pi d^2$:

$$\text{Brightness of sunlight} = \frac{L_{Sun}}{4\pi d^2}$$

A planet does not absorb *all* the sunlight that falls on it. **Albedo**, a, is the fraction of the sunlight that reflects from a planet. The corresponding fraction of the sunlight absorbed by the planet is 1 minus the albedo $(1 - a)$. A planet covered in snow would have a high albedo (close to 1), whereas a planet covered by black rocks would have a low albedo, close to 0:

$$\text{Fraction of sunlight absorbed} = 1 - a$$

We can now calculate the energy absorbed by the planet each second. Writing that relationship as an equation, we say that

$$\begin{pmatrix} \text{Energy absorbed} \\ \text{by the planet} \\ \text{each second} \end{pmatrix} = \begin{pmatrix} \text{Absorbing} \\ \text{area of} \\ \text{the planet} \end{pmatrix} \times \begin{pmatrix} \text{Brightness} \\ \text{of sunlight} \end{pmatrix} \times \begin{pmatrix} \text{Fraction} \\ \text{of sunlight} \\ \text{absorbed} \end{pmatrix}$$

$$= \pi R^2_{planet} \times \frac{L_{Sun}}{4\pi d^2} \times (1 - a)$$

Now let's turn to the other piece of the equilibrium: the amount of energy that the planet radiates away into space each second. We can calculate that amount by multiplying the number of square meters of the planet's total surface area by the energy radiated by each square meter each second. The surface area for the planet is given by $4\pi R^2_{planet}$. According to the Stefan-Boltzmann law, the energy radiated by each square meter each second is given by σT^4. Thus,

$$\begin{pmatrix} \text{Energy radiated} \\ \text{by the planet} \\ \text{each second} \end{pmatrix} = \begin{pmatrix} \text{Surface} \\ \text{area of} \\ \text{the planet} \end{pmatrix} \times \begin{pmatrix} \text{Energy radiated} \\ \text{per square meter} \\ \text{per second} \end{pmatrix}$$

$$= 4\pi R^2_{planet} \times \sigma T^4$$

If the planet's temperature is to remain stable—not heating up or cooling down—then each second the "energy radiated" must be equal to "energy absorbed":

$$\begin{pmatrix} \text{Energy radiated} \\ \text{by the planet} \\ \text{each second} \end{pmatrix} = \begin{pmatrix} \text{Energy absorbed} \\ \text{by the planet} \\ \text{each second} \end{pmatrix}$$

When we set those two quantities equal to each other, we arrive at the following expression:

$$4\pi R^2_{planet} \sigma T^4 = \pi R^2_{planet} \frac{L_{Sun}}{4\pi d^2}(1 - a)$$

Canceling out πR^2_{planet} on both sides, and rearranging the equation to put T on one side and everything else on the other, gives

$$T^4 = \frac{L_{Sun}(1 - a)}{16\sigma\pi d^2}$$

If we take the fourth root of each side, we get

$$T = \left(\frac{L_{Sun}(1 - a)}{16\sigma\pi d^2}\right)^{1/4}$$

Putting in the appropriate numbers for the known luminosity of the Sun, L_{Sun}, and the constants π and σ yields a simpler equation:

$$T = 279\,\text{K} \times \left(\frac{1 - a}{d^2_{AU}}\right)^{1/4}$$

where d_{AU} is the distance of the planet from the Sun in astronomical units.

To use that equation, we need to know a planet's distance from the Sun and its average albedo. For a blackbody ($a = 0$) at 1 AU from the Sun, the temperature is 279 K. For Earth, with an albedo of 0.3 and a distance from the Sun of 1 AU, the temperature is

$$T = 279\,\text{K} \times \left(\frac{1 - 0.3}{1^2}\right)^{1/4} = 255\,\text{K}$$

(Calculator hint: To take a fourth root, you can take the square root twice, or use the x^y button with $y = 0.25$.)

Earth is cooler than a blackbody at 1 AU from the Sun because its average albedo is greater than zero. If Earth's albedo changed or the Sun's luminosity changed, that would affect the result. When we examine planets around other stars, we must use the luminosity of the particular star in the equation, instead of the Sun's luminosity, so the temperature at 1 AU will be different from what it is for Earth.

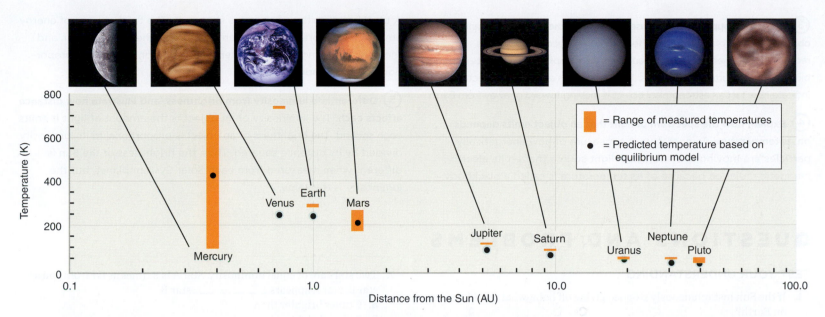

Figure 5.25 Predicted temperatures for the planets and dwarf planet Pluto are based on the equilibrium between absorbed sunlight and thermal radiation into space. Those temperatures are compared with ranges of observed surface temperatures.

temperatures they have is probably pretty good. The data for Mercury, Mars, and Pluto agree particularly well.

Sometimes, however, the predictions are wrong. For Earth, the actual measured temperature is a bit higher than the predicted temperature, and for Venus the actual surface temperature is much higher than the prediction.

The predicted values assume that the temperature of the planet is the same everywhere, but planets are likely to be hotter on the day side than on the night side. The predictions also assume that a planet's only source of energy is sunlight and that the fraction of sunlight reflected is constant over the surface of each planet. The values also incorporate the assumption that the planets absorb and radiate energy into space as blackbodies.

The discrepancies between the calculated and the measured temperatures of some of the planets indicate that for those planets, some or all of those assumptions are incorrect. For example, the planet may have its own source of energy besides sunlight, or it may have an atmosphere that traps more solar energy and increases the temperature of the planet. Understanding the temperatures of planets makes it possible to hypothesize which planets are more suitable for life.

SUMMARY

Light carries both information and energy throughout the universe. The speed of light in a vacuum is 300,000 km/s; nothing can travel faster. Visible light is only a tiny portion of the entire electromagnetic spectrum. Atoms absorb and emit radiation at unique wavelengths, giving them spectral fingerprints. A planet's temperature depends on its distance from its star, its albedo, and the luminosity of its star.

(1) Compare the wave and particle properties of light, and describe the electromagnetic spectrum. Light is simultaneously a stream of particles called photons and an electromagnetic wave. Different types of electromagnetic radiation, from gamma rays to visible light to radio waves, are electromagnetic waves that differ in frequency and wavelength.

(2) Describe how to measure the chemical composition of distant objects by using the unique spectral lines of different types of atoms. Light can reveal the identity of the chemical elements present in matter. The electron energy levels of each element have different (unique) spacings, and the wavelengths of the photons that each element emits correspond to the differences in those levels. As a result, we can identify different chemical elements and molecules in distant objects.

③ **Apply the Doppler effect and use it to measure the motion of distant objects.** Because of the Doppler effect, light from receding objects is redshifted to longer wavelengths, and light from approaching objects is blueshifted to shorter wavelengths. The wavelength shifts of the spectral lines indicate how fast an astronomical object is moving toward or away from Earth.

④ **Explain how the spectrum of light that an object emits depends on its temperature.** Temperature is a measure of how energetically particles are moving in an object. A light source that emits electromagnetic radiation because of its temperature is called a blackbody.

A blackbody emits a continuous spectrum. The total amount of energy emitted is proportional to the temperature to the fourth power, and the peak wavelength, which determines its color, is inversely proportional to the temperature.

⑤ **Differentiate luminosity from brightness, and illustrate how distance affects each.** The luminosity of an object is the amount of light it emits each second. The brightness of an object is proportional to its luminosity divided by its distance squared. Thus, the brightness of the Sun is different when measured from each Solar System planet, but the luminosity is the same.

QUESTIONS AND PROBLEMS

TEST YOUR UNDERSTANDING

1. If the Sun instantaneously stopped giving off light, what would happen on Earth?
   a. Earth would immediately get dark.
   b. Earth would get dark 8 minutes later.
   c. Earth would get dark 27 minutes later.
   d. Earth would get dark 1 hour later.

2. Why is an iron atom a different element from a sodium atom?
   a. A sodium atom has fewer neutrons in its nucleus than an iron atom has.
   b. An iron atom has more protons in its nucleus than a sodium atom has.
   c. A sodium atom is bigger than an iron atom.
   d. A sodium atom has more electrons.

3. Suppose an atom has three energy levels, specified in arbitrary units as 10, 7, and 5. In those units, which of the following energies might an emitted photon have? (Select all that apply.)
   a. 3
   b. 2
   c. 5
   d. 4

4. When a boat moves through the water, the waves in front of the boat bunch up, whereas the waves behind the boat spread out. That is an example of
   a. the Bohr model.
   b. the wave nature of light.
   c. emission and absorption.
   d. the Doppler effect.

5. As a blackbody becomes hotter, it also becomes _____ and _____.
   a. more luminous; redder
   b. more luminous; bluer
   c. less luminous; redder
   d. less luminous; bluer

6. Which of the following factors directly influences the temperature of a planet? (Choose all that apply.)
   a. the luminosity of the Sun
   b. the distance from the planet to the Sun
   c. the albedo of the planet
   d. the size of the planet

7. Two stars are of equal luminosity. Star A is 3 times as far from you as star B. Star A appears _____ star B.
   a. 9 times brighter than
   b. 3 times brighter than
   c. the same brightness as
   d. $\frac{1}{3}$ as bright as
   e. $\frac{1}{9}$ as bright as

8. When less energy radiates from a planet, its _____ increases until a new _____ is achieved.
   a. temperature; equilibrium
   b. size; temperature
   c. equilibrium; size
   d. temperature; size

9. How does the speed of light in a medium compare with the speed in a vacuum?
   a. The speed is the same in both a medium and a vacuum because the speed of light is a constant.
   b. The speed in the medium is always faster than the speed in a vacuum.
   c. The speed in the medium is always slower than the speed in a vacuum.
   d. The speed in the medium may be faster or slower, depending on the medium.

10. When an electron moves from a higher energy level in an atom to a lower energy level,
    a. a continuous spectrum is emitted.
    b. a photon is emitted.
    c. a photon is absorbed.
    d. a redshifted spectrum is emitted.

11. In Figure 5.16, the red photons come from the transition from E_3 to E_2. Those photons will have the _____ wavelengths because they have the _____ energy compared with that of the other photons. ✦
    a. shortest; least
    b. shortest; most
    c. longest; least
    d. longest; most

12. Star A and star B appear equally bright in the sky. Star A is twice as far away from Earth as star B. How do the luminosities of stars A and B compare?
    a. Star A is 4 times as luminous as star B.
    b. Star A is 2 times as luminous as star B.
    c. Star B is 2 times as luminous as star A.
    d. Star B is 4 times as luminous as star A.

13. What is the surface temperature of a star that has a peak wavelength of 290 nm?
    a. 1000 K
    b. 2000 K
    c. 5000 K
    d. 10,000 K
    e. 100,000 K

14. If a planet is in thermal equilibrium,
    a. no energy is leaving the planet.
    b. no energy is arriving on the planet.
    c. the amount of energy leaving equals the amount of energy arriving.
    d. the temperature is very low.

15. The temperature of an object has a specific meaning as it relates to the object's atoms. A high temperature means that the atoms
    a. are very large.
    b. are moving very fast.
    c. are all moving together.
    d. have a lot of energy.

THINKING ABOUT THE CONCEPTS

16. ★ **WHAT AN ASTRONOMER SEES**   Suppose that you read about a new observation of an object that combines data from the ultraviolet and the radio parts of the spectrum. Use Figure 5.6 to determine the approximate wavelength and frequency ranges included in this observation. Why would astronomers want to combine multiple parts of the spectrum when observing an object? 👁

17. The speed of light in a vacuum is 3×10^5 km/s. Can light travel at a lower speed? Explain your answer.

18. Is light a wave or a particle or both? Explain your answer.

19. If any of the experiments mentioned in the Process of Science Figure had *not* agreed with the others, what would that mean for the conclusion that light has a finite, constant speed? ★

20. If photons of blue light have more energy than photons of red light, how can a beam of red light carry as much energy as a beam of blue light?

21. Patterns of emission or absorption lines in spectra can uniquely identify individual atomic elements. How can the positive identification of atomic elements be used to test the validity of the cosmological principle discussed in Chapter 1?

22. An atom in an excited state can drop to a lower energy state by emitting a photon. Can we predict exactly how long the atom will remain in the higher energy state? Explain your answer.

23. Spectra of astronomical objects show both bright and dark lines. Describe what those lines indicate about the atoms responsible for the spectral lines.

24. Astronomers describe certain celestial objects as being *redshifted* or *blueshifted*. What do those terms indicate about the objects?

25. An object somewhere near you is emitting a pure tone at middle C on the octave scale (262 Hz). You, having perfect pitch, hear the tone as A above middle C (440 Hz). Describe the motion of that object relative to where you are standing.

26. During a popular art exhibition, the museum staff finds it necessary to protect the artwork by limiting the total number of viewers in the museum at any particular time. New viewers are admitted at the same rate that others leave. Is that system an example of static equilibrium or of dynamic equilibrium? Explain.

27. A favorite object for amateur astronomers is the double star Albireo, with one of its components a golden yellow and the other a bright blue. What do those colors tell you about the relative temperatures of the two stars?

28. The stars you see in the night sky cover a large range of brightness. What does that range tell you about the distances of the various stars? Explain your answer.

29. Why is it not surprising that sunlight peaks in the "visible"?

30. In Figure 5.25, why is the range of temperatures so much greater for Mercury than for the other planets? 👁

APPLYING THE CONCEPTS

31. Your microwave oven cooks by vibrating water molecules at a frequency of 2.45 gigahertz (GHz), or 2.45×10^9 Hz. What is the wavelength of the microwave's electromagnetic radiation? 🟡–🟢–🟢
    a. Make a prediction: Do you expect this wavelength to be longer or shorter than visible wavelengths (which are a few hundred nanometers)?
    b. Calculate: Follow Working It Out 5.1 to convert the given frequency into a wavelength.
    c. Check your work: Does your answer agree with your prediction? Does it lie in the microwave region of the spectrum?

32. Assume that an object emitting a pure tone of 440 Hz is on a vehicle approaching you at a speed of 25 m/s. If the speed of sound at this particular atmospheric temperature and pressure is 340 m/s, what will be the frequency of the sound that you hear? 🟡–🟢–🟢
    a. Make a prediction: Do you expect the wavelength of sound from an approaching vehicle to be longer or shorter? So then, do you expect the frequency to be higher or lower?
    b. Calculate: Follow Working It Out 5.2 to calculate the observed sound frequency.
    c. Check your answer: Check your units, and compare your answer to your prediction.

33. The Sun has a radius of 6.96×10^5 km and a blackbody temperature of 5780 K. Calculate the Sun's luminosity. 🟡–🟢–🟢
    a. Make a prediction: Do you expect the Sun's luminosity to be small, large, very large, or extremely large? Suppose that you calculate that the Sun's luminosity is 100 W; would you believe this answer?
    b. Calculate: Follow Working It Out 5.3 to calculate the Sun's luminosity.
    c. Check your work: Verify that you have the correct units for luminosity in your answer, and compare the luminosity you calculate to your prediction and to the known luminosity of the Sun.

34. Suppose our Sun had 10 times its current luminosity. What would the average blackbody surface temperature of Earth be if Earth had the same albedo? ⬤–⬤–⬤
    a. Make a prediction: Do you expect Earth to be hotter or colder than it currently is, if the Sun were 10 times as luminous?
    b. Calculate: Follow Working It Out 5.4 to calculate the surface temperature of Earth if the Sun were 10 times as luminous.
    c. Check your work: Verify that your answer has the correct units for temperature, and compare it to your prediction.

35. Some of the hottest stars known have a blackbody temperature of 100,000 K. What is the peak wavelength of their radiation? What type of radiation is it?
    a. Make a prediction: Given that these are the hottest stars, do you expect the peak wavelength to be longer than red wavelengths, or shorter than blue wavelengths?
    b. Calculate: Follow Working It Out 5.3 to calculate the peak wavelength of the hottest stars.
    c. Check your work: Verify that you have the correct units for wavelength in your answer, and compare your numerical value to your prediction.

36. You observe a spectral line of hydrogen at a wavelength of 502.3 nm in a distant galaxy. The rest wavelength of that line is 486.1 nm. What is the radial velocity of that galaxy? Is it moving toward you or away from you?

37. If half of the phosphorescent atoms in a glow-in-the-dark toy give up a photon every 30 minutes, how bright (relative to its original brightness) will the toy be after 2 hours?

38. How bright would the Sun appear to be from Neptune, 30 AU from the Sun, compared with its brightness as seen from Earth? The spacecraft *Voyager 1* is now about 140 AU from the Sun and heading out of the Solar System. Compare the brightness of the Sun as seen from *Voyager 1* with that seen from Earth.

39. You are tuned to 790 on AM radio. That station is broadcasting at a frequency of 790 kHz (7.90×10^5 Hz). You switch to 98.3 on FM radio. That station is broadcasting at a frequency of 98.3 MHz (9.83×10^7 Hz).
    a. What are the respective wavelengths of the AM and FM radio signals?
    b. Which broadcasts at higher frequencies, AM or FM?
    c. What are the respective photon energies of the two broadcasts?

40. On a dark night you notice that a distant lightbulb happens to have the same brightness as a firefly 5 meters away from you. If the lightbulb is a million times more luminous than the firefly, how far away is the lightbulb?

41. Two stars appear to have the same brightness, but one star is 3 times more distant than the other. How much more luminous is the more distant star?

42. A panel with an area of 1 square meter (m^2) is heated to a temperature of 500 K. How many watts is the panel radiating into its surroundings?

43. Your body emits radiation at a temperature of about 37°C.
    a. What is that temperature in kelvins? What is the peak wavelength, in microns, of your emitted radiation? In what region of the spectrum is this?
    b. If you assume an exposed body surface area of 0.25 m^2, how many watts of power do you radiate?

44. A planet with no atmosphere at 1 AU from the Sun would have an average blackbody surface temperature of 279 K if it absorbed all the Sun's electromagnetic energy falling on it (albedo = 0).
    a. What would the average temperature on that planet be if its albedo were 0.1, typical of a rock-covered surface?
    b. What would the average temperature be if its albedo were 0.9, typical of a snow-covered surface?

45. The orbit of Eris, a dwarf planet, carries it out to a maximum distance of 97.7 AU from the Sun. If you assume an albedo of 0.8, what is the average temperature of Eris when it is farthest from the Sun?

EXPLORATION Light as a Wave

digital.wwnorton.com/astro7

Visit the Digital Resources Page, and on the Student Site open the "Light as a Wave, Light as a Photon" AstroTour in Chapter 5. Watch the first section and then click through, using the "Play" button, until you reach "Section 2 of 3."

Here we explore the following questions: How many properties does a wave have? Are any of those properties related to each other?

Work your way to the experimental section, where you can adjust the properties of the wave. Watch the simulation for a moment to see how fast the frequency counter increases.

1 Increase the wavelength by pressing the arrow key. What happens to the rate of the frequency counter?

_____

_____

2 Reset the simulation and then decrease the wavelength. What happens to the rate of the frequency counter?

_____

_____

3 How are the wavelength and frequency related to each other?

_____

_____

4 Imagine that you increase the frequency instead of the wavelength. How should the wavelength change when you increase the frequency?

_____

_____

5 Reset the simulation, and increase the frequency. Did the wavelength change in the way you expected?

_____

_____

6 Reset the simulation, and increase the amplitude. What happens to the wavelength and the frequency counter?

_____

_____

_____

7 Decrease the amplitude. What happens to the wavelength and the frequency counter?

_____

_____

8 Is the amplitude related to the wavelength or frequency?

_____

_____

9 Why can't you change the speed of this wave?

_____

_____

The Tools of the Astronomer

Some telescopes use lenses to change the path of light. The properties of the lens determine the appearance of the image. A glass full of water may be used as a lens. Fold a piece of paper so that it stands up. Draw a horizontal arrow (pointing to the side) on the vertical part of the paper. Place the arrow about 3 inches behind a clear empty glass and observe the arrow through the glass. Make a prediction about what you will see if you fill the glass with water. As you fill the glass with water, watch the arrow through the glass. Write down your observations. Make a sketch, drawn from above, of the path of the light rays as they leave the two ends of the arrow and pass through the empty glass on their way to your eye. Repeat the sketch for the water-filled glass. Move the glass closer to and farther away from the arrow. Write down your observations of any changes that occur at the different distances.

EXPERIMENT SETUP

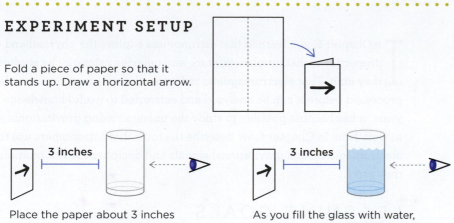

Fold a piece of paper so that it stands up. Draw a horizontal arrow.

3 inches

3 inches

Place the paper about 3 inches behind a clear empty glass and observe the arrow through the glass.

As you fill the glass with water, watch the arrow through the glass. Write down your observations.

Make a sketch, drawn from above, of the path of the light rays as they leave the two ends of the arrow and pass through the empty glass on their way to your eye. Repeat the sketch for the water-filled glass.

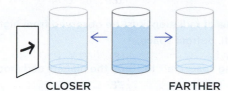

CLOSER   **FARTHER**

Move the glass closer to and farther away from the arrow. Write down your observations of any changes that occur at the different distances.

PREDICTION

I predict that when I view the arrow through the glass, that it will be:

_____

SKETCH OF RESULTS (in progress)

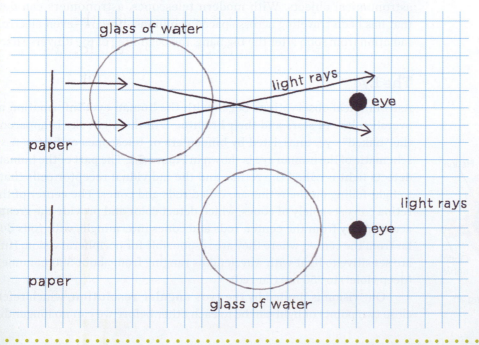

glass of water

light rays

eye

paper

light rays

eye

paper

glass of water

6

In Chapter 5, you learned that astronomers explore the physical and chemical properties of distant planets, stars, and galaxies primarily by studying the light they emit. That electromagnetic radiation, though, must first be collected and processed before it can be analyzed and converted to useful knowledge. In recent years, it has become possible to study the universe using gravitational waves as well. Here in Chapter 6, we describe the tools that astronomers use to collect electromagnetic and gravitational signals and decipher the information that they carry.

LEARNING GOALS

By the end of this chapter, you should be able to:

1. Compare how the two main types of optical telescopes gather and focus light.

2. Summarize the main types of detectors that are used on telescopes.

3. Explain why some wavelengths of radiation must be observed from high, dry, and remote observatories on Earth, or from space.

4. Explain the benefits of sending spacecraft to study the planets and moons of our Solar System.

5. Describe other astronomical tools, such as gravitational wave detectors, that contribute to the study of the universe.

6.1 The Optical Telescope Revolutionized Astronomy

The development of **telescopes**—devices for collecting and focusing light—in the 17th century greatly increased the amount of light that can be collected from astronomical objects. With modern telescopes, astronomers can detect light that has been traveling across space for billions of years—even electromagnetic radiation from soon after the Big Bang, the beginning of the universe itself.

The Eye

Astronomical observations began with the human eye. Information about the overall colors of stars and their brightness in the night sky is apparent even to the "naked" eye, unassisted by binoculars or telescopes or filters. A simplified schematic of the human eye is shown in **Figure 6.1**. The part of the human eye that detects light is called the retina, and the individual receptor cells that respond to light falling on the retina are called rods and cones. The center of the human retina consists solely of cones, which detect color and provide the greatest visual acuity. Away from the center, rods and cones intermingle, with rods dominating far from the center, where they are responsible for peripheral vision. Human eyes are sensitive to light with wavelengths ranging from about 380 nanometers (nm) (deep violet) to 750 nm (far red).

Our vision is limited by the eye's **angular resolution**, which refers to how close two points of light can be to each other before we can no longer distinguish

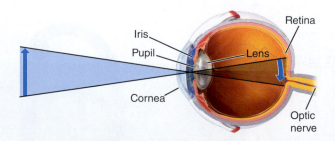

Figure 6.1 A schematic view of the human eye creating an image of an object (the blue arrow).

them. Unaided, the best human eyes can resolve objects separated by 1 **arcminute** (1/60 of a degree), an angular distance of about 1/30 the diameter of the full Moon. (A more in-depth description of angular units—radians, degrees, arcminutes, and arcseconds—can be found in Appendix 1.) That may seem small, but thousands of stars and galaxies may reside within a patch of sky with that diameter.

Refracting Telescopes

Optical telescopes come in two primary types: **refracting telescopes**, which use lenses, and **reflecting telescopes**, which use mirrors. For all telescopes, the "size" of the telescope refers to the diameter of the largest mirror (or lens), known as the **primary mirror** (or lens) which determines the light-collecting area. That diameter is called the **aperture**. The light-gathering power of a telescope is proportional to the area of its opening—that is, to the square of its diameter. The larger the aperture, the more light the telescope can collect. A "1-meter telescope" has a primary mirror (or lens) that is 1 meter in diameter. The aperture of the human eye is 6–7 millimeters (mm).

In the late 13th century, craftsmen in Venice were making small lentil-shaped disks of glass that could be mounted in frames and worn over the eyes to improve vision. More than 300 years later, Hans Lippershey (1570–1619), a spectacle maker living in the Netherlands, put two of his lenses together in a tube. With that new instrument, he saw distant objects magnified and could see farther. Galileo Galilei heard news of that invention and constructed one of his own. Recall from Chapter 3 that by the early 1600s, Galileo had become the first to see the phases of Venus and the moons of Jupiter and was among the first to see craters on the Moon. He also was the first to realize that the Milky Way is made up of many individual stars. The refracting telescope—one that uses lenses—quickly revolutionized the science of astronomy.

Recall from Chapter 5 that the speed of light is constant in a vacuum, but through a medium such as air or glass, the speed of light is always lower. As light enters a new medium, its speed changes. As shown in the diagram in **Figure 6.2a**, if the light traveling in the direction of the green line strikes a surface at an angle, some of the crest of the wave (red lines) arrives at the surface earlier and some arrives later. **Figure 6.2b** shows an actual light ray passing into and out of a medium (here, glass). The ray bends each time the medium changes. When the medium through which it travels changes, that bending of light is called **refraction**. Refraction is the basis for the refracting telescope.

The amount of refraction depends on the properties of the medium—what it's made of, its temperature, even the pressure that it's under. A medium's **index of refraction** (n) is equal to the ratio of the speed of light in a vacuum (c) to the speed of light in the medium (v). That relationship can be expressed by the equation $n = c/v$. For example, most glass has an index of refraction of approximately 1.5, so the speed of light in glass is 300,000 kilometers per second (km/s) divided by 1.5, or 200,000 km/s. The amount of refraction depends on the index of refraction of the materials involved and the angle at which the light strikes.

The primary lens in a refracting telescope is a simple convex lens, called the **objective lens (Figure 6.3)**, whose curved surfaces refract the light from a distant object. That refracted light forms an image on the telescope's **focal plane**, which is perpendicular to the *optical axis*—the path that light takes through the center of the lens. Because the telescope's glass lens is curved, light at the outer edges of the lens strikes the surface at a different angle than light near the center. As a result,

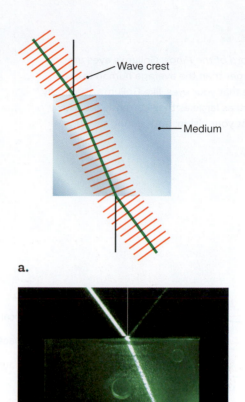

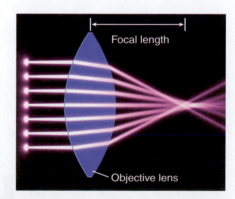

Figure 6.2 **a.** When wavefronts enter a new medium, they bend in a new direction relative to a line perpendicular to the surface (black lines). **b.** An actual light ray entering and leaving a medium.

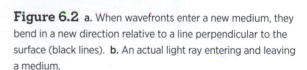

Figure 6.3 For a curved lens such as the one shown, refraction causes the light to focus to a point. That point is in a slightly different location for different wavelengths (colors) of light.

light at the outer edges of the lens is refracted more than light near its center. The lens concentrates the light rays entering the telescope, bringing them to a sharp focus at a distance called the **focal length**. Sometimes focal length is specified on a telescope as *focal ratio* (or "f-number"), which equals focal length divided by aperture; that term may be familiar to you from the lenses used in photography. Aperture and focal length are the two most important parameters of a telescope.

Figure 6.4 illustrates how focal length affects the image created by a refracting telescope. Figure 6.4a shows the light from two stars passing through a lens and converging at the focal plane of the lens. Figure 6.4b shows the same situation for a lens with a longer focal length. Longer focal lengths increase the size and separation of objects in the focal plane. In some telescopes, a second lens makes it possible for the user to view the image directly. The focal length of this interchangeable **eyepiece** determines the magnification (**Working It Out 6.1**). In modern research telescopes, however, the images are sent directly to a camera or other detector.

Refracting telescopes have two major shortcomings. First, physical limits constrain the size of refracting telescopes. The larger the area of the objective lens, the more light-gathering power it has and the fainter the stars we can observe. Larger objective lenses are heavier, however, and the weight of a massive lens can bend the tube that holds it. Refracting telescopes grew in size until the 1897 completion of the Yerkes 1-meter refractor (**Figure 6.5**), the world's largest refracting telescope. Located in Williams Bay, Wisconsin, the Yerkes telescope consists of a 450-kilogram (kg) objective lens mounted at the end of a 19.2-meter tube.

The second major shortcoming of refracting telescopes is **chromatic aberration** (**Figure 6.6**). Starlight is made up of all the colors of the rainbow, and each color refracts at a slightly different angle because the index of refraction depends on the wavelength of the light. Shorter (bluer) wavelengths are refracted more than longer (redder) wavelengths (Figure 6.6a). That wavelength-dependent difference in refraction, which spreads the white light out into its spectral colors, is called **dispersion**. Dispersion causes bluer light to come to a shorter focus than that of the longer visible wavelengths, creating chromatic aberration. In a refracting telescope with a simple convex lens, chromatic aberration produces haloed images around the star. That issue was addressed in early telescopes by increasing the focal length, but then the telescope requires a longer tube. Now manufacturers of quality cameras and telescopes use a **compound lens** composed of two types of glass to substantially correct for chromatic aberration (Figure 6.6b).

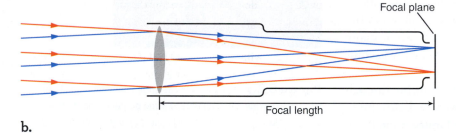

a.

b.

Figure 6.4 **a.** A refracting telescope uses a lens to collect and focus light from two stars, forming images of the stars on its focal plane. **b.** A telescope with a longer focal length produces larger images.

a.

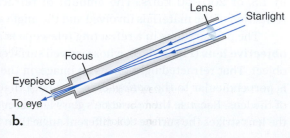

b.

Figure 6.5 **a.** The Yerkes 1-m refractor is the world's largest refracting telescope. **b.** The main parts of a refractor.

working it out 6.1

Telescope Aperture and Magnification

If you are shopping for a telescope, you need to consider its aperture and magnification.

Aperture

The light-gathering power of a telescope is proportional to the area of its lens or mirror, $\pi \times (D/2)^2$, and thus to the square of the aperture (D). The amount of light a telescope collects increases as the aperture increases. How does the light-gathering power of a telescope with a diameter of 200 mm, or 8 inches, compare with that of the pupil of your eye, which is about 6 mm in the dark?

$$\text{Light-gathering power of telescope} = \frac{\pi}{4} \times (200\,\text{mm})^2$$

and

$$\text{Light-gathering power of eye} = \frac{\pi}{4} \times (6\,\text{mm})^2$$

So, to compare:

$$\frac{\text{Light-gathering power of telescope}}{\text{Light-gathering power of eye}} = \frac{\frac{\pi}{4}(200\,\text{mm})^2}{\frac{\pi}{4}(6\,\text{mm})^2} = \left(\frac{200}{6}\right)^2 = 1111$$

Thus, an 8-inch telescope has more than 1000 times the light-gathering power of your eye.

Comparing that 8-inch telescope with the Keck 10-meter telescope shows why bigger is better: 200 mm = 0.2 meter, and we cancel out the $\pi/4$ again to obtain

$$\frac{\text{Light-gathering power of Keck}}{\text{Light-gathering power of 8-inch telescope}} = \left(\frac{10\,\text{m}}{0.2\,\text{m}}\right)^2 = 2500$$

Even larger telescopes, 25–40 meters in diameter, are under construction.

Magnification

Most telescopes have a set focal length and come with a collection of eyepieces. The magnification of the image in the telescope is given by

$$\text{Magnification} = \frac{\text{Telescope focal length}}{\text{Eyepiece focal length}}$$

Suppose the focal length of the 200-mm telescope in the preceding example is 2000 mm. Combined with the focal length of a standard eyepiece, 25 mm, that telescope will give the following magnification:

$$\text{Magnification} = \frac{2000\,\text{mm}}{25\,\text{mm}} = 80$$

That telescope and eyepiece combination has a magnifying power of 80, meaning that a crater on the Moon will appear 80 times larger in the telescope's eyepiece than it does when viewed by the naked eye. An eyepiece that has a focal length of 8 mm will have about 3 times more magnifying power, a magnification of 250.

A higher magnification will not necessarily let you see an object better. A faint and fuzzy image will not look clearer when magnified. Although a larger aperture collects more light so that you can observe fainter objects, magnification spreads that light out again, making faint objects more difficult to see; there is an inherent trade-off between aperture and magnification. If you are buying a telescope, opt for the larger aperture, and a collection of eyepieces that let you change the magnification to suit your observation target.

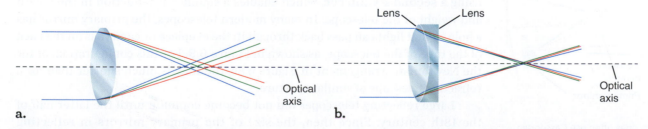

a. **b.**

Figure 6.6 **a.** Different wavelengths of light come to a focus at different places along the optical axis of a simple lens, causing chromatic aberration. **b.** A compound lens using two types of glass with different indices of refraction can compensate for much of the chromatic aberration, so different colors of light all come to a focus at the same point.

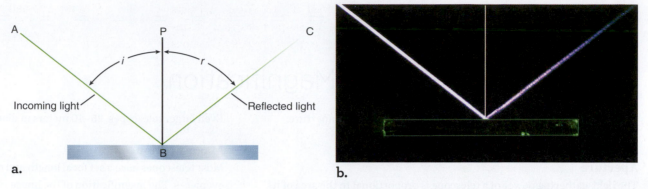

Figure 6.7 **a.** When a ray of incoming light (AB) shines on a flat surface, it reflects from the surface, becoming the reflected ray BC. The angle between AB and PB, the perpendicular to the surface, is the angle of incidence (*i*). The angle between BC and PB is the angle of reflection (*r*). The angles of incidence and reflection are always equal. **b.** Light from a laser beam is reflected from a flat glass surface.

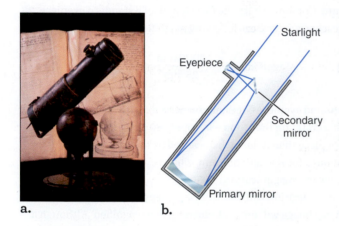

Figure 6.8 **a.** Newton's reflecting telescope. **b.** A Newtonian focus telescope has a curved primary and a flat secondary mirror, as shown in the sketch. The eyepiece is on the side of the tube.

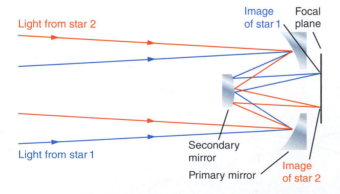

Figure 6.9 Large reflecting telescopes often use Cassegrain focus, in which a secondary mirror directs the light back through a hole in the concave parabolic primary mirror to an accessible focal plane behind the primary mirror.

Reflecting Telescopes

When light encounters a different medium, some of the light will bounce off of the surface, and remain in the original medium. This effect, called **reflection**, is the basis for reflecting telescopes. As shown in **Figure 6.7a**, the angle of the incoming light and the angle of outgoing light are always equal. You have experienced this when viewing an ordinary flat mirror (**Figure 6.7b**). A reflected image in a mirror looks like the original image, and is not stretched or distorted. Reflecting telescopes solve the two problems of refracting telescopes. Because mirrors can be supported from the back, they can be much heavier without affecting the telescope's performance. Because the angle of reflection does not depend on the wavelength of light, chromatic aberration is not a problem in reflecting telescopes.

In 1668, Isaac Newton designed a reflecting telescope, which uses mirrors instead of lenses (**Figure 6.8a**). The important parts of Newton's reflecting telescope are shown schematically in **Figure 6.8b**. To make that reflecting telescope, Newton cast a 2-inch primary mirror made of copper and tin and polished it to a special curvature. He then placed the primary mirror at the bottom of a tube with a secondary flat mirror mounted above it at a 45° angle. The second mirror directed the focused light to an eyepiece on the outside of the tube. The curve of the primary mirror of any reflecting telescope is shaped so that parallel rays of light will strike the mirror at different angles, and reflect them all so that they cross at the focal length of the mirror, as shown by the blue rays in Figure 6.8b.

The light path from the primary mirror to the focal plane can be "folded" by using a **secondary mirror**, which enables a significant reduction in the length and weight of the telescope. In many modern telescopes, the primary mirror has a hole so that light can pass back through it; the eyepiece or camera is on the back of the tube of the telescope, as shown in **Figure 6.9**. For this configuration, or for the Newtonian arrangement in Figure 6.8b, the tube is much shorter than for a refracting telescope of similar aperture.

Large reflecting telescopes did not become common until the latter half of the 18th century. Since then, the size of the primary mirrors in reflecting telescopes has grown larger every decade. Primary mirrors can be supported from the back, and they can be made thinner and therefore less massive than the objective lenses found in refracting telescopes. The limitation on the size of reflecting telescopes is the cost of their fabrication and support structure. **Table 6.1** lists the world's largest optical telescopes. All are reflecting telescopes. The largest single mirrors constructed today are 8 meters in diameter. Larger reflecting telescopes use an array of smaller segments. For example, the primary mirror of each of the 10-meter twin Keck telescopes has 36 hexagon-shaped segments that are each 1.8 meters in diameter (**Figure 6.10**). Located on 4205-meter-high Mauna Kea in Hawai'i, the Keck telescopes are among the world's largest reflecting telescopes. Each one can gather 4 million times more light than the human eye.

▶❚❚ **AstroTour:** Geometric Optics and Lenses

| Table 6.1 | The World's Largest Optical Telescopes |

Mirror Diameter (meters)	Telescope	Sponsor(s)	Location	Operational Date
39.3	Extremely Large Telescope (ELT)	European Southern Observatory (Europe, Chile, Brazil)	Cerro Armazones, Chile	Under construction
30.0	Thirty Meter Telescope (TMT)	International collaboration led by Caltech, U. of California, U. of Hawai'i, China, Japan, India, and Canada	Mauna Kea, Hawai'i, or Canary Islands, Spain	TBD
24.5	Giant Magellan Telescope (GMT)	Carnegie Institution, Arizona State U., Harvard U., Smithsonian Institution, U. of Arizona, U. of Texas, Texas A&M U., U. of Chicago, São Paulo Research Foundation (FAPESP), Australian National U., Astronomy Australia Ltd., and Korea Astronomy and Space Science Institute	Cerro Las Campanas, Chile	Under construction
10.4	Gran Telescopio Canarias (GTC)	Spain, Mexico, U. of Florida	Canary Islands	2007
10	Keck I	Caltech, U. of California, NASA	Mauna Kea, Hawai'i	1993
10	Keck II	Caltech, U. of California, NASA	Mauna Kea, Hawai'i	1996
~10	South African Large Telescope (SALT)	South Africa, USA, UK, Germany, Poland, New Zealand, India	Sutherland, South Africa	2005
10	Hobby-Eberly Telescope (HET)	U. of Texas, Penn State U., Stanford U., Germany	Mount Fowlkes, Texas	1999
8.4 × 2	Large Binocular Telescope (LBT)	U. of Arizona, Ohio State U., Italy, Germany, Arizona State, and others	Mount Graham, Arizona	2008
8.4	Large Synoptic Survey Telescope (LSST)	National Science Foundation, Dept. of Energy, and other partners through the private LSST corporation	Cerro Pachón, Chile	Under construction
8.2	Subaru Telescope	Japan	Mauna Kea, Hawai'i	1999
8.2 × 4	Very Large Telescope (VLT)	European Southern Observatory	Cerro Paranal, Chile	2000
8.1	Gemini North	USA, UK, Canada, Chile, Brazil, Argentina, Australia	Mauna Kea, Hawai'i	1999
8.1	Gemini South	USA, UK, Canada, Chile, Brazil, Argentina, Australia	Cerro Pachón, Chile	2000
6.5	MMT	Smithsonian Institution, U. of Arizona	Tucson, Arizona	2000
6.5	Magellan I	Carnegie Institution, U. of Arizona, Harvard U., U. of Michigan, MIT	Cerro Las Campanas, Chile	2000
6.5	Magellan II	Carnegie Institution, U. of Arizona, Harvard U., U. of Michigan, MIT	Cerro Las Campanas, Chile	2002

Figure 6.10 Each of the Keck 10-m reflectors uses an aligned group of 36 hexagonal mirrors to collect light.

—Segmented primary mirror

a.   **b.**

Figure 6.11 Angular resolution is the ability to separate two images that appear close together. When angular resolution is lower **a.**, the two images blend together. When angular resolution is higher **b.**, individual images can be distinguished.

CHECK YOUR UNDERSTANDING **6.1a**

Why are all large astronomical telescopes reflectors (choose all reasons that apply)? (a) chromatic aberration is minimized; (b) they are not as heavy; (c) they can be shorter; (d) only one surface of a mirror needs polishing.

Answers to Check Your Understanding questions are in the back of the book.

Optical and Atmospheric Limitations

The **angular resolution** (sometimes just "resolution") of a telescope determines how close two points of light can be to each other before they are indistinguishable (**Figure 6.11**). Review Figure 6.4a to see the path followed by rays of light from two distant stars as they pass through the lens of a refracting telescope. Figure 6.4b illustrated that increasing the focal length increases the size of and separation between the images that a telescope produces. That is one reason why telescopes provide a much clearer view of the stars than that obtained with the naked eye. The focal length of a human eye is typically about 20 mm, whereas telescopes used by professional astronomers often have focal lengths of tens or even hundreds of meters. Such telescopes make images that are far larger than those formed by the human eye, so they contain far more detail.

Focal length explains only one difference between the angular resolution of telescopes and that of the unaided eye. The other difference results from the wave nature of light. **Figure 6.12** shows what happens when light waves spread out from the edges of the lens or mirror as they pass through the aperture of a telescope. The distortion that occurs as light passes the edge of an opaque object is called **diffraction**. Diffraction diverts some of the light from its path, slightly blurring the image made by the telescope. Some of these light rays will overlap, and the waves will **interfere**, producing bright and dark "fringes." The degree of blurring depends on the wavelength of the light and the telescope's aperture. The larger the aperture, the smaller the problem diffraction poses. The best angular resolution that a given telescope can theoretically achieve is known as the **diffraction limit** (Working It Out 6.2).

Larger telescopes have better angular resolution and can distinguish objects that appear closer together. Theoretically, the 10-meter Keck telescopes have a

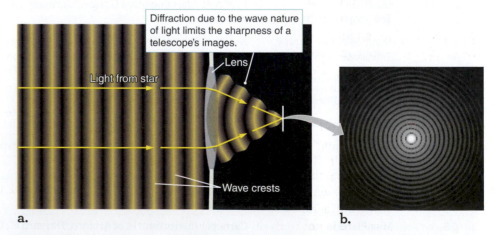

Diffraction due to the wave nature of light limits the sharpness of a telescope's images.

Lens

Light from star

Wave crests

a.   **b.**

Figure 6.12 a. Light waves from a star are diffracted by the edges of a telescope's lens or mirror. **b.** That diffraction causes the stellar image to be blurred, limiting a telescope's ability to resolve objects.

working it out 6.2

Diffraction Limit

The best possible angular resolution (θ) that can be obtained with a telescope is called the diffraction limit. That limit is determined by the ratio of the wavelength of light passing through the telescope (λ) to the diameter of the aperture (D):

$$\theta = 2.06 \times 10^5 \left(\frac{\lambda}{D}\right) \text{arcsec}$$

The constant, 2.06×10^5, has units of arcseconds (arcsec). An **arcsecond** is a tiny angular measure found by first dividing the sky into 360 degrees, then dividing each degree by 60 to get arcminutes, and then dividing each arcminute by 60 to get arcseconds. An arcsecond is 1/1800 of the size of the Moon in the sky, or about the size of a tennis ball if you could see it from 13 km (8 miles) away.

Both λ and D must be expressed in the same units—usually meters (m). The smaller the ratio of λ/D, the better the angular resolution. For example, the size of the human pupil (see Figure 6.1) ranges from about 2 mm in bright light to 8 mm (0.008 m) in the dark. Visible (green) light

has a wavelength (λ) of 550 nm—that is, 550×10^{-9} m, or 5.5×10^{-7} m. Using those values for the aperture and the wavelength gives

$$\theta = 2.06 \times 10^5 \left(\frac{5.5 \times 10^{-7}\,\text{m}}{0.008\,\text{m}}\right) \text{arcsec} = 14\,\text{arcsec}$$

or about $\frac{1}{4}$ arcmin for the theoretical best angular resolution. The typical angular resolution of the human eye is 1 arcmin. We do not achieve the best possible angular resolution with our eyes because the physical properties of our eyes are not perfect.

How does the angular resolution of the human eye compare with that of the Hubble Space Telescope in the visible part of the spectrum? The aperture of the Hubble Space Telescope is 2.4 m. Substituting that value for D and again using visible (green) light gives

$$\theta = 2.06 \times 10^5 \left(\frac{5.5 \times 10^{-7}\,\text{m}}{2.4\,\text{m}}\right) \text{arcsec} = 0.047\,\text{arcsec}$$

which is about 300 times better than the theoretical best angular resolution of the human eye.

---

diffraction-limited angular resolution of 0.0113 arcsec in visible light. If your eyes had this angular resolution, you could read this text 15 km away. But for telescopes with apertures larger than about a meter, Earth's atmosphere stands in the way of this ideal angular resolution. If you have ever looked out across a large asphalt parking lot on a summer day, you have probably seen the distant horizon shimmer as light is bent this way and that by turbulent bubbles of warm air rising off the hot pavement. The problem of the shimmering atmosphere is less pronounced when we look overhead, but the same phenomenon causes the twinkling of stars in the night sky. As telescopes magnify the angular diameter of an object, they also magnify the shimmering effects of the atmosphere. The limit on the angular resolution of a telescope on the surface of Earth caused by that atmospheric distortion is called **astronomical seeing**. One advantage of launching telescopes such as the Hubble Space Telescope into orbit around Earth is that placing them above Earth's atmosphere means they are no longer hindered by astronomical seeing.

To understand the effect Earth's atmosphere has on angular resolution, consider light from a distant star that arrives at the top of Earth's atmosphere with flat, parallel wave crests. If Earth's atmosphere were perfectly uniform, the crests would remain flat as they reached the objective lens or primary mirror of a ground-based telescope. After making their way through the telescope's optical system, the crests would produce a tiny diffraction disk in the focal plane, as shown in Figure 6.12b. But Earth's atmosphere is not uniform. It is filled with bubbles of air that have slightly different temperatures from those of their surroundings. Different temperatures mean different densities, and different densities mean different refractive properties, so each bubble bends light differently. Those air bubbles act as

unanswered questions

Will telescopes be placed on the Moon? The Moon has no atmosphere to make stars twinkle, cause weather, or block certain wavelengths of light from reaching its surface. All parts of the Moon have days and nights that last for 2 Earth weeks each. China had a small (15 cm) ultraviolet telescope operated from Earth on their *Chang'e 3* lander. One proposal is for placing a small array of radio telescopes on the side of the Moon facing Earth. These would study the Sun and the solar wind. Eventually, telescopes would go on the far side of the Moon, which faces away from the light and radio radiation of Earth. On the far side one might put an array of hundreds of radio telescopes that would be deployed to study the earliest formation of stars and galaxies. Another proposal is for the Lunar Liquid Mirror Telescope (LLMT), with a diameter of 20–100 m, to be located at one of the Moon's poles. Gravity would settle the rotating liquid into the necessary parabolic shape, and those liquid mirror telescopes are much simpler than are arrays of telescopes with large glass mirrors. Astronomers debate whether telescopes on the Moon would be easier to service and repair than those in space and whether problems caused by lunar dust would outweigh any advantages.

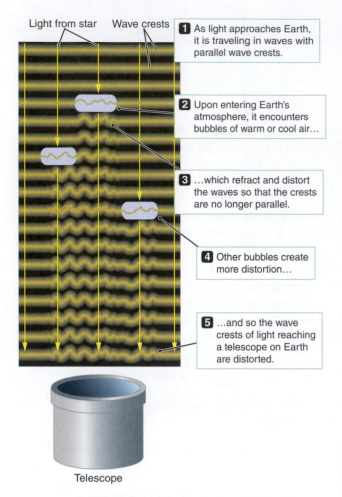

1 As light approaches Earth, it is traveling in waves with parallel wave crests.

2 Upon entering Earth's atmosphere, it encounters bubbles of warm or cool air...

3 ...which refract and distort the waves so that the crests are no longer parallel.

4 Other bubbles create more distortion...

5 ...and so the wave crests of light reaching a telescope on Earth are distorted.

Telescope

Figure 6.13 Bubbles of warmer or cooler air in Earth's atmosphere distort the wavefront of light from a distant object.

weak lenses, and by the time the waves reach the telescope they are far from flat, as shown in **Figure 6.13**. Instead of a tiny diffraction disk, the image in the telescope's focal plane is distorted and swollen, degrading the angular resolution.

Modern technology has improved ground-based telescopes with computer-controlled **adaptive optics** that compensate for much of the atmosphere's distortion. First, an optical device within the telescope constantly measures the wave crests. To calibrate those systems, a laser beam is shone into the atmosphere to create an artificial "star" in the same direction as the telescope is pointing. Then, before reaching the telescope's focal plane, the light from a target star is reflected off yet another mirror, which has a flexible surface. A computer analyzes the light and bends the flexible mirror so that it accurately corrects for the distortion of the artificial star caused by the air bubbles. **Figure 6.14** shows an example of an image corrected by adaptive optics. The widespread use of adaptive optics has made the image quality of ground-based telescopes competitive with the quality of Hubble images from space at some wavelengths. Telescopes also can have active optics, in which actuators on the rear of the mirror correct for factors such as wind, temperature, and sagging from gravity.

Observatory Locations

What makes a good location for a telescope on Earth? As you could deduce from Table 6.1, astronomers look for sites that are high, dry, and dark. The best sites are far away from the lights of cities; in locations with little moisture, humidity,

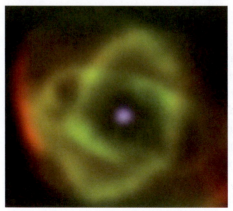

Figure 6.14 These images of the Cat's Eye Nebula from the Palomar Observatory telescope without (left) and with (right) adaptive optics show the benefit of the technique.

★ **WHAT AN ASTRONOMER SEES** An astronomer, looking at this image, would instantly identify it as a nebula (a "cloud"), because it is extended rather than pointlike, and even in the sharper image, is a little bit "fuzzy." This indicates that the light she is seeing comes from an extended cloud of dust and gas. She will also take particular note of the white dot, which indicates a central star. Because of the color, she will know that this is a relatively hot star (recall Wien's law from Chapter 5). An astronomer will notice the green and red colors of the nebula, and take note of them. An astronomer will be pleased that the image on the right allows her to see fine detail regarding the distribution of material in the nebula, including where it is red and where it is green. Without further information, she will not know for certain whether these colors are "true" colors that indicate composition (that there is line emission from particular elements in those particular colors), or whether the color represents some other aspect, like the spectral region of the observations or the velocity of the dust and gas. While it is likely that, for a nebula of this type, the colors indicate composition, even a professional astronomer cannot be certain of it without more information.

or rain; and where the atmosphere is relatively still. Telescopes are located as high as possible, above a significant part of Earth's atmosphere, which distorts images and blocks infrared and microwave light. Many telescopes are situated on remote, high mountaintops surrounded by desert or ocean. Recall from Chapter 2 that the stars that can be seen throughout the year depend on latitude, and only at the equator would a telescope have access to all the stars in the sky. But because equatorial latitudes have tropical weather—wet, humid, and stormy—they are poor locations for telescopes. To cover the entire sky, astronomers have built telescopes in both northern and southern locations. In the United States, large telescopes are located in California, Arizona, New Mexico, Texas, and Hawai'i. The largest southern-sky observatories are found in Chile, South Africa, and Australia. The twin Gemini telescopes, designed to be a matched pair, are located in Hawai'i in the Northern Hemisphere and in Chile in the Southern Hemisphere.

Newer and larger telescopes are planned for many of the same locations listed in Table 6.1. The 8-m Vera C. Rubin Observatory (formerly LSST) is under construction in Cerro Pachón in Chile, current site of the Gemini South telescope. The Giant Magellan Telescope (GMT), consisting of seven 8.4-m mirrors in a pattern equivalent to a 24.5-m mirror, is being constructed at Cerro Las Campanas in Chile. The Thirty Meter Telescope (TMT) is planned for Mauna Kea in Hawai'i or the Canary Islands, and the European Southern Observatory (ESO) is building the 39-m Extremely Large Telescope (ELT) at Cerro Armazones in Chile (**Figure 6.15**). As telescopes get larger—and more expensive—international collaboration becomes increasingly important.

Today's professional astronomers rarely look through the eyepiece of a telescope because they learn much more and make better use of observing time by permanently recording an object's image at a variety of wavelengths or seeing its light spread out into a revealing spectrum. Some astronomers no longer travel to telescopes at all, instead observing remotely from the base of the mountain or far away at their own institutions.

Professional and amateur astronomers alike are concerned about loss of the dark sky. As cities and suburbs around the world grow and expand, the use of outdoor artificial light becomes more widespread. Pictures from space show how bright many areas of Earth are at night (**Figure 6.16**). In the United States, two-thirds of the population resides in an area too bright to see the Milky Way in the sky at night, and it has been estimated that by 2025 the continental United States will have almost no dark skies. Increased air pollution also dims the view of the night sky in many locations. The U.S. National Park Service now advertises evening astronomy programs in natural, unpolluted dark skies as one of the reasons to visit some parks. Several international astronomy associations are working with UNESCO (the United Nations Educational, Scientific and Cultural Organization) to promote the "right to starlight," arguing that for historical, cultural, and scientific reasons, it would be a huge loss if humanity could no longer view the stars. Those organizations are encouraging countries to create starlight reserves and starlight parks where people can experience increasingly rare dark skies and a natural nocturnal environment.

Satellites are becoming an increasing hazard to dark skies. Many are bright enough to be seen with the naked eye. As more and more of them are launched, astronomers have to work hard to figure out how to avoid having their images of astronomical objects ruined by satellite tracks across the sky. Famously, SpaceX's Starlink satellites have already impacted astronomical observations

Figure 6.15 An artist's rendering of the European Extremely Large Telescope, a 39-m reflecting telescope under construction in Chile. The telescope will become operational in 2025. Credit: ESO/L. Calçada, https://www.eso.org/public/images/eso1440e/?lang=no. https://creativecommons.org/licenses/by/4.0/.

Figure 6.16 This satellite image of Earth at night shows that few populated areas are free from light pollution.

Figure 6.17 This image of the sky taken from the Lowell Observatory shows some stars and distant galaxies. The most noticeable features, however, are the bright streaks of more than two dozen of SpaceX's Starlink satellites passing overhead, a few weeks after they launched. Satellites such as these increasingly cause problems for astronomical observations.

(see **Figure 6.17**). By the end of 2020, SpaceX had launched more than 800 of these satellites into orbit around Earth, but it eventually plans to launch more than 10,000. Astronomers are working with SpaceX on plans to mitigate the effect by painting the satellites black, for example.

CHECK YOUR UNDERSTANDING **6.1b**

In practice, the smallest angular size that you can resolve with a 10-inch telescope is governed by the: (a) blurring caused by Earth's atmosphere; (b) diffraction limit of the telescope; (c) size of the primary mirror; (d) magnification of the telescope.

6.2 Optical Detectors and Instruments Used with Telescopes

Beginning in the 1800s, the development of film photography, and later digital photography, revolutionized astronomy, allowing astronomers to detect fainter and more-distant objects than it's possible to detect with the eye alone. Here in Section 6.2, we examine some of the more common types of detectors.

Integration Time and Quantum Efficiency

Originally, the retina of the human eye was the only astronomical detector. The limit of the faintest stars we can see with our unaided eyes is determined in part by two factors that are characteristic of all detectors: *integration time* and *quantum efficiency*.

Integration time is the limited time interval during which the eye can add up photons. It is analogous to the time the shutter is left open on a camera when taking a picture. The brain "reads out" the information gathered by the eye about every 100 milliseconds (ms). When anything happens faster than that, the human eye cannot detect the faster speed. If two images on a computer screen appear 30 ms apart, you will see them as a single image because your eyes will add up (or integrate) whatever they see over an interval of 100 ms or less. If the images occur 200 ms apart, however, you will see them as separate images. That relatively brief integration time is the most important factor limiting our nighttime vision. Stars too faint to be seen with the unaided eye are those from which you receive too few photons for your eyes to process in 100 ms.

Quantum efficiency measures how likely it is that any photon will produce a signal. For the human eye, 10 photons must strike a cone within 100 ms to activate a single response. So, the quantum efficiency of our eyes is about 10 percent. For every 10 events, then, the eye sends one signal to the brain. Together, integration time and quantum efficiency determine the rate at which photons must arrive at the retina before the brain says, "I see something." Astronomers seek to use detectors with longer integration times and higher quantum efficiency than our eyes can achieve.

From Photographic Plates to Charge-Coupled Devices

For more than two centuries after the invention of the telescope, astronomers struggled with the problem of **surface brightness**. Only *point sources* such as stars appear brighter in a telescope. Extended astronomical objects, such as the

Moon, appear bigger in the eyepiece, but the light is spread over that larger image, so their surfaces are no brighter than they appear to the unaided eye. Even when astronomers built larger telescopes, nebulae and galaxies appeared larger, but the details of these faint objects remained elusive. The problem was not with the telescopes but with the limitations of optics and the human eye. Only with the longer exposure times made possible by the invention of photography and the later development of electronic cameras could astronomers finally discern intricate details in faint objects.

In 1840, John W. Draper (1811–1882), a New York chemistry professor, created the earliest known astronomical photograph (**Figure 6.18**). By the late 1800s, astronomers had filled thousands of photographic plates with permanent images of planets, nebulae, and galaxies. The quantum efficiency of most photographic systems used in astronomy was poorer than that of the human eye—typically 1–3 percent. But unlike the eye, photography can overcome poor quantum efficiency by leaving the shutter open on the camera, increasing the integration time to many hours of exposure. Photography enabled astronomers to record and study objects invisible to the human eye. However, the response of photography to light is not linear, especially at long exposures, so if you doubled the exposure time, you did not get an image twice as bright. By the middle of the 20th century, the search was on for electronic detectors that would overcome photography's problems in sensitivity, spectral range, and nonlinearity.

In 1969, scientists at Bell Laboratories invented a detector called a **charge-coupled device (CCD)**. By the late 1970s, the CCD had become the detector of choice in almost all astronomical-imaging applications. CCDs are linear, so doubling the exposure means you record twice as much light. As a result, they are good for measuring objects that vary in brightness, as well as for faint objects that require long exposures. CCDs have a quantum efficiency far superior to that of either photography or the eye—up to 80 percent at some wavelengths. That improvement dramatically increases our ability to view faint objects with short exposure times.

A CCD is an ultrathin wafer of silicon—less than the thickness of a human hair—that is divided into a two-dimensional array of picture elements, or **pixels** (**Figure 6.19a**). When a photon strikes a pixel, it liberates an electron from an atom and creates a small electric charge within the silicon. At very low light levels, this response can be dwarfed by thermal "noise"—electrons that are moving freely because the camera is warmer than absolute zero. Liquid nitrogen or helium is used to cool some CCDs down to very low temperatures to reduce this thermal noise. Conversely, at very high light levels, it's possible to basically run out of electrons to liberate with another photon. Any further photons that arrive will not register. Just as a sponge can become saturated with water, so that it can absorb no more water droplets, a CCD camera can become saturated with light so that it can register no more photons. Between these two limits, the digital signal that flows to the computer is proportional to the accumulated charge, so the CCD is a linear device.

The output from a CCD is a digital signal that can either be sent directly from the telescope to image-processing software or stored electronically for later analysis. Nearly every spectacular astronomical image in ultraviolet (UV), visible, or infrared (IR) wavelength that you find online was recorded by a CCD in a telescope either on the ground or in space. The first astronomical CCDs were small arrays containing a few hundred thousand pixels. The larger CCDs used in astronomy today may contain arrays totaling more than 100 million pixels (**Figure 6.19b**). Still larger arrays are under development as ever-faster computing

Figure 6.18 A photograph of the Moon taken by John W. Draper in 1840. This is one of the first known photographs of an astronomical object.

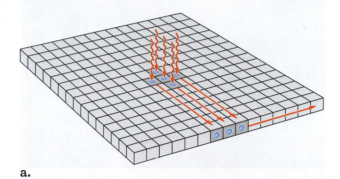

a.

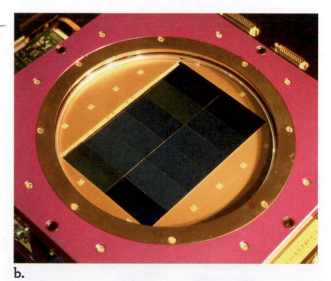

b.

Figure 6.19 a. In this simplified diagram of a charge-coupled device (CCD), photons from a star land on pixels (represented by gray squares) and produce free electrons within the silicon. The electron charges are electronically moved sequentially to the collecting register at the bottom. Each row is then moved out to the right to an electronic amplifier, which converts the electric charge of each pixel into a digital signal. **b.** This large CCD (about 6 inches across) contains 12,288 × 8192 pixels.

a.

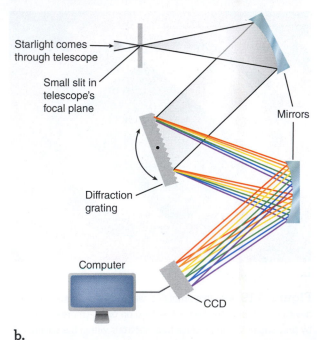

Starlight comes through telescope

Small slit in telescope's focal plane

Mirrors

Diffraction grating

Computer

CCD

b.

Figure 6.20 a. A spectrum is created by the reflection of light from the closely spaced tracks of a CD. **b.** In a grating spectrograph, light goes through the telescope and then a slit, where it is reflected to the diffraction grating and split into components. The spectrum is recorded on the CCD.

power keeps up with image-processing demands. CCDs are found in many everyday devices, too, such as digital cameras, digital video cameras, and camera phones.

Your cell phone takes color pictures by using a grid of CCD pixels arranged in groups of three. Each pixel in a group is constructed to respond to only a particular range of colors—to only red light, for example. That also is true for digital image displays. You can see for yourself if you place a small drop of water on the screen of your smartphone or tablet and turn it on. The water droplet magnifies the grid of pixels so that you can see them individually. In a camera, this grid degrades the angular resolution because each spot in the final image requires three pixels of information. Astronomers are motivated to maximize angular resolution, so instead of making a camera that takes color pictures, they use a camera that takes grayscale images at higher angular resolution. They put filters in front of the camera to allow light of only particular colors to pass through. Color pictures are constructed by taking multiple pictures in different filters, using software to color each one, and then carefully aligning and overlapping them to produce beautiful and informative images. Sometimes the colors are "true"; that is, they are close to the colors you would see if you were actually looking at the object with your eyes. Other times "false" colors represent different portions of the electromagnetic spectrum and tell you the temperature or composition of different parts of the object. Using changeable filters instead of designated color pixels gives astronomers greater flexibility and better angular resolution.

Spectrographs

Spectroscopy is the study of an object's *spectrum* (plural: *spectra*)—its electromagnetic radiation split into component wavelengths. **Spectrographs** (sometimes called **spectrometers**) are instruments that take the spectrum of an object and then record it. The first spectrographs used prisms to disperse the light. Modern spectrographs use a **diffraction grating**, which is made by engraving closely spaced lines on glass to disperse incoming light into a spectrum. **Figure 6.20a** shows the light reflected from a CD or DVD; the closely spaced tracks act as a grating and create a spectrum. **Figure 6.20b** shows a schematic of a grating spectrograph: Light from an astronomical object enters a telescope and passes through a slit. The light is reflected onto the diffraction grating, which creates a spectrum like the one shown in Figure 5.14. A CCD captures the spectrum for later analysis. Some modern spectrographs use bundles of optical fibers, or masks with multiple slits, to obtain spectra simultaneously from multiple objects in the field of view of the telescope.

CHECK YOUR UNDERSTANDING 6.2

CCD cameras have much higher quantum efficiency than other detectors. Therefore, CCD cameras: (a) can collect photons for longer times; (b) can collect photons of different energies; (c) can generate a signal from fewer photons; (d) can split light into different colors.

6.3 Astronomers Observe in Wavelengths Beyond the Visible

Astronomers use telescopes that observe at all the wavelengths of the electromagnetic spectrum. Recall from Section 5.4 that Wien's law states that an object's temperature can be found from the peak wavelength of its continuous spectrum.

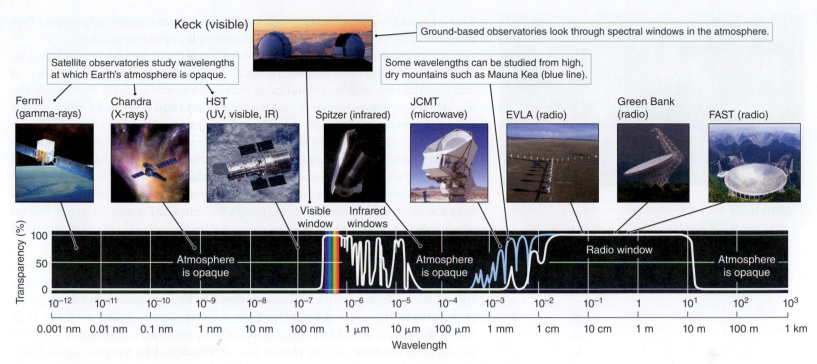

Figure 6.21 Earth's atmosphere blocks most electromagnetic radiation. Fermi = Fermi Gamma-ray Space Telescope (orbiting); Chandra = Chandra X-ray Observatory (orbiting); HST = Hubble Space Telescope (orbiting); Keck = Keck Observatory (Hawai'i); Spitzer = Spitzer Space Telescope (orbiting); JCMT = James Clerk Maxwell Telescope (Hawai'i); EVLA = Expanded Very Large Array (New Mexico); Green Bank = Robert C. Byrd Green Bank Telescope (West Virginia); FAST = Five-hundred-meter Aperture Spherical Telescope (Guizhou, China).

Hot objects emit the most light in the X-ray or gamma-ray regions of the spectrum, whereas cool objects emit the most light in the radio or infrared regions of the spectrum. Radio or infrared telescopes are used to study cool objects, such as clouds of dust, whereas X-ray or gamma-ray telescopes are used to study violently hot gas. Not all of these spectral regions are accessible from the ground. **Figure 6.21** shows Earth's **atmospheric windows** and which parts of the spectrum they let in. The largest window is in radio wavelengths, including microwaves at the short-wavelength end of the radio window. Radio telescopes can be built on the ground. However, gamma-ray, X-ray, ultraviolet, and most infrared light from astronomical objects fails to reach the ground because it is partially or completely absorbed by ozone, water vapor, carbon dioxide, and other molecules in Earth's atmosphere. Light at those wavelengths has to be observed from space.

Radio Telescopes

Karl Jansky (1905–1950), a young physicist working for Bell Laboratories in the early 1930s, identified a radio source in the Milky Way in the direction of the galactic center, in the constellation Sagittarius. Jansky's discovery marked the birth of radio astronomy, and in his honor, the basic unit for the strength of a radio source is called the **jansky** (Jy). A few years later, Grote Reber (1911–2002), a radio engineer and ham radio operator, built his own radio telescope and conducted the first survey of the sky at radio frequencies. Reber was largely responsible for the rapid advancement of radio astronomy in the post–World War II era.

Most radio telescopes are large, steerable dishes, typically tens of meters in diameter (**Figure 6.22a**). The world's largest single-dish radio telescope is the FAST (Five-hundred-meter Aperture Spherical Telescope) dish in China (**Figure 6.22b**).

a.

b.

Figure 6.22 a. The Green Bank telescope in West Virginia is the world's largest steerable radio telescope. **b.** The FAST radio telescope in China is the world's largest single-dish telescope.

Figure 6.23 The EVLA in New Mexico combines signals from 27 telescopes so that they act as one "very large" telescope.

Figure 6.24 The new Atacama Large Millimeter/submillimeter Array (ALMA) telescope in the Atacama Desert in northern Chile has many international partners.

This large single-dish telescope is not steerable, so it can observe only sources that pass within 20° of the zenith as Earth's rotation carries them overhead.

As large as radio telescopes are, they have relatively poor angular resolution. Recall that a telescope's angular resolution (θ) is determined by the ratio λ/D and that the best angular resolution is obtained when θ is small. As a result, the angular resolution gets poorer and poorer as that ratio increases (which occurs when λ increases and/or D decreases). Radio telescopes have diameters (D) much larger than the apertures of most optical telescopes. However, the wavelengths (λ) of radio waves range from about 1 centimeter (cm) to 10 meters, or up to several hundred thousand times greater than the wavelengths of visible light, which makes the ratio larger. Radio telescopes are thus limited by the very long wavelengths they are designed to receive. Even though China's FAST radio telescope has an effective aperture of 300 m, its angular resolution is similar to that of the unaided human eye—about 1 arcmin.

Radio astronomers have developed a clever way to improve angular resolution. Mathematically combining the signals from two radio telescopes turns them into a telescope with a diameter equal to the separation between them. For example, if two 10-m telescopes are located 1000 m apart, then the D in λ/D is 1000, not 10. That combination of two (or more) telescopes is called an **interferometer**, and it makes use of the wavelike properties of light. Usually, several telescopes are used in an arrangement called an **interferometric array**. Through the use of very large arrays, radio astronomers can better observe bright sources and exceed the angular resolution possible with optical telescopes.

The Expanded Very Large Array (EVLA) in New Mexico (**Figure 6.23**) is an interferometric array made up of 27 movable dishes spread out in a Y-shaped configuration up to 36 km across. At a wavelength of 10 cm, that array reaches angular resolutions of less than 1 arcsec. The Very Long Baseline Array (VLBA) uses 10 radio telescopes spread out over more than 8000 km from the Virgin Islands in the Caribbean to Hawai'i in the Pacific. At a wavelength of 10 cm, that array can attain angular resolutions of better than 0.003 arcsec. A radio telescope put into near-Earth orbit as part of a Space Very Long Baseline Interferometer (SVLBI) overcomes even that limit. The Event Horizon Telescope combines many of the most advanced existing radio telescopes, from Greenland to the South Pole, to make an *Earth-sized* interferometer. For a few nights each year, all the telescopes observe the same object, with a combined angular resolution that may be good enough to image objects near the center of the Milky Way.

Some radio telescopes use many small dishes. The Atacama Large Millimeter/submillimeter Array (ALMA; **Figure 6.24**), located at an elevation of 5000 m in the Atacama Desert in Chile, was completed in 2013. That project, an international collaboration of astronomers from Europe, North America, East Asia, and Chile, consists of 66 dishes with diameters of 12 m and 7 m for observations in the 0.3- to 9.6-mm-wavelength range. The Square Kilometre Array (SKA) is designed to have *thousands* of small radio dishes, which together will act as one dish with a collecting area of 1 square kilometer (km^2). Twenty countries are supporting that telescope, which will be located in Australia and South Africa.

Optical telescopes also can be combined in an array to yield angular resolutions greater than those of single telescopes, although for technical reasons the individual units cannot be spread as far apart as radio telescopes. The Very Large Telescope Interferometer (VLTI) in Chile, operated by ESO, combines four

8-m telescopes with four movable 1.8-m auxiliary telescopes. It has a baseline of up to 200 m, yielding angular resolution of about 0.001 arcsec. The six-telescope Center for High Angular Resolution Astronomy (CHARA) array in California works in the visible and near-infrared regions. It has a baseline of 330 m with angular resolution of 0.0003 arcsec.

Infrared Telescopes

Molecules such as water vapor in Earth's atmosphere block infrared photons from reaching astronomical telescopes on the ground, so telescopes that observe in the infrared [0.75–30 microns (µm)] are at the highest locations. Mauna Kea, a dormant volcano and home of the Mauna Kea Observatories (MKO), rises 4205 m above the Pacific Ocean. At that altitude, the MKO telescopes sit above 40 percent of Earth's atmosphere; but more important, 90 percent of Earth's atmospheric water vapor lies below. Still, for the infrared astronomer, the remaining 10 percent is troublesome.

Airborne observatories overcome atmospheric absorption of infrared light by placing telescopes above most of the water vapor in the atmosphere. NASA's Stratospheric Observatory for Infrared Astronomy (SOFIA) (**Figure 6.25**), a joint project with the German Aerospace Center (DLR), is a modified 747 airplane that carries a 2.5-m telescope and works in the far-infrared region of the spectrum, from 1 to 650 µm. It flies in the stratosphere at an altitude of about 12 km, above 99 percent of the water vapor in Earth's lower atmosphere. Because airplanes are highly mobile, SOFIA can observe in both the Northern and Southern Hemispheres. Other infrared wavelengths must be observed from space.

Orbiting Observatories

Gaining full access to the complete electromagnetic spectrum requires getting completely above Earth's atmosphere. The first astronomical satellite was the British Ariel 1, launched in 1962 to study solar UV and X-ray radiation. Today, many orbiting astronomical telescopes cover the electromagnetic spectrum from gamma rays to microwaves, with more in the planning stage (**Table 6.2**). Optical telescopes, such as the 2.4-m Hubble Space Telescope (HST; see Figure 6.21), operate successfully at low Earth orbit, 600 km above Earth's surface. Launched in 1990, HST has been the workhorse for UV, visible, and IR space astronomy for more than 30 years. Low Earth orbit is also the region where the International Space Station and many scientific satellites orbit.

For certain other satellites and space telescopes, however, 600 km is not high enough. The Chandra X-ray Observatory (see Figure 6.21), NASA's X-ray telescope, cannot see through even the tiniest traces of atmosphere and therefore orbits more than 16,000 km above Earth's surface. NASA's Spitzer Space Telescope, an infrared telescope, is so sensitive that it needs to be completely free from Earth's own infrared radiation. The solution was to put it into a *solar* orbit, trailing tens of millions of kilometers behind Earth. The James Webb Space Telescope (JWST), scheduled to succeed the HST, will observe primarily in infrared wavelengths. It will be located 1.5 million miles away from Earth, orbiting the Sun at a fixed distance from the Sun and Earth.

Orbiting telescopes located above the atmosphere are unaffected by atmospheric image distortions, weather, or brightening night skies. But space observatories are much more expensive than ground-based observatories and can be difficult or

what if . . .

What if human life had developed on a planet containing almost no atmosphere? How might astronomy have developed differently?

a.

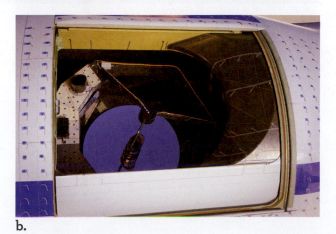

b.

Figure 6.25 SOFIA is a 2.5-m infrared telescope mounted in a Boeing 747 aircraft.

Table 6.2	Selected Space Observatories			
Telescope	**Sponsor(s)**	**Description**		**Year**
Hubble Space Telescope (HST)	NASA, ESA	Optical, infrared, ultraviolet observations		1990–
Chandra X-ray Observatory	NASA	X-ray imaging and spectroscopy		1999–
X-ray Multi-Mirror Mission (XMM-Newton)	ESA	X-ray spectroscopy		1999–
Galaxy Evolution Explorer (GALEX)	NASA	Ultraviolet observations		2003–2012
Spitzer Space Telescope	NASA	Infrared observations		2004–2020
Neil Gehrels Swift Observatory	NASA	Gamma-ray bursts		2004–
Convection Rotation and Planetary Transits (COROT) space telescope	CNES (France)	Planet finder		2006–2013
Fermi Gamma-ray Space Telescope	NASA, European partners	Gamma-ray imaging and gamma-ray bursts		2008–
Planck telescope	ESA	Cosmic microwave background radiation		2009–2013
Herschel Space Observatory	ESA	Far-infrared and submillimeter observations		2009–2013
Kepler telescope	NASA	Planet finder		2009–2018
Solar Dynamics Observatory (SDO)	NASA	Sun, solar weather		2010–
RadioAstron	Russia, international collaborators	Very-long-baseline interferometry in space		2011–
Nuclear Spectroscopic Telescopic Array (NuSTAR)	NASA	High-energy X-ray		2012–
Gaia	ESA	Optical, digital 3D space camera		2013–
Transiting Exoplanet Survey Satellite (TESS)	NASA, MIT, SAO	Survey entire sky for transiting planets around nearby and bright stars		2018–
CHEOPS (CHaracterising ExOPlanet Satellite)	ESA	High-precision brightness measurements of stars known to have planets		2019–
James Webb Space Telescope (JWST)	NASA, ESA, Canadian Space Agency	Primarily infrared		2021 (scheduled)

impossible to repair. The HST required several servicing missions, but such missions are impossible for the observatories in more distant Earth orbits. Ground-based telescopes at even the most remote mountaintop locations can receive shipments of replacement parts in a few days; space telescopes cannot. Some wavelengths can be observed from space only, but issues of cost and repair are why ground-based telescopes remain more prevalent for wavelengths that can be observed from the ground.

CHECK YOUR UNDERSTANDING 6.3

Which of the following is the biggest disadvantage of putting a telescope in space? (a) Astronomers don't have as much control in choosing what to observe. (b) Astronomers have to wait until the telescopes come back to Earth to get their images. (c) Space telescopes can observe only in certain parts of the electromagnetic spectrum. (d) Space telescopes are much more expensive and riskier than similar ground-based telescopes.

6.4 Planetary Spacecraft Explore the Solar System

Recall from Chapter 2 that everyone always sees the same face of the Moon from Earth because the Moon's orbital and rotational periods are equal. The first view of the "far" side was in 1959, when the Soviet flyby mission *Luna 3* sent back pictures showing that the far side of the Moon was very different from its Earth-facing half. No matter how powerful our ground-based or Earth-orbiting telescopes, sometimes we need to send a spacecraft for a different view.

Spacecraft have now visited all of our Solar System's planets and some of their moons, as well as some comets and asteroids, providing the first close-up views of those distant worlds. The study of the Solar System from space is an international collaboration involving NASA, the European Space Agency (ESA), the Russian Federal Space Agency (Roscosmos), the Japan Aerospace Exploration Agency (JAXA), the China National Space Administration (CNSA), and the Indian Space Research Organisation (ISRO). Other countries may soon join the endeavor. In this section, we look at the types of spacecraft used to explore our Solar System.

Flybys and Orbiters

Exploration of the Solar System began with a reconnaissance phase, using spacecraft to fly by or orbit a planet or other body. A **flyby** is a spacecraft that first approaches and then continues flying past the target. As those spacecraft speed by, instruments aboard them briefly probe the physical and chemical properties of the target and its environment.

Flyby missions are the most common first phase of exploration. They cost less than orbiters or landers and are easier to design and execute. Flyby spacecraft, such as *Voyager 1* and *Voyager 2,* can sometimes visit several worlds during their travels (**Figure 6.26**). The downside of flyby missions is that the spacecraft must

Figure 6.26 The two *Voyager* spacecraft **a.** flew past the outer planets and **b.** have recently passed the boundary of our Solar System.

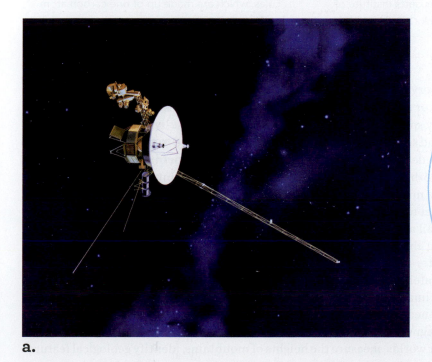

a.

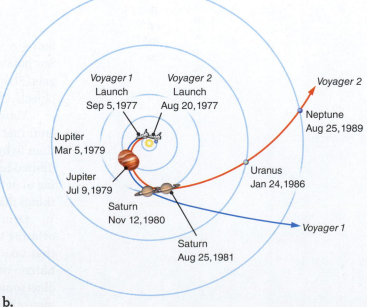

Voyager 1
Launch
Sep 5, 1977

Voyager 2
Launch
Aug 20, 1977

Voyager 2

Neptune
Aug 25, 1989

Jupiter
Mar 5, 1979

Jupiter
Jul 9, 1979

Uranus
Jan 24, 1986

Saturn
Nov 12, 1980

Voyager 1

Saturn
Aug 25, 1981

b.

Reading Astronomy News

NASA's *Perseverance* Mars Rover Extracts First Oxygen from Red Planet

NASA, *April 21, 2021*

In the 21st Century, missions to other planets do more than observe and take data; they are beginning to explore resource extraction and utilization.

The growing list of "firsts" for *Perseverance*, NASA's newest six-wheeled robot on the Martian surface, includes converting some of the Red Planet's thin, carbon dioxide–rich atmosphere into oxygen. A toaster-size, experimental instrument aboard *Perseverance* called the Mars Oxygen In-Situ Resource Utilization Experiment (MOXIE) accomplished the task. The test took place April 20, the 60th Martian day, or sol, since the mission landed Feb. 18.

While the technology demonstration is just getting started, it could pave the way for science fiction to become science fact—isolating and storing oxygen on Mars to help power rockets that could lift astronauts off the planet's surface. Such devices also might one day provide breathable air for astronauts themselves. MOXIE is an exploration technology investigation—as is the Mars Environmental Dynamics Analyzer (MEDA) weather station—and is sponsored by NASA's Space Technology Mission Directorate (STMD) and Human Exploration and Operations Mission Directorate.

"This is a critical first step at converting carbon dioxide to oxygen on Mars," said Jim Reuter, associate administrator for STMD. "MOXIE has more work to do, but the results from this technology demonstration are full of promise as we move toward our goal of one day seeing humans on Mars. Oxygen isn't just the stuff we breathe. Rocket propellant depends on oxygen, and future explorers will depend on producing propellant on Mars to make the trip home."

For rockets or astronauts, oxygen is key, said MOXIE's principal investigator, Michael Hecht of the Massachusetts Institute of Technology's Haystack Observatory.

To burn its fuel, a rocket must have more oxygen by weight. Getting four astronauts off the Martian surface on a future mission would require approximately 15,000 pounds (7 metric tons) of rocket fuel and 55,000 pounds (25 metric tons) of oxygen. In contrast, astronauts living and working on Mars would require far less oxygen to breathe. "The astronauts who spend a year on the surface will maybe use one metric ton between them," Hecht said.

Hauling 25 metric tons of oxygen from Earth to Mars would be an arduous task. Transporting a one-ton oxygen converter—a larger, more powerful descendant of MOXIE that could produce those 25 tons—would be far more economical and practical.

Mars' atmosphere is 96 percent carbon dioxide. MOXIE works by separating oxygen atoms from carbon dioxide molecules, which are made up of one carbon atom and two oxygen atoms. A waste product,

---

move by very swiftly due to the physics of orbits. Thus, they are limited to just a few hours or at most a few days in which to conduct close-up studies of their targets. Flyby spacecraft give astronomers their first close-up views of Solar System objects, and sometimes the data obtained are then used to plan follow-up studies.

More detailed reconnaissance work is done by **orbiters**, which are spacecraft that orbit around their target. Such missions are intrinsically more difficult than flyby missions because they have to make risky maneuvers and use their limited fuel to change their speed to enter an orbit. But orbiters can linger, looking in detail at more of the surfaces of the objects they are orbiting and studying things that change with time, such as planetary weather.

Orbiters use remote-sensing instrumentation like that used by Earth-orbiting satellites to study our own planet. Those instruments include cameras that take images at different wavelength ranges, radar that can map surfaces hidden beneath obscuring layers of clouds, and spectrographs that analyze the electromagnetic spectrum. Those instruments enable planetary scientists to map other worlds, measure the heights of mountains, identify geological features

carbon monoxide, is emitted into the Martian atmosphere.

The conversion process requires high levels of heat to reach a temperature of approximately 1470 degrees Fahrenheit (800 Celsius). To accommodate this, the MOXIE unit is made with heat-tolerant materials. These include 3D-printed nickel alloy parts, which heat and cool the gases flowing through it, and a lightweight aerogel that helps hold in the heat. A thin gold coating on the outside of MOXIE reflects infrared heat, keeping it from radiating outward and potentially damaging other parts of *Perseverance*.

In this first operation, MOXIE's oxygen production was quite modest—about 5 grams, equivalent to about 10 minutes worth of breathable oxygen for an astronaut. MOXIE is designed to generate up to 10 grams of oxygen per hour.

This technology demonstration was designed to ensure the instrument survived the launch from Earth, a nearly seven-month journey through deep space, and touch-down with *Perseverance* on Feb. 18. MOXIE is expected to extract oxygen at least nine more times over the course of a Martian year (nearly two years on Earth).

These oxygen-production runs will come in three phases. The first phase will check out and characterize the instrument's function, while the second phase will run the instrument in varying atmospheric conditions, such as different times of day and seasons. In the third phase, Hecht said, "we'll push the envelope"—trying new operating modes, or introducing "new wrinkles, such as a run where we compare operations at three or more different temperatures."

"MOXIE isn't just the first instrument to produce oxygen on another world," said Trudy Kortes, director of technology demonstrations within STMD. It's the first technology of its kind that will help future missions "live off the land," using elements of another world's environment, also known as in-situ resource utilization.

"It's taking regolith, the substance you find on the ground, and putting it through a processing plant, making it into a large structure, or taking carbon dioxide—the bulk of the atmosphere—and converting it into oxygen," she said. "This process allows us to convert these abundant materials into useable things: propellant, breathable air, or, combined with hydrogen, water."

QUESTIONS

1. What does MOXIE do?

2. In this test, MOXIE produced 5.4 grams of oxygen in two hours; enough to sustain an astronaut for ten minutes. How much oxygen is needed to sustain an astronaut for one hour?

3. In future tests, MOXIE will produce 10 grams of oxygen per hour. Will MOXIE produce oxygen fast enough to sustain an astronaut?

4. In order for humans to travel to Mars, what other activities (besides sustaining breathing astronauts) require oxygen?

5. NASA plans to have MOXIE run nine other tests on Mars. Why is it important that MOXIE conduct these other tests?

Source: https://www.nasa.gov/press-release/nasa-s-perseverance-mars-rover-extracts-first-oxygen-from-red-planet.

and rock types, watch weather patterns develop, measure the composition of atmospheres, and get a general sense of the place. Additional instruments make measurements of the extended atmospheres and space environment through which they travel.

Landers, Rovers, and Atmospheric Probes

Reconnaissance spacecraft provide a wealth of information about a planet, but no better way exists to explore a planet than doing so within a planet's atmosphere or on solid ground. Spacecraft have landed on the Moon, Mars, Venus, Saturn's large moon Titan, and several asteroids and comets. Those spacecraft have taken pictures of planetary surfaces, measured surface chemistry, and conducted experiments to determine the physical properties of the surface rocks and soils.

Using **landers**—spacecraft that touch down and remain on the surface—has several disadvantages. Because of the expense, only a few landings in limited areas are practical. With that limitation, the results may apply only to the small area

Figure 6.27 The robotic rover *Curiosity* took this selfie on the surface of Mars.

Figure 6.28 This photograph of "Earthrise" over the limb of the Moon was taken by the astronauts on *Apollo 8*.

unanswered questions

Will further human exploration of the Solar System occur within your lifetime? Since *Apollo 17*, in 1972, humans have not returned to the Moon or traveled to other planets or moons in the Solar System. Sending humans to the worlds of the Solar System is much more complicated, risky, and expensive than sending robotic spacecraft. Humans need life support such as air, water, and food. Radiation in space can be dangerous. Furthermore, human explorers would expect to return to Earth, whereas most spacecraft do not come back. Astronomers and space scientists have heated debates about human spaceflight versus robotic exploration. Some argue that true exploration requires that human eyes and brains actually go there; others argue that the costs and risks are too high for the potential additional scientific knowledge. Beyond basic exploration, we also do not know whether humans will ever permanently colonize space.

around the landing site. Imagine, for example, what a different picture of Earth you might get from a spacecraft that landed in Antarctica, as opposed to a spacecraft that landed in a volcano or on the floor of a dry riverbed. Sites to be explored with landed spacecraft must be carefully chosen on the basis of reconnaissance data. Some landers have wheels and can explore the vicinity of the landing site. Such remote-controlled vehicles, called **rovers**, were used first by the Soviet Union on the Moon four decades ago and more recently by the United States and China on Mars. **Figure 6.27** shows a self-portrait of the *Curiosity* rover on Mars.

Atmospheric probes descend into the atmospheres of planets and continually measure and send back data on temperature, pressure, and wind speed, along with other properties, such as chemical composition. Atmospheric probes have survived all the way to the solid surfaces of Venus and of Saturn's moon Titan, sending back streams of data during their descent. An atmospheric probe sent into Jupiter's atmosphere never reached that planet's surface because, as you will learn later in the book, Jupiter does not have a solid surface in the same sense that terrestrial planets and moons do. After sending back its data, the Jupiter probe eventually melted and vaporized as it descended into the hotter layers of the planet's atmosphere.

Sample Returns

If you pick up a rock from the side of a road, you might learn a lot from the rock by using tools that you could carry in your pocket or in your car. Much better, though, would be to pick up a few samples and carry them back to a laboratory equipped with a full range of state-of-the-art instruments that can measure chemical composition, mineral type, age, and other information needed to reconstruct the story of your rock sample's origin and evolution. The same is true of Solar System exploration. One of the most powerful methods for investigating remote objects is to collect samples of the objects and bring them back to Earth for study. So far, only samples of the Moon, a comet, and streams of charged particles from the Sun have been collected and returned to Earth. Scientists have found meteorites on Earth that are pieces of Mars that were blasted loose when objects crashed into that world. In the next 10 years, robotic "sample and return" missions may go to Mars.

The missions discussed so far in this section have all been conducted with robotic spacecraft. The only spacecraft that took people to another world were the *Apollo* missions to the Moon. That program ran from 1961 to 1972 and included several missions before the first Moon landing on July 20, 1969. The *Apollo 8* astronauts took the magnificent picture of Earth viewed over the surface of the Moon (**Figure 6.28**). Each mission from *Apollo 11* through *Apollo 17* had three astronauts—two to land on the Moon and one to remain in orbit. *Apollo 13* did not reach the Moon but returned to Earth safely. Twelve American astronauts walked on the Moon between 1969 and 1972 and brought back a total of 382 kg of rocks and other material.

The return of extraterrestrial samples to Earth is governed by international treaties and standards to ensure that the samples do not contaminate our world. For example, before the lunar samples that the *Apollo* missions brought back could be studied, they (and the astronauts) had to be quarantined and tested for the presence of alien life-forms. The same international standards apply to spacecraft landing elsewhere. The goal of those standards is to avoid transporting life-forms from Earth to another planet; we do not want to "discover" life that we, in fact, introduced. In addition, we do not want to potentially harm life that may exist on other planets.

With many missions under way and others on the horizon, robotic exploration of the Solar System is an ongoing, dynamic activity. Appendix 5 summarizes some recent and current missions. We discuss some of those missions in the relevant chapter. Information on the latest discoveries can be found on mission websites and in science news sources.

CHECK YOUR UNDERSTANDING 6.4

Spacecraft are the most effective way to study planets in our Solar System because: (a) planets move too fast across the sky for us to image them well from Earth; (b) planets cannot be imaged from Earth; (c) spacecraft can collect more information than is available just from images from Earth; (d) space missions are easier than long observing campaigns.

6.5 Other Tools Contribute to the Study of the Universe

High-profile space missions have sent back stunning images and data from across the electromagnetic spectrum. But astronomers use other tools as well, including particle accelerators and colliders, neutrino and gravitational-wave detectors, and super-computers.

Particle Accelerators

Ever since the early years of the 20th century, physicists have been peering into the structure of the atom by observing what happens when small particles collide. By the 1930s, physicists had developed the technology to accelerate charged subatomic particles such as protons to very high speeds and then observe what happens when they slam into a target. From such experiments, physicists have discovered many kinds of subatomic particles and learned about their physical properties. High-energy particle colliders have proved to be an essential tool for physicists studying the basic building blocks of matter.

Astronomers have realized that to understand the very largest structures seen in the universe, it is important to understand the physics that took place during the earliest moments in the universe, when everything was extremely hot and dense. High-energy particle colliders that physicists use today are designed to approach the energies of the early universe. The effectiveness of particle accelerators is determined by the energy they can achieve and the number of particles they can accelerate. Modern particle colliders such as the Large Hadron Collider near Geneva, Switzerland (**Figure 6.29**), reach very high energies. Particles also can be studied from space. The Alpha Magnetic Spectrometer, installed on the International Space Station in 2011, searches for some of the most exotic forms of matter, such as dark matter, antimatter, and high-energy particles called cosmic rays.

Neutrinos and Gravitational Waves

The **neutrino** is an elusive elementary particle that plays a major role in the physics of the interiors of stars. Neutrinos are extremely difficult to detect. In less time than you take to read this sentence, a thousand trillion (10^{15}) solar neutrinos from

what if . . .

What if it proves possible to use powerful lasers to accelerate tiny space probes to speeds that can reach nearby solar systems in a few decades? What kind of information would you like to learn from such a flyby?

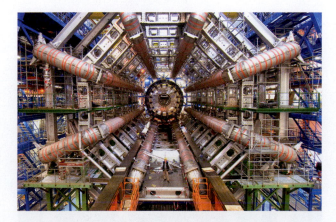

Figure 6.29 The ATLAS particle detector at CERN's Large Hadron Collider near Geneva, Switzerland. The instrument's enormous size is evident from the person standing near the bottom center of the picture.

Credit: Photograph: Maximilien Brice, © 2005-2021 CERN, http://cds.cern.ch/record/910381. https://creativecommons.org/licenses/by/4.0/.

what if . . .

What if, instead of detecting several black hole or neutron star merger events per year, LIGO had seen only one such event during its three years of observing: What might you reasonably conclude from that result?

the Sun are passing through your body, even during the night. Neutrinos are so nonreactive with matter that they can pass right through Earth (and you) as though it (or you) weren't there at all. To be observed, a neutrino has to interact with a detector. Neutrino detectors typically record only one of every 10^{22} (10 billion trillion) neutrinos passing through them, but that's enough to reveal processes deep within the Sun or the violent death of a star 160,000 light-years away.

Experiments designed to look for neutrinos originating outside Earth are buried deep underground in mines or caverns or under the ocean or ice to ensure that only neutrinos are detected. For example, the ANTARES experiment uses the Mediterranean Sea as a neutrino telescope. Detectors located 2.5 km under the sea, off the coast of France, observe neutrinos that originated in objects visible in southern skies and passed through Earth. In the IceCube neutrino observatory located at the South Pole in Antarctica, the neutrino detectors are 1.5–2.5 km under the ice, and they observe neutrinos that originated in objects visible in northern skies (**Figure 6.30**).

Another elusive phenomenon is the **gravitational wave**. Gravitational waves are disturbances in a gravitational field, similar to the waves that spread out from the disturbance you create when you toss a pebble onto the quiet surface of a pond. Several facilities, including the Laser Interferometer Gravitational-Wave Observatory (LIGO) and the European VIRGO interferometer, have been constructed to detect gravitational waves (**Process of Science Figure**). Recently, LIGO has detected several pairs of coalescing binary black holes and neutron stars, as we explore in Chapter 18.

Computers

Astronomers use powerful computers to gather, analyze, and interpret data. A single CCD image may contain millions of pixels, with each pixel displaying roughly 30,000 levels of brightness. That adds up to several trillion pieces of information in each image. To analyze their data, astronomers typically do calculations for *every single pixel* of an image to remove unwanted contributions from Earth's atmosphere or to correct for instrumental effects. Astronomers conduct many types of sky surveys—in which one or more telescopes survey a specific part of the sky—yielding thousands of images that need to be analyzed.

High-performance computers also play an essential role in generating and testing theoretical models of astronomical objects. Even when we completely understand the underlying physical laws that govern the behavior of a particular object, often the object is so complex that calculating its properties and behavior would be impossible without the assistance of high-performance computers. As discussed in Chapter 4, for example, you can use Newton's laws to compute the orbits of two stars that are gravitationally bound to each other because their orbits take the form of simple ellipses. However, understanding the orbits of the several hundred billion stars that make up the Milky Way Galaxy is not so easy, even though the underlying physical laws are the same.

Computer modeling is used to determine the interior properties of stars and planets, including Earth. Although astronomers cannot see beneath the surfaces of those bodies, they have a surprisingly good understanding of their interiors, which we describe in later chapters. Astronomers start a model by assigning well-understood physical properties to tiny volumes within a planet or star. The computer assembles an enormous number of those individual elements into an overall representation. The result is a rather good picture of what the interior of the star or planet is like.

Figure 6.30 The IceCube neutrino telescope at the South Pole, Antarctica.

Technology and Science Are Symbiotic

Scientists have been searching for waves that carry gravitational information for nearly 100 years, but the accuracy of their measurements has been limited by the available technology.

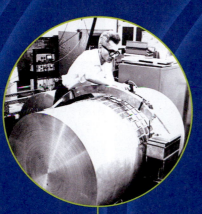

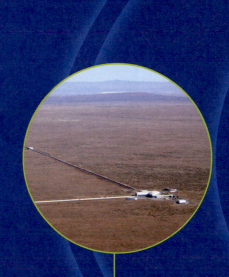

Einstein predicts that changing gravitational fields should produce gravitational waves. He calculates the size of the effect, and concludes that it will be impossible to detect them.

1916

c. 1970

Joseph Weber constructs precision-machined bars of metal that should "ring" as a gravitational wave passes by. His results were never confirmed.

1990s

Construction begins on LIGO, an observatory which uses lasers to precisely measure the change in the length of long tubes as gravitational waves pass by.

2015

LIGO detects gravitational waves for the first time, announcing the discovery in early 2016, 100 years after Einstein's prediction. This first detection has been followed by many more, so many that detections are now nearly routine.

Future (2034?)

LISA, a gravitational wave observatory in space, will be more sensitive; it will be able to detect much weaker events than LIGO.

Technology and science develop together. New technologies enable humans to ask new scientific questions. Asking new scientific questions pushes the development of better instrumentation and new technologies.

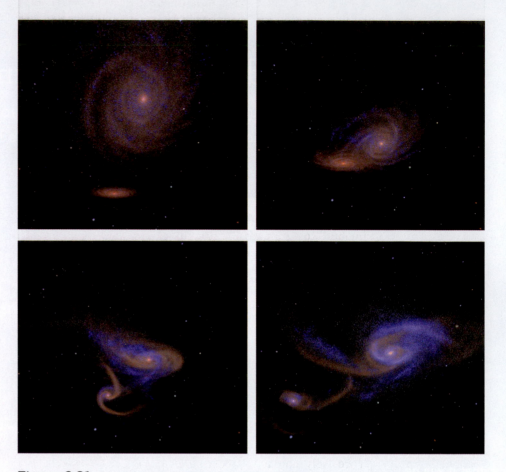

Figure 6.31 These images show supercomputer simulations of the collision of two galaxies. Astronomers compare simulations such as these with telescopic observations.

Astronomers also use high-performance computers to study how astronomical objects, systems of objects, and the universe as a whole evolve. For example, astronomers create models of galaxies and then run computer simulations to study how those galaxies might change over billions of years. **Figure 6.31** shows a simulation of the collision of two galaxies. The results of the computer simulations are then compared with telescopic observations. If the simulations do not match the observations, the model is adjusted and the simulations are run again until general agreement exists between them.

CHECK YOUR UNDERSTANDING **6.5**

High-performance computers have become one of an astronomer's most important tools. Which of the following require the use of that type of computer? (Choose all that apply.) (a) analyzing images taken with very large CCDs; (b) generating and testing theoretical models; (c) pointing a telescope from object to object; (d) studying how astronomical objects or systems evolve

Origins: Microwave Telescopes Detect Radiation from the Big Bang

In this chapter, we explored the tools of the astronomer, from basic optical telescopes to instruments that observe in different wavelengths. Now let's examine in more detail one type of telescope that has aided the study of the history of the universe. Recall from Chapter 1 that astronomers think the universe originated with a hot Big Bang. The multiple strands of evidence for that conclusion are discussed in Chapter 21. Here, we look at one piece: the observation of faint microwave radiation left over from the early hot universe. Two Bell Laboratories physicists, Arno Penzias (1933–) and Robert Wilson (1936–), were working on satellite communications when they first detected that radiation in 1964 with a microwave antenna in New Jersey. Today, we routinely use cell phones and handheld GPS devices that communicate directly with satellites, but at the time, that capability was at the limit of technology.

Penzias and Wilson needed a very sensitive microwave telescope for the work they were doing for Bell Labs because any spurious signals coming from the telescope itself might wash out the faint signals bounced off a satellite. To that end, they were working hard to eliminate all possible sources of microwave radiation originating from within their instrument, including keeping the telescope free of bird droppings. No matter how carefully they tried to eliminate sources of extraneous noise, they always still detected a faint signal at microwave wavelengths. That faint signal was the same in every direction and turned out to be from the Big Bang. Penzias and Wilson shared the 1978 Nobel Prize in Physics for discovering the **cosmic microwave background radiation (CMB)** left over from the Big Bang itself.

Since 1964, astronomers from around the world have designed increasingly precise instruments to measure that radiation from the ground, from high-altitude balloons, from rockets, and from satellites. The Russian experiment RELIKT-1, launched in 1983, found some limits on the variation of the CMB. The COBE (Cosmic Background Explorer) satellite, launched in 1989, showed that the spectrum of that radiation precisely matched that of a blackbody with a temperature of 2.73 K—exactly what was predicted for the radiation left over from the Big Bang. (Compare **Figure 6.32** with the curves in Figure 5.23.) The data also showed some slight differences in temperature—small fractions of a degree—over the map of the sky. Those slight variations tell us about how the universe evolved from one dominated by radiation to one that contains structures such as galaxies, stars, planets, and us. John Mather and George Smoot shared the 2006 Nobel Prize in Physics for that work.

In 1998 and 2003, a high-altitude balloon experiment called BOOMERANG (short for "balloon observations of millimetric extragalactic radiation and geophysics") flew over Antarctica at an altitude of 42 km to study CMB variations and estimate the overall geometry of the universe. The *WMAP* (Wilkinson Microwave Anisotropy Probe) satellite, launched in 2001, created an even more detailed map of the temperature variations in that radiation, yielding more precise values for the age and shape of the universe and the presence of dark matter and dark energy. The Planck space telescope, operated from 2009 to 2013 by the European Space Agency, was much more sensitive than *WMAP* and studied those CMB variations in even more precise detail. The Atacama Cosmology Telescope and Simons Array (Chile) and the South Pole Telescope (Antarctica) study that radiation to look for evidence of when galaxy clusters formed. Those experiments and observations have opened up the current era of precision cosmology, in which astronomers can make detailed models of how the universe was born, eventually leading to stars, planets, and us.

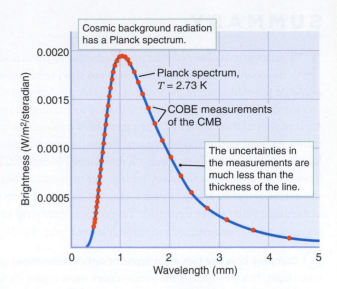

Figure 6.32 This graph shows the spectrum of the cosmic microwave background radiation (CMB) as measured by the COBE satellite (red dots). A steradian is a unit of solid angle. The uncertainty in the measurement at each wavelength is much less than the size of a dot. The line running through the data is a Planck blackbody spectrum with a temperature of 2.73 K.

SUMMARY

Earth's atmosphere blocks many spectral regions and distorts telescopic images, so telescopes are sited to be above as much of the atmosphere as possible. Telescopes are matched to the wavelengths of observation, with different technologies required for each region of the spectrum. The aperture of a telescope both determines its light-gathering power and limits its angular resolution; larger telescopes are better in both measures. Modern CCD cameras have better quantum efficiency and longer integration times, allowing astronomers to study fainter and more distant objects than were observable with earlier detectors. Telescopes observing at microwave wavelengths have detected radiation left over from the Big Bang.

① Compare how the two main types of optical telescopes gather and focus light. The telescope is the astronomer's most important tool. Ground-based telescopes that observe in visible wavelengths are either refractors (lenses) or reflectors (mirrors). All large astronomical telescopes are reflectors. Large telescopes collect more light and have better angular resolution. The diffraction limit is the limiting angular resolution of a telescope.

② Summarize the main types of detectors that are used on telescopes. Photography improved the ability of astronomers to record details of faint objects seen in telescopes. CCDs are today's astronomical detector

of choice because they are much more linear, have a broader spectral response, and can send electronic images directly to a computer. Spectrographs are specialized instruments that take the spectrum of an object to reveal what the object is made of and many other physical properties.

③ Explain why some wavelengths of radiation must be observed from high, dry, and remote observatories on Earth, or from space. Radio, near-infrared, and optical telescopes can see through our atmosphere. Those types of telescopes can be arrayed to greatly increase angular resolution. Putting telescopes in space solves problems created by Earth's atmosphere.

④ Explain the benefits of sending spacecraft to study the planets and moons of our Solar System. Most of what is known about the planets and their moons comes from observations by spacecraft. Flyby and orbiting missions obtain data from space, and landers and rovers collect data from the ground.

⑤ Describe other astronomical tools, such as gravitational wave detectors, that contribute to the study of the universe. Astronomers also use particle accelerators, neutrino detectors, and gravitational-wave detectors to study the universe. High-performance computers are essential to acquiring, analyzing, and interpreting astronomical data.

QUESTIONS AND PROBLEMS

TEST YOUR UNDERSTANDING

1. If one telescope has an aperture of 20 cm, and another has an aperture of 30 cm, and if aperture size is the only difference, then which should you choose, and why?
    a. The 20 cm, because the light-gathering power will be better.
    b. The 20 cm, because the image size will be larger.
    c. The 30 cm, because the light-gathering power will be better.
    d. The 30 cm, because the image size will be larger.

2. Study Figure 6.21. Which of the following can be observed from Earth's surface? (Choose all that apply.) 👁★
    a. radio waves
    b. gamma radiation
    c. far UV light
    d. X-ray light
    e. visible light

3. Match the following properties of telescopes (lettered) with their corresponding definitions (numbered).
    a. aperture
    b. angular resolution
    c. focal length
    d. chromatic aberration
    e. diffraction
    f. interferometer
    g. adaptive optics

    (1) two or more telescopes connected to act as one
    (2) distance from lens to focal plane
    (3) diameter
    (4) ability to distinguish close objects
    (5) computer-controlled correction for atmospheric distortion
    (6) color-separating effect
    (7) smearing effect due to sharp edge

4. Two 10-m telescopes, separated by 85 m, can operate as an interferometer. What is its angular resolution when it observes in the infrared at a wavelength of 2 microns?
    a. 0.01 arcsec
    b. 0.005 arcsec
    c. 0.2 arcsec
    d. 0.05 arcsec

5. Arrays of radio telescopes can produce much better angular resolution than single-dish telescopes can because they work based on the principle of
    a. reflection.
    b. refraction.
    c. diffraction.
    d. interference.

6. Refraction is caused by
    a. light bouncing off a surface.
    b. light changing colors as it enters a new medium.
    c. light changing speed as it enters a new medium.
    d. two light beams interfering.

7. The light-gathering power of a 4-m telescope is _____ than that of a 2-m telescope.
    a. 4 times larger
    b. 8 times larger
    c. 16 times smaller
    d. 2 times smaller

8. Improved angular resolution is helpful to astronomers because
   a. they often want to look in detail at small features of an object.
   b. they often want to look at very distant objects.
   c. they often want to look at many objects close together.
   d. all of the above

9. The part of the human eye that acts as the detector is the
   a. retina.
   b. pupil.
   c. lens.
   d. iris.

10. Cameras that use adaptive optics provide images with better angular resolution primarily because
    a. they operate above Earth's atmosphere.
    b. deformable mirrors are used to correct the blurring due to Earth's atmosphere.
    c. composite lenses correct for chromatic aberration.
    d. they simulate a much larger telescope.

11. The advantage of an interferometer is that
    a. the angular resolution is dramatically improved.
    b. the focal length is dramatically increased.
    c. the light-gathering power is dramatically increased.
    d. diffraction effects are dramatically decreased.
    e. chromatic aberration is dramatically decreased.

12. The angular resolution of a ground-based telescope is usually determined by
    a. diffraction.
    b. the focal length.
    c. refraction.
    d. atmospheric seeing.

13. A grating can spread white light out into a spectrum of colors because of the property of
    a. reflection.
    b. interference.
    c. dispersion.
    d. diffraction.

14. Why would astronomers put telescopes in airplanes?
    a. to get the telescopes closer to the stars
    b. to get the telescopes above most of the water vapor in Earth's atmosphere
    c. to be able to observe one object for more than 24 hours without stopping
    d. to allow the telescopes to observe the full spectrum of light

15. If we could increase the quantum efficiency of the human eye, doing so would
    a. allow humans to see a larger range of wavelengths.
    b. allow humans to see better at night or in other low-light conditions.
    c. increase the angular resolution of the human eye.
    d. decrease the angular resolution of the human eye.

THINKING ABOUT THE CONCEPTS

16. Galileo's telescope used simple lenses. What is the primary disadvantage of using a simple lens in a refracting telescope?

17. The largest astronomical refractor has an aperture of 1 m. List several reasons why building a larger refractor with twice that aperture would be impractical.

18. Your camera may have a zoom lens, ranging between wide angle (short focal length) and telephoto (long focal length). How does the size of an object in the camera's focal plane differ between wide angle and telephoto?

19. Optical telescopes reveal much about the nature of astronomical objects. Why do astronomers also need information provided by gamma-ray, X-ray, infrared, and radio telescopes?

20. For light reflecting from a flat surface, the angles of incidence and reflection are the same. That is also true for light reflecting from the curved surface of a reflecting telescope's primary mirror. Sketch a curved mirror and several of those reflecting rays.

21. Consider two optically perfect telescopes having different diameters but the same focal length. Is the image of a star larger or smaller in the focal plane of the larger telescope? Explain your answer.

22. Study the Process of Science Figure. Make a flowchart for the symbiosis between technology and science that led to the detection of gravitational waves. ⊙

23. ★ WHAT AN ASTRONOMER SEES   Use Figure 6.14 as an example to explain adaptive optics and describe how they improve a telescope's image quality. ⊙

24. Explain integration time and quantum efficiency and how each contributes to the detection of faint astronomical objects.

25. Some people believe that we put astronomical telescopes on high mountaintops or in orbit because doing so gets them closer to the objects they are observing. Explain what is wrong with that popular misconception, and give the actual reason telescopes are located in those places.

26. Humans have sent various kinds of spacecraft—including flybys, orbiters, and landers—to all the planets in our Solar System. Explain the advantages and disadvantages of each of those types of spacecraft.

27. If Earth has meteorites that are pieces of Mars, why is going to Mars and bringing back samples of the martian surface so important?

28. Humans had a first look at the far side of the Moon as recently as 1959. Why had we not seen it earlier—when Galileo first observed the Moon with his telescope in 1610?

29. Where are neutrino detectors located? Why are neutrinos so difficult to detect?

30. Why do telescopes in space give a better picture of the leftover radiation from the Big Bang?

APPLYING THE CONCEPTS

31. Many amateur astronomers start out with a 4-inch (aperture) telescope and then graduate to a 16-inch telescope. By what factor does the light-gathering power of the telescope increase with that upgrade? ●━●━●
    a. Make a prediction: Do you expect the 16-inch telescope to have a larger or smaller light-gathering power than the 4-inch telescope? Do you expect the "factor" that you are solving for to be more or less than one? Study Working It Out 6.1. Will you need to change the units of inches to meters in order to find this factor?
    b. Calculate: Find the ratio of the light-gathering power of the 16-inch telescope to the 4-inch telescope.
    c. Check your work: Verify that the factor you have calculated is dimensionless (the units have canceled out), and compare it to your prediction.

32. Assume that you have a telescope with an aperture of 1 m. What is the telescope's theoretical angular resolution when you are observing in the near-infrared region of the spectrum (λ = 1000 nm)? ●━●━●
    a. Make a prediction: Study Working It Out 6.2. Is this wavelength in the near-infrared region longer or shorter than the wavelength of green light used there? Is the diameter of this telescope more or less than the diameter of Hubble Space Telescope? Which term do you expect to dominate in this calculation; that is, do you expect the angular resolution you calculate to be smaller or larger than that of Hubble Space Telescope?
    b. Calculate: Use the angular resolution equation to calculate the angular resolution of this telescope at this wavelength.
    c. Check Your Work: Verify that the units of your answer are correct, and compare your result to your prediction.

33. Compare the light-gathering power of the Thirty Meter Telescope with that of the dark-adapted human eye (aperture, 8 mm).

34. The diameter of the full Moon in the focal plane of an average amateur's telescope (focal length, 1.5 m) is 13.8 mm. How big would the Moon be in the focal plane of a very large astronomical telescope (focal length, 250 m)?
    a. Make a prediction: Study the magnification equation in Working It Out 6.1. Do telescopes with longer focal lengths magnify more or less than telescopes with shorter focal lengths? Do you expect the Moon to be larger or smaller in the focal plane of the very large astronomical telescope?
    b. Calculate: Set up a ratio of the magnification of the Moon in the larger telescope to the magnification in the smaller telescope (assume the eyepiece used is the same in both cases). Then use that ratio of magnifications to find the size of the Moon in the larger telescope.
    c. Check your work: Verify that the unit of your answer is a unit of length, and compare your answer to your prediction.

35. The angular resolution of the human eye is about 1.5 arcmin. What would the aperture of a radio telescope (observing at 21 cm) have to be to have that angular resolution? Even though the atmosphere is transparent at radio wavelengths, humans do not see light in the radio range. Using your calculations and logic, explain why.

36. Assume that the maximum aperture of the human eye, D, is approximately 8 mm and the average wavelength of visible light, λ, is 5.5×10^{-4} mm.
    a. Calculate the diffraction limit of the human eye in visible light.
    b. How does the diffraction limit compare with the actual angular resolution of 1–2 arcmin (60–120 arcsec)?
    c. To what do you attribute the difference?

37. Study the photograph of light entering and leaving a block of refractive material in Figure 6.2b. Use a protractor to measure the angles of the green light as it enters the block and as it leaves the block. How are those angles related? ◉

38. One of the earliest astronomical CCDs had 160,000 pixels, each recording 8 bits (256 levels of brightness). A new generation of astronomical CCDs may contain a billion pixels, each recording 15 bits (32,768 levels of brightness). Compare the number of bits of data that each of those two CCD types produces in a single image.

39. Consider a CCD with a quantum efficiency of 80 percent and a photographic plate with a quantum efficiency of 1 percent. If an exposure time of 1 hour is required to photograph a celestial object with a given telescope, how much observing time would be saved by substituting a CCD for the photographic plate?

40. The VLBA uses an array of radio telescopes ranging across 8000 km of Earth's surface from the Virgin Islands to Hawai'i.
    a. Calculate the angular resolution of the array when radio astronomers are observing interstellar water molecules at a microwave wavelength of 1.35 cm.
    b. How does that angular resolution compare with the angular resolution of two large optical telescopes separated by 100 m and operating as an interferometer at a visible wavelength of 550 nm?

41. When operational, the Space VLBI may have a baseline of 100,000 km. What will be the angular resolution when studying interstellar molecules emitting at a wavelength of 17 mm from a distant galaxy?

42. The *Mars Reconnaissance Orbiter* (*MRO*) flies at an average altitude of 280 km above the martian surface. If its cameras have an angular resolution of 0.2 arcsec, what is the size of the smallest objects that the *MRO* can detect on the martian surface?

43. At this writing, *Voyager 1* is about 140 astronomical units (AU) from Earth, continuing to record its environment as it departed the boundary of our Solar System.
    a. How far away (in kilometers) is *Voyager 1*?
    b. How long do observational data take to come back to us from *Voyager 1*?
    c. How does *Voyager 1*'s distance from Earth compare with that of the nearest star (other than the Sun)?

44. Gravitational waves travel at the speed of light. Their speed, wavelength, and frequency are related as $c = \lambda \times f$. If we were to observe a gravitational wave from a distant cosmic event with a frequency of 10 hertz (Hz), what would be the wavelength of the gravitational wave?

45. Compute the peak of the blackbody spectrum with a temperature of 2.73 K. What region of the spectrum is this?

EXPLORATION Geometric Optics and Lenses

digital.wwnorton.com/astro7

Visit the Digital Resources Page and on the Student Site open the "Geometric Optics and Lenses" AstroTour in Chapter 6. Read through the animation until you reach the optics simulation, pictured in **Figure 6.33**. The simulator shows a converging lens and a pencil. Rays come from the pencil on the left of the converging lens, pass through the lens, and make an image to the right of the lens. The view that would be seen by an observer at the position of the eye is shown in the circle at upper right. Initially, when the pencil is at position 2.3 and the eye is at position 2.0, the pencil is out of focus and blurry.

1 Is the eraser at the top or the bottom of the actual pencil? (That answer becomes important later.)

_____

Using the red slider in the upper left of the window, try moving the pencil to the right. Pause when the observer's eye sees a recognizable pencil (even if it's still blurry).

2 Does the eye see the pencil right side up or upside down?

_____

3 This is somewhat analogous to the view through a telescope. The objects are very far from the lenses, and the observer sees things upside down in the telescope. If an object in your field of view is at the top of the field and you want it in the center, should you move the telescope up or down?

_____

Now return the pencil to position 2.3. Use the red slider in the lower right of the window to move the eye closer to the lens (to the left).

4 At what distance does the image of the pencil first become crisp and clear?

_____

5 Is the pencil right side up or upside down?

_____

6 In practice at the telescope, we do not move the observer back (away from the eyepiece) to bring the image into focus. Why not?

_____

7 Instead of moving the observer, we use a focusing knob to move the lens in the eyepiece, which brings the image into focus. Imagine that you are looking through the eyepiece of a telescope and the image is blurry. You turn the focusing knob and things get blurrier! What should you try next?

_____

8 Now imagine that you get the image focused just right, so it is crisp and sharp. The next person to use the telescope wears glasses and insists that the image is blurry. When you look through the telescope again, though, the image is still crisp. Explain why your experiences differ.

_____

Step through the animation to the next picture. Carefully study the two telescopes shown and the path the light takes through them.

9 Which telescope has a longer focal length: the top one or the bottom one?

_____

10 Which telescope produces an image with the red and the blue stars more separated: the top one or the bottom one?

_____

11 A longer focal length is an advantage in one sense, but it's not the entire story. What are some disadvantages of a telescope with a very long focal length?

_____

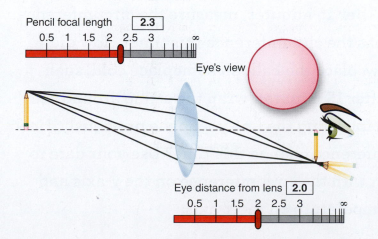

Pencil focal length   2.3

Eye's view

Eye distance from lens   2.0

Figure 6.33 Use this simulation to change the position of the object of the eye to explore what will be seen for various configurations.

The Formation of Planetary Systems

Hydrostatic equilibrium is a balance between forces. The stability of clouds, planets, and stars depends on it. So does the stability of an inflated balloon. Blow up a balloon and tie it off. Wrap a string around the widest part of the balloon, and use a marker to trace the string on the balloon. Mark on the string where it overlaps; this indicates the balloon's circumference at room temperature. Use a ruler to measure the circumference. Predict how the circumference of the balloon will change if it is warmed above room temperature and if it is cooled below room temperature. Next, place the balloon in a sink, pot, or bucket and cover it completely with hot water (but not boiling!). After 15 minutes, measure the circumference again. What was the circumference this time? How did it change? Now, place the balloon someplace cold, such as a freezer. After 15 minutes, wrap the string around the balloon again, following the line you made previously. How did the circumference change this time? Use your data to sketch a graph with the circumference on the *y*-axis and the relative temperature on the *x*-axis.

EXPERIMENT SETUP

Blow up a balloon and tie it off. Wrap a string around the widest part of the balloon.

Use a marker to trace the string on the balloon.

Mark on the string where it overlaps.

00:15

Submerge the balloon in hot water (but not boiling!) for 15 minutes. Measure the circumference of the balloon.

00:15

Place the balloon someplace cold for 15 minutes. Measure the circumference of the balloon.

PREDICTION

The circumference will be
☐ larger ☐ smaller ☐ the same
when the balloon is cooled. It will be
☐ larger ☐ smaller ☐ the same
when the balloon is heated.

SKETCH OF RESULTS (in progress)

CIRCUMFERENCE

large

medium

small

cold   room   hot

TEMPERATURE

7

The planetary system containing Earth—our Solar System—is a by-product of the birth of the Sun, but the physical processes that shaped the formation of the Solar System are not unique to it. The same processes have formed many other multiplanet systems. Here in Chapter 7, we examine how planetary systems are born and evolve.

LEARNING GOALS

By the end of this chapter, you should be able to:

(1) Describe how our understanding of planetary system formation developed from the work of both planetary and stellar scientists.

(2) Discuss the role of gravity and angular momentum in explaining why planets orbit the Sun in a plane and why they revolve in the same direction that the Sun rotates.

(3) Explain how temperature at different locations in the protoplanetary disk affects the composition of planets, moons, and other bodies.

(4) Discuss the processes that resulted in the formation of planets and other objects in our Solar System.

(5) Describe how astronomers both find planets around other stars and determine exoplanet properties.

▶❚❚ **AstroTour:** Solar System Formation

7.1 Planetary Systems Form around a Star

Earth is part of a collection of **planets**—large, round isolated bodies that orbit a star. Astronomers call a system of planets surrounding a star a **planetary system**. The Solar System, shown in **Figure 7.1**, is the planetary system that includes Earth, seven other planets, and the Sun. The system also includes moons that orbit planets and small bodies that occupy particular regions of the Solar System, such as the asteroid belt or the Kuiper Belt. Our Solar System is a tiny part of our galaxy, which is a tiny part of the universe. Review Figure 1.3 to remind yourself of the size scales involved. Light takes about 4 hours to travel to Earth from Neptune, the outermost known planet in the Solar System, but light from the most distant galaxies has taken more than 13 *billion* years to reach Earth.

Until the latter part of the 20th century, the origin of the Solar System remained speculative. Over the past century, with the aid of spectroscopy, astronomers have determined that the Sun is an ordinary star, one of hundreds of billions in its galaxy, the Milky Way, and that the Milky Way is an ordinary galaxy, one of hundreds of billions in the universe. In the past few decades, stellar astronomers studying the formation of stars and planetary scientists analyzing clues about the history of the Solar System have arrived at the same picture of the early Solar System—but from two very different directions. That unified understanding provides the foundation for the way astronomers now think about the Sun and the objects that orbit it. In this section, we look at how the work of stellar and planetary scientists converged to inform our understanding of planetary system formation.

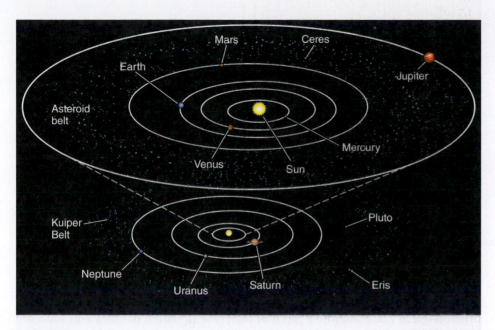

Figure 7.1 Our Solar System includes planets, moons, and other small bodies. Sizes and distances shown are not to scale.

The Nebular Hypothesis

The first plausible theory for the formation of the Solar System, the **nebular hypothesis**, was proposed in 1755 by the German philosopher Immanuel Kant (1724–1804) and conceived independently in 1796 by the French astronomer Pierre-Simon Laplace (1749–1827). Kant and Laplace argued that a rotating cloud of interstellar gas, or **nebula** (Latin for "cloud"), gradually collapsed and flattened to form a disk with the Sun at its center. Surrounding the Sun were rings of material from which the planets formed. That configuration would explain why the planets orbit the Sun in the same direction in the same plane. The nebular hypothesis remained popular throughout the 19th century, and those basic principles of the hypothesis are still retained today.

Our modern theory of planetary system formation calculates the conditions required for a cloud of interstellar gas to collapse under the force of its own self-gravity to form stars. Recall from Chapter 4 that self-gravity is the gravitational attraction between the parts of an object, such as a planet or star, that pulls all the parts toward the object's center. That inward force is opposed by either structural strength (for rocks that make up terrestrial planets) or the outward force resulting from gas pressure and radiation pressure within a star. If the outward force is less than self-gravity, the object contracts; if it is greater, the object expands. In a stable object, the inward and outward forces are balanced.

In support of the nebular hypothesis, disks of gas and dust have been observed surrounding young stellar objects (**Figure 7.2**). From that observational evidence, stellar astronomers have shown that, much as a spinning ball of pizza dough spreads out to form a flat crust, the cloud that produces a star—the Sun, for

Figure 7.2 a. Hubble Space Telescope image of a disk around newly formed stars. The dark band is the silhouette of the disk seen edge on. Bright regions are dust illuminated by the star's light. The jets are shown in green. **b.** The Atacama Large Millimeter /submillimeter Array (ALMA) obtained this image of a protoplanetary disk around the star HL Tau. This image shows substructures and possibly planets in the system's dark patches.

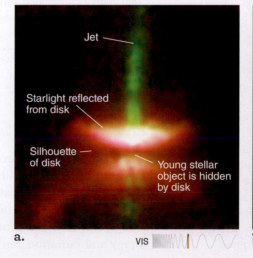

Jet
Starlight reflected from disk
Silhouette of disk
Young stellar object is hidden by disk

a.   VIS

b.   RADIO

What if astronomers find another planetary system where the planets orbit the central star but align in two separate planes which are oblique to each other? What would such an observation tell you about the nebular hypothesis?

Figure 7.3 Meteorites are the surviving pieces of Solar System fragments that land on planets. This meteorite formed from many smaller components that stuck together.

unanswered questions

How typical is the Solar System? Only within the past decade have astronomers found other systems containing four or more planets, and so far, the observed distributions of large and small planets in those multiplanet systems have looked different from those of the Solar System. Computer simulations of planetary system formation suggest that a system with stable orbits and a planetary distribution like those of the Solar System may develop only rarely. Improved supercomputers can run more complex simulations, which can be compared with the observations to better understand how solar systems are configured.

example—collapses first into a rotating disk. Material in the disk eventually suffers one of three fates: it travels inward onto the forming star at its center, it remains in the disk itself to form planets and other objects, or it is ejected back into interstellar space.

Planetary Scientists and the Convergence of Evidence

While astronomers were working to understand star formation, other groups of scientists with very different backgrounds were piecing together the history of the Solar System. Planetary scientists, geochemists, and geologists looking at the current structure of the Solar System inferred what some of its early characteristics must have been. The orbits of all the planets lie very close to a single plane, so the early Solar System must have been flat. In addition, all the planets orbit the Sun in the same direction, so the material from which the planets formed must have been orbiting the Sun in the same direction as well.

To find out more, scientists study samples of the very early Solar System. Rocks that fall to Earth from space, known as **meteorites**, include pieces of material left over from the Solar System's youth. Many meteorites, such as the one in **Figure 7.3**, resemble a piece of concrete in which pebbles and sand are mixed with a much finer filler, suggesting that the larger bodies in the Solar System must have grown from the aggregation of smaller bodies. That finding suggests an early Solar System in which the young Sun was surrounded by a flattened disk of both gaseous and solid material. Our Solar System formed from that swirling disk of gas and dust.

As astronomers and planetary scientists compared notes, they realized they had arrived at the same picture of the early Solar System from two completely different directions. The rotating disk from which the planets formed was the remains of the disk that had accompanied the formation of the Sun. Earth, along with all the other orbiting bodies that make up the Solar System, formed from the remnants of an *interstellar cloud* that collapsed to form the local star, the Sun. The connection between the formation of stars and the origin and later evolution of the Solar System is one of the cornerstones of both astronomy and planetary science—a central theme of our understanding of our Solar System (see the **Process of Science Figure**).

CHECK YOUR UNDERSTANDING 7.1

Which of the following pieces of evidence support the nebular hypothesis? (Choose all that apply.) (a) Planets orbit the Sun in the same direction. (b) The Solar System is relatively flat. (c) Earth has a large Moon. (d) We observe disks of gas and dust around other stars.

Answers to Check Your Understanding questions are in the back of the book.

7.2 The Solar System Began with a Disk

Planets form in a disk around young stars, but what are some of the specifics of the process? **Figure 7.4** illustrates the young Solar System as it appeared roughly 5 billion years ago. At that time, the Sun was still a **protostar**—a large ball of gas but not yet hot enough in its center to be a star. As the cloud of interstellar gas collapsed to form the protostar, its gravitational energy was converted into heat energy and radiation. Surrounding the protostellar Sun was a flat, orbiting disk

Converging Lines of Inquiry

Why is the Solar System a disk, with all planets orbiting in the same direction? Scientists from different disciplines often contribute to the solution to a problem. These scientists approach the problem from different directions and perspectives, so when they arrive at the same conclusion, that is compelling evidence that they are all on the right track.

Mathematicians:

Suggested the nebular hypothesis—that a collapsing rotating cloud formed the Solar System.

Stellar Astronomers:

Tested the nebular hypothesis, seeking evidence for or against.

Found dust and gas around young stars.

Observed this gas and dust to be in the shape of disks.

Planetary Scientists:

Tested the nebular hypothesis, seeking evidence for or against.

Studied meteorites, which showed that the planets formed from many smaller bodies.

Beginning from the same fundamental observations about the shape of the Solar System, theorists (the mathematicians), stellar astronomers, and planetary scientists **converged on the nebular theory** that stars and planets form together from a collapsing cloud of gas and dust.

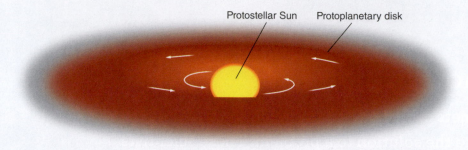

Protostellar Sun   Protoplanetary disk

Figure 7.4 Think of the young Sun as being surrounded by a flat, rotating disk of gas and dust that was flared at its outer edge.

📹 **Astronomy in Action:** Angular Momentum

of gas and dust. Each bit of the material in that thin disk orbited the Sun in accordance with the same laws of motion and gravitation that govern the orbits of the planets. The disk around the Sun—like the disks that astronomers see today surrounding protostars elsewhere in our galaxy—is called a **protoplanetary disk**. The disk is so "fluffy" and low-density compared to the protostar that it probably contained less than 1 percent of the mass of the star forming at its center, but that amount was more than enough to account for the bodies that make up the Solar System today.

The Collapsing Cloud and Angular Momentum

Conservation of *angular momentum* causes protoplanetary disks to form. **Angular momentum** is a conserved quantity associated with a revolving or rotating system and depends on both the velocity and distribution of the system's mass. The angular momentum of an isolated system is always conserved; that is, it remains unchanged unless acted on by an external force. A figure-skater spinning on the ice (**Figure 7.5**), like any other rotating object, has some amount of angular momentum. Unless frictional forces act to reduce her angular momentum, she will always have the same amount of angular momentum.

The amount of angular momentum depends on three factors:

1. How fast the object is rotating. The faster an object is rotating, the more angular momentum it has.

2. The mass of the object. If a similarly sized bowling ball and a basketball are spinning at the same speed, the bowling ball has more angular momentum because it has more mass.

3. How the mass of the object is distributed relative to the spin axis—that is, how spread out the object is. For an object of a given mass and rate of rotation, the more spread out it is, the more angular momentum it has. A spread-out object rotating slowly might have the same angular momentum as a compact object rotating rapidly.

Both an ice-skater and a collapsing interstellar cloud are subject to the **conservation of angular momentum**: the angular momentum must remain the same in the absence of an external force. For angular momentum to be conserved, a change in one of the three quantities (the rate of spin, mass, or distribution of mass) must be accompanied by a compensating change in another quantity. Because an ice-skater's mass doesn't change, for example, she can control how rapidly she spins by pulling in or extending her arms or legs. As she pulls in her arms to become more compact, she changes her distribution of mass and must spin faster to maintain the same angular momentum. When her arms are held tightly in front of her and one leg is wrapped around the other, the skater's spin becomes a blur. She finishes with a flourish by throwing her arms and leg out—an action that abruptly slows her spin by spreading out her mass. The skater's angular momentum remains constant throughout the maneuver. Similarly (see **Figure 7.6**), the cloud that formed our Sun rotated faster and faster as it collapsed, just as the ice-skater speeds up when she pulls in her arms.

That description, however, presents a puzzle. Suppose the Sun formed from a typical cloud—one about a light-year across and rotating so slowly that completing one rotation took a million years. By the time the cloud collapsed to the size of

Figure 7.5 A figure-skater relies on the principle of conservation of angular momentum to change the speed at which she spins.

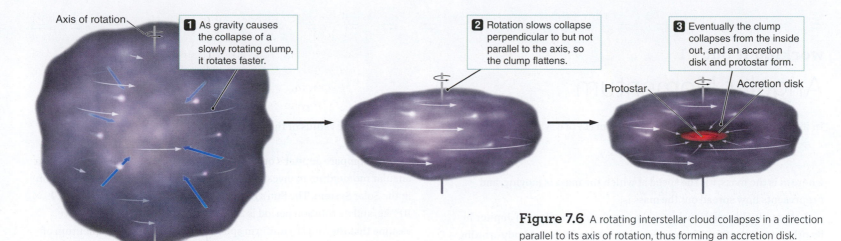

Axis of rotation

1 As gravity causes the collapse of a slowly rotating clump, it rotates faster.

2 Rotation slows collapse perpendicular to but not parallel to the axis, so the clump flattens.

3 Eventually the clump collapses from the inside out, and an accretion disk and protostar form.

Protostar   Accretion disk

Figure 7.6 A rotating interstellar cloud collapses in a direction parallel to its axis of rotation, thus forming an accretion disk.

the Sun today, it would have been spinning so fast that one rotation would occur every 0.6 second. That rate is more than 3 million times faster than our Sun actually spins. At that rate of rotation, the Sun would tear itself apart. It *appears* that angular momentum was *not* conserved in the actual formation of the Sun—but that can't be right because angular momentum *must* be conserved. We must be missing something. Where did the angular momentum go?

The Formation of an Accretion Disk

To understand how angular momentum is conserved in disk formation, we must think in three dimensions. Imagine that the ice-skater bends her knees, compressing herself downward instead of bringing her arms toward her body. As she does so, she again makes herself less spread out, but her rate of spin does not change because no part of her body has become any closer to the axis of spin. Similarly, as shown in Figure 7.6, a clump of a molecular cloud can flatten out without speeding up by collapsing parallel to its axis of rotation. Instead of collapsing into a ball, the interstellar cloud flattens into a disk. As the cloud collapses, its self-gravity increases and the inner parts begin to fall freely inward, raining down on the growing disk at the center. The outer portions of the cloud lose the support of the collapsed inner portion, and they start falling inward, too. As that material makes its final inward plunge, it lands on a thin, rotating disk—called an **accretion disk**—that forms from the accretion of material around a massive object.

The formation of accretion disks, shown in **Figure 7.7**, is common in the universe. As material falls onto the disk at an angle, it impacts material coming up to the disk from below. Over time, as more collisions occur, these perpendicular motions gradually cancel out. But the part of the motion parallel to the disk remains unchanged. This is much like two football players approaching from opposite sides and colliding as they jump for the ball. Their motion across the field cancels out, but their motion down the field remains, so they fall to the ground closer to the goal. In the case of accretion disks, the material flattens out and continues to rotate, which conserves angular momentum.

Thus, the angular momentum of the infalling material is transferred to the accretion disk. The rotating accretion disk has a radius of hundreds of astronomical units, and that is *thousands* of times greater than the radius of the star that will

▶❙❙ **AstroTour:** Traffic Circle Analogy

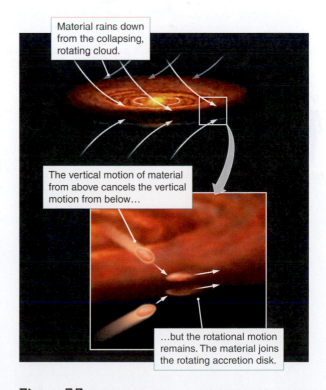

Material rains down from the collapsing, rotating cloud.

The vertical motion of material from above cancels the vertical motion from below…

…but the rotational motion remains. The material joins the rotating accretion disk.

Figure 7.7 Gas from a rotating cloud falls inward from opposite sides, piling up onto a rotating disk.

working it out 7.1

Angular Momentum

In its simplest form, the angular momentum (L) of a system is given by

$$L = m \times v \times r$$

where m is the mass, v is the speed at which the mass is moving, and r represents how spread out the mass is.

Let's apply this relationship to the angular momentum of Jupiter in its orbit about the Sun. The angular momentum from one body orbiting another is called *orbital* angular momentum, $L_{orbital}$. The mass (m) of Jupiter is 1.90×10^{27} kilograms (kg), the speed of Jupiter in orbit (v) is 1.31×10^4 meters per second (m/s), and the radius of Jupiter's orbit (r) is 7.79×10^{11} meters. Putting all that together gives

$$L_{orbital} = (1.90 \times 10^{27} \text{ kg}) \times (1.31 \times 10^4 \text{ m/s}) \times (7.79 \times 10^{11} \text{ m})$$

$$L_{orbital} = 1.94 \times 10^{43} \text{ kg m}^2/\text{s}$$

Calculating the *spin* angular momentum of a spinning object, such as a skater, a planet, a star, or an interstellar cloud, is more complicated. Here, we must add up the individual angular momenta of *every tiny mass element* within the object. For a uniform sphere, the spin angular momentum is

$$L_{spin} = \frac{4\pi mR^2}{5P}$$

where R is the radius of the sphere and P is the rotation period of its spin.

Let's compare Jupiter's orbital angular momentum with the Sun's spin angular momentum to investigate the distribution of angular momentum in the Solar System. The Sun's radius is 6.96×10^8 meters, its mass is 1.99×10^{30} kg, and its rotation period is 24.5 days = 2.12×10^6 seconds. If we assume that the Sun is a uniform sphere, the spin angular momentum of the Sun is

$$L_{spin} = \frac{4 \times \pi \times (1.99 \times 10^{30} \text{ kg}) \times (6.96 \times 10^8 \text{ m})^2}{5 \times (2.12 \times 10^6 \text{ s})}$$

$$L_{spin} = 1.14 \times 10^{42} \text{ kg m}^2/\text{s}$$

$L_{orbital}$ of Jupiter is about 17 times greater than L_{spin} of the Sun. Thus, most of the angular momentum of the Solar System now resides in the orbits of its major planets.

For a collapsing sphere to conserve L_{spin}, its rotation period P must be proportional to R^2. As with the skater, when a sphere decreases in radius, its rotation period decreases; that is, it spins faster.

what if . . .

What if you observe a close pair of stars forming a binary system? How would you expect the disk around such a pair of stars to differ from the disk around a single star?

eventually form at its center. Therefore, most of the angular momentum in the original interstellar cloud ends up in the accretion disk rather than in the central protostar (see **Working It Out 7.1** for an example of the relevant calculation).

Most of the matter that lands on the accretion disk either becomes part of the star or is ejected back into interstellar space in the form of jets or other outflows (**Figure 7.8**). Those jets are bipolar—they come in pairs aligned along an axis. Material swirling in the bipolar jets carries angular momentum away from the accretion disk in the general direction of the poles of the rotation axis. However, a small amount of material is left behind in the disk. The objects in that leftover disk—the dregs of the process of star formation—form planets and other objects that orbit the star.

Formation of Large Objects

Random motions of the gas within the protoplanetary disk eventually push the smaller grains of solid material toward larger grains. As that happens, the smaller grains stick to the larger grains. The "sticking" process among smaller grains is due to the same static electricity that causes dust bunnies to grow under your bed. Starting out at only a few microns (μm) across—about the size of particles

Figure 7.8 ★ WHAT AN ASTRONOMER SEES This gorgeous image from the Hubble Space Telescope shows a portion of the Carina Nebula. An astronomer looking at this image will immediately notice the colors, because color often indicates where different atoms or molecules are present. She will not necessarily know which atoms or molecules are represented by these colors because different astronomers will use a different palette. But even without reading any background on the image, she will know that there is something different about the hazy blue areas and the hazy pink or green ones. An astronomer will recognize and then mostly ignore the diffraction spikes (mentioned in Chapter 6) that form an X around each bright star. She will notice the brown clumps of material that are too dense to see through. These high-density regions indicate that star formation might be happening in this nebula, and this will be confirmed by the small oval protostar in the upper right corner and by the dense "fingers" that stick out in various places. These fingers point the way toward a source of interstellar wind outside the image to the upper right. That wind has eroded away the less dense material around these denser regions and may have triggered star formation by compressing the material at the top of each finger. Each finger is a dense blob of material that creates a wind "shadow" behind it. A new star has just formed at the top of the finger near the middle of the image. An astronomer will identify this new star because of the thin jets of material that are being ejected in opposite directions; new stars sometimes create such jets.

in smoke—the slightly larger bits of dust grow to the size of pebbles and then to clumps the size of boulders, which are not as easily pushed around by gas (**Figure 7.9**). When clumps grow to about 100 meters across, the objects are so far apart that they collide less often, and their growth rate slows down but does not stop. Within a protoplanetary disk, the larger dust grains become larger at the expense of the smaller grains.

For two large clumps to stick together rather than explode into many small pieces, they must bump into each other very gently: collision speeds must be about 0.1 m/s or less. If your stride is about a meter long, a collision speed of 0.1 m/s would correspond to only one step every 10 seconds. The process does not yield larger and larger bodies with every collision. When violent collisions occur in an accretion disk, larger clumps break back into smaller pieces. But over a long time, large bodies do form.

Objects continue to grow by "sweeping up" smaller objects that get in their way. Those objects can eventually measure up to several hundred meters across. As the clumps reach the size of about a kilometer, a different process becomes important. Those kilometer-sized objects are massive enough that their gravity pulls

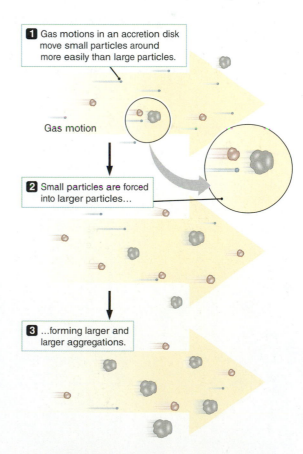

1 Gas motions in an accretion disk move small particles around more easily than large particles.

Gas motion

2 Small particles are forced into larger particles...

3 ...forming larger and larger aggregations.

Figure 7.9 Motions of gas in a protoplanetary disk blow smaller particles of dust into larger particles, making the larger particles larger still. That process continues, eventually creating objects many meters in size.

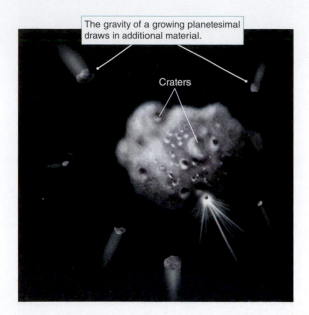

The gravity of a growing planetesimal draws in additional material.

Craters

Figure 7.10 The gravity of a planetesimal is strong enough to attract surrounding material, which causes the planetesimal to grow more rapidly. Some of this additional material impacts with enough energy to make craters on the planetesimal.

on nearby bodies, as shown in **Figure 7.10**. Those bodies of rock and ice are known as **planetesimals** ("tiny planets") and eventually combine with one another to form planets. Besides chance collisions with other objects, a planetesimal's gravity can now pull in and capture small objects outside its direct path. That process speeds up the growth of planetesimals, so larger planetesimals quickly consume most of the remaining bodies near their orbits. The final survivors of that process are large enough to be called planets. As with the major bodies in orbit around the Sun, some of the planets may be relatively small and others quite large.

CHECK YOUR UNDERSTANDING 7.2

Where does most of the angular momentum of the original cloud go? It (a) goes into the orbital angular momentum of planets; (b) goes into the star; (c) goes into the spin of the planets; (d) is lost along the jets from the star.

7.3 The Inner Disk and Outer Disk Formed at Different Temperatures

In the Solar System, the inner planets are small and mostly rocky, whereas the outer planets are very large and mostly gaseous. That distinct difference between the inner and outer Solar System can be explained by how the local disk environment affects the formation process. In this section, we examine these differences.

Energy in the Disk

The accretion disks surrounding young stars form from interstellar material that may have a temperature of only a few kelvins, but the disks themselves reach temperatures of hundreds of kelvins or more. Astronomers want to understand what heats up the disk around a forming star so that we can calculate how hot those disks get.

According to the law of **conservation of energy**, the total amount of energy in a system must remain constant unless energy is added to or taken away from the system from the outside. The form the energy takes, however, can change. With that concept in mind, why does gas falling on the disk make the disk hot? Imagine you are working against gravity by lifting a heavy object, such as a bowling ball. Lifting the bowling ball takes energy, and the law of conservation of energy states that energy is never lost. Where does that energy go? The energy is stored and changed into a form called **gravitational potential energy**. If you drop the bowling ball, it falls, and as it falls, it speeds up. The gravitational potential energy that was stored is converted to energy of motion, which is called **kinetic energy**. When the bowling ball hits the floor, it stops suddenly. The bowling ball loses its energy of motion, so what form does that energy take now? Some of the energy is converted into the sound the bowling ball makes when it hits the floor, and some goes into heating and distorting the floor. But much of the energy is converted into **thermal energy**. The atoms and molecules that make up the bowling ball are moving around within the bowling ball a bit faster than they were before the bowling ball hit, so the bowling ball and its surroundings, including the floor, grow a tiny bit warmer. Similarly, as gas falls toward the disk surrounding a protostar, gravitational potential energy is converted first to kinetic energy, causing the gas to pick up speed. When the gas hits the disk and stops suddenly, that kinetic energy turns into thermal energy.

Material falling onto the accretion disk around a forming star causes the disk to heat up, too (**Figure 7.11**). The amount of heating depends on *where* the material hits the disk. Material hitting the inner part of the disk (the *inner disk*) has fallen farther and picked up greater speed within the gravitational field of the forming star than material hitting the disk farther out. Like a brick dropped from a tall building, material striking the inner disk is moving quite rapidly when it hits, so it heats the inner disk to high temperatures. In contrast, material falling onto the outer part of the disk (the *outer disk*) is moving much more slowly, like a brick dropped from just a foot or so above the ground. As a result, the temperature at the outermost parts of the disk is not much higher than that of the original interstellar cloud. Stated another way: material falling onto the inner disk has more gravitational potential energy to convert into thermal energy than does material falling onto the outer disk.

The energy released as material falls onto the disk is not the only source of thermal energy in the disk. Even before the nuclear reactions that will one day power the new star have ignited, the conversion of gravitational energy into thermal energy drives the temperature at the surface of the protostar to several thousand kelvins, and it drives the luminosity of the huge ball of glowing gas to many times the luminosity of the present-day Sun. For the same reasons why Mercury is hot whereas Pluto is not (see Chapter 5), the radiation streaming outward from the protostar at the center of the disk drives the temperature in the inner parts of the disk even higher, increasing the difference in temperature between the inner and outer parts of the disk.

The Compositions of Planets

Temperature affects whether a material exists as a solid, a liquid, or a gas. On a hot summer day, ice melts and water quickly evaporates; on a cold winter night, water in your breath freezes into tiny ice crystals. Metals and rocky materials, such as iron, **silicates** (minerals containing silicon and oxygen), and carbon, remain solid even at high temperatures. Substances that can withstand high temperatures without melting or being vaporized are called **refractory materials**. Other materials, such as water, ammonia, and methane, remain in a solid form only if their temperature is very low. Those materials, which become gases at moderate temperatures, are called **volatile materials** (or *volatiles* for short). Astronomers generally call the solid form of any volatile material an **ice**.

Differences in temperature from place to place within the protoplanetary disk significantly affect the makeup of the dust grains in the disk. **Figure 7.12** illustrates that only refractory substances exist in the hottest parts of the disk—the area closest to the protostar. In the inner disk, dust grains are composed

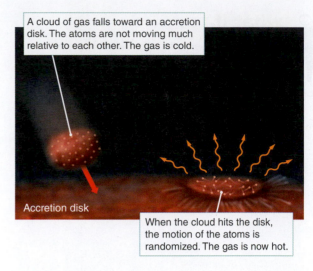

Figure 7.11 Atoms in a gas fall together until they hit the accretion disk, at which point their motions become randomized, raising the temperature of the gas.

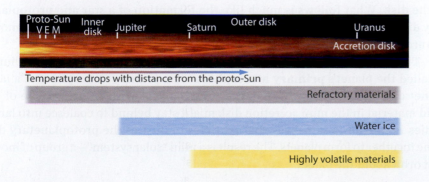

Figure 7.12 Differences in temperature within a protoplanetary disk determine the composition of dust grains that then evolve into planetesimals and planets. The colored bars show that refractory materials are found throughout the disk, whereas water ice is found only outside Jupiter's orbit, and highly volatile materials are found only outside Saturn's orbit. Shown here are the proto-Sun and the orbits of Venus (V), Earth (E), Mars (M), Jupiter, Saturn, and Uranus.

what if . . .

What if you observed an Earth-mass planet that is far from its central star, but has almost no atmosphere at all? What could you conclude about this planet's formation history?

almost entirely of refractory materials. Some substances can survive in solid form somewhat farther out, including some hardier volatiles, such as water ice and certain chemical compounds that are **organic** (meaning that they contain molecules with a carbon–hydrogen bond). Those solids add to the materials that make up dust grains. In the coldest, outermost parts of the accretion disk, far from the central protostar, highly volatile components such as methane, ammonia, and carbon monoxide ices and other organic molecules survive only in solid form. The differences in composition of dust grains within the disk are reflected in the composition of the planets formed from that dust. Planets that form closer to the central star tend to be made up mostly of refractory materials, such as rock and metals, but are deficient in volatiles. Planets that form farther from the central star contain not only refractory materials but also large quantities of ices and organic materials.

In the Solar System, the inner planets are composed of rocky material surrounding metallic cores of iron and nickel. Objects in the outer Solar System, including moons, giant planets, and comets, are composed largely of ices of various types. But not all planetary systems are so neatly organized as our Solar System. When planets around other stars were first discovered, they appeared to be very different, with large planets close to their respective stars. Astronomers now think that **chaotic** encounters—in which a small change in the initial state of a system can lead to a large change in the final state of the system—may change the organization of planetary compositions. In a process called **planet migration**, the force of gravity from all the nearby objects can move some planets so that they end up far from the place of their birth. In our Solar System, for example, Uranus and Neptune originally may have formed nearer to the orbits of Jupiter and Saturn but were then driven outward to their current locations by gravitational encounters with Jupiter and Saturn. A planet also can migrate when it gives up some of its orbital angular momentum to the disk material that surrounds it. Such a loss of angular momentum causes the planet to slowly spiral inward toward the central star. Thus, the order of planets in a system can change.

Formation of an Atmosphere

Once a solid planet has formed, it may continue to grow by capturing gas from the protoplanetary disk. To do so, it must act quickly. Young stars and protostars emit fast-moving particles and intense radiation that can quickly disperse the gaseous remains of the accretion disk. Gaseous planets such as Jupiter probably have only about 10 million years to form and to grab whatever gas they can. Because of their strong gravitational fields, more massive young planets can capture more of the hydrogen and helium gases that makes up the bulk of the disk. What follows is much like the formation of a star and protoplanetary disk, but on a smaller scale—namely, gas from a mini accretion disk moves inward and falls onto the planet.

The gas that a planet captures when it forms—primarily hydrogen and helium—is called the planet's **primary atmosphere**. The primary atmosphere of a large planet can be more massive than the solid body, as with Jupiter. Some of the solid material in the mini accretion disk might stay behind to coalesce into larger bodies in much the same way that particles of dust in the protoplanetary disk came together to form planets. The result is a mini "solar system"—a group of moons that orbit about the planet.

A less massive planet also may capture some gas from the protoplanetary disk, only to lose it later. The gravity of small planets may be too weak to hold low-mass gases such as hydrogen or helium. Even if a small planet can gather some hydrogen and helium from its surroundings, that primary atmosphere will not last long. In the inner solar system, the temperatures are higher, so the hydrogen and helium atoms are moving faster than in the outer solar system and will escape from a small planet. The atmosphere that remains around a small planet such as Earth is a **secondary atmosphere**, which forms later in the life of a planet. Volcanism is one important source of a secondary atmosphere because it releases heavier and thus slower-moving gases, such as carbon dioxide, water vapor, and other gases trapped in the planet's interior. In addition, volatile-rich comets that formed in the outer parts of the disk fall inward toward the new star long after its planets have formed, and they sometimes collide with planets. **Comets** are icy planetesimals that survive planetary accretion. They may serve as a significant source of water, organic compounds, and other volatile materials on planets close to the central star.

CHECK YOUR UNDERSTANDING 7.3

In our Solar System, the inner planets are rocky because: (a) the original cloud had more rocky material near the center; (b) warm temperatures in the inner disk caused the inner planetesimals to be formed of only rocky material; (c) the inner disk filled a smaller volume, and so it was denser; (d) the hydrogen and helium atoms were too low mass to remain in the inner disk.

7.4 The Formation of Our Solar System

Nearly 5 billion years ago, the Sun was still a protostar surrounded by a protoplanetary disk of gas and dust. During the next few hundred thousand years, much of the dust in the disk had collected into planetesimals—clumps of rock and metal near the emerging Sun, and aggregates of rock, metal, ice, and organic materials farther from the Sun. In this section, we look at the formation of the types of planets in our own Solar System.

The Terrestrial Planets

Within the inner 5 astronomical units (AU) of the disk, several rock and metal planetesimals quickly grew larger to become the dominant masses in their orbits. With their ever-strengthening gravitational fields, they either captured most of the remaining planetesimals or ejected them from the inner part of the disk. **Figure 7.13** shows some results from a computer simulation of how that might have happened. The dominant planetesimals became planet-sized bodies with masses ranging between 5 percent and 100 percent of Earth's mass. Those dominant planetesimals evolved into the **terrestrial planets**, which are rocky, Earth-like planets. Today, the surviving terrestrial planets are Mercury, Venus, Earth, and Mars. Earth's Moon is often grouped with those terrestrial planets because of its similar physical and geological properties, even though it is not a planet itself and formed differently. It is possible that one or two other planets or large moons formed in the young Solar System but were later destroyed.

For several hundred million years after the four surviving terrestrial planets formed, leftover pieces of debris still in orbit around the Sun continued to rain

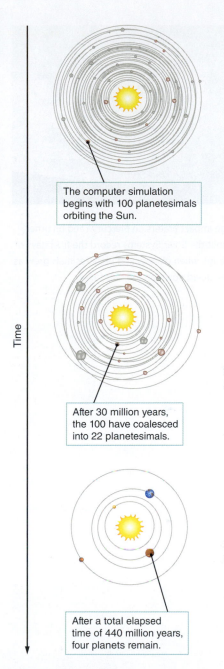

Time

The computer simulation begins with 100 planetesimals orbiting the Sun.

After 30 million years, the 100 have coalesced into 22 planetesimals.

After a total elapsed time of 440 million years, four planets remain.

Figure 7.13 Computer models simulate how material in the protoplanetary disk became clumped into the planets over time. Only a few planets remain at the end.

Figure 7.14 Large impact craters on Mercury (and on other solid bodies throughout the Solar System) record the final days of the Solar System's youth, when planets and planetesimals grew as smaller planetesimals rained down on their surfaces.

down on the surfaces of those planets. Today, we can still see the scars of those early impacts on the cratered surfaces of all the terrestrial planets (**Figure 7.14**). That rain of debris continues even today, but at a much lower rate.

Before the proto-Sun became a true star, gas in the inner part of the protoplanetary disk was still plentiful. During that early period the two larger terrestrial planets, Earth and Venus, may have held on to weak primary atmospheres of hydrogen and helium, but those thin atmospheres were soon lost to space. The terrestrial planets did not develop thick atmospheres until the formation of the secondary atmospheres that now surround Venus, Earth, and Mars. Mercury's size and proximity to the Sun and the Moon's small mass prevented those bodies from retaining significant secondary atmospheres.

The Giant Planets

Beyond 5 AU from the Sun, in a much colder part of the accretion disk, planetesimals combined to form several bodies with masses about 5–20 times that of Earth (5–20 M_{Earth}). Those planet-sized objects formed from planetesimals containing volatile ices and organic compounds in addition to rock and metal. In a process astronomers call **core accretion–gas capture**, mini accretion disks formed around those planetary cores, capturing massive amounts of hydrogen and helium and funneling that material onto the planets. Four such massive bodies became the cores of the **giant planets**—Jupiter, Saturn, Uranus, and Neptune. Those giant planets are many times the mass of any terrestrial planet.

Jupiter's massive solid core captured and retained the most gas—roughly 300 M_{Earth}. The other outer planetary cores captured less hydrogen and helium, perhaps because their cores were less massive or because less gas was available to them. Saturn ended up with less than 100 M_{Earth} of gas, and Uranus and Neptune grabbed less than 20 M_{Earth} of gas.

The core accretion model indicates that a Jupiter-like planet could take up to 10 million years to accumulate. Some planetary scientists think that our protoplanetary disk could not have survived long enough to form gas giants such as Jupiter through the general process of core accretion. All the gas may have dispersed in roughly half that time, cutting off Jupiter's supply of hydrogen and helium. An alternative explanation is a process called *disk instability*, in which the protoplanetary disk suddenly and quickly fragments into massive clumps equivalent to those of a large planet. Both core accretion and disk instability may have played a role in the formation of our own and other planetary systems.

During the formation of the planets, gravitational energy was converted into thermal energy as individual atoms and molecules moved faster. That conversion warmed the gas surrounding the cores of the giant planets. Proto-Jupiter and proto-Saturn probably became so hot that they glowed a deep red color, similar to the heating element on an electric stove. Their internal temperatures may have been even higher.

In the mini accretion disks surrounding the giant planets, some of the remaining material combined into small bodies, which became moons. A **moon** is any natural satellite in orbit about a planet or asteroid. The composition of the moons that formed around the giant planets followed the same trend as that of the planets that formed around the Sun: the innermost moons formed under the hottest conditions and therefore contained the smallest amounts of volatile material. For example, the closest of Jupiter's many moons may have experienced high temperatures from nearby Jupiter's glowing so intensely that it would have evaporated most of the volatile substances in the inner part of its mini accretion disk.

Remaining Planetesimals

Not all planetesimals in the disk became planets. For example, dwarf planets orbit the Sun but have not cleared other, smaller bodies from their orbits. Ceres and Pluto (Figure 7.1) are dwarf planets. More dwarf planets, along with many smaller bodies, are found in the Kuiper Belt, beyond Pluto's orbit. Asteroids are small bodies found inside Jupiter's orbit around the Sun; most are located in the main asteroid belt between the orbits of Mars and Jupiter. Jupiter's gravity kept the region between Jupiter and Mars so stirred up that most planetesimals there never formed a large planet.

Planetesimals persist to this day in the outermost part of the Solar System as well. Formed in a deep freeze, those objects have retained most of the highly volatile materials found in the grains present when the accretion disk formed. Unlike the crowded inner part of the disk, the outermost parts had planetesimals too sparsely distributed for large planets to grow. Icy planetesimals in the outer Solar System that survived planetary accretion remain today as **comet nuclei**. The frozen, distant dwarf planets Pluto and Eris are especially large examples of those residents of the outer Solar System.

Many Solar System objects show evidence of cataclysmic impacts that reshaped worlds, suggesting that the early Solar System must have been a remarkably violent and chaotic place. The dramatic difference in the terrain of the northern and southern hemispheres on Mars, for example, has been interpreted as the result of one or more colossal collisions. The leading theory for the origin of our Moon is that it resulted from the collision of an object with Earth. Mercury has a crater on its surface from an impact so devastating that it caused the crust to buckle on the opposite side of the planet. In the outer Solar System, one of Saturn's moons, Mimas, has a crater roughly one-third the diameter of the moon itself. Uranus suffered one or more collisions violent enough to knock it on its side. As a result, its equatorial plane is tilted at almost a right angle to its orbital plane. Other examples are discussed in later chapters.

CHECK YOUR UNDERSTANDING 7.4

Suppose that astronomers found a rocky, terrestrial planet beyond the orbit of Neptune. What is the most likely explanation for its origin? (a) It formed close to the Sun and migrated outward. (b) It formed in that location and was not disturbed by migration. (c) It formed later in the Sun's history than other planets. (d) It is a captured planet that formed around another star.

7.5 Planetary Systems Are Common

When astronomers turn their telescopes to young nearby stars, they see disks of the same type from which the Solar System formed. When the light from the central star is blocked (**Figure 7.15**), evidence of the planetary disk is observed. The physical processes that led to the formation of the Solar System should be commonplace wherever new stars are being born. Compared with stars, however, planets are small and dim objects. They shine primarily by reflection and therefore are millions to billions of times fainter than their host stars. Thus, they were difficult to find until advances in telescope detector technology in the 1990s enabled astronomers to discover them through indirect methods. In 1995, astronomers announced the first confirmed **exoplanet**—a planet orbiting around a star other than the Sun.

what if . . .

What if astronomers had observed that all other systems were similar to our own, with rocky planets closer to the star and gas giants farther away? What would that tell us about star formation in general?

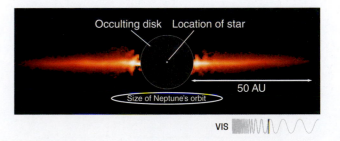

Figure 7.15 An edge-on dust disk around a star is seen extending outward to 60 AU from the young (12-million-year-old) star AU Microscopii. The star itself, whose brilliance would otherwise overpower the dust disk, is hidden behind an opaque mask (called an "occulting disk") placed in the telescope's focal plane. The star's position is represented by the dot.

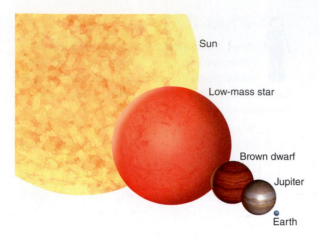

Figure 7.16 A comparison of the diameters of the Sun, a low-mass star, a brown dwarf, Jupiter, and Earth.

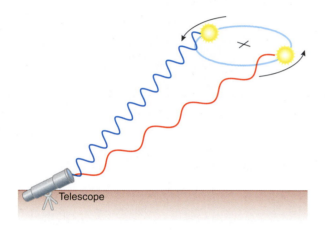

Figure 7.17 Doppler shifts observed in the spectrum of a star as it moves around its common center of gravity with its planet. The spectral lines are blueshifted as the star moves toward us and redshifted as it moves away from us.

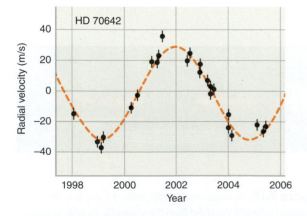

Figure 7.18 Radial velocity data for a star with a planet. A positive velocity is motion away from the observer, whereas a negative velocity is motion toward the observer. The plot repeats every 5.7 years as the planet completes another orbit around the star.

(These are sometimes called "extrasolar planets," although "exoplanet" is more accepted today.) Today, the number of known exoplanets has grown to the thousands, and new discoveries occur almost daily.

The International Astronomical Union (IAU) currently defines an exoplanet as an object that orbits a star other than the Sun and has a mass less than 10–13 Jupiter masses (10–13 M_{Jup}). Objects more massive than 10–13 M_{Jup} but less massive than 0.08 solar masses (0.08 M_{Sun}; about 80 M_{Jup}) are **brown dwarfs**. Objects more massive than 0.08 M_{Sun} are defined as stars. **Figure 7.16** compares the diameters of typical objects of each class.

The Search for Exoplanets

The first planets were discovered indirectly, by observing their gravitational tug on the central star. As technology has improved, other methods have become more productive. Astronomers now have direct images of planets orbiting stars and have taken spectra of some exoplanets to observe the composition of their atmospheres. Almost certainly, between the time we write this and the time you read it, new discoveries will have been made. The field is advancing extremely quickly: more than 100 projects are searching for exoplanets from the ground and from space. We will now look at each discovery method.

The Radial Velocity Method   As a planet orbits a star, the planet's gravity tugs the star around ever so slightly. If the star has a radial velocity (Chapter 5), an observable Doppler shift may appear in the spectrum of the star. **Figure 7.17** illustrates that motion. When the star is moving toward us (negative radial velocity), the light is blueshifted; when the star is moving away from us (positive radial velocity), the light is redshifted. That pattern of radial velocity repeats. After detecting those changes in the radial velocity (**Figure 7.18**), astronomers can infer the existence of the planet and find the planet's mass and distance from the star.

The smallest planet that can be found depends on the precision of the observations. For example, in our Solar System, Jupiter's mass is greater than the mass of all the other planets, asteroids, and comets combined, so Jupiter is the planet with the largest effect on the Sun. Both the Sun and Jupiter orbit a common center of gravity (sometimes called center of mass—the location where the effect of one mass balances the other) that lies just outside the surface of the Sun, as shown in **Figure 7.19**. Jupiter tugs the Sun around in a circle with a speed of 12 m/s. Alien astronomers would find that the Sun's radial velocity varies by ±12 m/s, with a period equal to Jupiter's orbital period of 11.86 years. From that information, the astronomers would rightly conclude that the Sun has at least one planet with a mass comparable to Jupiter's. Without greater precision, the observers would be unaware of the other, less massive planets. To detect Saturn, they would need to improve the precision of their measurements to 2.7 m/s. Earth would not be detectable unless the aliens could detect motions as small as 0.09 m/s.

The precision of radial velocity instruments has been about 0.3 m/s. The technique enabled astronomers to detect giant planets around solar-type stars, but not yet to find planets with masses similar to Earth's. Finding the signal of the Doppler shift in the noise of the observation requires the star to be quite bright in our sky. A new instrument at the Very Large Telescope is expected to improve precision and be able to detect smaller planets. **Working It Out 7.2** explains more about the spectroscopic radial velocity method.

working it out 7.2

Estimating the Size of a Planet's Orbit

In the spectroscopic radial velocity method, the star is moving about its center of mass, and its spectral lines are Doppler-shifted accordingly. Recall from Figure 7.19 that an alien astronomer looking toward the Solar System would observe a shift in the wavelengths of the Sun's spectral lines—caused by the presence of Jupiter—of ~12 m/s.

Figure 7.18 showed the radial velocity data for a star with a planet discovered with that method. How do astronomers use that method to estimate the distance (A) of the planet from the star? Recall from Chapter 4 that Newton generalized Kepler's law relating the period of an object's orbit to the orbital semimajor axis:

$$P^2 = \frac{4\pi^2}{G} \times \frac{A^3}{M}$$

where A is the semimajor axis of the orbit, P is its period, and M is the combined mass of the two objects. To find A, we rearrange the equation as follows:

$$A^3 = \frac{G}{4\pi^2} \times M \times P^2$$

According to the graph of radial velocity observations in Figure 7.18, the period of the orbit is 5.7 years. A year has 3.16×10^7 seconds, so

$$P = 5.7 \text{ yr} \times (3.16 \times 10^7 \text{ s/yr}) = 1.8 \times 10^8 \text{ s}$$

The mass of the star is much greater than the mass of the planet, so the combined masses of the star and the planet can be approximated as the mass of the star, which here is about equal to the mass of the Sun, 2×10^{30} kg. (Stellar masses can be estimated from their spectra.) The gravitational constant G is 6.67×10^{-20} km³/(kg s²). Plugging in the numbers gives

$$A^3 = \frac{6.67 \times 10^{-20} \dfrac{\text{km}^3}{\text{kg s}^2}}{4\pi^2} \times (2 \times 10^{30} \text{ kg}) \times (1.8 \times 10^8 \text{ s})^2$$

$$A^3 = 1.1 \times 10^{26} \text{ km}^3$$

Taking the cube root,

$$A = 4.8 \times 10^8 \text{ km}$$

If we convert that number into astronomical units (where 1 AU = 1.5 $\times$ 10⁸ km), the semimajor axis of the orbit of this planet is

$$A = \frac{4.8 \times 10^8 \text{ km}}{1.5 \times 10^8 \text{ km/AU}} = 3.2 \text{ AU}$$

The planet is 3.2 times farther from its star than Earth is from the Sun.

The Transit Method From Earth it is sometimes possible to see the inner planets Mercury and Venus transit, or pass in front of, the Sun. An alien located somewhere in the plane of Earth's orbit would see Earth pass in front of the Sun and could infer the existence of Earth by detecting the 0.009 percent drop in the Sun's brightness during the transit. Similarly, for astronomers on Earth to observe a planet passing in front of a star, Earth must lie nearly in the orbital

Astronomy in Action: Doppler Shift

Interactive Simulation: Radial Velocity

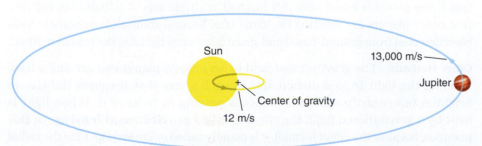

Figure 7.19 The Sun and Jupiter orbit around a common center of gravity (+), which lies just outside the Sun's surface. Spectroscopic measurements made by an extrasolar astronomer would reveal the Sun's radial velocity varying by ±12 m/s over 11.86 years, which is Jupiter's orbital period. Jupiter travels around its orbit at a speed of 13,000 m/s.

working it out 7.3

Estimating the Radius of an Exoplanet

The masses of exoplanets can often be estimated using Kepler's laws and the conservation of angular momentum. When planets are detected with the transit method, astronomers can estimate the radius of an exoplanet. In that method, astronomers look for planets that eclipse their stars and then observe how much the star's light decreases during that eclipse (see Figure 7.20). When Venus or Mercury transits the Sun, a black circular disk is visible on the face of the circular Sun. During the transit, the amount of light from the transited star is reduced by the area of the circular disk of the planet divided by the area of the circular disk of the star:

$$\text{fractional reduction in light} = \frac{\text{Area of disk of planet}}{\text{Area of disk of star}} = \frac{\pi R_{\text{planet}}^2}{\pi R_{\text{star}}^2} = \frac{R_{\text{planet}}^2}{R_{\text{star}}^2}$$

Then, to solve for the radius of the planet, astronomers need an estimate of the radius of the star and a measurement of the fractional reduction in light during the transit. The radius of a star is estimated from the surface temperature and the luminosity of the star.

Kepler-11, for example, is a system of at least six planets that transit a star. The radius of the star, R_{star}, is estimated to be 1.1 times the radius of the Sun, or $1.1 \times (7.0 \times 10^5 \text{ km}) = 7.7 \times 10^5$ km. The light from the Kepler-11 star is observed to decrease by 0.077 percent, or 0.00077 (see Figure 7.20), from planet Kepler-11c. What is the radius of Kepler-11c?

$$0.00077 = \frac{R_{\text{Kepler-11c}}^2}{R_{\text{star}}^2} = \frac{R_{\text{Kepler-11c}}^2}{(7.7 \times 10^5 \text{ km})^2}$$

$$R_{\text{Kepler-11c}}^2 = 4.6 \times 10^8 \text{ km}^2 \quad \text{thus} \quad R_{\text{Kepler-11c}} = 2.1 \times 10^4 \text{ km}$$

Dividing $R_{\text{Kepler-11c}}$ by R_{Earth} (6400 km) shows that $R_{\text{Kepler-11c}} = 3.3 \, R_{\text{Earth}}$.

---

plane of that planet. When an exoplanet passes in front of its parent star, the light from the star diminishes by a tiny amount (**Figure 7.20**). This is the **transit method** of detecting exoplanets, in which we observe the dimming of starlight as a planet passes in front of its parent star. Whereas the radial velocity method gives us the mass of the planet and its orbital distance from a star, the transit method provides the radius of a planet. **Working It Out 7.3** shows how the radii are estimated.

Current ground-based technology limits the sensitivity of the transit method to about 0.1 percent of a star's brightness. Telescopes in space improve the sensitivity because smaller dips in brightness can be measured. The small French COROT telescope (27 cm) discovered 32 planets during its 6 years of operation (2007–2013). NASA's 0.95-meter Kepler telescope has discovered many planets and has found thousands more candidates that are being investigated further. **Figure 7.21** shows that if one planet is found with this method, multiple sets of transits can indicate that other planets are orbiting the same star. Several thousand exoplanets have been detected from ground-based and space telescopes by using the transit method.

Other Methods   The gravitational field of an unseen planet can act like a lens, bending the light from a distant star in such a way that it causes the star to brighten temporarily while the planet is passing in front of it. When light is bent by a gravitational field, the effect is called **gravitational lensing**. In this instance, because the effect is small, it is usually called *microlensing*. Like the radial velocity method, microlensing provides an estimate of the mass of the planet. To date, about 60 exoplanets have been found with this technique.

Planets also may be detected by *astrometry*—precisely measuring the position of a star in the sky. If the system is viewed from "above," the star moves in a mini-orbit as the planet pulls it around. That motion is generally tiny and therefore very

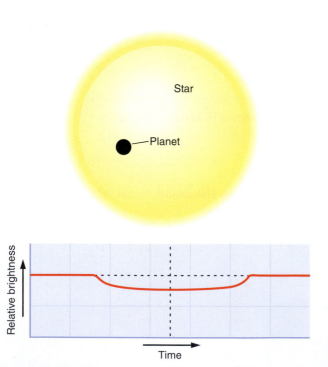

Figure 7.20 A planet that passes in front of a star blocks some of the light coming from the star's surface, causing the brightness of the star to decrease slightly. (The decrease in brightness is exaggerated here.)

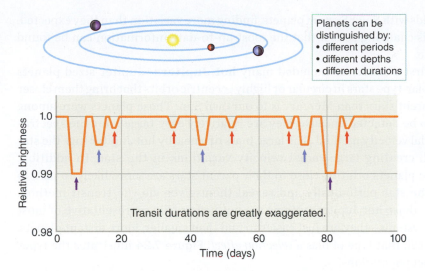

Planets can be distinguished by:
• different periods
• different depths
• different durations

Transit durations are greatly exaggerated.

Figure 7.21 Multiple planets can be detected through multiple transits with different changes in brightness. The colored arrows point to the changes in the total light as the three planets (red, blue, and purple) transit the star.

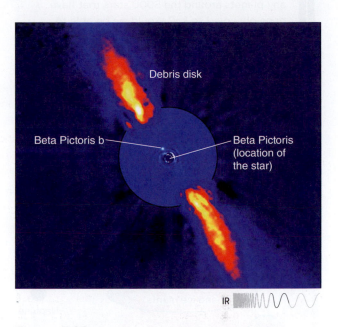

Figure 7.22 The planet Beta Pictoris b is seen orbiting within a dusty debris disk that surrounds the bright naked-eye star Beta Pictoris. The planet's estimated mass is 8 times that of Jupiter. The star is hidden behind an opaque mask, and the planet appears through a semitransparent mask used to subdue the brightness of the dusty disk.
Credit: ESO/A.-M. Lagrange et al., https://www.eso.org/public/images /eso0842b/. https://creativecommons.org/licenses/by/4.0/.

difficult to measure. For systems viewed from above the plane of the planet's orbit, however, none of the prior methods will work because the planet neither passes in front of the star nor causes a shift in its speed along the line of sight. Space missions such as the Gaia observatory, launched in 2013 by the European Space Agency, conduct such observations.

Direct imaging involves taking a picture of the planet directly. The technique is conceptually straightforward but is technically difficult because it involves searching for a relatively faint planet in the overpowering glare of a bright star—a challenge far more difficult than looking for a star in a clear, bright daytime sky. Even when an object is detected by direct imaging, an astronomer must still determine whether the observed object is actually a planet. Suppose we detect a faint object near a bright star. Could it be a more distant star that just happens to be in the line of sight? Future observations could tell whether the object shares the bright star's motion through space, but it also could be a brown dwarf rather than a true planet. An astronomer would need to make further observations to determine the object's mass.

Some planets have been discovered through that method with large ground-based telescopes operating in the infrared region of the spectrum, using adaptive optics. **Figure 7.22** is an infrared image of Beta Pictoris b. Hubble Space Telescope observations of Fomalhaut, a bright naked-eye star only 25 light-years away, revealed a 3 Jupiter–mass planet in the dusty debris ring about 17 billion km from the central star (**Figure 7.23**). A related form of direct observation involves separating the spectrum of a planet from the spectrum of its star to obtain information about the planet directly. Large ground-based telescopes have obtained spectra of the atmospheres of some exoplanets and have found, for example, carbon monoxide and water in those atmospheres.

Types of Exoplanets

Searches for exoplanets have been remarkably successful. Between the discovery of the first (in 1995) and this writing, more than 4000 more have been confirmed, and thousands more candidates are under investigation. As the number of observed systems with single and multiple planets increases, astronomers can compare

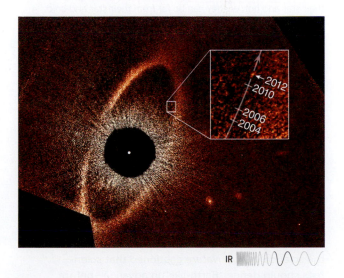

Figure 7.23 A Hubble Space Telescope image of Fomalhaut b, seen here in its 2012 position in the orbit. Prior positions are indicated by the tic marks on the white line overlaid on the orbit. The parent star, hidden by an obscuring mask, is about a billion times brighter than the planet.

what if . . .

What if astronomers had not yet discovered any planets around the 5000 stars that have been observed so far? Would we be able to conclude (with current technology) that there are no planets around the 100 billion stars in the Milky Way? Why or why not?

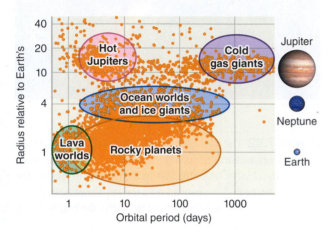

Figure 7.24 Different populations of exoplanets. (Although Kepler uses the transit method, its discoveries are shown separately, in orange.) Jupiter-sized planets can be close to their star and hot, or they can be far from their star and cold. Neptune-sized planets are probably composed of water and ice. Earth-sized planets have higher density, indicating that they are made of rock.

unanswered questions

How Earth-like must a planet be before scientists declare it to be "another Earth"? An editorial in the science journal *Nature* cautioned that scientists should define "Earth-like" in advance—before multiple discoveries of planets "similar" to Earth are announced and a media frenzy ensues. Must a planet be of similar size and mass, be located in the habitable zone, and have spectroscopic evidence of liquid water before we call it "Earth 2.0"?

those worlds with Solar System planets, finding more variation than they expected. The field is changing so fast that the most up-to-date information can be found only online.

The first discoveries included many **hot Jupiters**—Jupiter-sized planets orbiting solar-type stars in circular or highly eccentric orbits that bring them closer to their parent stars than Mercury is to our own Sun. Those planets were among the first to be detected because they are relatively easy targets for the spectroscopic radial velocity method. The large mass of a nearby hot Jupiter tugs the star very hard, creating large radial velocity variations in the star. In addition, those large planets orbiting close to their parent stars are more likely to pass in front of the star periodically and reveal themselves via the transit method. Therefore, those hot Jupiter systems are not necessarily representative of most planetary systems—they are just easier to find than smaller, more distant planets. Scientists call that type of bias a *selection effect*. **Figure 7.24** illustrates the types of exoplanet populations.

Astronomers were surprised by the hot Jupiters because, according to the theory of planet formation described earlier in the chapter, those giant, volatile-rich planets should not have been able to form so close to their parent stars. From theories based on the Solar System, astronomers expected that Jupiter-type planets should form in the more distant, cooler regions of the protoplanetary disk, where the volatiles that make up much of their composition can survive. Hot Jupiters may form much farther from their parent stars and later migrate inward to a closer orbit. That migration may be caused by an interaction with gas or planetesimals in which orbital angular momentum is transferred from the planet to its surroundings, allowing it to spiral inward.

The radial velocity method yields an estimate of the mass of a planet (see Working It Out 7.2), and the transit method yields an estimate of the size of a planet (see Working It Out 7.3). With both pieces of information, the density (mass divided by the volume) of the planet can be computed to determine whether the planet is mostly gaseous (low density) or mostly rock (high density). Many of the new planets discovered by Kepler are mini-Neptunes (gaseous planets with masses of 2–10 M_{Earth}) or **super-Earths** (rocky planets more massive than Earth). **Figure 7.25** illustrates the formation of those two types.

Planets with longer orbital periods, and therefore larger orbits, can be discovered only when the observations have gone on long enough to observe more than one complete orbit. Some of the exoplanets have highly elliptical orbits compared with those in the Solar System. Planets have been found with orbits that are highly tilted with respect to the plane of the rotation of their star, and some planets move in orbits whose direction is opposite that of their star's rotation. Multiple-planet systems have been observed in which the larger mini-Neptunes alternate with smaller super-Earths. The multiple-planet systems that have been found with the transit method reside in flat systems like our own, offering further evidence that the planets formed in a flat protoplanetary disk around a young star. But the current hypothesis to explain the Solar System's inner, small rocky planets and outer, large gaseous planets may not apply to all other planetary systems.

In addition, some planets, discovered through microlensing, don't have a star at all. Those planets may have been ejected from their solar systems after they formed and are no longer in gravitationally bound orbits around their stars. Others are "circumbinary"; they orbit two stars that form a binary star system. A scroll through NASA's exoplanet archive reveals that new planet discoveries are confirmed

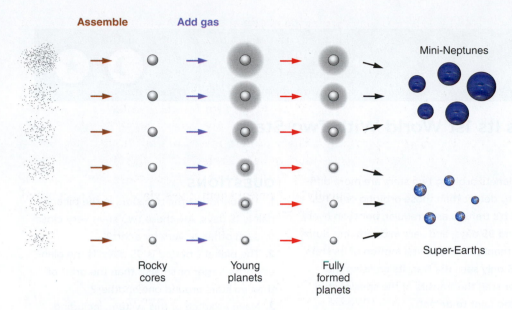

Assemble   Add gas

Mini-Neptunes

Super-Earths

Rocky cores   Young planets   Fully formed planets

Figure 7.25 Many exoplanets are between the sizes of Earth and Neptune (~$4R_E$). "Super-Earths" began with smaller rocky cores and less material, whereas "mini-Neptunes" had larger cores and collected more gas while forming.

nearly every week. The frequent new discoveries requiring revisions of existing theories make exoplanets one of the most exciting topics in astronomy today.

We return to exoplanets at several points in this book—namely, when we review the planets in our own Solar System, when we consider the types of stars, and when we discuss the search for Earth-like planets in Chapter 24.

CHECK YOUR UNDERSTANDING **7.5**

What is the most common method for discovering an Earth-mass planet around a star? (a) Doppler spectroscopy; (b) direct imaging; (c) transit; (d) astrometric

Origins: The Search for Earth-Sized Planets

The discovery of planetary systems, many different from the Solar System, shows us that the formation of planets often, and perhaps always, accompanies the formation of stars. The implications of that conclusion are profound. Planets are a common by-product of star formation. In a galaxy of 200 billion stars and a universe of hundreds of billions of galaxies, how many planets (and moons) might exist? And with all those planets in the universe, how many might have conditions suitable for the particular category of chemical reactions that we refer to as "life"? (We return to that point in Chapter 24.)

The Kepler Mission was developed by NASA to find Earth-sized and larger planets in orbit about a variety of stars. Kepler was a 0.95-meter telescope with 42 CCD detectors that observed approximately 150,000 stars in 100 square degrees of sky to look for planetary transits. To confirm a planetary detection, transits needed to be observed three times with repeatable changes in brightness, duration of transit times, and computed orbital period. Kepler could detect a dip of 0.01 percent in the brightness of a star—sensitive enough to detect an

Reading Astronomy News

NASA's TESS Mission Uncovers Its 1st World With Two Stars

NASA

Discovering more planets means that we discover more unusual ones; like this planet that orbits two stars at once

The TOI 1338 system lies 1,300 light-years away in the constellation Pictor. The two stars orbit each other every 15 days. One is about 10% more massive than our Sun, while the other is cooler, dimmer and only one-third the Sun's mass.

Newly discovered TOI 1338 b is the only known planet in this binary star system. It's around 6.9 times larger than Earth, or between the sizes of Neptune and Saturn. The planet orbits in almost exactly the same plane as the stars, so it experiences regular stellar eclipses.

Scientists use the observations from TESS to generate graphs of how the brightness of stars change over time. When a planet transits in front of its star from our perspective, its passage causes a dip in the star's brightness.

Planets orbiting two stars are more difficult to detect than those orbiting one. TOI 1338 b's transits are irregular, between every 93 and 95 days, and vary in depth and duration thanks to the orbital motion of its stars. TESS only sees the transits crossing the larger star; the transits of the smaller star are too faint to detect.

"These are the types of signals that algorithms really struggle with," said lead author Veselin Kostov, a research scientist at the SETI Institute and Goddard. "The human eye is extremely good at finding patterns in data, especially non-periodic patterns like those we see in transits from these systems."

After identifying TOI 1338 b, the research team used a software package called eleanor, named after Eleanor Arroway, the central character in Carl Sagan's novel "Contact," to confirm the transits were real.

TOI 1338 had already been studied from the ground by radial velocity surveys. Kostov's team used this archival data to analyze the system and confirm the planet.

QUESTIONS

1. The period of the two stars in the binary pair is 15 days. Are these two stars very close to each other, or very far apart?

2. The planet's period is 95 days. Is the planet's orbit larger or smaller than the orbit of the two stars around one another?

3. Make a sketch of this system, including the two stars and the planet. Draw circles to show how the orbits are related, and whether or not they cross.

4. The article states that the discovery was confirmed using archival data from ground-based telescopes. Speculate: Why was the planet not discovered prior to this, using that ground-based data?

5. This system provides a window into the naming conventions for stars and planets. Study the names of the stars and planet. How are the two stars of a binary star system designated? How is the first planet in a planetary system designated?

Source: https://www.nasa.gov/feature/goddard /2020/nasa-s-tess-mission-uncovers-its-1st-world -with-two-stars.

Earth-sized planet. Kepler identified the first Earth-sized planets in 2011. Stars with transiting planets detected by Kepler also are observed spectroscopically to obtain radial velocity measurements that can lead to an estimate of a planet's mass. If a planet's radius and mass are known, the planet's density (mass per volume) can be estimated, too. From the density, astronomers can get a sense of whether the planet is composed primarily of gas, rock, ice, water, or some mixture of those.

On Earth, liquid water was essential for life to form and evolve. Because life on Earth is the only example of life for which we have evidence, we do not know whether liquid water is a cosmic requirement, but it is a place to start. The Kepler Mission searched for rocky planets at the right distance from

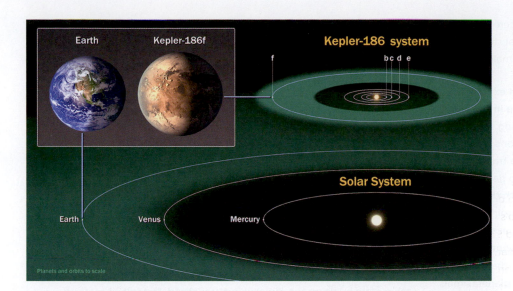

their stars to permit the existence of liquid water, a distance known as the **habitable zone**. The location and width of the zone varies depending on the temperature of the star. If a planet is too close to its star, water will exist only as a vapor; if too far, water will be frozen as ice. In the Solar System, Earth is the only planet currently in the habitable zone. Although announcements of new planets often state whether the planet is in the habitable zone, just being in the zone doesn't guarantee that the planet actually has liquid water—or that the planet is inhabited! An example of an Earth-sized planet in a habitable zone is shown in **Figure 7.26**.

Kepler identified thousands of planet candidates, some in the habitable zones of their respective stars. The candidates are confirmed by follow-up observations of more transits or of radial velocities before they are officially announced as planet detections. Citizen scientists contribute to that search through the Exoplanet Explorers project.

The Transiting Exoplanet Surveying Satellite (TESS) mission, launched in 2018, monitors 200,000 nearby stars for dimming caused by planetary transits. The mission is beginning to yield thousands more planetary candidates, with subsequent confirmations taking substantially longer. The James Webb Space Telescope (JWST), scheduled for launch in 2021, will observe in the infrared and should be able to detect gases in the atmospheres of exoplanets.

SUMMARY

Stars and their planetary systems form from collapsing interstellar clouds of gas and dust, following the laws of gravity and conservation of angular momentum. Conservation of angular momentum produces an accretion disk around a protostar that often fragments to form multiple planets, as well as smaller objects such as asteroids and dwarf planets, through the gradual accumulation of material into larger and larger objects. Multiple methods exist for finding planets around other stars, and such planets are now thought to be very common. This field of study is evolving very quickly due to advances in technology.

① **Describe how our understanding of planetary system formation developed from the work of both planetary and stellar scientists.**
Planetary scientists observed the motions of planets in our Solar System, as well as compositions of planets and meteorites. Stellar scientists observed the life cycle of stars and the nebula that precede and follow the time when a star fuses hydrogen. From these observations, the two groups developed the story of planetary system formation. Planets are a common by-product of star formation, and many stars are surrounded by planetary systems. Gravity pulls clumps of gas and dust together, causing them to shrink and heat up. Angular momentum must be

conserved, leading to both a spinning central star and an accretion disk that rotates and revolves in the same direction as the central star. Solar System meteorites show that larger objects build up from smaller objects.

(2) **Discuss the role of gravity and angular momentum in explaining why planets orbit the Sun in a plane and why they revolve in the same direction that the Sun rotates.** As particles orbit the forming star, gravity pulls the dust and gas inward toward the center. Collisions between particles that are traveling upward in their orbit with those traveling downward in their orbit cause the cloud of dust and gas to flatten into a plane. Conservation of angular momentum determines both the speed and the direction of the revolution of the objects in the forming system. Dust grains in the protoplanetary disk first stick together because of collisions and static electricity. As those objects grow, they eventually have enough mass to attract other objects gravitationally. Once that occurs, they begin emptying the space around them. Collisions of planetesimals lead to the formation of planets.

(3) **Explain how temperature at different locations in the protoplanetary disk affects the composition of planets, moons, and other bodies.** The temperature is higher near the central protostar, forcing volatile elements, such as water, to evaporate and leave the inner part of the disk. Planets in the inner part of the disk will have fewer volatiles than those in the outer part of the disk. The gas that a planet captures when it forms is the planet's primary atmosphere. Less massive planets lose their primary atmospheres and then form secondary atmospheres.

(4) **Discuss the processes that resulted in the formation of planets and other objects in our Solar System.** In the current model of the formation of the Solar System, solid terrestrial planets formed in the inner disk, where temperatures were high, whereas giant gaseous planets formed in the outer disk, where temperatures were low. Dwarf planets such as Pluto formed in the asteroid belt and in the region beyond the orbit of Neptune. Asteroids and comet nuclei remain today as leftover debris.

(5) **Describe how astronomers both find planets around other stars and determine exoplanet properties.** Astronomers find planets around other stars by using the radial velocity method, the transit method, microlensing, astrometry, and direct imaging. As technology has improved, the number and variety of known exoplanets has increased dramatically, with thousands of planets and planet candidates discovered orbiting other stars near the Sun within the Milky Way Galaxy in just the past few years.

QUESTIONS AND PROBLEMS

TEST YOUR UNDERSTANDING

1. Place the following events in the order that corresponds to the formation of a planetary system.
   a. Gravity collapses a cloud of interstellar gas.
   b. A rotating disk forms.
   c. Small bodies collide to form larger bodies.
   d. A stellar wind "turns on" and sweeps away gas and dust.
   e. Primary atmospheres form.
   f. Primary atmospheres are lost.
   g. Secondary atmospheres form.
   h. Dust grains stick together by static electricity.

2. If the radius of an object's orbit is halved, and angular momentum is conserved, what must happen to the object's speed?
   a. It must be halved.   c. It must be doubled.
   b. It must stay the same.   d. It must be squared.

3. Unlike the giant planets, the terrestrial planets formed when
   a. the inner Solar System was richer in heavy elements than the outer Solar System.
   b. the inner Solar System was hotter than the outer Solar System.
   c. the outer Solar System took up more volume than the inner Solar System, so more material was available to form planets.
   d. the inner Solar System was moving faster than the outer Solar System.

4. The terrestrial planets and the giant planets have different compositions because
   a. the giant planets are much larger.
   b. the terrestrial planets formed closer to the Sun.
   c. the giant planets are made mostly of solids.
   d. the terrestrial planets have few moons.

5. The spectroscopic radial velocity method preferentially detects
   a. large planets close to the central star.
   b. small planets close to the central star.
   c. large planets far from the central star.
   d. small planets far from the central star.
   e. The method detects all those planets equally well.

6. The concept of disk instability was developed to solve the problem that
   a. Jupiter-like planets migrate after formation.
   b. not enough gas was in the Solar System to form Jupiter.
   c. the early solar nebula probably dispersed too soon to form Jupiter.
   d. Jupiter consists mostly of volatiles.

7. Because angular momentum is conserved, an ice-skater who throws her arms out will
   a. rotate more slowly.   c. rotate at the same rate.
   b. rotate more quickly.   d. stop rotating.

8. Clumps grow into planetesimals by
   a. gravitationally pulling in other clumps.
   b. colliding with other clumps.
   c. attracting other clumps with opposite charge.
   d. conserving angular momentum.

9. The transit method preferentially detects
   a. large planets close to the central star.
   b. small planets close to the central star.
   c. large planets far from the central star.
   d. small planets far from the central star.
   e. The method detects all those planets equally well.

10. If the radius of a spherical object is halved, what must happen to the period so that the spin angular momentum is conserved?
    a. It must be divided by 4.
    d. It must double.
    b. It must be halved.
    e. It must be multiplied by 4.
    c. It must stay the same.

11. Which of the following affects the angular momentum of a spherical object? (Choose all that apply.)
    a. radius
    c. rotation speed
    b. mass
    d. temperature

12. The planets in the inner part of the Solar System are made primarily of refractory materials; the planets in the outer Solar System are made primarily of volatiles. That difference occurs because
    a. refractory materials are heavier than volatiles, so they sank farther into the nebula.
    b. no volatiles were in the inner part of the accretion disk.
    c. the volatiles on the inner planets were lost soon after the planets formed.
    d. the outer Solar System has gained more volatiles from space since formation.

13. If scientists want to find out about the composition of the early Solar System, the best objects to study are
    a. the terrestrial planets.
    b. the giant planets.
    c. the Sun.
    d. asteroids and comets.

14. The direction of revolution in the plane of the Solar System was determined by
    a. the plane of the galaxy in which the Solar System sits.
    b. the direction of the gravitational force within the original cloud.
    c. the direction of rotation of the original cloud.
    d. the amount of material in the original cloud.

15. A planet in the "habitable zone"
    a. is close to the central star.
    b. is far from the central star.
    c. is the same distance from its star as Earth is from the Sun.
    d. is at a distance where liquid water can exist on the surface.

THINKING ABOUT THE CONCEPTS

16. ★ **WHAT AN ASTRONOMER SEES**   In Figure 7.8, identify an unlabeled finger, jet, and diffraction spike. Make a sketch of the image, and label the locations of the features you have identified. 👁

17. What is the source of the material that now makes up the Sun and the rest of the Solar System?

18. Describe the different ways by which stellar astronomers and planetary scientists each came to the same conclusion about how planetary systems form.

19. What is a protoplanetary disk? What are two reasons why the inner part of the disk is hotter than the outer part?

20. Physicists describe certain properties, such as angular momentum and energy, as being *conserved*. Does this property mean conservation laws imply that an individual object can never lose or gain angular momentum or energy? Explain your reasoning.

21. The Process of Science Figure in this chapter makes the point that different areas of science must agree with one another. Suppose that a few new exoplanets are discovered that appear not to have formed from the collapse of a stellar nebula (for example, the planetary orbits might be in random orientations). What will scientists do with that new information? 👁

22. How does the law of conservation of angular momentum control a figure-skater's rate of spin?

23. What is an accretion disk?

24. Look under your bed, the refrigerator, or any similar place for dust bunnies. Once you find them, blow one toward another. Watch carefully and describe what happens as they meet. What happens if you repeat that action with additional dust bunnies? Will those dust bunnies ever have enough gravity to begin pulling themselves together? If they were in space instead of on the floor, might that happen? What force prevents their mutual gravity from drawing them together into a "bunny-tesimal" under your bed?

25. Why do we find rocky material everywhere in the Solar System but find large amounts of volatile material only in the outer regions?

26. Why could the four giant planets collect massive gaseous atmospheres, whereas the terrestrial planets could not? Explain the source of the secondary atmospheres surrounding the terrestrial planets.

27. Describe four methods that astronomers use to search for exoplanets. What are the limitations of each method; that is, what circumstances are necessary to detect a planet by each method?

28. Why is it difficult to obtain an image of an exoplanet?

29. Many of the first exoplanets that astronomers found orbiting other stars were giant planets with Jupiter-like masses and with orbits located very close to their parent stars. Explain why those characteristics are a selection effect of the discovery method.

30. How has the Kepler telescope found Earth-like planets, and what do astronomers mean by "Earth-like"?

APPLYING THE CONCEPTS

31. Compare Earth's orbital angular momentum with its spin angular momentum by using the following values: $m = 5.97 \times 10^{24}$ kg, $v = 29.8$ kilometers per second (km/s), $r = 1$ AU, $R = 6378$ km, and $P = 1$ day. Assume that Earth is a uniform body. What fraction does each component (orbital and spin) contribute to Earth's total angular momentum? Refer to Working It Out 7.1. ●–●–●
    a. Make a prediction: Study Working It Out 7.1. Do you expect Earth to have more angular momentum in its spin, or in its orbit?
    b. Calculate: Follow Working It Out 7.1 to calculate Earth's orbital angular momentum and its spin angular momentum. Add these together to find the total angular momentum. Then divide each component by the total to find the fraction of the angular momentum that is accounted for by that component.
    c. Check your work: Verify in both cases that you obtain the correct units for angular momentum (kg m²/s), and compare your results to your prediction.

32. A planet has been found to orbit a 1-M_{Sun} star in 200 days. What is the semimajor axis of this exoplanet's orbit? Compare the semimajor axis of this exoplanet with that of the planets around our own Sun. What temperatures must that planet experience? ●–■–●
   a. Make a prediction: Is 200 days longer or shorter than Earth's orbital period? Therefore, do you expect the semimajor axis to be larger or smaller than 1 AU?
   b. Calculate: Follow Working It Out 7.2 to find the semimajor axis of this exoplanet.
   c. Check your work: Verify that your semimajor axis has correct units, and compare your result to your prediction.
   d. Evaluate: Given the semimajor axis of this planet's orbit, what do you expect the temperature to be like on the surface?

33. The star Kepler-11 has an orbital radius of 1.1 R_{Sun}. If the planet has a radius of 4.5 R_{Earth}, by what percentage does the brightness of Kepler-11 decrease when that planet transits the star? ●–■–●
   a. Make a prediction: Study Working It Out 7.3. Do you expect a large or a small drop in brightness of the star when the planet transits?
   b. Calculate: Follow Working It Out 7.3 to calculate the percent reduction in brightness of Kepler-11.
   c. Check your work: Verify that all the units cancel out in your percentage, and compare your result to your prediction.

34. Jupiter has a mass equal to 318 times Earth's mass, an orbital radius of 5.2 AU, and an orbital velocity of 13.1 km/s. Earth's orbital velocity is 29.8 km/s. What is the ratio of Jupiter's orbital angular momentum to that of Earth?
   a. Make a prediction: Consider the equation for orbital angular momentum. Do you expect Jupiter to have more or less orbital angular momentum than Earth? Do you expect the ratio of Jupiter's orbital angular momentum to Earth's orbital angular momentum to be greater or smaller than one?
   b. Calculate: Set up the ratio of orbital angular momenta, using letters first. Divide out terms that appear in top and bottom. (Hint: Given the problem statement, does Earth's mass appear in both the top and bottom?) Then use your calculator to solve for the value of the ratio.
   c. Check your work: Verify that the final ratio is dimensionless (all the units should have divided out). Compare your result with your prediction.

35. The asteroid Vesta has a diameter of 530 km and a mass of 2.7×10^{20} kg. Calculate the density (mass/volume) of Vesta.
   a. Make a prediction: The density of water is 1000 kg/m3, whereas that of rock is about 2500 kg/m3. Do you expect the density of Vesta to be similar to the density of rock, or similar to the density of water?
   b. Calculate: Use the formula for density (mass/volume) to find the density of Vesta. You may need to recall that the volume of a sphere is (4/3)ϖR3.
   c. Check your work: Verify that your answer has correct units of density, and compare it to the densities of water and rock.

36. In Figure 7.18, what is the maximum radial velocity of HD 70642 in meters per second? Convert that number to miles per hour (mph). How does that value compare with the speed at which Earth orbits the Sun (67,000 mph)? ✦

37. Use Appendix 4 to answer the following:
   a. What is the total mass of all the planets in the Solar System, expressed in Earth masses (M_{Earth})?
   b. What fraction of that total planetary mass is Jupiter?
   c. What fraction does Earth represent?

38. Venus has a radius 0.950 times that of Earth and a mass 0.815 times that of Earth. Venus's rotation period is 243 days. What is the ratio of Venus's spin angular momentum to that of Earth? Assume that Venus and Earth are uniform spheres.

39. Suppose you can measure radial velocities of about 0.3 m/s. Suppose, too, you are observing a spectral line with a wavelength of 575 nanometers (nm). How large a shift in wavelength would a radial velocity of 0.3 m/s produce?

40. Using data in Appendix 4, compute the densities of Venus, Jupiter, and Neptune. Compare your answers with the densities of rock, water, and gas.

41. Recalling Kepler's laws, put the three planets in Figure 7.21 in order from fastest to slowest. Compare the duration of the transits in Figure 7.21. Why does the outermost planet have the longest duration? ✦

42. Earth tugs the Sun around as it orbits, but that effect (only 0.09 m/s) is much smaller than that of any known exoplanet. How large a shift in wavelength does that effect cause in the Sun's spectrum at 500 nm?

43. If an alien astronomer observed a plot of the light curve as Jupiter passed in front of the Sun, by how much would the Sun's brightness drop during the transit?

44. Kepler detected a planet with a diameter of 1.7 Earth (D_{Earth}).
   a. How much larger is the volume of that planet than Earth's?
   b. Assume that the density of the planet is the same as Earth's. How much more massive is it than Earth?

45. The planet COROT-11b was discovered using the transit method, and astronomers have followed up with radial velocity measurements, so both its radius (1.43 R_{Jup}) and its mass (2.33 M_{Jup}) are known. The density provides a clue about whether the object is gaseous or rocky.
   a. What is the mass of the planet in kilograms?
   b. What is the planet's radius in meters?
   c. What is the planet's volume?
   d. What is the planet's density? How does that density compare with the density of water (1000 kg/m³)? Is the planet likely to be rocky or gaseous?

EXPLORATION Exploring Exoplanets

digital.wwnorton.com/astro7

Visit the Student Site at the Digital Resources page and open the Radial Velocity Simulation in Chapter 7. This applet has a number of different panels that allow you to experiment with the variables that are important for measuring radial velocities. Compare the views shown in the various panels with the colored arrows in the first panel to see where an observer would stand to see the view shown. Start the animation (press "Play") and allow it to run while you watch the planet orbit its star from each of the views shown. Stop the animation, and in the "Sample Systems" panel select "Option A."

1. Is Earth's view of this system most nearly like the "side view" or most nearly like the "orbit view"?

2. Is the orbit of this planet circular or elongated?

3. Study the radial velocity graph in the upper right panel. The blue curve shows the radial velocity of the star over a full period. What is the maximum radial velocity of the star?

4. The horizontal axis of the graph shows the "phase," or fraction of the period. A phase of 0.5 is halfway through a period. The vertical red line indicates the phase shown in views in the upper left panel. Start the animation to see how the red line sweeps across the graph as the planet orbits the star. The period of this planet is 365 days. How many days pass between the minimum radial velocity and the maximum radial velocity?

5. When the planet moves away from Earth, the star moves toward Earth. The sign of the radial velocity tells the direction of the motion (toward or away). Is the radial velocity of the star positive or negative at this time in the orbit? If you could graph the radial velocity of the planet at this point in the orbit, would it be positive or negative?

In the "Sample Systems" window, select "Option B:"

6. What has changed about the orbit of the planet as shown in the views in the upper left panel?

7. When is the planet moving fastest—when it is close to the star or when it is far from the star?

8. When is the star moving fastest—when the planet is close to it or when it is far away?

9. Explain how an astronomer would determine, from a radial velocity graph of the star's motion, whether the orbit of the planet was in a circular or elongated orbit.

10. Study the "Earth View" panel. Would this planet be a good candidate for a transit observation? Why or why not?

The Terrestrial Planets and Earth's Moon

The Moon has many visible surface features that can be used to reveal its history. Use the Internet to find out the date of the next full Moon. Go outside on a clear night within 3 days of the full Moon (either before or after) and sketch the face of the Moon. You may want to look at the Moon through a cardboard tube, such as a toilet paper roll or paper towel roll, or a rolled-up piece of paper. The tube will block some of the ambient light and focus your attention on the surface of the Moon itself, thus helping you to see more detail. Use a map of the Moon to identify at least two maria and two craters and label them on your sketch.

EXPERIMENT SETUP

Use the internet to find out the date of the next full Moon.

Go outside on a clear night within 3 days of the full Moon (either before or after), and sketch the face of the Moon.

You may want to look at the Moon through a cardboard tube or a rolled up piece of paper to block some of the ambient light and focus your attention on the surface of the Moon itself, thus helping you see more detail.

Use a map of the Moon to identify at least two maria and two craters on the surface of the Moon. Label them on your sketch.

Mare Frigoris
Sinus Iridum
Mare Imbrium
Aristarchus
Oceanus Procellarum
Mare Serenitatis
Posidonius
Mare Crisium
Eratosthenes
Copernicus
Mare Tranquillitatis
Mare Fecunditatis
Ptolemaeus
Mare Nectaris
Mare Humorum
Mare Nubium
Schickard
Tycho

SKETCH OF RESULTS
(in progress)

Mare Crisium

Mare Fecunditatis

The objects that formed in the inner part of the protoplanetary disk around the Sun are relatively small, rocky worlds, one of which is Earth. A comparison of those worlds reveals the forces that shape a planet. Since the mid-20th century, humans have been collecting detailed information about objects in the Solar System, in order to make these comparisons. Robotic probes have visited every planet, and astronauts have walked on the Moon. In addition to discoveries from new space missions and telescopes, improved analytical techniques applied to the rocks and soil brought back from the Moon in the 1960s and 1970s have led to surprising new results. The information from those missions has revolutionized the understanding of the Solar System, offering insights into the current state of each of the neighboring planets and clues about their histories.

LEARNING GOALS

By the end of this chapter, you should be able to look at an image of a planet and identify which geological features occurred early in the history of that world and which occurred late. You also should be able to:

(1) Describe how impacts have affected the evolution of the terrestrial planets.

(2) Explain how radiometric dating is used to measure the ages of rocks and terrestrial planetary surfaces.

(3) Explain how scientists use both theory and observation to determine the structure of terrestrial planetary interiors.

(4) Compare tectonism and volcanism and the forms they take on different terrestrial planets.

(5) Summarize what is known about the presence of water on the terrestrial planets.

8.1 Impacts Help Shape the Evolution of the Planets

Four principal geological processes constantly reshape the terrestrial planets: *impact cratering, tectonism, volcanism,* and *erosion.* Some geological processes originate in a planet's interior, whereas other processes are external. The relative importance of each of those processes varies among the planets. Planetary scientists can learn about the evolutionary history of the Solar System by comparing those geological processes on the terrestrial planets and the Moon. In this section, we examine **impact cratering**, which occurs when large collisions by other Solar System objects leave distinctive scars on the outer layer of a planet.

Comparative Planetology

The four innermost planets in the Solar System—Mercury, Venus, Earth, and Mars—are collectively known as the *terrestrial planets.* Although the Moon is Earth's lone natural satellite, we include it in this chapter because of its similarity to the terrestrial planets.

When comparing planets, we first compare the basic physical characteristics, such as distance from the Sun, size and density, and gravitational pull at the surface. Those characteristics reveal what a planet is made of, what its surface temperature is likely to be, and how well it can hold an atmosphere (planetary atmospheres are discussed in Chapter 9). The correct explanation for a particular aspect of one planet must be consistent with what is known about the other planets. For example, an analysis of why the Moon is covered with craters must allow for the fact that preserved craters are rare on Earth. An explanation for why Venus has such a massive atmosphere should point to reasons why Earth and Mars do not. Such comparisons are the basis of an approach called **comparative planetology**. Some of the basic physical properties of the terrestrial planets are compared in **Table 8.1**.

Impacts and Craters

Of the four principle geological processes, impact cratering causes the most concentrated and sudden release of energy. Planets and other objects orbit the Sun at very high speeds. As seen in Table 8.1, for example, Earth orbits the Sun at an average speed of around 30 kilometers per second (km/s), or about 67,000 miles per hour (mph). Collisions between orbiting bodies traveling at such speeds can release huge amounts of energy and produce craters such as the one in **Figure 8.1**. When an object hits a planet, the object's kinetic energy heats and compresses the surface and throws material far from the resulting impact crater, as shown in **Figure 8.2**. Sometimes, material thrown from the crater, called **ejecta**, falls back to the surface of the planet with enough energy to cause **secondary craters**. The rebound of heated and compressed material can also lead to the formation

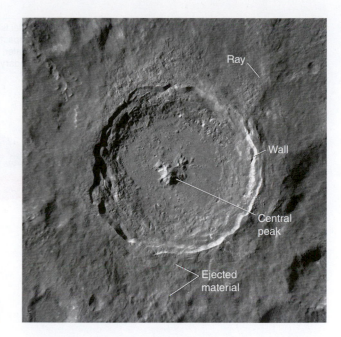

Figure 8.1 A lunar crater showing the crater wall and central peak surrounded by ejected material (*ejecta*) and rays—all typical features associated with impact craters.

Table 8.1	Comparison of Physical Properties of the Terrestrial Planets and the Moon				
Property	Mercury	Venus	Earth	Mars	Moon
Orbital radius	0.387 AU	0.723 AU	1.000 AU	1.524 AU	384,400 km
Orbital period	0.241 yr	0.615 yr	1.000 yr	1.881 yr	27.32 days
Orbital velocity (km/s)	47.4	35.0	29.8	24.1	1.02
Mass/M_{Earth}	0.055	0.815	1.000	0.107	0.012
Equatorial diameter (km)	4880	12,104	12,756	6794	3476
Equatorial diameter (D/D_{Earth})	0.383	0.950	1.000	0.532	0.273
Density/density of water	5.43	5.24	5.51	3.93	3.34
Sidereal rotation period*	58.65^d	243.02^d	23^h56^m	24^h37^m	27.32^d
Obliquity (degrees)†	0.03	177.36	23.44	25.19	6.68
Surface gravity (m/s²)	3.70	8.87	9.81	3.71	1.62
Escape velocity (km/s)	4.25	10.36	11.19	5.03	2.38

*The superscript letters *d*, *h*, and *m* stand for days, hours, and minutes, respectively.

†An obliquity greater than 90° indicates that the planet rotates in a retrograde, or backward, direction.

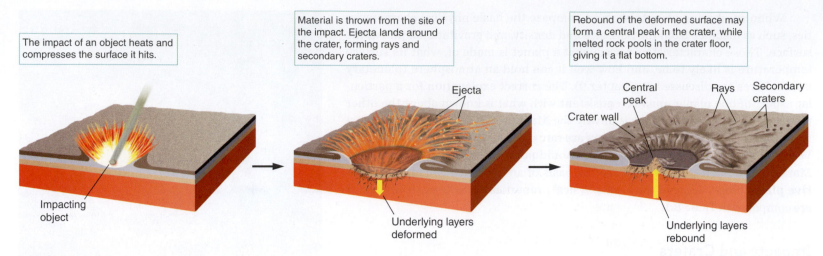

The impact of an object heats and compresses the surface it hits.

Material is thrown from the site of the impact. Ejecta lands around the crater, forming rays and secondary craters.

Rebound of the deformed surface may form a central peak in the crater, while melted rock pools in the crater floor, giving it a flat bottom.

Impacting object

Ejecta

Underlying layers deformed

Central peak

Rays

Secondary craters

Crater wall

Underlying layers rebound

Figure 8.2 Stages in the formation of an impact crater.

of a central peak or a ring of mountains on the crater floor, as shown in the lunar crater in Figure 8.1. A similar effect occurs when a drop lands in liquid, as caught by the slow-motion sequence in **Figure 8.3**. In the last panel of the sequence, a droplet rises from the surface of the liquid at the center of the splash.

The energy of an impact can be great enough to melt or even vaporize rock. The floors of some craters are the cooled surfaces of melted rock that flowed as lava.

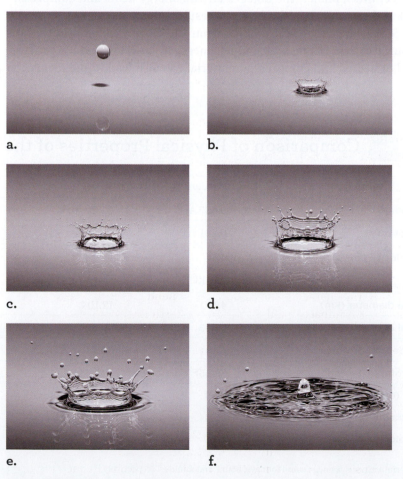

a.

b.

c.

d.

e.

f.

Figure 8.3 A drop **a.** hitting a pool of liquid illustrates the formation of features in an impact crater, including crater walls **b, c**, secondary craters **d, e**, and a central peak **f.**

The energy released in an impact also can lead to the formation of new minerals. Because some minerals form only during an impact, they are evidence of ancient impacts on Earth's surface. The space rocks that commonly cause such impacts are defined by three related terms: **Meteoroids** are small (less than 100 meters in diameter) cometary or asteroid fragments in space. A meteoroid that enters and burns up in a planetary atmosphere is called a **meteor**. Any meteoroids that survive to hit the ground are known as **meteorites** (see Figure 7.3).

One of the best-preserved impact structures on Earth is Meteor Crater in Arizona (**Figure 8.4**). That impact occurred about 50,000 years ago. From the crater's size and shape and from the remaining pieces of the impacting body, we know that the nickel–iron asteroid fragment was about 50 meters across, had a mass of about 300 million kilograms (kg), and was traveling at 13 km/s relative to Earth when it hit Earth's upper atmosphere. Approximately half of the original mass was vaporized in the atmosphere before the rest hit the ground. That collision released about 300 times as much total energy as the first atom bomb. At only 1.2 km in diameter, Meteor Crater is tiny in comparison with impact craters seen on the Moon or more ancient impact scars on Earth.

Impact craters cover the surfaces of Mercury, Mars, and the Moon. The Moon has hundreds of millions of craters of at least 10 meters in diameter and even more smaller ones. As seen outlined by yellow circles in **Figure 8.5**, craters of different sizes often exist one on top of another. Nearly all those craters are the result of impacts. Fewer than 200 impact craters have been identified on Earth, however, and only about 1000 have been found on Venus. It's not that Earth and Venus have been impacted significantly less than the Moon. Rather, geologic processes have obliterated most of the craters. On Earth, those processes (discussed later in the chapter) are plate tectonics in Earth's ocean basins and erosion on land. On Venus, it's lava flows.

The atmospheres of Earth and Venus provide another explanation for their low number of small craters. The surfaces of the Moon and Mercury are directly exposed to bombardment from space, whereas the surfaces of Earth and Venus are partly protected by their atmospheres. Rock samples from the Moon show craters smaller than a pinhead, formed by micrometeoroids. In contrast, most meteoroids smaller than 100 meters in diameter that enter Earth's atmosphere are either burned up or broken up by friction before they reach the surface. Small meteorites found on the ground on Earth are probably pieces of much larger bodies that broke apart on entering the atmosphere. With an atmosphere far thicker than that of Earth, Venus is even better protected.

Planetary scientists can tell a lot about the surface of a planet by studying its craters because the characteristics of a crater depend on the properties of the planetary surface. An impact in a deep ocean on Earth might create an impressive wave but leave no lasting crater. In contrast, an impact scar formed in an ancient rocky area can be preserved for billions of years. For example, craters on the Moon's pristine surface are often surrounded by strings of smaller secondary craters formed from material thrown out by the impact. Some examples are visible in Figure 8.1.

Some craters on Mars look very different. They are surrounded by structures that look like a fluid washed over the surface (**Figure 8.6**). The flows suggest that the martian surface rocks contained water or ice at the time of the impact. The energy released by an impact would have melted that ice, turning the surface material into a slurry with a consistency much like that of wet concrete. When thrown from the crater by the force of the impact, that slurry would hit the surrounding terrain

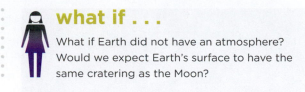

what if . . .

What if Earth did not have an atmosphere? Would we expect Earth's surface to have the same cratering as the Moon?

Figure 8.4 Meteor Crater (also known as Barringer Crater), located in northern Arizona, is an impact crater 1.2 km in diameter that formed some 50,000 years ago when a nickel-iron meteoroid collided with Earth.

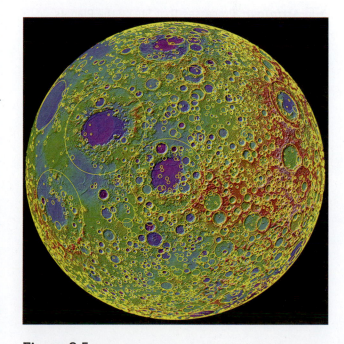

Figure 8.5 *Lunar Reconnaissance Orbiter* false-color map of craters on the Moon.

and slide across the surface, forming the mudlike craters we see today. Not all martian craters have that appearance, so the water or ice must have been concentrated in only some areas.

Features resembling canyons and dry riverbeds are further evidence that liquid water may have existed on the surface of Mars in the past. Today, however, the surface of Mars is dry in some regions and frozen in others, which suggests that water once on the surface has evaporated, or soaked into the ground, much like water frozen in the ground in Earth's polar regions.

Giant Impacts Reshape Planets

Because many planetesimals were roaming around the early Solar System, every young planet experienced frequent impacts early in its history. The last major bombardment is called *the late heavy bombardment*, which took place from 4.1 billion to 3.8 billion years ago. Observations that the thicknesses of the crusts of inner planets are uneven have led to theories that some or all of the terrestrial planets were disrupted by at least one giant impact—a collision with an object the size of a large asteroid. A giant impact with Mercury could have removed some of the lighter material of its outside layer, leaving behind an overall denser planet. A giant impact with Venus may have led to its retrograde (backward) rotation. A giant impact probably explains the large differences between the northern and southern hemispheres of Mars. The southern highlands have a thicker crust and formed early in martian history. The northern lowlands have a thinner and younger crust and may have formed when melting occurred after the impact. Early giant impacts on Mars also may be responsible for the loss of its magnetic field. Impacts with smaller icy comets from the outer Solar System brought water, atmospheric gases, and possibly organic molecules to the inner planets. As we discuss at the end of this chapter, an impact with Earth about 65 million years ago might have been a crucial event for the evolution of our species, *Homo sapiens.*

Although several ideas have been proposed to account for how the Moon formed, the leading theory involves a giant impact. According to this theory, a Mars-sized protoplanet collided with proto-Earth about 4.5 billion years ago, blasting off and vaporizing parts of proto-Earth's outer layers. The debris from the impact condensed into many "moonlets" in orbit around Earth, and those collected and formed our Moon. That theory accounts for the similarities in composition between the Moon and Earth's outer layers. It also explains the smaller amounts of volatiles on the Moon: during the vaporization stage of the collision, most gases were lost to space, leaving primarily the nonvolatiles to condense as the Moon. Earth, in contrast, was massive enough to keep more of its volatiles, which continued to be released from its interior after the collision. Because of Earth's stronger gravity, those gases were retained as part of Earth's atmosphere. If that account of the Moon's formation is correct, scientists expect to observe chemical evidence of material from the colliding protoplanet that was incorporated into the Moon's composition along with material from Earth. Scientists still debate whether the collision was glancing or head-on, how the Earth-Moon system lost angular momentum, and whether chemical evidence of the colliding object has been found (**Process of Science Figure**). In some models, the *inner* part of the Moon would still contain volatiles. As explained in Section 8.6, debate persists about evidence of water in the Moon's interior.

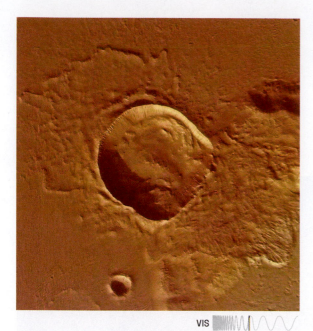

VIS

Figure 8.6 Some craters on Mars look like those formed by rocks thrown into mud, suggesting that material ejected from the crater contained large amounts of water. This crater is about 20 km across.

▶❚❚ **AstroTour:** Processes That Shape the Planets

Certainty Is Sometimes Out of Reach

There are several hypotheses for how the Moon formed. One of these fits the data better than the others, but none has been absolutely ruled out.

Did the Moon split off from Earth?

Was the Moon captured?

Did the Moon and Earth form together?

Did the Moon form from an impact of another object with Earth?

In some cases, hypotheses cannot be definitively falsified (at least not yet). The working hypothesis, then, is the one that best fits the data, but other ideas are kept in mind.

8.2 Radiometric Dating Tells Us the Age of the Moon and the Solar System

The number of visible craters on a planet is determined by the rate at which those craters are destroyed. Geological activity on Earth, Mars, and Venus erased most evidence of early impacts. By contrast, the Moon's surface still preserves the scars of craters dating from about 4 billion years ago. The lunar surface has remained essentially unchanged for more than a billion years because the Moon has no atmosphere or surface water and a cold, geologically dead interior. Mercury also has well-preserved craters, although recent evidence from the *Messenger* mission shows tilted crater floors that are higher on one side than the other—evidence that internal forces lifted the floors unevenly after the craters formed.

Planetary scientists use that cratering record to estimate the ages and geological histories of planetary surfaces: extensive cratering indicates an older planetary surface that remains relatively unchanged because of minimal geological activity. The amount of cratering can indicate the relative ages of surfaces—this area is older than that one. However, to determine the age of a surface—for example, this area is 1.3 billion years old—we need more information.

To assign real dates to the layers in rock, scientists use a technique called **radiometric dating** (occasionally called radioactive dating). Recall that the nuclei of atoms contain neutrons and protons. An **isotope** is an atom that has the same number of protons as other atoms of the same chemical element but has a different number of neutrons. Isotopes are often unstable, and undergo radioactive decay, turning from one element into another, and releasing energy in the process. These decaying isotopes are known as **radioisotopes**. The radioisotope is often called the **parent element**, and the decay products are called **daughter products**. Because it takes time for all of the atoms of the radioisotope to decay, a geologist can find the age of a rock by measuring the relative amounts of both a parent element and its daughter products. If no daughter element is present in the environment, then immediately after its formation, a rock containing radioisotopes would contain the parent element, but the daughter products would be absent because they would not have formed yet. As radioactive atoms decay, the amount of parent element decreases and the amount of daughter products builds up.

The time interval over which a radioactive isotope decays to half its original amount is called that isotope's **half-life**. With every half-life that passes, the remaining amount of the radioisotope decreases by a factor of 2. After 3 half-lives, for example, the remaining amount of a parent radioisotope will be $1/2 \times 1/2 \times 1/2 = 1/8$ of its original amount. In **Figure 8.7**, a sample is formed in which 100 percent of the material is parent element (in red), with no daughter products (in blue) yet. After 1 half-life has passed, half of the parent element has decayed, and there are equal numbers of atoms of parent element and daughter products. After another half-life has passed, the sample is now only 1/4 parent element and 3/4 daughter products, and so on. By comparing the percentages of parent and daughter

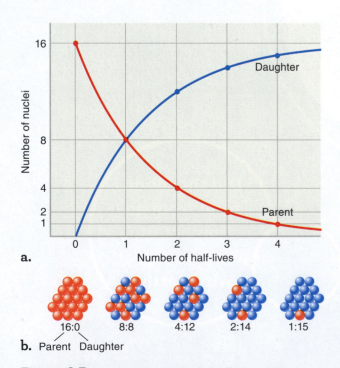

a.

b. Parent Daughter

Figure 8.7 A parent population of 16 radioactive nuclei decays over several half-lives. That information can be presented **a.** graphically or **b.** as collections of particles.

working it out 8.1

Computing the Ages of Rocks

With every half-life that passes, the remaining amount of radioactive isotope decreases by a factor of 2. If we express the number of half-lives more generally as n, then that relationship can be expressed mathematically as

$$\frac{P_F}{P_O} = \left(\frac{1}{2}\right)^n$$

where P_O and P_F are the original and final amounts, respectively, of a parent radioisotope, and n is the number of half-lives that have gone by, which equals the time interval of decay (its age) divided by the half-life of the isotope.

For example, the most abundant isotope of the element uranium (uranium-238, or ^{238}U—the parent) decays through a series of intermediate daughters to an isotope of the element lead (lead-206, or ^{206}Pb—its final, stable daughter). The half-life of ^{238}U is 4.5 billion years. Therefore, after 4.5 billion years, a sample that originally contained the uranium isotope (the parent) but no lead (its final daughter) would instead contain equal numbers of uranium and lead atoms. If we were to find a mineral with such composition, we would know that half the uranium atoms had turned to lead and that the mineral formed 4.5 billion years ago.

Another isotope of uranium (^{235}U) decays to a different lead isotope (^{207}Pb) with a half-life of 700 million years. Suppose that a lunar mineral brought back by astronauts has 15 times as much ^{207}Pb (the daughter product) as ^{235}U (the parent radioisotope). Therefore, 15/16 of the parent radioisotope (^{235}U) has decayed to the daughter product (^{207}Pb), leaving only 1/16 of the parent remaining in the mineral sample. By noting that 1/16 is $(1/2)^4$, we see that 4 half-lives have elapsed since the mineral was formed, in which case the lunar sample is 4×700 million years $=$ 2.8 billion years old.

Because the measured quantity of the isotope is not always a neat power of 2, looking at how we would solve the equation

mathematically is worthwhile. We do that by taking the logarithm on both sides:

$$\log_{10}\frac{P_F}{P_O} = \log_{10}\left(\frac{1}{2}\right)^n$$

$$\log_{10}\frac{P_F}{P_O} = n\log_{10}\left(\frac{1}{2}\right)$$

$$\log_{10}\frac{P_F}{P_O} = -0.3n$$

Putting that back into words, we can write the relationship as

$$\log_{10}\left(\frac{\text{Actual measured quantity of isotope}}{\text{Original quantity of isotope}}\right) = -0.3 \times \left(\frac{\text{Time it has been decaying (age)}}{\text{Half-life}}\right)$$

Solving for age,

$$\text{Age} = -3.3 \times \text{Half-life} \times \log_{10}\left(\frac{\text{Actual measured quantity of isotope}}{\text{Original quantity of isotope}}\right)$$

(Most calculators have a button called "log" or "$\log_{10}$" for calculating such numbers.)

For the ^{235}U that decays to ^{207}Pb with a half-life of 700 million years, the lunar mineral is measured to have 15 times as much lead as uranium, so the mineral now contains only 1/16 of the original quantity of uranium:

$$\text{Age of mineral} = -3.3 \times (700 \times 10^6\,\text{yr}) \times \log_{10}\left(\frac{1}{16}\right) = 2.8 \times 10^9\,\text{yr}$$

The mineral is 2.8 billion years old.

isotopes in a mineral, scientists can determine the number of half-lives that have passed and thus the age of the mineral. Some numerical examples are discussed in **Working It Out 8.1.**

The age of the Solar System is estimated from radiometric dating of meteorites found on Earth that are 4.5 billion to 4.6 billion years old. Earth may be as young as 4.4 billion years. The age of the Moon is determined from radiometric dating of lunar rocks. Between 1969 and 1976, *Apollo* astronauts and Soviet unmanned probes brought back samples from nine locations on the lunar surface. By measuring relative amounts of various radioactive elements and the elements into which they decay, scientists assigned ages to those lunar regions. The oldest, most heavily cratered regions on the Moon date back to about 4.4 billion years ago,

what if . . .

What if the crystalline structure of a rock required that at its formation, the rock had equal numbers of parent and daughter elements? Would it still be possible to determine the age of the rock from radiometric dating?

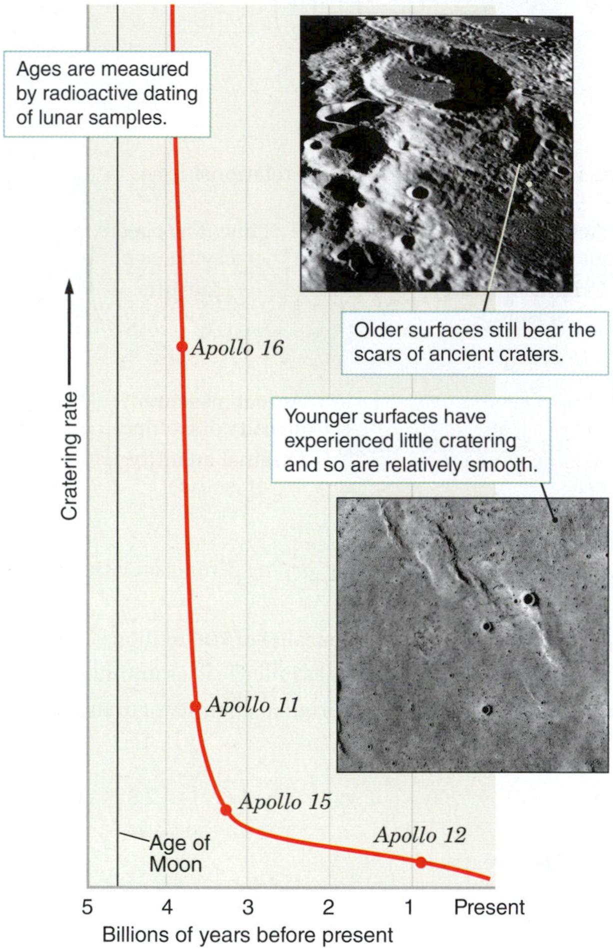

whereas most of the smoother parts of the lunar surface are typically 3.1 billion to 3.9 billion years old. That finding suggests that the Moon formed after Earth, and the heavy cratering suggests heavy bombardment at that time. As the graph in **Figure 8.8** indicates, almost all the major cratering in the Solar System took place within its first billion years.

CHECK YOUR UNDERSTANDING 8.2

If radioactive element A decays into radioactive element B with a half-life of 20 seconds, then after 40 seconds: (a) none of element A will remain; (b) none of element B will remain; (c) half of element A will remain; (d) one-quarter of element A will remain.

8.3 The Surface of a Terrestrial Planet Is Affected by Processes in the Interior

Whereas impact cratering is driven by forces outside a planet, tectonism and volcanism are determined by conditions in the interior of the planet. To understand those processes, we must understand the structure and composition of the interiors of planets. Scientists have determined a lot about the interior of Earth but less about the interiors of the other terrestrial planets.

Probing the Interior of Earth

How do we know what the interiors of planets are like? On Earth, the deepest holes drilled are about 12 km deep; that is tiny in comparison with Earth's radius of 6378 km. It is impossible to drill down into Earth's core to observe Earth's interior structure directly. The composition and structure of Earth's interior is found indirectly, instead.

In one approach, Kepler's or Newton's laws are used to find the mass of Earth—for example, by applying Kepler's third law to a satellite orbiting Earth. Dividing the mass by the volume of Earth gives an average density of 5500 kilograms per cubic meter (kg/m^3). Comparing this density to the density of other substances gives clues to Earth's overall composition. The density of Earth is 5.5 times the density of water, so Earth is certainly not made mostly of water. Rocky surface material averages only 2900 kg/m^3, so the interior of Earth must contain material denser than surface rocks. Because meteorites are left over from a time when the Earth was forming from similar materials, the overall composition of Earth should be similar to the composition of meteorite material. Meteorites include minerals with large amounts of iron, which has a density of nearly 8000 kg/m^3. From those considerations, planetary scientists determine that Earth's interior has a large percentage of iron.

Vibrations from earthquakes, known as **seismic waves**, are used to model the structure of Earth's interior. **Surface waves** travel across the surface of a planet, much like waves on the ocean. If conditions are right, surface waves from earthquakes can be seen rolling across the countryside like ripples on water. Those waves are responsible for much of the heaving of Earth's surface during an earthquake, causing damage such as the buckling of roadways.

Other types of seismic waves pass *through* the interior of Earth. These waves travel faster than surface waves. **Primary waves** (P waves) are a **longitudinal wave** that alternately compress and stretch the material they pass

Ages are measured by radioactive dating of lunar samples.

Older surfaces still bear the scars of ancient craters.

Younger surfaces have experienced little cratering and so are relatively smooth.

Apollo 16

Apollo 11

Apollo 15

Apollo 12

Age of Moon

Cratering rate

5   4   3   2   1   Present
Billions of years before present

Figure 8.8 Radiometric dating of lunar samples returned from *Apollo* missions that landed at different sites was used to determine how the cratering rate has changed. Cratering records can then be used to establish the age of other parts of the lunar surface.

through. To visualize a longitudinal wave, imagine a stretched-out spring. A quick push along its length will make a longitudinal wave, as shown in **Figure 8.9a**. **Secondary waves** (S waves) are **transverse waves** that result from the sideways motion of material, like the wave on a string shown in **Figure 8.9b**.

The progress of seismic waves through Earth's interior depends on the characteristics of the material through which they are moving. Seismic waves travel at different speeds, depending on the density and composition of the rocks they encounter. As a result, seismic waves that pass a boundary between rocks of various densities or composition are bent in much the same way that waves of light are bent when they enter or leave glass. Primary waves (yellow; **Figure 8.10**) can travel through either solids or liquids, but secondary waves (blue) can *only* travel through solids. The refraction of primary waves at the outer edge of Earth's liquid outer core and the inability of secondary waves to penetrate the liquid outer core create "shadows" on the side of Earth opposite an earthquake's epicenter (Figure 8.10), where direct S and P waves do not appear.

Scientists use instruments called **seismometers** to measure the distinctive patterns of seismic waves. For more than 100 years, thousands of seismometers scattered around the globe have measured the vibrations from countless earthquakes and other seismic events, such as volcanic eruptions and nuclear explosions. A single seismometer can record ground motion at only one place on Earth, but when combined with the recordings of many other seismometers placed all over Earth, scientists can measure the shadows of the core to find the size and shape of interior components.

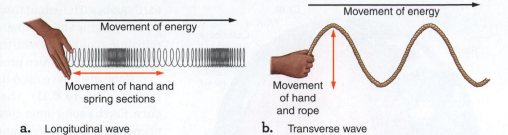

a. Longitudinal wave   **b.** Transverse wave

Figure 8.9 a. A longitudinal wave involves oscillations along the direction in which the wave travels. **b.** A transverse wave involves oscillations perpendicular to the direction in which the wave travels. Primary seismic waves are longitudinal; secondary seismic waves are transverse.

Figure 8.10 Primary and secondary seismic waves move through the interior of Earth in distinctive ways. Measurements of when and where different types of seismic waves arrive after an earthquake enable scientists to test predictions from detailed models of Earth's interior. Note the "shadow areas" caused by the refraction of primary waves (yellow) at the outer boundary of the liquid outer core and the inability of secondary waves (blue) to pass through the liquid outer core.

Building a Model of Earth's Interior

At any point in Earth's interior, the weight of all the material above the point must be balanced by the outward force resulting from the outward pressure of the compressed material below that point. If the outward pressure at some point within a planet were *less* than the weight per unit area of the overlying material, that material would further compress, and shrink. If the outward pressure at some point within a planet were *greater* than the weight per unit area of the overlying material, the material would expand, lifting the overlying material. The situation is stable only when the weight of matter above is precisely balanced by the pressure within the whole interior of the planet. This balance between pressure and weight—known as **hydrostatic equilibrium**—is important to the structure of planetary interiors, planetary atmospheres, and the structure and evolution of stars.

Earth is in hydrostatic equilibrium; it neither expands nor contracts. Combining this fact with seismic wave measurements, scientists construct a layered model of Earth's interior. They then test their model by comparing its predictions of how seismic waves would propagate through Earth with actual observations of seismic waves from real

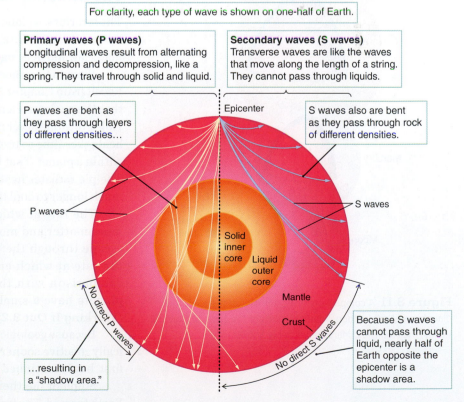

For clarity, each type of wave is shown on one-half of Earth.

Primary waves (P waves)
Longitudinal waves result from alternating compression and decompression, like a spring. They travel through solid and liquid.

Secondary waves (S waves)
Transverse waves are like the waves that move along the length of a string. They cannot pass through liquids.

P waves are bent as they pass through layers of different densities…

Epicenter

S waves also are bent as they pass through rock of different densities.

P waves

S waves

No direct P waves

Solid inner core

Liquid outer core

Mantle

Crust

No direct S waves

…resulting in a "shadow area."

Because S waves cannot pass through liquid, nearly half of Earth opposite the epicenter is a shadow area.

Figure 8.11 A comparison of the interiors of the terrestrial planets and Earth's Moon. Some fractions of the cores of Mercury, Venus, the Moon, and Mars are probably liquid.

earthquakes. The extent to which the predictions agree with observations identifies both strengths and weaknesses of the model. Geologists adjust the model—always remaining consistent with the known physical properties of materials—until a good match is found between prediction and observation.

Geologists have used this method to arrive at the current picture of Earth's interior (**Figure 8.11**). The innermost region of Earth's interior consists of a **core**. Earth's solid inner core is at a temperature of about 6000 K and is composed primarily of iron, nickel, and other dense metals. The liquid outer core is cooler, about 4000 K, and is composed of liquid metals. Outside the outer core is Earth's **mantle**, a rocky shell made of solid, medium-density materials such as silicates. That hot, rocky shell is under pressure, and over long timescales—hundreds of years— it flows. Covering the mantle is the **crust**, a thin, hard layer of lower-density materials that is chemically distinct from the interior.

Earth's interior is not uniform. The materials have been separated according to density, a process known as **differentiation.** When rocks of different types are mixed, they tend to stay mixed. Once that rock melts, however, the denser materials sink to the center and the less dense materials float toward the surface. Today, little of Earth's interior is molten, but the differentiated structure shows that Earth was much hotter right after formation, and its interior was liquid throughout. Similarly, the cores of all the terrestrial planets and the core of the Moon were molten when they formed. Using new, improved methods to reanalyze 8 years of data from seismometers left on the Moon by the *Apollo* astronauts, planetary scientists found that the Moon has a solid inner core, possibly a liquid outer core, and a partially melted layer between the core and the mantle. Cross sections of the interior structures of each terrestrial planet and Earth's Moon are compared in Figure 8.11.

The Evolution of Planetary Interiors

The interiors of planets evolve as their temperatures change. Factors that influence how the temperature changes include the size of the planet, the composition of the material, and heating from various sources.

Planets lose thermal energy from their surfaces primarily through radiation. Recall from Chapter 5 that the hotter an object is, the more energy it radiates; and that the peak wavelength of that radiation (whether in the infrared, optical, ultraviolet, or some other range) also depends on the temperature of the object. Because heat takes time to travel through rock, the temperature increases the deeper we go within a planet. That trend is similar to the effect of taking a hot pie out of an oven. The pie radiates heat from the surface and eventually cools, but the filling takes much longer to cool than the crust.

The rate at which a planet cools depends on its size. A larger planet has more matter and more thermal energy trapped inside. Thermal energy has to escape through the planet's surface, so the planet's surface area determines the rate at which energy is lost. Smaller planets have more surface area in comparison with their small volumes, so they cool faster, whereas larger planets have a smaller surface-area-to-volume ratio and cool more slowly (**Working It Out 8.2**).

Because geological activity is powered by heat, smaller objects become geologically inactive sooner. Major geological activity ended on Mercury and the Moon first, but it continued on the larger terrestrial planets—Venus, Earth, and Mars.

Some of the thermal energy in the interior of Earth is left over from when Earth formed. The tremendous energy of collisions and the energy from short-lived

working it out 8.2

How Planets Cool

If we assumed that all the terrestrial planets formed with the same percentage of radioactive materials and that those radioactive materials are their sole source of internal thermal energy, a planet's volume would determine the total amount of the thermal energy–producing material it contains. The volume of a spherical planet is proportional to the *cube* of the planet's radius (volume = $4/3 \times \pi R^3$).

A planet loses its internal energy by radiating it away at its surface, so a planet's surface area determines the rate at which it can get rid of its thermal energy. The surface area of the planet is proportional to the *square* of the radius (surface area = $4\pi R^2$). The ratio of the two—the volume divided by the surface area through which thermal energy can escape—is given by

$$\frac{\text{Volume}}{\text{Surface area}} = \frac{\frac{4}{3} \times \pi R^3}{4\pi R^2} = \frac{R}{3}$$

Thus, the time it takes to lose internal energy is proportional to the planet's radius.

A planet's ability to transfer internal energy from its hot core to its cooling surface also depends on some of its own internal properties. Nevertheless, all things being equal, planets with larger radii retain their internal energy longer than smaller planets do. For example, Mars has a radius about half that of Earth, so it has been losing its internal thermal energy to space about twice as fast as Earth has. That is one reason why Mars is now less geologically active than Earth.

radioactive elements melted the planet, leading to the differentiated structure. As the surface of Earth radiated energy into space, it cooled rapidly. A solid crust formed above a molten interior. Because a solid crust does not conduct thermal energy well, the remaining heat took longer to escape. Over a long time, energy from the interior of the planet continued to leak through the crust and radiate into space. As a result, the interior of the planet cooled slowly, and the mantle and the inner core solidified.

If the thermal energy from Earth's formation were the only source of heating in Earth's interior, Earth would have solidified completely long ago. Most of the rest of the thermal energy in Earth's interior comes from long-lived radioactive elements trapped in the mantle. As those radioactive elements decay, they release energy, which heats the planet's interior. Today, the temperature of Earth's interior is determined by a dynamic equilibrium between the radioactive heating of the interior and the loss of energy to space. As the radioactive elements decay, the amount of thermal energy generated declines, and Earth's interior cools as it ages. A small amount of additional heating of Earth's interior is due to friction generated by tidal effects of the Moon and Sun.

Although temperature plays an important role in a planet's interior structure, whether a material is solid or liquid also depends on pressure. Higher pressure forces atoms and molecules closer together and makes the material more likely to become a solid. Toward the center of Earth, the effects of temperature and pressure oppose each other: the higher temperatures make it more likely that material will melt, but the higher pressure favors a solid form. In Earth's outer core, the high temperature wins, allowing the material to exist in a molten state. At the center of Earth, the temperature is higher still, but the pressure is so great that Earth's inner core is solid.

CHECK YOUR UNDERSTANDING **8.3a**

Differentiation refers to materials that are separated on the basis of their: (a) weight; (b) mass; (c) volume; (d) density.

what if . . .

What if Earth's magnetic field showed signs that it would undergo a reversal within the next hundred years? Would we care whether the magnetic pole now in the north simply migrates to the South Pole or whether the magnetic field disappears entirely and reappears as a reversed field?

a.

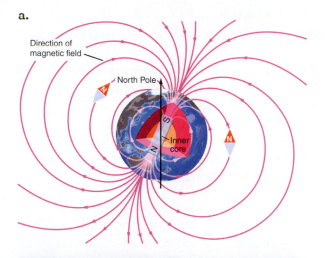

b.

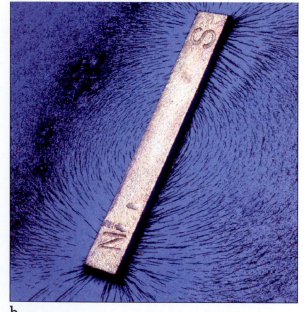

Figure 8.12 **a.** Earth's magnetic field can be visualized as though it were a giant bar magnet tilted relative to Earth's axis of rotation. Compass needles line up along magnetic-field lines and point toward Earth's North Pole.   **b.** Iron filings sprinkled around a bar magnet help us visualize such a magnetic field. A careful observer will notice that the south end of the bar magnet is at the top of the image. We've aligned it this way because Earth's south *magnetic* pole is near the Earth's north *geographic* north pole; that's why the north end of a compass needle points that way. Opposite magnetic poles attract.

Magnetic Fields

A magnetic field is created by moving charges and exerts a force on magnetically reactive objects, such as iron and other charged particles. In a navigational compass, for example, the compass needle lines up with Earth's magnetic field and points "north" and "south," as shown in **Figure 8.12a**. In the north, a compass needle points to a location in the Arctic Ocean off the coast of northern Canada, near but not quite at the geographic North Pole (about which Earth spins). In the south, a compass needle points to a location off the coast of Antarctica, 2800 km from the geographic South Pole. Earth behaves almost as though it contained a giant bar magnet that were slightly tilted with respect to the planet's rotation axis and had its two endpoints at the two magnetic poles (**Figure 8.12b**).

The general idea of how Earth's magnetic field (and those of other planets) originates is called the **dynamo theory**. In general, magnetic fields result from electric currents, which are moving electric charges. Earth's magnetic field is thought to be a side effect of three factors: Earth's rotation about its axis; an electrically conducting, liquid outer core; and fluid motions within the outer core. The theory suggests that any rotating planet with an internal heat source will have a magnetic field. A planet's magnetic field extends into space, filling the planet's **magnetosphere**: the region surrounding a planet that is filled with relatively intense magnetic fields and charged particles. Because the dynamo theory connects the internal structure of a planet to the magnetosphere, planetary scientists can find out about the internal structure of a planet by probing its magnetic field.

Earth's magnetic field is constantly changing. At the moment, the north magnetic pole is traveling several tens of kilometers per year toward the northwest. If that rate and direction continue, the north magnetic pole could be in Siberia before the end of the century. The magnetic pole tends to wander, constantly changing direction as a result of changes in the core.

Much more dramatic changes in the magnetic field have occurred over Earth's history. This is recorded in the geologic record of iron-bearing minerals. When a magnet made of material such as iron gets hot enough, it loses its magnetization. As the material cools, it again becomes magnetized by any magnetic field surrounding it. Thus, iron-bearing minerals record the direction of Earth's magnetic field at the time that they cooled. For example, lava extruded from a volcano carries a record of Earth's magnetic field at the moment the lava cooled. By using radiometric techniques to date those materials, geologists create a record of how Earth's magnetic field has changed. Earth's magnetic field has probably existed for at least 3.5 billion years, and the north and south magnetic poles switch from time to time—on average, about every half-million years.

What about the magnetic fields of other terrestrial worlds? During the *Apollo* program, astronauts measured the Moon's local magnetic fields, and small satellites have searched for global magnetism. The Moon has a very weak field, possibly none at all, because the Moon is very small and therefore has a very small, solid (not liquid and rotating) inner core. However, remnant magnetism is preserved in lunar rocks from an earlier time before the lunar surface and rocks solidified. Recent analysis of the oldest lunar rocks brought back on *Apollo 17* suggests that 4.2 billion years ago, the Moon had a liquid core that generated a magnetic field until about a billion years ago. Data from India's *Chandrayaan-1* spacecraft suggest that the Moon may have a very weak and localized mini-magnetosphere on its far side.

Other than Earth, Mercury is the only terrestrial planet with a significant global magnetic field today—although its field is only about 1 percent as strong as Earth's.

Reading Astronomy News

Moon Had a Magnetic Field for at Least a Billion Years Longer than Thought

Ashley Yeager, *ScienceNews*

The Moon had a magnetic field for at least 2 billion years, or maybe longer.

Analysis of a relatively young rock collected by *Apollo* astronauts reveals the Moon had a weak magnetic field until 1 billion to 2.5 billion years ago, at least a billion years later than previous data showed. Extending this lifetime offers insights into how small bodies generate magnetic fields, researchers report August 9 in *Science Advances*. The result may also suggest how life could survive on tiny planets or moons.

"A magnetic field protects the atmosphere of a planet or moon, and the atmosphere protects the surface," says study coauthor Sonia Tikoo, a planetary scientist at Rutgers University in New Brunswick, NJ. Together, the two protect the potential habitability of the planet or moon, possibly those far beyond our solar system.

The Moon does not currently have a global magnetic field. Whether one ever existed was a question debated for decades. On Earth, molten rock sloshes around the outer core of the planet over time, causing electrically conductive fluid moving inside to form a magnetic field. This setup is called a dynamo. At 1 percent of Earth's mass, the Moon would have cooled too quickly to generate a long-lived roiling interior.

Magnetized rocks brought back by *Apollo* astronauts, however, revealed that the Moon

must have had some magnetizing force. The rocks suggested that the magnetic field was strong at least 4.25 billion years ago, early on in the Moon's history, but then dwindled and maybe even got cut off about 3.1 billion years ago.

Tikoo and colleagues analyzed fragments of a lunar rock collected along the southern rim of the Moon's Dune Crater during the *Apollo 15* mission in 1971. The team determined the rock was 1 billion to 2.5 billion years old and found it was magnetized. The finding suggests the Moon had a magnetic field, albeit a weak one, when the rock formed, the researchers conclude.

A drop in the magnetic field strength suggests the dynamo driving it was generated in two distinct ways, Tikoo says. Early on, Earth and the Moon would have sat much closer together, allowing Earth's gravity to tug on and spin the rocky exterior of the Moon. That outer layer would have dragged against the liquid interior, generating friction and a very strong magnetic field.

Then slowly, starting about 3.5 billion years ago, the Moon moved away from Earth, weakening the dynamo. But by that point, the Moon would have started to cool, causing less dense, hotter material in the core to rise and denser, cooler material to sink, as in Earth's core. This roiling of material would have sustained a weak field that lasted for at least a billion years, until the Moon's interior cooled, causing the dynamo to die completely, the team suggests.

The two-pronged explanation for the Moon's dynamo is "an entirely plausible idea," says planetary scientist Ian Garrick-Bethell of the University of California, Santa Cruz. But researchers are just starting to create computer simulations of the strength of magnetic fields to understand how such weaker fields might arise. So it is hard to say exactly what generated the lunar dynamo, he says.

If the idea is correct, it may mean other small planets and moons could have similarly weak, long-lived magnetic fields. Having such an enduring shield could protect those bodies from harmful radiation, boosting the chances for life to survive.

QUESTIONS

1. Refer to Working It Out 8.2. Is the smaller *mass* of the Moon the primary reason it cooled more quickly? Explain.

2. What is the other cause of the dynamo's weakening?

3. How might the presence of a magnetic field affect the possibility of habitability?

4. Why did the Moon move away from Earth billions of years ago? (Hint: See Chapter 4.)

Source: https://www.sciencenews.org/article/moon-had-magnetic-field-least-billion-years-longer-thought.

Slow rotation and a large iron core, parts of which are molten and circulating, cause Mercury's magnetic field. Because the field is so weak, Mercury's magnetosphere is small, meeting the solar wind about 1700 km above its surface. At that boundary, twisted bundles of magnetic fields transfer magnetic energy from the planet to space.

Planetary scientists expected that Venus would have a magnetic field because the planet's mass and distance from the Sun imply an iron-rich core and partly molten interior like Earth's. Venus lacks a magnetic field, however, which might be due to its extremely slow rotation (see Table 8.1). It may be that Venus's core has settled into several stable layers, so that the convection needed to generate a magnetic field cannot occur; Earth may have had a similar fate, except that an early giant impact scrambled its inner layers so that convection occurs throughout the core. Or perhaps Venus's magnetic field is temporarily dormant—a condition that Earth is believed to have experienced at times of magnetic field reversals.

Mars has a weak magnetic field, presumably frozen in place early in its history. The magnetic signature occurs only in the ancient crustal rocks, showing that early in the history of Mars, some sort of an internally generated magnetic field must have existed. Geologically younger rocks lack that residual magnetism, so the planet's original magnetic field has long since disappeared. Determining how long the magnetic field lasted is difficult, because the impacts that formed young basins on Mars displaced the crust, which may have erased the magnetic signature. The lack of a strong magnetic field today on Mars might be the result of its small core. Impacts may also play a role here, if they heated the mantle of Mars enough to reduce the flow of heat out of the core to the mantle. The retained heat would disrupt the generation of the magnetic field in the core. The oldest large impact basins on Mars appear to be magnetized, whereas newer ones are not.

CHECK YOUR UNDERSTANDING **8.3b**

According to the dynamo theory, a planet will have a strong magnetic field if it has: (a) fast rotation and a solid core; (b) slow rotation and a liquid core; (c) fast rotation and a liquid core; (d) slow rotation and a solid core; (e) fast rotation and a gaseous core.

▶❚❚ **AstroTour:** Continental Drift

Figure 8.13 Tectonic processes fold and warp Earth's crust, as seen in these rocks along a roadside north of Denver, Colorado.

8.4 Planetary Surfaces Evolve through Tectonism

The interior conditions of a planet are connected to the processes that shape the surface. The crust and part of the upper mantle form the **lithosphere** of a planet. **Tectonism**, the deformation of a planet's lithosphere, warps, twists, and shifts the lithosphere to form visible surface features. If you have driven through mountainous or hilly terrain, you may have seen places like the one shown in **Figure 8.13**, where the roadway has been cut through rock. The exposed layers tell the story of Earth through the vast expanse of geological time. In this section, we look at tectonic processes that create those layers and play an important part in shaping the surface of a planet.

The Theory of Plate Tectonics

Early in the 20th century, some scientists recognized that Earth's continents could be fit together like pieces of a giant jigsaw puzzle. In addition, the layers in the rock and the fossil records they hold on the east coast of South America match those on the west coast of Africa. On the basis of that evidence, Alfred Wegener (1880–1930) proposed a hypothesis called **continental drift** that the continents were originally joined in one large landmass that broke apart as the continents began to "drift" away from one another over millions of years. That hypothesis was further developed into the theory known today as **plate tectonics**. Geologists now recognize that Earth's outer shell is composed of several relatively brittle segments, or **lithospheric plates**. About seven major plates and about a half dozen smaller plates are floating on the mantle. The motion of those plates is constantly changing the surface of Earth.

Originally, the idea of plate tectonics was met with great skepticism among geologists because they could not imagine a mechanism that could move such huge landmasses. In the late 1950s and early 1960s, however, studies of the ocean floor provided compelling evidence for plate tectonics. Those surveys showed surprising characteristics in bands of **basalt**—a type of rock formed from cooled lava—that were found on both sides of the ocean rifts. Ocean floor rifts such as the Mid-Atlantic Ridge are **spreading centers**. As **Figure 8.14** shows, hot material in those rifts rises toward Earth's surface, becoming new ocean floor. When that hot material cools, it becomes magnetized along the direction of Earth's magnetic field, thus recording the direction of Earth's magnetic field at that time. Greater distance from the rift indicates the ocean floor is older and formed earlier. Combined with radiometric dates for the rocks, that magnetic record proved that the spreading of the seafloor and the motions of the plates have continued over long geological time spans.

Precise surveying techniques and global positioning systems (GPS) can now determine locations on Earth to within a few centimeters. Those measurements confirm that Earth's lithospheric plates are moving. Some areas are being pulled apart by more than 15 centimeters (cm) each year. That may not sound very fast, but over 10 million years—a short time by geological standards—a location traveling at 15 cm/yr moves 1500 km.

The Role of Convection

Moving lithospheric plates requires immense forces. Those forces are the result of thermal energy escaping from the interior of Earth by a process called **convection**: the transport of thermal energy by the movement of packets of gas or liquid. If you have ever watched water in a heated pot on a stovetop, you have observed convection (see the sketch in **Figure 8.15a**). Thermal energy from the stove warms water at the bottom of the pot. The warm water expands slightly, becoming less dense than the cooler water above it. Thus, the cooler water with higher density sinks, displacing the warmer water upward. When the lower-density water reaches the surface, it gives up part of its energy to the air and cools; as the water cools, it then becomes denser and sinks back toward the bottom of the pot. Water rises in some locations and sinks in others, forming convection cells. Convection is also important in planetary atmospheres and inside stars.

Figure 8.15b shows how convection works in Earth's mantle, where radioactive decay supplies the heat to drive convection. The mantle is viscous, having the consistency of hot molten glass. That consistency allows convection to take place very slowly. Convection cells in Earth's mantle drive the plates, carrying both continents and ocean crust along with them. Crust forms along rift zones in the ocean basins, where mantle material rises up by convection, cools, and slowly spreads out.

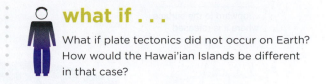

what if . . .

What if plate tectonics did not occur on Earth? How would the Hawai'ian Islands be different in that case?

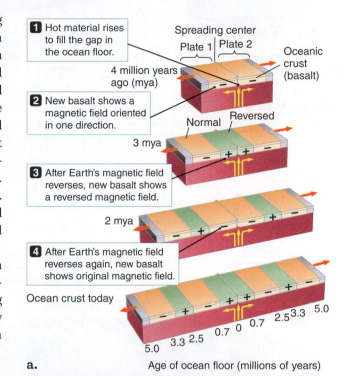

1 Hot material rises to fill the gap in the ocean floor.

Spreading center
Plate 1 Plate 2
Oceanic crust (basalt)

4 million years ago (mya)

2 New basalt shows a magnetic field oriented in one direction.

Normal   Reversed

3 mya

3 After Earth's magnetic field reverses, new basalt shows a reversed magnetic field.

2 mya

4 After Earth's magnetic field reverses again, new basalt shows original magnetic field.

Ocean crust today

5.0 3.3 2.5 0.7 0 0.7 2.5 3.3 5.0

a.   Age of ocean floor (millions of years)

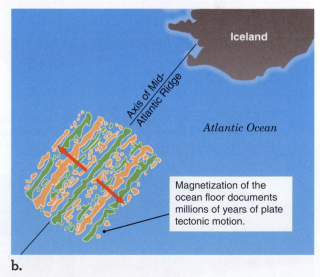

Iceland

Axis of Mid-Atlantic Ridge

Atlantic Ocean

Magnetization of the ocean floor documents millions of years of plate tectonic motion.

b.

Figure 8.14 a. New seafloor is formed at a spreading center. The cooling rock becomes magnetized and is then carried away by tectonic motions. **b.** Maps such as this one of banded magnetic structure in the seafloor near Iceland support the theory of plate tectonics.

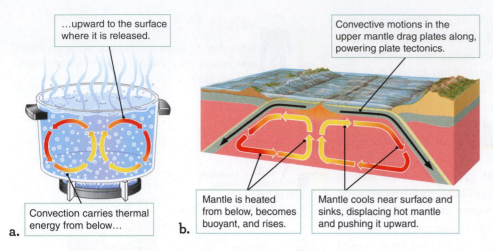

…upward to the surface where it is released.

Convective motions in the upper mantle drag plates along, powering plate tectonics.

Convection carries thermal energy from below…

Mantle is heated from below, becomes buoyant, and rises.

Mantle cools near surface and sinks, displacing hot mantle and pushing it upward.

a.   **b.**

Figure 8.15 a. Convection occurs when a fluid is heated from below. **b.** Convection in Earth's mantle drives plate tectonics.

Spreading

Where two continental plates meet, the crust can push up into high mountains.

Himalayan mountains

India

Fault

Indo-Australian Plate

Eurasian Plate

The continental crust can form cracks or faults as it deforms.

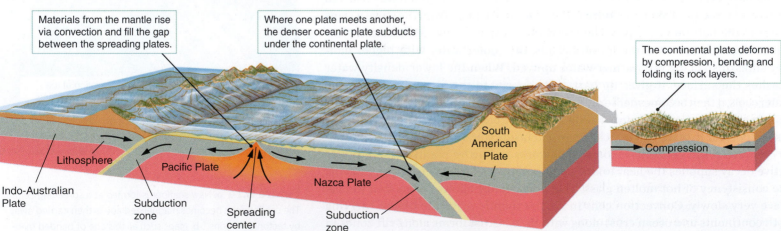

Materials from the mantle rise via convection and fill the gap between the spreading plates.

Where one plate meets another, the denser oceanic plate subducts under the continental plate.

The continental plate deforms by compression, bending and folding its rock layers.

Lithosphere

Pacific Plate

Nazca Plate

South American Plate

Compression

Indo-Australian Plate

Subduction zone

Spreading center

Subduction zone

Figure 8.16 Divergence and collision of tectonic plates create a variety of geological features.

Figure 8.16 illustrates some of the consequences of plate tectonics. If material rises and spreads in one location, it must converge and sink in another. Locations where plates converge and convection currents turn downward are called **subduction zones**. In a subduction zone, one plate slides beneath the other, and convection drags the submerged lithospheric material back down into the mantle. The Mariana Trench—the deepest part (11 km) of Earth's ocean floor—is a subduction zone. Much of the ocean floor lies between spreading centers and subduction zones, so the ocean floor is the youngest portion of Earth's crust. In fact, the *oldest* seafloor rocks are less than 200 million years old.

In some places, the plates are not sinking but instead are colliding and, consequently, shoved upward. The highest mountains on Earth, the Himalayas, grow a half-meter per century as the Indo-Australian subcontinental plate collides with the Eurasian Plate. In other places, plates meet at oblique angles and slide along past each other. A fracture in a planet's crust along which material can slide is called a **fault**. One such place is the San Andreas Fault in California, where the Pacific Plate slides past the North American Plate.

Where plates meet, enormous stresses build up, so these boundaries tend to be very active geologically. Earthquakes occur when a portion of the boundary between two plates suddenly slips, relieving the stress. Volcanoes are created when friction between plates melts rock, which is then pushed up through cracks to the surface. A map of where earthquakes and volcanism occur (**Figure 8.17**) reveals the outlines of Earth's plates. Earth also has many **hot spots**, where hot deep-mantle material rises, releasing thermal energy.

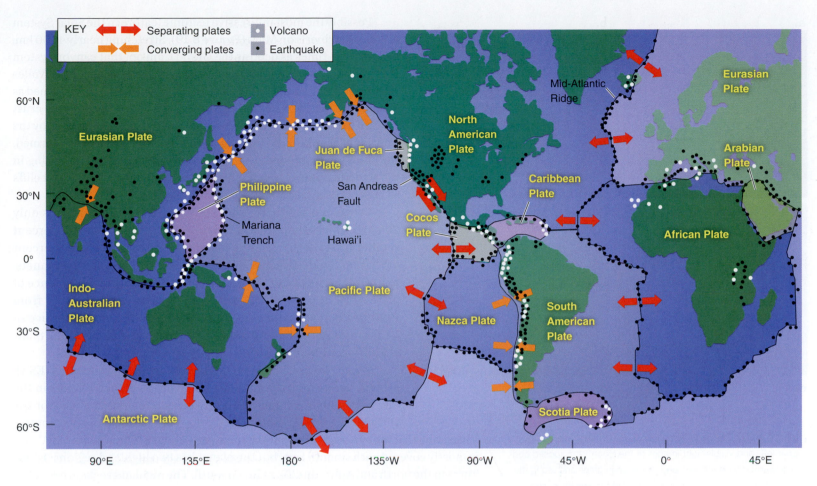

Figure 8.17 Major earthquakes and volcanic activity are often concentrated along the boundaries of Earth's principal tectonic plates.

As plates shift, some parts move more rapidly than others, causing the plates to stretch, buckle, or fracture. Those effects are seen on the surface as folded and faulted rocks. Mountain chains are common near converging plate boundaries, where plates buckle and break.

Tectonism on Other Planets

We have observed plate tectonics only on Earth, but all the terrestrial planets and some moons show evidence of tectonic disruptions. Many of these are caused by cooling: as the world cools, it shrinks, and the crust cracks and buckles, much like the skin of a grape wrinkles when it becomes a raisin.

NASA's *Lunar Reconnaissance Orbiter* (*LRO*) revealed a number of ridges with exposed bedrock (**Figure 8.18**). These could be evidence that not too long ago, tectonic activity disturbed the lunar regolith (fine dust) that lies across most of the surface. That regolith builds up relatively quickly, so these regions of exposed bedrock may be evidence of tectonic activity occurring even today. Further evidence comes from seismometers placed by the *Apollo* astronauts, which picked up numerous "moonquakes" caused by the Moon's long-term cooling, which causes the crust to crack and shake. On Mercury, cliffs that are hundreds of kilometers long appear to result from the shrinking of Mercury. Observations from the *Messenger* orbiter suggest that Mercury's diameter has shrunk by about 12 km, and continues to shrink even today.

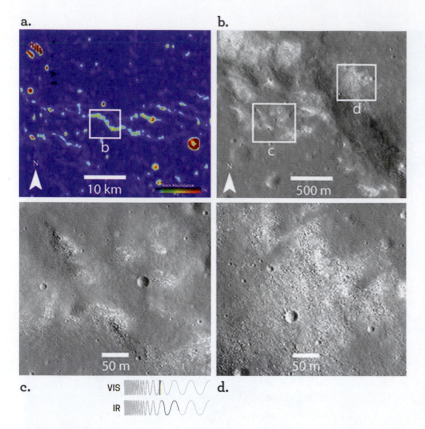

c.   VIS
      IR

d.

Figure 8.18 The *Lunar Reconnaissance Orbiter* took infrared (upper left) and visible-light images of the Moon.   **a.** Exposed bedrock is warmer than regolith, and shows up in green and red in the infrared image. Zooming in to the visible light image in   **b.** shows that the exposed bedrock is whiter than the surrounding region. Zooming in still further, in   **c.** and   **d.**, shows that the "smoothing" quality of lunar regolith is absent from this exposed bedrock surface.

Possibly the most impressive tectonic feature in the Solar System is Valles Marineris on Mars (**Figure 8.19**). Stretching nearly 4000 km, and nearly 4 times as deep as the Grand Canyon, that canyon system is as long as the distance between San Francisco and New York. Valles Marineris includes massive cracks in the crust of Mars that formed as local forces, perhaps related to mantle convection, pushed the crust upward from below. The surface could not be equally supported by the interior everywhere, so unsupported segments fell in. Once formed, the cracks were eroded by wind, water, and landslides, resulting in the structure we see today. Other parts of Mars have faults, but cliffs as high and long as those seen on Mercury are absent.

Venus and Earth are very similar; the mass of Venus is only 20 percent less than that of Earth, and its radius is just 5 percent smaller than Earth's, leading to a surface gravity that is 90 percent of that on Earth. Because of the similarities between the planets, many scientists predicted that Venus might also show evidence of plate tectonics. The NASA *Magellan* mission orbited Venus from 1990 to 1994, and *Magellan* mapped nearly the entire surface of Venus, providing the first high-resolution radar views of the planet's surface (**Figure 8.20**).

Venus Express, launched by the European Space Agency (ESA), orbited Venus from 2006 to 2014. The probe mapped Venus in the infrared, which can penetrate the clouds to enable a view of the surface. The impact craters on Venus seem to be evenly distributed, suggesting that the surface is all about the same age—about a billion years old. Venus is mostly covered with smooth lava, but the planet has two highland regions: Ishtar Terra in the north and Aphrodite Terra in the south. The highland rocks are rougher and older than those on the rest of Venus and may be similar to granite rocks on Earth. Because **granite** results from plate tectonics and water, the data hint at the possibility that those highlands on Venus are ancient continents, created by volcanic activity, on a planet that once had liquid water oceans.

Figure 8.19 a. A mosaic of *Viking Orbiter* images shows Valles Marineris, the major tectonic feature on Mars, stretching across the center of the image from left to right. This canyon system is more than 4000 km long. The dark spots on the left are huge shield volcanoes. **b.** This close-up perspective view of the canyon wall was photographed by the European Space Agency's *Mars Express* orbiting spacecraft.

a.   VIS

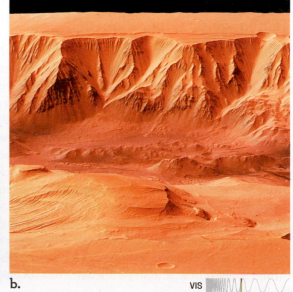

b.   VIS

Because of the similarities between Venus and Earth, the interior of Venus should be very much like the interior of Earth, and convection should be occurring in its mantle. On Earth, mantle convection and plate tectonism release the most thermal energy from the interior. On Venus, however, hot spots may be the principal way that thermal energy escapes from the planet's interior. Circular fractures called coronae on the surface of Venus, ranging from a few hundred kilometers to more than 2500 km across, may be the result of upwelling plumes of hot mantle that have fractured Venus's lithosphere. Alternatively, energy may build up in the interior until large chunks of the lithosphere melt and overturn, releasing an enormous amount of energy. Then, the surface cools and solidifies. Scientists remain uncertain about why Venus and Earth are so different with regard to plate tectonics.

CHECK YOUR UNDERSTANDING 8.4

On which of the following does plate tectonics occur now? (Select all that apply.) (a) Mercury; (b) Venus; (c) Earth; (d) the Moon; (e) Mars

8.5 Volcanism Signifies a Geologically Active Planet

The molten rock that emerges from a volcano is known as **lava**. It originates deep in the crust and in the upper mantle, where sources of thermal energy combine. Those sources include rising convection cells in the mantle, heating by friction from movement in the crust, and concentrations of radioactive elements. In this section, we look at the occurrence of volcanic activity on a planet or moon, called **volcanism**. Volcanism not only shapes planetary surfaces but also is a key indicator of a geologically active planet.

Terrestrial Volcanism Is Related to Tectonism

On Earth, volcanoes are usually located along plate boundaries and over hot spots. Maps of geological activity such as the one in Figure 8.17 leave little doubt that most terrestrial volcanism is linked to the same forces responsible for plate motions. A tremendous amount of friction is generated as plates slide under each other. That friction raises the temperature of rock toward its melting point. However, material at the base of a lithospheric plate is under a great deal of pressure because of the weight of the plate pushing down on it. That pressure increases the melting temperature of the material, so the material remains solid even at high temperature. As the material is forced up through the crust, its pressure drops and the material's melting temperature drops, too. Material that was solid at the base of a plate becomes molten as it nears the surface. Places where convection carries hot mantle material toward the surface are frequent sites of eruptions. Iceland, one of the most volcanically active regions in the world, sits astride one such spreading center—the Mid-Atlantic Ridge (see Figure 8.17).

Once magma reaches the surface of Earth, it can form many types of structures. Flows often form vast sheets, especially if the eruptions come from long fractures called **fissures**. If very fluid lava flows from a single *point source*, it spreads out over the surrounding terrain or ocean floor, forming a **shield volcano (Figure 8.21a)**. A

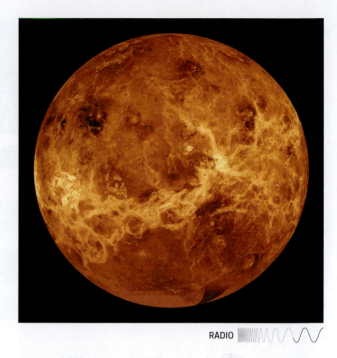

RADIO

Figure 8.20 The atmosphere of Venus blocks our view of the surface in visible light. This false-color view of Venus is a radar image made by the *Magellan* spacecraft. Bright yellow and white areas are mostly fractures and ridges in the crust. Some circular features seen in the image may be regions of mantle upwelling, or *hot spots*. Most of the surface is formed by lava flows, shown in orange.

▶❚❚ **AstroTour:** Hot Spot Creating a Chain of Islands

unanswered questions

Why is Venus so different from Earth? The two planets have similar size, mass, and composition, but they are very different geologically with respect to magnetic fields, plate tectonics, and recent activity, and it is not yet known why. In addition, how did Venus end up rotating in the direction opposite that of its revolution around the Sun? Did it form with a different orbit or rotation? Was it the result of an impact early in its history? Did it change very slowly because of tidal effects from other planets?

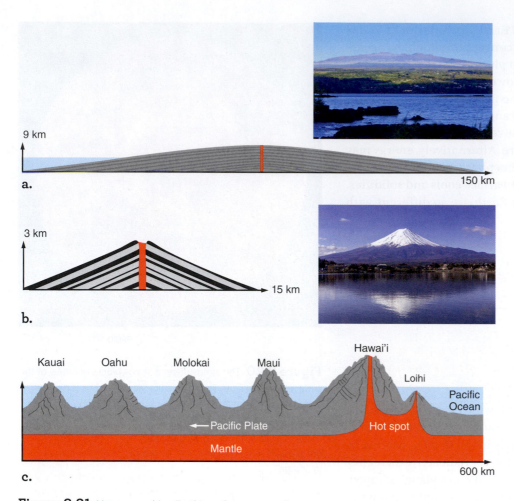

Figure 8.21 Magma reaching Earth's surface commonly forms **a.** shield volcanoes, such as Mauna Kea, which have gently sloped sides built up by fluid lava flows; and **b.** composite volcanoes, such as Mt. Fuji, which have steeply symmetric sides built up by viscous lava flows. **c.** Hot spots are convective plumes of lava that can form a successive series of volcanoes as the plate above them slides by.

Figure 8.22 This rock sample from the Moon, collected by the *Apollo 15* astronauts from a lunar lava flow, shows gas bubbles typical of gas-rich volcanic materials. The rock is about 6 × 12 cm.

composite volcano (Figure 8.21b) forms when thick lava flows alternating with explosively generated rock deposits build a steep-sided structure.

Terrestrial volcanism also occurs where convective plumes rise toward the surface in the interiors of lithospheric plates, creating local hot spots. Volcanism over hot spots works much like volcanism elsewhere at a spreading center, except that the convective upwelling occurs at a single spot rather than along the edge of a plate. Those hot spots force mantle and lithospheric material toward the surface, where the material emerges as liquid lava.

Earth has many hot spots, including the regions around Yellowstone Park and the Hawai'ian Islands (**Figure 8.21c**). The Hawai'ian Islands are a chain of shield volcanoes that formed as the lithospheric plate moved across a hot spot. Volcanoes erupt over a hot spot, building an island. The island ceases to grow as the plate motion carries the island away from the hot spot. Meanwhile, a new island grows over the hot spot. Today, the Hawai'ian hot spot is located off the southeast coast of the Big Island of Hawai'i, where it continues to power the active volcano forming the newest Hawai'ian island, Loihi. Loihi is already a massive shield volcano, rising more than 3 km above the ocean floor. In about 100,000 years, Loihi will break the surface of the ocean and merge with the Big Island of Hawai'i.

Volcanism in the Solar System

Evidence of volcanism appears throughout the Solar System, including several moons of the outer planets.

The Moon   Some of the first observers to use telescopes to view the Moon noted dark areas that looked like bodies of water—thus these areas were named **maria** (singular: *mare*), Latin for "seas." Maria are vast, hardened lava flows, with a composition similar to volcanic rocks known as basalts on Earth. Because the maria contain relatively few craters, those volcanic flows must have occurred after the period of heavy bombardment ceased.

Many of the rock samples that the *Apollo* astronauts brought back from the lunar maria contained gas bubbles typical of volcanic materials (**Figure 8.22**). The lava that flowed across the lunar surface must have been relatively fluid to form vast sheets that fill low-lying areas such as impact basins (**Figure 8.23**). The fluidity, due partly to the lava's chemical composition, partly explains the Moon's lack of classic volcanoes: the lava was too fluid to pile up—like motor oil that spreads out when poured from a container, the lava spread out across the lunar surface.

The lunar rock samples also showed that most of the lunar lava flows are older than 3 billion years. Samples from the heavily cratered terrain of the Moon also originated from magma, indicating that the young Moon went through a molten stage. Those rocks cooled from a "magma ocean" and are more than 4 billion years old, thus preserving the early history of the Solar System. Most of the sources of heating and volcanic activity on the Moon must have shut down some 3 billion years ago—unlike on Earth, where volcanism continues. That conclusion is consistent with the idea that smaller objects and planets cool more efficiently (see Working It Out 8.2) and are thus less active than larger planets.

Only in a few limited areas of the Moon are younger lavas thought to exist, but most of those have not been sampled directly. The *Lunar Reconnaissance Orbiter* observed volcanic cones that were probably built up from volcanic rocks erupting from the surface. Those volcanic rocks are far different from the mare basalt rocks and contain silica and thorium. Those domes could have been formed as recently as 800 million years ago, which would make them the result of the most recent volcanic activity found on the Moon.

Mercury   Mercury also shows evidence of past volcanism. The *Mariner 10* and *Messenger* missions revealed smooth plains on Mercury that looked similar to the lunar maria. Those sparsely cratered plains are the youngest areas on Mercury and, like those on the Moon, are almost certainly created when fluid lavas flowed into and filled huge impact basins. Many of the volcanic plains on Mercury also are associated with impact scars. The volcanic activity that created the plains probably ceased 3.8 billion years ago. Volcanic activity may have ceased because of the shrinking of the planet as it cooled. The ending of the late heavy bombardment might also have been a factor. High-resolution imaging by *Messenger* has also identified several volcanoes. Vents that could be from explosive volcanism have been found around the large, old impact basin Caloris (**Figure 8.24**) and may be as young as 1 billion to 2 billion years.

Mars   Mars also has been volcanically active. More than half the surface of Mars is covered with volcanic rocks. Lavas covered huge regions of Mars, flooding the older, cratered terrain. Most of the vents or long cracks that created those flows are buried under the lava that poured forth from them. Among the most impressive features on Mars are its enormous shield volcanoes—the largest mountains in the Solar System. Olympus Mons (**Figure 8.25**), standing 27 km high at its peak and 550 km wide at its base, would tower over Earth's largest mountains. Olympus Mons and its neighbors grew as the result of hundreds of thousands of individual eruptions. Because Mars does not have plate tectonics, its hot spots remained stationary for billions of years, so its volcanoes grew taller and broader in the lower surface gravity with each successive eruption.

Lava flows and other volcanic landforms span nearly the entire history of Mars and cover more than half the planet's surface. Volcanism on Mars began with the formation of the crust some 4.4 billion years ago and continued to geologically recent times, possibly even to present day. Although some "fresh appearing" lava flows have been identified on Mars, we will not know the age of those latest eruptions until rock samples are radiometrically dated. Mars could, in principle, experience eruptions today.

Venus   Of the terrestrial planets, Venus has the most volcanoes. Radar images reveal a variety of volcanic landforms. They include lava flows covering thousands of square kilometers, shield volcanoes approaching the size and complexity of those on Mars, and lava channels thousands of kilometers long. Those lavas must have been extremely hot and fluid to flow for such long distances. Some of the volcanic eruptions on Venus are thought to have been associated with the deformation of Venus's lithosphere above hot spots, such as the circular features mentioned previously.

what if . . .

What if we observed a large volcanic eruption on an Earth-like planet orbiting a nearby star? What might we conclude about the planet just from knowing about the eruption?

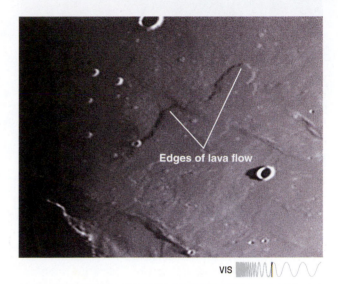

VIS

Figure 8.23 The lava flowing across the surface of Mare Imbrium on the Moon must have been extremely fluid to spread out for hundreds of kilometers in sheets only tens of meters thick.

Edges of lava flow

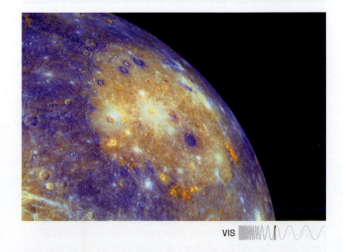

VIS

Figure 8.24 The Caloris Basin on Mercury (yellow) is one of the largest impact basins in the Solar System, with a span of about 1500 km. The orange regions may be volcanic vents. The false color is enhanced to show more detail.

Figure 8.25 The largest known volcano in the Solar System, Olympus Mons is a 27-km-high shield-type volcano on Mars, similar to but much larger than Hawai'i's Mauna Kea. This partial view of Olympus Mons was taken by the *Mars Global Surveyor*.

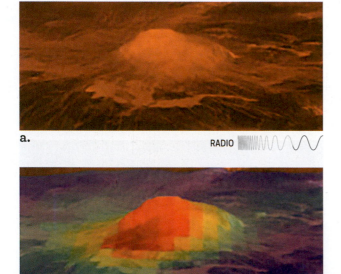

Figure 8.26 The volcanic peak Idunn Mons on Venus is 200 km in diameter and 2.5 km high. **a.** Radar data on top of the topographic data from NASA's *Magellan* mission. **b.** Visible and infrared data from ESA's *Venus Express* orbiter. The red-orange area at the summit is the warmest, suggesting the presence of some volcanic outflow activity.

Lavas on Venus are basalts, much like the lavas on Earth, the Moon, Mars, and possibly Mercury. There are nine Hawai'i-like hot spots on Venus. Each hot spot has several volcanoes, with altitudes of 500–1200 meters above the nearby plains. The *Venus Express* spacecraft imaged three of the nine Hawai'i-like hot spots in infrared wavelengths. One of these is shown in **Figure 8.26**. Some of the volcanic regions were hotter than the regions nearby. That finding suggests that those volcanic regions have younger material and that volcanic activity had taken place within the past 2.5 million years—perhaps as recently as a few thousand years ago. As we describe in Chapter 9, Venus's atmosphere has significant amounts of the volcanic gas sulfur dioxide. This further supports the idea that Venus remains volcanically active.

No rock samples have been retrieved from the surface of Venus to establish the age of the surface. Because of its relative lack of impact craters, most of the surface of Venus is probably less than 1 billion years old, and some of it may be much more recent. When volcanism began on Venus and how much active volcanism exists today remain unanswered questions.

CHECK YOUR UNDERSTANDING **8.5**

Which of the following is a reason for the large size of volcanoes on Mars in comparison with the smaller ones on Earth? (Choose all that apply.) (a) absence of plate tectonics; (b) distance from the Sun; (c) lower surface gravity than Earth's; (d) many repeated eruptions

8.6 The Geological Evidence for Water

Today, Earth is the only planet in the Solar System where the temperature and atmospheric conditions allow extensive liquid water to exist on the surface. The other inner planets do not have extensive liquid surface water, but evidence exists for water ice in deep craters or in the polar regions and for water below the surface in permafrost, subsurface glaciers, or possibly as liquid in the core.

Life, as we know it on Earth, requires water as a solvent and as a delivery mechanism for essential chemistry. In addition, humans would need a source of water to live on another terrestrial planet. Therefore, the search for water is central to the exploration of the Solar System. In this section, we explain how water modifies the surface of a planet and then we review the search for water in the Solar System.

Water and Erosion

Tectonism, volcanism, and impact cratering reshape Earth's surface by creating variations in the height of the surface. These variations form the surface's **topography**, as depicted on a topographic map of mountains or canyons. **Erosion** wears away a planet's surface by mechanical action through a variety of processes. Erosion, not only by running water but also by wind and by the actions of living organisms, wears down hills, mountains, and craters; the resulting debris fills in valleys, lakes, and canyons. If erosion were the only geological process operating, it would eventually smooth out the surface of the planet completely. Because Earth is a geologically and biologically active world, however, its surface is an ever-changing battleground between processes that build up topography and those that tear it down.

Weathering is the first step in erosion. During weathering, rocks are broken into smaller pieces and may be chemically altered. For example, rocks on Earth are physically weathered along shorelines, where the pounding waves break them into beach sand. Other weathering processes include chemical reactions, such as when oxygen in the air combines with iron in rocks to form a type of rust. One of the most efficient forms of weathering is caused by water: liquid water runs into crevices and then freezes. As the water freezes, it expands and shatters the rock.

After weathering, the resulting debris can be carried away by flowing water, glacial ice, or blowing wind and deposited in other areas as **sediment**. Where material is eroded, we can see features such as river valleys, wind-sculpted hills, or mountains carved by glaciers. Where eroded material is deposited, we see features such as river deltas, sand dunes, or piles of rock at the bases of mountains and cliffs. Erosion is most efficient on planets with water and wind. On Earth, where water and wind are prevalent, most impact craters on land are worn down and filled in.

Even though the Moon and Mercury have almost no atmosphere and no running water, a type of slow erosion is still at work. Radiation from the Sun and from deep space slowly decomposes some types of minerals, effectively weathering the rock. Such effects are only a few millimeters deep at most. Impacts of micrometeoroids also chip away at rocks. In addition, landslides can occur wherever gravity and differences in elevation are present. Although water enhances landslide activity, landslides also are seen on dry bodies such as Mercury and the Moon.

Earth, Mars, and Venus have significant atmospheres, and all three planets show the effects of windstorms. Images of Mars and Venus returned by spacecraft landers show surfaces that have been subjected to the forces of wind. Sand dunes are common on Earth and Mars (**Figure 8.27**), and some have been identified on Venus. Orbiting spacecraft also have found wind-eroded hills and surface patterns called wind streaks. Those surface patterns appear, disappear, and change in response to winds blowing sediments around hills, craters, and cliffs. The surface patterns serve as local weather vanes, telling planetary scientists about the direction of local prevailing surface winds. Planet-encompassing windy dust storms have been seen on Mars.

The Search for Water in the Solar System

A priority of recent planetary exploration missions is the search for water on the terrestrial planets and the Moon. Some of the evidence for past water comes from the geological processes discussed earlier in the chapter. Water was brought in by impacts in the early Solar System and is affected by geological and atmospheric activity on a planet. The search for water includes examination of images of the terrain obtained by flybys, orbiters, and landers. For the Moon, the search has included reanalyzing lunar rocks and soil brought back to Earth during the *Apollo* missions (1969–1972) and crashing spacecraft into the surface to analyze the resulting debris.

Mars The search for water on Mars goes back almost a century and a half. Even small telescopes show polar ice caps, which change with the martian seasons. In 1877, Italian astronomer Giovanni Schiaparelli (1835–1910) observed what appeared to be linear features on Mars and dubbed them *canali* (Italian for "channels"). Some other observers, including the U.S. astronomer Percival Lowell (1855–1916), incorrectly translated Schiaparelli's *canali* as "canals," concluding that they were artificially constructed by intelligent life rather than naturally formed as a result of geological processes. Lowell argued that those "canals" were built to move water around a drying planet Mars. Other observers at the time disputed that idea,

VIS

Figure 8.27 A *Mars Reconnaissance Orbiter* (*MRO*) image of the Nili Patera dune field on Mars. The dunes change over months because of winds on Mars.

what if . . .

What if Mars had large quantities of liquid water on its surface throughout its history? How might water erosion have changed the appearance of Olympus Mons or Valles Marineris?

VIS

Figure 8.28 A photograph of gully channels in a crater on Mars taken by the *MRO*. The gullies coming from the rocky cliffs near the crater's rim (out of the image, to the upper left) show meandering and braided patterns similar to those of water-carved channels on Earth.

arguing that the features were optical illusions seen in telescopes. Larger telescopes and astrophotographs did not show canals, astronomical spectroscopy did not find water vapor, and the idea of artificial canals faded away after Lowell died.

The geological evidence suggests that water once flowed across the surface of Mars in vast quantities. Canyons and huge, dry riverbeds attest to tremendous floods that poured across the martian surface. In addition, many regions on Mars show small networks of valleys that probably were carved by flowing water (**Figure 8.28**). Large deposits of ice have been detected under the surface. Mars may have contained oceans at one time, including one ocean that might have covered a third of the planet's surface. Scientists debate how recently significant liquid water has flowed on the surface of Mars.

In 2004, NASA sent two instrument-equipped roving vehicles, *Opportunity* and *Spirit*, to search for evidence of water on Mars. Each rover landed inside a crater, on opposite sides of the planet. For the first time, martian rocks were available for study in the original order in which they had been laid down. Previously, the only rocks that landers and rovers had come across were those that had been dislodged from their original settings by either impacts or river floods. *Spirit* encountered basaltic rocks with a structure that appears volcanic in nature, and on its travels found several rocks that may have been affected by water in the past. The layered rocks at the *Opportunity* site revealed that they had once been soaked in or transported by water. The form of the layers was typical of layered sandy deposits laid down by gentle currents of water. Magnified images of the rocks showed "blueberries"—small spheres a few millimeters across that probably formed in place among the layered rocks. Analysis of the spheres revealed abundant hematite, an iron-rich mineral that forms in the presence of water.

Observations by the ESA's *Mars Express* and NASA's *Mars Odyssey* and *Mars Reconnaissance Orbiter* (*MRO*) have shown the hematite signature and the presence of sulfur-rich compounds in a vast area surrounding *Opportunity*'s

VIS

Figure 8.29 The water-worn gravel embedded in sand shown in this Mars image from the *Curiosity* rover is evidence of an ancient streambed.

landing site. Those observations suggest the existence of an ancient martian sea larger than the combined area of the Great Lakes and as deep as 500 meters.

In 2012, NASA's Mars rover *Curiosity* landed in Gale Crater, a large (150 km), old (3.8 billion years) crater just south of the equator of Mars. *Curiosity* found evidence of a stream that flowed at a rate of about 1 meter per second and was as much as 2/3 meter deep. The streambed is identified by water-worn gravel (**Figure 8.29**). The rover, about the size of a car, includes cameras, a drill, and an instrument to measure chemical composition. When the rover drilled into a rock, it found sulfur, nitrogen, hydrogen, oxygen, phosphorus, and carbon, together with clay minerals that formed in a water-rich environment that was not very salty. That crater may well have once been a large lake that held water in the ground long after the lake evaporated.

Where did the water go? Some escaped into the thin atmosphere of Mars, and some is locked up as ice in the polar regions, just as the ice caps on Earth hold much of its water. Unlike Earth's polar caps of frozen water, those on Mars are a mixture of frozen carbon dioxide and frozen water. Water must be hiding elsewhere on Mars. Small amounts of water can be found on the surface, and in 2008, NASA's *Phoenix* lander found water ice just a centimeter or so beneath surface soils at high northern latitudes (**Figure 8.30**). However, most of the water on Mars appears to be trapped well below the surface. Radar imaging by *Mars Express* and the *MRO* indicates huge quantities of subsurface water ice, not only in the polar areas as expected but also at lower latitudes under craters, even as far south as the equator. In addition, *MRO* images suggest that seasonal salt water might flow on the surface far from the poles. Salt water freezes at a lower temperature, so some sites could be warm enough to have temporary liquid salt water. Other *MRO* images show underground ice exposed on a steep slope (**Figure 8.31**). Thus, Mars may have had conditions suitable to support Earth-like microbial life in the distant past.

Venus   Evidence for liquid water on Venus comes primarily from water vapor in its atmosphere, but some geological indications of past water are apparent, such as

VIS

Figure 8.30 Water ice appears a few centimeters below the surface in this trench dug by a robotic arm on the *Phoenix* lander. The trench measures about 20 × 30 cm.

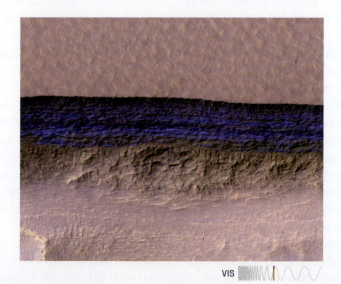

VIS

Figure 8.31 Color-enhanced *MRO* image of a steep slope on Mars in which underground water ice is exposed (and appears bright blue). *MRO* observed eight of these slopes, suggesting that Mars has places where water is accessible.

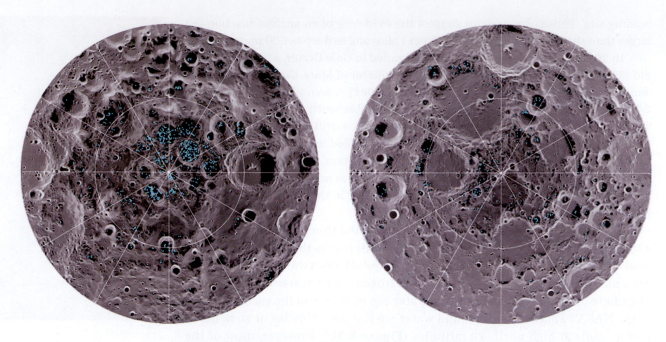

Figure 8.32 NASA's Moon Minerology Mapping instrument found surface ice at the poles of the Moon. In this image, the blue areas show the location of the ice, in the coldest and darkest locations at the South Pole (left) and North Pole.

unanswered questions

Will humans someday "live off the land" on the Moon? Recent space missions have provided evidence of some water ice on the Moon. If the Moon has water, that would certainly make living there more practical than if water had to be brought from Earth or synthesized from hydrogen and oxygen extracted from the lunar soil. Scientists and engineers have been studying several methods to see whether oxygen can be extracted from lunar rocks to make breathable air for people. Other researchers have looked at using lunar rock as a building material (to make concrete, for example). But the most valuable material on the Moon might turn out to be helium-3 (^{3}He), which is an isotope of helium with two protons and one neutron). On Earth, helium-3 exists only as a by-product of nuclear weapons, but up to a million tons of it may be on the Moon. Some scientists and engineers think that helium-3 could be used in a "clean" type of nuclear energy. Thus, that helium-3 could be brought back from the Moon for use on Earth or possibly even used in a power plant on the Moon to create energy for a lunar colony.

color differences between the highland and lowland regions. As noted previously, such a difference on Earth indicates the presence of granite, which requires water for its formation. We return to the subject of what happened to the water on Venus and Mars in Chapter 9, when we discuss their atmospheres.

The Moon   Infrared measurements of the Moon by the U.S. *Clementine* mission in 1994 supported the possibility of ice at the lunar poles. In 1998, NASA's *Lunar Prospector* observations suggested subsurface water ice in the polar regions. When the lunar orbiter's primary mission was completed, NASA crashed *Lunar Prospector* into a crater near the Moon's south pole while ground-based telescopes searched for evidence of water vapor above the impact site, but none was seen. Since 2007, several new missions have been sent to the Moon, including Japan's *Kaguya*, India's *Chandrayaan*, and NASA's *Lunar Reconnaissance Orbiter* (*LRO*) and its companion *Lunar Crater Observation and Sensing Satellite* (*LCROSS*).

In 2009, NASA crashed the *LCROSS* launch vehicle into the Cabeus Crater at the lunar south pole, which sent up dust and vapor that the *LCROSS* and *LRO* spacecraft analyzed. More than 5 percent of the resulting plume was water, which makes that part of the Moon wetter than many Earth deserts. Other volatiles also were detected. Measurements of hydrogen by the *LRO* suggest that a fair amount of buried water ice exists in the cold southern polar region. In mid-2018, data from an instrument on *Chandrayaan* definitively confirmed that there was water ice in the shadows of craters in the polar regions (**Figure 8.32**).

Those space observations of lunar ice sent planetary scientists back to the collections of lunar rocks and soil returned to Earth decades ago by the *Apollo* mission astronauts. New analysis of volcanic glass beads in lunar soil suggests that the interior of the Moon may have a much larger amount of volatiles than previously believed. Reanalysis of the *Apollo* lunar rocks also found evidence of water. One way to distinguish whether water on the Moon originated from its interior or from impacts is to look at the ratio of water molecules composed of regular hydrogen (with one proton) and oxygen to water molecules composed of oxygen, hydrogen, and an isotope of hydrogen called deuterium (hydrogen with one proton and one neutron). The ratio is higher in the water in lunar rocks than in the water on Earth. Water that originated in comets or in meteorites rich in water has a different ratio. Alternatively, protons from the solar wind or from high-energy cosmic rays in space could have combined with oxygen on the lunar surface, yielding a different ratio. Findings from a recent study suggest that the Moon's water came from the interactions between solar wind protons and oxygen.

Those results on lunar water are preliminary. Considerable debate continues among scientists about exactly how much water ice exists on the lunar surface and how much liquid water is in the interior, how the Moon acquired that water, and how the presence and origin of water affect the current theories of lunar formation. Several countries—and some private companies—are considering proposals to send robotic spacecraft to the Moon to collect additional lunar material and bring it back to Earth for analysis.

Mercury   Water ice also has been detected in the polar regions of Mercury. Some deep craters in the polar regions of Mercury have floors that are in perpetual shadow and thus receive no sunlight. Temperatures in those permanently shadowed areas remain very cold—below 180 K. For many years, planetary scientists had speculated that ice could be found in those polar craters, with a possible detection by radar occurring in the early 1990s. The *Messenger* spacecraft, which orbited Mercury from 2011 to 2015, found deposits of ice at craters at the planet's north pole. Other frozen volatiles also were seen in the polar craters. The icy areas have sharp boundaries, which indicates they are relatively recent, either from comet impacts or from some ongoing process on the planet.

In this chapter, you have used a strategy called "comparative planetology" to learn about the four processes that shape terrestrial worlds: impacts, tectonics, volcanism, and erosion. In December 1968, *Apollo 8* astronauts in orbit photographed Earth rising above the Moon (**Figure 8.33**). This image gives an opportunity for you to practice what you've learned about comparing terrestrial worlds.

Figure 8.33 ★ WHAT AN ASTRONOMER SEES   In December 1968, *Apollo 8* astronauts in orbit photographed Earth rising above the Moon. An astronomer viewing this image will be as awestruck as anyone by the simple fact that humans managed to figure out how to orbit the Moon and take such a picture. Then she will start to look at details. She will notice that the Moon has topography. There are elevation changes, particularly along the horizon, and large and small circular craters that vary in brightness. And she will notice that Earth is very different from the Moon. Studying Earth, she will notice blue water and brown land, as well as swirling cloud formations that indicate an atmosphere with active weather patterns. The Earth is only partially illuminated by the Sun, and that illumination shows that the Sun is located beyond the top of the image and slightly to the right. Because Earth is approximately in quarter phase, the Moon must likewise be approximately in quarter phase. An astronomer may spend considerable time trying to decipher which of the continents are visible on Earth, to figure out whether Earth's North Pole is on the right or the left, deciding eventually that north is on the right. From this information, she can make a mental sketch and figure out that the Moon was just about at first quarter, which is very satisfying.

Which of the following worlds show evidence of the current presence of liquid or frozen water? (Choose all that apply.) (a) Mercury; (b) Venus; (c) Earth; (d) the Moon; (e) Mars

Origins: The Death of the Dinosaurs

When large impacts happen on Earth, they can have far-reaching consequences for Earth's climate and for terrestrial life. One of the biggest and most significant impacts happened at the end of the Cretaceous Period, which lasted from 146 million years ago to 65 million years ago. At the end of the Cretaceous Period, more than half of all living species, including the dinosaurs, became extinct. That mass extinction is marked in Earth's fossil record by the **Cretaceous-Paleogene boundary**, or *K-Pg boundary* (the *K* comes from *Kreide*, German for "Cretaceous"). Fossils of dinosaurs and other now-extinct life-forms are found in older layers below the K-Pg boundary. Fossils in the newer rocks above the K-Pg boundary lack more than half of all previous species but contain a record of many other newly evolving species. Big winners in the new order were the mammals—distant ancestors of humans—that moved into the ecological niches that extinct species vacated.

How do scientists know that an impact was involved? The K-Pg boundary is marked in the fossil record in many areas by a layer of clay. Studies at more than 100 locations around the world have found large amounts of the element iridium in that layer, as well as traces of soot. Iridium is very rare in Earth's crust but is common in meteorites. The soot at the K-Pg boundary possibly indicates that widespread fires burned the world over. The thickness of the layer of clay at the K-Pg boundary and the concentration of iridium increases toward what is today the Yucatán Peninsula in Mexico. Although erosion has largely erased the original crater, geophysical surveys and rocks from drill holes in the area show a deeply deformed subsurface rock structure, similar to that seen at known impact sites. Those results provide compelling evidence that 65 million years ago an asteroid about 10 km in diameter struck the area, throwing great clouds of red-hot dust and other debris into the atmosphere (**Figure 8.34**) and possibly igniting a worldwide conflagration. The energy of the impact is estimated to have been more than that released by 5 billion nuclear bombs.

An impact of that energy would have devastated terrestrial life. In addition to the possible firestorm ignited by the impact, computer models suggest that earthquakes and tsunamis would have occurred. Dust from the collision and soot from the firestorms thrown into Earth's upper atmosphere would have remained there for years, blocking sunlight and plunging Earth into decades of a cold and dark "impact winter." Recent measurements of ancient microbes in ocean sediments suggest that Earth may have cooled by 7°C. The firestorms, temperature changes, and decreased food supplies could have led to a mass starvation that would have been especially hard on large animals such as the dinosaurs.

Not all paleontologists believe that mass extinction event was the result of an impact; some think volcanic activity was important as well. However, the evidence is compelling that a great impact did occur at the end of the Cretaceous Period. Life on our planet has had its course altered by sudden and cataclysmic events when asteroids and comets have slammed into Earth. We quite possibly owe our existence to the luck of our remote ancestors—small rodent-like mammals—that could live amid the destruction after such an impact 65 million years ago.

Figure 8.34 This artist's rendition depicts an asteroid or comet, perhaps 10 km across, striking Earth 65 million years ago in what is now the Yucatán Peninsula in Mexico. The lasting effects of the impact might have killed off most forms of terrestrial life, including the dinosaurs.

SUMMARY

The terrestrial planets in the Solar System are Mercury, Venus, Earth, and Mars, all of which have evidence of past or present water. The Moon is usually included in discussions of the terrestrial worlds because it is similar to them in many ways. Comparative planetology is the key to understanding the planets. Four geological processes—impacts, volcanism, tectonism, and erosion—are responsible for the topography on the terrestrial planets. Active volcanism and tectonics are the results of a "living" planetary interior: one that is still hot inside. Over time, the interiors cool, and tectonics and volcanism weaken. On Earth, radioactive decay and tidal effects from the Moon help heat the interior. Erosion is a surface phenomenon that results from weathering by wind or water. Surface features on the terrestrial planets, such as tectonic plates, volcanoes, mountain ranges, or canyons, are the result of the interplay among those four processes. Evolution on Earth may have been affected by impacts, such as the impact of an asteroid 65 million years ago that might have led to the death of the dinosaurs.

(1) Describe how impacts have affected the evolution of the terrestrial planets. Impact cratering is the result of a direct interaction of an astronomical object with the surface of a planet. The layering of craters gives their relative ages, with more recent craters found superimposed on older ones. Crater densities can be used to find the relative ages of regions on a surface; more heavily cratered regions are older than less cratered ones. Planets protected by atmospheres, such as Earth and Venus, have fewer small impact craters. The Moon was probably created when a Mars-sized protoplanet collided with Earth.

(2) Explain how radiometric dating is used to measure the ages of rocks and terrestrial planetary surfaces. Radioactive isotopes found in rocks can be used to measure their age. The oldest rocks measured, from the Moon and from meteorites, give the Solar System an age of 4.5 billion to 4.6 billion years.

(3) Explain how scientists use both theory and observation to determine the structure of terrestrial planetary interiors. Models of Earth's interior are used to predict how seismic waves should propagate through the interior, and those predictions are compared with observations of seismic waves. The interiors of other planets are modeled using physical principles, along with observational data of their magnetic fields. Earth has a strong magnetic field, but Venus and Mars do not. The cause for that difference between the terrestrial planets is still uncertain.

(4) Compare tectonism and volcanism and the forms they take on different terrestrial planets. Tectonism folds, twists, and cracks the outer surface of a planet. Plate tectonics is unique to Earth currently, although other types of tectonic disruptions are observed on the other terrestrial planets, such as cracking and buckling on the surface. Smooth areas on the Moon and Mercury are ancient lava flows. While Venus has the most volcanoes, the largest mountains in the Solar System are volcanoes on Mars. Earth's surface is still changing as volcanic hot-spot activity forms new members of island chains.

(5) **Summarize what is known about the presence of water on the terrestrial planets.** Geological features observed in orbiting and surface missions to Mars suggest that liquid water was once on the surface. Mars today has large amounts of subsurface water ice. Venus might have had liquid oceans early in its history. Space mission data indicate that water ice exists near the poles of the Moon and Mercury. The search for water is important to both the search for extraterrestrial life and the possibilities of human colonization of space.

QUESTIONS AND PROBLEMS

TEST YOUR UNDERSTANDING

1. _____, _____, and _____ build up structures on the terrestrial planets, whereas _____ tear(s) them down.
   a. impacts, erosion, volcanism; tectonism
   b. impacts, tectonism, volcanism; erosion
   c. tectonism, volcanism, erosion; impacts
   d. tectonism, impacts, erosion; volcanism

2. When geologists place features in order from top to bottom, this determines
   a. the relative size of features.
   b. the origin of features.
   c. the relative age of features.
   d. the actual age (in years) of features.

3. Scientists can learn about the interiors of the terrestrial planets from
   a. seismic waves.
   b. satellite observations of gravitational fields.
   c. physical arguments about cooling.
   d. satellite observations of magnetic fields.
   e. all of the above

4. Earth's interior is heated by
   a. angular momentum and gravity.
   b. radioactive decay and gravity.
   c. radioactive decay and tidal effects.
   d. angular momentum and tidal effects.
   e. gravity and tidal effects.

5. If a radioactive element has a half-life of 10,000 years, what fraction of it is left in a rock after 40,000 years?
   a. 1/2   c. 1/8   e. 1/32
   b. 1/4   d. 1/16

6. Lava flows on the Moon and Mercury created large, smooth plains. We don't see similar features on Earth because
   a. Earth has less lava.
   b. Earth had fewer large impacts in the past.
   c. Earth has plate tectonics that recycle the surface.
   d. Earth is large with respect to the size of those plains, so they are not as noticeable.
   e. Earth rotates much faster than either of those other worlds.

7. Scientists know the history of Earth's magnetic field because
   a. the magnetic field hasn't changed since Earth formed.
   b. they see today's changes and project backward in time.
   c. the magnetic field becomes frozen into rocks, and plate tectonics spreads those rocks apart.
   d. they compare the magnetic fields on other planets to Earth's.

8. Suppose an earthquake occurs on an imaginary planet. Scientists on the other side of the planet detect primary waves but not secondary waves after the quake. That suggests that
   a. part of the planet's interior is liquid.
   b. the planet's entire interior is solid.
   c. the planet has an iron core.
   d. the planet's interior consists entirely of rocky materials.
   e. the planet's mantle is liquid.

9. Geologists can determine the actual age of features on a planet by
   a. radiometric dating of rocks retrieved from the planet.
   b. comparing cratering rates on one planet with those on another.
   c. assuming that all features on a planetary surface are the same age.
   d. both a and b
   e. both b and c

10. Impacts on the terrestrial planets and the Moon
    a. are more common than they used to be.
    b. have occurred at approximately the same rate since the Solar System formed.
    c. are less common than they used to be.
    d. periodically become more common and then less common.
    e. never occur anymore.

11. Earth has fewer craters than Venus because
    a. Earth's atmosphere protects better than Venus's.
    b. Earth is a smaller target than Venus.
    c. Earth is closer to the asteroid belt.
    d. Earth's surface experiences more erosion.

12. Scientists propose an early period of heavy bombardment in the Solar System because
    a. the Moon is heavily cratered.
    b. all the craters on the Moon are old.
    c. the smooth part of the Moon is nearly as old as the heavily cratered part.
    d. all the craters on the Moon are young.

13. Scientists know that Earth was once completely molten because
    a. the surface is smooth.
    b. the interior layers are denser.
    c. the chemical composition indicates that.
    d. volcanoes exist today.

14. What is the main reason why Earth's outer core is liquid today?
    a. the tidal force of the Moon on Earth
    b. seismic waves that travel through Earth's interior
    c. the decay of radioactive elements
    d. convective motions in the mantle
    e. pressure on the core from Earth's outer layers

15. Mars has about half of Earth's diameter. If the interiors of those planets are heated by radioactive decays, how does the heating rate of the interior of Mars compare with that of Earth?
    a. The heating rate of Mars is 0.125 times that of Earth.
    b. The heating rate of Mars is 8 times that of Earth.
    c. The heating rate of Mars is 0.5 times that of Earth.
    d. The heating rate of Mars is 4 times that of Earth.
    e. The heating rates are about the same.

THINKING ABOUT THE CONCEPTS

16. When discussing the terrestrial planets, why do we include Earth's Moon?

17. Can all rocks be dated with radiometric methods? Explain.

18. Explain how scientists know that rock layers at the bottom of the Grand Canyon are older than those found on the rim.

19. Describe the sources of heating responsible for generating Earth's magma.

20. Explain why the Moon's core is cooler than Earth's.

21. Explain the difference between longitudinal waves and transverse waves.

22. How do we know that Earth's core includes a liquid zone?

23. Study the Process of Science Figure. What evidence makes the impactor theory the currently preferred explanation for the origin of the Moon? What evidence remains to be found to rule out the competing theories? ⊛

24. Explain plate tectonics and identify the only planet on which that process has been observed.

25. Volcanoes have been found on all the terrestrial planets. Where are the largest volcanoes in the inner Solar System?

26. Explain the criteria you would apply to images (assume adequate resolution) to distinguish between a crater formed by an impact and one formed by a volcanic eruption.

27. What are the primary reasons scientists have decided that the surfaces of Venus, Earth, and Mars are younger than those of Mercury and the Moon?

28. Explain some of the geological evidence suggesting that Mars once had liquid water on its surface.

29. What evidence supports the theory suggesting that a mass extinction occurred as a consequence of an enormous impact on Earth 65 million years ago?

30. ★ **WHAT AN ASTRONOMER SEES**   Study the surface of the Moon that is visible in the image in Figure 8.1. Is this surface old (dating to the period of early heavy bombardment) or young (formed much later)? How do you know? ⊙

APPLYING THE CONCEPTS

31. The glaze on a piece of ancient pottery contains radium, a radioactive element that decays to radon and has a half-life of 1620 years. Radon could not have been in the glaze when the pottery was being fired, but now it contains three atoms of radon for each atom of radium. How old is the pottery? ●–●–●
    a. Make a prediction: What do you already know about what the magnitude of a reasonable answer might be? Could the pottery be tens of years old? Hundreds? Thousands? Hundreds of thousands? Millions?
    b. Calculate: Follow Working It Out 8.1 to find the age of the pottery.
    c. Check your work: Verify that your answer has the correct units. Compare your answer to your prediction to check that it has approximately the right magnitude.

32. Assume that Earth is a perfect sphere with radius of 6371 km. What are the volume and the surface area of Earth? ●–●–●
    a. Make a prediction: What do you expect the units of volume and surface area to be? Which of these two numbers do you expect to be larger?
    b. Calculate: Follow Working It Out 8.2 to find the volume and the surface area of Earth.
    c. Check your work: Verify that your units are correct, and compare your two numerical results to your prediction.

33. Archaeological samples are often dated by using radiocarbon dating. The half-life of carbon-14 is 5700 years. How much time will have passed if a sample has only 1/64 as much carbon-14 as it originally contained?
    a. Make a prediction: Make an estimate by considering the number "64." About how many times would you have to multiply 2 by itself to get 64: 10 times? 100 times? 1000 times? Based on your prediction, do you expect the age to be about 10 times the half-life? 100 times the half-life? 1000 times the half-life? What kind of "ballpark" is reasonable?
    b. Calculate: Follow Working It Out 8.1 to determine how many half-lives have passed.
    c. Calculate: Multiply the number of half-lives by the half-life in years to determine how much time has passed.
    d. Check your work: Compare your result to your prediction to see if your answer "makes sense"; that is, is your answer reasonable or is it vastly too large or too small?
    e. Consider further: If the daughter product of carbon-14 is present in the sample when it forms (even before any radioactive decay happens), you cannot assume that every daughter product you see is the result of carbon-14 decay. If you did make that assumption, would you overestimate or underestimate the age of a sample?

34. The Moon's mean radius is 1738 km, and its mass is 7.2×10^{22} kg. Calculate the density of the Moon. What does this density tell you about the composition of the Moon?

   a. Make a prediction: In Section 8.3, you read about the density of water, rock, and iron. Given what you learned in this chapter about the formation of the Moon, which of these densities do you expect to be closest to the density of the Moon?

   b. Calculate: Use the mass of the Moon and the formula for the volume of a sphere given in Working It Out 8.2 to calculate the density of the Moon.

   c. Check your work: Verify that your units are correct, and compare your answer to the given densities for water, rock, and iron to check your intuition about what the density of the Moon should be!

35. Earth's mean radius is 6371 km, and its mass is 6.0×10^{24} kg. The Moon's mean radius is 1738 km, and its mass is 7.2×10^{22} kg. Which body cooled faster, and how many times faster did it cool?

   a. Make a prediction: Working It Out 8.2 shows that the rate at which a planet cools is proportional to its radius. Does Earth or the Moon have a larger radius, and by about how many times is it larger?

   b. Calculate: Set up a ratio of volume/surface area (V/A) for the larger body to V/A of the smaller body, using the equation for V/A shown in Working It Out 8.2. Calculate this ratio to determine how many times faster the smaller body cools than the larger one.

   c. Check your work: Verify that all the units cancel in your answer (because it's a ratio), and then compare your answer to your prediction.

36. Use Figure 8.8 to answer the following questions. ⊙

   a. How has the cratering rate changed? Has it fallen off gradually or abruptly?

   b. At present, what is the cratering rate compared with that about 4 billion years ago?

   c. Why does that falloff in cratering rate fit nicely into the theory of planet formation?

37. Are the vertical and horizontal axes in Figure 8.7 linear or logarithmic? After how many half-lives does the number of parent isotopes equal the number of daughter isotopes? Is the result unique to that example? Why or why not? ⊙

38. The destruction of the parent isotope is an example of exponential decay (Figure 8.7). Is the growth of the daughter isotope an example of exponential growth? How can you tell? ⊙

39. Study Figure 8.23. Is this region more similar in age to the region sampled by *Apollo 16* or *Apollo 12* (see Figure 8.8). How do you know? ⊙

40. Assume that Earth and Mars are perfect spheres with radii of 6371 km and 3390 km, respectively.

   a. Calculate the surface area of Earth.

   b. Calculate the surface area of Mars.

   c. If 0.72 (72 percent) of Earth's surface is covered with water, compare the amount of Earth's land area with the total surface area of Mars.

41. Compare the kinetic energy ($= 1/2 \, mv^2$) of a 1-gram piece of ice (about half the mass of a dime) entering Earth's atmosphere at a speed of 50 km/s with that of a 2-metric-ton SUV (mass $= 2 \times 10^3$ kg) speeding down the highway at 90 km/h.

42. The object that created Arizona's Meteor Crater (Figure 8.4) was estimated to have a radius of 25 meters and a mass of 300 million kg. Calculate the density of the impacting object, and explain what that may tell you about its composition.

43. Earth's mean radius is 6371 km, and its mass is 6.0×10^{24} kg. The Moon's mean radius is 1738 km, and its mass is 7.2×10^{22} kg.

   a. Calculate Earth's average density. Show your work; do not look the value up.

   b. The average density of Earth's crust is 2600 kg/m^3. What does that tell you about the composition of Earth's interior?

   c. Compute the Moon's average density. Show your work.

   d. Compare the average densities of the Moon, Earth, and Earth's crust. What do those values tell you about the Moon's composition in comparison with that of Earth and of Earth's crust?

44. Different radioisotopes have different half-lives. For example, the half-life of carbon-14 is 5700 years, the half-life of uranium-235 is 704 million years, the half-life of potassium-40 is 1.3 billion years, and the half-life of rubidium-87 is 49 billion years.

   a. Why would an isotope with a half-life like that of carbon-14 be a poor choice to get the age of the Solar System?

   b. The age of the universe is approximately 14 billion years. Does that mean that no rubidium-87 has decayed yet?

45. Assume that the east coast of South America and the west coast of Africa are separated by an average distance of 4500 km. Assume also that GPS measurements indicate that those continents are now moving apart at 3.75 cm/yr. If that rate has been constant over geological time, how long ago were the continents joined as part of a supercontinent?

EXPLORATION Exponential Behavior

Knowledge of the exponential function is critical to understanding life in modern times. That function shows up in contexts as diverse as economics, population studies, and climate change.

Radioactive decay exhibits exponential decay, where the amount of a radioactive material that remains after an elapsed time is proportional to the amount present at the beginning of the period. In this Exploration, you will use a small (1.69-ounce) bag of plain M&M's to investigate that behavior.

A 1.69-ounce bag of M&M's contains approximately 56 pieces of candy. Each piece has an *M* stamped on one side. Thus, each piece can land in two ways when dropped: *M* side up or *M* side down. That is much like what happens to a radioactive nucleus: for any given time, either it decays or it doesn't.

Instead of acquiring 10 bags of M&M's, you will use the same bag 10 times and then add your results together. That approach makes the sample large enough that the exponential behavior becomes apparent.

Step 1 For the first trial, shake all the M&M's into your hands and then pour them onto the table.

Step 2 Count how many land *M* side up, and record that number in the provided table under Trial 1, Time Step 1.

Step 3 Set the *M*-side-up candies to one side.

Step 4 Repeat steps 1–3 for time steps 2, 3, 4, and so on, until the last candy lands *M* side up.

Step 5 Repeat steps 1–4 for nine more trials (for a total of 10 trials).

Step 6 Add the results of the trials for each time step, and record them in the Sum column of the table.

Step 7 Making sense of numbers in tabular form like this is hard, so plot the results on a graph with the sum on the *y*-axis and the time step on the *x*-axis.

Time Step	Trial 1	Trial 2	Trial 3	Trial 4	Trial 5	Trial 6	Trial 7	Trial 8	Trial 9	Trial 10	Sum
1											
2											
3											
4											
5											
6											
7											
8											
9											
10											
11											
12											

1. Study your graph. At first, does the number of M&M's landing *M* side up decrease slowly or quickly? In later time steps (such as 8 or 9), does the number decrease slowly or quickly?

2. Generalize your answer to question 1 to the behavior of radioactive sources. Does the radioactivity fall off slowly or quickly at the beginning of an observation? How about at later times?

3. As time goes by, what happens to the number of M&M's that remain? What happens to the number of M&M's in the pile set aside?

4. Generalize your answer to question 3 to the behavior of radioactive sources. What happens to the number of radioactive isotopes over time? What happens to the number of daughter products?

5. Imagine that you walk into a room and observe another student performing this experiment. She pours the M&M's onto the table and counts 10 candies that landed *M* side up. About how many time steps have passed since she started the experiment?

6. Apply your answer to question 5 to the behavior of radioactive sources. When scientists study radioactive sources, they generally study the ratio of the number of radioactive isotopes to the number of daughter products. Explain how that method, though different from what you did in step 5, contains the same information about the amount of time that has passed.

Atmospheres of the Terrestrial Planets

Throughout the history of Earth, life has modified the atmosphere and the surface, as you will learn here in Chapter 9. In this experiment, you will see that even a small amount of microscopic life can significantly modify its environment. Stretch out a balloon by blowing it up and letting the air out a few times. Combine 1 cup of very warm water, 2 tablespoons of sugar, and 1 packet of active dry yeast in a clean soda bottle, and mix them together until bubbling starts, which is the sign that the yeast are producing CO_2. Attach the balloon to the top of the bottle. Measure the circumference of the balloon. After the first measurement, begin to sketch a graph with circumference on the y-axis and time on the x-axis. How do you expect the balloon's size to increase with time? Continue measuring the balloon's circumference until it stops expanding. Graph your results and compare them to your prediction. How long did it take for the balloon to stand upright? Why did the balloon eventually stop expanding?

EXPERIMENT SETUP

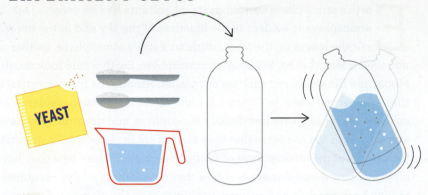

Stretch out a balloon by blowing it up and letting the air out a few times. Combine 1 cup of very warm water, 2 tablespoons of sugar, and 1 packet of active dry yeast in a clean soda bottle. Mix them together until bubbling starts.

Attach the balloon to the top of the bottle. Note the time, and wait for the balloon to stand upright above the bottle.

Note the time again, and measure the circumference of the balloon. At even time intervals (every few minutes), measure the circumference of the balloon, as you did in Chapter 7. Continue measuring the balloon's circumference until it stops expanding.

 00:00   00:10   00:12   00:14

PREDICTION

I predict that the circumference of the balloon will stop changing
☐ abruptly.   ☐ smoothly.

SKETCH OF RESULTS (in progress)

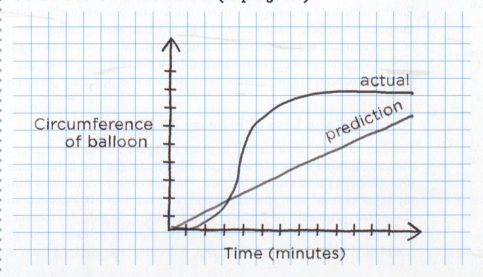

Circumference of balloon

actual

prediction

Time (minutes)

9

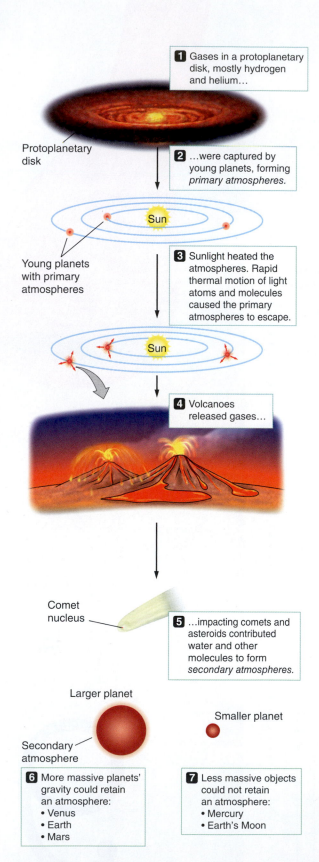

1 Gases in a protoplanetary disk, mostly hydrogen and helium…

Protoplanetary disk

2 …were captured by young planets, forming *primary atmospheres.*

Sun

Young planets with primary atmospheres

3 Sunlight heated the atmospheres. Rapid thermal motion of light atoms and molecules caused the primary atmospheres to escape.

Sun

4 Volcanoes released gases…

Comet nucleus

5 …impacting comets and asteroids contributed water and other molecules to form *secondary atmospheres.*

Larger planet

Smaller planet

Secondary atmosphere

6 More massive planets' gravity could retain an atmosphere:
• Venus
• Earth
• Mars

7 Less massive objects could not retain an atmosphere:
• Mercury
• Earth's Moon

Figure 9.1 Planetary atmospheres form and evolve in phases.

E arth's atmosphere surrounds its inhabitants like an ocean of air. The atmosphere is evident in the blueness of the sky and in the breezes that stir the leaves on the trees. Without Earth's atmosphere, neither clouds nor oceans would exist. Without an atmosphere, Earth would look much like the Moon, and life would not exist on our planet. Among the five terrestrial bodies that you learned about in Chapter 8, only Venus and Earth have dense atmospheres. Mars has a very low-density atmosphere, and the atmospheres of Mercury and the Moon are so sparse that they can hardly be detected. To understand the origins of the atmospheres of Earth, Venus, and Mars—how they have changed, how they compare, and how they are likely to evolve—requires us to look back nearly 5 billion years to a time when the planets were just completing their growth.

LEARNING GOALS

Here in Chapter 9, we compare the atmospheres of the terrestrial planets. By the end of this chapter, you should be able to:

1 Identify the processes that cause primary and secondary atmospheres to be formed, retained, and lost.

2 Compare the strength of the greenhouse effect on Earth, Venus, and Mars and how it contributes to differences among the atmospheres of those planets.

3 Describe the layers of the atmospheres on Earth, Venus, and Mars.

4 Explain how Earth's atmosphere has been reshaped by the presence of life.

5 Describe the evidence that shows Earth's climate is changing and how comparative planetology contributes to a better understanding of those changes.

9.1 Atmospheres Change over Time

An **atmosphere** is a layer of gas that sits above the surface of a body such as a planet, moon, or star. On Earth, a blanket of atmosphere warms and sustains Earth's temperate climate. On Venus, a thick atmosphere of carbon dioxide pushes the planet's surface temperature very high. A thin atmosphere leaves the surface of Mars unprotected and frozen. Mercury and the Moon have essentially no atmosphere. Why do some terrestrial planets have dense atmospheres, whereas others have little or none? In this section, you will learn how planetary atmospheres form.

Formation and Loss of Primary Atmospheres

Planetary atmospheres formed in phases (**Figure 9.1**). In the first phase, young planets captured some of the surrounding hydrogen and helium that filled the protoplanetary disk around the Sun. This gas capture continued until the supply of gas ran out, soon after the planets formed. This early gaseous atmosphere, composed primarily of hydrogen and helium, is called a planet's primary atmosphere. For terrestrial planets near the Sun, the primary atmospheres were temporary; the lightweight atoms and molecules later escaped from the planets' gravity. How can gas molecules escape from a planet? To understand that, you must learn more about how particles move within a planetary atmosphere.

Recall from Chapter 4 that any object can escape a planet if the object's speed exceeds the escape velocity and the object is traveling in the right direction. The escape velocity depends on the mass of the planet; terrestrial planets have small masses, so the escape velocity from them is relatively low. Fast-moving hydrogen and helium in the primary atmospheres of the terrestrial planets leaked back into space, even while the primary atmospheres were forming. In addition, giant impacts by large planetesimals early in the history of the Solar System may have imparted large amounts of energy all at once, and blasted away some of the particles in the primary atmospheres of the terrestrial planets. But the most important contributor to the loss of primary atmospheres was heating by the Sun.

After the protoplanetary disk dissipated, intense radiation from the Sun heated the planets' atmospheres, increasing the kinetic energy of the atmospheric particles. Particles of all masses would have had, on average, the same kinetic energy. The kinetic energy of a particle depends on its mass and its speed. Hydrogen and helium both have very low mass, so increasing their kinetic energy makes them move much faster than heavier particles. For example, in a mixture of hydrogen and oxygen at room temperature, hydrogen molecules will be rushing around at about 2000 meters per second (m/s) on average, whereas the much more massive oxygen molecules will be moving at only 500 m/s. Heating from the Sun increased the chances that some of the hydrogen and helium would be traveling faster than the planet's escape velocity, so as the atmosphere heated, more and more of it escaped into space. The ability of planets to hold on to their atmospheres is explored further in **Working It Out 9.1**.

The process accelerates over time. Deep within a planet's atmosphere, fast molecules near the ground will almost certainly collide with other molecules before the fast molecules can escape. These collisions tend to equalize the speeds, making slow particles move faster, and slowing down the faster particles, so that few are moving faster than escape speed. The upper levels of the atmosphere, however, contain fewer molecules. Fast molecules in the upper atmosphere are less likely to collide with other molecules before they escape. As particles escape into space, the atmosphere close to the ground becomes thinner, and acts more like the upper atmosphere. Over time, more and more particles are lost, due to heating by the Sun. In contrast, the giant planets, which are much more massive than the terrestrial planets and farther from the Sun, had both stronger gravity and lower temperatures so they were able to retain nearly all of their massive primary atmospheres.

▶❚❚ **AstroTour:** Atmospheres: Formation and Escape

The Formation of Secondary Atmospheres

If the primary atmospheres of the terrestrial planets were lost, why does Earth have an atmosphere today? The secondary atmospheres of Earth, Venus, and Mars that exist today come from accretion, volcanism, and impacts. During the planetary accretion process, minerals containing water, carbon dioxide, and other volatile matter collected in the planetary interiors. Later, as the interiors heated up, those gases were released from the minerals that had held them. Volcanism then brought the gases to the surface, where they accumulated and created a secondary atmosphere (see step 4 of Figure 9.1).

Impacts by comets and asteroids were another important source of gases. Huge numbers of comets formed near Jupiter and beyond, so they were rich in volatiles such as water (see Figure 7.12). As the giant planets grew to maturity or migrated

working it out 9.1

Atmosphere Retention

To estimate a planet's ability to retain its atmosphere, we compare the escape velocity from the planet (which depends on the planet's gravity, determined from its mass and radius) with the average speed of the molecules in a gas (which depends on the temperature of the gas and the mass of the molecules that make up the gas). The escape velocity is

$$v_{esc} = \sqrt{\frac{2GM}{R}}$$

and the values of v_{esc}, in kilometers per second (km/s), are given for the inner planets in Table 8.1.

The temperature, T, of a gas is proportional to the kinetic energy of the particles, $1/2\,mv^2$ (Chapter 5). Rearranging that relationship to solve for v and inserting the constants of proportionality gives the average speed of a molecule in a gas:

$$v_{molecule} = \sqrt{\frac{3kT}{m}}$$

In this expression, T is the temperature of the gas in kelvins, m is the mass of the molecule in kilograms (kg), and k is the Boltzmann constant. The atomic mass of a molecule is found by adding the atomic masses of its composite atoms as specified in the periodic table. (Atomic masses of atoms come from the total number of neutrons and protons; the weight of the electrons is negligible in comparison.) An oxygen molecule (O_2), for example, has an atomic weight of 32 (16 for each O atom), whereas a hydrogen molecule (H_2) has an atomic weight of 2 (1 for each H atom). Thus, O_2 has 16 times the mass of H_2.

If we put in the value of the Boltzmann constant ($k = 1.38 \times 10^{-23}$ joule per kelvin, or J/K) and the mass of the hydrogen atom ($m = 1.67 \times 10^{-27}$ kg), then $v_{molecule}$, in kilometers per second, is given by

$$v_{molecule} = 0.157 \text{ km/s} \times \sqrt{\frac{\text{Temperature of gas}}{\text{Atomic weight of molecule}}}$$

The higher the temperature, the higher the average kinetic energy of the individual particles. All the particles will have the same kinetic energy, on average, but particles that have smaller mass will have more of that kinetic energy apportioned to their speed. That difference explains why Earth can hold on to the oxygen in its atmosphere but loses hydrogen to space.

In our example of hydrogen and oxygen molecules, H_2 has an atomic weight of 2 (1 for each hydrogen atom), whereas O_2 has an atomic weight of 32 (16 for each oxygen atom, from 8 protons and 8 neutrons). Earth has an average temperature of 288 K. The average speeds of H_2 and O_2 are thus

$$\text{For } H_2: v_{molecule} = 0.157 \times \sqrt{\frac{288}{2}} = 1.88 \text{ km/s}$$

$$\text{For } O_2: v_{molecule} = 0.157 \times \sqrt{\frac{288}{32}} = 0.47 \text{ km/s}$$

Thus, in a gas containing both H_2 and O_2, the hydrogen molecule will, on average, be moving 4 times faster than the average oxygen molecule. At any given temperature, the lighter molecules will be moving faster. Not all molecules in a gas are moving at the average speed: some move faster and some move slower (**Figure 9.2**). The general rule is that over the age of the Solar System, a planet can keep its atmosphere if, for that type of gas molecule,

$$v_{molecule} \leq 1/6\, v_{esc}$$

The escape velocity from Earth is 11.2 km/s, and one-sixth of that is 1.87 km/s (that is, slightly less than the average speed of H_2). Those values explain why Earth has kept its O_2 but not its H_2. A similar analysis shows that on the Moon, with its lower v_{esc} value, both H_2 and O_2 escape. Jupiter, with its much colder temperatures and higher v_{esc} value, keeps both H_2 and O_2.

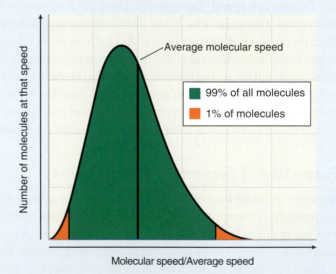

Figure 9.2 The distribution of the speeds of molecules in a gas. The shape of the curve and the exact numbers depend on the temperature of the gas and the masses of the molecules. In all cases, some of the speedier molecules may escape if they exceed the escape velocity.

their orbits, their gravity disturbed the orbits of nearby comets. Many of those icy bodies were flung outward by the giant planets to the outermost regions of the Solar System. Others were scattered into the inner Solar System. Upon impact with the terrestrial planets, those objects brought ices such as water, carbon monoxide, methane, and ammonia. On the terrestrial planets, cometary water mixed with the water that had been released into the atmosphere by volcanism (see step 5 of Figure 9.1). On Earth, and perhaps Mars and Venus, most of the water vapor then condensed as rain and flowed into the lower areas to form the earliest oceans. Rocky asteroids also brought materials to the terrestrial planets.

Sunlight altered the composition of secondary atmospheres. Ultraviolet (UV) light from the Sun easily breaks down molecules such as ammonia (NH_3) and methane (CH_4). Ammonia, for example, breaks down into hydrogen and nitrogen. When that happens, the lighter hydrogen atoms quickly escape to space, leaving behind the much heavier nitrogen atoms. Nitrogen atoms then combine in pairs to form more massive nitrogen molecules (N_2), and those more massive molecules are even less likely to escape into space. This process is the primary source of molecular nitrogen in the atmospheres of the terrestrial planets. On Earth, molecular nitrogen is about 78 percent of the atmosphere today.

Among the terrestrial planets, only Venus, Earth, and Mars have significant secondary atmospheres today. But both the Moon and Mercury have virtually no atmosphere today. Why? Some carbon dioxide and water must have accumulated on the Moon and Mercury during volcanic eruptions and comet impacts. But Mercury and the Moon are both less massive and smaller than the other terrestrial planets, so the escape velocity from their surface is lower. Each lost their secondary atmosphere through the same processes that cause the loss of a primary atmosphere. Even molecules as massive as carbon dioxide can escape from a small planet if the temperature is high enough, as it is on Mercury's sunlit side. Furthermore, intense UV radiation from the Sun can break molecules into less massive fragments, which are lost to space even more quickly. The Moon is much cooler than Mercury, but its mass is so small that molecules easily escaped even at relatively low temperatures. The weak magnetic fields of Mercury and the Moon are also a factor. With a weaker magnetic field, particles in the atmosphere are less protected from the **solar wind**—a constant stream of charged particles from the Sun. Particles in the solar wind can smash into atmospheric particles, causing some to reach escape velocity. This process is known as **sputtering**, and it speeds up the loss of the atmosphere to space.

CHECK YOUR UNDERSTANDING 9.1

Which are reasons Mercury has so little gas in its atmosphere? (Choose all that apply.) (a) Its mass is small. (b) It has a high temperature. (c) It is close to the Sun. (d) Its escape velocity is low. (e) It has no moons.

Answers to Check Your Understanding questions are in the back of the book.

what if . . .

What if the atmosphere of Mars were suddenly ripped away by some catastrophic event? What processes today would help restore the atmosphere?

9.2 Secondary Atmospheres Evolve

Venus, Earth, and Mars most likely started out with atmospheres of similar composition. All three planets must have shared the intense cometary showers of the early Solar System. All three planets have volcanism in their history: Earth is volcanically active, Venus might still be volcanically active, and Mars has been volcanically

Table 9.1	Atmospheres of the Terrestrial Planets		
Physical Properties and Composition			
	PLANET		
Property	Venus	Earth	Mars
Surface Pressure (bars)	92	1.0	0.006
Atmospheric Mass (kg)	4.8×10^{20}	5.1×10^{18}	2.5×10^{16}
Surface Temperature (K)	740	288	210
Carbon Dioxide (%)	96.5	0.041	95.3
Nitrogen (%)	3.5	78.1	2.7
Oxygen (%)	0.00	20.9	0.13
Water (%)	0.002	0.1 to 3	0.02
Argon (%)	0.007	0.93	1.6
Sulfur Dioxide (%)	0.015	0.02	0.00

Astronomy in Action: Changing Equilibrium

active in the recent past. These similar histories suggest that their early secondary atmospheres might also have been similar. However, today these atmospheres are very different from one another. To understand why, you need to learn a bit more about the effects of planetary mass and the greenhouse effect.

Planetary Mass Affects a Planet's Atmosphere

Table 9.1 shows that the *compositions* of the atmospheres of Venus and Mars today are nearly identical, but the composition of Earth's atmosphere is very different. The development of life on Earth increased the amount of oxygen in Earth's secondary atmosphere. Earth's atmosphere is now made up primarily of nitrogen and oxygen, with only a trace of carbon dioxide. The atmospheres of Venus and Mars, in contrast, are both nearly entirely carbon dioxide, with small amounts of nitrogen and virtually no oxygen. The total *amount* of atmosphere, however, differs widely among the three planets. The extremely dense atmosphere on Venus produces a pressure at the surface nearly 100 times greater than that on Earth. By contrast, the average pressure at the surface of Mars is less than a hundredth of that on Earth. Venus is nearly 8 times as massive as Mars, so Venus had roughly 8 times as much carbon within its interior to produce carbon dioxide, the principal secondary-atmosphere component of both planets.

The large difference in atmospheric mass comes from the relative strengths of each planet's **surface gravity**—the pull of gravity measured at the surface, which depends not only on the mass of the planet but also on its radius. Venus has strong enough surface gravity to hang on to its atmosphere; Mars does not (see Working It Out 9.1). As with the loss of a primary atmosphere, once a planet begins to lose its atmosphere, the process accelerates.

Atmospheric loss also arises from the solar wind's interaction with planetary atmospheres, especially in the absence of a planetary magnetic field. In 2017, program scientists with NASA's *Mars Atmosphere and Volatile EvolutioN* (*MAVEN*) mission measured argon levels in the atmosphere of Mars and confirmed that sputtering was responsible for most of the loss of the secondary atmosphere (**Figure 9.3**). This means that Mars formerly had a much thicker atmosphere, so it was warmer and wetter in the past.

The Atmospheric Greenhouse Effect

Venus, Earth, and Mars have dramatically different surface temperatures. In Chapter 5, we calculated the temperature of a planet by finding the equilibrium between the amount of energy it absorbs from sunlight and the amount of energy it radiates back into space. Recall that we found that this calculation gives a good result for planets without atmospheres: Mercury, the Moon, and Mars. Earth, on the other hand, is about 33°C warmer than expected, whereas Venus is hundreds of degrees hotter than that simple model predicted. When the predictions of a model fail, something was probably left out of the model. Here, that something was the *atmospheric greenhouse effect*, which traps solar radiation near the surface of a planet.

The atmospheric greenhouse effect in planetary atmospheres and the conventional **greenhouse effect** operate differently, although the results are much the same. Planetary atmospheres and the interiors of greenhouses are both heated by trapping the Sun's energy, but there the similarities end. The conventional greenhouse effect is the rise in temperature in a car on a sunny day when you leave the windows closed. The same effect keeps a literal greenhouse

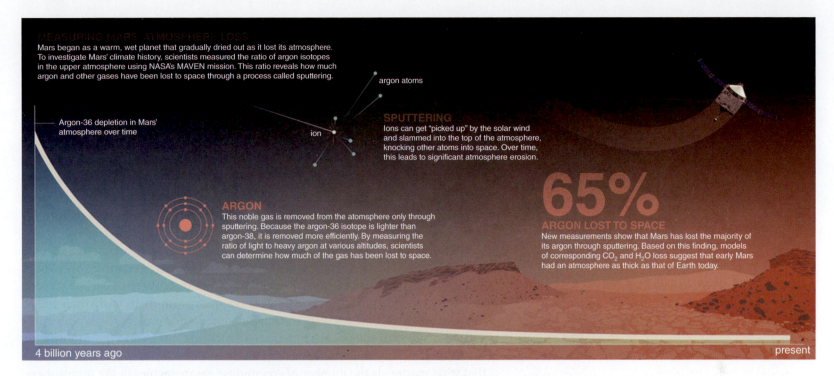

MEASURING MARS' ATMOSPHERE LOSS
Mars began as a warm, wet planet that gradually dried out as it lost its atmosphere. To investigate Mars' climate history, scientists measured the ratio of argon isotopes in the upper atmosphere using NASA's MAVEN mission. This ratio reveals how much argon and other gases have been lost to space through a process called sputtering.

argon atoms

ion

Argon-36 depletion in Mars' atmosphere over time

SPUTTERING
Ions can get "picked up" by the solar wind and slammed into the top of the atmosphere, knocking other atoms into space. Over time, this leads to significant atmosphere erosion.

ARGON
This noble gas is removed from the atmosphere only through sputtering. Because the argon-36 isotope is lighter than argon-38, it is removed more efficiently. By measuring the ratio of light to heavy argon at various altitudes, scientists can determine how much of the gas has been lost to space.

65%
ARGON LOST TO SPACE
New measurements show that Mars has lost the majority of its argon through sputtering. Based on this finding, models of corresponding CO_2 and H_2O loss suggest that early Mars had an atmosphere as thick as that of Earth today.

4 billion years ago

present

warm so that tender plants can survive in the winter. In both cases, sunlight pours through the glass, heating the interior and raising the internal air temperature. With the windows closed, hot air is trapped, and temperatures can climb as high as 80°C (about 180°F). Heating by solar radiation is most efficient when an enclosure is transparent, which is why the walls and roofs of real greenhouses are made primarily of glass or clear plastic.

The atmospheric greenhouse effect is illustrated in **Figure 9.4**. **Greenhouse gases** are those that transmit visible radiation but absorb infrared radiation. Examples include water vapor, carbon dioxide, methane, and nitrous oxide, as well as industrial chemicals such as halogens. Greenhouse gases freely allow visible light from the Sun to reach a planet's surface and warm it up. The warmed surface radiates the energy in the infrared region of the spectrum. Greenhouse gases strongly absorb that infrared radiation and reemit it in random directions. Some of the energy continues into space, but much of it returns to the ground, raising the planet's surface temperature. As a result of that redirection, the planet's surface receives energy from both the Sun and the atmosphere. The planet will be warmer than one without greenhouse gases in its atmosphere.

The temperature will not rise indefinitely. For a given number of greenhouse gas molecules in the atmosphere, the temperature increases until the surface becomes hot enough—and therefore radiates enough energy—that the infrared radiation leaking out through the atmosphere balances the absorbed sunlight, and equilibrium is reached. Convection also helps maintain equilibrium by transporting thermal energy to the top of the atmosphere, where it can be more easily radiated to space. The greenhouse effect from naturally occurring greenhouse gases in the atmosphere warms Earth by 33°C above the temperature predicted for a planet with no atmosphere.

If the number of greenhouse gas molecules decreases, less infrared radiation is trapped, and the temperature of the planet decreases. If the number of greenhouse gas molecules increases, more infrared radiation is trapped, and the

Figure 9.3 Interaction with the solar wind caused sputtering in the atmosphere of Mars. By measuring the argon atoms in the current atmosphere, scientists were able to show that sputtering was a major cause of the loss of atmosphere.

▶❚❚ **AstroTour:** Greenhouse Effect

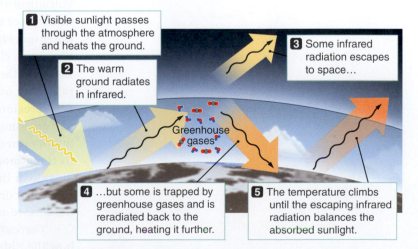

1 Visible sunlight passes through the atmosphere and heats the ground.

2 The warm ground radiates in infrared.

3 Some infrared radiation escapes to space…

Greenhouse gases

4 …but some is trapped by greenhouse gases and is reradiated back to the ground, heating it further.

5 The temperature climbs until the escaping infrared radiation balances the absorbed sunlight.

Figure 9.4 In the atmospheric greenhouse effect, greenhouse gases such as water vapor and carbon dioxide trap infrared radiation, increasing a planet's temperature.

temperature of the planet also increases. Even though the mechanisms are different, the conventional greenhouse effect and the atmospheric greenhouse effect produce the same result: the environment enclosed by the greenhouse is heated by trapped solar radiation.

Similarities and Differences among the Terrestrial Planets

How does the atmospheric greenhouse effect operate on Mars, Earth, and Venus? It is important to understand that what really matters is the actual number of greenhouse gas molecules in a planet's atmosphere, not the fraction they represent. For example, even though the atmosphere of Mars is composed almost entirely of carbon dioxide (see Table 9.1)—an effective greenhouse gas molecule— the atmosphere is very thin and contains fewer greenhouse gas molecules than the atmosphere of Venus or Earth. As a result, the atmospheric greenhouse effect is relatively weak on Mars and raises the average surface temperature by only about 5 K (9°F). At the other extreme, Venus's massive atmosphere of carbon dioxide and sulfur compounds raises its average surface temperature by more than 400 K, to about 740 K (467°C, 870°F). At such high temperatures, any remaining water and most carbon dioxide locked up in surface rocks are driven into the atmosphere, further enhancing the atmospheric greenhouse effect. Mars and Venus show that what matters is the number of greenhouse gas molecules in the atmosphere, not the fraction of the atmosphere that is greenhouse gas.

The atmospheric greenhouse effect on Earth is not as severe as it is on Venus—the average global temperature near Earth's surface is about 288 K (15°C, 59°F). Temperatures on Earth's surface are currently about 35 K (35°C, 63°F) warmer than they would be without an atmospheric greenhouse effect, mainly because of water vapor and carbon dioxide. Without that comparatively small greenhouse effect, though, Earth's average global temperature would be 255 K (−18°C, 0°F)—well below the freezing point of water, leaving us with a world of frozen oceans and ice-covered continents.

How has the atmospheric greenhouse effect made the composition of Earth's atmosphere so different from the high–carbon dioxide atmospheres of Venus and Mars? The answer lies in Earth's location in the Solar System. Earth and Venus have about the same mass, but Venus orbits the Sun closer than Earth, at 0.7 AU. Volcanism and cometary impacts produced large amounts of carbon dioxide and water vapor to form early secondary atmospheres on both planets. Most of Earth's water quickly rained out of the atmosphere to fill vast ocean basins. But because Venus was closer to the Sun, its surface temperatures were higher than those of Earth. As the Sun itself aged and brightened, Venus got warmer, and most of the rainwater on Venus immediately reevaporated, much as water does in Earth's desert regions. Venus was left with a surface that contained very little liquid water and an atmosphere filled with water vapor. The water vapor caused even higher temperatures, which led to the release of more carbon dioxide from the rocks to the atmosphere. The continuing buildup of both water vapor and carbon dioxide in the atmosphere of Venus led to a runaway atmospheric greenhouse effect that drove up the surface temperature of the planet even more. Ultimately, the surface of Venus became so hot that no liquid water could exist on it.

That early difference between a watery Earth and an arid Venus forever changed how their atmospheres and surfaces evolved. On Earth, water erosion caused by rain and rivers continually exposed fresh minerals, which then reacted chemically with atmospheric carbon dioxide to form solid carbonates. That reaction removed some of the atmospheric carbon dioxide, burying it within Earth's crust as a component of a rock called limestone. Later, the development of life in Earth's oceans accelerated the removal of atmospheric carbon dioxide. Tiny sea creatures built their protective shells out of carbonates, and as they died they built up massive beds of **limestone** on the ocean floors. Water erosion and the chemistry of life tied up all but a trace of Earth's total inventory of carbon dioxide in limestone beds. Earth's particular location in the Solar System seems to have spared it from the runaway atmospheric greenhouse effect. If those reactions had not locked up all the carbon dioxide now in limestone beds, Earth's atmosphere would be composed of about 98 percent carbon dioxide, similar to that of Venus or Mars. The atmospheric greenhouse effect would be much stronger, and Earth's temperature would be much higher.

Scientists study the details of the differences in the amount of water on Venus, Earth, and Mars. Geological evidence indicates that liquid water was once plentiful on the surface of Mars. Several of the spacecraft orbiting Mars have found evidence that significant amounts of water still exist on Mars as subsurface ice—far more than the atmospheric abundance indicated in Table 9.1. Earth's liquid and solid water supply is even greater, about 0.02 percent of its total mass. More than 97 percent of Earth's water is in the oceans, which have an average depth of about 4 km. Earth today has 100,000 times more water than Venus.

Some scientists think that Venus once had as much water as Earth—as liquid oceans or as more water vapor than is measured today. As the Sun aged and became brighter, and as the planets received more solar energy, water molecules high in the atmosphere of Venus were broken apart into hydrogen and oxygen by solar UV radiation. The low-mass hydrogen atoms were quickly lost to space. Oxygen escaped more slowly, so some eventually migrated downward to the planet's surface, where it was removed from the atmosphere by bonding with minerals on the surface. In support of that theory, the *Venus Express* spacecraft has measured hydrogen and some oxygen escaping from the upper levels of Venus's atmosphere.

CHECK YOUR UNDERSTANDING 9.2

The main greenhouse gases in the atmospheres of the terrestrial planets are: (a) oxygen and nitrogen; (b) methane and ammonia; (c) carbon dioxide and water vapor; (d) hydrogen and helium.

9.3 Earth's Atmosphere Has Detailed Structure

Now that we have considered some of the overall processes that have influenced how the terrestrial planet atmospheres evolved, we look in depth at each of them. We begin with the composition and structure of Earth's atmosphere, not only because we know it best but also because it helps us better understand the atmospheres of other worlds.

Life and the Composition of Earth's Atmosphere

Earth's atmosphere is about four-fifths nitrogen (N_2) and one-fifth oxygen (O_2) (see Table 9.1). Many important minor constituents such as water vapor and carbon dioxide (CO_2) are also present, the amounts of which vary depending on global location and season. The composition of Earth's atmosphere is relatively uniform on a global scale, but temperatures can vary widely. Atmospheric temperatures near Earth's surface can range from as high as 60°C (140°F) in the deserts to as low as −90°C (−130°F) in the polar regions. The mean global temperature is about 15°C (59°F).

○ **what if . . .**

What if just after the terrestrial planets formed, a passing star stripped away the Kuiper Belt, the Oort Cloud, and all of the comets in the outer solar system? How would that have affected the formation of atmospheres on the terrestrial planets?

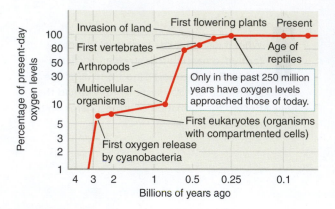

Figure 9.5 The amount of oxygen in Earth's atmosphere has built up as a result of photosynthesizing plant life.

Oxygen   Table 9.1 shows that Earth's atmosphere contains abundant amounts of oxygen (O_2), whereas the atmospheres of other planets do not. Oxygen is a highly reactive gas: it chemically combines with, or oxidizes, almost any material it touches. The rust (iron oxide) that forms on steel is an example. The reddish surface of Mars is coated with oxidized iron-bearing minerals—one reason the martian atmosphere is almost completely free of oxygen. A planet with significant amounts of oxygen in its atmosphere must be continuously replacing oxygen lost through oxidation. On Earth, plants perform that role by breaking CO_2 into oxygen and carbon. The oxygen is released to the atmosphere, and the carbon is used by the plants.

The oxygen concentration in Earth's atmosphere has changed over the history of the planet (**Figure 9.5**). When Earth's secondary atmosphere first formed about 4 billion years ago, it had very little oxygen because O_2 is not found in volcanic gases or comets. Studies of ancient sediments show that about 2.8 billion years ago, an ancestral form of cyanobacteria—single-celled organisms that contain chlorophyll, which enables them to obtain energy from sunlight—began releasing oxygen into Earth's atmosphere as a waste product of their metabolism. At first, that biologically generated oxygen combined with exposed metals and minerals in surface rocks and soils, so it was removed from the atmosphere as quickly as it formed. Ultimately, the explosive growth of cyanobacteria and then plant life accelerated the production of oxygen, building up atmospheric concentrations that settled at today's levels only about 250 million years ago.

All plants, from tiny green algae to giant redwoods, use the energy of sunlight to build carbon compounds out of carbon dioxide and produce oxygen as a metabolic waste product in a process called **photosynthesis**. Plants dramatically changed the very composition and appearance of Earth's surface—the first of many widespread modifications of Earth by living organisms. Earth's atmospheric oxygen content is held in a delicate balance primarily by plants. If plant life on the planet were to disappear, so, too, would nearly all Earth's atmospheric oxygen, and therefore all animal life—including us.

Ozone   Ozone (O_3) is another constituent in Earth's atmosphere. Ozone is formed when UV light from the Sun breaks molecular oxygen (O_2) into its individual atoms. Those oxygen atoms can then recombine with other oxygen molecules to form ozone (the net reaction is $O_2 + O \rightarrow O_3$). Most of Earth's natural ozone is concentrated in a layer in the upper atmosphere at altitudes between 10 and 50 km (the **stratosphere**). Ozone in the upper atmosphere absorbs UV sunlight very strongly. Without the ozone layer, that radiation would reach Earth's surface, where it would be lethal to nearly all forms of life. Ozone exists in the lower atmosphere, too, where it is primarily a by-product of power plants, factories, and automobiles. That ozone is a human-made pollutant and a health hazard that raises the risk of respiratory and heart problems.

In the mid-1980s, scientists began noticing that the measured amount of ozone in Earth's upper atmosphere had been decreasing seasonally since the 1970s, primarily over the polar latitudes during springtime in both the Northern and Southern hemispheres. Scientists called those depleted regions "ozone holes": the term indicates reduced concentrations of ozone and not an actual hole in the ozone layer. Ozone depletion is caused by a seasonal buildup of atmospheric halogens—mostly chlorine, fluorine, and bromine—such as those found in industrial refrigerants, especially chlorofluorocarbons (CFCs). Halogens diffuse upward into the stratosphere, where they destroy ozone without themselves being consumed. Such agents are called **catalysts**—materials that participate in and accelerate chemical reactions but are not themselves modified in the process.

Because they are not modified or used up, halogens may remain in Earth's upper atmosphere for decades or even centuries. Even though more of the chemicals originated in the north, the depletions are greater in the Southern Hemisphere because the colder temperatures in the southern polar regions produce a type of cloud that provides a surface on which the ozone-destroying chemical reactions can take place (**Figure 9.6**).

Scientists predicted that the continuing removal of ozone from the high atmosphere could cause trouble for terrestrial life as more and more UV radiation reached the ground. Measured increases in the levels of UV radiation correlate with increases in skin cancer in humans, and the mutating effects it may have on other life-forms are not completely understood. By the late 1980s, international agreements on phasing out the production of CFCs and other ozone-depleting chemicals were signed (the Montreal Protocol), and world production of these chemicals has steadily declined. The largest Southern Hemisphere ozone hole occurred over Antarctica in 2006. The southern polar ozone hole has mostly stabilized: the area of the hole has varied, but is following a downward trend, while the annual minimum ozone level is following a definite upward trend. The ozone hole in the north has been smaller than in the south but affects more populated areas when it grows, as it did in 2011. Ozone still seems to be decreasing slowly in the lower stratosphere in midlatitudes over more populated regions, perhaps from the release of other chemicals or from changes in the climate. Full recovery to 1980 levels is not expected before the late 21st century.

Carbon Dioxide Carbon dioxide (CO_2) is another variable component of Earth's atmosphere. How much carbon dioxide is in the atmosphere at any one time depends on a complex pattern of carbon dioxide *sources* (places where it originates) and *sinks* (places where it goes). Plants consume carbon dioxide in great quantities as part of their metabolic process. Coral reefs are colonies of tiny ocean organisms that build their protective shells with carbonates produced from dissolved carbon dioxide. Fires, decaying vegetation, and the human burning of fossil fuels all release carbon dioxide back into the atmosphere. That balance between carbon dioxide sources and sinks changes. As we describe later, the amount of carbon dioxide in the atmosphere has varied historically but has been increasing rapidly since the industrial revolution (since 1840 or so). That recent increase in carbon dioxide, in turn, has directly affected global temperature because carbon dioxide is a powerful greenhouse gas.

Water Vapor Water vapor (H_2O) in Earth's atmosphere also affects daily life and is a powerful greenhouse gas. Over the range of temperatures on Earth, the amount of water in the atmosphere varies from time to time and from place to place. In warm, moist climates, water vapor may account for as much as 3 percent of the total atmospheric composition. In cold, arid climates, it may be less than 0.1 percent. The continual process of condensation and evaporation of water involves the exchange of thermal and other forms of energy, making water vapor a major contributor to Earth's weather.

The Layers of Earth's Atmosphere

Earth's atmosphere is a blanket of gas several hundred kilometers thick. It has a total mass of approximately 5×10^{18} kg, less than one-millionth of Earth's total mass. The weight of Earth's atmosphere creates a force of approximately 100,000 newtons (N) acting on each square meter of the planet's surface, equivalent to about 14.7 pounds pressing on every square inch. That amount of pressure

what if . . .
What if you could use spectroscopy to detect different kinds of molecules in the atmospheres of planets orbiting other stars? What molecules would be suggestive of life on such a planet and why?

Figure 9.6 Polar stratospheric clouds form in the polar springtime and serve as the surface upon which ozone destruction takes place.

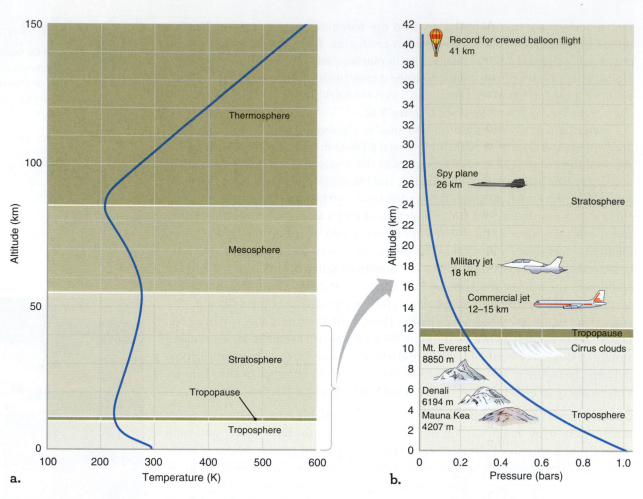

Figure 9.7 These graphs show **a.** temperature and **b.** pressure plotted for Earth's atmospheric layers as a function of altitude. Most human activities are confined to the bottom layers of Earth's atmosphere.

is called a **bar** (from the Greek *baros*, meaning "weight" or "heavy"). Earth's average atmospheric pressure at sea level is approximately 1 bar. A millibar (mb) is one-thousandth of 1 bar and is more commonly used in meteorology and in weather reports. Underwater, each depth of 10 meters adds 1 bar of pressure. We are largely unaware of Earth's atmospheric pressure because the same pressure exists both inside and outside our bodies, so the force pushing in is precisely balanced by the force pushing out.

Recall from Chapter 8 that the pressure at any point within a planet's interior must be great enough to balance the weight of the overlying layers. The same principle holds true in a planetary atmosphere. The atmospheric pressure on a planet's surface must be great enough to support the weight of the overlying atmosphere. Different forms of matter provide the pressure within a planet's interior and in its atmosphere. In the interior of a solid planet, solid materials exert pressure as they resist being compressed. In a planetary atmosphere, the motions of gas molecules exert enough pressure to support the atmosphere.

Earth's atmosphere consists of several distinct layers (**Figure 9.7**). Those layers are distinguished by the changes in temperature and pressure through the atmosphere. The lowest layer, the one in which humans live and breathe, is called the **troposphere**. It contains 90 percent of Earth's atmospheric mass and is the source of all our weather. At Earth's surface, usually called sea level, the troposphere has an average temperature of 15°C (288 K). Within the troposphere, atmospheric pressure, density, and temperature all decrease as altitude increases. For example, at an altitude of 5.5 km (18,000 feet, a few thousand feet below the summit of Denali in Alaska), the atmospheric pressure and density are only 50 percent of their sea-level values, and the average temperature has dropped to −20°C (253 K). In the lower stratosphere, at an altitude of 12–15 km, where commercial jets cruise, the temperature is −60°C (213 K), and the density and pressure are less than one-fifth what they are at sea level.

The atmosphere is warmer near Earth's surface because the air is closer to the sunlight-heated ground, which warms the air by infrared radiation. The atmosphere is cooler at very high altitudes because there the atmosphere freely radiates its thermal energy into space. **Figure 9.8** illustrates how convection carries thermal energy upward through Earth's atmosphere, keeping the upper layers warmer than they would otherwise be. At a given pressure, cold air is denser than warm air. The denser cold air falls below the less dense warm air, pushing the warm air upward. That convection sets up air circulation between the lower and upper levels of the atmosphere and tends to diminish the temperature extremes caused by heating at the bottom and cooling at the top.

Convection also affects the vertical distribution of atmospheric water vapor. Warm air can hold more water vapor than cold air. The amount of water vapor in the air relative to the maximum amount the air could hold at a particular temperature is called the **relative humidity**. Air saturated with water vapor has a relative humidity of 100 percent. As air is carried upward by convection, it cools. When the air temperature decreases to the point at which the air can no longer hold all its water vapor, water begins to condense to tiny droplets or ice crystals. In large numbers those become visible as clouds. When those droplets combine to form large drops, they fall as rain or snow, so most of the water vapor in Earth's atmosphere stays within 2 km of the surface. At an altitude of 4 km, the Mauna Kea Observatories are higher than approximately one-third of Earth's atmosphere, but they lie above nine-tenths of the atmospheric water vapor. That altitude is important for astronomers who observe in the infrared region of the spectrum because water vapor strongly absorbs infrared light.

The layer of the atmosphere above the troposphere and extending upward to an altitude of 50 km above sea level is the stratosphere. The boundary between the troposphere and stratosphere is called the **tropopause**. It varies between 10 and 15 km above sea level, depending on latitude, and is highest at the equator. Little convection takes place in the stratosphere because the temperature no longer decreases with increasing altitude. In fact, the temperature begins to *increase* with altitude because of the ozone layer, which warms the stratosphere by absorbing UV radiation from the Sun.

The region above the stratosphere is the **mesosphere**, which extends from an altitude of 50 km to about 90 km. The mesosphere has no ozone to absorb sunlight, so temperatures once again decrease with altitude. The base of the stratosphere and the upper boundary of the mesosphere are two of the coldest levels in Earth's atmosphere. Higher in Earth's atmosphere, interactions with space become important. At altitudes above 90 km, solar UV radiation and high-energy particles from the solar wind strip electrons from, or **ionize**, atmospheric molecules, causing the temperature once again to increase with altitude. That region, called the **thermosphere**, is the hottest part of the atmosphere. The temperature can reach 1000 K near the top of the thermosphere, at an altitude of 600 km.

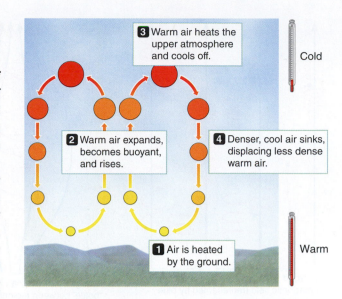

3 Warm air heats the upper atmosphere and cools off.

Cold

2 Warm air expands, becomes buoyant, and rises.

4 Denser, cool air sinks, displacing less dense warm air.

1 Air is heated by the ground.

Warm

Figure 9.8 Atmospheric convection carries thermal energy from the Sun-heated surface of Earth upward through the atmosphere.

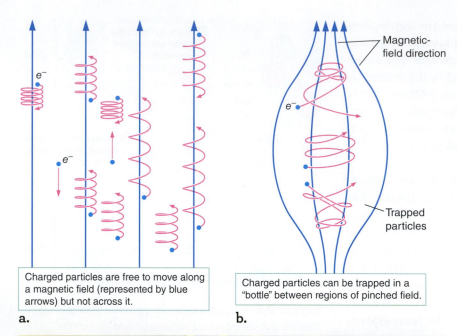

Charged particles are free to move along a magnetic field (represented by blue arrows) but not across it.

a.

Magnetic-field direction

Trapped particles

Charged particles can be trapped in a "bottle" between regions of pinched field.

b.

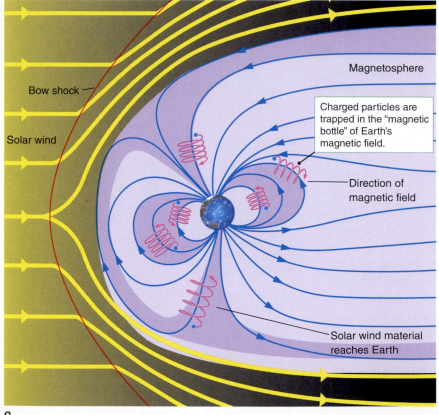

Bow shock

Solar wind

Magnetosphere

Charged particles are trapped in the "magnetic bottle" of Earth's magnetic field.

Direction of magnetic field

Solar wind material reaches Earth

c.

Figure 9.9 a. Charged particles, here electrons, spiral in a uniform magnetic field. **b.** When the field is pinched, charged particles can be trapped in a "magnetic bottle." **c.** Earth's magnetic field acts like a bundle of magnetic bottles, trapping particles in Earth's magnetosphere. The bow shock is where the solar wind slows as it meets the magnetosphere. In all these images, the radius of the helix that the charged particle follows is greatly exaggerated.

The atoms and molecules in the gases within and beyond the thermosphere are ionized by UV photons and high-energy particles from the Sun. That region of ionized atmosphere is called the **ionosphere**, and it not only overlaps the thermosphere but also extends farther into space. The ionosphere reflects certain frequencies of radio waves back to the ground. For example, the frequencies used by AM radio bounce back and forth between the ionosphere and the surface, enabling radio receivers to pick up stations at great distances from the transmitters. Amateur radio operators can communicate with one another around the world by bouncing their signals off the ionosphere.

CHECK YOUR UNDERSTANDING 9.3a

List the layers of the atmosphere in order from nearest to the surface to farthest from the surface: (a) stratosphere; (b) thermosphere; (c) troposphere; (d) ionosphere; (e) mesosphere.

Earth's Magnetosphere

Even farther out than the ionosphere is Earth's magnetosphere, a large region filled with electrons, protons, and other charged particles from the Sun that have been captured by the planet's magnetic field. That region has a radius approximately 10 times that of Earth. Earth and its magnetosphere are immersed in the solar wind. When the charged particles of the solar wind first encounter Earth's magnetic field, the smooth flow is interrupted. They are diverted by Earth's magnetic field like a river diverted around a boulder.

Magnetic fields affect only moving charges. Charged particles move freely along the direction of the magnetic field, parallel to the magnetic field line. If charged particles try to move across the field lines, they experience a force that is perpendicular both to the motion of the particle and to the direction of the magnetic field line. This force pushes the charged particles in a circle around the direction of the magnetic field line, as illustrated in **Figure 9.9a**. The motions along and around the magnetic field line combine to make a path that is coiled like a spring.

If the magnetic field lines are pinched together at some point, particles moving into the pinch will experience a magnetic force that reflects them back along the direction they came from. If there are two such pinch points, they form a "magnetic bottle" that contains the charged particles, as shown in **Figure 9.9b**. Earth's magnetic field is pinched together at the two magnetic poles and spreads out around the planet. A charged particle may bounce back and forth between the poles many times before it strikes another particle or is neutralized. As the charged particles of the solar wind flow past Earth's magnetosphere, some particles flow past, but some of those charged particles are trapped, as illustrated in **Figure 9.9c**.

The interaction between the solar wind and Earth's magnetosphere has serious implications for our technological society. Disturbances in Earth's magnetosphere caused by changes in the solar wind can lead to changes in Earth's magnetic field that are large enough to trip power grids, cause blackouts, and disrupt communications. In addition, regions in the magnetosphere that contain especially strong concentrations of energetic charged particles, called **radiation belts**, can be very damaging to both electronic equipment and astronauts.

Earth's magnetic field funnels energetic charged particles down into the ionosphere in two rings located around the magnetic poles; one at either end of the magnetic bottle. Those charged particles (mostly electrons) collide with atoms and molecules such as oxygen, nitrogen, and hydrogen in the upper atmosphere, causing them to glow like the gas in the tubes of a neon sign. Interactions with different atoms cause different colors. Those glowing rings, called **auroras**, can be seen from space (**Figure 9.10a**). When viewed from the ground (**Figure 9.10b**), auroras appear as eerie, shifting curtains of multicolored light. People living far from the equator are often treated to spectacular displays of the aurora borealis (the "northern lights") in the Northern Hemisphere or the aurora australis (the "southern lights") in the Southern Hemisphere. When the solar wind is particularly strong, auroras can be seen at lower latitudes, sometimes as far south as Texas. Auroras also have been seen on Venus, Mars, all the giant planets, and some moons, when the solar wind interacts with the magnetic fields of those Solar System bodies.

The general structure we have described here is not limited to Earth's atmosphere. The major vertical structural components—troposphere, tropopause, stratosphere, and ionosphere—also exist in the atmospheres of Venus and Mars,

what if . . .

What if astronauts travel from Earth to Mars and back? What are the risks to the astronauts associated with leaving the protection of Earth's magnetic field?

Astronomy in Action: Charged Particles and Magnetic Forces

Figure 9.10 Auroras result when particles trapped in Earth's magnetosphere collide with molecules in the upper atmosphere. **a.** An auroral ring around Earth's south magnetic pole, as seen from space. **b.** Aurora borealis—the "northern lights"—viewed from the ground in Alaska.

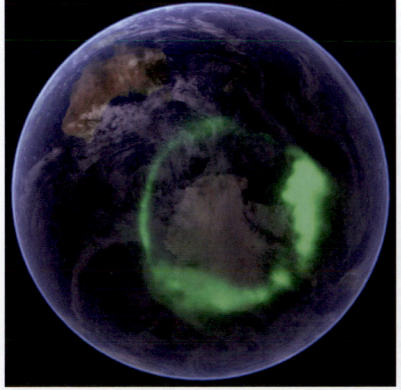

a.   VIS

b.   VIS

as well as in the atmospheres of Titan and the giant planets. The magnetospheres of the giant planets are among the largest structures in the Solar System.

Weather and Wind

Weather is the local day-to-day state of the atmosphere. Local weather is caused by winds and convection. Winds are the natural movement of air, both locally and on a global scale, in response to variations in temperature from place to place. Recall from Chapter 5 that heating a gas increases its pressure, which causes it to push into its surroundings. Winds are caused by pressure differences of this type. The air is usually warmer in the daytime than at night, warmer in the summer than in winter, and warmer at the equator than in the polar regions. Large bodies of water, such as oceans, also affect atmospheric temperatures. The strength of the winds is governed by the size of the temperature difference from place to place.

Recall from Chapter 2 that the effect of Earth's rotation on winds—and on the motion of any object—is called the Coriolis effect (see Figure 2.12). As air in Earth's equatorial regions is heated by the warm surface, convection causes that air to rise. The warmed surface air displaces the air above it, which then has nowhere to go but toward the poles. That air becomes cooler and denser as it moves toward the poles, so it sinks back down through the atmosphere. In the process, it displaces the surface polar air, which is forced back toward the equator, completing the circulation. As a result, the equatorial regions remain cooler and the polar regions remain warmer than they otherwise would be. Air moves between the equator and poles of a planet in a pattern known as **Hadley circulation** (**Figure 9.11a**).

On Earth, other factors break up the planetwide flow into smaller Hadley cells. Most Solar System planets and their atmospheres rotate rapidly enough that the Coriolis effect strongly interferes with Hadley circulation by redirecting the horizontal flow (**Figure 9.11b**). The Coriolis effect creates winds that blow predominantly in an east–west direction and are often confined to relatively narrow bands of latitude. Meteorologists call those **zonal winds**. Planets that rotate faster have a stronger Coriolis effect and stronger zonal winds than planets that rotate more slowly. Between the equator and the poles in most planetary atmospheres, the zonal winds alternate between winds blowing from the east toward the west (easterlies) and winds blowing from the west toward the east (westerlies).

In Earth's atmosphere, several bands of alternating zonal winds lie between the equator and each hemisphere's pole. That zonal pattern is called Earth's **global circulation** because its extent is planetwide. The best-known zonal currents are the subtropical trade winds—more or less easterly winds that once carried sailing ships from Europe westward to the Americas—and the midlatitude prevailing westerlies that carried them home again.

Embedded within Earth's global circulation pattern are systems of winds associated with large high-pressure and low-pressure regions. A combination of a low-pressure region and the Coriolis effect produces a circulating pattern called **cyclonic motion** (**Figure 9.12**). Cyclonic motion is associated with stormy weather, including hurricanes. Similarly, high-pressure systems are localized regions where the air pressure is higher than average. Owing to the Coriolis effect, high-pressure regions rotate in a direction opposite to that of low-pressure

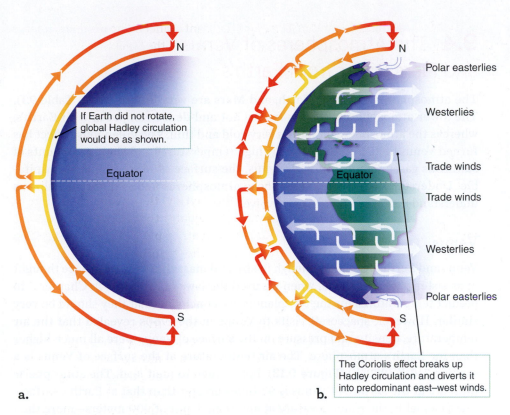

Figure 9.11 **a.** Hadley circulation covers an entire hemisphere. **b.** On Earth, Hadley circulation breaks up into smaller circulation cells because of the Coriolis effect, which diverts the north–south flow into east–west zonal flow.

regions. Those high-pressure circulating systems experience **anticyclonic motion** and are generally associated with fair weather.

Earth has a **water cycle** in which water from the surface enters the air and later returns to the oceans. When liquid water in Earth's oceans, lakes, and rivers absorbs enough thermal energy from sunlight, it turns to water vapor. The water vapor carries that thermal energy as it circulates throughout the atmosphere, releasing the energy to its surroundings when the water vapor condenses back into rain or snow. That process powers rainstorms, thunderstorms, hurricanes, and other dramatic weather.

For example, Coriolis forces acting on air rushing into regions of low atmospheric pressure create huge circulating systems that result in **hurricanes**. The conditions must be just right: warm tropical seawater, light winds, and a region of low pressure in which air spirals inward. Sustained winds near the center of the storm can exceed 300 kilometers per hour (km/h), causing widespread damage and fatalities. **Tornadoes** are small but violent circulations of air associated with storm systems. **Dust devils** are similar in structure to tornadoes, but generally smaller and less intense. They usually occur in fair weather in places such as the deserts of the American Southwest. The lifetime of a typical tornado or dust devil is brief—usually a dozen or so minutes.

CHECK YOUR UNDERSTANDING 9.3b

All weather and wind on Earth are a result of convection in the: (a) troposphere; (b) stratosphere; (c) mesosphere; (d) ionosphere; (e) thermosphere.

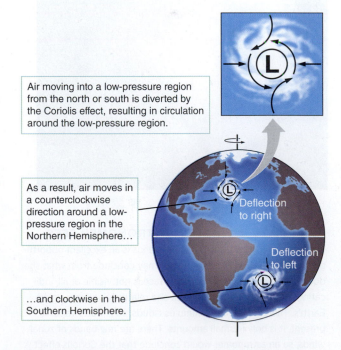

Air moving into a low-pressure region from the north or south is diverted by the Coriolis effect, resulting in circulation around the low-pressure region.

As a result, air moves in a counterclockwise direction around a low-pressure region in the Northern Hemisphere…

…and clockwise in the Southern Hemisphere.

Figure 9.12 As a result of the Coriolis effect, air circulates around regions of low pressure on the rotating Earth.

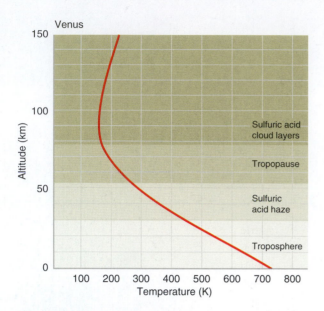

Figure 9.13 The temperature of the atmosphere on Venus primarily decreases as altitude increases, unlike temperatures in Earth's atmosphere, which fall and rise and fall again through the troposphere, stratosphere, and mesosphere (compare Figure 9.7a).

VIS

Figure 9.14 ★ WHAT AN ASTRONOMER SEES
This enhanced visible-light image of Venus is an excellent opportunity to describe what an astronomer may conclude from what she does NOT see. In this image, the surface is not visible at all, indicating that the cloud layer is thick and relatively uniform, unlike Earth's. There are no fluffy cumulus clouds, so if water vapor is present, it is only in small amounts. There are few bands of zonal winds, so an astronomer would conclude that the Coriolis effect is weak on this planet, and thus the planet probably rotates slowly. Phenomena that are absent in astronomical images are sometimes as important as those that are present.

9.4 The Atmospheres of Venus and Mars Differ from Earth's

The atmospheres of Venus, Earth, and Mars are very different (see Table 9.1). The lower atmosphere of Venus is very hot and dense compared with Earth's, whereas the atmosphere of Mars is very cold and thin. The greenhouse effect has turned Venus hellish, with extremely high temperatures and choking amounts of sulfurous gases. Compared with Venus, the surface of Mars is almost hospitable. Understanding why and how those atmospheres are so different helps us understand how atmospheres may evolve.

Venus

Venus and Earth are similar enough in size and mass that they were once thought of as sister planets. Indeed, when we used the laws of radiation in Chapter 5 to predict temperatures for the two planets, we concluded that they should be very similar. However, spacecraft visits to Venus in the 1960s revealed that the air temperature, density, and pressure on the surface of Venus were all much higher than for Earth's atmosphere. The air temperature at the surface of Venus is a sizzling 740 K (872°F) (**Figure 9.13**), hot enough to melt lead. The atmospheric pressure at the surface of Venus is 92 times greater than that at Earth's surface: that is equal to the water pressure at an ocean depth of 900 meters—more than enough to crush the hull of a submarine. Ninety-six percent of Venus's massive atmosphere is carbon dioxide, with only 3.5 percent nitrogen and lesser amounts of other gases. The greenhouse effect of this carbon dioxide is responsible for the extreme atmospheric properties measured by the spacecraft.

The graph in Figure 9.13 shows that, as on Earth, the atmospheric temperature of Venus drops with altitude throughout the planet's troposphere, reaching a low of about 160 K at the tropopause. At an altitude of approximately 50 km, Venus's atmosphere has an average temperature and pressure similar to those of Earth's atmosphere at sea level. At altitudes between 50 and 80 km, the atmosphere is cool enough for sulfurous oxide vapors to react with water vapor to form dense clouds of concentrated droplets of sulfuric acid (H_2SO_4). Those dense clouds block the view of the surface of Venus (**Figure 9.14**). Large variations in the observed amounts of sulfurous compounds in the high atmosphere of Venus suggest that the sulfur arises from sporadic episodes of volcanic activity. That finding, along with some hot spots seen near a large shield volcano, strengthens the possibility that Venus is currently volcanically active.

In the 1960s, radio telescopes and spacecraft with cloud-penetrating radar provided low-resolution views of the surface of Venus. Not until 1975, when the Soviet Union landed cameras there, did scientists get a clear picture of the surface. Those images showed fields of rocks 30–40 centimeters (cm) across and basalt-like slabs surrounded by weathered material. Soviet landers in the 1980s revealed similar landscapes (**Figure 9.15**). Radar images taken by the *Magellan* spacecraft in the early 1990s (see Figure 8.20) produced a global map of the surface of Venus. The high atmospheric temperatures on Venus also mean that neither liquid water nor liquid sulfurous compounds can exist on its surface, leaving an extremely dry lower atmosphere with only 0.01 percent water and sulfur dioxide vapor.

Imagine standing on the surface of Venus. Because sunlight cannot easily penetrate the dense clouds above you, noontime is no brighter than a very

Reading Astronomy News

Phosphine on Venus

The detection of phosphine on Venus is a tantalizing hint that life might exist on Venus. But extraordinary claims require . . . more study.

An international team of astronomers detected phosphine (PH_3) in the atmosphere of Venus. They studied the origin of phosphine, but no inorganic processes, including supply from volcanos and atmospheric photochemistry can explain the detected amount of phosphine. The phosphine is believed to originate from unknown photochemistry or geochemistry, but the team does not completely reject the possibility of biological origin. This discovery is crucial to examine the validity of phosphine as a biomarker.

"When we got the first hints of phosphine in Venus's spectrum, it was a shock!," says team leader Jane Greaves of Cardiff University in the UK, who first spotted signs of phosphine in observations from the James Clerk Maxwell Telescope (JCMT), operated by the East Asian Observatory, in Hawai'i. Confirming their discovery required the Atacama Large Millimeter/submillimeter Array (ALMA) in Chile, a more sensitive telescope. The reason why she was so shocked is that phosphine can be produced by microbes on the Earth [1], although the research team does not think that they found life on Venus.

How do you find life on a planet from quite far away? One way is to study its atmosphere and find a biomarker that can be evidence of the presence of living forms. If a molecule in the atmosphere is mainly produced by living organisms and the contribution from abiotic origins is negligibly small, it can be a good biomarker.

The international team led by Greaves, including Hideo Sagawa at Kyoto Sangyo University, studied the signal of phosphine in the radio spectra and found that the amount of the molecule is about 20 parts per billion in the atmospheric molecules. This is quite a small amount, but enough to astonish the researchers. This is because researchers have supposed that most of the phosphorus, if it existed in the first place, would bind with oxygen atoms because the Venusian atmosphere has a huge amount of oxygen atoms, although most of them are in the form of carbon dioxide (CO_2).

The team carefully examined the possible origins of the phosphine: production by chemical reaction in the atmosphere driven by strong sunlight or lightning, supply from volcanic activity, and delivery by meteorites. The team found that all of these known processes failed to produce the observed amount of phosphine. The amount of phosphine molecules produced by those processes is 10,000 times smaller than the amount detected with the radio telescopes.

The researchers supposed that phosphine is produced by unknown photochemistry or geochemistry, but they also considered the possibility of biological origin. On Earth, some microbes produce and egest phosphine. If similar living organisms were in the Venusian atmosphere, they could produce the detected amount of phosphine.

"Although we concluded that known chemical processes cannot produce enough phosphine, there remains the possibility that some hitherto unknown abiotic process exists on Venus," says Sagawa. "We have a lot of homework to do before reaching an exotic conclusion, including re-observation of Venus to verify the present result itself."

Venus is Earth's twin, in terms of size. However, the atmospheres of the two planets are quite different. Venus has a very thick atmosphere and the devastating greenhouse effect raises the surface temperature as high as 460 degrees Celsius. Some researchers argue that the upper atmosphere is much milder and possibly habitable, but the extremely dry and deadly acidic atmosphere would make it difficult for a life form similar to the ones on Earth to survive on Venus.

Further observations with large telescopes on the Earth, including ALMA, and ultimately on-site observations and a sample return of the Venusian atmosphere by space probes will provide crucial information to understand the mysterious origin of the phosphine.

[1] Phosphine has been detected in the atmospheres of Jupiter and Saturn. The molecule is formed deep inside the giant planets and transported to the upper layers by atmospheric circulation. On the other hand, Venus is a rocky planet so that a similar chemistry can not be used to produce phosphine.

QUESTIONS

1. The chemical formula for phosphine is PH_3. How many different elements make up this molecule?

2. What part of the electromagnetic spectrum contains the signature of phosphine?

3. The team examined several known mechanisms to see if they might produce phosphine in Venus's atmosphere: sunlight, lightning, volcanoes, and others. Why were all of these other nonbiological processes ruled out?

4. Ruling out all of those known processes means that the only remaining known process is biological in origin. Yet the researchers "supposed that phosphine is produced by unknown photochemistry or geochemistry." Do you see this as a reasonable supposition, or is it unnecessarily cautious? Explain.

5. What will astronomers do to follow up on this observation and determine the origin of phosphine on Venus?

Source: https://www.eurekalert.org/pub_releases /2020-09/nion-pov091420.php.

what if . . .

What if humans were to "terraform" Mars, by creating a much thicker atmosphere, among other things? How would this change wind patterns on Mars? Would Mars have zonal winds, like Earth, or be more like Venus? How do you know?

VIS

Figure 9.15 This image of Venus is from the 1982 Soviet *Venera 14* mission. The spacecraft is in the foreground. Note the rocky ground and the orange sky.

Figure 9.16 This true-color image of the surface of Mars was taken by the rover *Spirit*. Without dust, the sky's thin atmosphere would appear deep blue. In this image, windblown dust turns the sky pinkish.

unanswered questions

Was Venus once hospitable to life? Earlier in the history of the Solar System, the young Sun was fainter and Venus was cooler. One research group, using three-dimensional climate simulations to model conditions on a young Venus, concluded that Venus could have had moderate temperatures if it was rotating slowly but in a prograde direction. Another group modeled the conditions on early Venus and concluded that Venus may have had an ocean of liquid water early in its history, if it had the right balance of cloud cover and the same amount of carbon dioxide as today. Understanding the differences in the evolution of Venus and Earth is of great interest to astronomers as they detect exoplanets of that size near the edge of their stars' habitable zone.

cloudy day on Earth. High temperatures and very light winds keep the lower atmosphere free of clouds and hazes. The local horizon can be seen clearly, but molecules in the dense atmosphere **scatter** light, bouncing it in all directions. This scattering effect would greatly soften any view you might have of distant mountains.

Unlike the other Solar System planets, Venus rotates on its axis in a direction opposite to its motion around the Sun. Astronomers call that *retrograde rotation*. Relative to the stars, Venus rotates on its axis once every 243 Earth days. However, a solar day on Venus—the time it takes for the Sun to return to the same place in the sky—is only 117 Earth days. The slow rotation means that Coriolis effects on the atmosphere are small. Because of its slow rotation, Venus is the only planet with global circulation close to the classic Hadley pattern shown in Figure 9.11a. Other planets have strong enough Coriolis effect to break up those cells, as shown in Figure 9.11b.

The massive atmosphere on Venus transports thermal energy around the planet very efficiently, so the polar regions are only a few degrees cooler than the equatorial regions, with almost no temperature difference between day and night. Because Venus's equator is nearly in the plane of the planet's orbit, seasonal effects are small, producing only negligible changes in surface temperature. Such small temperature variations also mean that wind speeds near the surface of Venus are low, typically about a meter per second, so wind erosion is weaker than that on Earth and Mars. High up in the atmosphere (at about 70 km), temperature differences are larger, contributing to superhurricane-force winds that reach speeds of 110 m/s (400 km/h), circling the planet in only 4 days. The variation of that high-altitude wind speed with latitude can be seen in the V-shaped cloud patterns.

When the *Pioneer Venus* spacecraft was orbiting Venus during the 1980s, its radio receiver picked up many bursts of lightning static—so many that Venus appears to have a rate of lightning activity comparable to that of Earth. On Venus, as on Earth, lightning is created in the clouds; but Venus's clouds are so high—typically 55 km above the surface of the planet—that the lightning bolts never hit the ground. The Japanese Space Agency mission *Akatsuki* has been orbiting Venus since late 2015 to study its weather and look for lightning and signs of volcanic activity. So far, its observations have revealed a more complex atmospheric structure than expected, and the mission continues as of this writing.

Mars

Mars has a stark landscape, colored reddish by the oxidation of iron-bearing surface minerals. The sky is sometimes dark blue but more often has a pinkish color due to windblown dust (**Figure 9.16**). The lower density of the martian atmosphere makes it more responsive than Earth's atmosphere to heating and cooling, so Mars has greater temperature extremes. The surface near the equator at noontime is a comfortable 20°C—a cool room temperature (68°F) on Earth. However, nighttime temperatures typically drop to a frigid −100°C, and during the polar night the air temperature can reach −150°C—cold enough to freeze carbon dioxide out of the air in the form of a dry-ice frost. The temperature profile of the atmosphere of Mars (**Figure 9.17**) has a range of only 100 degrees up to about 125 km. Above that the temperature rises because sunlight is absorbed in the upper atmosphere. The temperature profile of Mars is more similar to that of Earth than that of Venus.

The average atmospheric surface pressure of Mars is equivalent to the pressure at 35 km above sea level on Earth (well into the stratosphere; see Figure 9.7). The highest surface pressures on Mars are found in the lowest impact basins; pressure there is only 11.5 mb (1.1 percent of Earth's pressure at sea level). At the summit of Olympus Mons, the highest mountain on Mars, this pressure drops to 0.3 mb. Like Earth, Mars has some water vapor in its atmosphere, in the form of clouds of ice crystals because the atmospheric temperature is so low. Mars can have early-morning ice fog in the lowlands (**Figure 9.18**) and clouds hanging over the mountains.

Like Venus, the atmosphere of Mars is composed almost entirely of carbon dioxide (95 percent) and a small amount of nitrogen (2.7 percent). In the absence of plants, Mars has only a tiny trace of oxygen, which is crucial to life on Earth. The near absence of oxygen means that Mars has very little ozone, so solar UV radiation reaches the surface. Those UV rays could be lethal to any surface life-forms, so any life on Mars would either need to develop protective layers or be located away from direct exposure on the surface of the planet, such as in caves or below the surface.

The tilt of the rotation axis of Mars is similar to Earth's at present, so the two planets have similar seasons. Seasonal effects on Mars, though, are larger for two reasons. First, the elliptical orbit of Mars has a higher eccentricity, so the annual orbital distance of Mars from the Sun varies more than Earth's does. Second, the low density of the martian atmosphere makes it more responsive to seasonal change. The large daily, seasonal, and latitudinal surface temperature differences on Mars often create locally strong winds—some estimated to be greater than 100 m/s (360 km/h). High winds can stir up huge quantities of dust and distribute it around the planet's surface. For more than a century, astronomers have watched the development of dust storms on Mars during its spring season. The stronger storms spread quickly and can envelop the planet in a shroud of dust within a few weeks (**Figure 9.19**). Such large amounts of windblown dust can take many months to settle out of the atmosphere. Seasonal movement of dust from one area to another alternately exposes and covers large areas of dark, rocky surface. That phenomenon led some astronomers of the late 19th and early 20th centuries to believe (wrongly) that they were witnessing the seasonal growth and decay of vegetation on Mars.

The *Viking* landers first noticed dust devils on Mars in 1976. More recently, *Mars Reconnaissance Orbiter* spotted many dust devils that were visible because of the shadows they cast on the martian surface. **Figure 9.20** shows a 20-km-high, 70-meter-wide dust devil. Most martian dust devils leave dark meandering trails behind them where they have lifted bright surface dust, revealing the dark surface rock that lies beneath. Dust devils on Mars are typically higher, wider, and stronger than those on Earth.

Mars probably had a more massive secondary atmosphere in the distant past. Geological evidence strongly suggests that liquid water once flowed across its surface (Chapter 8), but the low incidence (or possibly cessation) of volcanism and the planet's low gravity, and perhaps the decrease of its magnetic field, was responsible for the loss of much of that earlier atmosphere. Scientists have not yet reached a consensus on how massive the martian atmosphere was in the past.

Recent analysis of five years of data from instruments on the Mars *Curiosity* Rover have indicated the presence of organic (carbon-based) compounds. These could have formed on the surface with or without life, or been brought to Mars on comets or meteors. One of these compounds, methane gas, has been detected in Mars' atmosphere, and varies with the seasons. The methane is thought to have originated on or below the surface of Mars.

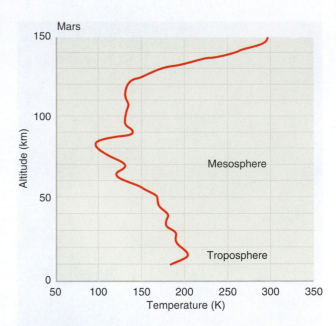

Figure 9.17 The temperature profile of the atmosphere of Mars. Note the differences in temperature and structure between this profile and the profile of the atmosphere of Venus in Figure 9.13.

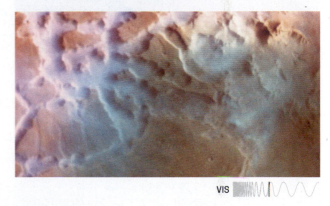

VIS

Figure 9.18 Patches of early-morning water vapor fog forming in canyons on Mars.

June 26, 2001   September 4, 2001

VIS

Figure 9.19 Hubble Space Telescope images show the development of a global dust storm that enshrouded Mars in September 2001. The same region of the planet is shown in both images; surface features are obscured by the thick layer of dust.

VIS

Figure 9.20 This dust devil on Mars was imaged by the *Mars Reconnaissance Orbiter.*

Mercury and the Moon

The ultrathin atmospheres of Mercury and the Moon are known as **exospheres**, and they are less than a million-billionth (10^{-15}) as dense as Earth's atmosphere. The recent NASA *Lunar Atmosphere and Dust Environment Explorer* (*LADEE*) mission found helium, argon, and dust in the Moon's exosphere. Other atoms, such as sodium, calcium, and even water-related ions, were seen in Mercury's exosphere by the *Messenger* spacecraft, and they may have been blasted loose from Mercury's surface by the solar wind or micrometeoroids. The exospheres of Mercury and the Moon probably vary with the strength of the solar wind and the atoms of hydrogen and helium they capture from it. Exospheres do not affect local surface temperatures, but astronomers study how exospheres interact with the solar wind.

CHECK YOUR UNDERSTANDING 9.4

Rank, from greatest to smallest, the seasonal variations on (a) Mercury, (b) Venus, (c) Earth, and (d) Mars.

9.5 Greenhouse Gases Affect Global Climates

Climate is the *average* state of an atmosphere, including its temperature, humidity, winds, and so on. Climate describes the planet as a whole, over timescales of years or decades. That is an important distinction from weather, which is the state of an atmosphere at a particular time and place. The study of climate change on Earth and Mars is not new to the 21st century. Nineteenth-century scientists found evidence of past ice ages and knew that Earth's climate had been very different earlier in its history. Observations of changes in the martian ice caps led to speculation about whether Mars also had ice ages. In this section, we look at the natural factors that can cause climates to change on planets, and we then examine the additional factors that affect Earth.

Factors That Can Cause Climate Change on a Planet

Scientists of all kinds study the astronomical, geological, and (on Earth) biological mechanisms controlling climate on the planets. Astronomical mechanisms that influence changes in planetary temperature include changes in the Sun's energy output, which has increased very slowly as the Sun ages, and possibly changes in the galactic environment as the Sun travels in its orbit around the center of the Milky Way. Scientists have suggested that sporadic bursts of gamma rays or cosmic rays (fast-moving protons) from distant exploding stars could interact with planetary atmospheres. Those mechanisms would affect *all* planets in the Solar System at the same time.

Other astronomical mechanisms relevant to climate change and specific to each planet are the Milankovitch cycles, named for geophysicist Milutin Milanković (1879–1958). Milankovitch cycles arise because a planet's energy balance is related to a planet's motion. Periodic changes in its orbital eccentricity, the tilt of its rotational axis, and its precession can affect the climate. If a planet's orbit becomes more eccentric, the amount of energy it receives from the Sun will vary more over its year. If the tilt of a planet increases, its seasons will become more extreme, and its temperature variation during the year will increase.

The precession cycle affects which hemisphere is pointed toward the Sun at different times of the elliptical orbit, so that one hemisphere may have longer winters and the other, longer summers.

The tilt of Earth's axis varies from 22.1° to 24.5°, and Earth's relatively large Moon keeps that tilt from changing more than that. In contrast, the moons of Mars are small, and the gravitational influence of Jupiter is a greater factor for Mars, so the tilt of Mars is thought to vary from 13° to 40° or possibly more. Given the precession of Mars and its more eccentric orbit, which creates differences in season length between its northern and southern hemispheres, Mars may have had very large swings in its climate throughout its history as the tilt changed.

A second set of factors that can affect climate is the geological activity of a planet. Volcanic eruptions can produce dust, aerosol particles, clouds, or hazes that block sunlight over the entire globe and lower the temperature. Impacts by large objects can kick up sunlight-blocking particles. Tectonic activity also can affect climate. On Earth, for example, the shifting of the plates has led to different configurations of the oceans and the shifting continents, thus affecting global atmospheric and oceanic circulatory patterns. The albedo of a planet can increase if it has more clouds and ice or can decrease if ice melts or is covered by volcanic ash. Changes also may arise from variations in carbon cycles. On Earth, long-term interactions of the oceans, land, and atmosphere affect the levels of greenhouse gases such as carbon dioxide and water vapor.

A third set of mechanisms that trigger climate changes is biological. Over billions of years on Earth, photosynthesis by bacteria and later by plants removed carbon dioxide from the atmosphere and replaced it with oxygen (Section 9.3). Biological (and geological) activity on Earth can produce methane, a strong greenhouse gas. Certain microorganisms produce methane as a metabolic by-product. For example, bubbles rising to the surface of a stagnant pond—swamp gas—contain biologically produced methane. Methane also is emitted from the guts of grain-fed livestock (and, in the past, from some large dinosaurs) as well as from termites. Another biological effect on climate could come from phytoplankton. If the oceans get more solar energy and warm up because of one of the astronomical mechanisms, the phytoplankton in the ocean may grow faster, leading to the release of more aerosols from sea spray and the formation of more clouds, which increases albedo. Finally, human activities are triggering some major changes, as discussed in the next subsection.

In short, many factors affect the temperature of a planet. Earth's climate is the most complicated of those of the terrestrial planets because Earth is the most geologically and biologically active. How do scientists sort out all those factors? They use the scientific method. Scientists create mathematical models to simulate the general circulation and energy balance of a planet, incorporating all the appropriate factors. Their goal is to create a global climate model that reproduces the data from observations of a planet. Once the model correctly predicts past and present climate, it can be used to predict future climate. The first simple climate models for Earth were run on the earliest computers in the 1950s and 1960s. (Fifty years later, scientists are impressed with how well a climate model from 1967 quantified the factors affecting Earth's climate.) One set of models developed at NASA's Goddard Institute for Space Studies in the 1970s was a spinoff of a program originally designed to study Venus. The insights from comparative planetology are important for producing better models that will aid in scientific predictions of climate change on Earth.

Figure 9.21 Global variations in carbon dioxide (CO_2), temperature, and methane (CH_4) concentrations over the past 800,000 years of Earth's history. Notice the multiple *y*-axes on this graph. The axis on the right relates to the temperature data (black), whereas the axes on the left relate to the CO_2 (blue) and CH_4 (red) data. These data sets have been plotted on the same graph to make the similarities and differences easier to see. The low points correspond to ice ages. ppb = parts per billion; ppm = parts per million. 2017 values for CH_4 (red) and CO_2 (blue), indicated by the arrows, are larger than at any time in the past 800,000 years.

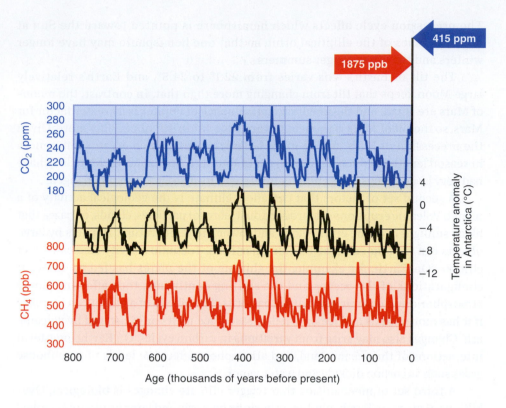

Climate Change on Earth

Paleoclimatology is the study of changes in Earth's climate throughout history. Scientists use evidence from geology and paleontology, such as sediments, ice sheets, rocks, tree rings, coral, shells, and fossils, to get data on Earth's past climate. Researchers have found that Earth's climate has lengthy temperature cycles, some lasting hundreds of thousands of years and some tens of thousands of years. As you can see in the middle plot in **Figure 9.21**, Earth has had periods of colder temperatures, known as ice ages. Earth's atmosphere is so sensitive to even small temperature changes that a drop of only a few degrees in the mean global temperature can plunge the planet's climate into an ice age. Those oscillations in the mean global temperature are far smaller than typical geographic or seasonal temperature variations.

Periodic Milankovitch cycle changes in Earth's orbit correspond to some temperature cycles. As shown in **Figure 9.22**, Earth's axial tilt varies in cycles of 41,000 years, the eccentricity of Earth's orbit varies with two cycles of about 100,000 and 413,000 years, and the time of the year when Earth is closest to the Sun varies in cycles of about 21,000 years. Global climate models using those Milankovitch cycles have replicated much of the observed paleoclimatology data. Temperature changes that are not periodic may have been triggered by volcanic eruptions or long-term interactions between Earth's oceans and its atmosphere or by other factors already mentioned.

Earth is currently experiencing the sixth known mass extinction. A mass extinction is a period of dramatic die-off of plant and animal species. At least four of the five mass extinction events in Earth's history probably arose from changes in the level of greenhouse gases. Those mass extinctions occurred 439 million, 364 million, 250 million, and 200 million years ago. The increase in carbon dioxide emitted by active volcanoes, along with methane released from the ground, led to ocean acidification and acid rain and perhaps reduced oxygen in the ocean. Most species

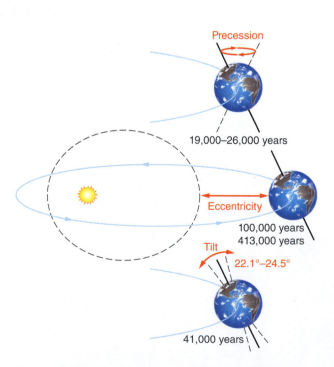

Figure 9.22 The Milankovitch cycles for Earth. The precession of Earth and the rotation of its elliptical orbit combine to yield a cyclic variation of about 21,000 years, its eccentricity varies with two cycles of about 100,000 and 413,000 years, and its tilt varies in cycles of 41,000 years. (Not to scale.)

died off. An increase in CO_2 from volcanic activity may also have been a factor, along with the impacting asteroid discussed in Chapter 8, for the fifth major extinction, which led to the death of the dinosaurs 65 million years ago. The current mass extinction results from a combination of human-caused factors, from deforestation to pollution to climate change.

If Earth's climate has been changing naturally for most of its history, why are scientists especially concerned about the current trend in global climate? Figure 9.21 shows the carbon dioxide levels (top), methane levels (bottom), and temperature (middle) of Earth's atmosphere over the past 800,000 years, obtained from measuring deep ice cores in Antarctica. Notice that those three factors are correlated: when one rises, so do the others. Those data show the naturally occurring ranges since before the first humans existed. The temperature difference between ice ages and interglacial periods is only 10°C–15°C, and those changes are gradual, occurring over tens of thousands of years.

Two major changes have taken place on Earth during the past 150 years. First, the industrial revolution led to an increase in the production of greenhouse gases, especially from the burning of fossil fuels, which releases carbon dioxide into the atmosphere. In 1896, Svante Arrhenius (1859–1927), a Nobel Prize–winning chemist from Sweden, calculated that CO_2 released from burning fossil fuels could increase the greenhouse effect and raise Earth's surface temperature. The CO_2 produced by these fossil fuels has a different carbon isotope than the naturally occurring CO_2 already in the atmosphere, so it has been possible to verify that the measured CO_2 increase is due to burning fossil fuels, and not some other source.

The second change has been the rapid growth in human population. A larger population means more deforestation to clear land for agriculture and industry, reducing the amount of CO_2 absorbed by photosynthesizing plants, increasing CO_2 emissions, and locally changing Earth's albedo. A larger population also means more agricultural soil that releases nitrous oxide, more livestock that releases methane, and more people who release carbon dioxide by burning fossil fuels. **Figure 9.23** shows that concentrations of methane, nitrous oxide, and CO_2 have all been increasing since the industrial revolution. The CO_2 level has risen and is higher than any of the historic levels seen in Figure 9.21. Zooming in to more recent times, **Figure 9.24** shows the rise in the level of CO_2 plotted with the average global

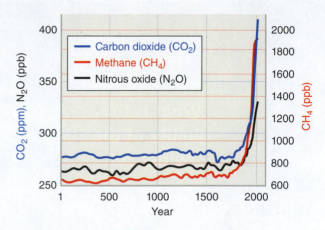

Figure 9.23 Concentrations of greenhouse gases from the year 1 to 2005, showing the increases beginning at the time of industrialization.

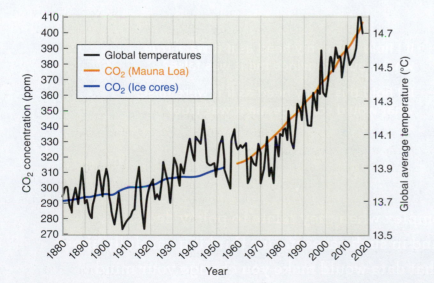

Figure 9.24 Earth's global average temperature and CO_2 concentrations since 1880. This graph shows that global temperatures are climbing along with concentrations of carbon dioxide. Annual variations in atmospheric CO_2 arise from seasonal variations in plant life and fossil fuel use, whereas the overall steady climb is due to human activities.

Thinking about Complexity

Climate change is a complex scientific issue. When confronted with complex science, there are several questions you should ask.

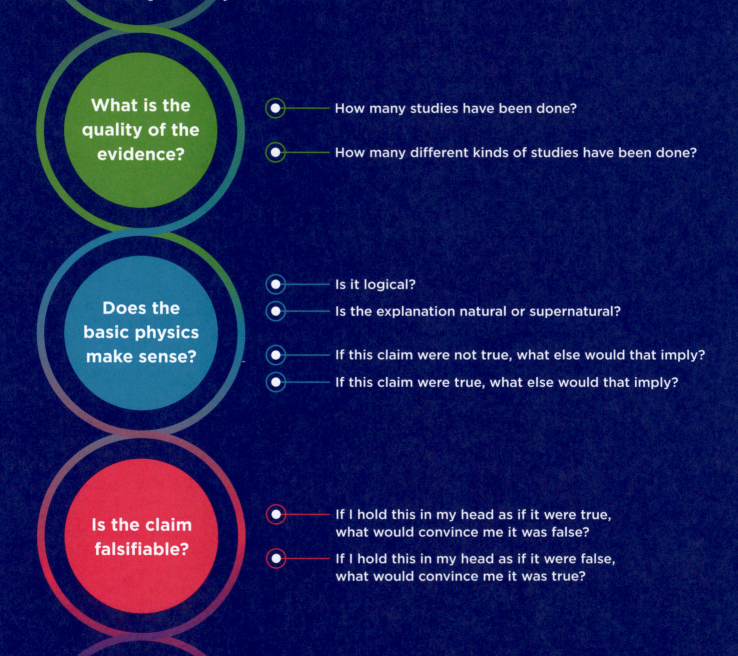

What is the quality of the evidence?
- How many studies have been done?
- How many different kinds of studies have been done?

Does the basic physics make sense?
- Is it logical?
- Is the explanation natural or supernatural?
- If this claim were not true, what else would that imply?
- If this claim were true, what else would that imply?

Is the claim falsifiable?
- If I hold this in my head as if it were true, what would convince me it was false?
- If I hold this in my head as if it were false, what would convince me it was true?

Scientific issues become remarkably complex when they relate to policy decisions with a broad reach. Keeping an open mind in such cases means thinking carefully about the quality of the evidence, as well as what data would make you change your mind. If your mind cannot be changed, then you are not making decisions based on evidence.

temperature on Earth to show that they rise together. **Figure 9.25** shows that levels of carbon dioxide, nitrous oxide, and methane have continued to steadily increase over the past few decades.

Nearly all climatologists accept the computer models indicating that this trend represents the beginning of a long-term change in temperature caused by the buildup of human-produced greenhouse gases. That anthropogenic (human-caused) change is happening much faster than the historic changes seen in Figure 9.21. Earth's atmosphere is a delicately balanced mechanism, and its climate is a complex system within which tiny changes can produce enormous and often unexpected results. Even an average increase of a few degrees can change the climate from an ice age to an interglacial period.

To add to the complexity, Earth's climate is intimately tied to ocean temperatures and currents. Ocean currents are critical in transporting energy from one part of Earth to another, and it is unclear how increased temperatures may affect those systems. Warmer oceans evaporate more, leading to wetter air, which can mean more intense summer and winter storms (including more snow). We see examples of that connection in the periodic El Niño and La Niña conditions, in which small shifts in ocean temperature cause much larger global changes in air temperature and rainfall. **Figure 9.26** shows that in the past several decades, however, natural factors have been overwhelmed by the contribution from **anthropogenic climate change**: the release of greenhouse gases into the atmosphere from the burning of fossil fuels and other human activities. Models including only natural causes fail to fit the data, whereas models that include both human activities and natural factors fit the data well. The **Process of Science Figure** discusses how scientists think about such complex issues.

Changes in climate affect where plants and animals can live, so scientists look for evidence of climate change in ecosystems around the world. Scientists have observed that the dates and locations of breeding, migration, hibernation, and so on, have changed. Agricultural growing seasons and pollination also are affected, as is the availability of freshwater. Rainfall patterns are shifting, causing long-term droughts that led to the historic wildfire seasons in Australia and the United States in 2020. The melting of mountain glaciers, Antarctic glaciers, and polar sea ice due to higher global average temperature is already being observed. The levels of the oceans have measurably risen and are predicted to rise further—not only from melted ice in Greenland and Antarctica but also from the thermal expansion of the water as the ocean temperature increases. That is a serious problem for the many people who live in coastal or low-lying regions, and it is already an issue in Miami and Venice. A warmer Arctic can lead to a greater release of methane from the permafrost. Less ice and snow can decrease Earth's albedo, allowing more sunlight to reach the surface, although albedo may rise because of an increase in cloud cover, caused by more water in the atmosphere.

In a real sense, we are *experimenting* with Earth. We are asking the question: What happens to Earth's climate if we steadily increase the number of greenhouse molecules in its atmosphere? We do not yet know the full answer, but we are already seeing some of the consequences. These consequences allow us to further test the models that scientists use to predict climate, which in turn improves those models. The processes are very complex, but over time, the accuracy and precision of scientists' predictions have continuously improved. The most recent studies are able to make confident predictions about the effects of climate change at size scales as small as counties within states in the United States. Anthropogenic

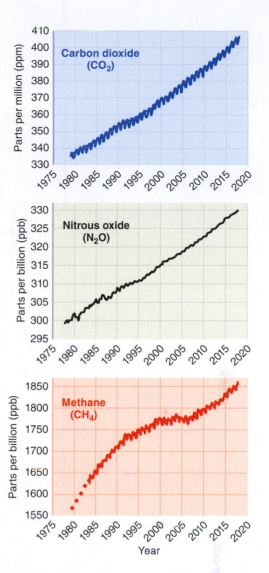

Figure 9.25 National Oceanic and Atmospheric Administration (NOAA) plots of global average amounts of the major greenhouse gases.

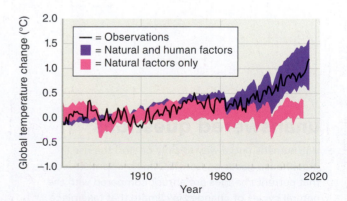

Figure 9.26 Global temperature variations on Earth since 1860, overlaid with model results that include only natural causes of temperature change (pink) and those that include both natural and human factors (purple).

climate change is the best explanation for the range of changes that are occurring on Earth today.

CHECK YOUR UNDERSTANDING **9.5**

Over the past 800,000 years, according to ice-core data, which of the following are correlated? (Choose all that apply.) (a) temperature; (b) CO_2 levels; (c) methane levels; (d) the size of the ozone hole

Origins: Our Special Planet

Why are we here on Earth instead of elsewhere in the Solar System? The habitable zone of a planetary system (Chapter 7) is the range of distances from its star within which a planet could have a surface temperature such that large amounts of water could exist in liquid form. The temperature of a planet, though, can depend on more than just distance from its star. Temperature can change due to orbital variations, the atmospheric greenhouse effect, and the planet's ability to hold on to its atmosphere. Mercury and the Moon were too small to hold on to either their primary or their secondary atmospheres, leaving them as essentially airless rocks. Venus, Earth, and Mars all had some liquid water at one time, so early in the history of the Solar System all three might have been classified as habitable. Their atmospheres were probably similar then, too, before the atmosphere of Mars escaped, the atmosphere of Venus was heated by the greenhouse effect, and the atmosphere of Earth was oxygenated. Primitive life may even have developed on Venus and Mars at the same time it developed on Earth, about a billion years after the Solar System formed.

Ultimately, the three planets evolved differently, and only Earth now has the liquid water vital for our biology. Venus receives more energy from the Sun and was slightly warmer than the young Earth. The young Venus might have been too hot for liquid water to form oceans, or any liquid water might have quickly evaporated. Because Venus had a lot of water vapor in its atmosphere and lacked a liquid ocean to store carbon dioxide, the greenhouse effect created a thicker atmosphere. That thicker atmosphere fed back to a stronger greenhouse effect, which created an even thicker atmosphere. With that resulting runaway greenhouse effect, Venus became just too hot: the evaporated water molecules broke apart, the hydrogen escaped to space, and any water cycle was destroyed. Primitive bacteria may have lingered in water vapor in the clouds on Venus, but life did not evolve into anything more complex.

Mars is smaller and less massive than Venus or Earth (**Figure 9.27**), so its gravity is weaker. The red planet has a larger orbit than Venus and Earth and so receives less energy from the Sun. Over time, much of its atmosphere escaped and was not replaced by volcanic emissions, in which case the atmospheric pressure became too low to maintain liquid water. Whereas Venus was too hot, the young Mars was too cold. Back then it had a thicker atmosphere and liquid rain. Images of the surface of Mars show flood basins, indicating huge rivers in the past. The process that prevented Earth from becoming a Venus-like hothouse continued further on Mars, and the temperature fell until the water froze. Hints of subsurface water exist on Mars, and perhaps some form of martian bacteria will be found beneath the ground.

Thus, our Solar System contains astronomical evidence of how the greenhouse effect can influence planetary atmospheres, including Earth's. Planetary scientists view those other planets as a cautionary tale, showing the results of varied "doses"

unanswered questions

Will humans find a way to slow or stop the rise in greenhouse gases on Earth? Climate scientists worry that current changes are abrupt compared with the natural cycles of changes in climate that take place gradually over thousands of years. Will nations reach an agreement to reduce the production of those gases, as they did to reduce the use of chemicals that created the ozone hole?

Figure 9.27 Physical and orbital differences among Venus, Earth, and Mars cause the planets to have very different atmospheres.
Credit (Mars): Phil James (Univ. Toledo), Todd Clancy (Space Science Inst., Boulder, CO), Steve Lee (Univ. Colorado), and NASA/ESA, https://esahubble.org/images/opo9715c/. https://creativecommons.org/licenses/by/4.0/.

of greenhouse gases. Only Earth stayed "just right" and could retain the liquid oceans in which more complex life evolved. We owe our lives to the blanket of atmosphere that covers the planet. The study of the Solar System reveals that Earth is maintained by the most delicate of balances. Over billions of years, life has shaped Earth's atmosphere, and today, through the activities of humans, life is reshaping the planet's atmosphere once again. Human civilization is much younger than Earth and has been brief in comparison with the cycles of climate on the planet. The past 10,000 years has been a relatively stable period of Earth's climate, and many people argue that this stability enabled agriculture and civilization to develop. Scientists are uncertain what will happen to agriculture, civilization, or the planet itself if the climate undergoes fast, intense changes. The other planets in the Solar System are not places where billions of people from our planet can live. Earth is the only planet suitable for human life.

SUMMARY

Earth, Venus, and Mars are warmer than they would be from solar illumination alone. Earth's atmosphere is thick enough to warm the surface to life-sustaining temperatures, but not so thick that Earth becomes overheated. Earth, Venus, and Mars all have significant atmospheres that are different from the original atmospheres they captured when they formed. Their atmospheres are complex, both in chemical composition and in physical characteristics such as temperature and pressure. The climates of Earth, Venus, and Mars are all determined by their individual atmospheres. The atmospheres of Earth, Venus, and Mars have different chemical compositions. They led, in turn, to dramatic differences in temperature and pressure. Life has altered Earth's atmosphere several times, most notably in the distant past from an increase in the amount of oxygen in the atmosphere and in modern times from an increase in greenhouse gases. Mars and Venus might have been habitable early in the history of the Solar System, but now only Earth has substantial liquid water on its surface.

(1) Identify the processes that cause primary and secondary atmospheres to be formed, retained, and lost. Planetary atmospheres evolve.

The primary atmospheres consisted mainly of hydrogen and helium captured from the protoplanetary disk. The terrestrial planets lost their primary atmospheres soon after the planets formed. Secondary atmospheres were created by volcanic gases and from volatiles brought in by impacting comets and asteroids. Planetary bodies must have enough mass to hold on to their atmospheres.

(2) Compare the strength of the greenhouse effect on Earth, Venus, and Mars and how it contributes to differences among the atmospheres of those planets. Earth, Venus, and Mars have naturally occurring greenhouse gases, which increase the average surface temperature of each planet. The amount by which those greenhouse gases raise the temperature of a planet depends on the number of greenhouse gas molecules in the atmosphere. The differences in global temperatures among those planets can be explained in part by their distances from the Sun. However, their different compositions and atmospheric densities are highly significant factors in determining their global temperatures. The atmospheric greenhouse effect keeps Earth from freezing, but it turns Venus into an inferno.

③ Describe the layers of the atmospheres on Earth, Venus, and Mars. The atmospheres of Earth, Venus, and Mars have different temperatures, pressures, and compositions. Earth's atmosphere, in particular, has many layers. The layers are determined by the vertical variations in temperature and absorption of solar radiation throughout the atmosphere. Temperature and pressure decrease with altitude in the tropospheres of Earth, Venus, and Mars. Earth's magnetosphere shields the planet from the solar wind. Venus has a massive, hot atmosphere of carbon dioxide and sulfur compounds. Venus has surprisingly fast winds for a slowly rotating planet. Mars has a thin, cold, carbon dioxide atmosphere that may have been much thicker in the past. Earth's atmosphere is thicker than that of Mars but thinner than that of Venus and is composed primarily of nitrogen.

④ Explain how Earth's atmosphere has been reshaped by the presence of life. The oxygen levels in Earth's atmosphere have been enhanced through photosynthesis by bacteria and then by plants. Increased oxygen levels led to the formation of the ozone layer and to the development of more advanced forms of life.

⑤ Describe the evidence that shows Earth's climate is changing and how comparative planetology contributes to a better understanding of those changes. Astronomical, geological, and biological processes can lead to large changes in a planet's climate. The study of climate on the terrestrial planets expands scientists' knowledge of Earth's past, present, and future conditions. Large variations in global temperature over the past 800,000 years correlate strongly with the number of greenhouse molecules in the atmosphere. The current level of greenhouse gases in Earth's atmosphere is higher than any seen during that period and correlates with a later increase in temperature.

QUESTIONS AND PROBLEMS

TEST YOUR UNDERSTANDING

1. Place in chronological order the following steps in the formation and evolution of Earth's atmosphere.
   a. Plant life converts carbon dioxide (CO_2) to oxygen.
   b. Hydrogen and helium are lost from the atmosphere.
   c. Volcanoes, comets, and asteroids increase the inventory of volatile matter.
   d. Hydrogen and helium are captured from the protoplanetary disk.
   e. Oxygen enables the growth of new life-forms.
   f. Life releases CO_2 from the subsurface into the atmosphere.

2. On which of the following planets is the atmospheric greenhouse effect strongest?
   a. Venus
   b. Earth
   c. Mars
   d. Mercury

3. The oxygen molecules in Earth's atmosphere
   a. were part of the primary atmosphere.
   b. arose when the secondary atmosphere formed.
   c. are the result of life.
   d. are being rapidly depleted by the burning of fossil fuels.

4. The differences in the climates of Venus, Earth, and Mars are caused primarily by
   a. the composition of their atmospheres.
   b. their relative distances from the Sun.
   c. the thickness of their atmospheres.
   d. the time at which their atmospheres formed.

5. The words *weather* and *climate*
   a. mean essentially the same thing.
   b. refer to very different timescales.
   c. refer to very different-sized scales.
   d. refer to both b and c.

6. Less massive molecules tend to escape from an atmosphere more often than more massive molecules because
   a. the gravitational force on them is less.
   b. they are moving faster.
   c. they are more buoyant.
   d. they are smaller, so they experience fewer collisions on their way out.

7. Venus is hot and Mars is cold primarily because
   a. Venus is closer to the Sun.
   b. Venus has a much thicker atmosphere.
   c. the atmosphere of Venus is dominated by CO_2, but the atmosphere of Mars is not.
   d. Venus has stronger winds.

8. Studying climate on other planets is important to understanding climate on Earth because (select all that apply)
   a. the underlying physical processes are the same on every planet.
   b. other planets offer a variety of extremes to which Earth can be compared.
   c. comparing climates on other planets helps scientists understand which factors are important.
   d. other planets can be used to test atmospheric models.

9. The atmosphere of Mars is often pink-orange because
   a. it is dominated by carbon dioxide.
   b. the Sun is at a low angle in the sky.
   c. Mars has no oceans to reflect blue light to the sky.
   d. winds lift dust into the atmosphere.

10. Auroras are the result of
    a. the interaction of particles from the Sun and Earth's atmosphere and magnetic field.
    b. upper-atmosphere lightning strikes.
    c. the destruction of stratospheric ozone, which leaves a hole.
    d. the interaction of Earth's magnetic field with Earth's atmosphere.

11. The stratospheric ozone layer protects life on Earth from
    a. high-energy particles from the solar wind.
    b. micrometeorites.
    c. ultraviolet radiation.
    d. charged particles trapped in Earth's magnetic field.

12. Hadley circulation is broken into zonal winds by
    a. convection from solar heating.
    b. hurricanes and other storms.
    c. interactions with the solar wind.
    d. the planet's rapid rotation.

13. The _____ of greenhouse gas molecules affects the temperature of an atmosphere.
    a. percentage
    b. fraction
    c. number
    d. mass

14. Over the past 800,000 years, Earth's temperature has closely tracked with
    a. solar luminosity.
    b. oxygen levels in the atmosphere.
    c. the size of the ozone hole.
    d. carbon dioxide levels in the atmosphere.

15. Convection in the _____ causes weather on Earth.
    a. stratosphere
    b. mesosphere
    c. troposphere
    d. ionosphere

THINKING ABOUT THE CONCEPTS

16. ★ WHAT AN ASTRONOMER SEES In Figure 9.14, an astronomer would see evidence that Venus is a slowly rotating planet. Make a sketch of what the clouds in this figure would look like if Venus rotated quickly. Annotate your sketch with arrows to show how the clouds might move relative to one another. ◉

17. Primary atmospheres of the terrestrial planets were composed almost entirely of hydrogen and helium. Explain why they contained only those gases and not others.

18. How were the secondary atmospheres of the terrestrial planets created?

19. Nitrogen, the principal gas in Earth's atmosphere, was not a significant component of the protostellar disk from which the Sun and planets formed. Where did Earth's nitrogen come from?

20. What are the likely sources of Earth's water?

21. In what ways does plant life affect the composition of Earth's atmosphere?

22. What is the difference between ozone in the stratosphere and ozone in the troposphere? Which is a pollutant, and which protects terrestrial life?

23. What is the principal cause of winds in the atmospheres of the terrestrial planets?

24. Global warming appears to be responsible for increased melting of the ice in Earth's polar regions.
    a. Why does the melting of Arctic ice, which floats on the Arctic Ocean, *not* affect the level of the oceans?
    b. How is the melting of glaciers in Greenland and Antarctica affecting the level of the oceans?

25. Why can we *not* get a clear view of the surface of Venus, as we have so successfully done with the surface of Mars?

26. What is the evidence that the greenhouse effect exists on Earth, Venus, and Mars?

27. Explain why surface temperatures on Venus hardly vary between day and night and between the equator and the poles.

28. Why do scientists think that Mars and Venus were once more habitable but no longer are?

29. The last step in the Process of Science Figure is one that anyone can carry out about any complex issue. Write down your current take on the issue of anthropogenic climate change: Do you accept the evidence? If yes, what evidence persuaded you? Whether or not you accept climate change, write down a piece of evidence that you would need to see in order to change your mind about this important issue. ◉

30. Given the current conditions on Venus and Mars, which planet might be easier to engineer to make it habitable to humans? Explain.

APPLYING THE CONCEPTS

31. On average, will helium atoms be lost or retained in Earth's atmosphere? ●—●—●
    a. Make a prediction: Study Working It Out 9.1. Is a helium atom more or less massive than a hydrogen atom? As a result, will the speed of a helium atom be larger or smaller than the speed of a hydrogen molecule? Will the difference between the two speeds be large or small?
    b. Calculate: Find the average speed of a helium atom in Earth's atmosphere. Compare your answer to the general rule for retaining particles in an atmosphere. Determine whether helium atoms will be retained or lost.
    c. Check your work: Verify that your answer has units of km/s, and compare your answer to your prediction.

32. Calculate the average speed of a hydrogen molecule (H_2) in the atmosphere of Mars.
    a. Make a prediction: Think about what you know about the composition of the atmosphere of Mars. Do you expect the average speed of a hydrogen molecule to be greater than or less than the escape speed from Mars?
    b. Calculate: Use the equation for the average speed of a molecule from Working It Out 9.1 to find the average speed of a hydrogen molecule on Mars.
    c. Check your work: Verify that the units of your answer are in km/s, and compare your answer to your prediction.

33. Calculate the escape speed from Mars.
    a. Make a prediction: Do you expect this escape speed to be greater than or less than the escape speed from Earth?
    b. Calculate: Follow Working It Out 9.1 to find the escape speed from Mars.
    c. Check your work: Verify that the units of your answer are in km/s, and compare your answer to your prediction.

34. Suppose that your research team has discovered an Earth-like planet around a Sun-like star. The Earth-like planet has the same mass and radius as Earth, but it is significantly closer to its star than Earth is to the Sun. The average temperature is twice as high as the temperature on Earth. If this planet had H_2 and O_2 in its atmosphere, would you expect either of them to be retained?
    a. Make a prediction: This planet is twice as warm as Earth. Studying the formula for the velocity of a molecule shows that the temperature is under the square root; so, the velocity of the molecules will be 1.414 times faster than on Earth, because 1.414 is the square root of 2. The velocities of these molecules on Earth are given in Working It Out 9.1. Do you expect that multiplying by 1.414 will push these average speeds over the boundary set by the "general rule" that if $v_{mol} \leq \frac{1}{6} v_{esc}$, the planet will retain that gas?
    b. Calculate: Calculate the speed of each molecule, and compare it to 1.87 km/s, which is 1/6 of Earth's escape velocity.
    c. Check your work: Verify that your answer has units of km/s, and compare your answer to your prediction.

35. Oxygen molecules (O_2) are 16 times more massive than hydrogen molecules (H_2). Carbon dioxide molecules (CO_2) are 22 times more massive than H_2. Compare the average speed of O_2 and CO_2 molecules in a volume of air.
    a. Make a Prediction: When you take a ratio, you divide the two numbers. Study the third equation in Working It Out 9.1. When you divide that equation (calculated for O_2) by that equation (calculated for CO_2), what terms will divide out? Does O_2 have a larger or smaller mass than CO_2? Which one will move faster? Will the ratio of the speed of O_2 to the speed of CO_2 be larger or smaller than 1?
    b. Calculate: Set up the ratio, cancel terms, and solve for the numerical ratio of the speed of O_2 molecules to the speed of CO_2 molecules.
    c. Check Your Work: Verify that all the units have canceled, so that your answer has no units. Compare your answer to your prediction.

36. Calculate the average speed of a carbon dioxide molecule in the atmospheres of Earth and Mars. Compare those speeds with their respective escape velocities. What does that tell you about each planet's hold on its atmosphere?

37. The average surface pressure on Mars is 6.4 mb. Using Figure 9.7, estimate how high you would have to go in Earth's atmosphere to experience the same atmospheric pressure that you would experience if you were standing on Mars. ⊙

38. Water pressure in Earth's oceans increases by 1 bar for every 10 meters of depth. Compute how deep you would have to go to experience pressure equal to the atmospheric surface pressure on Venus.

39. Study Figure 9.21. ⊙
    a. Are the axes linear or logarithmic?
    b. Compare the CO_2 levels (top) with the temperature relative to present (middle). How would you describe the relationship between the graphs?
    c. How much higher is the current CO_2 level than the previous highest value?

40. Study Figure 9.23. ⊙
    a. According to the figure, when (approximately) did greenhouse gases begin rising exponentially?
    b. These graphs show that several greenhouse gases have behaved similarly in recent times. Was that true in the past (in general)?
    c. Speculate on possible causes for the common behavior of greenhouse gases in modern times.

41. Why do commercial jet planes fly at the altitudes shown in Figure 9.7b? ⊙

42. Commercial jets are pressurized. If you take an unopened bag of chips on a commercial jet airplane, the bag puffs up as you travel to the cruising altitude of 14 km.
    a. Is the pressure in the cabin higher or lower than the pressure on the ground?
    b. If a second bag of chips were attached to the outside of the plane, which bag would puff up more? About how much more (assume an unbreakable bag)? (See Figure 9.7b.)

43. Atmospheric pressure is caused by the weight of a column of air above you pushing down. At sea level on Earth, that pressure is equal to 10^5 newtons per square meter (N/m^2).
    a. Estimate the total force on the top of your head from that pressure. (Note: Force = pressure × area.)
    b. Recall that the acceleration due to gravity is 9.8 m/s^2. If the force in part (a) were caused by a kangaroo sitting on your head, what would the mass of the kangaroo be?
    c. Assume that a typical kangaroo has a mass of 60 kg. How many kangaroos would have to be sitting on your head to equal the mass of the extremely massive kangaroo in part (b)?
    d. Why are you *not* crushed by that astonishing force on your head?

44. Increasing the temperature of a gas inside a closed, rigid container increases the pressure. (That is why you should not put an unopened can of soup directly on the stove!) Explain how Figure 9.7b shows that phenomenon at work in Earth's atmosphere. ⊙

45. The total mass of Earth's atmosphere is 5×10^{18} kg. Carbon dioxide (CO_2) makes up about 0.06 percent of Earth's atmospheric mass.
    a. What is the mass of CO_2 (in kilograms) in Earth's atmosphere?
    b. The annual global production of CO_2 is now estimated to be 3×10^{13} kg. What annual fractional increase does that value represent?
    c. The mass of a molecule of CO_2 is 7.31×10^{-26} kg. How many molecules of CO_2 are added to the atmosphere each year?
    d. Why does an increase in CO_2 have such a big effect, even though it represents a small fraction of the atmosphere?

EXPLORATION Climate Change

One prediction about climate change is that as the planet warms, ice in the polar caps and in glaciers will melt. Such melting certainly seems to be occurring in almost all glaciers and ice sheets around the planet. Does that actually matter, and if so, why? In this Exploration, we explore several consequences of the melting ice on Earth.

Experiment 1: Floating Ice

For this experiment, you will need a permanent marker, a translucent plastic cup, water, and ice cubes. Place a few ice cubes in the cup and add water until the cubes float (so they don't touch the bottom). Mark the water level on the outside of the cup with the marker, and label that mark "initial water level."

1. As the ice melts, what do you expect to happen to the water level in the cup?

   _____

   _____

Wait for the ice to melt completely, and then mark the cup again.

2. What happened to the water level in the cup when the ice melted?

   _____

   _____

3. From the results of your experiment, what do you predict will happen to global sea levels when the Arctic ice sheet, which floats on the ocean, melts?

   _____

   _____

Experiment 2: Ice on Land

For this experiment, you will need the same materials as in experiment 1, plus a paper or plastic bowl. Fill the cup about halfway with water and then mark the water level, labeling the mark "initial water level." Poke a hole in the bottom of the bowl and set the bowl over the cup. Add some ice cubes to the bowl.

4. As the ice melts, what do you expect will happen to the water level in the cup?

   _____

   _____

Wait for the ice to melt completely, and then mark the cup again.

5. What happened to the water level in the cup when the ice melted?

   _____

   _____

6. In this experiment, the water in the cup is analogous to the ocean, and the ice in the bowl is analogous to ice on land. From the results of your experiment, what do you predict will happen to global sea levels when the Antarctic ice sheet, which sits on land, melts?

   _____

   _____

Experiment 3: Why Does It Matter?

Search online for the phrase "Earth at night" to find a satellite picture of Earth taken at night. The bright spots on the image trace out population centers. In general, the brighter a spot, the more populous the area (although a confounding factor relates to technological advancement).

7. Where do humans tend to live—near coasts or inland? Coastal regions are, by definition, near sea level. If both the Greenland and Antarctic ice sheets melted completely, sea levels would rise by as much as 80 meters. How would a sea-level rise of a few meters (in the range of reasonable predictions) over the next few decades affect the global population? (Note that each story of a building is about 3 meters.)

   _____

   _____

Worlds of Gas and Liquid— The Giant Planets

The giant planets Jupiter and Saturn, along with terrestrial planets Venus and Mars, are bright enough that even city-dwellers can see them at night when the planets are in the right part of the sky. Use the Internet or an app to find out which planets are visible during the term. Go outside on a clear night and sketch the positions of the planets that are visible. If you can identify stars or constellations near the planets (star maps like those at the end of this book, or ones you can obtain online, will help!), include them in your sketch. Be sure to record your location, the date, the time, and a note about the sky conditions.

EXPERIMENT SETUP

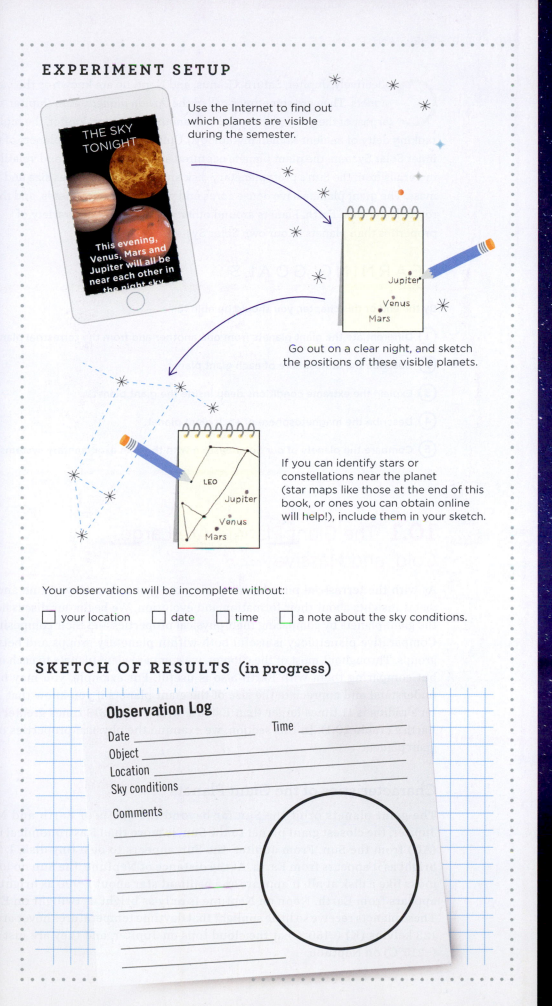

THE SKY TONIGHT

This evening, Venus, Mars and Jupiter will all be near each other in the night sky.

Use the Internet to find out which planets are visible during the semester.

Jupiter
Venus
Mars

Go out on a clear night, and sketch the positions of these visible planets.

LEO
Jupiter
Venus
Mars

If you can identify stars or constellations near the planet (star maps like those at the end of this book, or ones you can obtain online will help!), include them in your sketch.

Your observations will be incomplete without:

☐ your location ☐ date ☐ time ☐ a note about the sky conditions.

SKETCH OF RESULTS (in progress)

Observation Log

Date _____ Time _____

Object _____

Location _____

Sky conditions _____

Comments

_____

_____

_____

_____

_____

_____

_____

10

Collectively, Jupiter, Saturn, Uranus, and Neptune are known as the *giant planets*. They are sometimes called the *Jovian planets*, after Jupiter, the largest of the giant planets. (*Jove* is another name for Jupiter, the highest-ranking deity of ancient Roman mythology.) Unlike the terrestrial planets of the inner Solar System, the giant planets captured and retained gases and volatile materials from the Sun's protoplanetary disk and grew to enormous size and mass. The giant planets have dense cores and very large atmospheres, and they rotate faster than Earth. Planets around other stars have a wider variety of properties than planets in our own Solar System.

LEARNING GOALS

By the end of this chapter, you should be able to:

(1) Differentiate the giant planets from one another and from the terrestrial planets.

(2) Describe the atmosphere of each giant planet.

(3) Explain the extreme conditions deep inside the giant planets.

(4) Describe the magnetosphere of each giant planet.

(5) Compare the planets of our Solar System with those in exoplanetary systems.

10.1 The Giant Planets Are Large, Cold, and Massive

As with the terrestrial planets, comparing the giant planets with one another yields insights about their formation and evolution. We begin our discussion of the giant planets by comparing their physical properties and their compositions. Comparative planetology is useful both within planetary groups and between groups. Throughout most of the chapter, we compare giant planets to each other, but comparing them with Earth also is useful. For example, you may better understand and appreciate the size of the giant planets if you know that Jupiter's radius is 11 times larger than Earth's, with a mass 318 times greater than Earth's (**Table 10.1**). In this section, we examine the physical properties of the giant planets.

Characteristics of the Giant Planets

The giant planets orbit the Sun far beyond the orbits of Earth and Mars. Jupiter, the closest giant planet to the Sun, is more than 5 astronomical units (AU) from the Sun. From Jupiter, the Sun appears to be a tiny disk, 1/27 as bright as it appears from Earth. At the distance of Neptune, the Sun no longer looks like a disk at all: it appears as a brilliant star about 1/900 as bright as it appears from Earth. Noon on Neptune is only as bright as twilight on Earth. These planets receive so little sunlight that daytime temperatures hover around 123 kelvins (K) (−150°C) at the cloud tops on Jupiter, and they are just 58 K (−215°C) on Neptune.

Table 10.1	Physical Properties of the Giant Planets, Compared to Earth				

Property	Earth	Jupiter	Saturn	Uranus	Neptune
Orbital semimajor axis (AU)	1	5.20	9.6	19.2	30
Orbital period (Earth years)	1	11.9	29.5	84.0	164.8
Orbital velocity (km/s)	29.29	13.1	9.7	6.8	5.4
Mass (Earth masses)	1	317.8	95	14.5	17.2
Equatorial radius (km)	6378	71,490	60,270	25,560	24,300
Equatorial radius (Earth radii)	1	11.2	9.4	4.0	3.8
Oblateness	0.003	0.065	0.098	0.023	0.017
Density (water = 1)	5.5	1.33	0.69	1.27	1.64
Rotation period (hours)	24	9.9	10.7	17.2	16.1
Tilt (degrees)	23.5	3.13	26.7	97.8	28.3
Surface gravity ($g_{Earth} = 1$)	1	2.53	1.07	0.91	1.14
Escape speed (km/s)	11.2	59.5	35.5	21.3	23.5

Jupiter, Saturn, Uranus, and Neptune are enormous in comparison with the rocky terrestrial planets. Jupiter is the largest of the eight planets and is more than 1/10 the diameter of the Sun. Saturn is slightly smaller than Jupiter, with a diameter of 9.5 Earths. Uranus and Neptune are each about 4 Earth diameters across. **Figure 10.1a** shows the relative diameters of the giant planets. Because they are at different distances, their angular diameters as seen from Earth vary even more, as shown in **Figure 10.1b**.

The most accurate method of finding the diameter of a giant planet is to observe a **stellar occultation**, which occurs when the planet eclipses a star in the sky. The star disappears behind the planet and then reappears a short while later. Because we know the relative orbital speeds of Earth and the giant planets, we can calculate the size of the eclipsing giant planet from the length of time the star is eclipsed, as shown in **Figure 10.2**. Occultations of the radio signals transmitted from orbiting spacecraft and images taken by spacecraft cameras also provide accurate measurements of the diameters and shapes of planets and their moons.

Although far away, the giant planets are so large that all but Neptune can be seen with the unaided eye (although Uranus is so far away that it is very faint; you must have excellent vision and be in an extremely clear, dark location to see Uranus). Jupiter and Saturn were known to the ancients, but Uranus was not discovered until 1781, when William Herschel accidentally noticed a tiny disk in the eyepiece of his 6-inch telescope. At first he thought he had found a comet, but the object's slow nightly motion soon convinced him that it was a planet beyond the orbit of Saturn. During the

a.

Jupiter   Saturn   Uranus   Neptune

VIS

Figure 10.1 a. Relative sizes of the giant planets, shown to the same physical scale. **b.** (next page) The same images, scaled according to the relative sizes as they appear from Earth. The full Moon is 1800 arcseconds across in our sky

decades that followed Herschel's discovery, astronomers found that Uranus's position differed from the path predicted from Newton's laws of motion and suggested that the gravitational pull of an unknown planet caused Uranus's surprising behavior. Using the astronomers' measured positions of Uranus, two young mathematicians—Urbain-Jean-Joseph Le Verrier (1811–1877) in France and John Couch Adams (1819–1892) in England—independently predicted the location of the hypothetical

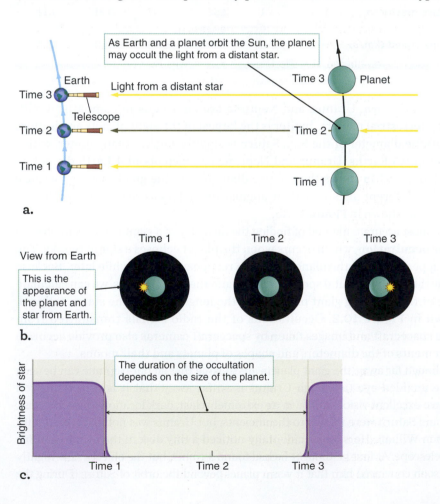

Figure 10.2 a. Occultations occur when a planet, moon, or ring passes in front of a star. **b.** As the planet moves (from right to left as seen from Earth), the starlight is blocked. **c.** The amount of time that the star is hidden, combined with information about how fast the planet is moving, gives the size of the planet.

Jupiter   Saturn   Uranus   Neptune

b.   VIS

planet. The German astronomer Johann Gottfried Galle (1812–1910) found the planet on his first observing night, just where Le Verrier and Adams had predicted it would be. Thus, Galle's discovery of Neptune in 1846 became a triumph for mathematical prediction based on physical law—and for the later confirmation of theory by observation (see the **Process of Science Figure**). Neptune remains the outermost known planet in the Solar System and cannot be seen without binoculars.

The Sun contains the vast majority of the mass in the Solar System. It is about 1000 times as massive as Jupiter. Jupiter contains more than twice the mass of all the other planets combined; it is 318 times as massive as Earth, 3.3 times as massive as Saturn, and about 20 times as massive as either Uranus or Neptune. Together, the giant planets contain 99.5 percent of all the mass in the Solar System, apart from the Sun. All other Solar System objects—terrestrial planets, dwarf planets, moons, asteroids, and comets—make up the remaining 0.5 percent.

Originally, astronomers combined Newton's universal law of gravitation and Kepler's third law to find the mass of a planet from the period and semimajor axis of a moon of that planet. Planetary spacecraft now make it possible to measure the masses of planets even more accurately. As a spacecraft flies by a planet, the planet's gravity tugs on it, changing its path. Astronomers at multiple locations on Earth track and compare the spacecraft's radio signals to detect these tiny changes in the spacecraft's path. From those changes, astronomers calculate the gravitational force on the spacecraft and accurately measure the planet's mass.

Composition of the Giant Planets

In the preceding chapters, you learned that the terrestrial planets are composed mostly of rocky materials, with atmospheres forming a thin layer of very little or no mass. The giant planets, in contrast, are made up primarily of gases and liquids. Jupiter and Saturn are composed of hydrogen and helium and are therefore known as **gas giants**. Uranus and Neptune, on the other hand, are known as **ice giants** because they both contain much larger amounts of ices, such as water ice, than Jupiter and Saturn. On a giant planet, the atmosphere merges seamlessly into a liquid hydrogen ocean, which merges smoothly into a denser liquid or solid core. No abrupt transition exists from atmosphere to solid ground, as is found on

Scientific Laws Make Testable Predictions

How did astronomers predict that Neptune existed, before it was discovered?
Newton's laws of motion and gravity do more than describe what we see.
They also enable us to predict the existence of things yet unseen.

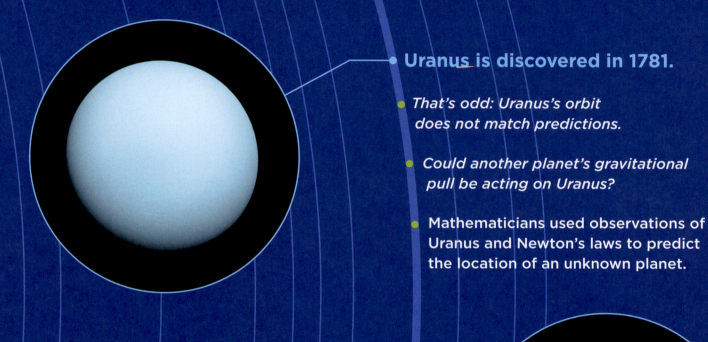

Uranus is discovered in 1781.

That's odd: Uranus's orbit does not match predictions.

Could another planet's gravitational pull be acting on Uranus?

Mathematicians used observations of Uranus and Newton's laws to predict the location of an unknown planet.

Neptune is discovered in 1846.

Newton's laws pass another test!

Laws make predictions that can be tested in order to verify or falsify them. Each test that does not falsify a scientific law strengthens scientists' confidence in its predictions.

the terrestrial planets. The atmospheres of giant planets are much deeper than the atmospheres of the terrestrial planets—thousands of kilometers rather than hundreds. In visible-light photos of Jupiter and Saturn, we see the top of a layer of thick clouds that obscures lower layers (see **Figure 10.3a**). Only a few thin clouds are visible on Uranus, although atmospheric models suggest that thick cloud layers must lie below. Neptune displays a few high clouds with a deep, clear atmosphere showing between them (**Figure 10.3b**).

The giant planets have lower densities than the terrestrial planets because they are composed almost entirely of lighter materials, such as hydrogen, helium, and water. Among the giant planets, Saturn has the lowest mean density, only 0.7 times that of water. Therefore, Saturn would float in water—if you had a large enough swimming pool—with 70 percent of its volume submerged. Neptune has the highest density, about 1.6 times that of water. (Earth is nearly 3.5 times denser.) The densities of Jupiter and Uranus are between those of Neptune and Saturn.

Jupiter's chemical composition is similar to that of the Sun—mostly hydrogen and helium. (Recall Figure 5.17, the astronomer's periodic table.) Only a small percentage of Jupiter's mass is made up of **heavy elements**, which astronomers define as all elements more massive than helium. In Jupiter's atmosphere, many of those heavy elements combine chemically with hydrogen (H). For example, atoms of oxygen (O), carbon (C), and nitrogen (N) combine with hydrogen to form molecules of water (H_2O), methane (CH_4), and ammonia (NH_3), respectively. More complex combinations, such as ammonium hydrosulfide (NH_4HS), also form there. Jupiter's core, which contains most of the planet's iron and silicates, remains from the original rocky planetesimals around which Jupiter grew.

The four giant planets differ in the amounts of hydrogen and helium that each one contains. Because of its larger mass, Jupiter accumulated a higher percentage of hydrogen and helium when it formed than the other planets did. Saturn contains more heavy elements and less hydrogen and helium than Jupiter. Heavy elements are significant components of Uranus and Neptune. Methane is a particularly important molecule in the atmospheres of those two planets, giving them their characteristic blue-green color. Those differences in the fraction of hydrogen and helium are the principal compositional differences among the giant planets, and they contain important clues to understanding how the giant planets formed.

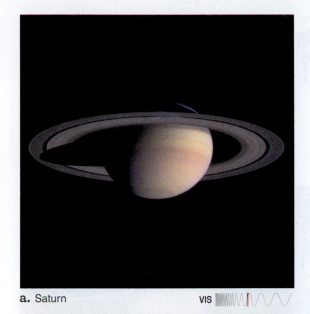

a. Saturn   VIS

b. Neptune   VIS

Figure 10.3 a. Saturn, imaged in visible light by *Cassini*. **b.** Neptune, imaged in visible light by *Voyager*.

Days and Seasons on the Giant Planets

The giant planets rotate rapidly (Table 10.1), so their days are short (ranging from 10 to 17 hours) and their shapes are distorted from perfectly spherical to **oblate**—they bulge at their equators and have an overall flattened appearance. Saturn's oblateness is noticeable. In **Figure 10.4**, we have drawn a circle around Saturn, and you can see that the planet does not completely fill the circle at the top and bottom. Saturn's equatorial diameter is almost 10 percent greater than its polar diameter. In contrast, Earth's polar and equatorial diameters differ by only 0.3 percent.

The intensity of a planet's seasons is determined by the tilt of its axis (Chapter 2). With a tilt of only 3°, Jupiter has almost no seasons at all. The tilts of Saturn (27°) and Neptune (28°), which are slightly larger than those of Earth (23.5°) and Mars (25°), cause moderate but well-defined seasons. Uranus is tilted about 98°, so it spins on an axis that is nearly in the plane of its orbit. Each polar region alternately experiences 42 years of continuous sunshine followed by 42 years of total darkness, causing Uranus's seasons to be extreme. Averaged over an entire orbit,

what if . . .

What if Earth's axis were tilted by 98°? How might that affect weather and life on Earth?

VIS

Figure 10.4 This Hubble Space Telescope image of Saturn was taken in 2009. The oblateness of the planet is apparent when compared with the white circle. The large orange moon Titan appears near the top of the disk of Saturn, along with its black shadow.

the poles receive more sunlight than the equator—a situation markedly different from that of any other Solar System planet.

A tilt greater than 90° indicates that the planet rotates clockwise when seen from above its orbital plane. This orientation is rare in the Solar System. Only Venus, Pluto, Pluto's moon Charon, and Neptune's moon Triton also have tilts greater than 90°. Why is Uranus tilted so differently from most other planets? One possible explanation is that Uranus was "knocked over" by the impact of one huge or several large planetesimals near the end of its accretion phase.

CHECK YOUR UNDERSTANDING 10.1

How are Uranus and Neptune different from Jupiter and Saturn? (a) Uranus and Neptune have a higher percentage of ices in their interiors; (b) Uranus and Neptune have more hydrogen; (c) Uranus and Neptune have no storms; (d) Uranus and Neptune are closer to the Sun.

Answers to Check Your Understanding questions are in the back of the book.

10.2 The Giant Planets Have Clouds and Weather

When we observe the giant planets through a telescope or in visible images from a spacecraft, only the top layers of the atmosphere are visible. Sometimes we can see a bit deeper into the clouds, but in essence, we are seeing a two-dimensional view of the cloud tops. The existence of deeper cloud layers on those giant planets is inferred from physical models of how the temperature changes with depth. In this section, we explore the atmospheres of the giant planets.

Viewing the Cloud Tops

Even when viewed through small telescopes, Jupiter is very colorful. Parallel bands—ranging in hue from bluish gray to various shades of orange, reddish brown, and pink—stretch out across its large, pale yellow disk. Astronomers call the darker bands *belts* and the lighter bands *zones*. Many clouds—some dark and some bright, some circular, and others more elliptical—appear along the edges of, or within, the belts. Whirlpool-like, swirling features are known as vortices (the singular is "vortex"). Those vortices are familiar to us on Earth as high- and low-pressure systems, hurricanes, and supercell thunderstorms.

The most prominent cloud structure is a large, red, oval feature in Jupiter's southern hemisphere known as the **Great Red Spot (Figure 10.5)**. The Great Red Spot was first observed more than three centuries ago, shortly after the telescope was invented. Since then, it has varied unpredictably in size, shape, color, and motion as it drifts among Jupiter's clouds. In the 1800s, the Great Red Spot was greater than twice the diameter of Earth, but now it has shrunk to just over one Earth diameter. Observations of small clouds circling the perimeter of the Great Red Spot show that it rotates counterclockwise with a period of about a week. Its cloud pattern looks a lot like that of a terrestrial hurricane, but it rotates in the opposite direction—exhibiting *anticyclonic* rather than cyclonic flow, indicating a high-pressure system.

Vortices are evident in many of the smaller elliptical clouds found elsewhere in Jupiter's atmosphere and in similar clouds observed in the atmospheres

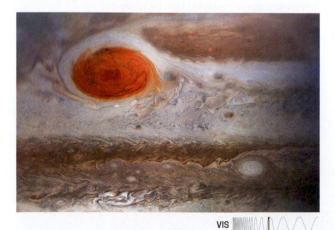

VIS

Figure 10.5 A digitally enhanced image of Jupiter taken by *Juno*. The Great Red Spot is a hurricane about 1.3 times the diameter of Earth.

Reading Astronomy News

Juno Spacecraft Spots Jupiter's Great Red Spot

jpl.nasa.gov

The newest mission to Jupiter observes its clouds and atmosphere.

Images of Jupiter's Great Red Spot (**Figure 10.6**) reveal a tangle of dark, veinous clouds weaving their way through a massive crimson oval. The JunoCam imager aboard NASA's *Juno* mission snapped pics of the most iconic feature of the solar system's largest planetary inhabitant during its Monday (July 10) flyby. The images of the Great Red Spot were downlinked from the spacecraft's memory on Tuesday and placed on the mission's Juno-Cam website Wednesday morning.

"For hundreds of years scientists have been observing, wondering and theorizing about Jupiter's Great Red Spot," said Scott Bolton, *Juno* principal investigator from the Southwest Research Institute in San Antonio. "Now we have the best pictures ever of this iconic storm. It will take us some time to analyze all the data from not only JunoCam, but Juno's eight science instruments, to shed some new light on the past, present and future of the Great Red Spot."

As planned by the *Juno* team, citizen scientists took the raw images of the flyby

Figure 10.6 An enhanced-color image of Jupiter's Great Red Spot.

from the JunoCam site and processed them, providing a higher level of detail than available in their raw form. The citizen-scientist images, as well as the raw images they used for image processing, can be found at: https://www.missionjuno.swri.edu/junocam /processing.

"I have been following the *Juno* mission since it launched," said Jason Major, a JunoCam citizen scientist and a graphic designer from Warwick, Rhode Island. "It is always exciting to see these new raw images of Jupiter as they arrive. But it is even more thrilling to take the raw images and turn them into something that people can appreciate. That is what I live for."

Measuring in at 10,159 miles (16,350 kilometers) in width (as of April 3, 2017) Jupiter's Great Red Spot is 1.3 times as wide as Earth. The storm has been monitored since 1830 and has possibly existed for more than 350 years. In modern times, the Great Red Spot has appeared to be shrinking.

All of *Juno*'s science instruments and the spacecraft's JunoCam were operating during the flyby, collecting data that are now being returned to Earth. *Juno*'s next close flyby of Jupiter will occur on Sept. 1.

Juno reached perijove (the point at which an orbit comes closest to Jupiter's center) on July 10 at 6:55 p.m. PDT (9:55 p.m. EDT). At the time of perijove, *Juno* was about 2200 miles (3500 kilometers) above the planet's cloud tops. Eleven minutes and 33 seconds later, *Juno* had covered another 24,713 miles (39,771 kilometers), and was passing directly above the coiling, crimson cloud tops of the Great Red Spot. The spacecraft passed about 5600 miles (9000 kilometers) above the clouds of this iconic feature.

Juno launched on Aug. 5, 2011, from Cape Canaveral, Florida. During its mission of exploration, *Juno* soars low over the planet's cloud tops—as close as about 2100 miles (3400 kilometers). During these flybys, *Juno* is probing beneath the obscuring cloud cover of Jupiter and studying its auroras to learn more about the planet's origins, structure, atmosphere and magnetosphere.

Early science results from NASA's *Juno* mission portray the largest planet in our solar system as a turbulent world, with an intriguingly complex interior structure, energetic polar aurora, and huge polar cyclones.

"These highly anticipated images of Jupiter's Great Red Spot are the 'perfect storm' of art and science. With data from *Voyager*, *Galileo*, *New Horizons*, Hubble and now *Juno*, we have a better understanding of the composition and evolution of this iconic feature," said Jim Green, NASA's director of planetary science. "We are pleased to share the beauty and excitement of space science with everyone."

QUESTIONS

1. Why has the Great Red Spot been seen for only ~350 years?

2. Why is it difficult for astronomers to understand what is happening below the cloud tops?

3. Do a search to see whether the Great Red Spot is still shrinking.

4. Go to the website mentioned in the article. Why is *Juno* using "citizen scientists" to process the images?

5. What new observational capabilities does the *Juno* mission bring to understanding the Jovian system?

Source: https://www.nasa.gov/feature/jpl/nasa -s-juno-spacecraft-spots-jupiter-s-great-red-spot.

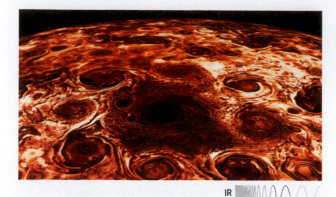

Figure 10.7 *Juno* image of cyclones at the north pole of Jupiter. The colors represent temperature: the thinner yellow clouds are about −13°C, whereas the dark red thicker clouds are about −118°C.

of Saturn and Neptune. Jupiter's vortices are so complex that scientists still do not fully understand how they interact with one another. In 1979, *Voyager 2* observed several Alaska-sized clouds being swept into the Great Red Spot. Some of those clouds were carried around the vortex a few times and then ejected, whereas others were swallowed up and never seen again. Other smaller clouds with structure and behavior similar to that of the Great Red Spot are seen in Jupiter's middle latitudes.

The *Juno* mission entered an elongated orbit around Jupiter in 2016. The spacecraft uses infrared and microwave instruments to further analyze the atmosphere. When closest to Jupiter, *Juno* is only a few thousand kilometers above the cloud tops. *Juno* has shown that the clouds are present even in Jupiter's polar regions and that the depths of the clouds vary greatly. For example, data from the *Juno* spacecraft suggest that the roots of the Great Red Spot go at least 300 kilometers (km) into the atmosphere, making the storm at least 20 times taller than the largest storms on Earth. In a particularly striking observation, *Juno* captured a 4000-km-wide vortex near the north pole of Jupiter, with a ring of eight smaller vortices surrounding it (**Figure 10.7**). At that time, in the southern polar region, five smaller vortices circled another central, larger vortex.

Saturn also displays atmospheric bands, but they tend to be wider than those on Jupiter, with much more subdued colors and contrasts. The largest atmospheric features on Saturn are about the size of the continental United States, but many are smaller than terrestrial hurricanes. Close-up views from the *Cassini* spacecraft show immense lightning-producing storms in a region of Saturn's southern hemisphere known as "storm alley." In December 2010, a large storm appeared in Saturn's northern hemisphere (**Figure 10.8a**) that was visible in even small amateur telescopes. Individual clouds are not seen often on Saturn, so that was an unusual event. That large storm eventually wrapped itself around the planet. *Cassini* also discovered a spinning vortex at Saturn's northern polar region (**Figure 10.8b**), but it is not known how long it has been active.

Through most telescopes on Earth, Uranus and Neptune look like tiny, featureless, pale bluish green disks. With the largest ground-based telescopes (using adaptive optics systems) or the telescopes in space, however, optical and infrared imaging reveals several individual clouds and belts suggestive of those

Figure 10.8 Saturn images from *Cassini*. **a.** A northern hemisphere 2010 storm in Saturn's atmosphere is shown in true color. **b.** The eye of this storm in Saturn's north polar region is about 2000 km across and has wind speeds up to 150 meters per second. The false colors represent three near-infrared wavelengths; red shows low clouds and green indicates higher clouds.

a.

b.

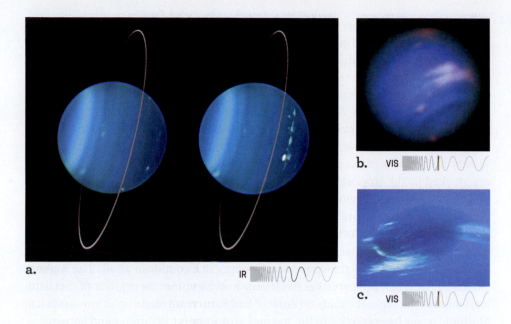

Figure 10.9 a. The ground-based Keck telescope image of Uranus and **b.** a 2016 Hubble Space Telescope (HST) image of Neptune were taken at wavelengths of light that are strongly absorbed by methane. The visible clouds are high in the atmosphere. **c.** The Great Dark Spot on Neptune disappeared between the time *Voyager 2* flew by Neptune in 1989 and the time previous HST images were obtained in 1994, but a new spot has now appeared.

seen on Jupiter and Saturn, but more subdued (**Figure 10.9a**). Methane's strong absorption of reflected sunlight causes the atmospheres of Uranus and Neptune to appear dark in the near infrared, allowing the highest clouds and bands to stand out against the dark background.

Hubble Space Telescope (HST) images show several bright cloud bands in Neptune's atmosphere (**Figure 10.9b**). Located near the planet's tropopause, those cloud bands cast their shadows downward through the clear upper atmosphere onto a dense cloud layer 50 km below. A large, dark, oval feature seen in Neptune's southern hemisphere was first observed in images taken by *Voyager 2* in 1989. This feature reminded astronomers of Jupiter's Great Red Spot, so they called it the Great Dark Spot (**Figure 10.9c**). However, Neptune's great spot was gray rather than red, and it changed in length and shape more rapidly than the Great Red Spot. When HST observed Neptune in 1994, the Great Dark Spot had disappeared. A different spot of comparable size was first seen in 2016. This spot has continued to grow, indicating that such storms come and go.

The Structure below the Cloud Tops

As on the terrestrial planets, atmospheric temperature, density, pressure, and even chemical composition vary with altitude and vary over horizontal distances on the giant planets. As a rule, atmospheric temperature, density, and pressure all decrease with increasing altitude, although temperature is sometimes higher at very high altitudes, as in Earth's thermosphere. The stratospheres above the cloud tops of the giant planets appear relatively clear, but closer inspection shows that they contain layers of thin haze that show up best when seen in profile above the edges of the planets. The composition of the haze particles remains unknown, but they may be smoglike products created when ultraviolet sunlight acts on hydrocarbon gases such as methane.

Volatiles are materials that become gases at moderate temperatures; at lower temperatures, they are liquids or ices. At the temperatures and pressures that exist

what if . . .

What if tidal forces caused Jupiter's orbit to shrink, so that it ended up in the same orbit as Venus? How would Jupiter's appearance change as a result of this?

in Earth's lower atmosphere, water is the only volatile that can condense into clouds. The atmospheres of the giant planets have a much larger range of temperatures and pressures, so more kinds of volatiles can condense and form clouds. Each kind of volatile forms clouds at a different altitude because each kind of volatile, such as water or ammonia, condenses at a different combination of temperature and pressure. Convection carries volatile materials upward along with other atmospheric gases. When each particular volatile reaches an altitude with its condensation temperature, most of that volatile condenses and separates from the other gases, so very little of it is carried higher. **Figure 10.10** shows how the layers of volatiles are stacked in the tropospheres of the giant planets. Volatiles form dense layers of cloud separated by regions of relatively clear atmosphere.

Planets that are more distant from the Sun have colder tropospheres. Therefore, the distance from the Sun determines the altitude at which a particular volatile, such as ammonia or water, will condense to form a cloud layer on each planet. If temperatures are too high, some volatiles may not condense at all. The highest clouds in the frigid atmospheres of Uranus and Neptune are crystals of methane ice, whereas the highest clouds on Jupiter and Saturn are made up of ammonia ice. Methane never freezes to ice in the warmer atmospheres of Jupiter and Saturn.

In 1995, an atmospheric probe on the *Galileo* spacecraft descended slowly via parachute into the atmosphere of Jupiter. Near the top of Jupiter's troposphere at a temperature of about 130 K (about −140°C), the probe found that ammonia had condensed. Next it found a layer of ammonium hydrosulfide clouds at a temperature of about 190 K (about −80°C). Soon after descending to an atmospheric pressure of 22 bars and a temperature of about 373 K (100°C), the *Galileo* probe failed, presumably because its transmitter got too hot. These direct observations confirm the existence of volatile layers predicted by models of Jupiter's atmosphere.

Why are some clouds so colorful, especially Jupiter's? In their purest form, the ices that make up the clouds of the giant planets are all white, similar to snow on Earth. The colors come from impurities in the ice crystals. Those impurities are elemental sulfur and phosphorus, as well as various organic materials produced when ultraviolet sunlight breaks up hydrocarbons such as methane, acetylene, and ethane. The molecular fragments can then recombine to form complex organic compounds that condense into solid particles, many of which are colorful. Such reactions also occur in Earth's atmosphere, often producing smog.

Similarly, Uranus and Neptune are bluish-green because of the composition of their atmospheres. The upper tropospheres of Uranus and Neptune are relatively clear, with only a few white clouds that are probably composed of methane ice crystals. Methane gas is much more abundant in the atmospheres of Uranus and Neptune than in those of Jupiter and Saturn. Like water, methane gas tends to selectively absorb the longer wavelengths of light—yellow, orange, and red. Absorption of the longer wavelengths leaves only the shorter wavelengths—green and blue—to be scattered from the atmospheres of Uranus and Neptune.

Winds and Weather

On the giant planets, the thermal energy that drives convection comes both from the Sun and from the hot interiors of the planets themselves. Convection results from vertical temperature differences (Chapter 9). As heating drives air up and down, the Coriolis effect shapes that convection into atmospheric vortices, visible as isolated circular or elliptical cloud structures, such as the Great Red Spot on Jupiter and the Great Dark Spot on Neptune. As a packet of gas in the atmosphere rises near

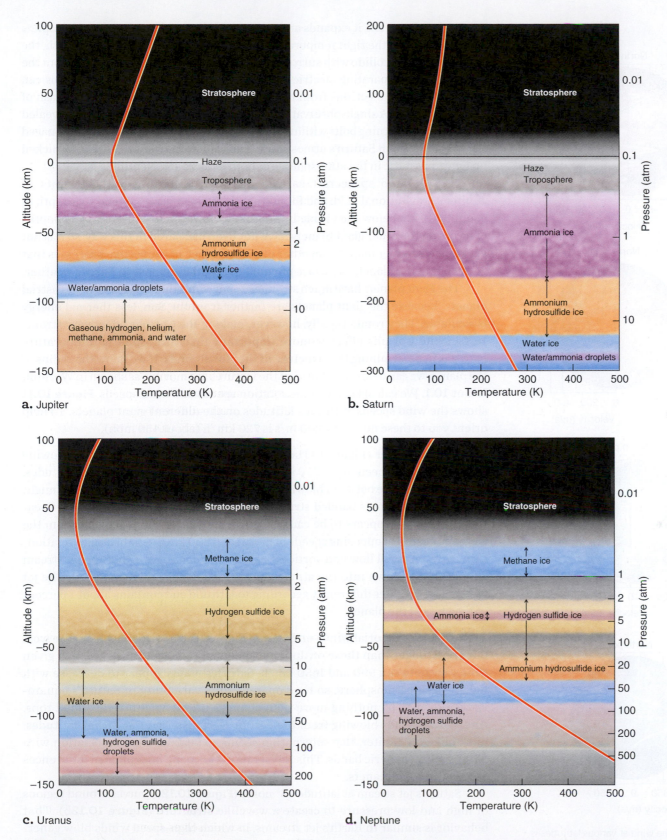

a. Jupiter

b. Saturn

c. Uranus

d. Neptune

Figure 10.10 Volatile materials condense at different levels in the atmospheres of the giant planets, leading to chemically different types of clouds at different depths in the atmospheres. The red graph line in each diagram shows how atmospheric temperature changes with height. Because those planets have no solid surface, the zero point of altitude is arbitrary. Here, the arbitrary zero points of altitude are at 0.1 atmosphere (atm) for **a.** Jupiter and **b.** Saturn and at 1.0 atm for **c.** Uranus and **d.** Neptune. The value of 1.0 atm corresponds to the atmospheric pressure at sea level on Earth. Saturn's altitude scale is compressed to better show the layered structure.

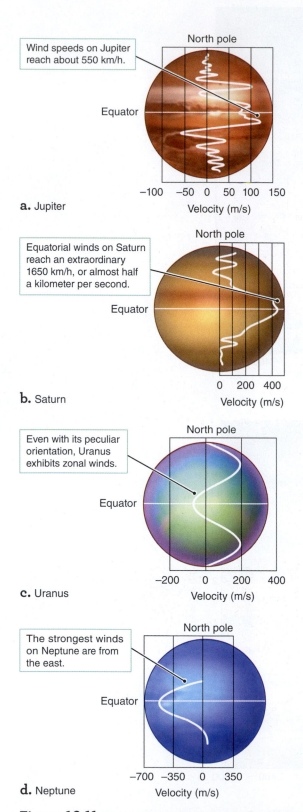

Figure 10.11 The speed of the wind at various latitudes is shown by the white graph line on **a.** Jupiter, **b.** Saturn, **c.** Uranus, and **d.** Neptune. When the speed is positive (to the right of zero), the wind is from the west. When the speed is negative (to the left of zero), the wind is from the east.

the center of a vortex, it expands and cools. Volatiles condense into liquid droplets when the gas is at the right temperature, and then they fall as rain. As on Earth, the falling raindrops collide with surrounding molecules, stripping electrons from the molecules and separating electric charges in the air. Many falling raindrops can separate so many electrons from molecules that a surge of current and a flash of lightning results. A single observation of Jupiter's night side by *Voyager 1* revealed several dozen lightning bolts within an interval of 3 minutes. *Cassini* has also imaged lightning flashes in Saturn's atmosphere, and radio receivers on *Voyager 2* picked up lightning static in the atmospheres of both Uranus and Neptune.

NASA's *Juno* spacecraft carried out high-precision measurements of Jupiter's gravitational field, and found that the striped bands extend to a depth of about 3000 km below the visible layers that form Jupiter's "surface." The depths of layers on the other giant planets have not been measured in this way, so it is not yet known whether features on other planets are part of convective processes that go deep into the planet, or surface phenomena similar to those in Earth's weather.

The giant planets have much stronger zonal winds than those of the terrestrial planets. Because the giant planets are farther from the Sun, less thermal energy is available. They rotate rapidly, however, making the Coriolis effect very strong. In fact, the Coriolis effect is more important than atmospheric temperature patterns in determining the structure of the global winds. If we know the radius of the planet, we can determine how fast the features are moving, as shown in **Working It Out 10.1**. We can also determine rotation speeds and wind speeds. **Figure 10.11** shows the wind speeds at various latitudes on the different giant planets. To help orient you to these numbers, 200 m/s is 720 km/h (about 450 mph).

jupiter   On Jupiter (Figure 10.11a), the strongest winds are equatorial, blowing from the west, at speeds up to 550 kilometers per hour (km/h). At higher latitudes, the winds alternate between blowing from the west or east in a pattern that might be related to Jupiter's banded structure. Near a latitude of 20° south, the Great Red Spot vortex appears to be caught between a pair of winds blowing from the west and east with opposing speeds of more than 200 km/h—indicating a relationship between zonal flow and vortices. Data from *Juno* suggest that the jet stream goes 3000 km down into the atmosphere, much deeper than expected. The jet stream layer rotates differently according to latitude and contains about 1 percent of the mass of the planet.

Saturn   The equatorial winds on Saturn (Figure 10.11b) also blow from the west but are stronger than those on Jupiter. The maximum wind speeds at any given time vary between 990 and 1650 km/h. Saturn's winds appear to decrease with height in the atmosphere, so the apparent time variability of Saturn's equatorial winds may be nothing more than changes in the height of the cloud tops. Alternating winds blowing from the east or west also occur at higher latitudes; but unlike on Jupiter, that alternation seems to bear no clear association with Saturn's atmospheric bands. This is just one of the many unexplained differences among the giant planets.

Saturn's jet stream at latitude 45° north (**Figure 10.12a**) curves around regions of high and low pressure to create a wavelike structure (**Figure 10.12b**). That behavior is similar to Earth's jet streams, in which high-speed winds blow generally from west to east but wander toward and away from the poles. Nested within the crests and troughs of Saturn's jet stream are anticyclonic and cyclonic vortices. They appear similar in both form and size to terrestrial high- and low-pressure systems, which bring alternating periods of fair and stormy weather.

working it out 10.1

Measuring Wind Speeds on Different Planets

How do astronomers measure wind speeds on planets so far away? If we can see individual clouds in their atmospheres, we can measure their winds. As on Earth, clouds are carried by local winds. The local wind speed can be calculated by measuring the positions of individual clouds and noting how much they move during an interval of a day or so. We also need to know how fast the planet is rotating so we can measure the speed of the winds with respect to the planet's rotating surface. For the giant planets, no solid surface exists against which to measure the winds. Scientists must instead assume a hypothetical surface—one that rotates as though it were somehow "connected" to the planet's deep interior. Periodic bursts of radio energy caused by the rotation of a planet's magnetic field tell us how fast the interior of the planet is rotating.

If we know the radius of the planet, we can find the speed of the features. Consider a small white cloud in Neptune's atmosphere. The cloud, on Neptune's equator, is observed to be at longitude 73.0° west on a given day. (That longitude system is anchored in the planet's deep interior and already takes into account Neptune's rotation.) The spot is then seen at longitude 153.0° west exactly 24 hours later. Neptune's equatorial winds have carried the white spot 80.0° in longitude in 24 hours.

The circumference (C) of a planet is given by $2\pi r$, where r is the equatorial radius. From Table 10.1, the equatorial radius of Neptune is 24,300 km, so Neptune's circumference is

$$C = 2\pi r = 2\pi \times 24{,}300 \text{ km} = 152{,}700 \text{ km}$$

The full circle represented by the circumference has 360° of longitude. The spot has moved 80°/360° of the circumference and thus has traveled

$$\frac{80}{360} \times 152{,}700 \text{ km} = 33{,}930 \text{ km}$$

in 24 hours. Therefore, the wind speed is given by

$$\text{Speed} = \frac{\text{Distance}}{\text{Time}} = \frac{33{,}930 \text{ km}}{24 \text{ h}} = 1410 \frac{\text{km}}{\text{h}} = 392 \text{ m/s}$$

The equatorial winds are very strong and are blowing in a direction opposite to the planet's rotation. On Earth, the much slower equivalents of those winds are called *trade winds* (see Figure 9.11).

Uranus Less is known about global winds on Uranus (Figure 10.11c) than about those on the other giant planets. When *Voyager 2* flew by Uranus in 1986, the few visible clouds were in the southern hemisphere because the northern hemisphere was in complete darkness at the time. The strongest winds observed were 650 km/h from the west in the middle to high southern latitudes, and no winds from the east were seen. Because Uranus's peculiar orientation makes its poles warmer than its equator, some astronomers had predicted that the global wind system of Uranus might differ greatly from that of the other giant planets. But *Voyager 2* observed that the Coriolis forces dominate on Uranus as they do on other planets, so the dominant winds on Uranus are zonal, just as on the other giant planets.

As Uranus has moved along in its orbit, regions previously unseen by modern telescopes have become visible (**Figure 10.13**). Observations by HST and ground-based telescopes showed bright cloud bands in the far north extending more than 18,000 km and revealed wind speeds of up to 900 km/h. As Uranus approaches northern summer solstice in the year 2027, much more about its northern hemisphere will be learned.

Neptune On Neptune (Figure 10.11d), the southern hemisphere's summer solstice occurred in 2005, so much of the north is still in darkness. Observers will have to wait until Neptune's equinox in 2045 to see its northern hemisphere. The strongest winds on Neptune occur in the tropics, similar to winds on Jupiter and Saturn. On Neptune, however, the winds are from the east rather than from the west, with speeds in excess of 2000 km/h. Winds from the west with speeds higher than

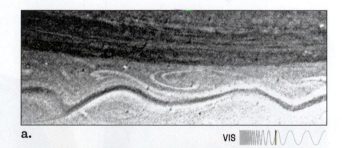

a.   VIS

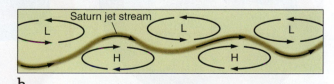

b.

Figure 10.12 a. This jet stream in Saturn's northern hemisphere, imaged by *Voyager 2*, is similar to jet streams in Earth's atmosphere. **b.** The jet stream dips toward the equator below regions of low pressure and is forced toward the pole above regions of high pressure.

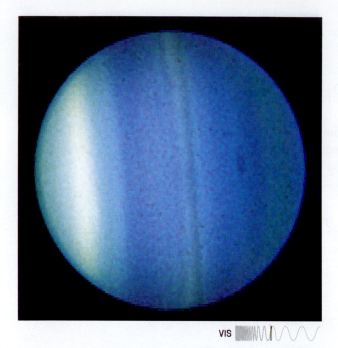

Figure 10.13 Uranus is approaching equinox in this 2006 Hubble Space Telescope image. Much of its northern hemisphere is becoming visible. The dark spot in the northern hemisphere (to the right) is similar to but smaller than the Great Dark Spot seen on Neptune in 1989.

900 km/h have been seen in Neptune's south polar regions. With wind speeds 5 times greater than those of the fiercest hurricanes on Earth, Neptune and Saturn are the windiest planets known.

CHECK YOUR UNDERSTANDING 10.2

Why does Jupiter appear reddish in color? (a) It is very hot; (b) its atmosphere consists of reddish chemical compounds; (c) it is moving very quickly; (d) it is rusty, like Mars.

10.3 The Interiors of the Giant Planets Are Hot and Dense

At the center of each giant planet is a dense, liquid core consisting of a very hot mixture of heavier materials such as water, molten rock, and metals. Based on the interior structure of the giant planets shown in **Figure 10.14**, the gas giants differ from the ice giants in their amounts of hydrogen and ices. We look at each in turn.

The Cores of Jupiter and Saturn

The overlying layers of Jupiter and Saturn press down on the liquid core, raising its temperature. For example, the pressure at Jupiter's core is about 45 million bars, and a pressure that high heats the fluid to 35,000 K. Central temperatures and pressures of the other giant planets are lower because they are less massive than Jupiter. In the extremely hot core of the giant planets, water is still liquid, because the extremely high pressures prevent liquid water from evaporating to steam.

Unlike the terrestrial planets, the giant planets are still contracting. This contraction converts gravitational potential energy into thermal energy that heats the interiors of the planets. The contraction necessary to sustain their internal temperature is only a tiny fraction of their radius each year. Jupiter, for example, contracts by a bit less than 2 centimeters per year. The thermal energy from the core is carried outward by convection in the atmosphere and eventually escapes to space as radiation (**Working It Out 10.2**). For Saturn, moreover, and perhaps for Jupiter, liquid helium separates from a hydrogen–helium mixture under the right conditions

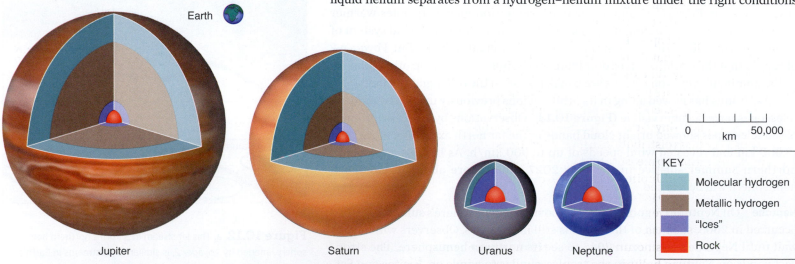

Figure 10.14 The interiors of the giant planets have central cores and outer liquid shells. Only Jupiter and Saturn have significant amounts of the molecular and metallic forms of liquid hydrogen surrounding their cores.

working it out 10.2

Internal Thermal Energy Heats the Giant Planets

The equilibrium between the absorption of sunlight and the radiation of infrared light into space was introduced in Chapter 5. In Chapter 9, you learned how the equilibrium temperature responds to the greenhouse effect on Venus, Earth, and Mars. When that equilibrium is calculated for the giant planets, however, it doesn't match the measurements. According to those calculations, the equilibrium temperature for Jupiter should be 109 K, but the average temperature turns out to be 124 K. A difference of 15 K might not seem like much, but the energy radiated by an object depends on its temperature raised to the fourth power (the Stefan-Boltzmann law; Chapter 5). Applying that relationship to Jupiter, we get:

$$\left(\frac{T_{actual}}{T_{expected}}\right)^4 = \left(\frac{124\ K}{109\ K}\right)^4 = 1.67$$

That result implies that Jupiter is radiating roughly two-thirds more energy into space than it absorbs from sunlight. Similarly, Neptune emits 2.6 times more energy than it absorbs from the Sun. Those planets, then, are *not* in equilibrium; instead, they are slowly contracting and thus generating more heat. The internal energy escaping from Uranus turns out to be smaller than the solar energy the planet absorbs, but this does not mean Uranus is in thermal equilibrium.

and rains downward toward the core. As the droplets of liquid helium sink, they release their gravitational potential energy as thermal energy. Planetary physicists think that most of Saturn's internal energy, and perhaps some of Jupiter's internal energy, comes from that separation of liquid helium. The continuous production of thermal energy is enough to replace the energy escaping from their interiors.

The pressure within the atmospheres of Jupiter and Saturn increases with depth because overlying layers of atmosphere press down on lower layers. At depths of a few thousand kilometers, the atmospheric gases of Jupiter and Saturn are so compressed by the weight of the overlying atmosphere that they condense to liquid. That depth roughly marks the lower boundary of the atmosphere. The difference between a liquid and a highly compressed, very dense gas is subtle, so Jupiter and Saturn have no clear boundary between the atmosphere and the ocean of liquid that lies below. Jupiter's atmosphere is about 20,000 km deep, whereas Saturn's atmosphere is about 30,000 km deep; at those depths, the pressure climbs to 2 million bars and the temperature reaches 10,000 K (hotter than the surface of the Sun: 5780 K; see Chapter 14). Under those conditions, hydrogen molecules are battered so violently that their electrons are stripped free, and the hydrogen acts like a liquid metal. In that state, it is called metallic hydrogen. Those oceans of hydrogen and helium are tens of thousands of kilometers deep.

NASA's *Juno* spacecraft is orbiting Jupiter and has made very careful measurements that show tiny variations in Jupiter's gravitational field. These variations indicate that the mass distribution inside of Jupiter is not perfectly spherical. From repeated observations to measure how the mass rotates, scientists determined that although the interior of Jupiter is composed of an extremely dense gas of hydrogen and helium, the inner 96 percent of Jupiter rotates as if it were solid. There may be a small hard core very, very deep within the interior, but even there, the material is still just a very dense gas.

Differentiation has occurred and is still occurring in Saturn and perhaps in Jupiter. On Saturn, helium condenses out of the hydrogen–helium oceans. Helium also can be compressed to a metal, but it does not reach that metallic state under the physical conditions of the interiors of the giant planets. Because

what if . . .

What if the diameter of Jupiter were to shrink at a rate of about 1 mm/year, instead of the current 2 cm/year? If Jupiter's mass were larger, how would that affect this contraction?

those droplets of helium are more dense than the hydrogen–helium liquid in which they condense, they sink toward the center of the planet, converting gravitational energy to thermal energy. That process heats the planet and enriches the helium concentration in the core while depleting it in the upper layers. In Jupiter's hotter interior, by contrast, the liquid helium is mostly dissolved together with the liquid hydrogen.

The heavy-element components of the cores of Jupiter and Saturn have masses of about 5–20 Earth masses. Jupiter and Saturn have total masses of 318 and 95 Earth masses, respectively. The heavy materials in their cores contribute little to their average chemical composition. Therefore, Jupiter and Saturn have approximately the same composition as the Sun and the rest of the universe: about 98 percent hydrogen and helium, leaving only 2 percent for everything else.

The Cores of Uranus and Neptune

Uranus and Neptune are considerably less massive than Jupiter and Saturn (see Table 10.1), have lower interior pressures, and contain smaller fractions of hydrogen—their interiors probably contain only a small amount of liquid hydrogen, with little or none in a metallic state. Uranus and Neptune are made of denser material than Saturn and Jupiter. Neptune, the most dense of the giant planets, is about 1.6 times denser than water and only about half as dense as rock. Uranus is less dense than Neptune (again, see Table 10.1). These observations tell us that water and other low-density ices, such as ammonia and methane, must be the major compositional components of Uranus and Neptune, along with lesser amounts of silicates and metals. The total amount of gaseous hydrogen and helium in those planets is probably limited to no more than 1 or 2 Earth masses, and most of those gases reside in the relatively shallow atmospheres of the planets. Computer simulations suggest that under their conditions of high pressure and temperature, the water that makes up so much of Uranus and Neptune might be *super-ionic*—a state between a liquid and a solid.

Why do Jupiter and Saturn have so much more hydrogen and helium than Uranus and Neptune? Although each giant planet formed around cores of rock and metal, they turned out differently. Those differences are an important clue to their origins. The variation may be due to both the time those planets took to form and the distribution of material from which they formed. The cores of Uranus and Neptune were smaller and formed much later than those of Jupiter and Saturn, when most of the gas in the protoplanetary disk had been blown away by the emerging Sun. The icy planetesimals from which they formed were more widely dispersed at their greater distances from the Sun. With more space between planetesimals, their cores would have taken longer to build up. Saturn may have captured less gas than Jupiter both because its core formed a bit later and also because less gas was available at its greater distance from the Sun. Some astronomers hypothesize that Uranus and Neptune might have formed in a location different from where they are now (see the "Origins" section).

CHECK YOUR UNDERSTANDING 10.3

The interiors of the giant planets are heated by gravitational contraction. We know that because: (a) the cores are very hot; (b) most of the giant planets radiate more energy than they receive from the Sun; (c) the giant planets have strong winds; (d) the giant planets are mostly atmosphere.

unanswered questions

What are the mass and size of the core of each giant planet? Is there a rocky core underneath the thick atmosphere? Is the core the size of a terrestrial planet, or larger? The NASA *Juno* mission is measuring Jupiter's gravitational and magnetic fields to map the amount and distribution of mass in its core and atmosphere. Preliminary results from *Juno* suggest that Jupiter's core is larger and "fuzzier" than previously thought, so that question might be answered for Jupiter in a few years.

10.4 The Giant Planets Are Magnetic Powerhouses

In Jupiter and Saturn, circulating currents within deep layers of metallic hydrogen generate magnetic fields. In Uranus and Neptune, magnetic fields arise within deep oceans of water and ammonia made electrically conductive by dissolved salts or ionized molecules. The magnetospheres of the giant planets are very large and interact with not only the solar wind (as Earth's does) but also with the rings and moons that orbit the giant planets.

The Size and Shape of the Magnetospheres

The magnetic field strengths of the giant planets range from 50 to 20,000 times stronger than Earth's, at the same radius. Field strength drops precipitously with distance, however, so the fields at the cloud tops of Saturn, Uranus, and Neptune are comparable in strength to Earth's surface field. Even for Jupiter's exceptionally strong field, the field strength at the cloud tops is only about 15 times stronger than Earth's surface field. Just as Earth's magnetic field extends past the surface to trap energetic charged particles in Earth's magnetosphere, the magnetic fields of the giant planets extend past the cloud tops to trap energetic particles in magnetospheres of their own. The most colossal of those by far is Jupiter's magnetosphere. Its diameter is 100 times that of the planet itself, roughly 10 times the diameter of the Sun. Even the relatively weak magnetic fields of Uranus and Neptune form magnetospheres that are comparable in size to the Sun. Evidence of the giant planets' magnetospheres comes from spacecraft in the outer Solar System, from telescopes orbiting Earth, and from radio emissions received on Earth.

Every magnetic field has a north pole and a south pole, and we often picture them as being formed by bar magnets, with the magnet aligned along the magnetic field axis, as sketched in **Figure 10.15**. Jupiter's magnetic axis is inclined 10° with respect to its rotation axis—an orientation similar to Earth's—but it is offset about a tenth of a radius from the planet's center. Saturn's magnetic axis is located almost precisely at the planet's center and is almost perfectly aligned with the rotation axis. The magnetic axis of Uranus is inclined nearly 60° to its rotation axis and is offset by a third of a radius from the planet's center. Neptune's rotation axis is tilted by about the same amount as those of Earth, Mars, and Saturn. However, Neptune's magnetic-field axis is inclined 47° to that rotation axis, and the center of that magnetic field is displaced from the planet's center by more than half the radius—an offset even greater than that of Uranus. The field is displaced primarily toward Neptune's southern hemisphere, so the field at the cloud tops is 20 times stronger in the south than in the north. The reasons for the differences in the orientations of the magnetic field axes of the giant planets are not well understood. In particular, the reason for the unusual geometries of the magnetic fields of Uranus and Neptune remains uncertain, but it is not related to the orientations of their rotation axes.

The solar wind also influences the magnetosphere. The solar wind supplies charged particles, and the pressure of the solar wind pushes on and compresses a magnetosphere. The size and shape of a planet's magnetosphere depends on how the solar wind is blowing at any particular time. Conversely, planetary magnetic fields divert the solar wind, which flows around magnetospheres the way a stream flows around boulders. Just as a rock in a river creates a wake that extends downstream

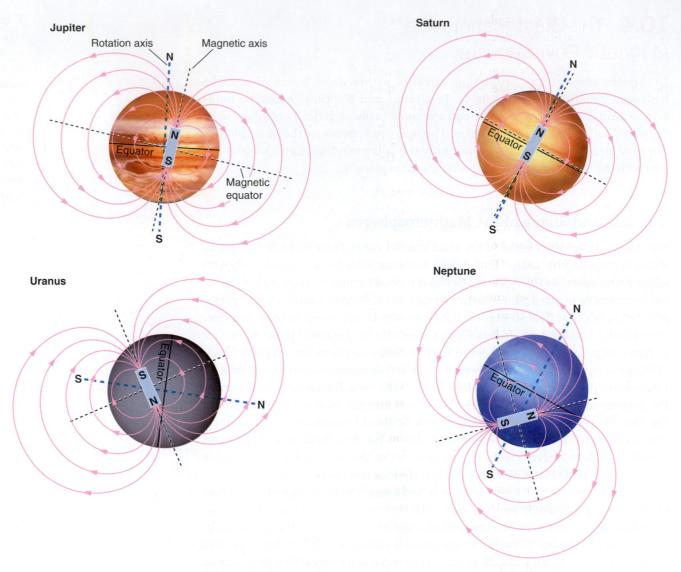

Figure 10.15 The magnetic fields of the giant planets can be approximated by the fields from bar magnets offset and tilted with respect to the planets' axes. Compare these magnetic fields with Earth's (see Figure 8.12).

(**Figure 10.16a**), the magnetosphere of a planet produces a wake that can extend for great distances. **Figure 10.16b** shows that the wake of Jupiter's magnetosphere extends well past the orbit of Saturn. Jupiter's magnetosphere is the largest permanent structure in the Solar System, surpassed in size only by the tail of an occasional comet.

In addition to protons and electrons from the solar wind, the magnetospheres of the giant planets contain large amounts of various elements (some ionized), including sodium, sulfur, oxygen, nitrogen, and carbon. Those elements come from several sources, including the planets' extended atmospheres and the moons that orbit within them.

Rapidly moving electrons in planetary magnetospheres spiral around the magnetic field lines, and as they do so they emit a type of low-energy radiation, known as **synchrotron radiation**, concentrated in the radio part of the spectrum. Precise measurement of periodic variations in the radio signals "broadcast" by the giant planets indicates the planets' true rotation periods. The magnetic field of each planet is locked to the conducting liquid layers deep within the planet's interior, so the magnetic field rotates with the same period as the deep interior of the planet. With the fast and highly variable winds that push around clouds in the atmospheres of the giant planets, measuring the radio emission is the only way to determine their true rotation periods.

If your eyes were sensitive to radio waves, the second-brightest object in the sky would be Jupiter's magnetosphere. The Sun would still be brighter, but even at a distance from Earth of 4.2–6.2 AU, Jupiter's magnetosphere would appear roughly twice as large as the Sun in the sky. Saturn's magnetosphere also would be large enough to see, if we could see radio waves, but it would be much fainter than Jupiter's. Even though Saturn has a strong magnetic field, pieces of rock, ice, and dust in Saturn's spectacular rings act like sponges, soaking up charged particles soon after they enter the magnetosphere. With far fewer charged particles in the magnetosphere, Saturn has much less radio emission. The magnetic tails of Uranus and Neptune have a curious structure. For both Uranus and Neptune, the tilt and the large displacement of the magnetic field from the center of each planet cause the magnetosphere to wobble as the planet rotates. That wobble causes the tail of the magnetosphere to twist like a corkscrew as it stretches away.

Auroras and Radiation Belts of the Giant Planets

Charged particles spiral along the magnetic-field lines of the giant planets, bouncing back and forth between each planet's two magnetic poles, just as they do around Earth. As with Earth, those energetic particles collide with atoms and molecules in a planet's atmosphere, knocking them into excited energy states that decay and emit light. The results are bright auroral rings (**Figure 10.17**). Those auroral rings surround the magnetic poles of the giant planets, just as the aurora borealis and aurora australis ring the north and south magnetic poles, respectively, of Earth (see Figure 9.10).

As a planet rotates on its axis, it drags its magnetosphere around with it, and charged particles are swept around at high speeds. Those fast-moving charged particles slam into neutral atoms, and the energy released in the resulting high-speed collisions heats the **plasma** to extreme temperatures. (A plasma is a gas consisting of electrically charged particles.) In 1979, while passing through Jupiter's magnetosphere, *Voyager 1* encountered a region of plasma with a temperature

a.

Figure 10.16 **a.** Water flowing past a rock sweeps the algae against the rock and into a "tail" pointing in the direction of the water's flow. **b.** The solar wind compresses Jupiter's (or any other planet's) magnetosphere on the side toward the Sun and draws it out into a magnetic tail away from the Sun. Jupiter's tail stretches beyond the orbit of Saturn. (Not to scale.)

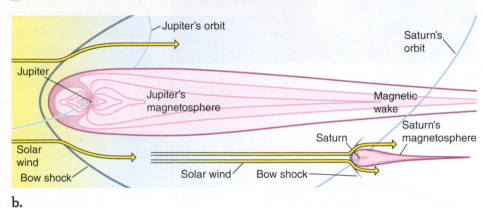

b.

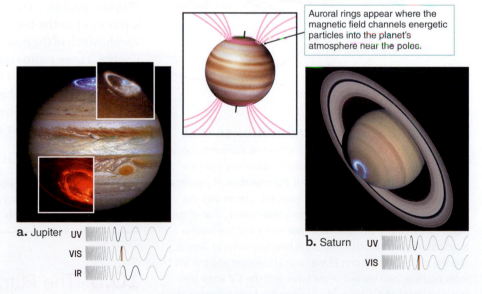

a. Jupiter UV VIS IR

b. Saturn UV VIS

Auroral rings appear where the magnetic field channels energetic particles into the planet's atmosphere near the poles.

Figure 10.17 The Hubble Space Telescope took images of auroral rings around the poles of **a.** Jupiter and **b.** Saturn. The auroral images (bright rings near the north pole) were taken in ultraviolet light and then superimposed on visible-light images. Credit (part a): NASA/ESA, https://esahubble.org/images/heic1613a/. https://creativecommons.org/licenses/by/4.0/; (part a, top inset): John Clarke (University of Michigan), and NASA/ESA, https://esahubble.org /images/opo9804b/. https://creativecommons.org/licenses/by/4.0/.

what if . . .

What if you searched for planets around other stars by looking for radio emissions from their magnetospheres? Would this constitute a good way to detect other large planets?

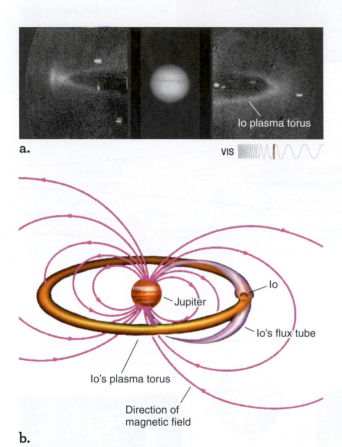

a.

VIS

b.

Figure 10.18 ★ WHAT AN ASTRONOMER SEES

a. In this image, a faintly glowing torus of plasma surrounds Jupiter (center). The torus is made up of atoms knocked free from the surface of Io by charged particles. Io is the innermost of Jupiter's Galilean moons. (A semitransparent mask was placed over the disk of Jupiter to prevent its light from overwhelming that of the much fainter torus.) An astronomer will notice that the plasma torus has a radius several times larger than the radius of Jupiter. She will notice that the part of the torus at the right and the left looks brighter than the rest of the torus, and she will know that this is an effect of the geometry; because the torus is viewed at an angle, she is seeing more material there, so it appears brighter. This is the same reason that the glass of an empty wine glass is most visible at the edges. An astronomer will also notice that the camera was tracking the motion of Jupiter across the sky for a considerable amount of time, because the background stars in the image are not points of light but have become smeared into bright streaks. b. The geometry of Io's plasma torus and flux tube.

of more than 300 million K—20 times the temperature at the center of the Sun. *Voyager 1* did not melt when passing through that region, though, because the plasma was so tenuous that the plasma's particles were very far apart in space. Each particle was extraordinarily energetic, but too few of them were present to harm the spacecraft.

Charged particles trapped in planetary magnetospheres are concentrated in *radiation belts*. Although Earth's radiation belts are severe enough to worry astronauts, the radiation belts that surround Jupiter are searing by comparison. In 1974, the *Pioneer 11* spacecraft passed through the radiation belts of Jupiter. Several of the instruments onboard were permanently damaged as a result, and the spacecraft itself barely survived to continue its journey to Saturn.

The most intense radiation belt in the Solar System is a doughnut-shaped ring, or **torus**, of plasma associated with Io, the innermost of Jupiter's four Galilean moons. As we discuss in more detail in Chapter 11, Io has low surface gravity and violent volcanic activity. Some of the gases erupting from Io's interior escape and become part of Jupiter's radiation belt. As charged particles are slammed into Io by the rotation of Jupiter's magnetosphere, even more material is knocked free of its surface and ejected into space. Images of the region around Jupiter, taken in the light of emission lines from atoms of sulfur or sodium, show a faintly glowing ring of plasma supplied by Io (**Figure 10.18a**). Other moons also influence the magnetospheres of the planets they orbit. *Cassini* found that Saturn's moon Enceladus leaks ionized molecules (including nitrogen), water vapor, and ice grains from icy geysers and provides most of the torus of plasma in Saturn's magnetosphere.

As Jupiter's magnetic field sweeps past Io, electrons spiral along Jupiter's magnetic-field lines. The result is a magnetic channel, called a **flux tube**, that connects Io with Jupiter's atmosphere near the planet's magnetic poles (**Figure 10.18b**). The electrical power contained in Io's flux tube is roughly equivalent to the total power produced by all electrical generating stations on Earth. Much of the power generated within the flux tube is radiated away as radio energy. Those radio signals are received on Earth as intense bursts. However, a substantial fraction of the energy of the particles in the flux tube is deposited into Jupiter's atmosphere, causing intense auroral activity where Io's flux tube intercepts Jupiter's atmosphere. As Jupiter rotates, that spot leaves behind an auroral trail in Jupiter's atmosphere.

CHECK YOUR UNDERSTANDING 10.4

Why are the radiation belts around Jupiter much stronger than those around Earth? (a) Jupiter has larger storms than Earth; (b) Jupiter is colder than Earth; (c) Jupiter rotates faster than Earth; (d) Jupiter has a stronger magnetic field than Earth.

10.5 The Planets of Our Solar System Might Not Be Typical

Chapters 8–10, which discuss the planets of our Solar System, divide them into two categories: inner terrestrial rocky planets versus outer gas and ice giants. The first planets discovered around other stars were completely different from those in our Solar System. Those exoplanets were massive, like the giant planets, but located close to their central star, like our terrestrial planets. As the number of

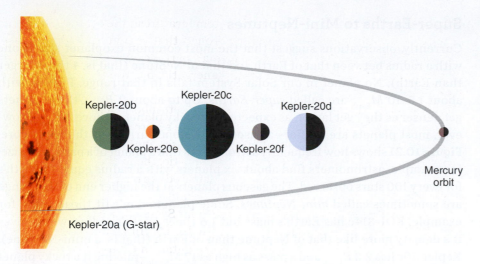

Figure 10.19 Illustration of the Kepler-20 system. Large and small planets alternate, but they are all closer to their star than Mercury is to the Sun.

confirmed exoplanets increases, astronomers compare them with those of our own system. Exoplanet systems differ in all kinds of ways from our own Solar System. In the Kepler-20 system, for example, the large planets alternate with the smaller ones, but all five known planets in that system have orbits smaller than the orbit of Mercury (**Figure 10.19**). In this section, we describe types of planets that are unlike those in our Solar System.

Different Types of "Jupiters"

The first exoplanets were discovered using the radial velocity method (Chapter 7), which measures the wobble of a star caused by the gravity of its planet. That method is most successful at finding a massive planet located close to its star, where the planet's gravitational tug on the star is stronger than if the planet were of smaller mass and/or farther away. Those gas giants, discovered within a few percent of an AU to half an AU from their respective stars, with correspondingly short and sometimes highly elliptical orbits, are known as *hot Jupiters*. Some of those planets may be *puffy Jupiters*, which have a larger radius and a lower density (that is, a density closer to that of Saturn than to that of Jupiter). The larger radius is a result of the heated, and thus expanding, gaseous atmosphere. Current models of planetary formation suggest that not enough excess hydrogen would have existed that close to their stars for those planets to form there. These planets have almost certainly moved from their original orbits.

Astronomers have identified several hundred exoplanets known as *super-Jupiters* because their masses are 2–13 times that of Jupiter. Some are also hot Jupiters, but most are not. The greater mass of these planets gives them stronger gravitational contraction than that in planets with Jupiter's mass, so they shrink. As a result, most have a higher density than Jupiter. The gases may be compressed by self-gravity so much that a super-Jupiter could be denser than Earth. The stronger gravitational contraction would create hotter cores, so they might have more intense winds and weather than Jupiter has. An artist's depiction of the super-Jupiter planet Kappa Andromedae b, with about 13 times the mass of Jupiter, is shown in **Figure 10.20**.

Figure 10.20 Artist's depiction of the super-Jupiter planet Kappa Andromedae b.

what if . . .

What if, early in the Solar System's history, Jupiter had migrated inward to an orbit half the radius of Mercury's orbit? How might that have affected the development and evolution of life on Earth?

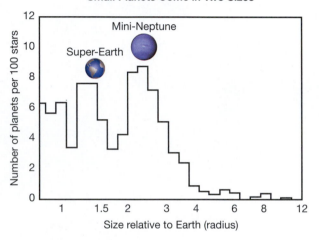

Small Planets Come in Two Sizes

Figure 10.21 Smaller planets tend to be super-Earths or mini-Neptunes. Data on nearby super-Earths suggest that at about 2 R_{Earth}, mean planet density decreases with size because the planet has more gas accumulated on its rocky core.

unanswered questions

Did the Solar System start with more planets? In the same types of computer models of the early Solar System that we discuss in "Origins," astronomers can run simulations with different initial configurations of planets to see which configurations evolve and then compare those with what is observed today. In one set of models, astronomers found that starting with five giant planets best reproduced the current outer Solar System. The fifth planet, which would have been kicked out of the Solar System after a close encounter with Jupiter, may still be wandering through the Milky Way.

Super-Earths to Mini-Neptunes

Currently, observations suggest that the most common exoplanet size is one with a radius between that of Earth and that of Neptune (that is, 4 times larger than Earth). No planet in our Solar System falls in that range. Planets with about 1.5–10 M_{Earth} are called *super-Earths*. Up to about 2 R_{Earth}, those planets get denser as they get larger, as expected for rocky planets. Above 2 R_{Earth}, however, most planets are puffier—a gaseous envelope surrounds the rocky core. **Figure 10.21** shows how frequently astronomers find a planet of a particular size; for example, astronomers find about six planets with a radius equal to Earth's for every 100 stars observed. The gaseous planets at the higher end of that range are sometimes called *mini-Neptunes*. Some planets don't fit those rules; for example, KOI-314c has Earth's mass but 1.6 times Earth's radius, which gives it a density more like that of Neptune than of Earth (that is, a mini-Neptune). Kepler-10c has 2.3 R_{Earth} and a mass as high as 17 M_{Earth}—making it a rocky planet with a surprisingly high mass.

In Chapter 7, we presented a scenario in which the gaseous giant planets formed in the cold outer Solar System and the small rocky terrestrial planets formed in the warm inner Solar System. That model seemed to make sense chemically and physically, so astronomers were shocked when the hot Jupiters were first discovered in the mid-1990s. To explain hot Jupiters, astronomers built computer models to more precisely explore gravitational interactions, either between a planet and the protoplanetary disk or among the planets in the system. These models showed that these interactions could cause planets to migrate to different orbital distances from where they had formed. Migration is also necessary to explain the locations of super-Earths close to their respective stars.

Several hundred multiplanet systems have been discovered around other stars. Observations and computer models show that many combinations of planetary sizes, masses, and compositions can exist and persist within planetary systems. To date, the mix of planet types in our Solar System—outer giant gaseous planets and inner small rocky planets, all with nearly circular orbits—has not been seen in any other system. It is not yet known whether planetary systems like ours are rare or if they just haven't been discovered yet.

CHECK YOUR UNDERSTANDING **10.5**

Place in order of increasing diameter the following types of exoplanets: (a) super-Earths; (b) puffy Jupiters; (c) super-Jupiters; (d) mini-Neptunes.

Origins: Giant Planet Migration and the Inner Solar System

As long as only two objects are involved, the motions that result from Newton's law of gravitation are simple. Kepler's laws describe the regular, repeating elliptical orbit of a planet around the Sun. When more than two objects are involved, however, the resulting motions may be anything but simple and regular. Each planet in the Solar System moves under the gravitational influence of the Sun combined with that of all the other planets. Although those extra influences are small, they are not negligible. Over millions of years, they lead to significant differences in the locations of planets in their orbits. In many possible exoplanetary systems,

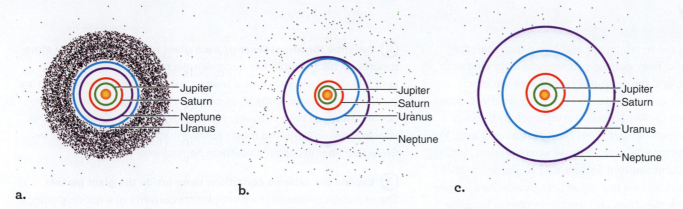

a.   b.   c.

Figure 10.22 Snapshots from computer simulations of planet migration show how Jupiter and Saturn affect the outer Solar System. The four circles are the orbits of the Jovian planets, and the black dots are planetesimals. **a.** The orbital period of Saturn becomes twice that of Jupiter. **b.** Planetesimals are scattered as the giant planets change orbits. **c.** In some models, Neptune and Uranus change places. The inner Solar System is left more stable.

such interactions among planets might cause planets to dramatically change their orbits or even be ejected from the system entirely.

Computer models developed to understand the exoplanet systems have been applied to the Solar System as well, and the results are intriguing. The giant planets may not have formed in their current locations and could have migrated quite far from the orbits in which they formed. A key point seems to be the gravitational influence of Saturn on Jupiter, especially the ratio of their orbital periods. **Figure 10.22** shows snapshots from one such migration model. When Saturn's orbital period became twice that of Jupiter, their respective orbits became more elongated. The result was an outward migration of Uranus and Neptune, whose orbits grew larger. Uranus and Neptune may actually have switched places. That shuffling cleared away nearby planetesimals, sending some to the inner Solar System, and made the planetary orbits more stable. In another set of models, Jupiter migrated inward to 1.5 AU—the current orbital distance of Mars. Then Saturn migrated inward even faster to a point at which its orbital period was 1.5 times that of Jupiter, and finally they both migrated outward, pushing Uranus and Neptune into larger orbits.

In some exoplanetary systems, gas giants are observed at the right distance from their respective stars to be in the habitable zone (see Chapter 7), but it is not known whether gas giants can support life. Super-Earths are sometimes found in the habitable zone of their stars, too. In the Solar System, migrations of the giant planets could have helped confine and stabilize the orbits of the inner planets, so that they reside in or near the habitable zone. Migration of Jupiter may have destabilized the orbits of close-in hot planets, so that they crashed into the Sun. Planetesimals closer to 1.5 AU may have scattered, keeping Mars from growing so that it stayed small. A small Mars could not hold on to its atmosphere and its liquid water, making it a less likely starting point for life. In addition, the shuffling about of the outer planets may be responsible for the period of late heavy bombardment, which brought at least some of the atmospheric gases, water, and possibly organic molecules needed for life to form on Earth.

SUMMARY

The giant planets are much larger and less dense than the terrestrial planets. They consist primarily of light elements rather than rock, and their outer atmospheres are much colder. Because of their rapid rotation and the Coriolis effect, zonal winds are very strong on those planets. Storms, such as those on Saturn, tend to be larger and longer-lived than storms on Earth. Volatiles become ices at various heights in those atmospheres, leading to a layered cloud structure. Jupiter, Saturn, and Neptune are still shrinking, and their gravitational energy is being converted to thermal energy, heating both the cores and the atmospheres from the inside. Uranus does not seem to have as large a heat source inside. Temperatures and pressures in the cores of the giant planets are very high, leading to unusual states of matter, such as metallic hydrogen and super-ionic water. The current locations of those planets might be different from where they formed: models suggest that their positions may have migrated.

① **Differentiate the giant planets from one another and from the terrestrial planets.** Uranus and Neptune were discovered by telescope, unlike all the other planets, which have been known since ancient times. Jupiter and Saturn are gas giants, whereas Uranus and Neptune are ice giants. Jupiter and Saturn are made up mostly of hydrogen and helium (a composition similar to that of the Sun), whereas Uranus and Neptune contain larger amounts of ices, such as water, ammonia, and methane, than are found in Jupiter and Saturn. Those compositions set them apart from the terrestrial planets. In addition, all four giant planets are much larger than Earth.

② **Describe the atmosphere of each giant planet.** We see only atmospheres on the giant planets because solid or liquid surfaces, if they exist, are deep below the cloud layers. Clouds on Jupiter and Saturn are composed of various kinds of ice crystals colored by impurities. Uranus and Neptune have relatively few clouds, and so their atmospheres appear more uniform. The most prominent atmospheric feature is the Great Red Spot in Jupiter's southern hemisphere.

③ **Explain the extreme conditions deep inside the giant planets.** The ongoing collapse of the giant planets converts gravitational energy to thermal energy. That process heats most of the giant planets from within, producing convection. Powerful convection and the Coriolis effect drive high-speed winds in the upper atmospheres of all the giant planets. The interiors of the giant planets are very hot and very dense because of the high pressures exerted by their overlying atmospheres.

④ **Describe the magnetosphere of each giant planet.** The giant planets have enormous magnetospheres that emit synchrotron radiation and interact with their moons. The rotation speed of a gas giant is found from periodic bursts of radio waves generated as the planet's magnetic field rotates.

⑤ **Compare the planets of our Solar System with those in exoplanetary systems.** Systems of exoplanets found to date do not contain our Solar System's distribution of small, rocky inner planets and large, giant outer planets. Most planets found so far fall between Earth and Neptune in size. Our understanding of how solar systems form is incomplete.

QUESTIONS AND PROBLEMS

TEST YOUR UNDERSTANDING

1. The following steps lead to convection in the atmospheres of giant planets. After (a), place (b)–(f) in order.
   a. Gravity pulls particles toward the center.
   b. Warm material rises and expands.
   c. Particles fall toward the center, converting gravitational energy to kinetic energy.
   d. Expanding material cools.
   e. Thermal energy heats the material.
   f. Friction converts kinetic energy to thermal energy.

2. Deep inside the giant planets, water is still a liquid even though the temperatures are tens of thousands of degrees above the boiling point of water. That can happen because
   a. the density inside the giant planets is so high.
   b. the pressure inside the giant planets is so high.
   c. the outer Solar System is so cold.
   d. space has very low pressure.

3. Assume you want to deduce the radius of a Solar System planet as it occults a background star when the relative velocity between the planet and Earth is 30 km/s. If the star crosses through the middle of the planet and disappears for 26 minutes, what is the planet's radius?
   a. 3000 km   c. 15,000 km
   b. 23,000 km   d. 5000 km

4. Neptune's existence was predicted because
   a. Uranus did not seem to obey Newton's laws of motion.
   b. Uranus wobbled on its axis.
   c. Uranus became brighter and fainter in an unusual way.
   d. some of the solar nebula's mass was unaccounted for.

5. Which giant planet has the most extreme seasons?
   a. Jupiter   c. Uranus
   b. Saturn   d. Neptune

6. The magnetic fields of the giant planets
   a. align closely with the planet's rotation axis.
   b. extend far into space.
   c. are thousands of times stronger at the cloud tops than at Earth's surface field.
   d. have an axis that passes through the planet's center.

7. An occultation occurs when
   a. a star passes between Earth and a planet.
   b. a planet passes between Earth and a star.
   c. a planet passes between Earth and the Sun.
   d. Earth passes between the Sun and a planet.

8. Occultations directly determine a planet's
   a. diameter.   c. density.
   b. mass.   d. orbital speed.

9. The chemical compositions of Jupiter and Saturn are most similar to the chemical compositions of
   a. Uranus and Neptune.   c. their moons.
   b. the terrestrial planets. d. the Sun.

10. Individual cloud layers in the giant planets have different compositions because
    a. the winds are all in the outermost layer.
    b. the Coriolis effect occurs only close to the "surface" of the inner core.
    c. convection does not occur on the giant planets.
    d. different volatiles freeze out at different temperatures.

11. The Great Red Spot on Jupiter is
    a. a surface feature.
    b. a storm that has been raging for more than 300 years.
    c. caused by the interaction between the magnetosphere and Io.
    d. currently about the size of North America.

12. How are Uranus and Neptune different from Jupiter and Saturn?
    a. Uranus and Neptune have a higher percentage of ices in their interiors.
    b. Uranus and Neptune have no rings.
    c. Uranus and Neptune have no magnetic field.
    d. Uranus and Neptune are closer to the Sun.

13. What could have caused the planets to migrate through the Solar System?
    a. gravitational pull from the Sun
    b. interaction with the solar wind
    c. accreting gas from the solar nebula
    d. gravitational pull from other planets

14. Zonal winds on the giant planets are stronger than those on the terrestrial planets because
    a. they have more thermal energy.
    b. the giant planets rotate faster.
    c. the moons of giant planets provide additional pull.
    d. the moons feed energy to the planet through the magnetosphere.

15. A "hot Jupiter" gets its name from the fact that
    a. its temperature has been measured to be higher than Jupiter's.
    b. it is located around a much hotter star than the Sun.
    c. it has very high density, so its temperature is high.
    d. it orbits very close to its central star.

THINKING ABOUT THE CONCEPTS

16. Describe how the giant planets differ from the terrestrial planets.

17. Jupiter's chemical composition is more like that of the Sun than Earth's is, yet both planets formed from the same protoplanetary disk. Explain why they are different today.

18. Figure 10.2 shows the occultation of a star as a planet passes in front of it. In this figure, the center of the planet passes directly in front of the star. Suppose that the planet were shifted "down" in the "View from Earth," so that the center of the planet does not pass directly in front of the star. Would the time of the occultation be shorter or longer in this case? Explain how astronomers could still determine the diameter of the planet, even if the center of the planet does not pass directly in front of the star. 👁

19. What drives the zonal winds in the atmospheres of the giant planets?

20. Compare the sequence of events in the Process of Science Figure in this chapter with the flowchart of the Process of Science Figure in Chapter 1. Redraw the flowchart, incorporating each event leading to the discovery of Uranus as examples in the appropriate boxes. 👁

21. None of the giant planets is perfectly spherical. Explain why they have a slightly squashed shape.

22. What is the source of color in Jupiter's clouds? Uranus and Neptune, when viewed through a telescope, appear distinctly bluish-green. What are the two reasons for their striking appearance?

23. Which giant planets have seasons similar to Earth's, and which one experiences extreme seasons?

24. The bands on Jupiter penetrate to 3000 km below the surface. Study Figure 10.14. Do the bands penetrate down to the metallic hydrogen layer? Jupiter's core is thought to consist of rocky material and ices, all in a liquid state at a temperature of 35,000 K. How can materials such as water be liquid at such high temperatures? 👁

25. What is the Great Red Spot? How has it changed over the centuries it has been observed?

26. Jupiter, Saturn, and Neptune radiate more energy into space than they receive from the Sun. What is the source of the additional energy?

27. When viewed by radio telescopes, Jupiter is the second-brightest object in the sky. What is the source of its radiation?

28. ★ WHAT AN ASTRONOMER SEES   In Figure 10.18a, the astronomer concludes that the camera tracked Jupiter's motion across the sky while the image was being taken. The stars, therefore, formed bright streaks in the background as the camera moved. Did Jupiter move in an up-down direction or in a right-left direction, according to the orientation of this photo? 👁

29. What creates auroras in the polar regions of Jupiter and Saturn?

30. How might migration of the outer giant planets affect the sizes and orbits of the inner planets?

APPLYING THE CONCEPTS

31. A small cloud in Jupiter's equatorial region is observed to be at longitude 122.0° west in a coordinate system and to be rotating at the same rate as the deep interior of the planet. Another observation, made exactly 10 Earth hours later, finds the cloud at longitude 118.0° west. Jupiter's equatorial radius is 71,500 km. What is the observed equatorial wind speed in kilometers per hour? ●—●—●
 a. Make a prediction: The graphs in Figure 10.11 show average wind speeds on the giant planets at varying latitudes. About how fast are the equatorial winds on Jupiter, on average? ★
 b. Calculate: Follow Working It Out 10.1 to calculate the wind speed from the observations stated in the problem.
 c. Check your work: Make sure your answer has units of km/h, and compare your answer to your prediction.

32. The equilibrium temperature for Saturn should be 82 K, but the observed temperature is 95 K. How much more energy does Saturn radiate than it absorbs? ●—●—●
 a. Make a prediction: Working It Out 10.2 carries out this calculation for Jupiter. Study that calculation. Do you expect Saturn to have a larger or smaller ratio of energies than Jupiter? Do you expect the ratio to be "around 1" or "around 100"?
 b. Calculate: Follow Working It Out 10.2 to calculate the ratio of emitted to absorbed energies for Saturn.
 c. Check your work: Verify that your answer has no units and that the magnitude of the answer is "a few" rather than "a few hundred."

33. Suppose that you observe a storm on Saturn, just as it becomes visible around the edge in the equatorial region. How much time will pass before the storm disappears around the opposite side of Saturn's disk?
 a. Make a prediction: Use reason and the answer in Working It Out 10.1 to think about the order of magnitude that you expect for these storms to orbit the planet. Will this number be a few seconds? A few minutes? A few hours? Days? Months? Years?
 b. Calculate: Follow Working It Out 10.1 in reverse to find the time it takes a storm to travel half the circumference of Saturn. (Refer to Table 10.1 to find the radius of Saturn.)
 c. Check your work: Verify that your answer has units of time and that it is sensible when compared to your prediction.

34. Neptune radiates 2.6 times as much energy into space as it absorbs from the Sun. Its equilibrium temperature (see Chapter 5) is 47 K. What is its true temperature?
 a. Make a prediction: Working It Out 10.2 shows the reverse calculation for Jupiter—that is, it finds the ratio of energies from the temperature, rather than the temperature from the ratio. Study that calculation. Because the ratio of energies for Neptune is larger than the ratio for Jupiter, do you expect the difference between the temperatures to be larger or smaller than for Jupiter?
 b. Calculate: Algebraically solve the equation in Working It Out 10.2 for the actual temperature, and then plug in the values in the problem statement to find the actual temperature.
 c. Check your work: Verify that your answer is in kelvins, and compare your answer to your prediction to check that the magnitude of your answer is sensible.

35. Uranus occults a star when the relative motion between Uranus and Earth is 23.0 km/s. An observer on Earth sees the star disappear for 37 minutes, 2 seconds and notes that the center of Uranus passed directly in front of the star. On the basis of those observations, what value would the observer calculate for the diameter of Uranus?
 a. Make a prediction: Table 10.1 gives the radius of Uranus. Accounting for measurement error, roughly what do you expect to calculate for the diameter of Uranus?
 b. Calculate: This problem recalls the fundamental relationship: distance = rate × time. Calculate the distance Uranus travels during the time the star is blocked.
 c. Check your work: Verify that your answer has units of km, and compare it to the given radius in Table 10.1 to check that the size of your answer is sensible.

36. Figure 10.1 shows two sets of pictures of the outer planets. What is the difference between Figure 10.1a and Figure 10.1b? ★

37. Figure 10.11d graphs the winds on Neptune, but the graph does not cover the full planet. Is that likely to mean that the wind speed is zero where no white line is present or that the wind speed is unknown where no white line is present? Explain your reasoning. ★

38. Use Figure 10.18a to estimate the radius of Io's plasma torus in terms of the radius of Jupiter. Convert that value to kilometers and then look up the answer on the Internet. How close did you get with your simple measurement? ★

39. The Sun appears 400,000 times brighter than the full Moon in Earth's sky. How far from the Sun (in astronomical units) would you have to go for the Sun to appear only as bright as the full Moon appears in Earth's nighttime sky? How does the distance you would have to travel compare with the semimajor axis of Neptune's orbit?

40. Jupiter's equatorial radius (R_{Jup}) is 71,500 km, and its oblateness is 0.065. What is Jupiter's polar radius (R_{polar})? Oblateness is given by $(R_{Jup} - R_{polar})/R_{Jup}$.

41. Ammonium hydrosulfide (NH_4HS) is a molecule in Jupiter's atmosphere responsible for many of its clouds. Using the periodic table in Appendix 3, calculate the molecular weight of an ammonium hydrosulfide molecule, where the atomic weight of a hydrogen atom is 1. (Recall from Working It Out 9.1 that the weight of a molecule is equal to the sum of the weights of its component atoms.)

42. Jupiter is an oblate planet with an average radius of 69,900 km. Earth's average radius is 6370 km.
 a. Given that volume is proportional to the cube of the radius, how many Earth volumes could fit inside Jupiter?
 b. Jupiter is 318 times as massive as Earth. Show that Jupiter's average density is about one-fourth that of Earth.

43. The tilt of Uranus is 98°. From one of the planet's poles, how far from the zenith would the Sun appear on summer solstice?

44. Compare the graphs in Figures 10.10a and 10.10b. On which planet, Jupiter or Saturn, does atmospheric pressure increase more rapidly with depth? Compare the graphs in Figures 10.10c and 10.10d. Does pressure increase more rapidly with depth on Uranus or on Neptune? Of the four giant planets, which has the fastest pressure rise with depth? The slowest? ★

45. Use Figure 10.10 to find the temperature at an altitude of 100 km on each of the four giant planets. ★

EXPLORATION Estimating Rotation Periods of the Giant Planets

Study the two images of Neptune in **Figure 10.23**. The image on the right was taken 17.6 hours after the image on the left. During that time, the Great Dark Spot completed *less* than one full rotation. The small storm at the bottom of the image completed slightly *more* than one rotation. You would be very surprised to see that result for locations on Earth.

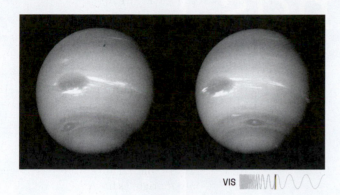

VIS

Figure 10.23 Images of Neptune, taken 17.6 hours apart by *Voyager 2*.

1 What do those observations tell you about the rotation of visible cloud tops of Neptune?

_____

You can find the rotation period of the smaller storm by equating two ratios. First, use a ruler to find the distance (in millimeters) from the left edge of the planet to the small storm in each image (**Figure 10.24a**). The right edge of the planet is not illuminated, so you will have to estimate the radius of the circles traveled by the storms. You can do that by measuring from the edge of the planet to a line through the planet's center (**Figure 10.24b**). Because the small storm travels along a line of latitude close to a pole, the distance it travels is significantly less than the circumference of the planet.

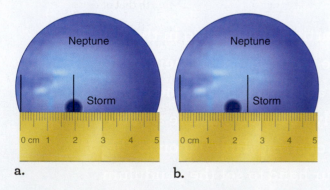

a.   **b.**

Figure 10.24 **a.** How to measure the position of the storm. **b.** How to measure the radius of the circle that the storm traveled.

2 Estimate the radius of the circle (in millimeters) that this small storm makes around Neptune by measuring from the edge of the disk to the line through the center of the planet.

_____

3 Find the circumference of that circle (in millimeters).

_____

Because the small storm rotated *more* than one time, the total distance it traveled is the circumference of the circle plus the distance between its locations in the two images.

4 Add the numbers you obtained in steps 2 and 3 to get the total distance traveled (in millimeters) between those images.

_____

Now take a ratio and find the rotation period. The ratio of the rotation period (T) to the time elapsed (t) must be equal to the ratio of the circumference of the circle around which it travels (C, in millimeters) to the total distance traveled (D, in millimeters): $T/t = C/D$.

5 You have all the numbers you need to solve for T. What value do you calculate for the small storm's rotation period? (To check your work, note that your answer should be less than 17.6 hours. Why?)

_____

Why does that calculation work at all? The actual distance the small storm traveled is *not* a few millimeters, nor is the circumference of the circle around which it travels. To find the actual distance or circumference, you would multiply both values by the same constant of proportionality. Because you are taking a ratio, however, that constant cancels out, so you might as well leave it out from the beginning.

6 Perform the corresponding measurements and calculations for the Great Dark Spot. What is its rotation period? Think carefully about how to find the total distance traveled because the Great Dark Spot has rotated around the planet *less* than once. (To check your work, note that your answer should be more than 17.6 hours. Why?)

_____

7 How similar are the rotation periods for those two storms?

_____

8 What does that comparison tell you about using that method to determine the rotation periods of the giant planets?

_____

9 What method do astronomers use instead?

_____

Planetary Moons and Rings

The Cassini Division is a gap in Saturn's rings caused by the gravity of the moon Mimas. A moon can cause a gap in the rings if the moon and the ring particles repeatedly line up so that the moon gives the particles a little tug in the same direction over and over again, shoving the particles out of their orbit. To see how a lot of little tugs can add up if they always happen in the right direction at the right place, hang any moderately heavy object, like a water bottle or a balled-up pair of socks, from a piece of string. Holding on to the loose end of the string, use small motions of your hand to set the pendulum swinging. Make it swing more, then make it swing less. Stop the motion of the pendulum entirely with small motions of your hand at the end of the string, and then set it swinging again.

EXPERIMENT SETUP

To see how a lot of little tugs can add up if they always happen in the right direction at the right place, hang any moderately heavy object, like a water bottle or a balled-up pair of socks, from a piece of string.

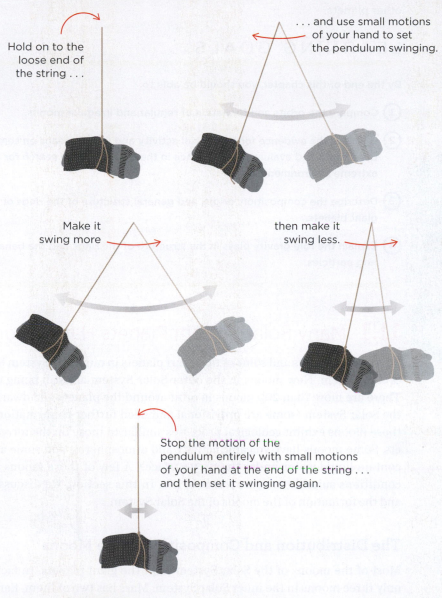

Hold on to the loose end of the string . . .

. . . and use small motions of your hand to set the pendulum swinging.

Make it swing more

then make it swing less.

Stop the motion of the pendulum entirely with small motions of your hand at the end of the string . . . and then set it swinging again.

PREDICTION

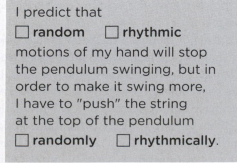

I predict that
☐ **random**   ☐ **rhythmic**
motions of my hand will stop the pendulum swinging, but in order to make it swing more, I have to "push" the string at the top of the pendulum
☐ **randomly**   ☐ **rhythmically**.

11

For centuries, Saturn's rings and the Galilean moons of Jupiter have delighted people who looked through telescopes. Since the dawn of the space age, robotic explorers traveling through the Solar System have revealed even more of the diverse collection of moons and rings orbiting other planets.

LEARNING GOALS

By the end of this chapter, you should be able to:

(1) Compare the orbits and formation of regular and irregular moons.

(2) Describe the evidence for geological activity and liquid oceans on some of the moons, and evaluate this evidence in the context of the search for life in extreme environments.

(3) Describe the composition, origin, and general structure of the rings of the giant planets.

(4) Explain the role gravity plays in the structure of the rings and the behavior of ring particles.

11.1 Many Solar System Planets Have Moons

Most of the planets and some of the dwarf planets in our Solar System have moons orbiting them. New moons in the outer Solar System are still being discovered. There are more than 200 moons in orbit around the planets and dwarf planets of the Solar System (some are provisional and need further confirmation). Many of those moons exhibit geological processes similar to those on the terrestrial planets. Some moons have volcanic activity and atmospheres, and some are likely to contain liquid water under their icy surfaces. A few of those moons could have conditions suitable to some forms of life. In this section, we discuss the orbits and the formation of the moons of the Solar System.

The Distribution and Composition of the Moons

Most of the moons of the Solar System orbit the giant planets. In fact, there are only three moons in the inner Solar System: Mars has two of them, Earth has one, and Mercury and Venus have none. Among the dwarf planets, Pluto has five known moons, Haumea has two, and Eris and Makemake each have one. While the giant planets were forming, they had greater attracting mass and more debris around them, so they have more moons.

Figure 11.1 shows the major moons in the Solar System. (Some properties of other notable planetary moons are listed in Appendix 4, and moons around asteroids are discussed in Chapter 12.) Some of these major moons, like Earth's Moon, are made of rock. Others, especially in the outer Solar System, are mixtures of rock and water ice. A few are made almost entirely of ice. The smallest known moons are only a kilometer in diameter, but some are comparable in size to planets. Two moons, Jupiter's Ganymede and Saturn's Titan, are larger in diameter than Mercury. Although most moons have no atmosphere, Titan has an atmosphere denser than Earth's, and several have very low-density atmospheres. Moons probably accreted

Figure 11.1 Various spacecraft have imaged the major moons of the Solar System. The images are to scale, with the planet Mercury and dwarf planet Pluto shown for comparison. At this scale, the martian moons, Phobos and Deimos, are too small to be shown.

from smaller bodies in much the same way that planets accreted from planetesimals, although some may be the product of collisions.

The Orbits of the Moons

One way to classify moons is by comparing their orbits. A **regular moon** has an orbit that lies near its planet's equatorial plane, is close to its planet, and is nearly circular (**Figure 11.2a**). A regular moon also orbits in the same sense (clockwise or counterclockwise) that the planet rotates. About one-third of the moons in the Solar System are regular. Most of those moons probably formed from an accretion disk around their planet at around the time the planet was forming. Our Moon, the Galilean moons of Jupiter, and Saturn's Titan are large, regular moons. Most regular moons are tidally locked to their parent planets. Recall from Chapter 4 that tidal locking causes a body to have a rotation period equal to its orbital period, as Earth's Moon does. When a moon is tidally locked to its planet, the leading hemisphere permanently faces "forward," in the direction in which the moon is traveling in its orbit around the planet. The trailing hemisphere faces backward. The leading hemisphere is always flying directly into any local debris surrounding the planet, so that hemisphere may have more impact craters on its surface than the trailing hemisphere.

what if . . .

What if a sixth moon was found in a circular orbit around Pluto, and what if it was tidally locked to Pluto but was not in orbital resonance with the other five moons? What conclusions might we draw about this moon?

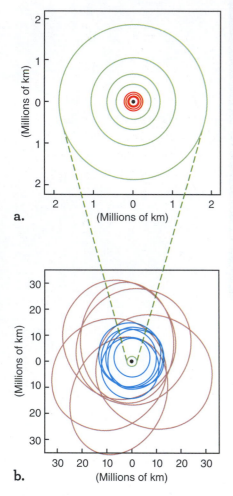

Figure 11.2 These diagrams illustrate the view from "above" the orbits of some of Jupiter's moons. **a.** Closer moons, including the Galilean moons, are regular, with nearly circular orbits in Jupiter's equatorial plane. **b.** Most of the more distant moons are irregular, with more elliptical, retrograde orbits not in the equatorial plane.

Compared with a regular moon, an **irregular moon** has a more elliptical and more inclined orbit and generally is farther away from its planet (**Figure 11.2b**). Most irregular moons orbit in a direction *opposite* to the rotation of their respective planets; that is, they have retrograde orbits. (Recall from Chapter 3 that apparent backward motion of a planet in the sky is called retrograde motion.) The largest irregular moons are Neptune's Triton and Saturn's Phoebe. Most of the recently discovered moons of the outer planets are irregular, and many are only a few kilometers across. Those moons are almost certainly bodies that formed elsewhere and were later captured by the planets.

Some of the regular moons have strange orbital characteristics. For example, Phobos, one of the two small moons of Mars (**Figure 11.3**), is so close to Mars that it actually orbits Mars faster than Mars rotates. As seen from Mars, Phobos rises in the west and sets in the east twice a day. Phobos is closer to its planet than any known moon. It is not known whether Phobos and Mars's other moon, Deimos, were captured from the nearby asteroid belt or whether they evolved together with Mars, possibly after a collision early in the history of Mars.

Another strange regular moon is Pluto's Charon, which has a radius one-half that of Pluto. The two are mutually tidally locked, so each has one hemisphere that always faces the other body and another hemisphere that never faces the other body. The small human figures on Pluto and Charon in **Figure 11.4** will always see each other. Pluto's highly compact moon system may have been created by a massive collision between Pluto and another planetesimal, producing a cloud of debris that coalesced to form Charon and the four smaller moons of Pluto, perhaps similar to the way Earth's Moon formed (Chapter 8).

Some sets of moons are in synchronized orbits, called **orbital resonances**, in which the orbital period of one moon is a multiple of the orbital period of another moon. Jupiter's moons Ganymede, Europa, and Io are in a resonance of 1:2:4; this means that for every one orbit of Ganymede, Europa orbits twice and Io orbits four times. When the moons are aligned, gravitational effects elongate Io's orbit, which is also affected by the tidal forces of Jupiter on Io. Pluto's five moons appear to be in a 1:3:4:5:6 sequence of near resonances. Pairs of some of Saturn's moons are in resonance, and resonances also exist between its moons and gaps in its rings.

The orbits of the moons not only indicate something about their origin but also can be used to estimate the masses of the planets by using Kepler's laws, just as the

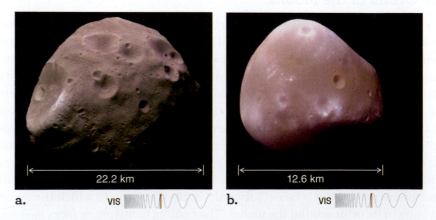

Figure 11.3 The *Mars Reconnaissance Orbiter* photographed the two tiny moons of Mars: **a.** Phobos and **b.** Deimos.

Credit (part a): ESA/DLR/FU Berlin (G. Neukum), https://www.esa.int/ESA_Multimedia /Images/2004/11/Phobos_in_colour_close-up2. https://creativecommons.org/licenses/by/4.0/.

working it out 11.1

Using Moons to Compute the Mass of a Planet

Recall from Working It Out 4.3 that Newton's version of Kepler's law for planets orbiting the Sun can be used to estimate the mass of the Sun. In Chapter 4, we used the following equation to calculate the mass (M) of the Sun:

$$M = \frac{4\pi^2}{G} \times \frac{A^3}{P^2}$$

where A is the semimajor axis of the orbit, P is the orbital period of a planet, and G is the universal gravitational constant.

For moons orbiting a planet, the same equation applies, as long as the moon is much less massive than the planet. Thus, we can use the orbital motion of the moons to estimate the mass of the planet. For example, Jupiter's moon Io has an orbital semimajor axis of $A = 422,000$ kilometers (km) and an orbital period of $P = 1.77$ days. To match the units in G, we need to convert P into seconds:

$$1.77 \text{ days} = 1.77 \text{ days} \times 24\frac{\text{h}}{\text{day}} \times 60\frac{\text{min}}{\text{h}} \times 60\frac{\text{s}}{\text{min}} = 152,928 \text{ s}$$

$G = 6.67 \times 10^{-20}$ km^3/(kg s^2), so the mass of Jupiter (M_{Jup}) is given by

$$M_{\text{Jup}} = \frac{4\pi^2}{G} \times \frac{A^3}{P^2} = \frac{4\pi^2}{6.67 \times 10^{-20} \text{ km}^3/(\text{kg s})^2} \times \frac{(422,000 \text{ km})^3}{(152,928 \text{ s})^2}$$

$$M_{\text{Jup}} = 1.90 \times 10^{27} \text{ kg}$$

Using any other moon of Jupiter, you would get the same answer.

Back before Newton published his law of gravity and before any measured value of the gravitational constant G was possible, Galileo and Kepler showed that P^2/A^3 was the same for each of the four Galilean moons of Jupiter. That finding demonstrated that Kepler's law applied to systems other than planets orbiting the Sun.

---

orbital properties of the planets can be used to find the mass of the Sun. An example of that calculation is shown in **Working It Out 11.1.**

CHECK YOUR UNDERSTANDING 11.1

Which of the following are characteristics of regular moons? (Choose all that apply.)
(a) They revolve around their planets in the same direction as the planets rotate.
(b) They have orbits that lie nearly in the equatorial planes of their planets. (c) They are usually tidally locked to their parent planets. (d) They are much smaller than all the known planets.

Answers to Check Your Understanding questions are in the back of the book.

11.2 Some Moons Have Geological Activity and Water

In this section, we organize our discussion of the moons by considering the history of their geological activity and the presence (or absence) of water and an atmosphere.

As with the terrestrial planets and Earth's Moon, surface features provide critical clues to a moon's geological history. For example, water ice is a common surface material among the moons of the outer Solar System, and the freshness of that ice indicates the age of those surfaces. Meteorite dust darkens the icy surfaces of moons just as dirt darkens snow in urban areas. A bright surface often means a fresh surface. The size and number of impact craters indicate the relative timing of events such as volcanism, and that timing enables scientists to gauge whether and

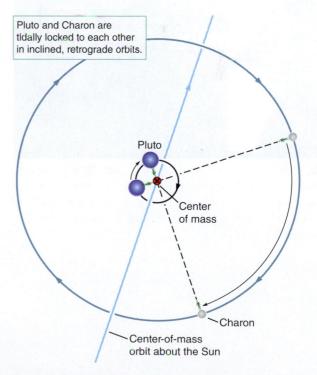

Figure 11.4 The doubly synchronous rotation and revolution in the Pluto-Charon system. The two bodies permanently face each other.

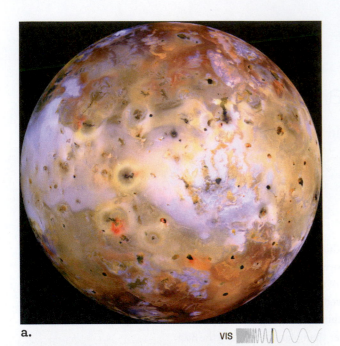

a.   VIS

Figure 11.5 ★ WHAT AN ASTRONOMER SEES An astronomer will be fascinated by this pair of images of Jupiter's moon Io. In the composite image **a**, she will apply several concepts from previous chapters. Noticing the smoothness of the surface, she will conclude that the surface is extremely young. Searching across the surface for impact craters, and finding none, she will become convinced that this world is unusually volcanically active. The colors will attract her attention, because she knows that color indicates composition; some areas of the moon have slightly different composition than others. The bright reflective patches imply that a sort of "snow," or crystallized volatile, has fallen there. She will puzzle over the dark squiggly structures emanating from the centers of a few volcanoes, particularly because they seem to point in various directions. Shifting to the second image **b**. will solve that mystery, because it shows how the volcanic plumes cast shadows. From the fact that the plumes in the first image point in different directions, she can guess that the first image is a composite of many images, taken over time. This is true; the image is a composite of pictures obtained by *Galileo*. Knowing the diameter of Io, she would be able to determine that the plume from the crater Pillan Patera rises 140 km above the edge of the moon on the left.

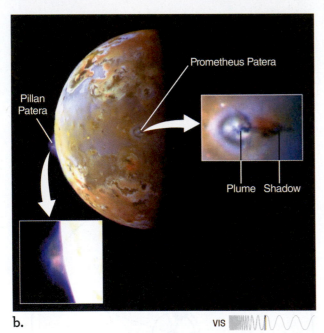

Pillan Patera

Prometheus Patera

Plume   Shadow

b.   VIS

when a moon may have been active in the past. Older surfaces have more craters. Observations of erupting volcanoes, which are found on Io and Enceladus, for example, are direct evidence that some moons are geologically active today. These types of observations show that some moons in the Solar System have been geologically dead since their formation during the early history of the Solar System, whereas others are even more geologically active than Earth.

Io, the Most Geologically Active Moon

One surprise in Solar System exploration was the discovery of active volcanoes on Io, the innermost of the four large moons of Jupiter. Because Io's orbit is elliptical, Jupiter's gravitational pull flexes Io's crust and generates enough thermal energy to melt parts of it. Have you ever taken a piece of metal and bent it back and forth, eventually breaking it in half? Touching the crease line can burn your fingers. Just as bending metal in your hands creates heat, the continual flexing of Io generates enough energy to melt parts of its mantle. In that way, Jupiter's gravitational energy is converted into thermal energy, powering the most active volcanism in the Solar System.

Io is just slightly larger than our Moon. Its surface is covered with volcanic features, including vast lava flows, volcanoes, and volcanic craters (**Figure 11.5a**). Lava flows and volcanic ash bury impact craters as quickly as they form, so no impact craters have been observed on the surface. The *Voyager*, *Galileo*, and *New Horizons* spacecraft and the Keck telescope have observed hundreds of volcanic vents and active volcanoes on Io. The most vigorous eruptions spray sulfurous gases, ash, and lava hundreds of kilometers above the surface. Some of that material escapes from Io. Ash and other particles rain onto the surface as far as 600 km from the vents. The moon is so active that several huge eruptions often occur at the same time. **Figure 11.5b** shows the volcanic activity on Io—the source of the material supplying Io's plasma torus and flux tube (recall Figure 10.18).

The surface of Io displays pale shades of red, yellow, orange, and brown. Mixtures of sulfur, sulfur dioxide frost, and sulfurous salts of sodium and potassium are responsible for that variety of colors. Bright patches may be fields of sulfur dioxide snow. Liquid sulfur dioxide flows beneath Io's surface, held at high pressure by the weight of the overlying material. That pressure forces sulfur dioxide out through fractures in the crust. The sulfur dioxide crystallizes in the abrupt transition to the lower pressure environment, much like liquid carbon dioxide contained at high pressure immediately turns to "dry ice" snow as it leaves the nozzle of a fire extinguisher. Sprays of sulfur dioxide snow crystals travel for up to hundreds of kilometers before settling back to the moon's surface.

Spacecraft images reveal the plains, irregular volcanic craters, and flows, all related to the eruption of mostly silicate magmas onto the surface of Io. The images also show tall mountains, some nearly twice the height of Mount Everest. Huge structures have multiple summit craters showing a long history of repeated eruptions followed by the collapse of partially emptied magma chambers. Many of the chamber floors are very hot (**Figure 11.6**) and might still contain molten material similar to the magnesium-rich lavas that erupted on Earth more than 1.5 billion years ago. Volcanoes on Io are spread much more randomly than those on Earth; there is no evidence of the plate boundaries that accompany plate tectonics.

Because of its active volcanism, Io's mantle has turned inside-out more than once, leading to chemical differentiation. Volatiles such as water and carbon dioxide escaped into space long ago, whereas most of the heavier materials sank to the interior to form a core. Sulfur and various sulfur compounds, as well as silicate magmas, are constantly being recycled to form the complex surface we see today.

Evidence of Liquid Oceans on Europa and Enceladus

Jupiter's moon Europa is slightly smaller than our Moon; it is made of rock and ice and it has an iron core. *Voyager* observed an outer shell of water ice with surface cracks and creases. Few impact craters are present, so the surface must be young. Regions of chaotic terrain (**Figure 11.7**) are places where the icy crust has been broken into slabs that have shifted into new positions. In other areas, the crust has split apart, and the gaps have filled in with new dark material rising from the interior. The young surface implies geological activity, which varies as Europa orbits Jupiter. As on Io, that activity is probably powered by continually changing tidal forces from Jupiter. However, the forces are not as strong on Europa as they are on Io because Europa is farther from Jupiter (**Working It Out 11.2**).

The *Galileo* spacecraft measured Europa's magnetic field and found that it varies, indicating an internal electrically conducting fluid. Detailed computer models of the interior of Europa suggest that it has a global ocean 100 km deep that contains more water than is held in all of Earth's oceans combined. That ocean might be salty as a result of dissolved minerals. The lightly cratered and thus geologically young surface indicates that energy exchange occurs between the icy crust and liquid water, but it is not known whether the icy crust is tens or hundreds of kilometers thick. It also is not yet known whether Europa has volcanic activity at the seafloor.

The Hubble Space Telescope may have detected two transient plumes of water vapor erupting from the icy surface, but they were not seen in later observations. The surface is too cold for liquid water, but lakes might lie a few kilometers underneath the surface. Scientists have reanalyzed observations of Europa in light of what has been learned about the many subsurface lakes in Antarctica, subglacial volcanoes in Iceland, and ice sheets at both poles of Earth. Many shallow "great lakes" may exist underneath the ice, and those would be prime targets for future exploration. **Figure 11.8** is an artist's schematic of Europa showing how some of the water from the ocean may leak out at the surface, making it easier to study.

Enceladus, one of Saturn's icy moons, shows a variety of ridges, faults, and smooth plains. That evidence of tectonic processes is unexpected for

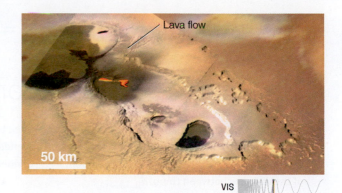

VIS

Figure 11.6 This *Galileo* image of Io shows regions where lava has erupted within the hollow at the top of a volcanic crater. The molten lava flow is shown in false color to make it more visible.

VIS

Figure 11.7 This high-resolution *Galileo* image of Jupiter's moon Europa shows where the icy crust has been broken into slabs that have shifted into new positions. Those areas of chaotic terrain are characteristic of a thin, brittle crust of ice floating atop a liquid or slushy ocean.

Figure 11.8 Artist's conception of Europa, showing liquid bubbling up from the liquid ocean underneath the icy surface. Jupiter and Io are visible in the Europan sky.

working it out 11.2

Tidal Forces on the Moons

Recall from Chapter 4 that the tidal force of a planet, such as Jupiter, acting on a small mass, m, on the surface of a moon depends on the mass of the planet, M_{Jup}, and the radius of the moon, R_{moon}, divided by the cube of the distance between them, $d_{Jup-moon}$:

$$F_{tidal} = \frac{2GM_{Jup}R_{moon}m}{d_{Jup-moon}^3}$$

We can use this equation to compare the tidal forces on a mass located on Io to the tidal forces on the same mass located on Europa by taking a ratio (as we did in Working It Out 4.4 to compare tidal forces between Earth and the Sun and Moon):

$$\frac{F_{tidal-Io}}{F_{tidal-Europa}} = \frac{\dfrac{2GM_{Jup}R_{Io}m}{d_{Jup-Io}^3}}{\dfrac{2GM_{Jup}R_{Europa}m}{d_{Jup-Europa}^3}} = \frac{R_{Io}d_{Jup-Europa}^3}{R_{Europa}d_{Jup-Io}^3}$$

Using data on Io and Europa from Appendix 4—namely, $R_{Io} = 1820$ km, $R_{Europa} = 1560$ km, $d_{Jup-Io} = 422{,}000$ km, and $d_{Jup-Europa} = 671{,}000$ km—we have

$$\frac{F_{tidal-Io}}{F_{tidal-Europa}} = \frac{R_{Io}d_{Jup-Europa}^3}{R_{Europa}d_{Jup-Io}^3} = \frac{(1820)(671{,}000)^3}{(1560)(422{,}000)^3} = 4.69$$

Thus, the tidal forces on Io are 4.69 times stronger than those on Europa.

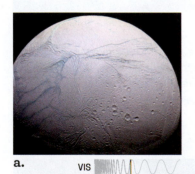

a.   VIS

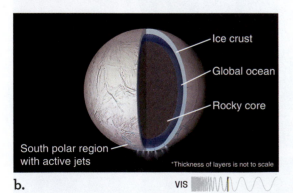

Ice crust

Global ocean

Rocky core

South polar region with active jets

*Thickness of layers is not to scale

b.   VIS

Figure 11.9 Images of Enceladus taken by *Cassini*. **a.** The deformed ice cracks (shown blue in false color) were found to be the sources of cryovolcanism. **b.** Cryovolcanic plumes in the south polar region are seen spewing ice particles into space.

a small (diameter = 500 km) body. The activity on Enceladus is an example of **cryovolcanism**, which is similar to terrestrial volcanism but is driven by subsurface low-temperature liquids such as water and hydrogen rather than by molten rock. Some impact craters appear softened, perhaps by the viscous flow of ice, such as the flow that occurs in the bottom layers of glaciers on Earth. Parts of the moon have no craters, indicating recent resurfacing. Terrain near the south pole of Enceladus is cracked and twisted (**Figure 11.9a**). The cracks are warmer than their surroundings, suggesting that tidal heating and radioactive decay within the moon's rocky core heat the surrounding ice and drive it to the surface.

Buried beneath 30–40 km of Enceladus's ice crust lies a liquid ocean that is 10 km deep (**Figure 11.9b**). Active cryovolcanic plumes expel water vapor, tiny ice crystals, and salts. Some of the crystals fall back onto the surface as an extremely fine, powdery snow. The rate of accumulation is very low—a fraction of a millimeter per year—but over time the snow builds up. *Cassini* scientists estimate that the snow may be 100 meters thick in one area near the south pole of Enceladus, indicating that the plume activity has continued on and off for at least tens of millions of years. Planetary scientists are very interested in sending a mission to Enceladus to explore this ocean.

Flexing from tidal forces is the likely source of the heat energy coming from Enceladus, as on Io and Europa. Enceladus has an orbital resonance with Saturn's moon Dione. A moon made completely of ice would be too stiff for tidal heating to be effective; tidal heating works more effectively with ice over liquid water or cracked ice with some liquid. It remains a mystery why Enceladus is so active, whereas Mimas, a neighboring moon of about the same size—but closer to Saturn and also subject to tidal heating—appears geologically dead.

Reading Astronomy News

NASA Killed *Cassini* to Avoid Contaminating Saturn's Moons

Nicole Mortillaro, *CBC News*

Spacecraft's 13 years in orbit have produced suggestions of potentially habitable worlds.

NASA scientists killed the hard-working *Cassini* spacecraft to avoid contaminating Saturn's moons with Earth microbes because they may have the potential to support life.

On Friday at 7:55 a.m. ET [September 15, 2017], the world said goodbye to *Cassini* after 20 years in space and 13 years orbiting Saturn and its moons, providing incredibly detailed, high-resolution photos of what many consider the jewel of our solar system.

Though *Cassini*'s life was extended twice with new missions, the spacecraft had a limited fuel supply. So NASA announced in April that it would carry out one final mission, dubbed The Grand Finale.

Scientists and engineers altered the orbit of the spacecraft to run into Saturn, where it would safely burn up in its atmosphere.

Cassini's fate arose directly from its success in shedding a light on the possibility of life on other worlds.

On Wednesday, NASA's planetary science division director Jim Green spoke about Enceladus, a small, icy moon, spewing organic material that likely originates from a subsurface ocean.

"What we thought was an icy ball, when we observed the southern hemisphere, and geysers of water spewing out into the Saturn system, it amazed us," he told reporters.

"And it began changing the way we view the habitability or potential habitability of moons in the outer part of our solar system."

Allowing *Cassini* to run out of fuel would leave NASA no way to control the spacecraft. One day it could crash into Enceladus or even Titan, Saturn's largest moon, which is also being studied for potential habitability.

While there is no guarantee that there is microbial life on any of these moons, there is a chance of contaminating their surface with Earth microbes.

"Now because of the importance of Enceladus that *Cassini* has shown us, and of Titan, another potential world that could be habitable for life, perhaps not like we know it, but perhaps completely different than ours, we had to make decisions on how to dispose of the spacecraft," Green said.

"And that led us, inevitably, to the plan of taking *Cassini* and plunging into Saturn."

On the plus side, *Cassini*'s Grande Finale has given us unprecedented views of the planet's north pole as well as its intricate ring system. For the first time in its mission, the spacecraft flew between the planet and the rings.

And, with its success and valuable insight into the potential habitability of other worlds, it's likely there could be new missions.

Missions have been suggested to explore Enceladus, including its interior and the subsurface liquid body, Larry Soderblom, an interdisciplinary scientist on the Cassini mission, told CBC News.

"We might find in fact a higher probability than I'm willing to admit right now that life exists in the interior of Enceladus."

While astronomers and planetary enthusiasts may miss the data and photos gathered by Saturn, NASA's *Juno* spacecraft is also orbiting Jupiter. And there are plans for the *Europa Clipper*, which will conduct flybys of another icy, potentially habitable moon, Europa.

If there is a chance for life to be found in the Saturn system, NASA is eager to protect it.

"Because of planetary protection and our desire to go back to Enceladus and go back to Titan and go back to the Saturn system, we must protect those bodies for future exploration," Green said.

QUESTIONS

1. How could *Cassini* "contaminate" a moon?

2. Do a search on YouTube to see the end of *Cassini*. What did the final dives reveal about Saturn's rings?

3. Do a search to see whether new missions to Saturn's moons have been approved. Which moons will be explored?

4. Go to the Web page for the NASA Office of Planetary Protection (https://sma.nasa.gov/sma-disciplines/planetary-protection). Which moons are affected by that policy? How might that policy affect future plans for a lander on Europa? What was done to protect Mars from contamination during the NASA Mars lander missions?

Source: https://www.cbc.ca/news/science/cassini-saturn-contaminate-moons-enceladus-titan-1.4287818.

Figure 11.10 Images of Saturn's largest moon, Titan, taken by *Cassini*. **a.** Titan's orange atmosphere is caused by organic, smoglike particles. **b.** Infrared-light imaging penetrates Titan's smoggy atmosphere and reveals surface features.

Titan's Atmosphere, Lakes, and Ocean

Saturn's moon Titan is slightly larger than Mercury and has a composition of about 45 percent water ice and 55 percent rocky material. What makes Titan especially remarkable is its thick atmosphere and the presence of liquid methane on its surface. Whereas Mercury's secondary atmosphere has escaped to space due to Mercury's proximity to the Sun, Titan's greater mass and distance from the Sun have allowed it to retain an atmosphere that is 30 percent denser than Earth's. Titan's atmosphere, like Earth's, is mostly nitrogen. As Titan differentiated, various ices, including methane (CH_4) and ammonia (NH_3), emerged from the interior to form an early atmosphere. Ultraviolet photons from the Sun have enough energy to break apart ammonia and methane molecules—a process called **photodissociation**. Photodissociation of ammonia is the likely source of Titan's atmospheric nitrogen. Methane breaks into fragments that recombine to form organic compounds, including complex hydrocarbons such as ethane. Those compounds tend to cluster in tiny particles, creating organic smog much like the air over Los Angeles on a bad day. This process gives Titan's atmosphere its characteristic orange hue (**Figure 11.10a**).

The *Cassini* spacecraft obtained close-up views of Titan's surface. Haze-penetrating infrared imaging showed broad regions of dark and bright terrain (**Figure 11.10b**). Radar imaging of Titan, moreover, revealed irregularly shaped features in its northern hemisphere that appear to be widespread lakes and seas of methane, ethane, and other hydrocarbons (**Figure 11.11**). The photodissociative process by sunlight should have destroyed all atmospheric methane within a geologically brief period of about 50 million years, so a process must be renewing the methane being destroyed by solar radiation. That, along with the near absence of impact craters on the surface of Titan, suggests recent methane-producing activity. Radar views also indicate an active surface, showing features that resemble terrestrial sand dunes and channels. Heat supplied by radioactive decay could cause cryovolcanism that releases "new" methane from underground. The evidence of active cryovolcanism on Titan is indirect—the

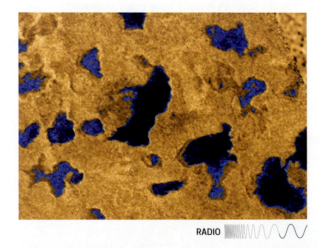

Figure 11.11 Radar imaging (false color) near Titan's north pole shows lakes of liquid hydrocarbons covering 100,000 square kilometers (km^2) of the moon's surface. Features such as islands, bays, and inlets are visible in many of those radar images.

presence of abundant atmospheric methane and of methane lakes strongly suggests that Titan has some geological activity.

Titan has terrains reminiscent of those on Earth, with networks of channels, ridges, hills, and flat areas that may be dry lake basins. Those terrains suggest a methane cycle (analogous to Earth's water cycle) in which methane rain falls to the surface, washes the ridges free of the dark hydrocarbons, and then collects into drainage systems that empty into low-lying, liquid methane pools. Stubby, dark channels appear to be springs where liquid methane emerges from the subsurface; bright, curving streaks could be water ice that has oozed to the surface to feed glaciers. An infrared camera photographed a reflection of the Sun from such a lake surface. The type of reflection observed proves that the lake contains a liquid and is not frozen or dry. Recent observations might indicate waves on one of those lakes.

Titan is the only moon (besides Earth's) that has been landed upon. In 2005, *Cassini* released the *Huygens* probe, which plunged through Titan's atmosphere, taking pictures and measuring the moon's composition, temperature, pressure, and wind speeds. *Huygens* confirmed the presence of nitrogen-bearing organic compounds in the clouds. During its descent, *Huygens* encountered 120-m/s winds and temperatures as low as 88 kelvins (K). As it reached the surface, though, winds died down to less than 1 m/s and the temperature warmed to 112 K. The pictures taken by *Huygens* showed that the surface was wet with liquid methane, which evaporated as the probe—heated during its passage through the atmosphere—landed in the frigid soil. The surface also was rich with other organic (carbon-bearing) compounds, such as cyanogen and ethane. **Figure 11.12** shows that the surface around the landing site is relatively flat and littered with rounded "rocks" of water ice. The dark "soil" is probably a mixture of water and hydrocarbon ices.

As with Saturn's moon Enceladus and Jupiter's moon Europa, gravitational mapping provides indirect evidence that Titan also has an ocean buried beneath its surface. Some of Titan's surface features move by as much as 35 km, which suggests that the crust is sliding on an underlying liquid layer. The current model of Titan (**Figure 11.13**) is that its rigid ice shell varies in thickness and surrounds an ocean 100 km below the surface. That ocean would be made of water mixed with dissolved salts—possibly saltier than Earth's Dead Sea. In that model, methane outgassing would occur in hot spots.

Titan is the only moon with a significant atmosphere and the only Solar System body other than Earth that has standing liquid on the surface and a cycle of liquid rain and evaporation. In many ways, Titan resembles a primordial Earth, albeit at much lower temperatures. The presence of liquids and of organic compounds that could be biological precursors for life in the right environment makes Titan another high-priority target for continued exploration.

CHECK YOUR UNDERSTANDING 11.2a

Which of the following moons is thought to have an ocean of water beneath its surface? (Choose all that apply.) (a) Io; (b) Europa; (c) Enceladus; (d) Titan

Cryovolcanism on Triton

Cryovolcanism also occurs on Triton, Neptune's largest moon. Triton is an irregular moon; its retrograde orbit suggests that Triton was captured by Neptune after the planet's formation. As Triton achieved its current circular, synchronous orbit,

Figure 11.12 The two water-ice "rocks" on Titan just below the center of this *Huygens* image are about 85 centimeters (cm) from the camera and roughly 15 and 4 cm across, respectively. Credit: © ESA/NASA/JPL /University of Arizona, https:// nssdc.gsfc.nasa.gov/planetary /titan_images.html. https:// creativecommons.org/licenses /by/4.0/.

VIS

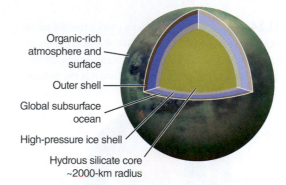

Organic-rich atmosphere and surface

Outer shell

Global subsurface ocean

High-pressure ice shell

Hydrous silicate core ~2000-km radius

Figure 11.13 Artist's conception of Titan's internal structure showing how Titan is differentiated, with a core of water-bearing rocks and a subsurface ocean of liquid water. A layer of high-pressure ice surrounds the core, with an outer ice shell on top of the subsurface ocean.

unanswered questions

What is the source of Titan's nitrogen atmosphere? One group of experimenters tried to answer that question by blasting a laser at water–ammonia (H_2O–NH_3) ice to simulate cometary impacts and see whether nitrogen gas (N_2) forms. They concluded that the observed amount of N_2 in Titan's atmosphere could have been created from ammonia ice in that way. Another idea is that those gases were accreted during the formation of the moon. Using data from the *Huygens* probe, other researchers concluded that if Titan had differentiated like Ganymede, hydrothermal activity released the gases from a hot core. Astronomers want to understand why Titan has an atmosphere and the other larger moons do not.

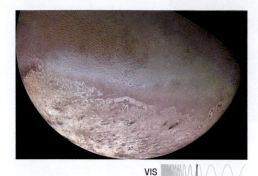

VIS

Figure 11.14 *Voyager 2* mosaic showing various terrains on the Neptune-facing hemisphere of Triton. The lack of impact craters in the "cantaloupe terrain," visible at the top, indicates a geologically younger age than that of the bright, cratered terrain at the bottom.

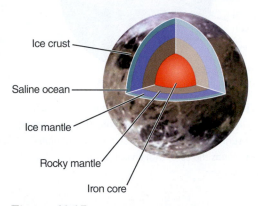

Ice crust

Saline ocean

Ice mantle

Rocky mantle

Iron core

Figure 11.15 Artist's conception of the interior of Jupiter's moon Ganymede, showing that the ocean and ice may be stacked in multiple layers.

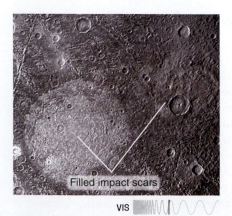

Filled impact scars

VIS

Figure 11.16 *Voyager* image showing filled impact scars on Jupiter's moon Ganymede.

it experienced extreme tidal stresses from Neptune, generating large amounts of thermal energy. The interior may have melted, allowing Triton to become chemically differentiated.

Triton has a thin atmosphere and a surface composed mostly of ices and frosts of methane and nitrogen at a temperature of about 38 K. The relative absence of craters tells us the surface is geologically young. Part of Triton is covered with terrain that looks like the skin of a cantaloupe (**Figure 11.14**), with irregular pits and hills that may be caused by slushy ice emerging onto the surface from the interior. Veinlike features include grooves and ridges that could result from ice oozing out along fractures. The rest of Triton is covered with smooth volcanic plains. Irregularly shaped depressions as wide as 200 km formed when mixtures of water, methane, and nitrogen ice melted in the interior of Triton and erupted onto the surface, much as rocky magmas erupted onto the lunar surface and filled impact basins on Earth's Moon.

Colorless nitrogen ice creates a localized greenhouse effect, in which solar energy trapped beneath the ice raises the temperature at the base of the ice layer. A temperature increase of only 4 K vaporizes the nitrogen ice. As that gas is formed, the expanding vapor exerts very high pressures beneath the ice cap. Eventually, the ice ruptures and vents the gas explosively into the low-density atmosphere. *Voyager 2* found four of those active geyserlike cryovolcanoes on Triton. Each consisted of a plume of gas and dust as much as 1 km wide rising 8 km above the surface, where the plume was caught by upper atmospheric winds and carried for hundreds of kilometers downwind. Dark material, perhaps silicate dust or radiation-darkened methane ice grains, is carried along with the expanding vapor into the atmosphere, from which it later settles to the surface, forming dark patches streaked out by local winds, as seen near the lower right of Figure 11.14.

Formerly Active Moons

Some moons show clear evidence of past ice volcanism and tectonic deformation, but no current geological activity. Jupiter's moon Ganymede, the largest moon in the Solar System, is larger than the planet Mercury. When Jupiter was forming, low temperatures enabled grains of water ice to survive and coalesce along with dust grains into larger bodies at the distance of Ganymede's orbit. In less than half a million years, those bodies accreted to form Ganymede. Heating from accretion melted parts of Ganymede so that it is fully differentiated (**Figure 11.15**), with outer water layers, an inner silicate zone, and an iron-rich liquid core. As the moon cooled, much of the outer water layer froze, forming a dirty ice crust. Most of the denser materials sank to the central core, leaving an intermediate ice–silicate zone. Ganymede might also have a large, salty ocean underneath its icy surface, maybe 800 km deep, containing 25 times the volume of Earth's oceans.

Its surface is composed of two prominent terrains: a dark, heavily cratered (and therefore ancient) terrain, and a bright terrain characterized by ridges and grooves. The abundance of impact craters on Ganymede's dark terrain reflects the period of intense bombardment during the early history of the Solar System. The largest region of ancient dark terrain includes a semicircular area more than 3200 km across on the leading hemisphere. Furrowlike depressions occurring in many dark areas are among Ganymede's oldest surface features. They may represent surface deformation from internal processes or may be relics of impact-cratering processes.

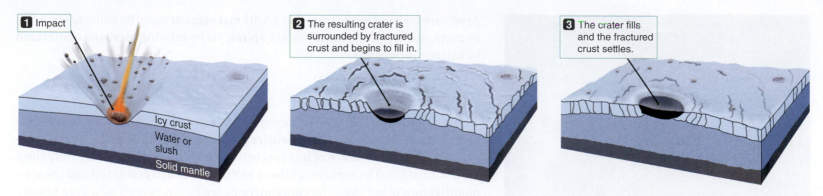

1 Impact

icy crust
Water or slush
Solid mantle

2 The resulting crater is surrounded by fractured crust and begins to fill in.

3 The crater fills and the fractured crust settles.

Figure 11.17 Filled impact scars form as viscous flow smooths out structures left by impacts on icy surfaces.

Impact craters on Ganymede range up to hundreds of kilometers in diameter, and the larger craters are proportionately shallower. The icy crater rims slowly slump, like a lump of soft clay. They are seen as bright, flat, circular patches found principally in the moon's dark terrain (**Figure 11.16**) and are thought to be scars left by early impacts onto a thin, icy crust overlying water or slush (**Figure 11.17**). In Chapter 8, we discussed how planetary surfaces can be fractured by faults or folded by compression resulting from movements initiated in the mantle. On Ganymede, the tectonic processes have been so intense that the fracturing and faulting have completely deformed the icy crust, destroying all signs of older features, such as impact craters, and creating the bright terrain. The energy that powered Ganymede's early activity was liberated during a period of differentiation when the moon was very young. After differentiation was complete, that source of internal energy ran out, and geological activity ceased.

Many other moons show evidence that they experienced an early period of geological activity that resulted in a dazzling array of terrains. A 400-km impact crater scars Saturn's moon Tethys, covering 40 percent of its diameter, and an enormous canyonland wraps around at least three-fourths of the moon's equator. Saturn's moon Dione shows bright ice cliffs up to several hundred meters high, created by tectonic fracturing. The trailing hemisphere of Saturn's Iapetus is bright, reflecting half the light that falls on it, whereas much of the leading hemisphere is as black as tar. Those dark deposits appear *only* in the leading hemisphere of Iapetus, suggesting that they might be debris that was blasted off small retrograde moons of Saturn by micrometeoritic impacts and swept up by Iapetus as it moved along in its prograde orbit around Saturn.

Saturn's moon Mimas, no larger than the state of Ohio, is heavily cratered with deep, bowl-shaped depressions. The most striking feature on Mimas is a huge impact crater in the leading hemisphere (**Figure 11.18**). Named "Herschel" after astronomer Sir William Herschel, who discovered many of Saturn's moons, the crater is 130 km across—a third the size of Mimas itself. It is doubtful that Mimas could have survived the impact of a body much larger than the one that created Herschel. Some astronomers think that Mimas (and perhaps other small, icy moons as well) was hit many times in the past by objects so large as to fragment the moon into many small pieces. Each time that happened, the individual pieces still in Mimas's orbit would coalesce to re-form the moon, perhaps in much the same way that Earth's Moon coalesced from fragments that remained in orbit around Earth after a large planetesimal impacted Earth early in its history.

Areas on Uranus's small moon Miranda have been resurfaced by eruptions of icy slush or glacierlike flows. Other moons of Uranus—Oberon, Titania, and

VIS

Figure 11.18 *Cassini* image showing Saturn's moon Mimas and the crater Herschel.

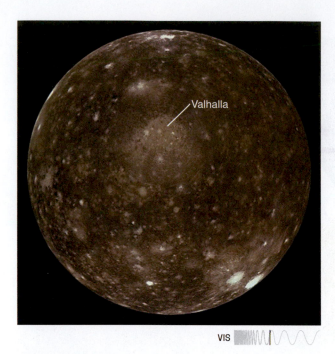

VIS

Figure 11.19 *Galileo* image showing Jupiter's second-largest moon, Callisto. The moon's ancient surface is dominated by impact craters and shows no sign of early internal activity.

VIS

Figure 11.20 Saturn's moon Hyperion rotates chaotically, with its rotation period and spin axis constantly changing. The 250-km moon's low density and spongelike texture, seen in this *Cassini* image, suggest that its interior houses a vast system of caverns.

Ariel—are covered with faults and additional signs of early tectonism. On Ariel, in particular, very old, large craters appear to be missing, perhaps obliterated by earlier volcanism.

Geologically Dead Moons

Jupiter's Callisto, Saturn's Hyperion, Uranus's Umbriel, and a large assortment of irregular moons are considered to be geologically dead because there exists little or no evidence of internal activity having occurred at any time since their formation. The surfaces of those moons are heavily cratered and show no modification other than the cumulative degradation caused by a long history of impacts.

Callisto is the third-largest moon in the Solar System, just slightly smaller than Mercury. Callisto also is the darkest of the Galilean moons of Jupiter, yet it is still twice as reflective as Earth's Moon. That brightness indicates that Callisto is rich in water ice, but with a mixture of dark, rocky materials. Except in areas that experienced large impact events, the surface is essentially uniform, consisting of relatively dark, heavily cratered terrain. Callisto's most prominent feature is a 2000-km, multi-ringed structure of impact origin named Valhalla (the largest bright feature visible in **Figure 11.19**). *Galileo* results suggest that a liquid ocean containing water or water mixed with ammonia could exist beneath the heavily cratered surface. Callisto may have partially differentiated, with rocky material separating from ices and sinking deeper into the interior.

Saturn's Hyperion (**Figure 11.20**) is one of the largest irregularly shaped moons and could be the remnant of an impact. The extensive craters look almost like sponges. In its chaotic orbit, Hyperion crosses Saturn's magnetosphere, which seems to have left the moon with some electric charge. Umbriel, the darkest and third largest of Uranus's moons, appears uniform in color, reflectivity, and general surface features, indicating an ancient surface. The real puzzle posed by Umbriel is why it is geologically dead, whereas the surrounding large moons of Uranus have been active at least at some time in their past.

CHECK YOUR UNDERSTANDING 11.2b

Rank the following moons according to the density of impact craters (from most to least) that you would expect to observe on the surface. (a) Callisto; (b) Titan; (c) Io; (d) Ganymede

11.3 Rings Surround the Giant Planets

A planetary **ring** is a collection of particles—ranging in size from tiny grains to house-sized boulders—that orbit individually around a planet, forming a flat disk. Ring systems do not occur around the terrestrial planets but are found around each of the giant planets. **Figure 11.21** shows how the ring system of each giant planet varies in size and complexity: some systems extend for hundreds of thousands of kilometers, and some have a detailed structure that includes many small rings. In this section, we discuss ring formation, composition, and evolution.

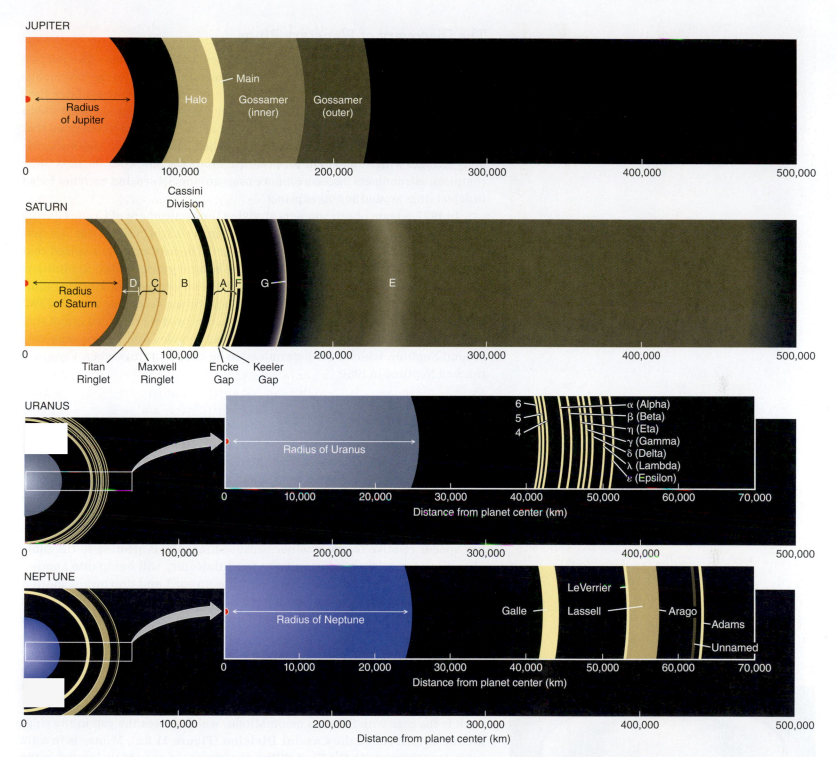

Figure 11.21 The ring systems of the four giant planets vary in size and complexity. Saturn's system, with its broad E Ring, is by far the largest and has the most complex structure in its inner rings.

The Discovery of Planetary Rings

Saturn's rings have been observed for centuries. In 1610, Galileo observed two small objects next to Saturn and thought they might be similar to the four moons orbiting Jupiter. But Saturn's "moons" did not move, and 2 years later they disappeared. In 1655, Dutch instrument maker Christiaan Huygens (1629–1695) pointed a superior telescope of his own design at Saturn. Huygens observed that an apparently continuous flat ring surrounds the planet and that the ring's visibility changes with its apparent tilt as Saturn orbits the Sun. Over the next three centuries, astronomers discovered more rings around Saturn, but searches failed to detect rings around any other planet.

In 1977, a team of astronomers studying the atmosphere of Uranus during stellar occultations saw brief, minute changes in the brightness of a star as it first approached and then receded from the planet. The astronomers realized that meant that Uranus has rings. Over the next several years, stellar occultations revealed nine rings surrounding the planet. In 1986, *Voyager 2* imaged two additional rings of Uranus, and in 2005 the Hubble Space Telescope recorded two more, bringing the total to 13. In 1979, cameras on *Voyager 1* recorded a faint ring around Jupiter. The occultation technique also revealed arclike ring segments around Neptune, which were determined to be complete rings when *Voyager 2* reached Neptune in 1989.

The Orbits of Ring Particles

Ring particles follow Kepler's laws, so the speed and orbital period of each particle depends upon its distance from the planet. The closest particles move the fastest and have the shortest orbital periods (see Working It Out 11.1). The orbital periods of particles in Saturn's bright rings, for example, range from 5 hours 45 minutes at the inner edge of the innermost bright ring to 14 hours 20 minutes at the outer edge of the outermost bright ring. Ring particles have low speeds relative to one another because they are all orbiting in the same direction. A particle moving on an upward trajectory will bump into another particle on a downward trajectory, and the upward and downward motion will cancel—leaving the particles moving in the same plane. A similar process occurs for particles moving inward and outward, leaving the particles moving at a constant radius.

The orbits of ring particles can also be influenced by the planet's larger moons. If the moon is massive enough, it exerts a gravitational tug on the ring particles as it passes by. If that happens over and over through many orbits, the particles are pulled out of the area, leaving a lower-density gap (see Figure 11.21). Such is the case with Saturn's moon Mimas, which causes the gap in the rings around Saturn called the **Cassini Division** (**Figure 11.22**). Mimas is in a 1:2 orbital resonance with the Cassini Division, giving a ring particle located in the Cassini Division an orbital period about Saturn that is equal to half the orbital period of Mimas. Such resonances with other moons produce some of the gaps that appear in Saturn's bright rings. One of the gaps is caused by a 4:1 resonance between the ring particles and Mimas.

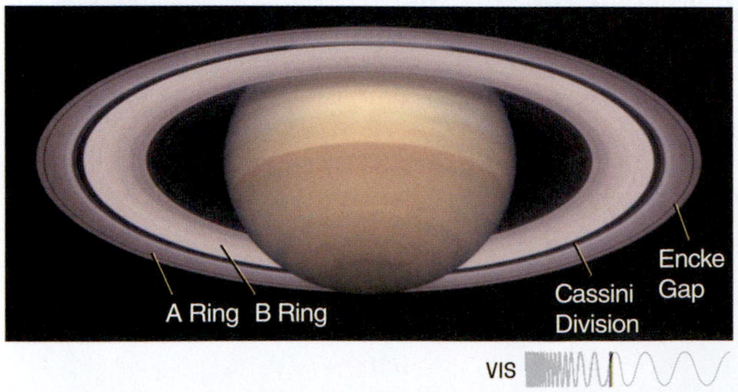

Other kinds of orbital resonances also are possible. For example, most narrow rings are caught up in a periodic gravitational tug-of-war with nearby moons, known as **shepherd moons** for how they "herd" the flock of ring particles. Shepherd moons are usually small and often come in pairs, with one orbiting just

VIS

Figure 11.22 Hubble Space Telescope image showing Saturn and its A Ring, B Ring, Cassini Division, and Encke Gap. The C Ring is too dim to be seen clearly.

inside and the other just outside a narrow ring. A shepherd moon just outside a ring robs orbital energy from any particles that drift outward beyond the edge of the ring, causing the particles to move back inward. A shepherd moon just inside a ring gives up orbital energy to a ring particle that has drifted too far in, nudging it back in line with the rest of the ring. Sometimes narrow rings are trapped between two shepherd moons in slightly different orbits.

Ring Formation and Evolution

Much of the material found in planetary rings is thought to be the result of tidal stresses. If a moon (or other planetesimal) orbits a large planet, the force of gravity will be stronger on the side of the moon close to the planet and weaker on the side farther away. That difference in gravitational force stretches out the moon, as you saw in the discussion of tidal forces in Chapter 4. If the tidal stresses are greater than the self-gravity that holds the moon together, the moon will be torn apart. The distance at which the tidal stresses equal the self-gravity is known as the Roche limit. The Roche limit applies only to objects held together by their own gravity; it does not apply to objects, such as people or cars, that are held together by other forces. If a moon or planetesimal comes within a planet's Roche limit, the object is pulled apart by tidal stresses, leaving many small pieces orbiting the planet. Those pieces gradually spread out, and their orbits are circularized and flattened out by collisions. The fragmented pieces of the disrupted body are then distributed around the planet in the form of a ring.

Planetary rings do not have the long-term stability of most Solar System objects. Ring particles are constantly colliding with one another in their tightly packed environment, either gaining or losing orbital energy. That redistribution of orbital energy can cause particles at the ring edges to leave the rings and drift away, aided by nongravitational influences such as the pressure of sunlight. Although moons may help guide the orbits of ring particles and delay the dissipation of the rings themselves, at best that condition can be only temporary. Saturn's brightest rings might be nearly as old as Saturn, but most planetary rings eventually disperse.

Even Earth may have had several short-lived rings at various times during its long history. Many comets or asteroids must have passed within Earth's Roche limit (about 25,000 km for rocky bodies and more than twice that for icy bodies) to disintegrate into a swarm of small fragments to create a temporary ring. Unlike the giant planets, however, Earth lacks shepherd moons to provide orbital stability to rings.

The Composition of Ring Material

Because much of the material in the rings of the giant planets comes from their moons, the composition of the rings is similar to that of the moons. Saturn's bright rings probably formed when a moon or planetesimal came within the Roche limit of Saturn. Those rings reflect about 60 percent of the sunlight falling on them. They are made of water ice, though a slight reddish tint indicates that they must contain small amounts of other materials, such as silicates. The icy moons around Saturn or the frozen comets of the outer Solar System could easily supply that material.

Saturn's rings are the brightest in the Solar System and are the only ones known to be composed of water ice. In stark contrast, the rings of Uranus and Neptune are among the darkest objects known in the Solar System. Only 2 percent of the

unanswered questions

What is the origin of Saturn's brightest rings? One early hypothesis was that the rings come from a moon that approached too close to the planet, but Saturn's moons are composed of rock and ice, and the rings are solely ice. Some recent computer models start with a differentiated moon the size of Titan that had a rock and iron core and a large, icy mantle. As the moon in the model slowly migrates toward Saturn and crosses the Roche limit, tidal forces rip away its water ice, but not the rocky core. According to that model, the core might have continued migrating inward until it fell into Saturn, with the ice forming Saturn's ring. As time went on, the ring spread, and as material crossed the Roche limit outward, Saturn's small moonlets formed. Computers are only now getting fast enough to test those models, which suggest a unified origin for many of the moons and rings.

what if . . .

What if Earth's Moon had active volcanoes that were strong enough to eject particles from the lunar surface? How would the resulting ring be oriented in the sky? Would the ring persist for a long time, or would it be short-lived? Describe how the night sky would be different when viewed from a dark site.

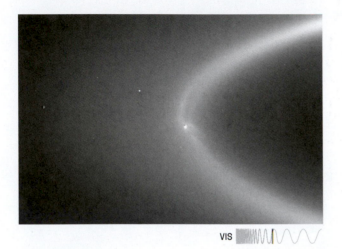

VIS

Figure 11.23 Saturn's moon Enceladus (the large bright spot appearing to be on the ring) is the source of material in Saturn's E Ring. Note the distortion in the distribution of ring material around the moon. That distortion is caused by the moon's gravitational influence on the orbits of ring particles. Other bright objects in this *Cassini* image are distant stars.

sunlight falling on them is reflected back into space, implying that the ring particles are blacker than coal or soot. No silicates or similar rocky materials are that dark, so those rings are likely to be composed of organic materials and ices that have been radiation darkened by high-energy, charged particles in the magnetospheres of those planets. Radiation blackens organic ices such as methane by releasing carbon from the ice molecules. Hydrogen is lost to space, while the carbon remains behind, making these particles as black as soot. Jupiter's rings are of intermediate brightness, suggesting that they may be rich in silicate materials, like the innermost of Jupiter's small moons.

The jumble of fragments that make up Saturn's rings is understood to be a product of tidal disruption of a moon or planetesimal, but moons can contribute material to rings in other ways. The brightest of Jupiter's rings is a relatively narrow strand only 6500 km across, consisting of material from the moons Metis and Adrastea. Those two moons orbit in Jupiter's equatorial plane, and the ring they form is narrow. Beyond that main ring, however, are the very different wispy rings called gossamer rings. The gossamer rings are supplied with dust by the moons Amalthea and Thebe. The innermost ring in Jupiter's system, called the halo ring, consists mostly of material from the main ring. As the dust particles in the main ring drift slowly inward toward the planet, they pick up an electric charge and are pulled into that thick torus by electromagnetic forces associated with Jupiter's powerful magnetic field.

Finally, moons may contribute ring material through volcanism. Volcanoes on Jupiter's moon Io continually eject sulfur particles into space, many of which are pushed inward by sunlight and find their way into a ring. The particles in Saturn's E Ring are ice crystals ejected from icy geysers on the moon Enceladus, in the very densest part of the E Ring (**Working It Out 11.3**). Ice particles ejected into space replace particles continually lost from Saturn's E Ring (**Figure 11.23**). The E Ring will survive for as long as Enceladus remains geologically active.

CHECK YOUR UNDERSTANDING 11.3

If rings are observed around a planet, that means: (a) a recent source of ring material is present; (b) the planet is newly formed; (c) the rings formed with the planet; (d) the rings are made of fine dust.

11.4 Ring Systems Have a Complex Structure

Huygens, with his mid-17th-century understanding of physics, thought that Saturn's ring was a solid disk surrounding the planet. Not until the mid-19th century did the brilliant Scottish mathematical physicist James Clerk Maxwell show that solid rings would be unstable and would quickly break apart. In this section, we examine the details of the rings of each outer planet.

Saturn's Magnificent Rings—A Closer Look

Saturn is adorned by a magnificent and complex system of rings, unmatched by any other planet in the Solar System. Figure 11.21 shows the components of Saturn's ring system and its major divisions and gaps. Among the four giant planets, Saturn's rings are the widest and brightest. The outermost bright ring, the A Ring, is the narrowest of the three bright rings. It has a sharp outer edge and contains several narrow gaps.

In 1675, the Italian-French astronomer Jean-Dominique Cassini (1625–1712) found a gap in the planet's seemingly solid ring. Saturn appeared to have two rings rather than one, and the gap that separated them became known as

working it out 11.3

Feeding the Rings

The moons of the giant planets have a low surface gravity and a much lower escape velocity than the 11.2 km/s of Earth. Thus, volcanic emissions from some of those small moons can escape and supply material to a ring. Recall the equation from Working It Out 4.2 for the escape velocity from a spherical object of mass M and radius R:

$$v_{esc} = \sqrt{\frac{2GM}{R}}$$

Saturn's moon Enceladus has a mass of 1.08×10^{20} kg and a radius of 250 km. The escape velocity from Enceladus is given by

$$v_{esc} = \sqrt{\frac{2 \times [6.67 \times 10^{-20}\,\text{km}^3/(\text{kg s}^2)] \times (1.08 \times 10^{20}\,\text{kg})}{250\,\text{km}}}$$

$$v_{esc} = 0.24\,\frac{\text{km}}{\text{s}};\ \text{or multiply by } 3600\frac{\text{s}}{\text{h}} \text{ to get } 864\,\frac{\text{km}}{\text{h}}$$

That escape velocity is much lower than the speed of the volcanic plumes on Enceladus, which is nearly 2200 km/h. The icy particles from the plumes supply particles to Saturn's E Ring.

the Cassini Division. The Cassini Division is so wide (4700 km) that the planet Mercury would almost fit within it. Astronomers once thought that the gap was empty, but images taken by *Voyager 1* show that the Cassini Division is filled with material, albeit less dense than the material in the bright rings.

The B Ring, whose width is roughly twice Earth's diameter, is the brightest of Saturn's rings and has no internal gaps on the scale of those seen in the other bright rings. The C Ring is so much fainter than neighboring rings that it often fails to show up in normally exposed photographs. Through the eyepiece of a telescope, that ring appears like delicate gauze. No known gap exists between the C Ring and either adjacent ring; only an abrupt change in brightness marks the boundary between them. The cause of that sharp change in the amount of ring material remains unknown. Too dim to be seen next to Saturn's bright disk, the D Ring is a fourth wide ring that was unknown until it was imaged by *Voyager 1*. The D Ring, which is the closest to the planet, shows less structure than any of the bright rings, and it does not appear to have a definable inner edge. The D Ring may extend all the way down to the top of Saturn's atmosphere, where its ring particles would burn up as meteors.

Saturn's bright rings are not uniform. The A and C rings contain hundreds, and the B Ring, thousands, of individual **ringlets**, some only a few kilometers wide (**Figure 11.24a**). Each ringlet is a narrowly confined concentration of ring

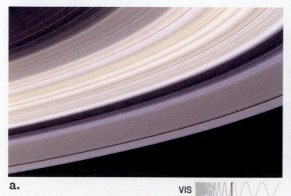

a.   VIS

b.   VIS

Figure 11.24 a. *Cassini* image of the rings of Saturn showing many ringlets and minigaps. The cause of most of that structure has yet to be explained in detail. **b.** The Hubble Space Telescope took these images of Saturn over 6 years, showing how the rings appear to change their shape as Saturn moves through its orbit around the Sun. That effect is caused by changes in the position of Earth relative to the rings.

Figure 11.25 a. Backlighting of hair creates a halo effect. **b.** This backlit image of Saturn shows the G Ring.

a.

b.

particles bounded on both sides by regions of relatively little material. About every 15 years, the plane of Saturn's rings lines up with Earth, and we view them edge on (**Figure 11.24b**). The rings are so thin that they almost vanish for a day or so in even the largest telescopes. While the glare of the rings is absent, astronomers search for undiscovered moons or other faint objects close to Saturn. In 1966, an astronomer looking for moons found weak but compelling evidence for a faint ring near the orbit of Saturn's moon Enceladus. In 1980, *Voyager 1* confirmed the existence of that faint ring, now called the E Ring, and found another, closer one known as the G Ring.

The E and G rings are examples of diffuse rings. In a diffuse ring, particles are far apart, and rare collisions between them can cause their individual orbits to become eccentric, inclined, or both. Because collisions are rare, the particles tend to remain in those disturbed orbits. Diffuse rings spread out horizontally and thicken vertically, sometimes without any obvious boundaries.

Diffuse rings contain tiny particles that show up best when the viewer is looking at those rings *into* the light—that is, in the direction of the Sun. In contrast, larger objects such as pebbles and boulders are easiest to see when the light illuminating them is coming from behind the viewer. For example, dust particles on your windshield appear brightest when you are driving toward the Sun. Photographers call that effect **backlighting** and often place their subjects in front of a bright light to highlight hair (**Figure 11.25a**). Backlighting happens when light falls on very small objects—those with dimensions a few times to several dozen times the wavelength of light. Cat hair and human hair are near the upper end of that range. Light falling on strands of hair tends to continue in the direction away from the source of illumination. Very little light is scattered off to the side, and almost none is scattered back toward the source.

Some of the dustier planetary rings are filled with particles just a few times larger than the wavelength of visible light. To a spacecraft approaching from the direction of the Sun, such rings may be difficult or even impossible to see. Those tiny ring particles scatter very little sunlight back toward the Sun and the approaching spacecraft. However, when the spacecraft passes by the planet and looks backward toward the Sun, those same dusty rings suddenly appear as a circular blaze of light, much like a halo surrounding the nighttime hemisphere of the planet. Many planetary rings are best seen with backlighting, and some, such as Saturn's G Ring (**Figure 11.25b**), have been observed only under those conditions. In 2009, astronomers using the infrared Spitzer Space Telescope discovered another diffuse ring around Saturn (**Figure 11.26**).

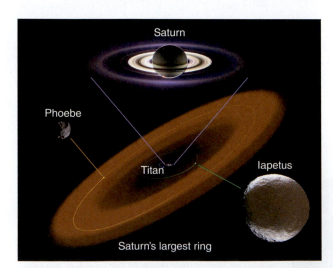

Figure 11.26 Artist's conception showing the highly inclined giant dust ring recently discovered around Saturn. That ring is so large that the rest of Saturn appears as a speck in the center (magnified in the inset).

That dusty ring is thicker than other rings, about 20 times larger than Saturn from top to bottom, and is tilted 27° with respect to the plane of the rest of the rings.

Although Saturn's bright rings are very wide—more than 62,000 km from the inner edge of the C Ring to the outer edge of the A Ring—they are extremely thin. Saturn's bright rings are no more than 100 meters thick and probably only a few tens of meters from their lower to upper surfaces. The diameter of Saturn's bright ring system is 10 million times the thickness of the rings. If the bright rings of Saturn were the thickness of a page in a book, their diameter would equal six football fields laid end to end.

Voyager 1 images showed that Saturn's F Ring is separated into several strands that appear to be intertwined as well as displaying what appear to be several knots and kinks. Saturn's F Ring is now understood to be a dramatic example of the action of a pair of shepherd moons. The F Ring is flanked by Prometheus, orbiting 860 km inside the ring, and Pandora, orbiting 1490 km on the outside, as seen in a more recent *Cassini* image (**Figure 11.27**). Both moons are irregular in shape, with average diameters of 85 and 80 km, respectively. Because of their relatively large size and their proximity, the moons exert significant gravitational forces on nearby ring particles. The resulting tug-of-war between Prometheus pulling nearby ring particles into larger orbits and Pandora drawing its neighboring particles into smaller orbits causes the bizarre structure in the F Ring (see the **Process of Science Figure**).

The F Ring is not an isolated case. The 360-km-wide Encke Gap in the outer part of Saturn's A Ring contains two narrow rings that show bright knots and dark gaps—a structure that must be related to a 20-km moon named Pan that orbits within the gap. Small moons orbiting within ring gaps can also disturb ring particles along the edges of the gaps. **Figure 11.28** shows the scalloped pattern caused by Pan that is found along the inner edge of the Encke Gap. Similarly, the 7-km moon Daphnis disrupts the inner and outer edges of Saturn's 35-km Keeler Gap, located near the outer edge of the A Ring.

Voyager 1 and then *Cassini* observed dozens of dark features in the outer part of Saturn's B Ring. Those temporary features, known as **spokes**, grow in a radial direction and are seasonal, lasting for less than half an orbit around Saturn, indicating that the particles in the spokes must be suspended above the ring plane, probably by electrostatic forces. One explanation is that when the charged particles interact with Saturn's magnetic field, the spokes rotate as the planet spins.

Rings around the Other Outer Planets

Ring structure among the other giant planets is not as diverse as Saturn's. Most rings other than Saturn's are narrow, although a few are diffuse. When *Voyager 1* scientists looked at Jupiter's ring system with the Sun behind the camera, they saw only a narrow, faint strand. But when *Voyager 2* looked back toward the Sun while in the shadow of the planet, Jupiter's rings suddenly blazed into prominence. **Figure 11.29** shows Jupiter's moons orbiting among that ring system. Most of the material in Jupiter's rings is made up of fine dust dislodged by meteoritic impacts on the surfaces of Jupiter's small inner moons.

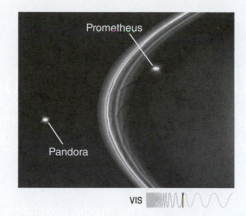

VIS

Figure 11.27 *Cassini* image showing Saturn's F Ring and its shepherd moons, Pandora and Prometheus. Also visible are some "kinks" in the inner ring.

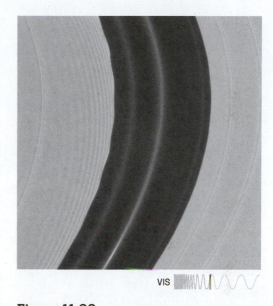

VIS

Figure 11.28 In this high-resolution view from *Cassini*, Saturn's Encke Gap reveals a scalloped pattern, caused by the moon Pan, along its inner edge.

Figure 11.29 A diagram of the Jupiter ring system and the small moons that form the rings.

Following Up on the Unexpected

Scientists expected to find dust particles in the rings of Saturn moving on undisturbed orbits. Instead, the F Ring particles seemed to disobey the laws of physics!

The media proclaimed:

" The laws of physics are wrong! "

Observers found more examples of deformed rings and the "shepherd moons" that cause them.

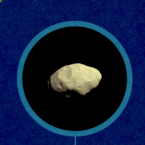

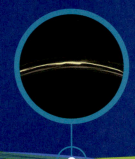

The *Voyager 1* spacecraft discovered multiple intertwining strands, knots, and kinks in Saturn's F Ring.

Further observations revealed previously unobserved moons that produced the unexpected behavior of F Ring particles.

The theory that rings can be distorted by the gravitational influence of nearby moons became widely accepted because it repeatedly passed tests of observation and simulation.

The scientists proclaimed:

" We must not have accounted for everything. "

Theorists verified gravitational effects with simulations.

Scientists are excited by apparent violations of well-supported theories because they may lead to new discoveries. One consequence is that unexpected or contradictory results often receive more attention than confirming results.

Of the 13 rings of Uranus, nine are very narrow and widely spaced relative to their widths (see Figure 11.21). Most are only a few kilometers wide, but they are many hundreds of kilometers apart. The two rings discovered by the Hubble Space Telescope in 2005 (**Figure 11.30**) are much wider and more distant than the narrow rings. The most prominent ring of Uranus, the Epsilon Ring, is eccentric and the widest of the planet's inner narrow rings, ranging between 20 and 100 km. The innermost ring is wide and diffuse, with an undefined inner edge. As with Saturn's D Ring, material in that ring may be spiraling into the top of the planetary atmosphere. When viewed under backlit conditions by *Voyager 2*, the space *between* the rings of Uranus turned out to be filled with dust, much as in Jupiter's ring system.

The rings in Saturn's Encke Gap are unusual but not unique. If shepherd moons are in eccentric or inclined orbits, they cause the confined ring to be eccentric or inclined, too. That is the case for the Epsilon Ring of Uranus. Because shepherd moons can be so small, they often escape detection. According to current theories of ring dynamics, several still-unknown shepherd moons must be interspersed among the ring systems of the outer Solar System.

For a while, Neptune seemed to be the only giant planet devoid of rings. Then, in the early to mid-1980s, occultation searches by teams of astronomers began yielding confusing results. Several occultation events that appeared to be due to rings were seen on only one side of the planet. The astronomers concluded that Neptune was surrounded not by complete rings but rather by several arclike ring segments. When *Voyager 2* reached Neptune in 1989, it was determined that Neptune's rings were complete. The **ring arcs** are high-density segments within one of its narrow rings. All of Neptune's rings are faint and, except for the ring arcs, contain too little material to be detected through the stellar occultation technique.

Four of Neptune's six rings are very narrow, similar to the 13 narrow rings surrounding Uranus. The other two are a few thousand kilometers wide (see Figure 11.21). Neptune's rings are named for 19th century astronomers who made major contributions to Neptune's discovery. Of those, the Adams Ring attracts the greatest attention. Much of the material in the Adams Ring is clumped into several ring arcs. Those high-density ring segments extend 4000–10,000 km yet are only about 15 km wide. When first discovered, the ring arcs were a puzzle because mutual collisions among their particles should cause the particles to be spread more or less uniformly around their orbits. Most astronomers now attribute that clumping to orbital resonances with the moon Galatea, which orbits just inside the Adams Ring (**Figure 11.31**). Images obtained by the Hubble Space Telescope in 2004 and 2005 compared with those taken by *Voyager 2* in 1989 show that some parts of the Neptune ring arcs are unstable. Slow decay is evident in two of the arcs, Liberté and Courage, suggesting that they may disappear before the end of the century. Uranus's Lambda Ring and Saturn's G Ring also show ring arcs.

Moons and Rings around Exoplanets

Large and small planets, many in multiplanet systems, have been detected within our galaxy, demonstrating that exoplanets are common. If our Solar System is typical, we might expect that other planetary systems contain planets with rings and planets with large moons, perhaps with geological activity and water. However, identifying exoplanet moons and rings is at the limit of current astronomical

VIS

Figure 11.30 The appearance of rings depends dramatically on lighting conditions and the angle from which they are seen. Earth's view of the rings of Uranus changes over several years, as shown.

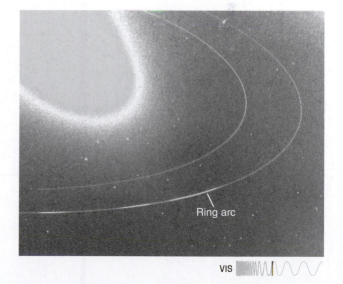

VIS

Figure 11.31 *Voyager 2* image showing the three brightest arcs in Neptune's Adams Ring. Neptune itself is greatly overexposed.

Figure 11.32 Artist's conception showing a hypothetical Earth-like moon in the foreground, orbiting around a Saturn-like exoplanet.

instrumentation, and while there are about a dozen candidate exoplanet moons, none have yet been confirmed.

The proposed methods for detecting *exomoons* are similar to those for detecting exoplanets. Recall that *Kepler* detects planets through the transit method (see Figures 7.20 and 7.21). The presence of a moon near the planet can slightly alter the depth and duration of the change in the light curve. The moon is probably in a different place in its orbit each time the planet orbits the star, so those alterations in the light curve will be different for each cycle. Over several cycles, the signature of a moon might be detected. It is also possible that astronomers could detect a large exomoon as the moon itself transits its star. Or a large moon could cause its planet's orbit to "wobble," which could be detected in the transit signal. NASA supercomputers are being used to analyze the large database from *Kepler* to look for signatures of exomoons.

Similarly, a large exoplanet with an extensive ring system, especially if the ring system has gaps, might be detectable through changes in the transit signal. The depth and length of the changes in the star's light curve could indicate the presence of such a system. **Figure 11.32** is an artist's conception of an exoplanet with rings and a large exomoon.

CHECK YOUR UNDERSTANDING **11.4**

If you wanted to search for faint rings around a giant planet by sending a spacecraft on a flyby, you could make your best observations: (a) as the spacecraft approached the planet; (b) after the spacecraft passed the planet; (c) while the spacecraft is orbiting the planet; (d) during the closest flyby; (e) while the spacecraft is orbiting one of its moons.

Origins: Extreme Environments

During the 1980s and 1990s, as the *Voyager* spacecraft were exploring the outer Solar System, biologists back on Earth were identifying strange forms of life. Off the coast of the Galápagos Islands, 2500 meters beneath the ocean's surface, plates grind against each other, creating friction, high temperatures, and seafloor volcanism. Mineral-rich, superheated water pours out of hydrothermal vents. The surrounding water contains very little dissolved oxygen. No sunlight reaches those depths, yet in the total darkness of the ocean bottom, life abounds. From tiny bacteria to shrimp to giant clams and tube worms, sea life thrives in that severe environment. Being without sunlight, the small, single-celled organisms at the bottom of the local food chain get their energy from *chemosynthesis*, a process by which inorganic materials are converted into food by using chemical energy. Biologists call such life-forms *extremophiles*.

Similarly, robust types of bacteria are found flourishing in the scalding waters of Yellowstone's hot springs; in the bone-dry oxidizing environment of Chile's Atacama Desert; and in the Dead Sea, where salt concentrations run as high as 33 percent. Bacteria have even been found in core samples of ancient ice 3600 meters below the East Antarctic ice sheet. When it comes to harsh habitats, life is amazingly adaptable. If life can exist under such extreme conditions on Earth, might it also exist on those moons of the giant planets that have the ingredients necessary for life on Earth: liquid water, an energy source, and the presence of organic compounds?

When scientists realized that Mars and Venus were not as Earth-like as once imagined, prospects for finding life elsewhere in the Solar System seemed dim. Now astrobiologists are turning their attention to some of the small worlds around the giant planets far from the Sun. Those moons may supply clues about the history of life in the Solar System. Their environments may be similar to some of the ecological niches on Earth that support extremophiles. The conditions necessary to create and support life on Earth—liquid water, heat, and organic material—could all be present in oceans on some of those moons.

Enceladus, Saturn's geologically active ice moon, spews salty water-ice grains, indicating liquid water below the surface. Perhaps its south polar region is habitable. The fractured ice floes on Jupiter's Europa may cover an ocean warmed and enriched by geothermal vents similar to those that dot the floors of Earth's oceans. Like Europa, Callisto also shows magnetic variability, possibly indicating a salty ocean. Saturn's Titan has an atmosphere, liquid methane lakes, and a possible subsurface ocean. Titan is similar in several ways to a much earlier Earth.

The presence of comet-borne organic material in large bodies of water on Europa, Callisto, Enceladus, or Titan cannot yet be confirmed. Methane can arise from biological processes or can come from chemical or geochemical processes, so its detection on Enceladus and Titan is tantalizing but is not definitive evidence of life. In addition to the presence of methane, spectroscopy reveals organic gases in Titan's massive atmosphere. Titan's nitrogen atmosphere contains compounds of biological interest. For example, five molecules of hydrogen cyanide (HCN) will spontaneously combine to form adenine, one of the four primary components of DNA and RNA. HCN is a building block of amino acids, too, which combine, in turn, to form proteins. Photodissociation and **recombination** of those various gases produce complex organic molecules that then rain out onto Titan's surface as a frozen tarry sludge. Biochemists think that many of those substances are biological precursors, similar to the organic molecules that preceded the development of life on Earth.

Astronomers anticipate future exploration of those moons, which may yield fascinating clues to the origins of terrestrial life.

SUMMARY

The moons of the outer Solar System are composed of rock and ice. A few moons are geologically active, but most are geologically dead. All four giant planets have ring systems, which are temporary, created from and maintained by moons also in orbit around those planets. Scientists are excited about the evidence for oceans on several of the moons and the possibility that some form of life may exist in those oceans.

(1) **Compare the orbits and formation of regular and irregular moons.** Most of the regular moons were formed along with their parent planets and have short and nearly circular orbits. Irregular moons were captured later, have more elongated orbits, and often orbit in the opposite direction of the planet's rotation. Observations of the orbits of moons can be used to find the masses of their host planets.

(2) **Describe the evidence for geological activity and liquid oceans on some of the moons, and evaluate this evidence in the context of the search for life in extreme environments.** Observations indicate that Jupiter's Io is the most volcanically active body in the Solar System. Jupiter's moon Europa contains an enormous subsurface ocean and probably has some geological activity; these characteristics make Europa a promising candidate in the search for life elsewhere in the Solar System. Saturn's moon Titan has lakes of liquid methane and perhaps a deep, salty ocean. Saturn's moon Enceladus and Neptune's moon Triton have cryovolcanoes. The large Galilean moons of Jupiter—Ganymede and Callisto—also may have subsurface oceans. Some moons were geologically active in the past, as indicated by crater scars and areas smoothed by flowing fluids. Moons that have always been geologically dead show nothing but impact craters on their surfaces.

③ **Describe the composition, origin, and general structure of the rings of the giant planets.** Rings are formed of countless numbers of particles all in the same plane, held to the host planet by gravity. Some rings form when moons cross a planet's Roche limit. The composition of those moons determines the composition of the rings that form from them. Shepherd moons often maintain and shape rings by gravitationally pulling and pushing those particles as they pass by. Ring particles also interact gravitationally with one another. Some rings may be temporary features held in place by moons. Saturn's bright rings and its

E Ring are made primarily of water ice: the rings of the other planets are composed of darker materials.

④ **Explain the role gravity plays in the structure of the rings and the behavior of ring particles.** Saturn's ring system is the most complex, and it is the best laboratory for understanding gravitational interactions between moons and rings, interactions between rings, and ring formation and dissipation. Gravity holds the ring particles in orbit around the planet, and gravitational interactions with moons determine the size and shape of rings.

QUESTIONS AND PROBLEMS

TEST YOUR UNDERSTANDING

1. Categorizing moons by geological activity is helpful because
   a. comparing them reveals underlying physical processes.
   b. geological activity levels decrease as the distance from the Sun increases.
   c. the size and composition of the moons depend on geological activity.
   d. most moons are very similar to one another.

2. Why are Ganymede and Callisto geologically dead, whereas the other two Galilean moons of Jupiter are geologically active?
   a. Ganymede and Callisto are larger.
   b. Ganymede and Callisto are farther from Jupiter.
   c. Ganymede and Callisto are more massive.
   d. Ganymede and Callisto have retrograde orbits.

3. Moons of outer planets may serve as a home for life because
   a. some have liquid water.
   b. some have organic molecules.
   c. some have an interior source of energy.
   d. all of the above are correct.

4. Io has the most volcanic activity in the Solar System because
   a. it is continually being bombarded with material in Saturn's E Ring.
   b. it is one of the largest moons and its interior is heated by radioactive decays.
   c. of gravitational friction caused by the moon Enceladus.
   d. its interior is tidally heated as it orbits around Jupiter.
   e. the ice on the surface creates a large pressure on the water below.

5. Gravitational interactions with moons produce
   a. fine structure within rings.
   b. short-lived rings.
   c. smoothed-out rings.
   d. rings with spokes.

6. Saturn's bright rings are located within the planet's Roche limit, supporting the theory that those rings (select all that apply)
   a. formed of moons torn apart by tidal stresses.
   b. formed when Saturn formed.
   c. are relatively recent.
   d. are temporary.

7. The story of the F Ring of Saturn is an example of
   a. an unexplained phenomenon.
   b. media bias.
   c. the self-correcting nature of science.
   d. a violation of causality.

8. If a moon revolves opposite to its planet's rotation, it probably
   a. was captured after the planet formed.
   b. had its orbit altered by a collision.
   c. has a different composition from that of other moons.
   d. formed very recently in the Solar System's history.

9. Under what lighting conditions are the tiny dust particles found in some planetary rings best observed?
   a. viewed from the shadowed side of the planet, looking toward deep space
   b. viewed from the shadowed side of the planet, looking toward the Sun
   c. viewed from the near side of the planet, looking toward deep space
   d. viewed from the near side of the planet, looking toward the Sun

10. Planets in the outer Solar System have more moons than those in the inner Solar System because
    a. the solar wind was weaker there.
    b. more debris was present around the outer planets when they were forming.
    c. the outer planets captured most of their moons.
    d. more planetesimal collisions occurred far from the Sun.

11. The difference between a moon and a planet is that
    a. moons orbit planets, whereas planets orbit stars.
    b. moons are smaller.
    c. moons and planets have different compositions.
    d. moons and planets formed differently.

12. Scientists determine the geological history of the moons of the outer planets from
    a. seismic probing.
    b. radioactive dating.
    c. surface features.
    d. time-lapse photography.

13. The energy that keeps Io's core molten comes from
  a. the Sun.
  b. radioactivity in the core.
  c. residual heat from the collapse.
  d. Jupiter's gravity.

14. We classify moons as formerly active if they
  a. are covered in craters.
  b. have no young craters.
  c. have regions with few craters.
  d. have regions with no craters.

15. Volcanoes on Enceladus affect the E Ring of Saturn by
  a. pushing the ring around.
  b. stirring the ring particles.
  c. supplying ring particles.
  d. dissipating the ring.

THINKING ABOUT THE CONCEPTS

16. ★ **WHAT AN ASTRONOMER SEES** An astronomer studying Figure 11.5 would determine that Io is remarkably volcanically active. List at least three pieces of evidence for that conclusion, and explain the process that drives volcanism on Jupiter's moon Io. 👁

17. Describe cryovolcanism and explain its similarities and differences with respect to terrestrial volcanism. Which moons show evidence of cryovolcanism?

18. Discuss evidence supporting the idea that Europa has a subsurface ocean of liquid water.

19. Titan contains abundant amounts of methane. What process destroys methane in Titan's atmosphere?

20. In certain ways, Titan resembles a frigid version of the early Earth. Explain the similarities.

21. Some moons display signs of geological activity in the past. Identify some of the evidence for past activity.

22. Why do the outer planets but not the inner planets have rings? Describe a ground-based technique that led to the discovery of rings around the outer planets.

23. What are ring arcs, and where are they found?

24. Identify and explain two possible mechanisms that can produce planetary ring material.

25. Explain two mechanisms that create gaps in Saturn's bright-ring system.

26. Describe ways in which diffuse rings differ from other planetary rings.

27. In Chapter 1, we stated that "all scientific knowledge is provisional." Explain how the discovery of the detailed structure of Saturn's F Ring (as described in the Process of Science Figure) challenged a scientific theory and how that apparent conflict was ultimately resolved. 👁

28. Astronomers think that most planetary rings eventually dissipate. Explain why the rings do not last forever. Describe and explain a mechanism that keeps planetary rings from dissipating.

29. Name one ring that might continue to exist indefinitely, and explain why it could survive when others might not.

30. Make a case for sending a space mission to one of the moons. Which moon would you choose, and what observations would you try to obtain?

APPLYING THE CONCEPTS

31. Use the value of P^2/A^3 for Europa, as in Working It Out 11.1, to estimate the mass of Jupiter. 🟢–🟢–🟢
  a. Make a prediction: How do you expect your answer to relate to the mass of Jupiter found from another moon's orbit in Working It Out 11.1?
  b. Calculate: Practice applying Newton's version of Kepler's law to find the mass of Jupiter. (You will likely do this again in other instances where the answer is not known.)
  c. Check your work: Verify that your answer has units of kg, and compare it with your prediction.

32. Use Working It Out 11.2 to compare the tidal force between Jupiter and Io with the tidal force between Earth and the Moon. 🟢–🟢–🟢
  a. Make a prediction: Do you expect the tidal force between Jupiter and Io to be greater or less than the tidal force between Earth and the Moon?
  b. Calculate: Follow Working It Out 11.2 to compare the tidal forces.
  c. Check your work: Verify that your answer has no units, and compare your answer to your prediction.

33. Io has a mass of 8.9×10^{22} kg and a radius of 1820 km. What is Io's escape velocity? 🟢–🟢–🟢
  a. Make a prediction: Does material from the volcanoes escape from Io? Therefore, how will Io's escape velocity compare with the vent velocities of 1 km/s from its volcanoes?
  b. Calculate: Using the formula provided in Working It Out 11.3, calculate Io's escape velocity.
  c. Check your work: Verify that your answer has units of km/s, and compare your answer to your prediction.

34. A 60-kg, 1.9-m tall astronaut is spacewalking outside the International Space Station, 380 km above Earth. Use Working It Out 11.2 (and Working It Out 4.4) to find the ratio of the tidal forces on the part of the astronaut closest to Earth versus the part of the astronaut farthest from Earth. Assume that her feet point toward the center of Earth.
  a. Make a prediction: Do you expect the tidal force between Earth and the astronaut to be greater or less than the gravitational force between Earth and the astronaut? The gravitational force on the astronaut in orbit will be less than on the ground, which could be estimated at about 600 N.
  b. Calculate: Follow Working It Out 11.2 to calculate the ratio of the tidal forces on the astronaut.
  c. Check your work: Verify that your answer is dimensionless (all the units have canceled out), and compare your answer to your prediction.

35. Find the escape velocity from Saturn's moon Mimas.

  a. Make a prediction: What units do you expect to find for escape velocity? See the relative size of Mimas in Figure 11.1; you can find its mass and radius in Appendix 4. Do you expect the escape velocity of Mimas to be larger or smaller than Earth's escape velocity of 11,000 m/s? ⬤

  b. Calculate: Follow Working It Out 11.3 to find the escape velocity.

  c. Check your work: Verify that your answer has units of m/s, and compare your answer to your prediction.

36. Use one of Saturn's moons and the calculation in Working It Out 11.1 to estimate the mass of Saturn.

37. Study Figure 11.2. ⬤

  a. Are the scales on Figure 11.2a and 11.2b linear or logarithmic?

  b. About how much larger is the space shown in Figure 11.2b than in Figure 11.2a?

38. Planetary scientists have estimated that extensive volcanism could be covering Io's surface with lava and ash to an average depth of up to 3 millimeters (mm) per year.

  a. If Io is a sphere with a radius of 1820 km, what are its surface area and volume?

  b. What is the volume of volcanic material deposited on Io's surface each year?

  c. How many years would it take for volcanism to perform the equivalent of depositing Io's entire volume on its surface?

  d. How many times might Io have "turned inside-out" over the age of the Solar System?

39. Use the formula for tidal forces in Working It Out 11.2 to answer the following questions:

  a. If the radius of a moon increases but its mass stays the same, does the tidal force increase, decrease, or stay the same?

  b. If the radius of a moon's orbit decreases, does the tidal force increase, decrease, or stay the same?

  c. If the mass of a central planet increases, does the tidal force increase, decrease, or stay the same?

40. Assuming that all other numbers are held constant, make a graph of the tidal force versus the distance between a planet and its moon. On the same graph, plot the gravitational force (which falls off as $1/d^2$). Compare the two graphs to determine the relative importance of tidal forces and gravitational forces at various distances.

41. Particles at the very outer edge of Saturn's A Ring are in a 7:6 orbital resonance with the moon Janus. If the orbital period of Janus is 16 hours 41 minutes (16^h41^m), what is the orbital period of the outer edge of Ring A?

42. The inner and outer diameters of Saturn's B Ring are 184,000 and 235,000 km, respectively. If the average thickness of the ring is 10 meters and the average density is 150 kilograms per cubic meter (kg/m^3), what is the mass of Saturn's B Ring?

43. The mass of Saturn's small, icy moon Mimas is 3.8×10^{19} kg. How does that mass compare with the mass of Saturn's B Ring, as calculated in Question 42? Why is that comparison meaningful?

44. The inner and outer diameters of Saturn's B Ring are 184,000 and 235,000 km, respectively. Use that information to find the ratio of the periods of particles at those two diameters. Does the B Ring orbit like a solid disk or like a collection of separate particles?

45. Use the escape velocity equation in Working It Out 11.3 to answer the following questions:

  a. For more massive planets, is the escape velocity higher or lower?

  b. For larger planets, is the escape velocity higher or lower?

  c. If you know the escape velocity of a planet, what other piece of information do you need to find the planet's mass?

EXPLORATION Measuring Features on Io

Part A: Finding the Image Scale

Finding the scale of an image is like finding the scale on a map. On a map, each inch or centimeter represents miles or kilometers of actual space. The same thing is true in an image. If you take a picture of a meterstick and then measure the meterstick in the picture to be 10 cm long, you know that 10 cm in the picture represents 1 meter of actual space, and 1 cm in the picture represents 10 cm of actual space.

To find the scale, you must compare the size of something in the image with its actual size in space. In **Figure 11.33**, the moon Io is the known object.

1 Use a ruler to measure the diameter of that image of Io in millimeters.

_____

2 Estimate the error in your measurement. (How far off could your measurement be?)

_____

3 Find the radius from that diameter.

_____

4 Look up the actual radius of Io (in kilometers) in Appendix 4 or online.

_____

5 Find the image scale (*s*) as follows:

$$s = \frac{\text{Actual size of object}}{\text{Size of object on the image}}$$

6 What are the units of that image scale?

_____

Part B: Finding the Sizes of Features on Io

7 Near the center of Io is a geyser, surrounded by a black circle and a white ring. That is Prometheus Patera. What is the diameter of the black ring around Prometheus in that image (in millimeters)? (Do not use the inset image.)

_____

8 Multiply the measured diameter by the image scale to find the actual size of the circle around Prometheus. Identify something on Earth that is about the same size.

_____

9 On the edge of the image of the moon is a purple plume from an erupting sulfur geyser. Follow steps 7 and 8 to find the height of that plume. Identify something on Earth that is about the same height.

_____

That measurement has at least two sources of error. One is the error of measurement—how well you use a ruler and how accurately you determined the top and the bottom of the plume. The other source of error is uncertainty about whether the plume is *exactly* at the edge of the image of the moon. If the eruption occurred on the far side of Io, would your calculation be an overestimate or underestimate of the plume height?

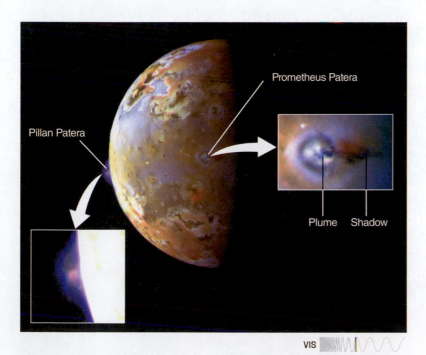

Pillan Patera

Prometheus Patera

Plume   Shadow

VIS

Figure 11.33 A *Galileo* image of Jupiter's moon Io, a very active moon. This image, which appeared previously in the chapter as Figure 11.5b, offers an opportunity to make some measurements.

Dwarf Planets and Small Solar System Bodies

Meteors occur when small pieces of space dust collide with Earth's atmosphere. Several times a year, many meteors occur at once, as you will find out here in Chapter 12. On these nights, the meteors appear to all come from the same location. Use the Internet to find out the date of the next meteor shower. Prepare an observation log page and take it with you when you go outside at night on that date. Sketch objects on the horizon onto the observation log, so that you fix different directions in your mind. If you can see the constellation for which the shower is named, sketch that, too! Spend a few hours watching for meteors. (Be sure to dress warmly!) Each time you see one, sketch it onto your observation log. Once you have observed about 10 meteors, trace their paths back with dotted lines, to see that they all intersect.

EXPERIMENT SETUP

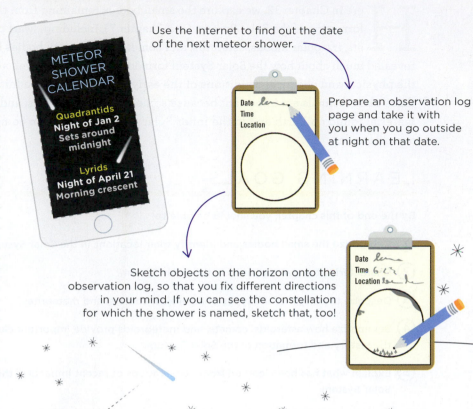

Use the Internet to find out the date of the next meteor shower.

METEOR SHOWER CALENDAR

Quadrantids
Night of Jan 2
Sets around midnight

Lyrids
Night of April 21
Morning crescent

Date
Time
Location

Prepare an observation log page and take it with you when you go outside at night on that date.

Date
Time
Location

Sketch objects on the horizon onto the observation log, so that you fix different directions in your mind. If you can see the constellation for which the shower is named, sketch that, too!

Spend a few hours watching for meteors. (Be sure to dress warmly!) Each time you see one, sketch it onto your observation log. Once you have observed about 10 meteors, trace their paths back with dotted lines, to see that they all intersect.

SKETCH OF RESULTS (in progress)

Observation Log

Date _____ Time _____
Object _____
Location _____
Sky conditions _____
Comments
_____
_____
_____
_____
_____
_____

12

Here in Chapter 12, we explore the small bodies remaining from the formation of the Solar System (see Chapter 7), including dwarf planets, irregular moons, asteroids, and comets. Those small bodies have revealed much about how the Solar System formed and evolved, as well as about the physical and chemical conditions of the early Solar System. In addition, these planetesimals are important because some of the water, gases, and organic material found on Earth and in the inner Solar System came from comets and asteroids.

LEARNING GOALS

By the end of this chapter, you should be able to:

(1) Categorize the small bodies and identify their locations in the Solar System.

(2) Characterize the dwarf planets in the Solar System.

(3) Describe the origin of the types of asteroids, comets, and meteorites.

(4) Summarize how asteroids, comets, and meteoroids provide important clues about the history and formation of the Solar System.

(5) Explain what has been learned from observations of recent impacts in the Solar System.

12.1 Dwarf Planets May Outnumber Planets

Recall from Chapter 7 that very early in the history of the Solar System—when the Sun was becoming a star—tiny grains of primitive material stuck together to produce swarms of small bodies called *planetesimals*. Those that formed in the hotter, inner part of the Solar System were composed mostly of rock and metal, whereas those in the colder, outer part were composed of ice, organic compounds, and rock. Some of the objects collided to become planets and moons. Many are still present, however, orbiting the Sun. They remain a scientifically important component of the present-day Solar System.

The **dwarf planets** orbit the Sun and have round shapes, but because they have relatively small mass, they have not cleared other small bodies from the area around their orbits. As of this writing, the Solar System has five officially recognized dwarf planets: Pluto, Eris, Haumea, Makemake, and Ceres (their properties are tabulated in Appendix 4). Ceres is a large object in the **main asteroid belt**—the region between the orbits of Mars and Jupiter that contains most of the asteroids in the Solar System. The other dwarf planets are found in the **Kuiper Belt**—a disk-shaped population of comet nuclei extending from Neptune's orbit (30 astronomical units) to about 50 AU, shown in **Figure 12.1**.

Ceres

In 1801, Sicilian astronomer Giuseppe Piazzi found a bright object between the orbits of Mars and Jupiter. He named the new object Ceres. Piazzi thought he might have found a hypothetical "missing planet," but as more objects were

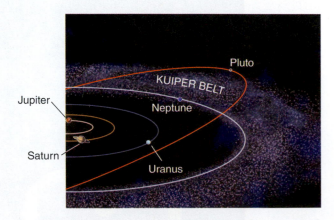

Figure 12.1 This sketch of the outer Solar System is shown in perspective. If viewed from "above," it would appear that Pluto and Neptune could someday collide, because their orbits would appear to cross in two places. The orbits do not cross, so Pluto and Neptune can never be in the same place at the same time. The Kuiper Belt contains a large number of icy bodies, many of which are probably as large or larger than Pluto.

discovered orbiting in the region between Mars and Jupiter, astronomers classified Ceres as belonging to a new category of Solar System objects called **asteroids**. The region between Mars and Jupiter in which they are found is called the main **asteroid belt**. Ceres is the largest body in the main asteroid belt. It also is now called a dwarf planet, because although it is round (**Figure 12.2a**), it has not cleared other small objects from the area around its orbit. With a diameter of about 940 km, Ceres is larger than most moons but smaller than any planet. It contains about a third of the total mass in the asteroid belt but only about 1.3 percent of the mass of Earth's Moon. Ceres rotates on its axis with a period of about 9 hours, typical of many asteroids.

Ceres was explored by the *Dawn* mission, which first orbited the asteroid Vesta for a little over a year, then traveled to Ceres. *Dawn* went into orbit around Ceres in 2015 and remained in orbit until its fuel ran out in 2018; it is now in an uncontrolled orbit around Ceres. The spacecraft found many features on Ceres similar to those seen on the inner planets. For example, the 34-km-wide Haulani Crater (**Figure 12.2b**) has landslides on the crater's limb and blue streaks radiating outward from the central crater peak. The color in that image is enhanced, and blue indicates young surfaces on Ceres. Those color variations indicate that the surface of Ceres has a different composition from that of the subsurface layer. Ceres has a large mountain, Ahuna Mons, that is 4–5 km (13,000–16,000 feet) high. It seems to be a salty-mud cryovolcano that may have been active in the last few hundred million years. Evidence also indicates that Ceres may have had cryovolcanoes that were flattened out by gravity over hundreds of millions of years. Because those volcanoes are made of ice rather than rock, they flow downward slowly, much like a blob of honey that gradually spreads out in the bottom of a mug.

Through gravity mapping, astronomers have confirmed that Ceres is much less dense than the other rocky bodies in the Solar System. Ceres is at least partly differentiated, with a rocky core and a lower-density, 40-km-thick crust of ice and salt (**Figure 12.2c**). Water ice has been detected at about a dozen sites, including within craters. Because hydrogen has been detected on the surface, scientists think that water probably exists only 1–2 meters beneath. That water is thought to have been with Ceres since it formed instead of having come later from comets and asteroids. Minerals that contain water were seen on the surface, and the icy crust might be the remains of an ancient ocean. As Ceres moved in its orbit around the Sun, crater ice released water vapor, which condensed and caused more landslides.

Pluto

Throughout the 19th century, astronomers found discrepancies between the observed and predicted orbital positions of Uranus and Neptune. Early in the 20th century, astronomers hypothesized that an unseen body was perturbing the orbits of those planets. Astronomers called that body Planet X and estimated that it had 6 times Earth's mass and was located beyond Neptune's orbit. Astronomer Clyde W. Tombaugh (1906–1997) discovered Planet X in 1930, not far from its predicted position. It became the Solar System's ninth planet and was named Pluto for the Roman god of the underworld. However, observational evidence soon indicated that the mass of Pluto was far too small to have produced the perturbations in the orbits of Uranus and Neptune. When astronomers reanalyzed the 19th century observations, they found that the orbital "discrepancies" were a mistake. Pluto's discovery thus turned out to be a coincidence.

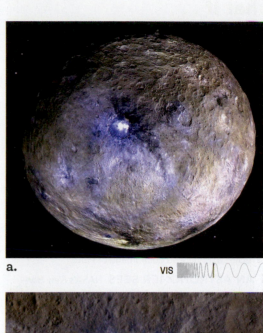

a.   VIS

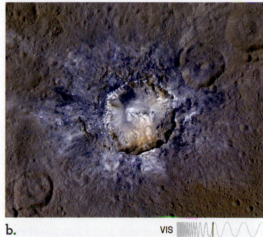

b.   VIS

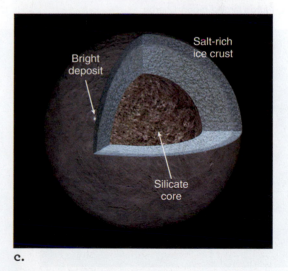

c.

Figure 12.2 **a.** Dwarf planet Ceres photographed by the *Dawn* mission. **b.** The 34-km-wide Haulani Crater shows landslides on the crater's limb, as well as smooth material and a central peak on the crater floor. Blue indicates a younger surface on Ceres. The color in this image is enhanced. **c.** Diagram of the interior of Ceres.

what if . . .

What if, in 2006, instead of demoting Pluto to a dwarf planet, the International Astronomical Union had redefined a planet to be any round astronomical object smaller than a star? What kinds of objects would then qualify as planets?

Figure 12.3 All images are from the *New Horizons* spacecraft flyby in 2015. **a.** Enhanced color image of Pluto.

★ **WHAT AN ASTRONOMER SEES** NASA's *New Horizons* visited Pluto briefly in 2015, taking images that surprised astronomers around the world. Astronomers expected to see a cratered ice world that had long been geologically dead. Instead, the pale heart-shaped region toward the lower right is very smooth, indicating to an astronomer that Pluto's core has been hot in the very recent past. The enhanced colors in this image indicate regions of different composition, further indicating that the surface has been modified over time. If Pluto had differentiated and then frozen in place, it should have the same composition over the whole surface. An astronomer would zoom in and notice ice mountains and other features that show that Pluto has had active geology. This realization about Pluto was so surprising that the *New Horizons* mission was extended and the spacecraft was repurposed to visit another distant Solar System object. **b.** Dark highlands shown in the lower right border a section of icy plains. **c.** Pluto's largest moon, Charon.

It takes 248 Earth years for Pluto to orbit the Sun just once. Pluto's orbit is so elliptical that it periodically crosses inside the nearly circular orbit of Neptune. Thus, from 1979 to 1999, Pluto was closer to the Sun than Neptune. As shown in Figure 12.1, however, Pluto's orbit is also tilted with respect to the plane of the Solar System, so the two orbits do not actually cross; Pluto and Neptune could never collide.

Pluto's diameter is only two-thirds the diameter of Earth's Moon. The largest of Pluto's five moons is Charon, about half the size of Pluto, 1/3 the size of Earth's Moon. The total mass of the Pluto-Charon system is 1/400 that of Earth, or 1/5 the mass of Earth's Moon. Similar to Uranus, Pluto rotates nearly on its side—that is, its rotation axis is nearly in its orbital plane. Pluto and Charon are a tidally locked pair: each has one hemisphere that always faces the other.

In 2005, astronomers identified an object more distant than Pluto, later named Eris, and then Eris's moon, Dysnomia. Observations of Dysnomia's orbit yielded a mass for Eris, which turned out to be about 28 percent greater than Pluto's mass. Pluto and Eris have similar nitrogen and methane abundances and each has a relatively large moon. So the inevitable question emerged: Should astronomers consider Eris the Solar System's tenth planet? Or should neither Pluto nor Eris be called a planet? The members of the International Astronomical Union (IAU) decided in August 2006: Pluto is round like the classical planets, but it cannot clear its neighborhood, so it was reclassified as a dwarf planet (see the **Process of Science Figure**).

Pluto and Charon were not on *Voyager*'s route through the Solar System, so scientists had to wait until NASA's *New Horizons* flyby spacecraft passed within 12,500 km of Pluto in July 2015 to find out about their surface properties and geological history. Astronomers were surprised to find that Pluto and Charon had geological activity. *New Horizons* images of Pluto showed varied surface features, including a large bright region whose western half is a basin containing nitrogen, methane, and carbon monoxide ices, possibly from a large impact (**Figure 12.3a**). The formation of this massive glacier reoriented the dwarf planet so that the basin now faces Charon. **Figure 12.3b** shows highlands and icy plains. Pluto's surface contains an icy mixture of frozen water, carbon dioxide, methane, and carbon monoxide, with

a.

b.

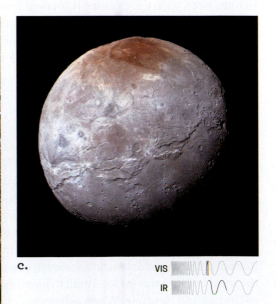

c.

VIS

IR

How to Classify Pluto

Pluto's reclassification from planet to dwarf planet in 2006 received a lot of publicity. The reclassification of Pluto is a clear example of the scientific method in practice.

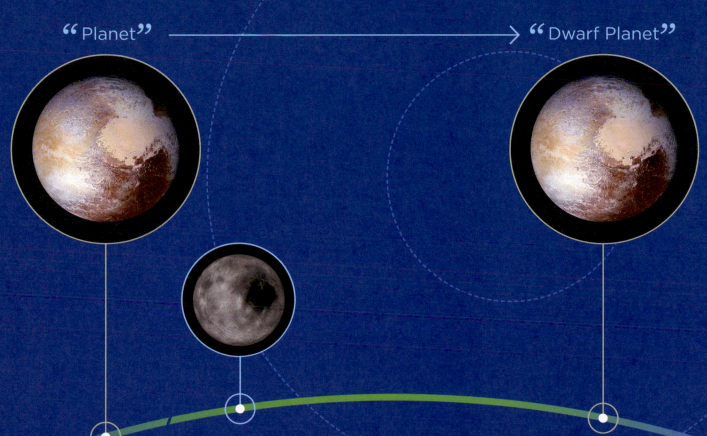

"Planet" ⟶ "Dwarf Planet"

1930
Pluto was discovered. It was classified as a planet because it orbited the Sun.

1978
Pluto's moon Charon was discovered. As information about Pluto accumulated, it became clear that Pluto was different from the other planets:

- Pluto's composition is not like the other planets.
- Pluto crosses Neptune's orbit.
- Pluto has a moon nearly as large as itself.

2006
In light of recent discoveries of new Pluto-like objects, the International Astronomical Union redefined the word *planet*. Pluto fell outside the new definition and was classified as a "dwarf planet."

Scientific decision making must follow the weight of the evidence, not historical precedent.

working it out 12.1

Eccentric Orbits

Many of the objects discussed in this chapter are far from the Sun. Their complete orbits take many years, but observing a complete orbit is not necessary to determine an object's semimajor axis and eccentricity: those values can be obtained from watching how the object moves in just a fraction of its orbit. Astronomers can calculate the orbits of distant objects as they approach the Sun in a highly elliptical orbit and determine whether they will come near Earth.

Kepler's third law for objects orbiting the Sun is $P^2 = A^3$, where P is the period of the orbit and A is the semimajor axis (Chapter 3). **Figure 12.4**

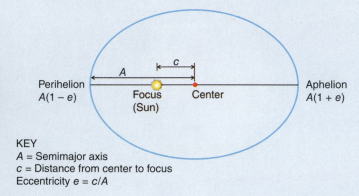

KEY
A = Semimajor axis
c = Distance from center to focus
Eccentricity $e = c/A$

Figure 12.4 The relationships between eccentricity, aphelion, and perihelion in an elliptical orbit.

shows how eccentricity (e) is defined mathematically as the distance from the center of the orbit to one focus (the Sun) divided by the semimajor axis (A). (The eccentricities of the orbits of the planets in the Solar System range from 0.007 for Venus, which is nearly circular, to 0.2 for Mercury.) The types of objects discussed here in Chapter 12 generally have higher eccentricities.

The eccentricity can be related to the closest approach and the farthest distance in the orbit (Figure 12.4). That is, the object's closest approach to the Sun, its **perihelion**, equals $A \times (1 - e)$, and the object's farthest distance from the Sun, its **aphelion**, equals $A \times (1 + e)$. So, if we know the semimajor axis and eccentricity of an orbit, we can calculate how close to and how far away from the Sun an object's orbit takes it.

What are perihelion and aphelion for Apollo asteroid 2005 YU55, which has a semimajor axis of 1.14 AU and an orbital eccentricity of 0.43?

$$\text{Perihelion} = A \times (1 - e) = 1.14 \times (1 - 0.43) = 1.14 \times 0.57 = 0.65\ \text{AU}$$

$$\text{Aphelion} = A \times (1 + e) = 1.14 \times (1 + 0.43) = 1.14 \times 1.43 = 1.63\ \text{AU}$$

Those results indicate that the orbit of 2005 YU55 crosses the orbits of Earth and Mars. In November 2011, it passed 324,900 km from Earth—about 85 percent of the distance to the Moon.

flowing nitrogen ice. Pluto has a subsurface ocean of liquid water; these features imply that Pluto formed from a collision of smaller bodies. The liquid ocean indicates that heat from that collision may be keeping Pluto tectonically active; there may even be active cryovolcanoes on Pluto. Pluto has a thin atmosphere of nitrogen, methane, ethane, and carbon monoxide: those gases freeze out of the atmosphere when Pluto is more distant from the Sun and therefore colder. In contrast, Charon has no atmosphere. Its surface has deep canyons, which might have formed as an ancient ocean froze and pushed the surface outward (**Figure 12.3c**).

Eris, Haumea, and Makemake

Eris, Haumea, and Makemake were discovered in the 21st century. Eris is about the same size as Pluto but has more mass. The highly eccentric orbit of Eris carries it from 37.8 AU out to 97.6 AU away from the Sun, with an orbital period of 557 years (**Working It Out 12.1**). Eris is near the most distant point in its orbit, making it the most remote known object in the Solar System. At that distance it is about 100 times fainter than Pluto. The eccentric orbits of other Solar System bodies will eventually carry them farther away than Eris, however, so Eris will not always be the most distant object known.

When astronomers combine the observed brightness of Eris with its diameter, they find it has a surprisingly high albedo of 0.96. (Recall from Chapter 5 that albedo is the percentage of light an object reflects.) The surface of Eris is more reflective than that of any other major Solar System body except Enceladus, so Eris too must have a coating of pristine ice. The surface of Enceladus is water ice, whereas Eris is covered with methane ice. At its present location, the average surface temperature on Eris is cold enough to freeze out any atmospheric methane, but it will probably develop a methane atmosphere when it comes closest to the Sun in the year 2257.

Haumea and Makemake are both smaller than Pluto (**Figure 12.5**) and have slightly larger orbits. Haumea has two moons—Hi'iaka and Namaka—enabling astronomers to calculate the system's mass. Although Haumea has enough mass to pull itself into a spherical shape, it spins so rapidly on its axis that its shape is flattened, with an equatorial radius approximately twice its polar radius. That difference between the equatorial and polar radii gives Haumea an oblateness (a measure of how far an object is from being perfectly round) of 0.5, giving it the most distorted shape of any of the planets or dwarf planets. Hubble Space Telescope (HST) infrared imaging indicates that Haumea and its two moons are covered in water ice. Astronomers think that those three objects and some smaller debris were left over after a larger body broke up after a collision. Makemake is about 2/3 the diameter of Pluto and has a surface covered in nitrogen, ethane, and methane ices. In 2016, HST discovered a moon orbiting Makemake. The moon has been nicknamed MK 2.

Figure 12.5 This NASA illustration shows the five dwarf planets compared with our Moon (Luna) and Earth.

Other Notable Kuiper Belt Objects

The innermost part of the Kuiper Belt contains tens of thousands of icy planetesimals known as **Kuiper Belt objects (KBOs)**. With a few exceptions, the sizes of KBOs are difficult to determine because, although brightness and approximate distance are known, their albedos are uncertain. Reasonable limits for the albedos can set maximum and minimum values for their size; the largest KBOs are likely similar in size to Pluto and Eris. Hundreds of dwarf planet candidates exist in the Kuiper Belt but, in general, their shapes have not yet been measured well enough to classify them definitively. Like asteroids, some KBOs have moons, and at least one has three moons. We know very little of the chemical and physical properties of most KBOs because of their great distance from us. Some KBOs, such as Arrokoth and Quaoar, are too small to be classified as dwarf planets, but are still notable for other reasons.

After encountering Pluto in 2015, the *New Horizons* spacecraft continued more than a billion kilometers past Pluto to fly by Arrokoth (**Figure 12.6**) in early 2019. Like Pluto, Arrokoth was a surprise. This fascinating binary object probably formed when two planetesimals formed close together, fell toward one another at low velocity, then orbited and gently merged. This "particle-cloud collapse" model is different from the hierarchical accretion model for other planetesimals in the early Solar System. Whether this model applies only to a few objects here and there or dominated planet formation throughout the Solar System is still uncertain. Since this flyby, *New Horizons* has continued outward through the Kuiper Belt, and will eventually leave the Solar System.

Quaoar (pronounced "kwa-whar") is one of the larger known KBOs. It is also one of the few whose diameter astronomers have independently measured—about

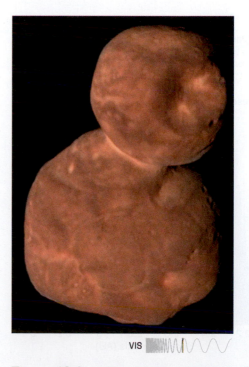

VIS

Figure 12.6 The *New Horizons* mission was redirected toward the Kuiper Belt Object "Arrokoth" after it passed Pluto. Learning more about Kuiper Belt objects helps astronomers better understand the history of the Solar System.

1100 km across. From its apparent brightness, distance, and size, astronomers calculate Quaoar's albedo to be 0.20, making it more reflective than the nuclei of those comets that have entered the inner Solar System but far less reflective than Pluto. Quaoar's remote location and pristine condition have allowed some volatile ices to survive on its surface, including crystalline water ice, methane, and ethane. Quaoar has a nearly circular orbit about the Sun and has a small moon, which enables astronomers to estimate the larger body's mass.

CHECK YOUR UNDERSTANDING 12.1

Why are these objects called dwarf *planets* even though they are actually smaller than some moons?

Answers to Check Your Understanding questions are in the back of the book.

12.2 Asteroids Are Pieces of the Past

After Piazzi found Ceres in 1801, several similar objects were discovered between the orbits of Mars and Jupiter. Because those new objects appeared in astronomers' eyepieces as nothing more than faint points of light, William and Caroline Herschel (the brother–sister pair of astronomers who discovered Uranus) named them *asteroids*, a Greek word meaning "starlike." As the years went by, more asteroids were discovered. Now, it's thought that an estimated 750,000 asteroids larger than 1 km exist, as well as many more that are smaller. Although a great many asteroids exist, they account for only a tiny fraction of the matter in the Solar System. Some of the asteroids are bound to another asteroid in a double system, and more than 200 asteroids have moons, some similar in size to the asteroids themselves. At least one asteroid has a ring.

The planetesimals that formed our Solar System's planets and moons have been so severely modified by planetary processes that nearly all information about their original physical condition and chemical composition has been lost. Asteroids are composed of the same types of rocky and metallic materials that formed the inner planets, but they did not take part in the accretion and differentiation process that formed planets. Therefore, asteroids constitute an ancient and far more pristine record of conditions in the inner early Solar System. Planetary scientists study asteroids to learn about the inner planets and their formation. In this section, we discuss the orbits and composition of the asteroids.

Where Are Asteroids Found?

Most asteroids orbit the Sun in several distinct zones, with many residing between the orbits of Mars and Jupiter in the main asteroid belt. The main belt contains at least 1000 objects larger than 30 km, of which about 200 are larger than 100 km. Many asteroids move quickly enough across the sky that their motion is noticeable over a few hours, making them interesting targets for amateur astronomers.

Asteroids are not distributed uniformly throughout the main asteroid belt: several regions exist that have very few asteroids in them, as shown in **Figure 12.7**. Those "gaps" in the asteroid belt are called **Kirkwood gaps**, after Daniel Kirkwood (1814–1895), the astronomer who first recognized them. All of the Kirkwood gaps in the asteroid belt correspond to resonances: asteroid orbits related to the orbital period of Jupiter by the ratio of two small integers. The orbital periods of

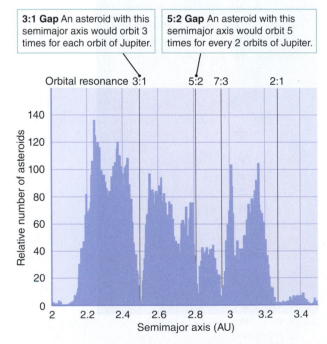

3:1 Gap An asteroid with this semimajor axis would orbit 3 times for each orbit of Jupiter.

5:2 Gap An asteroid with this semimajor axis would orbit 5 times for every 2 orbits of Jupiter.

Figure 12.7 A bar graph showing the relative number of asteroids in the main belt with a given orbital semimajor axis. The gaps in the distribution of asteroids, called *Kirkwood gaps*, are caused by orbital resonances with Jupiter.

some moons around their planets are similarly numerically related, as described in Chapter 11. The boundaries of the asteroid belt are also set by some of those resonances. The inner boundary of the asteroid belt, at 1.8 AU, corresponds to the 5:1 orbital resonance of Jupiter (that is, the asteroids make five orbits for every orbit of Jupiter). The outer boundary, at 3.3 AU, corresponds to the 2:1 orbital resonance of Jupiter.

To see why some orbits are occupied, but others are not, imagine an asteroid orbiting the Sun. While crossing the line between Jupiter and the Sun, that asteroid passes closest to Jupiter, in which case the asteroid receives an outward tug from Jupiter's gravity. At that point, the gravitational force of the Sun on the asteroid is more than 360 times stronger than the gravitational force of Jupiter on the asteroid, so a single close pass of the asteroid between Jupiter and the Sun does very little to the asteroid's orbit. For the majority of possible orbits, the next time the asteroid crosses the line between Jupiter and the Sun, Jupiter has moved. The asteroid is at a different place in its orbit, so it receives a tug in a different direction. The effects of those random tugs average out; even multiple passes close to Jupiter have little overall effect.

Now consider an asteroid starting with an orbital period exactly half that of Jupiter, a 2:1 orbital resonance. After 11.86 years—two complete asteroid orbits and one Jupiter orbit—the asteroid, Jupiter, and the Sun line up in the same location. Jupiter tugs the asteroid a second time in exactly the same direction. The repeated tugs from Jupiter at the same location in the asteroid's orbit add together, and after many small tugs, the asteroid moves out of that orbit, creating a gap in the distribution of asteroids there. As a result, no asteroids are in a 2:1 resonance with Jupiter. Asteroids in other orbital resonances, such as a 3:1 resonance, are similarly moved from their orbits. Asteroids are less likely to be found in the Kirkwood gaps because their gravitational interaction with Jupiter prevents them from staying there.

Several groups of asteroids exist outside the main asteroid belt. They are divided according to their orbital characteristics (**Figure 12.8**). **Trojan** asteroids share Jupiter's orbit and are held in place by interactions with Jupiter's gravitational field. Three other groups are defined by their relationship to the orbits of Earth and Mars: **Apollo** asteroids cross the orbits of Earth and Mars, **Aten** asteroids cross Earth's orbit but not that of Mars, and **Amor** asteroids cross the orbit of Mars but not Earth's. All three groups are named for a prototype asteroid that represents the group.

Asteroids whose orbits bring them within 1.3 AU of the Sun are called **near-Earth asteroids** because their orbits bring them close to Earth's orbit, at 1 AU. Those asteroids, along with a few comet nuclei, are known collectively as **near-Earth objects (NEOs)**. NEOs occasionally collide with Earth or the Moon. About 500–1000 NEOs have diameters larger than a kilometer. Collisions with NEOs are geologically important and have dramatically altered the history of Earth and life on Earth (see Chapter 8).

Part of NASA's mission is to identify and track NEOs. NASA's Wide-field Infrared Survey Explorer (WISE), an infrared telescope in space, surveyed the entire sky during 2010. The data suggest that about 20,000 mid-sized asteroids (100 meters to 1 km) exist near Earth. WISE also observed more than 150,000 asteroids in the main belt, including 34,000 new ones, as well as 2000 Trojans in Jupiter's orbit. The spacecraft, which was reactivated in late 2013 as NEOWISE, has discovered asteroids and comets and is now searching for NEOs. Observations continue in 2021, with NEOWISE's 15th survey of the sky.

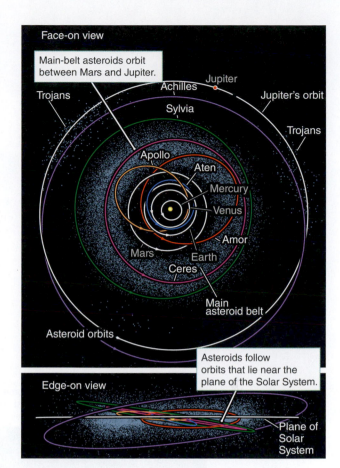

Figure 12.8 Face-on and edge-on views of asteroid orbits. Blue dots show the locations of known asteroids at a single point in time. The orbits of Aten, Amor, and Apollo (prototype members of some groups of asteroids) are shown. Most asteroids, such as Sylvia, are main-belt asteroids. Achilles was the first Trojan asteroid discovered.

unanswered questions

What can asteroids and comets reveal about the migration of the giant planets in the early Solar System? It is not accidental that the main asteroid belt and the Kuiper Belt are at the boundaries of the orbits of the giant planets. As noted in Chapter 10, the giant planets may have moved around quite a bit in the early Solar System. The migration of the giant planets may explain the spread in the orbits of the objects in the main asteroid belt. Migration may also have brought icy objects out of the Kuiper Belt and into the main asteroid belt. The Kuiper Belt would have been closer to the Sun originally, and Jupiter and Saturn would have pushed it outward. And, finally, the migration might have sent objects from both belts into the inner Solar System, creating the heavy bombardment of 4 billion years ago.

The Composition and Classification of Asteroids

Most asteroids are relics of rocky or metallic planetesimals that originated between the orbits of Mars and Jupiter. Although early collisions between those planetesimals created several bodies large enough to differentiate, Jupiter's tidal disruption and possible orbital migration prevented them from forming a single Moon-sized planet. As they orbit the Sun, asteroids continue to collide with one another, producing small fragments of rock and metal. Most meteorites are pieces of those fragments that have found their way to Earth and crashed to its surface.

Studies of meteorites led to a scheme for classifying asteroids by composition. Meteorites found on Earth come from asteroids, which come from planetesimals, as shown in **Figure 12.9**. As larger planetesimals accreted smaller objects, thermal energy from impacts and the decay of radioactive elements heated them. Despite that heating, some planetesimals never reached the high temperatures needed to melt their interiors: they simply cooled. The planetesimals look like rubble piles, pretty much as they were when they formed. Those planetesimals, the most common type of asteroid in the main belt, are called **C-type** (carbon-type) **asteroids**. They are composed of primitive material that has largely been unmodified since the origin of the Solar System almost 4.6 billion years ago.

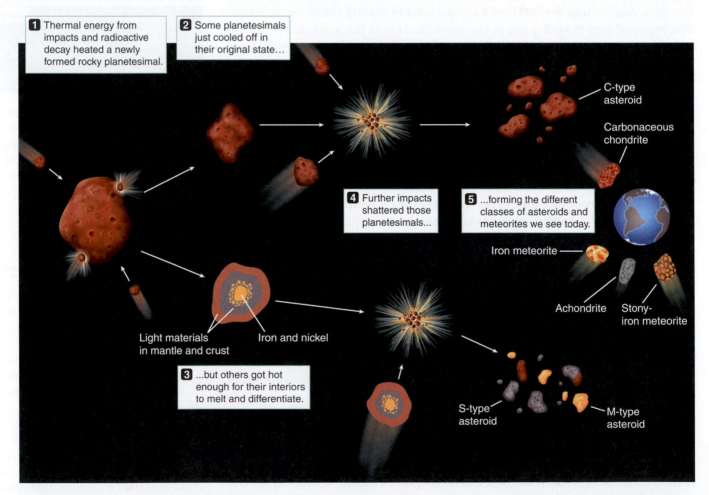

1 Thermal energy from impacts and radioactive decay heated a newly formed rocky planetesimal.

2 Some planetesimals just cooled off in their original state…

C-type asteroid

Carbonaceous chondrite

4 Further impacts shattered those planetesimals…

5 …forming the different classes of asteroids and meteorites we see today.

Iron meteorite

Light materials in mantle and crust

Iron and nickel

Achondrite   Stony-iron meteorite

3 …but others got hot enough for their interiors to melt and differentiate.

S-type asteroid

M-type asteroid

Figure 12.9 The fate of a rocky planetesimal in the young Solar System depends on whether it gets large and hot enough to melt and differentiate, as well as on the impacts it experiences. Different histories led to the varieties of asteroids and meteorites found today. (Images not to scale.)

In contrast, some planetesimals were heated enough by impacts and radio-active decay to cause them to melt and differentiate, with denser matter such as iron sinking to their centers. Lower-density material—such as compounds of calcium, silicon, and oxygen—floated toward the surfaces of those planetesimals and formed mantles and crusts of silicate rocks. **S-type** (stony) **asteroids** may be pieces of the mantles and crusts of such differentiated planetesimals and are chemically similar to volcanic rocks found on Earth. S-type asteroids were hot enough at some point to lose their carbon compounds and other volatile materials to space. **M-type** (metal) **asteroids** are fragments of the iron- and nickel-rich cores of one or more differentiated planetesimals that shattered into small pieces during collisions with other planetesimals.

Recently, some asteroids have been shown to have ice on their surface. Using a ground-based infrared telescope, astronomers found ice on 24 Themis, one of the largest main-belt asteroids (diameter 200 km) that orbits the Sun at the outer edge of the asteroid belt. Water ice covers its surface, and organic molecules also were found there. This was the first direct detection of water ice on an asteroid. Prior measurements, such as the identification of hydrated minerals on meteorites thought to have come from outer-main-belt asteroids, were indirect. The discovery of water ice on 24 Themis may indicate that a continuum rather than a strict boundary exists between icy comets and rocky asteroids. The observations support the idea that both asteroids and comets brought water and organic material to the early Earth.

With a few exceptions, the mass of an asteroid is too small for self-gravity to have pulled it into a spherical shape. Some asteroids have highly elongated irregular shapes, like potatoes, suggesting objects that either are fragments of larger bodies or were created haphazardly from collisions between smaller bodies. Astronomers have measured the masses of several asteroids by noting the effect of their gravity on Mars, on passing spacecraft, on one another, or by the orbits of their moons. The total mass of the asteroids in the main belt is estimated to be about 3 times the mass of Ceres, or 4 percent of the mass of the Moon. Their densities can be found from their mass and size, and they range between 1.3 and 3.5 times the density of water. The lower-density asteroids are shattered heaps of rubble, with large voids between the fragments.

Asteroids rotate just as planets and moons do, although irregularly shaped asteroids can wobble a lot as they spin. Their rotation periods range from 2 hours to longer than 40 Earth days. Rotation periods for asteroids are measured by watching changes in their brightness as they alternately present their broad and narrow faces to Earth. Different groups of asteroids have different average rotations.

Asteroids Viewed Up Close

Spacecraft have visited several asteroids, in missions that not only provide data about asteroids, but also practice maneuvers such as targeting, orbiting, landing, and taking off from these small bodies. In 1991, the *Galileo* mission passed by two S-type asteroids while on its way to Jupiter. The small asteroid Gaspra is cratered and irregular in shape (**Figure 12.10a**). Faint groovelike patterns may be fractures from the impact that chipped Gaspra from a larger planetesimal. Distinctive colors imply that Gaspra is covered with a variety of rock types. Later, *Galileo* passed close to asteroid Ida in the outer part of the main asteroid belt (**Figure 12.10b**). *Galileo* flew so close to Ida that the probe's cameras could see details as small as 10 meters across. Ida's surface is about a billion years old, twice the age estimated for Gaspra. Like Gaspra's, Ida's surface is fractured, indicating that both asteroids must be made

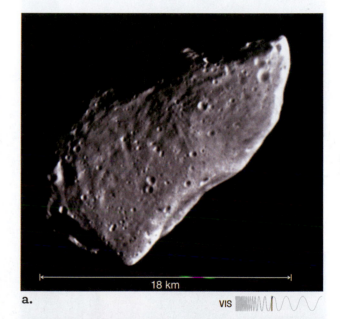

18 km

a.   VIS

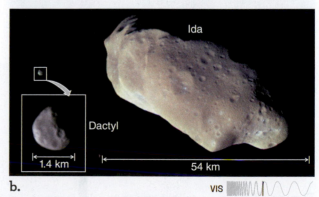

Ida

Dactyl

1.4 km

54 km

b.   VIS

Figure 12.10 a. The *Galileo* spacecraft passed near enough to the asteroid Gaspra to image its surface. **b.** Later, *Galileo* passed by the asteroid Ida and its tiny moon, Dactyl (shown enlarged in the inset).

what if . . .

What if you were an astronaut about to embark on a trip to a 200-km-diameter asteroid in the asteroid belt? What conditions do you think you will experience as you explore the asteroid's surface?

VIS

Figure 12.11 This image of Vesta was taken by *Dawn* in 2012. Its north pole is in the middle of the image.

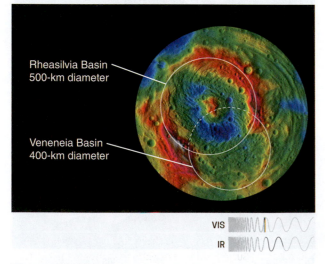

Rheasilvia Basin
500-km diameter

Veneneia Basin
400-km diameter

VIS

IR

Figure 12.12 Impact basins at the south pole of Vesta. By counting the craters on top of it, astronomers estimate Rheasilvia to be 1 billion years old. Veneneia is partly beneath Rheasilvia and is estimated to be 2 billion years old. Red indicates higher elevation.

of relatively solid rock. That finding supports the idea that some asteroids are chips from larger, solid objects. The *Galileo* images also revealed a tiny moon orbiting Ida, called Dactyl, which is itself cratered from impacts.

The first spacecraft to land on an asteroid was *NEAR Shoemaker*, which was gently crash-landed onto asteroid Eros in 2002 after a year of taking observations. Chemical analyses confirmed that the composition of Eros is like that of primitive meteorites. In November 2005, the Japanese spacecraft *Hayabusa* made contact with the small (less than 0.5 km) S-type asteroid Itokawa. *Hayabusa* collected small samples of dust that were returned to Earth in 2010—the first sample-return mission from an asteroid. Chemical analysis showed that such S-type asteroids are the parents of a type of meteorite found on Earth. The findings also suggest that Itokawa had been much larger, more than 20 km, when it formed.

In 2011, NASA's *Dawn* spacecraft went into orbit around Vesta (**Figure 12.11**), the second-most-massive body in the asteroid belt (after Ceres). Vesta is smaller (diameter 525 km) than the terrestrial planets but larger than the other visited asteroids. The data from *Dawn* indicate that Vesta is a leftover intact protoplanet that formed within the first 2 million years of the aggregation of the first solid bodies in the Solar System. Vesta has an iron core and is differentiated, so it is more like the terrestrial planets than it is like other asteroids.

Vesta's spectrum matches the reflection spectrum of a peculiar group of meteorites that look like rocks taken from iron-rich lava flows on Earth and the Moon. A collision—or two—that created the two large impact basins in the south polar region of Vesta (**Figure 12.12**) blasted material into space that then landed on Earth as those meteorites. Those basins are only 1 billion to 2 billion years old. The younger basin is 500 km across and 19 km deep—a depth greater than the height of Mauna Kea in Hawai'i (more than 10 km, measured from the ocean floor).

Two additional spacecraft have visited asteroids, with the aim of bringing samples to Earth for analysis. The Japanese space agency mission *Hayabusa 2* landed on asteroid 162173 Ryugu and, after studying the asteroid for 1.5 years, *Hayabusa 2* left Ryugu and returned a sample to Earth in December 2020. This successful mission has been extended to use up the rest of its xenon propellant. In 2026, the spacecraft will fly by the asteroid 2001 CC21, and in 2031, *Hayabusa 2* will arrive at the asteroid 1998 KY26. NASA's *OSIRIS-REx* (Origins, Spectral Interpretation, Resource Identification, Security-Regolith Explorer) mission arrived at the near-Earth asteroid 101955 Bennu in late 2018 and will bring a sample back to Earth in 2023. For both missions, the goal is to look for water and organic compounds from these relics of the early Solar System.

CHECK YOUR UNDERSTANDING 12.2

Remnants of volcanic activity on the asteroid Vesta indicate that members of the asteroid belt: (a) were once part of a single protoplanet that was shattered by collisions; (b) have all undergone significant chemical evolution since formation; (c) occasionally grow large enough to become differentiated and geologically active; (d) used to be volcanic moons orbiting other planets.

12.3 Comets Are Clumps of Ice

A *comet* is a complex object consisting of a small, solid, icy nucleus; an atmospheric halo; and a tail of dust and gas. The *comet nucleus* is the "heart" of a comet and contains most of the comet's mass. When very distant from the Sun, a comet consists entirely of its frozen nucleus. As a comet nucleus comes near enough to the Sun to heat up, the comet becomes active, emitting gas and dust that form an atmospheric halo. This gas and dust may form a tail, or sometimes two. Estimates of the number of comet nuclei in our Solar System go as high as a trillion (10^{12})—more than the number of stars in the Milky Way Galaxy—but astronomers have seen only several thousand, because they are small and faint. In this section, we discuss the orbits and composition of comets.

The earliest records of comets date from as long ago as the 23rd century BCE. Early cultures viewed the sudden and unexpected appearance of a bright comet as an omen. Comets were often seen as dire warnings of disease, destruction, and death, or sometimes as portents of victory in battle or as heavenly messengers announcing the impending birth of a great leader. Until the end of the Middle Ages, comets were regarded as atmospheric phenomena, like clouds or meteors, rather than as astronomical objects. In the 16th century, Tycho Brahe reasoned that if comets were atmospheric phenomena like clouds, their appearance and location in the sky should be very different to observers located many miles apart. But when Tycho compared sightings of comets made by observers at several sites, he found no evidence of such differences and concluded that comets must be at least as far away as the Moon.

Cometary Reservoirs and Orbits

We can extrapolate where comets come from by observing their orbits as they pass through the inner Solar System. Comets come from two distinct locations in the Solar System named for scientists Gerard Kuiper (1905–1973) and Jan Oort (1900–1992). Comets that come from the Kuiper Belt have different orbital characteristics, such as period and orientation, compared to those that come from the Oort Cloud.

Kuiper Belt Recall from Section 12.1 that the Kuiper Belt is a disk-shaped population of comet nuclei that begins about 30 AU from the Sun, near the orbit of Neptune, and extends outward to about 55 AU (**Figure 12.13**). The icy planetesimals in the Kuiper Belt are packed closely enough to interact gravitationally from time to time. When that happens, one object gains energy, whereas the other loses it. The "winner" may gain enough energy to be sent into an orbit that reaches far beyond the boundary of the Kuiper Belt. The "loser" may fall inward toward the Sun and may become an **active comet**.

Often these comets pass close enough to a planet for its gravity to change the comet's orbit about the Sun. One class of comet, the **short-period comets**, likely originated in the Kuiper Belt, but as they fell in toward the Sun, gravitational encounters with Jupiter forced them into their current short-period orbits relatively close to the Sun. Comets from the Kuiper Belt have orbits roughly aligned with the ecliptic, which are usually prograde; they orbit in the same direction as the planets. About 400 short-period comets are known, which by definition have periods less than 200 years.

Comet Halley is the most famous short-period comet. In 1705, Edmund Halley, using the gravitational laws of his colleague Isaac Newton, noted that a bright comet from 1682 had an orbit remarkably similar to those of comets seen in 1531 and 1607. He concluded that all three were the same comet and predicted that it would return in 1758. When it reappeared, astronomers quickly named it Halley's Comet and heralded it as a triumph for the genius of both Newton and Halley. Comet Halley's

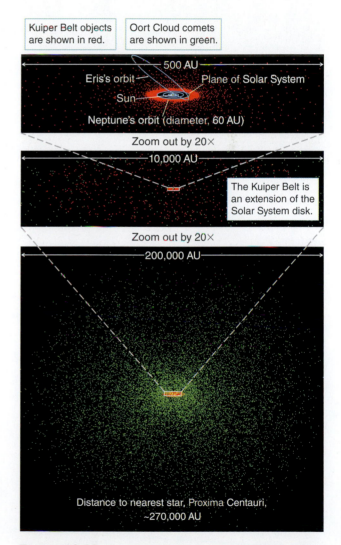

Figure 12.13 The top image shows that most comets within 500 AU of the Sun populate an extension to the disk of the Solar System called the Kuiper Belt (red). The orbit of dwarf planet Eris is highly inclined with respect to the disk. The middle image zooms out to show the expanded Kuiper Belt. The bottom image zooms out even more to illustrate the spherical Oort Cloud, which is far larger and contains many more comet nuclei (green).

highly elongated orbit takes it from perihelion, about halfway between the orbits of Mercury and Venus, out to aphelion beyond the orbit of Neptune. Astronomers and historians have now identified possible sightings of the comet that go back at least to 240 BCE. Comet Halley has an average period of 76 years. Its most recent appearance was in 1986, and it was not especially spectacular in comparison with its appearance in 1910 because in 1986, Comet Halley and Earth were on opposite sides of the Sun. Comet Halley will return and become visible to the naked eye once again in summer 2061.

Oort Cloud   Unlike the flat disk of the Kuiper Belt, the **Oort Cloud** is a spherical distribution of icy planetesimals (Figure 12.13) much too distant to be seen by even the most powerful telescopes. Astronomers determine the size and shape of the Oort Cloud from the orbits of its comets within the inner Solar System; these approach the Sun from all directions and from as far away as 100,000 AU—nearly halfway to the nearest stars.

Sedna is an object in the inner Oort Cloud whose highly elliptical orbit around the Sun takes it from 76 AU out to 937 AU. With such an extended orbit, Sedna requires more than 11,000 years to make a single trip around the Sun. When discovered in 2003, Sedna was about 90 AU from the Sun and getting closer. It will reach its perihelion in 2076. Herschel Space Observatory data suggest an albedo of 0.30 and a size of 1000 km. Sedna has no known moon, so its mass is hard to estimate. Water and methane ices have been detected in its spectrum. Like dwarf planet Eris, Sedna has a highly eccentric orbit. A second object in the inner Oort Cloud, 2012 VP113, was recently detected. Its distance ranges from 80 to 452 AU from the Sun, and it is thought to be about half the size of Sedna.

Objects in the Kuiper Belt are close enough to the Sun that the gravitational force of the Sun is far greater than the pull of any object outside the Solar System. In the distant Oort Cloud, however, the Sun's gravitational force on comet nuclei is so weak that they are barely bound to the Sun at all. The tug of a slowly passing star or interstellar cloud can compete with the Sun's gravity, significantly stirring up the Oort Cloud and changing the orbits of its objects. If the interaction adds to the orbital energy of a comet nucleus, the comet may move outward to an even more distant orbit or perhaps escape from the Sun. A comet nucleus that loses orbital energy as a result of that type of interaction will fall inward. Some of those comet nuclei come all the way into the inner Solar System, where they may appear briefly in Earth's skies before returning once again to the Oort Cloud.

Comets from the Oort Cloud may have orbital periods greater than 200 years and as long as hundreds of thousands or even millions of years. Hundreds of these **long-period comets** have well-determined orbits. Almost all their time is spent in the Oort Cloud in the frigid, outermost regions of the Solar System. Long-period comets reveal the existence of the Oort Cloud. Because of their very long orbital periods, those comets have been observed only once throughout recorded history. Each year, astronomers discover about six new long-period comets, and the total number observed to date is about 3000. Long-period comets were scattered to the outer Solar System by gravitational interactions, so they come into the inner Solar System from all directions. Some orbit the Sun in the same direction that the planets orbit (prograde), whereas others orbit in the opposite direction (retrograde).

Figure 12.14 shows the orbits of several comets, both long period and short period. Nearly all these orbits are highly elliptical, with one end of the orbit close to the Sun and the other in the distant parts of the Solar System. Most comets

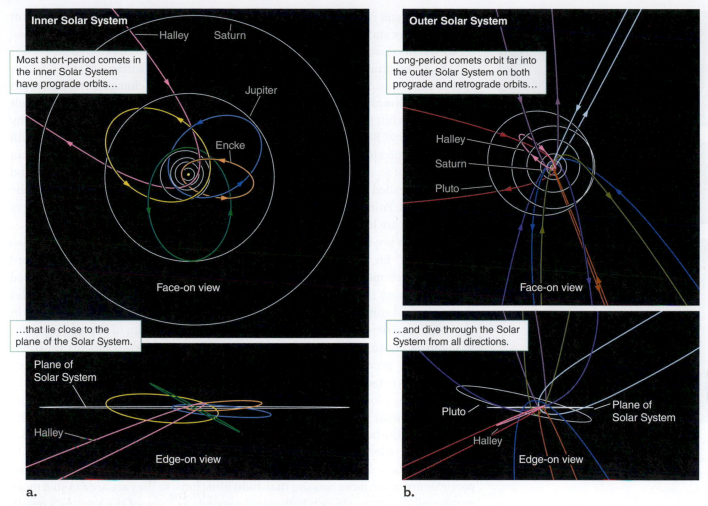

Inner Solar System

Halley   Saturn

Most short-period comets in the inner Solar System have prograde orbits…

Jupiter

Encke

Face-on view

…that lie close to the plane of the Solar System.

Plane of Solar System

Halley

Edge-on view

a.

Outer Solar System

Long-period comets orbit far into the outer Solar System on both prograde and retrograde orbits…

Halley

Saturn

Pluto

Face-on view

…and dive through the Solar System from all directions.

Pluto   Plane of Solar System

Halley

Edge-on view

b.

Figure 12.14 The orbits of several comets in face-on and edge-on views of the Solar System. Populations of **a.** short-period comets and **b.** long-period comets have very different orbital properties. Comet Halley, which appears in both diagrams for comparison, is a short-period comet.

passing through the inner Solar System have long orbital periods that carry them back to the Oort Cloud or the Kuiper Belt.

CHECK YOUR UNDERSTANDING **12.3a**

Oort Cloud objects will pass close to the Sun, and become comets, only if their orbits are (a) very large; (b) very small; (c) highly elliptical; (d) highly inclined.

Anatomy of an Active Comet

Unlike most asteroids, which have been through a host of chemical and physical changes as a result of collisions, heating, and differentiation, most comet nuclei have been preserved over the past 4.6 billion years by the "deep freeze" of the outer Solar System. Comet nuclei are made of nearly pristine material remaining from the formation of the Solar System.

The comet nucleus at the center is the smallest component of a comet, but it is the source of all the material that we see stretched across the sky as the comet nears the Sun (**Figure 12.15**). Comet nuclei range in size from a few dozen meters to several hundred kilometers. These "dirty snowballs" are composed of ice, organic

unanswered questions

Does a "Planet Nine" exist? That is, does a planet exist in the far Solar System that can explain some irregularities in the orbits of objects in the Kuiper Belt and beyond? The planet would need to be about 10 times the mass of Earth, about 2–4 times the size of Earth—similar to Uranus and Neptune. It would have an elongated orbit, with a distance from the Sun of about 200–1200 AU, and would take 10,000–20,000 years to orbit the Sun. Originally it might have either migrated outward in the early Solar System or been captured from another star. Scientists have extensively searched through existing data and made new observations, but nothing has been found. Because the planet would be cold, it would radiate in the infrared and might be detected by the James Webb Space Telescope.

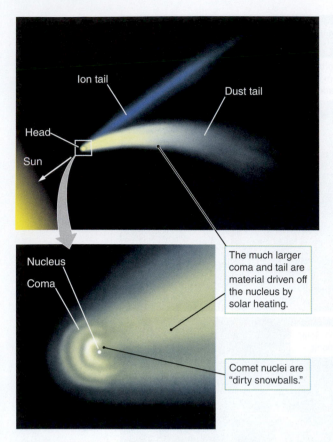

Figure 12.15 The principal components of a fully developed active comet are the nucleus, the coma, and two types of tails, called the dust tail and the ion tail. Together, the nucleus and the coma are called the head.

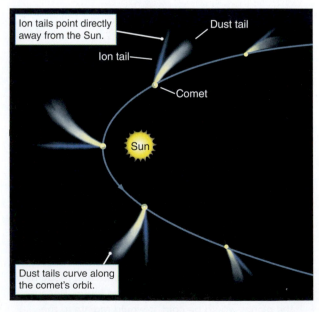

Figure 12.16 The orientation of the dust and ion tails at several points in a comet's orbit. The ion tail points directly away from the Sun, whereas the dust tail curves along the comet's orbit.

compounds, and dust grains. They are similar to deep-fried ice cream, with a soft porous interior surrounded by a crunchy crust of hardened water-ice crystals, topped off with sooty dust and organic molecules.

As a comet nucleus nears the Sun, sunlight heats its surface, vaporizing ices that stream away, carrying dust particles along. That direct change from solid to gas is called **sublimation**. For example, dry ice (frozen carbon dioxide) does not melt like water ice but instead sublimates directly into carbon dioxide gas—that is why it is called "dry." Set a piece of dry ice out in the Sun on a summer day, and you will get a pretty good idea of what happens to a comet. The gases and dust driven from the nucleus of an active comet form a nearly spherical atmospheric cloud around the nucleus called the **coma**. The nucleus and the inner part of the coma form the comet's **head**. Pointing from the head of the comet in a direction more or less away from the Sun are long streamers of dust, gas, and ions called **tails**.

The tails are the largest and most spectacular part of a comet. The tails also are the "hair" for which comets are named. (*Comet* comes from the Greek word *kometes*, which means "hairy one.") Active comets have two types of tails, called the **ion tail** and the **dust tail** (Figure 12.15). Many of the atoms and molecules that make up a comet's coma are ions. Because they are electrically charged, ions in the coma experience the effect of the solar wind—the stream of charged particles that blows continually away from the Sun. The solar wind pushes on those ions, rapidly accelerating them to speeds of more than 100 kilometers per second (km/s)—far greater than the orbital velocity of the comet itself—and sweeps them out into a long wispy structure. Because the particles that make up the ion tail are so quickly picked up by the solar wind, an ion tail is usually very straight: beginning at the head of the comet, an ion tail points directly away from the Sun.

Dust particles in the coma can also have a net electric charge and are subject to the force of the solar wind. Sunlight also exerts a force on cometary dust. But dust particles are much more massive than individual ions, so they are accelerated more gently and do not reach such high relative speeds as those of the ions. As a result, the dust particles cannot keep up with the comet, and the dust tail often curves away from the head of the comet as the dust particles are gradually pushed from the comet's orbit in the direction away from the Sun (Figure 12.15).

Figure 12.16 shows the tails of a comet at various points in its orbit. Both types of tails always point *away* from the Sun, regardless of which direction the comet is moving. As the comet approaches the Sun, then, its two tails trail *behind* its nucleus, but the tails extend *ahead* of the nucleus as the comet moves away from the Sun. Tails vary greatly from one comet to another. Some comets display both types of tails simultaneously; others, for reasons not yet understood, produce no tails at all. A tail often forms as a comet crosses the orbit of Mars, where the increase in solar heating drives gas and dust away from the nucleus.

The gas in a comet's tail is even more tenuous than the gas in its coma, with densities of no more than a few hundred particles per cubic centimeter. That is much, much less than the density of Earth's atmosphere, which at sea level contains more than 10^{19} molecules per cubic centimeter. Dust particles in the tail are typically about 1 micron (μm) in diameter, roughly the size of smoke particles.

The lifetime of a comet nucleus depends on how often it passes by the Sun and how close it gets. The nuclei of short-period comets have been badly worn out by repeated exposure to heating by the Sun. As the volatile ices are driven from a nucleus, some of the dust and organics are left behind on the surface. (Similarly, as a pile of dirty snow melts, the dirt left behind is concentrated on the surface of the snow.) The buildup of that covering slows down cometary activity. In contrast, long-period comets are usually relatively pristine. More of their supply of volatile ices still remains near the surface of the nucleus, and they can produce a truly magnificent show.

Most naked-eye comets first develop a coma and then an extended tail as they approach the inner Solar System. Comet McNaught in 2007 is such a comet, the brightest to appear in nearly 50 years. The comet's nucleus and coma were visible in broad daylight as its orbit carried it within 25 million km of the Sun. When Comet McNaught had passed behind the Sun and next appeared in the evening skies to observers in the Southern Hemisphere, its tail had grown to more than 160 million km and stretched 35° across the sky (**Figure 12.17**). Comet McNaught came into the inner Solar System from the Oort Cloud, but it left on a path that will carry it out of the Solar System.

Comet Hale-Bopp in 1997 was observed as a spectacular long-period comet with a long, beautiful tail (**Figure 12.18**). Hale-Bopp is a large comet, with a nucleus estimated at 60 km in diameter. It was discovered far from the Sun, near Jupiter's orbit, 2 years before its perihelion passage. That early discovery extended the total time available to study its development and plan observations as it approached the Sun. Warmed by the Sun, the nucleus produced large quantities of gas and dust and as much as 300 tons of water per second, with lesser amounts of carbon monoxide, sulfur dioxide, cyanogen, and other gases. Comet Hale-Bopp will continue its outward journey for more than 1000 years, and it will not return to the inner Solar System until sometime around the year 4530.

Comet Ikeya-Seki is a member of a family of comets called **sungrazers**, because the perihelia of these comets are located very close to the surface of the Sun. Many sungrazers fail to survive even a single orbit of the Sun. Ikeya-Seki became so bright as it neared perihelion in 1965 that it was visible in broad daylight, close to the Sun in the sky. Sungrazers generally come in groups, with successive comets following in nearly identical orbits. Each member of such a group started as part of a single larger nucleus that broke into pieces during an earlier perihelion passage.

A half dozen or so long-period comets arrive each year. Most pass through the inner Solar System at relatively large distances from Earth or the Sun and never become bright enough to attract much public attention. On average, a spectacular comet appears about once per decade.

Visits to Comets

Comets pose an engineering challenge for spacecraft designers. Researchers seldom have enough advance knowledge of a comet's visit or its orbit to mount a successful mission to intercept it. The relative speed between an Earth-launched spacecraft and a comet can be extremely high. Observations must be made very quickly, and debris from the nucleus is often moving at high speed, which poses a collision danger for a visiting spacecraft. About a dozen spacecraft have been sent to rendezvous with comets, including five spacecraft sent to Comet Halley by the Soviet, European, and Japanese space agencies in 1986. Much of what we know about comet nuclei and the innermost parts of the coma comes from data sent back by those missions. The spacecraft observed gas and dust jets, impact craters, and ice and dust on the comet nuclei.

The Soviet *Vega 1* and *Vega 2* and the European *Giotto* spacecraft entered the coma of Comet Halley when they were still nearly 300,000 km from its nucleus. The dust from Comet Halley was a mixture of light organic substances and

what if . . .

What if a short-period comet were to fracture in two as it approached and was heated by the Sun? As the two pieces begin to separate, what would happen to the tail, and would you expect to next see both pieces at the regular time?

VIS

Figure 12.17 Comet McNaught in 2007 was the brightest comet to appear in decades, but its true splendor was visible only to observers in the Southern Hemisphere.

VIS

Figure 12.18 Comet Hale-Bopp was a great comet in 1997. The ion tail is blue in this image, and the dust tail is white.

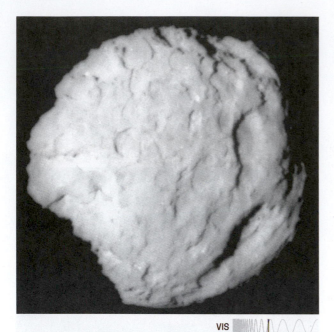

Figure 12.19 The nucleus of Comet Wild 2 was imaged by the *Stardust* spacecraft, which also sampled the comet's tail.

heavier rocky material, and the gas was about 80 percent water and 10 percent carbon monoxide, with smaller amounts of other organic molecules. The surface of Comet Halley's nucleus is among the darkest known objects in the Solar System. That means it is rich in complex organic matter that must have been present as dust in the disk around the young Sun—perhaps even in the interstellar cloud from which the Solar System formed. As the three spacecraft passed near Halley's nucleus, they observed jets of gas and dust moving away from its surface at speeds of up to 1 km/s, far above the escape velocity. By observing the jets of material streaming away from the nucleus of Halley, planetary scientists estimated that Comet Halley must have lost one-tenth of 1 percent of its mass as it went around the Sun.

Several space missions have visited short-period comets. In 2004, NASA's *Stardust* spacecraft flew within 235 km of the nucleus of Comet Wild 2 (**Figure 12.19**). Comet Wild 2 had previously resided between the orbits of Jupiter and Uranus, but a close encounter with Jupiter in 1974 had perturbed its orbit, bringing the relatively pristine body closer to the Sun as it traveled between the orbits of Jupiter and Earth. At the time of *Stardust*'s encounter with Wild 2, the comet had made only five trips around the Sun in its new orbit. Wild 2's nearly spherical nucleus is about 5 km across. At least 10 gas jets were active, some of which carried large chunks of surface material. The surface of Wild 2 is covered with features that may be impact craters modified by ice sublimation, small landslides, and erosion by jetting gas. Some craters show flat floors, suggesting a relatively solid interior beneath a porous surface layer.

The *Stardust* mission collected dust samples from Wild 2, which were returned to Earth in 2006. The samples revealed new kinds of organic substances unlike any seen before in materials from space. They were more primitive than those observed in meteorites and may have formed before the Solar System itself. Those grains can be used to investigate the conditions under which the Sun and planets formed. Minerals that form at high temperature also have been found, supporting the idea that the solar wind blew material out of the inner Solar System very early in the system's history. Scientists will be studying the particles from that mission for many years.

In 2005, NASA's *Deep Impact* spacecraft launched a 370-kilogram impacting projectile into the nucleus of Comet Tempel 1 at a speed of more than 10 km/s (**Figure 12.20**). The impact sent 10,000 tons of water and dust flying off into space

Figure 12.20 **a.** The surface of the nucleus of Comet Tempel 1 is shown just before impact by the *Deep Impact* projectile. The impact occurred between the two 370-meter craters located near the bottom of the image. The smallest features appearing in this image are about 5 meters across. **b.** Sixteen seconds after the impactor struck the comet, the parent spacecraft took this image of the initial ejecta.

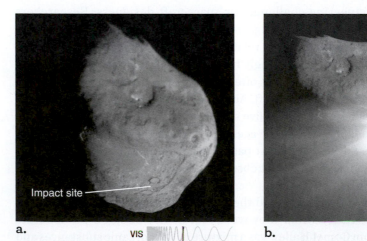
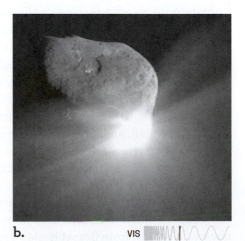

at speeds of 50 m/s. A camera mounted on the projectile snapped photos of its target until the projectile was vaporized by the impact. Observations of the event were made both locally by *Deep Impact* and back on Earth by orbiting and ground-based telescopes. Water, carbon dioxide, hydrogen cyanide, iron-bearing minerals, and a host of complex organic molecules were identified in the Comet Tempel 1 impact. The comet's outer layer is composed of fine dust with the consistency of talcum powder. Beneath the dust are layers made up of water ice and organic materials. Well-formed impact craters, which had been absent in previous images of Comet Wild 2, were also seen.

In 2010, the *EPOXI* spacecraft flew past Comet Hartley 2 (**Figure 12.21**), imaging not only jets of dust and gas, indicating a remarkably active surface, but also an unusual separation of rough and smooth areas. Carbon dioxide jets shoot out from the rough areas. The narrow part at the middle of Figure 12.21 is a smooth, inactive area where ejected material has fallen back onto the cometary nucleus. Further observations by the Herschel Space Observatory showed that the water on Hartley 2 has the same ratio of hydrogen isotopes as that of the water in Earth's oceans. That finding suggests that some of Earth's water could have originated in the Kuiper Belt. Measurements of water on comets from the Oort Cloud have a different ratio, and so they have been ruled out as the source of Earth's water.

The European Space Agency spacecraft *Rosetta* visited Comet 67P/Churyumov-Gerasimenko in 2014. The comet is "binary"—it consists of two pieces that bumped together billions of years ago. Organic compounds such as glycine (an amino acid), ethanol, and complex carbons were identified in the comet's dust. A small, separate spacecraft named *Philae* landed on the comet and sent back data for 2.5 days before running out of power. The landing site was dust-covered solid ice, too thick to drill. The coma is composed of water vapor, carbon monoxide (CO), carbon dioxide (CO_2), hydrogen sulfide, and ammonia. *Rosetta* orbited the comet as it approached the Sun and observed changes as the comet heated up. Dust and gas were released from the comet, including a bright jet. About half that dust is made up of organic molecules. The water on that comet did not have the same isotope ratio as Earth's water (see Reading Astronomy News). After 2 years of observations, *Rosetta* was crashed into 67P in 2016.

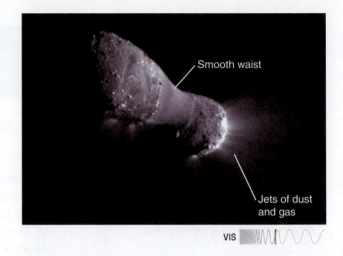

VIS

Figure 12.21 This image of Comet Hartley 2 taken by the *EPOXI* spacecraft reveals two distinct surface types. Water seeps through the dust at the smooth "waist" of the comet's nucleus, whereas carbon dioxide jets shoot gas, dust, and chunks of ice from the rough areas.

CHECK YOUR UNDERSTANDING 12.3b

The nucleus of a comet is mostly: (a) solid ice; (b) solid rock; (c) a porous mix of ice and dust; (d) frozen carbon dioxide.

12.4 Meteorites Are Remnants of the Early Solar System

Comet nuclei that enter the inner Solar System generally disintegrate within a few hundred thousand years as a result of passing repeatedly near the Sun. Asteroids have much longer lives but still are slowly broken into pieces from occasional collisions with one another. The disintegration of comet nuclei and collisions between asteroids create most of the debris that fills the inner Solar System. As Earth and other planets move along in their orbits, they continually sweep up that fine debris. Most of the meteoroids Earth encounters come from the disintegration of comets or asteroids. Meteoroids are small solid bodies ranging in size from

Reading Astronomy News

Rosetta Spacecraft Finds Water on Earth Didn't Come from Comets

Rebecca Jacobson, *PBS NewsHour*

Scientists report on the water observed on Comet 67P/Churyumov-Gerasimenko.

It's a mystery that has baffled scientists for decades: Where did Earth's water come from?

Some scientists believed comets might have been the original source of the Earth's oceans. But a study published this week in the journal *Science* is sending scientists back to the drawing board. In its first published scientific data, the ROSINA mass spectrometer on board the *Rosetta* probe found that water on Comet 67P/Churyumov-Gerasimenko doesn't match the water on Earth.

The result is surprising, says Kathrin Altwegg, principal investigator for ROSINA at the University of Bern and one of the authors of the study. For decades, scientists had ruled out comets from the Oort Cloud at the very edge of our Solar System as the source of Earth's water.

But three years ago, an analysis of water on the Hartley 2 comet near Jupiter found a perfect match to the Earth's oceans. That finding led scientists to believe that Earth's water could have come from much closer comets, either near Jupiter or in the Kuiper Belt just beyond Neptune. Comet 67P/Churyumov-Gerasimenko is one of those Jupiter family comets, which scientists believe originated in the Kuiper Belt.

"That was a big surprise, but now we are back to what I expected," she said. "I think it's very nice to see the diversity we have in Kuiper Belt, to see that not everything is as simple as it seemed."

To find the origin of Earth's water, scientists analyze the water's "fingerprint," says Claudia Alexander, project scientist of the U.S. *Rosetta* Project at NASA's Jet Propulsion Laboratory. Water has a chemical isotopic signature, which works just like a fingerprint. Planets, comets, even minerals all have a fingerprint, Alexander says, and scientists are looking for a match to Earth's.

On Earth, water is mostly two parts hydrogen and one part oxygen—H_2O. But there's also "heavy" water, Alexander explained, which is made with deuterium—a hydrogen atom with a neutron. That heavy water is what the *Rosetta* spacecraft found on Comet 67P/Churyumov-Gerasimenko. It's also a closer match to the water scientists have found on other comets, ruling them out as Earth's water source, Alexander said.

"The clues don't quite all add up," she said.

Altwegg agrees, saying that it's not likely the other Kuiper Belt comets have a match to Earth's water, but further studies would be helpful.

"You would have to assume that 67P is the exception in the Kuiper Belt," she said. "We need more missions to Kuiper Belt comets, which would be fabulous."

There are several ideas to explain the origin of Earth's water, Alexander said. Some believe that water has been on Earth since its formation, that it was beaten out of other minerals as the planet formed. Others think "wet planetesimals" near Jupiter—which were like planetary Silly Putty, loose sticky blobs of rock and ice, Alexander said—collided with Earth in the early formation of the Solar System.

Alexander believes the answer could be a combination of any of these ideas. The *Dawn* mission in 2015 [is studying] the water on the asteroid Ceres, near Jupiter. If it's a match for Earth's water, it may be another clue, Alexander said. But the *Rosetta* finding is a huge step in solving the mystery, she said.

"I think this is a big deal. . . . For me, I've not always been a believer in the story that comets brought the water," Alexander said. "In some respects, I'm somewhat relieved (this finding) doesn't confirm it. It's more complicated than that. I think we need more forensic evidence to settle the score."

QUESTIONS

1. This article uses the word *fingerprint* when discussing isotope ratios. Earlier in this book, we used that word in our discussion of spectral lines. How are the two phenomena similar?

2. What might explain why the water on Comet Hartley 2 matched Earth's water, but the water on Comet 67P/Churyumov-Gerasimenko did not?

3. Is the water on other comets more like the water on Comet Hartley 2 or on Comet 67P/Churyumov-Gerasimenko?

4. What are other possible sources of Earth's water?

5. Do an Internet search to see whether any results from the *Dawn* mission indicate the type of water observed on Ceres.

Source: https://www.pbs.org/newshour/science /rosetta-spacecraft-finds-water-earth-didnt-come -comets.

10 millimeters (mm) to 100 meters. When a meteoroid enters Earth's atmosphere, heat from the meteoroid's interaction with the air causes the air to glow, producing an atmospheric phenomenon called a *meteor*. If a meteoroid survives to reach the planet's surface, we call it a *meteorite*. Earth sweeps up about 100,000 kg of meteoritic debris every day, and particles smaller than 100 mm eventually settle to the ground as fine dust. In this section, we look more closely at meteorites and what can be learned about the early Solar System from them.

Observations of Meteors

If you stand outside for a few minutes on a moonless, starry night, away from bright city lights, you will almost certainly see a meteor, commonly known as a *shooting star*. The larger pieces that survive the plunge through Earth's atmosphere are usually fragments of asteroids. Most of the smaller pieces that burn up in the atmosphere before reaching the ground are cometary fragments. Typically less than a centimeter across, they have about the same density as cigarette ash.

A 1-gram meteoroid (about half the mass of a dime) entering Earth's atmosphere at 50 km/s has a kinetic energy comparable to that of an automobile cruising along at the fastest highway speeds. Radar measurements show that the altitudes of meteors are between 50 and 150 km. Most meteoroids are so small and fragile that they burn up before reaching Earth's surface. A meteor may streak across 100 km of Earth's atmosphere and last at most a few seconds. Meteorites probably litter the surfaces of all solid planets and moons.

Nearly all ancient cultures were fascinated by those rocks from the sky. Iron from meteorites was used to make the earliest tools. Early Egyptians preserved meteorites along with the remains of their pharaohs; Japanese placed them in Shinto shrines and ancient Greeks worshipped them. Despite many eyewitness accounts of meteorite falls, however, many people were slow to accept that those peculiar rocks actually come from far beyond Earth. By the early 1800s, scientists had documented so many meteorite falls that their true origin was indisputable. Today, hardly a year passes without a recorded meteorite fall, including some that have caused damage.

Fragments of asteroids are much denser than cometary meteoroids. If an asteroid fragment is large enough—about the size of your fist—it can survive all the way to the ground to become a meteorite. The fall of a 10-kg meteoroid can produce a fireball so bright that it lights up the night sky more brilliantly than the full Moon. Such a large meteoroid, traveling many times faster than the speed of sound, may create a sonic boom heard hundreds of kilometers away. The meteoroid may even explode into multiple fragments as it nears the end of its flight. Some fireballs glow with a brilliant green color, caused by elements in the meteoroid that created them.

Meteor showers occur when Earth's orbit crosses the orbit of a comet or asteroid and passes through a concentration of cometary or asteroid debris. During a shower, many meteors can be observed in just a few hours. More than a dozen comets and at least two asteroids have orbits that come close enough to Earth's orbit to produce annual meteor showers—see the list in **Table 12.1**. Because the meteoroids in a shower are in similar orbits, they all enter Earth's atmosphere moving in the same direction—the paths through the sky are parallel to one another. Therefore, all the meteors appear to originate from the same point in

Table 12.1	Selected Meteor Showers	
Shower	**Approximate Date**	**Parent Object**
Quadrantids	January 3–4	Asteroid 2003 EH1
Lyrids	April 21–22	Comet Thatcher
Eta Aquariids	May 5–6	Comet Halley
Perseids	August 12–13	Comet Swift-Tuttle
Draconids	October 8–9	Comet Giacobini-Zinner
Orionids	October 21–22	Comet Halley
Taurids	November 5–6	Comet Encke
Leonids	November 17–18	Comet Tempel-Tuttle
Geminids	December 13–14	Asteroid Phaethon
Ursids	December 22–23	Comet Tuttle

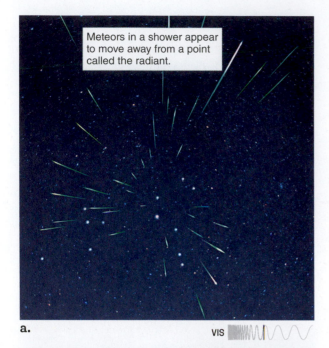

Meteors in a shower appear to move away from a point called the radiant.

a.   VIS

However, the meteor paths are actually parallel.

b.

Figure 12.22 **a.** Meteors appear to stream away from the radiant of the Leonid meteor shower. **b.** Such streaks are actually parallel paths that appear to emerge from a vanishing point, as in our view of these railroad tracks.

the sky (**Figure 12.22a**), just as the parallel rails of a railroad track appear to vanish to a single point in the distance (**Figure 12.22b**). That point is called the shower's **radiant**.

The **Perseid** shower occurs in August because that's when Earth crosses the orbit of Comet Swift-Tuttle. Although spread out along the comet's orbit, the debris is more concentrated near the comet itself. In 1992, Comet Swift-Tuttle returned to the inner Solar System for the first time since its discovery in 1862, resulting in an exceptional Perseid meteor shower with counts of up to 500 meteors per hour.

In mid-November each year, Earth passes almost directly through the orbit of Comet Tempel-Tuttle, a short-period comet with an orbital period of 33.2 years. That crossing produces the **Leonid** meteor shower, which is usually weak because Comet Tempel-Tuttle distributes little of its debris around its orbit in most years. In 1833 and 1866, however, Tempel-Tuttle was not far away when Earth passed through its orbit, and the Leonid showers were so intense that meteors filled the sky with as many as 100,000 meteors per hour. Further perturbations of the comet's orbit caused a spectacular Leonid shower in 1966 that may have produced as many as a half-million meteors per hour. The Leonid shower put on less spectacular but still impressive shows between 1999 and 2003, when several thousand meteors per hour were seen.

Types of Meteoroids

As asteroids orbit the Sun, they occasionally collide with one another, chipping off smaller rocks and bits of dust. Sometimes, one of those fragments is captured by Earth's gravity and survives its fiery descent through Earth's atmosphere as a meteor. Thousands of meteorites reach the surface of Earth every day, but only a tiny fraction of those are ever found and identified. Antarctica offers the best meteorite hunting in the world because in many places, the only stones to be found on the ice are meteorites. Because Antarctica has little precipitation, Antarctic meteorites also tend to show little weathering or contamination from terrestrial dust or organic compounds, making them excellent specimens for study.

Meteorites are categorized as stony, iron, or stony-iron on the basis of their materials and the degree of differentiation they experienced within their parent bodies (**Figure 12.23**). More than 90 percent of meteorites are **stony meteorites**, which are similar to terrestrial silicate rocks. A stony meteorite is characterized by the thin coating of melted rock that forms as it passes through the atmosphere. Many stony meteorites contain small, round spherules called **chondrules**, once-molten droplets that rapidly cooled to form crystallized spheres ranging in size from that of sand grains to that of marbles. Stony meteorites containing chondrules are called **chondrites** (Figure 12.23a); those without chondrules are known as **achondrites** (Figure 12.23b). The most primitive meteorites are called **carbonaceous chondrites** because they are chondrites rich in carbon. Indirect measurements suggest that those meteorites are about 4.56 billion years old—consistent with all other measurements of the time that has passed since the Solar System formed.

Iron meteorites (Figure 12.23c) come from M-type asteroids. Iron meteorites can be recognized by their melted and pitted appearance generated by frictional heating as they streaked through the atmosphere. Like most meteorites, many iron meteorites are never found, either because they land in water or because they are not recognized as meteorites. The Mars *Opportunity* rover discovered a few iron meteorites on the martian surface (**Figure 12.24**). Both their appearance—typical of iron meteorites found on Earth—and their position on the smooth, featureless plains made them instantly recognizable.

Stony-iron meteorites consist of a mixture of rocky material and iron-nickel alloys (Figure 12.23d). They are relatively rare.

what if . . .

What if a meteorite lands in your backyard? What might you do to determine the type of object from which that meteorite originated?

Meteorites and the History of the Solar System

Meteorites are extremely valuable because they are samples of the same relatively pristine material that makes up asteroids. Astronomers can take meteorites into the laboratory and thus study them more easily than sending spacecraft to comets or asteroids. Scientists compare meteorites to rocks found on Earth and the Moon and contrast their structure and chemical makeup with those of rocks studied by spacecraft that have landed on Mars and Venus. Comparing the spectra of meteorites with those of asteroids and planets reveals their origin.

Meteorites come from asteroids, which originate from stony-iron planetesimals. A few planetesimals between the orbits of Mars and Jupiter evolved toward becoming tiny planets before being shattered by collisions. Some became volcanically active, with lava erupting onto their surfaces. But instead of forming planets, those planetesimals broke into pieces in collisions with other planetesimals.

Some types of meteorites fail to follow the patterns discussed so far. Whereas most achondrites have ages of 4.5 billion to 4.6 billion years, some are less than 1.3 billion years old. Other achondrites are chemically and physically similar to the soil and the atmospheric gases that NASA's lander instruments have measured on Mars. The similarities are so strong that most planetary scientists think those meteorites are pieces of Mars knocked into space by large asteroidal impacts—as a result, researchers can study pieces of Mars in laboratories here on Earth. In 1996, a NASA research team announced that the meteorite ALH84001, found in Antarctica, showed possible physical and chemical evidence of past life on Mars, but the claim is still debated (see Chapter 24).

Another group of meteorites bear striking similarities to samples returned from the Moon. Like the meteorites from Mars, they are chunks of the Moon blasted into space by impacts and they later fell to Earth. Therefore, meteorites from Earth may have fallen on the Moon. If they are ever collected from the unchanging Moon, they could tell us about what conditions were like on the early Earth.

Figure 12.23 Cross sections of several kinds of meteorites: **a.** a chondrite (a stony meteorite with chondrules); **b.** an achondrite (a stony meteorite without chondrules); **c.** an iron meteorite; **d.** a stony-iron meteorite.

Zodiacal Dust

Like meteoroids, **zodiacal dust** is a mixture of cometary debris and ground-up asteroid material. Just as you can "see" sunlight streaming through an open window by observing its reflection from dust drifting in the air, you can see the sunlight reflected off tiny zodiacal dust particles that fill the inner Solar System close to the plane of the ecliptic. On a clear, moonless night, not long after the western sky has grown dark, that dust is visible as a faint column of light slanting upward from the western horizon along the path of the ecliptic. That band, called

Figure 12.24 This 2-meter-long iron meteorite lying on the surface of Mars was imaged by the Mars exploration rover *Curiosity*.

Figure 12.25 Zodiacal light shines in the western sky after sunset as seen from the La Silla Observatory in Chile. Credit: ESO/Y. Beletsky, https://www.eso.org/public/hungary /images/zodiacal_beletsky_potw/. https://creativecommons .org/licenses/by/4.0/.

the **zodiacal light**, also can be seen in the eastern sky just before dawn (**Figure 12.25**). With good eyes and an especially dark night, you may be able to follow the zodiacal dust band all the way across the sky. In its brightest parts, the zodiacal light can be several times brighter than the Milky Way, for which it is sometimes mistaken.

The dust grains are roughly a millionth of a meter in diameter—the size of smoke particles. Near Earth, each cubic kilometer of space contains only a few particles of zodiacal dust. The total amount of zodiacal dust in the entire Solar System is estimated to be 10^{16} kg, equivalent to a solid body about 25 km across, or roughly the size of a large comet nucleus. Grains of zodiacal dust are constantly being lost as they are swept up by planets or pushed out of the Solar System by the pressure of sunlight. Such interplanetary dust grains have been recovered from Earth's upper atmosphere by aircraft flying very high. If not replaced by new dust from comets, all zodiacal dust would be gone within the brief span of 50,000 years.

In the infrared region of the spectrum, thermal emission from the band of warm zodiacal dust makes it one of the brightest features in the sky. It is so bright that astronomers wanting to observe faint infrared sources are often hindered by its foreground glow.

CHECK YOUR UNDERSTANDING 12.4

Meteorites contain clues to which of the following? (Choose all that apply.) (a) the age of the Solar System; (b) the temperature in the early solar nebula; (c) changes in the composition of the primitive Solar System; (d) changes in the rate of cratering in the early Solar System; (e) the physical processes that controlled the formation of the Solar System.

unanswered questions

Does a small, dim star exist in the neighborhood of the Sun that comes close periodically and stirs up the Oort Cloud, sending a much larger than average number of comets into the inner Solar System? Some scientists think that fossil data show periodic mass extinctions on Earth and have investigated astronomical causes of the extinctions. One hypothesis is that a distant companion to the Sun has a wide orbit and periodically has come closer to the Sun as they both have traveled around the Milky Way Galaxy. That companion may have stirred up the Oort Cloud as a result. That Oort Cloud disturbance sent many comets to the inner Solar System, which could have caused many impacts on Earth, leading to a mass extinction similar to the one that wiped out the dinosaurs. But some scientists dispute that Earth's extinctions have occurred at regular intervals, and surveys with infrared telescopes have found no evidence of a small companion star.

12.5 Comet and Asteroid Collisions Still Happen Today

Almost all hard-surfaced objects in the Solar System still bear the scars of a time when tremendous impact events were common. In previous chapters, we explained how impacts from long ago may have been responsible for the formation of our Moon, the variation in hemispheric features on Mars, and the death of the dinosaurs. Although such impacts are far less frequent today than they once were, they still happen. In this section, we examine a few examples of recent collisions.

Comet Shoemaker-Levy 9 Collided with Jupiter

Early in the 20th century, the orbit of a comet nucleus called Shoemaker-Levy 9 from the Kuiper Belt was perturbed, and the comet's new orbit carried it close to Jupiter. Eventually, it was captured by Jupiter and orbited the planet. In 1992, that comet passed so close to Jupiter that tidal stresses broke it into two dozen major fragments, which then spread out along its orbit. The fragments took one more 2-year orbit around the planet, and throughout a week in 1994, the entire string of fragments crashed into Jupiter. The impacts occurred just behind the limb of the planet, so they were not visible on Earth until Jupiter's rotation put the impact points in view. Astronomers using ground-based telescopes and the HST could see immense plumes rising from the impacts to heights of more than 3000 km above the cloud tops at the limb. The debris in those plumes then rained back onto Jupiter's stratosphere, causing ripples like pebbles thrown into a pond. **Figure 12.26** shows some HST images of the impact features. Sulfur and carbon compounds released by the impacts formed Earth-sized scars in the atmosphere that persisted for months.

Collisions with Earth

In summer 1908, a remote region of western Siberia was blasted with the energy equivalent of 2000 times the energy of the atomic bomb dropped on Hiroshima in 1945. Eyewitness accounts detailed the destruction of dwellings, the incineration of reindeer (including one herd of 700), and the deaths of at least three people. Although trees were burned or flattened over more than 2150 square kilometers (km²)—an area greater than metropolitan New York City—no crater was left behind. The Tunguska event (named for the nearby river) was the result of a tremendous high-altitude explosion that occurred when a small body hit Earth's atmosphere, ripped apart, and formed a fireball before reaching Earth's surface. Recent expeditions to the Tunguska area have recovered resin from the trees blasted by the event. Chemical traces in the resin suggest that the impacting object may have been a stony asteroid.

In February 2013, a known near-Earth object about half the size of an American football field passed so close to Earth that it came within the orbit of artificial satellites. That near miss was uneventful, and the object simply continued on its way. In an unrelated event on the same day, however, a previously unknown meteoroid estimated to have a radius of about 20 meters exploded over Chelyabinsk, Russia. The shock waves from that explosion damaged thousands of buildings in six cities and injured more than 1000 people. That was likely to have been the largest impact on Earth since the Tunguska event, and many observations were recorded of the effect on that less remote location.

From car dashboard, cell phone, and security camera video and images (**Figure 12.27a**), as well as reports of the time between the brighter-than-the-Sun

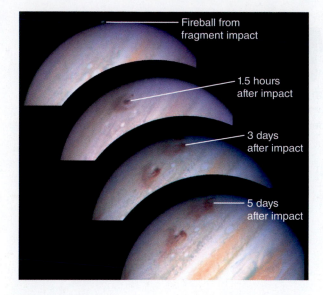

Figure 12.26 HST images of the evolution of the scar left by one fragment of Comet Shoemaker-Levy 9 when it impacted Jupiter in 1994.

Figure 12.27 a. In February 2013, a meteoroid entered the atmosphere over Russia, creating a fireball that eyewitnesses said was brighter than the Sun. **b.** A 600-kg piece of the meteorite on display.

a.

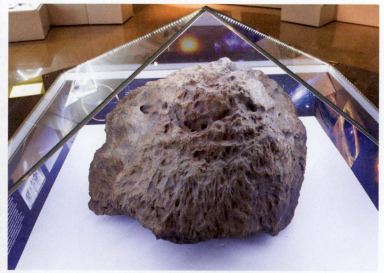

b.

flash and the sonic boom that followed, scientists determined the trajectory and speed of the incoming object as it traveled through the atmosphere. They estimate a preimpact orbit of the object in the inner asteroid belt and think it originally broke off from a known 2-km asteroid. From small pieces collected over a wide area and from a large, 600-kg chunk found in a frozen-over lake (**Figure 12.27b**), scientists could analyze the object's composition and density. It seems to be similar in composition to that of Itokawa, the S-type asteroid whose dust was collected and returned to Earth. It is estimated that only a few hundredths of 1 percent of the original 10-million-kg mass has been found on the surface of Earth. The energy of the explosion was about 30 times the energy released by the Hiroshima atomic bomb. Most of that energy went into the atmosphere, however, heating and breaking up the meteoroid at a much higher altitude than that at which bombs are detonated, so the effects on the ground were less than those of a bomb.

Those impacts are sobering events. The distribution of relatively large asteroids in the inner Solar System indicates it is highly improbable that an asteroid will impact a populated area on Earth during your lifetime. However, many comets and smaller asteroids have unknown orbits, and several previously unknown long-period comets enter the inner Solar System each year. If on a collision course with Earth, even a large comet might not be noticed until just a few weeks or months before impact. For example, Comet Hyakutake was discovered only 2 months before it passed near Earth, and a potentially destructive asteroid that just missed Earth in 2002 was not discovered until 3 days *after* its closest approach. The Chelyabinsk meteoroid was in an orbit such that it wasn't detectable at all. Earth's geological and historical record suggests that actual impacts by large bodies are infrequent events.

As many as 10 million asteroids larger than a kilometer across may exist, but only about 130,000 have well-known orbits, and most of the unknown asteroids are too small to see until they come very close to Earth. The U.S. government—along with the governments of several other nations—is aware of the risk posed by near-Earth objects. Although the probability of a collision between a small asteroid and Earth is small, the consequences could be catastrophic (**Working It Out 12.2**), so NASA has been given a congressional mandate to catalog 90 percent of all NEOs larger than 140 meters and to scan the skies for those that remain undiscovered.

CHECK YOUR UNDERSTANDING 12.5

Of the following, which is the most likely origin of an icy meteorite that reaches Earth? (a) the asteroid belt; (b) a prior impact on Mars; (c) the Oort Cloud; (d) the Kuiper Belt

Origins: Comets, Asteroids, Meteoroids, and Life

Water is essential to life on Earth, so where did it all come from? Scientists have thought that some of Earth's water was contributed during impacts of icy planetesimals early in the history of the Solar System. The icy planetesimals condensed from the protoplanetary disk surrounding the young Sun and grew to their current size near the orbits of the giant planets. Those planetesimals later interacted with the giant planets, which may themselves have been migrating to and from the inner Solar System. In such interactions, some of the planetesimals were flung outward to form the Kuiper Belt and Oort Cloud, and some were

working it out 12.2

Impact Energy

How much energy can be released from the impact of a comet nucleus? The kinetic energy of a moving object is given by

$$E_K = \frac{1}{2}mv^2$$

where E_K is the kinetic energy in joules (J), m is the mass in kilograms, and v is the speed in meters per second (m/s).

Suppose an asteroid or comet nucleus that is 10 km in diameter with a mass of 5×10^{14} kg hits Earth at a speed of 20 km/s = 20×10^3 m/s. Putting those values into the preceding equation gives us

$$E_K = \frac{1}{2} \times (5 \times 10^{14}\,\text{kg}) \times (20 \times 10^3\,\text{m/s})^2$$

$$E_K = 1.0 \times 10^{23}\,\text{J}$$

How much energy is that? A 1-megaton hydrogen bomb (67 times as energetic as the Hiroshima atomic bomb) releases 4.2×10^{15} J. If we divide the energy from our comet impact by that number, we have

$$\frac{1.0 \times 10^{23}\,\text{J}}{4.2 \times 10^{15}\,\text{J/megaton H-bomb}} = 2.4 \times 10^7$$

$$= 24\ \text{million 1-megaton H-bombs}$$

That's a lot of energy, and it shows why impacts have been so important in the history of the Solar System.

---

thrown inward toward the Sun, possibly hitting Earth. Because much of the mass in comet nuclei and some in asteroids appears to be in the form of water ice, some of Earth's current water supply may have come from that early bombardment. Spacecraft have measured the type of water in several comets and asteroids, and the water on Earth best matches that of the primitive carbonaceous chondrites and a few, but not most, comets.

Comets and asteroids can threaten life on Earth. Occasional collisions of comet nuclei and asteroids with Earth have probably resulted in widespread devastation of Earth's ecosystem and in the extinction of many species. Passing stars or the periodic passage of the Sun through giant gas clouds located in denser regions of the Milky Way Galaxy may have resulted in showers of comet nuclei into the inner Solar System, possibly changing the climate and contributing to mass extinctions. Although certainly qualifying as global disasters for the plants and animals alive at the time, such events also represent global opportunities for new life-forms to evolve and fill the niches left by species that did not survive. As noted in Chapter 8, such a collision with an asteroid or comet probably played a central role in ending the 180-million-year reign of dinosaurs and provided an opportunity for the evolution of mammals.

In studying comets, astronomers may also have found a key to the chemical origins of life on Earth. Comets are rich in complex organic material—the chemical basis for terrestrial life—and cometary impacts on the young Earth may have played a role in chemically seeding the planet. If comets are pristine samples of the material from which the Sun and planets formed, organic material must be widely distributed throughout interstellar space. Radio telescope observations of vast interstellar clouds throughout the Milky Way confirm the presence of organic material. The fact that asteroid belts and storms of comets have been observed in distant solar systems could have significant implications as astronomers consider the possibility of life elsewhere in the universe.

SUMMARY

The story of how planetesimals, asteroids, and meteorites are related is a great success of planetary science. Scientists have assembled a wealth of information about that diverse collection of objects to piece together a picture of how planetesimals grow, differentiate, and then shatter in collisions. That story fits well with the even larger story of how most planetesimals were accreted into the planets and their moons (Chapter 7). Comets, asteroids, and meteoroids may have supplied Earth with water, volatiles, and organic material in the early history of the Solar System. Impacts with large asteroids or comet nuclei may have led to mass extinctions on Earth that eventually enabled mammals (for example) to evolve and flourish. Recent spacecraft missions to comets and asteroids have begun to reveal details about the composition of those bodies.

(1) **Categorize the small bodies and identify their locations in the Solar System.** Small bodies in the Solar System that orbit the Sun include dwarf planets, asteroids, comets, Kuiper Belt objects, and meteoroids. Most asteroids are located in the main asteroid belt between the orbits of Mars and Jupiter. Comets are small, icy planetesimals that reside in the frigid regions of the Kuiper Belt and the Oort Cloud, beyond the planets. The orbits of nearly all comets are highly elliptical.

(2) **Characterize the dwarf planets in the Solar System.** Pluto, Eris, Haumea, Makemake, and Ceres are massive enough to have pulled themselves into spherical shapes. But they are classified as dwarf planets, not planets, because they are not massive enough to have cleared their surroundings of other bodies.

(3) **Describe the origin of the types of asteroids, comets, and meteorites.** Asteroids are small Solar System bodies made of rock and metal. Although early collisions between those planetesimals created several bodies large enough to differentiate, Jupiter's tidal disruption (and possible migration) prevented them from forming a single planet. Comets that venture into the inner Solar System are warmed by the Sun, which produces an atmospheric coma and a tail. Meteoroids are small fragments of asteroids and comets. When a meteoroid enters Earth's atmosphere, the energy released causes the air to glow, producing a meteor. Meteor showers occur when Earth passes through a trail of cometary debris. A meteoroid that survives to a planet's surface is called a meteorite. The various types of meteoroids that are formed depend on the differentiation of the parent body.

(4) **Summarize how asteroids, comets, and meteoroids provide important clues about the history and formation of the Solar System.** Asteroids, comets, and meteoroids are leftover debris from the formation of the Solar System. Asteroids are composed of the same type of material that became the inner planets, and comets are composed of the same type of material that became the outer planets. They provide samples of the initial composition and properties of the Solar System and furnish samples of material from its entire history.

(5) **Explain what has been learned from observations of recent impacts in the Solar System.** Impacts by comets and meteoroids have been observed recently on Jupiter and Earth. The debris from those impacts helps astronomers understand the conditions in the early Solar System when impacts were much more frequent.

QUESTIONS AND PROBLEMS

TEST YOUR UNDERSTANDING

1. This chapter has focused on leftover planetesimals. What became of most of the other planetesimals?
   a. They evaporated.
   b. They left the Solar System.
   c. They became part of larger bodies.
   d. They fragmented into smaller pieces.

2. The three types of meteorites come from different parts of their parent bodies. Stony-iron meteorites are rare because
   a. they are hard to find.
   b. the volume of a differentiated body that has both stone and iron is small.
   c. the Solar System has very little iron.
   d. the magnetic field of the Sun attracts the iron.

3. As a comet leaves the inner Solar System, the ion tail points
   a. back along the orbit.
   b. forward along the orbit.
   c. toward the Sun.
   d. away from the Sun.

4. Congress tasked NASA with searching for near-Earth objects because
   a. they might impact Earth, as others have.
   b. they are close by and easy to study.
   c. they are moving fast.
   d. they are scientifically interesting.

5. Meteorites can provide information about which of the following? (Choose all that apply.)
   a. the early composition of the Solar System.
   b. the composition of asteroids.
   c. the composition of comets.
   d. the Oort Cloud.

6. Perihelion is the point in an orbit _____ the Sun; aphelion is the point in an orbit _____ the Sun.
   a. closest to; farthest from
   b. farthest from; closest to
   c. at one focus of; at the other focus of

7. Kuiper Belt objects (KBOs) are mostly comet nuclei. Why doesn't a KBO have a coma and a tail?
   a. Most of the material has already been stripped from the objects.
   b. They are too far from the Sun.
   c. They are too close to the Sun.
   d. The coma and tail are pointing away from Earth, behind the object.

8. Asteroids are small
   a. rock and metal objects orbiting the Sun.
   b. icy objects orbiting the Sun.
   c. rock and metal objects found only between the orbits of Mars and Jupiter.
   d. icy bodies found only in the outer Solar System.

9. Aside from their periods, short-period and long-period comets differ in that
   a. short-period comets orbit prograde, whereas long-period comets orbit retrograde.
   b. short-period comets contain less ice, whereas long-period comets contain more.
   c. short-period comets do not develop ion tails, whereas long-period comets do.
   d. short-period comets come closer to the Sun at closest approach than long-period comets.

10. On average, a bright comet appears about once each decade. Statistically, that means that
   a. one will definitely be observed every tenth year.
   b. one will definitely be observed in each 10-year period.
   c. exactly 10 comets will be observed in a century.
   d. about 10 comets will be observed in a century.

11. Most asteroids are located between the orbits of
   a. Earth and Mars.
   b. Mars and Jupiter.
   c. Jupiter and Saturn.
   d. the Kuiper Belt and the Oort Cloud.

12. Comets, asteroids, and meteoroids may be responsible for delivering a significant fraction of the current supply of _____ to Earth.
   a. mass   c. oxygen
   b. water   d. carbon

13. An iron meteorite most likely came from
   a. an undifferentiated asteroid.
   b. a differentiated asteroid.
   c. a planet.
   d. a comet.

14. Meteor showers occur because Earth passes through the path of
   a. another planet.   c. a comet.
   b. a planetesimal.   d. the Moon.

15. Dwarf planets differ from the other planets in that they
   a. have no atmosphere.   d. have lower mass.
   b. have no moons.   e. are covered in ice.
   c. are all very far from the Sun.

THINKING ABOUT THE CONCEPTS

16. Describe ways in which Pluto differs significantly from the Solar System planets.

17. Consider the Process of Science Figure, and the accompanying text. By what criteria did Pluto fail to be considered a planet under the new IAU definition? Explain how that decision demonstrates the self-correcting nature of science. ★

18. How does the composition of an asteroid differ from that of a comet nucleus?

19. ★ WHAT AN ASTRONOMER SEES   In Figure 12.3a, the large heart-shaped feature is observed to be smooth, and astronomers take this to be evidence that Pluto must have recently been geologically active. Explain why a smooth surface implies recent geologic activity. 👁

20. What are the differences between a comet and a meteor in size, distance, and how long they remain visible?

21. Most meteorites are 4.54 billion years old. Carbonaceous chondrites, however, are 4.56 billion years old (20 million years older). What determines the time of "birth" of those pieces of rock? What does that information tell you about the history of their parent bodies?

22. Most asteroids are found between the orbits of Mars and Jupiter, but astronomers are especially interested in the relative few whose orbits cross that of Earth. Why?

23. How could you and a friend, armed only with your cell phones and knowledge of the night sky, prove conclusively that meteors are an atmospheric phenomenon?

24. Suppose you find a rock that has all the characteristics of a meteorite. You take it to a physicist friend, who confirms that it is a meteorite but says that radioisotope dating indicates an age of only a billion years. What might be the origin of that meteorite?

25. Describe differences between the Kuiper Belt and the Oort Cloud as sources of comets. What is the ultimate fate of a comet from each of those reservoirs?

26. What are the three parts of a comet? Which part is the smallest in radius? Which is the most massive?

27. In 1910, Earth passed directly through the tail of Comet Halley. Among the various gases in the tail was hydrogen cyanide (HCN). HCN is deadly to humans, yet nobody became ill from that event. Why?

28. Comets have two types of tails. Describe them and explain why they sometimes point in different directions.

29. What is zodiacal light, and what is its source?

30. How might comets and asteroids have contributed to the origin of life on Earth?

APPLYING THE CONCEPTS

31. Use the information in Appendix 4 to determine the perihelion and aphelion distances for Pluto. ●–●–●
  a. Make a prediction: Considering the size of Pluto's orbit, do you expect your distances to come out to be a few AU, a few tens of AU, or a few hundreds of AU? Which do you expect to be larger—the perihelion distance or the aphelion distance?
  b. Calculate: Follow Working It Out 12.1 to find the perihelion and aphelion distances for Pluto.
  c. Check your work: Verify that the units of your answer are in AU, and compare your answer to your predictions.

32. Some near-Earth objects are in binary systems, so it is possible to estimate their mass. How much energy would be released if a near-Earth asteroid with mass $m = 4.6 \times 10^{11}$ kg hit Earth at a velocity $v = 5$ km/s? ●–●–●
  a. Make a prediction: What units do you expect to have for the impact energy? Looking back at Working It Out 12.2, about how large do you expect your answer to be?
  b. Calculate: Find the impact energy of this asteroid.
  c. Check your work: Verify that your answer has the correct units, and compare your answer to your prediction.

33. Use the information in Appendix 4 to determine the perihelion and aphelion distances for Earth.
  a. Make a prediction: Considering the size of Earth's orbit, do you expect your distances to come out to be a few AU, a few tens of AU, or a few hundreds of AU? Which do you expect to be larger—the perihelion distance or the aphelion distance?
  b. Calculate: Follow Working It Out 12.1 to find the perihelion and aphelion distances for Earth.
  c. Check your work: Verify that the units of your answer are in AU, and compare your answer to your predictions.

34. What is the impact energy of an asteroid with a mass of 4.6×10^{11} kg traveling at 40 km/s? What is the equivalent in 1-megaton hydrogen bombs?
  a. Make a prediction: What units do you expect to have for the impact energy? Looking back at Working It Out 12.2, do you expect your answer to be equivalent to a few thousand, a few million, or a few billion hydrogen bombs?
  b. Calculate: Find the impact energy of this asteroid.
  c. Check your work: Verify that your answer has the correct units, and compare the size of your answer with your prediction.

35. Comet 67P/Churyumov-Gerasimenko has a diameter of 4 km and a mass of 1013 kg. What is the density of the comet? What is the escape velocity from the surface of that comet?
  a. Make a prediction: Considering the composition of a comet, do you expect the density to be more or less than the density of water? Do you expect the escape velocity from a comet to be more or less than the escape velocity of Earth?
  b. Calculate: Find the density of the comet and the escape velocity.
  c. Check your work: Verify that the density you calculated has units of mass per volume. Verify that the escape velocity you calculated has units of distance per time. Compare your answers to your predictions to test whether the size of the answer is sensible.

36. Ceres has a diameter of 975 km and a period of about 9 hours. What is the rotational speed of a point on the surface of that dwarf planet?

37. Figure 12.13 shows the scale of the Solar System out to the Oort Cloud. Judging from that figure, what fraction of the distance between the Sun and Proxima Centauri is occupied by the Oort Cloud, a part of the Solar System? ◉

38. The Moon has a diameter of 3474 km and orbits Earth at an average distance of 384,400 km. At that distance, it subtends an angle just slightly larger than half a degree in Earth's sky. Pluto's moon Charon has a diameter of 1186 km and orbits at a distance of 19,600 km from the dwarf planet.
  a. Compare the appearance of Charon in Pluto's skies with the Moon in Earth's skies.
  b. Describe where in the sky Charon would appear as seen from various locations on Pluto.

39. One recent estimate concludes that nearly 800 meteorites with mass greater than 100 grams (massive enough to cause personal injury) strike Earth each day. If you present a target of 0.25 square meter (m^2) to a falling meteorite, what is the probability that you will be struck by a meteorite during your 100-year lifetime? (The surface area of Earth is approximately 5×10^{14} m^2.)

40. Electra is a 182-km asteroid accompanied by a small moon orbiting at a distance of 1350 km in a circular orbit with a period of 3.92 days.
  a. What is the mass of Electra?
  b. What is Electra's density?

41. Calculate the orbital radius of the Kirkwood gap that is in a 3:1 orbital resonance with Jupiter.

42. The orbital periods of Comets Encke, Halley, and Hale-Bopp are 3.3, 76, and 2530 years, respectively. Their orbital eccentricities are 0.847, 0.967, and 0.995, respectively.
  a. What are the semimajor axes (in astronomical units) of the orbits of those comets?
  b. What are the minimum and maximum distances from the Sun (in astronomical units) reached by Comets Halley and Hale-Bopp in their respective orbits?
  c. Which region of the Solar System did each probably come from?
  d. Which would you guess is the most pristine comet among the three? Which is the least? Explain your reasoning.

43. Comet Halley has a mass of approximately 2.2×10^{14} kg. It loses about 3×10^{11} kg each time it passes the Sun.
  a. The first confirmed observation of the comet was made in 240 BCE. If you assume a constant period of 76.4 years, how many times has it reappeared since that early sighting?
  b. How much mass has the comet lost since 240 BCE?
  c. What percentage of the comet's total mass today does that amount represent?

44. A cubic centimeter of the air you breathe contains about 10^{19} molecules. A cubic centimeter of a comet's tail may typically contain 200 molecules. Calculate the volume of comet tail material that would hold 10^{19} molecules.

45. The estimated amount of zodiacal dust in the Solar System remains constant at approximately 10^{16} kg, yet zodiacal dust is constantly being swept up by planets or removed by the pressure of sunlight.
  a. If all the dust disappeared (at a constant rate) over a span of 30,000 years, what would the average production rate have to be, in kilograms per second, to maintain the current content?
  b. Is that an example of static or dynamic equilibrium? Explain your answer.

EXPLORATION Asteroid Discovery

Astronomers often discover asteroids and other small Solar System objects by comparing two (or more) images of the same star field and looking for bright spots that have moved between the images. The four images in **Figure 12.28** are all "negative images." That is, every dark spot would actually be bright in the sky, and all the white space is dark sky. A negative image sometimes helps the observer pick out faint details, and it is preferable for printing and photocopying. Study those four images. Can you find an asteroid that moves across the field in those images? That's the hard way

to do it. A much easier method is to use a "blink comparison," which lets you look at one image and then another very quickly. Make three photocopies of each image, cut out each one, and align them all carefully on top of one another in sequence, so that the stars overlap. You should have 12 pieces of paper, in this order: Image 1, Image 2, Image 3, Image 4; Image 1, Image 2, . . . , and so on. Staple the top edge and flip the pages with your thumb, looking carefully at the images. Can you find the asteroid now?

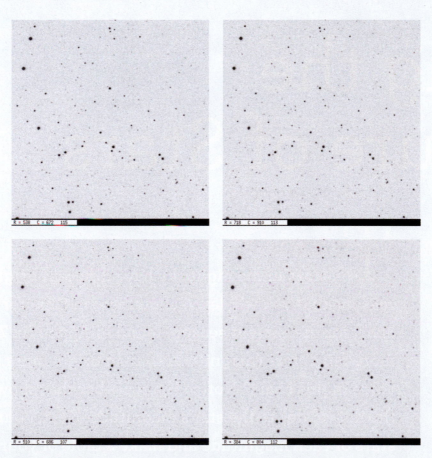

Figure 12.28 These four images of a star field were taken at different times. They contain a moving asteroid.

1 Circle the asteroid in each image.

The "blink comparison" method takes advantage of a feature of the human eye-brain connection. Humans are much better at noticing things that move than things that are stationary.

2 Why might that feature be a helpful evolutionary adaptation?

A third method is available, which sometimes makes things easier to see but requires high-quality digital images. In that method, one image is subtracted from another.

3 If you used the subtraction method with two of those images, what would you expect to see in the resulting image?

Taking the Measure of Stars

The stars in the night sky have different brightnesses for reasons that you will learn about here in Chapter 13. Study the star chart for the current season, found in Appendix 7. About how many more faint stars (magnitude 4) than bright stars (magnitude 1) are there in this chart? Notice the scale bar that shows how the size of the dot relates to the star's magnitude. Next, count the number of stars of each magnitude and make a table of your results. Does this more careful analysis agree with your estimate? Take this star chart outside on a clear night, and find all the stars with magnitude 1 that are above the horizon. This will orient you to the sky. Now find some stars with magnitude 2, 3, 4, and so on. If you can find no stars fainter than magnitude 3, then the "limiting magnitude" of your observing site is 3. What is the limiting magnitude of your observing site on this date? What sources of light are making it difficult to see stars fainter than this?

EXPERIMENT SETUP

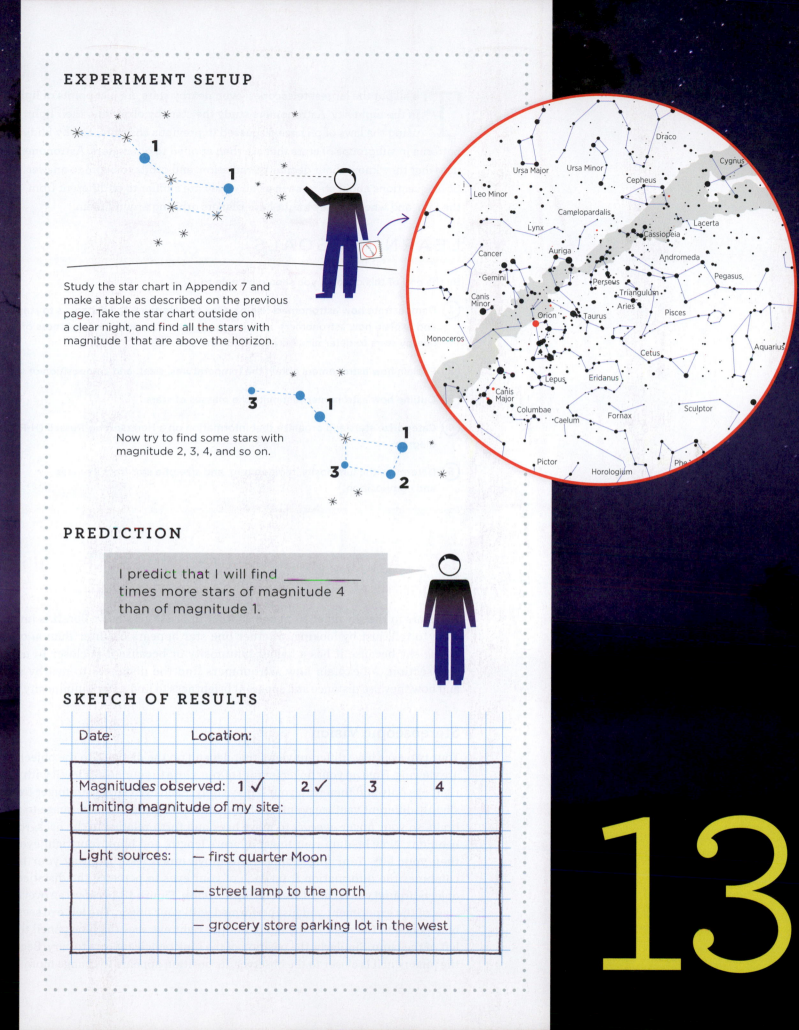

Study the star chart in Appendix 7 and make a table as described on the previous page. Take the star chart outside on a clear night, and find all the stars with magnitude 1 that are above the horizon.

Now try to find some stars with magnitude 2, 3, 4, and so on.

PREDICTION

I predict that I will find _____ times more stars of magnitude 4 than of magnitude 1.

SKETCH OF RESULTS

Date: Location:

Magnitudes observed: 1 ✓   2 ✓   3   4
Limiting magnitude of my site:

Light sources: — first quarter Moon

— street lamp to the north

— grocery store parking lot in the west

13

To all but the largest telescopes, even nearby stars are just points of light in the night sky. Astronomers study the stars by observing their light, by using the laws of physics discussed in previous chapters, and by finding patterns in subgroups of stars that are then applied to other stars. Astronomers use what they know about geometry, radiation, and orbits to begin to answer basic questions about stars, such as how they are similar to or different from the Sun, and whether they might have planets orbiting around them.

LEARNING GOALS

By the end of this chapter, you should be able to:

(1) Demonstrate how astronomers use parallax to determine the distances to stars, and explain how astronomers combine these distances with the brightness of nearby stars to determine how luminous the stars are.

(2) Explain how astronomers obtain the temperatures, sizes, and composition of stars.

(3) Outline how astronomers estimate the masses of stars.

(4) Categorize stars and organize that information on a Hertzsprung-Russell (H-R) diagram.

(5) Determine the luminosity, temperature, and size of a star from its mass and composition.

13.1 Astronomers Measure the Distance, Brightness, and Luminosity of Stars

The stars in the sky differ from one another in brightness and color. It is impossible to tell, just by looking, whether one star appears brighter than another in the sky because it has a higher luminosity or because it is closer to us. In this section, we explain how astronomers find the distances to nearby stars and how they use distance and apparent brightness to find a star's luminosity.

Stereoscopic Vision

Your two eyes have different views that depend on the distance to the object you are viewing. Hold up your finger in front of you, close to your nose. View it with your right eye only and then with your left eye only. Each eye views your finger from a slightly different vantage point. Each eye sends a slightly different image to your brain, so your finger *appears* to move back and forth relative to the background behind it. Now hold up your finger at arm's length, and blink your right eye and then your left. Your finger appears to move much less. The way your brain combines the information from each of your eyes to perceive the distances to objects around you is called **stereoscopic vision**. **Figure 13.1a** shows an overhead view of the experiment you just performed with your finger. The left eye sees the blue pencil to the right of the lamp. But the right eye sees the blue pencil to the left of the lamp. Similarly, the position of the pink pencil appears to vary. Because the pink pencil is closer to the observer, its position appears to change more than

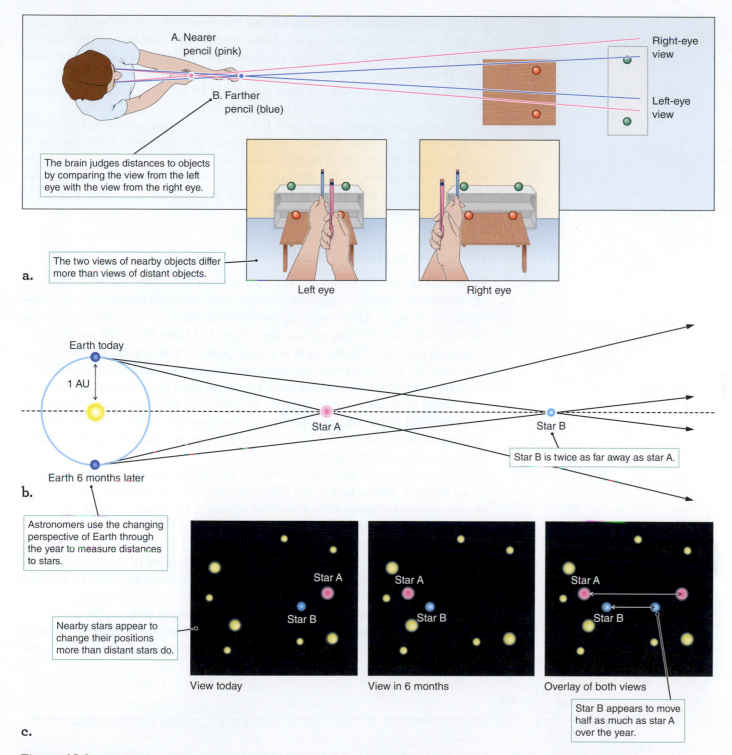

Figure 13.1 a. Stereoscopic vision enables you to determine the distance to an object by comparing the view from each eye. **b.** Similarly, comparing views from different places in Earth's orbit enables astronomers to determine the distances to stars. **c.** As Earth moves around the Sun, the apparent positions of nearby stars change more than the apparent positions of more distant stars. (Diagram not to scale.)

the position of the blue pencil—seeming to move from the right of the blue pencil to the left of the blue pencil.

Stereoscopic vision enables you to judge the distances of objects as far away as a few hundred meters, but beyond that it is of little use. Your right eye's view of a mountain several kilometers away is indistinguishable from the view seen by your left eye—all you can determine is that the mountain is too far away for you to judge its distance stereoscopically. The distance over which your stereoscopic vision works is limited by the relatively small separation between your eyes—about 6 centimeters (cm). If you could increase the distance between your eyes by several meters, the view from each eye would be different enough for you to determine the distances to objects kilometers away.

You cannot separate your eyes, but you can compare pictures taken with a camera from two widely separated locations. These images don't need to be simultaneous. If you take a picture of the sky tonight and then wait 6 months and take another picture, the distance between the two locations is the diameter of Earth's orbit (2 astronomical units [AU]). Without leaving Earth, this is the greatest possible separation obtainable.

📹 **Astronomy in Action:** Parallax

Distances to Nearby Stars

Figure 13.1b, which depicts Earth's orbit from far above the Solar System, shows how astronomers apply the concept of stereoscopic vision to measure the distances to stars. The change in Earth's position over 6 months is like the distance between the right eye and the left eye in Figure 13.1a. The nearby (pink and blue) stars are like the pink and blue pencils, whereas the distant yellow stars are like the objects on the bookcase. Because of the shift in perspective as Earth orbits the Sun, nearby stars appear to shift their positions. The pink star which is closer, appears to move farther than the more distant blue star (**Figure 13.1c**). Over 1 full year, the nearby star appears to move one way and then back again with respect to distant background stars, returning to its original position at the end of that year.

Astronomers use geometry to convert the shift in a star's apparent position into a distance. **Figure 13.2** shows Earth, the Sun, and two stars in a configuration similar to that in Figure 13.1b. Look first at star A, the closest star. When at the top of the figure, Earth forms a right triangle with the Sun and star A at the other corners. (A right triangle is one with a 90° angle in it.) The short leg of the triangle is the distance from Earth to the Sun, 1 AU. The long leg of the triangle is the distance from the Sun to star A. The small angle near star A is called the *parallax angle*, or simply **parallax**, of the star.

More distant stars make longer and skinnier triangles with smaller parallaxes. Star B is twice as far away as star A, and so star B's parallax is only half that of star A. The parallax (*p*) of a star is inversely proportional to its distance (*d*). A star

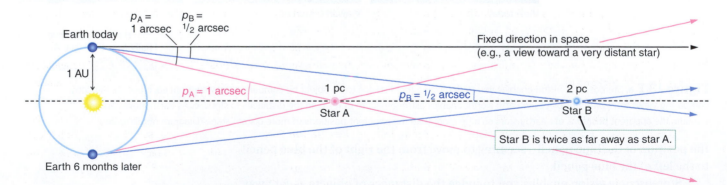

Figure 13.2 The parallax (*p*) of a star is inversely proportional to its distance. More distant stars have smaller parallaxes. (Diagram not to scale.)

Reading Astronomy News

NASA's *New Horizons* Conducts the First Interstellar Parallax Experiment

June 11, 2020

To make the most precise parallax measurements, scientists need the longest possible baseline. In 2020, they took advantage of a rare opportunity to use an unusually long baseline.

For the first time, a spacecraft has sent back pictures of the sky from so far away that some stars appear to be in different positions than we see from Earth.

More than four billion miles from home and speeding toward interstellar space, NASA's *New Horizons* has traveled so far that it now has a unique view of the nearest stars.

On April 22–23, the spacecraft turned its long-range telescopic camera to a pair of the closest stars, Proxima Centauri and Wolf 359, showing just how they appear in different places than we see from Earth. Scientists have long used this "parallax effect"—how a star appears to shift against its background when seen from different locations—to measure distances to stars.

"No human eye can detect these shifts," Stern said.

But when *New Horizons* images are paired with pictures of the same stars taken on the same dates by telescopes on Earth, the parallax shift is instantly visible. The combination yields a 3-D view of the stars "floating" in front of their background star fields.

"The *New Horizons* experiment provides the largest parallax baseline ever made—over 4 billion miles—and is the first demonstration of an easily observable stellar parallax," said Tod Lauer, New Horizons science team member from the National Science Foundation's

National Optical-Infrared Astronomy Research Laboratory who coordinated the parallax demonstration.

Working in Stereo

Lauer, *New Horizons* Deputy Project Scientist John Spencer, of SwRI, and science team collaborator, astrophysicist, Queen guitarist and stereo imaging enthusiast Brian May created the images that clearly show the effect of the vast distance between Earth and the two nearby stars.

"It could be argued that in astro-stereoscopy—3-D images of astronomical objects—NASA's *New Horizons* team already leads the field, having delivered astounding stereoscopic images of both Pluto and the remote Kuiper Belt object Arrokoth," May said. "But the latest *New Horizons* stereoscopic experiment breaks all records. These photographs of Proxima Centauri and Wolf 359—stars that are well-known to amateur astronomers and science fiction aficionados alike—employ the largest distance between viewpoints ever achieved in 180 years of stereoscopy!"

The companion Earth-based images of Proxima Centauri and Wolf 359 were provided by Edward Gomez of the Las Cumbres Observatory, operating a remote telescope at Siding Spring Observatory in Australia, and astronomers John Kielkopf, University of Louisville, and Karen Collins, Harvard and Smithsonian Center for Astrophysics, operating a remote telescope at Mt. Lemmon Observatory in Arizona.

An Interstellar Navigation First

Throughout history, navigators have used measurements of the stars to establish their position on Earth. Interstellar navigators can

do the same to establish their position in the galaxy, using a technique that *New Horizons* has demonstrated for the first time.

At the time of the observations, New Horizons was more than 4.3 billion miles (about 7 billion kilometers) from Earth, where a radio signal, traveling at the speed of light, needed just under 6 hours and 30 minutes to reach home.

Launched in 2006, *New Horizons* is the first mission to Pluto and the Kuiper Belt. It explored Pluto and its moons in July 2015—completing the space-age reconnaissance of the planets that started 50 years earlier—and continued on its unparalleled voyage of exploration with the close flyby of Kuiper Belt object Arrokoth in January 2019. *New Horizons* will eventually leave the Solar System, joining the *Voyagers* and *Pioneers* on their paths to the stars.

QUESTIONS

1. How many times longer was this baseline than the usual baseline, which is about the diameter of Earth's orbit?
2. Why did the team choose to measure the parallax of nearby stars, whose distances are already well known?
3. Go online to the source of the article to find the stereoscopic images of Proxima Centauri, which have been made into a "video." Where, in the image, is the shifting star?
4. How does improving the accuracy of the distances to nearby stars from trigonometric parallax affect astronomers' estimates of farther stars' distances, measured using spectroscopic parallax?

Source: http://pluto.jhuapl.edu/News-Center/News-Article.php?.

what if . . .

What if you measured the distances to all the visible stars in the constellation Sagittarius? Would you expect the distances to be similar or very different for these stars, and what would that imply?

3 times farther away than star A has a parallax 1/3 of star A's parallax. A star 10 times farther away has a parallax 1/10 of star A's parallax.

The parallaxes of real stars are tiny. Recall from Chapter 2 that the full circle of the sky can be divided into 360 degrees. The apparent diameter of the full Moon in the sky averages about half a degree. Just as an hour on the clock is divided into minutes and seconds, a degree of sky can be divided into arcminutes and arcseconds. An **arcminute** (abbreviated **arcmin**) is 1/60 of a degree. An **arcsecond** (abbreviated **arcsec**) is 1/60 of an arcminute and 1/3600 of a degree. To give you a better sense of these scales, an arcsecond is approximately the angle formed by the diameter of a golf ball viewed 9 km away from you.

We often use units of light-years to indicate distances to stars (Chapter 1). One light-year is the distance that light travels in 1 year—about 9.5×10^{12} (9.5 trillion) kilometers (km). Another common unit for discussing distances to stars and galaxies is the **parsec (pc)**, which is equal to 3.26 light-years (or 206,265 AU). The term is short for *parallax second*—a star at a distance of 1 parsec has a parallax of 1 arcsecond.

When astronomers began to measure the parallax angles of stars, they discovered that even the closest stars are very far away (**Working It Out 13.1**). F. W. Bessel (1784–1846) made the first successful measurement of stellar parallax in 1838, reporting a parallax of 0.314 arcsec for the star 61 Cygni. That finding indicated that 61 Cygni was 3.2 pc away, or 660,000 times as far away as the Sun. With that one measurement, Bessel increased the known volume of the universe by a factor of 10,000. Today, astronomers know of at least 60 stars, some in multiple-star systems, within 5 pc (16.3 light-years) of the Sun. In the neighborhood of the Sun, each star or star system has on average a volume of about 300 cubic light-years of space to itself.

Most stars are so far away that the parallax angle is too small to measure using ground-based telescopes, which are limited by Earth's atmosphere. In the 1990s, the European Space Agency's Hipparcos satellite measured the positions and parallaxes of over 100,000 stars with an uncertainty of about ±0.001 arcsec. Then, in 2013, the European Space Agency launched *Gaia*, a space mission that has been studying stellar parallaxes. *Gaia*'s second data release in 2018 included parallaxes of more than 1 billion stars, with uncertainties of ±0.00004 to ±0.0007 arcsec, depending on the star's brightness.

An uncertainty in the parallax will mean a corresponding uncertainty in the distance. For example, a star measured by Hipparcos to have a parallax of 0.004 ± 0.001 arcsec has a parallax between 0.003 and 0.005 arcsec. That range of parallaxes gives a corresponding distance range of 200–333 pc from Earth; a range of 133 pc. If the parallax measured by *Gaia* is 0.004 ± 0.0001 arcsec, then the distance range is 244–256 parsecs; a range of only 12 pc. Astronomers are very excited about these recent data from *Gaia*, which yield very small uncertainties compared to previously measured distances. Nevertheless, nearly all stars are too far away to find their distances by parallax. Other methods of measuring distance to stars too remote for parallax are discussed later in this chapter.

Luminosity, Brightness, and Distance

The stars in Earth's sky have noticeably different brightnesses, where brightness corresponds to the amount of energy falling on a square meter of area each second in the form of electromagnetic radiation (Chapter 5). Although a star's brightness can be measured directly, it does not immediately give much information about

working it out 13.1

Parallax and Distance

If star B in Figure 13.2 is twice as far away from us as star A, then star B will have half the parallax of star A. That is, the parallax of a star (*p*) is inversely proportional to its distance (*d*):

$$p \propto \frac{1}{d} \quad \text{or} \quad d \propto \frac{1}{p}$$

A star with a parallax of 1 arcsec is at a distance of 1 pc (Figure 13.2). The inverse proportionality between distance and parallax becomes

$$\left(\begin{array}{c}\text{Distance measured} \\ \text{in parsecs}\end{array}\right) = \frac{1}{\left(\begin{array}{c}\text{Parallax measured} \\ \text{in arcseconds}\end{array}\right)}$$

or

$$d(\text{pc}) = \frac{1}{p(\text{arcsec})}$$

Suppose that the parallax of a star is measured to be 0.5 arcsec. The distance can be found by

$$d(\text{pc}) = \frac{1}{0.5} = 2\,\text{parsecs}$$

Similarly, a star with a measured parallax of 0.01 arcsec is located at a distance of 1/0.01 = 100 pc.

Astronomers use the unit parsecs because doing so makes the relationship between distance and parallax easier to manage than when the distance is measured in light-years. One parsec equals 206,265 AU (see Appendix 1), which is 3.09×10^{13} km or 3.26 light-years.

After the Sun, the next-closest star to Earth is Proxima Centauri. Located at a distance of 4.24 light-years, Proxima Centauri is a faint member of a system of three stars called Alpha Centauri. What is that star's parallax? First, we convert the distance to parsecs:

$$d = 4.24\,\text{light-years} \times \frac{1\,\text{parsec}}{3.26\,\text{light-years}} = 1.30\,\text{parsecs}$$

Then,

$$p(\text{arcsec}) = \frac{1}{1.30\,\text{pc}} = 0.77\,\text{arcsec}$$

Even the closest star to the Sun has a parallax of only about 3/4 arcsec.

---

the star itself. As illustrated in **Figure 13.3**, a star may appear bright only because it is nearby. Conversely, a faint star may actually be very luminous but appear faint because it is far away.

Astronomers measure the brightness of stars by comparing them with one another. The system they use dates back 2100 years, when the Greek astronomer Hipparchus classified stars according to their brightness: the brightest stars he could see were magnitude 1, and the faintest were magnitude 6. The details of his system, still in use today, are discussed in **Working It Out 13.2** and Appendix 7.

To learn about the actual properties of a star, astronomers need to know the total energy radiated by a star each second—the star's luminosity. Recall from Chapter 5 that the brightness of an object that has a known luminosity *L* and is located at a distance *d* is given by

$$\text{Brightness} = \frac{\text{Total light emitted per second}}{\text{Area of sphere of radius } d} = \frac{\text{Luminosity}}{4\pi d^2}$$

You can rearrange that equation, moving the quantities you can measure (distance and brightness) to the right-hand side and the quantity you would like to know (luminosity) to the left, to get

$$\text{Luminosity} = 4\pi d^2 \times \text{Brightness}$$

That equation is used to find how luminous a star must be to appear as bright as it does when seen from Earth.

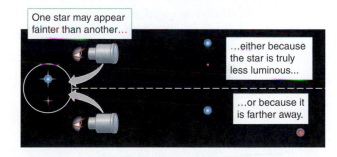

Figure 13.3 The brightness of a star visible in our sky depends on both its luminosity—how much light it emits—and its distance.

working it out 13.2

The Magnitude System

The **magnitude** system of brightness for celestial objects has been adjusted since Hipparchus. Astronomers have defined 1st magnitude stars to be exactly 100 times brighter than 6th magnitude stars. The system has been extended to stars brighter than 1st magnitude by using zero and negative numbers. Traditionally, Vega, a bright star in the constellation Lyra, is considered to have a magnitude of 0. A negative magnitude signifies that an object is *brighter* than Vega. For example, Sirius, the brightest star in the night sky, has a magnitude of −1.46. Venus can be as bright as magnitude −4.4, or about 15 times brighter than Sirius and bright enough to cast a shadow. The full Moon's magnitude is −12.6, and the Sun's is −26.7 (**Figure 13.4**). Hipparchus must have had typical eyesight because an average person under dark skies can see stars only as faint as 6th magnitude. Today, telescopes extend our vision to very faint objects. The Hubble Space Telescope (HST) can take long exposures and detect stars as faint as 30th magnitude.

With five steps between the 1st and 6th magnitudes, each step is equal to the fifth root of 100, or $100^{1/5}$, approximately 2.512. The magnitude system is logarithmic, but instead of the usual base 10, it is base 2.512. Fifth magnitude stars are 2.512 times brighter than 6th magnitude stars, and 4th magnitude stars are $2.512 \times 2.512 = 6.310$ times brighter than 6th magnitude stars. The brightness ratio between any two stars is equal to $(2.512)^N$, where N is the magnitude difference between them.

The limit of the Hubble Space Telescope is magnitude 30, whereas the limit of the human eye is magnitude 6. The difference is 24 magnitudes, so the HST can detect stars that are $(2.512)^{24} = 4 \times 10^9$, or 4 billion, times fainter than the magnitude 6 that the naked eye can see. Or, if we compare the Sun and the Moon, the Sun is 14 magnitudes brighter, or $(2.512)^{14} = 4 \times 10^5$; the Sun is 400,000 times brighter than the full Moon. (For more detailed calculations and a table of magnitudes and brightness differences, see Appendix 7.)

The magnitude of a star, as we have discussed it so far, is called the star's **apparent magnitude** because it is the star's brightness as it *appears* in Earth's sky. If all stars were located exactly 10 pc (32.6 light-years) away from Earth, the brightness of each star would be proportional to its luminosity. If the distance from Earth to a star is known, astronomers compute how bright the star would appear if it were located at 10 pc. A star's **absolute magnitude**— its apparent magnitude at a distance of 10 pc—specifies the star's luminosity.

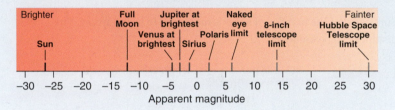

Figure 13.4 Apparent magnitude indicates the apparent brightness of an object in our sky. The brightest objects have a negative apparent magnitude, while telescopes have extended the observable range to fainter objects with higher magnitudes.

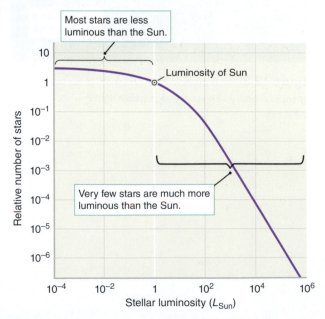

Figure 13.5 The distribution of the luminosities of stars is plotted logarithmically, with increments in powers of 10.

Stars have a wide range of luminosities—the most luminous are more than 10 billion (10^{10}) times more luminous than the least luminous. The Sun provides a convenient unit when measuring the properties of other stars, including their luminosity. The measured luminosity of the Sun is $L_{Sun} = 3.8 \times 10^{26}$ watts (W). The most luminous stars are a million times more luminous than the Sun ($10^6 L_{Sun}$). Very few stars are this luminous. The least luminous stars have luminosities less than 1/10,000 that of the Sun ($10^{-4} L_{Sun}$). Most stars are less luminous than the Sun. The graph in **Figure 13.5** shows how many stars of a particular luminosity exist for every star like the Sun. For every star like the Sun, there are about 3 stars that are 1/100 as luminous as the Sun. For every star like the Sun, there is only roughly 0.1 star that is 90 times as luminous. It's awkward to think about 0.1 star, so we often turn this around, to say that for every star that is 90 times as luminous as the Sun, there are 10 stars like the Sun. (Distances for the nearest stars are obtained from their parallaxes; other methods—discussed later in this chapter—are used for the more distant stars.) Two everyday concepts—stereoscopic vision and the fact that objects appear brighter when closer—have provided the tools needed to measure the distance and luminosity of the closest stars. In Section 13.2, we explain how the laws of radiation described in Chapter 5 reveal still more about stars.

CHECK YOUR UNDERSTANDING 13.1

Stars A and B appear equally bright, but star A is twice as far away from us as star B. Which of the following is true? (a) Star A is twice as luminous as star B. (b) Star A is 4 times as luminous as star B. (c) Star B is twice as luminous as star A. (d) Star B is 4 times as luminous as star A. (e) Star A and star B have the same luminosity because they have the same brightness.

Answers to Check Your Understanding questions are in the back of the book.

13.2 Astronomers Can Determine the Temperature, Size, and Composition of Stars

The gas in the outer layers of stars is dense enough that the radiation from a star comes close to obeying the same laws as the radiation from solid objects, such as the heating element on an electric stove. We can therefore use what you already learned about blackbody radiation (Chapter 5) to understand the radiation from stars. According to the Stefan-Boltzmann law, if two objects are the same size, then the hotter object is more luminous. And, according to Wien's law, the hotter object is also bluer. In this section, we explain how astronomers use those two laws to measure the temperatures and sizes of stars. We also provide a more detailed discussion of the absorption and emission of spectral lines mentioned in Chapter 5 to learn about the composition of stars.

Wien's Law Revisited: The Color and Surface Temperature of Stars

Wien's law relates an object's temperature to the peak wavelength of its spectrum (see Working It Out 5.3 for details); as the object's surface temperature increases, the light that it emits gets bluer. Stars with especially hot surfaces (tens of thousands of kelvins) are blue, stars with especially cool surfaces (a few thousand kelvins) are red, and the stars in between, such as the Sun (about 6000 K), are yellow-white. A more precise measurement of a star's spectrum can be used to find the wavelength at which the spectrum peaks, and then Wien's law can tell you the temperature of the star's surface. A star's color tells you about the temperature only at the surface because that layer is giving off most of the radiation that you see. (Stellar interiors are far hotter, as we discuss in Chapter 14.)

▶❚❚ **AstroTour:** Stellar Spectrum

   Astronomers can determine the temperature of a star without a spectrum. They often measure the colors of stars by comparing the brightness at two specific wavelengths. The brightness is usually measured through an optical **filter**—sometimes just a piece of colored glass—that lets through only a small range of wavelengths. Two of the most common are a blue filter, which allows light with wavelengths of about 440 nanometers (nm) to pass through, and a "visual" (yellow-green) filter, which allows light with wavelengths of about 550 nm to pass through. The ratio of brightness between the blue and visual filters is called the **color index** of the star (see Appendix 7 for more details). From a pair of pictures of a group of stars, each taken through a different filter, astronomers can find an approximate value of the surface temperature of every star in the picture—perhaps hundreds or even thousands—all at once. Based on this type of analysis, most stars have surface temperatures lower than that of the Sun.

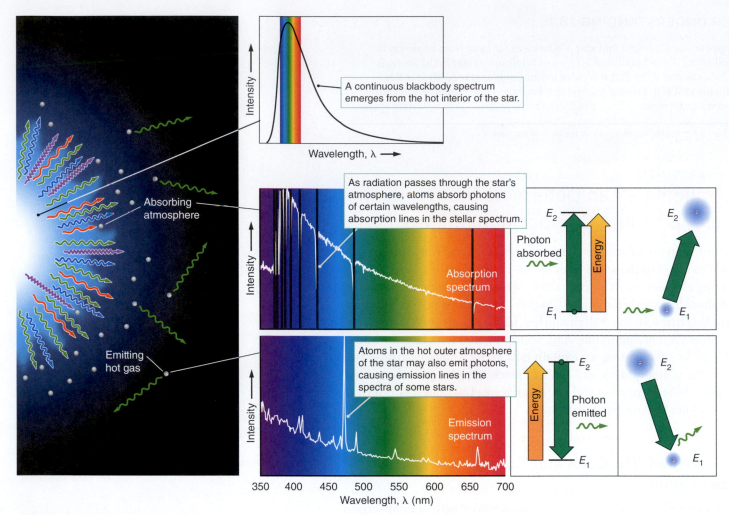

A continuous blackbody spectrum emerges from the hot interior of the star.

As radiation passes through the star's atmosphere, atoms absorb photons of certain wavelengths, causing absorption lines in the stellar spectrum.

Atoms in the hot outer atmosphere of the star may also emit photons, causing emission lines in the spectra of some stars.

Figure 13.6 Absorption and emission lines both appear in the spectra of stars. The Planck blackbody spectrum is the light emitted from a hot object just because it is hot. As that light passes through a gas, some of it is absorbed, producing an absorption spectrum. Hot gas also emits light and produces emission lines in the spectra of some stars.

Classifying Stars by Surface Temperature

Although the hot "surface" of a star emits radiation with a spectrum very close to a smooth Planck blackbody curve, that light must then escape through the outer layers of the star's atmosphere. The atoms and molecules in the cooler layers of the star's atmosphere leave their absorption line fingerprints in the escaping light (**Figure 13.6**). Under some circumstances, the atoms and molecules in the star's atmosphere, along with any gas that might be found near the star, can also produce emission lines in stellar spectra. Absorption and emission lines complicate how astronomers use the laws of Planck blackbody radiation to interpret light from stars, but spectral lines provide a wealth of information about the state and composition of the gas in a star's atmosphere.

The spectra of stars were first classified during the late 1800s, long before stars, atoms, and radiation were well understood. Stars were classified by the appearance of the dark bands (now known to be absorption lines) seen in their spectra. The original ordering of that classification was arbitrarily based on the prominence of particular absorption lines known to be associated with the element hydrogen. Stars with the strongest hydrogen lines were denoted *A stars*, stars with weaker hydrogen lines were denoted *B stars*, and so on.

Annie Jump Cannon (1863–1941) led an effort at the Harvard College Observatory to examine and classify the spectra of hundreds of thousands of stars systematically. She dropped many of the earlier spectral types, keeping only seven that were later reordered on the basis of surface temperatures. Spectra of stars of several spectral types are shown in **Figure 13.7**. The hottest stars, with surface temperatures above 30,000 K, are denoted *O stars*. O stars have only weak absorption lines from hydrogen and helium. The coolest stars—*M stars*—have temperatures as low as about 2800 K. M stars have absorption lines from many types of atoms and molecules. The sequence of **spectral types** of stars, from hottest to coolest, is O, B, A, F, G, K, M. That sequence has undergone several modifications, most recently to add cooler objects known as brown dwarfs with spectral types L, T, and Y.

Astronomers divide the main spectral types into a finer sequence of subclasses by adding numbers to the letter designations. For example, the hottest B stars are B0 stars, slightly cooler B stars are B1 stars, and so on. The coolest B stars are B9 stars, which are only slightly hotter than A0 stars. The boundaries between spectral types are not always easy to determine. A hotter-than-average G star is very similar to a cooler-than-average F star. The Sun is a G2 star.

In Figure 13.7, notice that hot stars emit more blue light than cool stars, as described by Wien's law. Notice also that the absorption lines in their spectra are different. The temperature of the gas in the atmosphere of a star affects the state of the atoms in that gas, affecting in turn the energy level transitions available to absorb radiation (see Section 5.2 to review atomic energy levels). In O stars, the temperature is so high that most atoms have had one or more electrons stripped from them by energetic collisions within the hot gas. As a result, few transitions are available in the visible part of the electromagnetic spectrum, making the visible spectrum of an O star relatively featureless. At lower temperatures, more atoms are in energy states that can absorb light in the visible part of the spectrum, so the visible spectra of cooler stars are far more complex than the spectra of O stars.

All absorption lines have a temperature at which they are strongest. For example, absorption lines from hydrogen are most prominent at temperatures of about 10,000 K, the surface temperature of an A star. At the very lowest stellar temperatures, atoms in a star's atmosphere react with one another, forming molecules. Molecules such as titanium oxide (TiO) are responsible for much of the absorption in the atmospheres of cool M stars.

Because different spectral lines are formed at different temperatures, astronomers can use those absorption lines to measure a star's temperature. The surface temperatures of stars measured in that way agree extremely well with the surface temperatures of stars measured by using Wien's law, confirming once again that the physical laws that apply on Earth apply to stars as well.

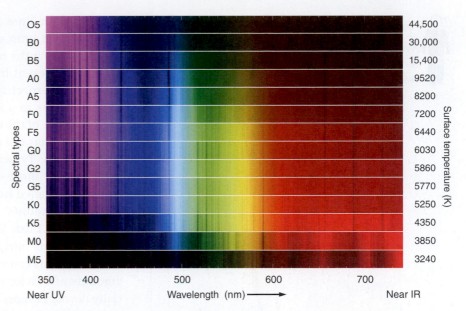

Figure 13.7 Visible light spectra of stars with different spectral types, ranging from hot, blue O stars to cool, red M stars. Hotter stars are more luminous at shorter wavelengths. The dark lines are absorption lines. These spectra have been adjusted to show only the brightest part of each spectrum. Comparing the O stars to the M stars makes clear that as the temperature decreases, the brightest part of the spectrum moves from the left (blue end of the spectrum) to the right (red end of the spectrum).

The Composition of Stars

Most variations in the lines of a particular chemical element seen in stellar spectra are due to temperature, but the details of the absorption and emission lines found in starlight also reveal a star's composition.

Each type of atom has different energy levels and thus different spectral lines (Chapter 5). The spectral line patterns are measured in laboratories on Earth and then used to identify the atoms (or molecules) present in stars. For example, if a star

⦿ **what if . . .**

What if you observe two stars that have very similar peak wavelengths for their spectra but very different spectral line strengths? What can you conclude from your observations?

Table 13.1	Relative Amounts of Chemical Elements in the Atmosphere of the Sun	
Element	Percentage of Atoms in the Sun	Percentage of Sun's Mass
Hydrogen	92.5	74.5
Helium	7.4	23.7
Oxygen	0.064	0.82
Carbon	0.039	0.37
Neon	0.012	0.19
Nitrogen	0.008	0.09
Silicon	0.004	0.09
Magnesium	0.003	0.06
Iron	0.003	0.16
Sulfur	0.001	0.04
Total of others	0.001	0.03

has absorption lines that correspond to the energy difference between two levels in the calcium atom, we know that calcium is present in the star's atmosphere.

The strengths of various absorption lines tell us not only what kinds of atoms are present in the gas but also the amount of each. However, we must carefully interpret spectra to account properly for the temperature and density of the gas in a star's atmosphere. Recall Figure 5.17—the "astronomer's periodic table of the elements." In 1925, Cecilia Payne-Gaposchkin (1900–1979) discovered that the composition of stars is very different from the composition of Earth. Typically, stars are composed of 90 percent hydrogen. Helium accounts for most of what remains, and all the other chemical elements, collectively called **heavy elements** or **massive elements**, are present in only very small amounts.

Table 13.1 lists the elements that make up the Sun's atmosphere. The Sun's composition is fairly typical for stars nearby, but the percentages of specific heavy elements can vary tremendously from star to star. Some stars have smaller amounts of heavier elements than the Sun. The existence of such stars, essentially devoid of more massive elements, provides important clues about the origin of chemical elements and the chemical evolution of the universe. Many of the atoms that make up Earth and its atmosphere (such as iron, silicon, nitrogen, oxygen, and carbon) exist as only a small percentage of the Sun.

By applying the physics of atoms and molecules to the study of stellar absorption lines, astronomers can accurately determine not only surface temperatures of stars but also pressures, chemical compositions, magnetic-field strengths, and other physical properties. In addition, by using the Doppler shift of emission and absorption lines, astronomers can measure rotation rates, atmospheric motions, expansion and contraction, "winds" driven away from stars, and other dynamic properties of stars. Stellar spectroscopy is one of the most powerful tools astronomers have for exploring the universe.

The Stefan-Boltzmann Law and Finding the Sizes of Stars

Stars are so far away that most cannot be imaged as more than point sources. Finding a star's radius, then, involves putting together other observable measurements. The Stefan-Boltzmann law relates an object's temperature to the amount of light that it emits from every square meter of its surface. The "from every square meter" part is important: if two stars have the same temperature and are the same distance from us, the larger one will be brighter—namely, it emits more light because it has more surface area. Unfortunately, stars are rarely at the same distance from us, or at the same temperature, so typically a few more steps are necessary to figure out which star is larger.

First, astronomers find a star's temperature directly, either from its color through Wien's law (**Figure 13.8a**) or from the strength of its spectral lines. The temperature of a star's surface is one factor that influences its luminosity. Second, astronomers find the distance to the star, perhaps from parallax. Third, astronomers find the luminosity from its brightness (easily measured) and its distance. Once an astronomer has determined both the temperature and the luminosity of the star, she is able to calculate the star's radius using the *luminosity-temperature-radius relationship*, which connects the three quantities, as shown in **Working It Out 13.3**.

As the Stefan-Boltzmann law states, if a large star and a small star are the same temperature, they will emit the same energy from every patch of surface. However,

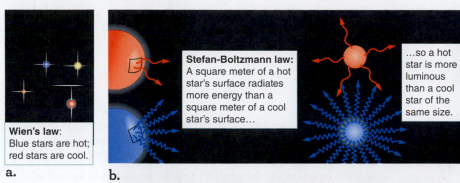

Figure 13.8 a. The temperature of a star can be found from its color using Wien's law. **b.** The luminosity depends on both the temperature and the size of the star. Blue stars are hotter; red stars are cooler. **c.** Once the temperature and the luminosity are known, the size of the star can be calculated.

the large star has more patches, so it will be more luminous altogether. Conversely, if two stars are the same size, but one is hotter than the other, the hot star will emit more light (**Figure 13.8b**). A small, hot star might even be more luminous than a larger cool star. Astronomers combine the temperature and the luminosity to find the star's radius (**Figure 13.8c**). A low-luminosity hot star must be small, whereas a high-luminosity cool star must be large.

Astronomers have used the luminosity-temperature-radius relationship to estimate the radii of many thousands of stars. The range of stellar sizes is smaller than the range of stellar luminosities: the radius of the largest star is 100,000 times larger than the radius of the smallest star. As with stellar luminosities, it's convenient to use the Sun as a unit of measure for size. The Sun's radius, designated R_{Sun}, is 696,000 km. One of the smallest types of stars, called white dwarfs, have radii only about 1 percent of the Sun's radius ($0.01\,R_{Sun}$)—about the size of Earth. The largest stars, called red supergiants, can have radii more than 1000 times that of the Sun ($1000\,R_{Sun}$). Many more stars are toward the small end of that range—smaller than the Sun—than giant stars.

CHECK YOUR UNDERSTANDING **13.2**

If star A has twice the surface temperature of the Sun but has the same luminosity, the radius of star A must be _____ the radius of the Sun. (a) 16 times; (b) 4 times; (c) 1/2; (d) 1/4

working it out 13.3

Estimating the Sizes of Stars

According to the Stefan-Boltzmann law (Chapter 5), the amount of energy radiated each second by each square meter of a star's surface is equal to the constant σ multiplied by the star's surface temperature raised to the fourth power:

$$\left(\begin{array}{c} \text{Energy radiated each} \\ \text{second by } 1\,\text{m}^2 \text{ of surface} \end{array} \right) = \sigma T^4$$

To find the total amount of light radiated each second by a star, we need to multiply the radiation emitted per second from each square meter by the number of square meters of the star's surface:

$$\left(\begin{array}{c} \text{Energy radiated} \\ \text{each second} \end{array} \right) = \left(\begin{array}{c} \text{Energy radiated each} \\ \text{second by } 1\,\text{m}^2 \text{of surface} \end{array} \right) \times \left(\begin{array}{c} \text{Surface} \\ \text{area} \end{array} \right)$$

The left-hand term in that equation—the total energy emitted by the star per second (in units of joules per second [J/s] = watts [W])—is the star's luminosity, L. The middle term—the energy radiated by each square meter of the star per second (in units of joules per square meter per second [J/m²/s])—can be replaced with the σT^4 factor from the Stefan-Boltzmann law. The remaining term—the number of square meters covering the star's surface—is the surface area of a sphere, $A_{\text{sphere}} = 4\pi R^2$ (in units of square meters [m²]), where R is the star's radius.

If we replace the words in the equation with the appropriate mathematical expressions for the Stefan-Boltzmann law and the area of a sphere, the equation for the luminosity becomes

$$L = \sigma T^4 \times 4\pi R^2$$

Combining gives

$$L = 4\pi R^2 \sigma T^4 \text{ J/s (W)}$$

That last equation is called the **luminosity-temperature-radius relationship** for stars. Because the constants (4, π, and σ) do not change, the star's luminosity is proportional only to $R^2 T^4$. Make a star 3 times as large, and its surface area becomes $3^2 = 9$ times as large. Because 9 times as much area is present to radiate, 9 times as much radiation is emitted. Make a star twice as hot, and each square meter of the star's surface radiates $2^4 = 16$ times as much energy. Thus, larger, hotter stars are more luminous than smaller, cooler stars.

Now, how large must a star of a given temperature be to have a total luminosity of L? The star's luminosity (L) and temperature (T) are measurable quantities, and the star's radius (R) is what we want to know. We can rearrange the previous equation, moving the properties that we know how to measure (temperature and luminosity) to the right-hand side of the equation and the property that we would like to know (the star's radius) to the left-hand side. After doing some algebra, we find that

$$R = \sqrt{\frac{L}{4\pi\sigma T^4}} = \frac{1}{T^2} \times \sqrt{\frac{L}{4\pi\sigma}}$$

Again, the right-hand side of the equation contains only things that we know or can measure. The constants 4, π, and σ are always the same. We can find L, the star's luminosity, from the measurements of the star's brightness and parallax (although only for nearby stars with known parallax). The star's surface temperature is T, which can be measured from its color. From the relationship of those measurements, we now know something new: the size of the star.

If we compare two stars, the constants all cancel out, leaving L, T, and R:

$$\frac{L_{\text{star 1}}}{L_{\text{star 2}}} = \frac{R_{\text{star 1}}^2}{R_{\text{star 2}}^2} \times \frac{T_{\text{star 1}}^4}{T_{\text{star 2}}^4}$$

Rearranging the terms, and taking the square root on both sides:

$$\frac{R_{\text{star 1}}}{R_{\text{star 2}}} = \sqrt{\frac{L_{\text{star 1}}}{L_{\text{star 2}}}} \times \frac{T_{\text{star 2}}^2}{T_{\text{star 1}}^2}$$

How does the Sun compare with the second-brightest star in the constellation Orion, a red star called Betelgeuse? From its spectrum, we know that Betelgeuse's surface temperature T is about 3500 K. Its distance is about 200 pc, and from that and its brightness, its luminosity is estimated to be 140,000 times that of the Sun. What can we say about the size of Betelgeuse? Using the preceding equation, we can determine the following:

$$\frac{R_{\text{Betelgeuse}}}{R_{\text{Sun}}} = \sqrt{\frac{L_{\text{Betelgeuse}}}{L_{\text{Sun}}}} \times \frac{T_{\text{Sun}}^2}{T_{\text{Betelgeuse}}^2}$$

$$\frac{R_{\text{Betelgeuse}}}{R_{\text{Sun}}} = \sqrt{\frac{140,000}{1}} \times \frac{5,800^2}{3,500^2} = \approx 370 \times 2.7 \approx 1,000$$

Thus, Betelgeuse has a radius about 1000 times larger than that of the Sun, making it a *supergiant*.

13.3 Measuring the Masses of Stars in Binary Systems

Determining a star's mass is difficult. Astronomers cannot use the amount of light from a star or the star's size as a measure of its mass, because stars can swell, contract, heat up, or cool down as they age. However, more massive stars *always* have stronger gravity. When astronomers are trying to determine the masses of astronomical objects, they almost always wind up looking for the effects of gravity. In Chapter 4, you learned that Kepler's laws of planetary motion are the result of gravity and that the properties of a planet's orbit can be used to measure the mass of the Sun. Similarly, astronomers can study two stars that orbit each other to determine their masses.

Many stars in our galaxy are in systems consisting of several stars moving about under the influence of their mutual gravity. Most of those systems are **binary stars**, in which two stars orbit each other in elliptical orbits with many of the same properties as the elliptical orbits described by Kepler's laws. In fact, Newton's version of Kepler's third law can be used to find the masses of the stars in a binary system, as we will demonstrate. However, most low-mass stars are single, and most stars are low-mass stars, so most stars are single and their mass cannot be calculated. We next explain how astronomers find the masses of stars in binary systems.

Binary Star Orbits

The **center of mass** is the balance point of a system. If two objects were sitting on a seesaw in a gravitational field, the support of the seesaw would have to be directly under the center of mass for the objects to balance (**Figure 13.9**). When Newton applied his laws of motion to the problem of orbits, he found that two objects must move in elliptical orbits around each other and that their common center of mass lies at one focus shared by both ellipses (**Figure 13.10**). The center of mass, which lies along the line between the two objects, remains stationary. The two objects will always be found on exactly opposite sides of the center of mass.

Imagine a binary star, as shown in **Figure 13.11a**. As seen from above, two stars orbit the common center of mass. Star 1, which is less massive, must complete its orbit in the same time as star 2, which is more massive. Because the less massive star has farther to go around the center of mass, it must be moving *faster* than the more massive star. If an astronomer can find the stars' velocities in their orbits, she can find each star's mass. However, in this face-on view, the Doppler shift (Chapter 5) cannot be measured because all the motion is in the plane of the sky, and none is toward or away from the observer.

When a system is edge-on to the observer, however, the astronomer can take advantage of the Doppler shift to find out about the motion. **Figure 13.11b** shows observations of the combined system's spectrum for each position in Figure 13.11a, if it were edge-on instead of face-on. Because the two stars are always on exactly opposite sides of their center of mass, they always move in opposite directions. Thus, when star 2 approaches, star 1 recedes. Because of the Doppler effect, the light coming from star 2 will be shifted to *shorter* wavelengths as it approaches, so the light will be *blue*shifted, and the light coming from star 1 will be shifted to *longer* wavelengths as it recedes, so the light will be *red*shifted. Half an orbital period later, the situation is reversed: lines from star 1 are blueshifted, and lines from star 2 are redshifted.

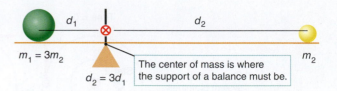

Figure 13.9 The center of mass of two objects is the "balance" point on a line joining the centers of two masses.

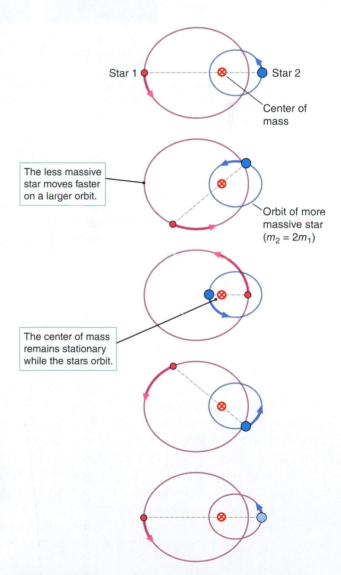

Figure 13.10 In a binary star system, the two stars orbit on elliptical paths about their common center of mass. Here, the blue star has twice the mass of the red one. The eccentricity of the orbits is 0.5. The time steps between the frames are equal.

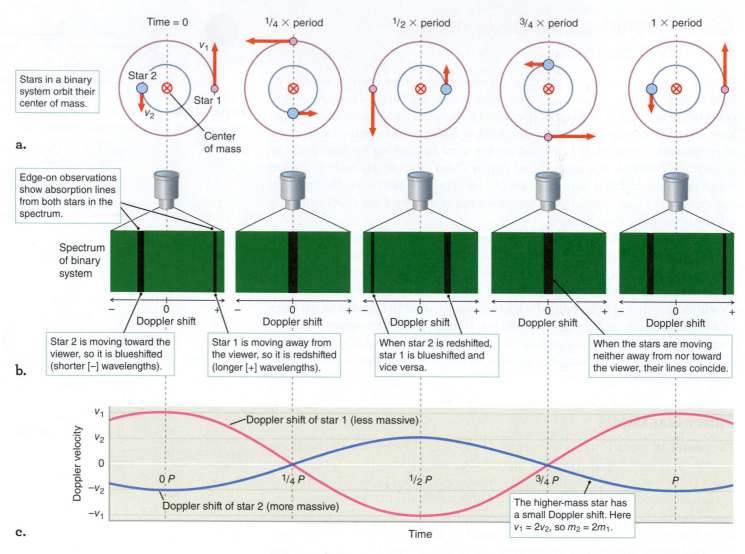

Figure 13.11 a. The view from "above" the binary system shows that both stars are on circular orbits around a common center of mass. b. The spectrum of the combined system (seen edge-on) shows that the spectral lines of each star shift back and forth. c. Graphing the Doppler shift of star 1 with star 2 versus time reveals that star 2 has half the maximum Doppler shift, so star 2 is twice as massive as star 1. P is the period of the orbit.

The less massive star, star 1, has a larger orbit, but both stars must orbit in the same amount of time in order to stay on opposite sides of the center of mass. Therefore, star 1 must move more quickly in order to travel around this larger orbit in the same amount of time that star 2 takes to complete a smaller orbit: there is an inverse proportion between velocity and mass. The velocities can be determined by finding the maximum value of the Doppler shift in a series of spectra taken throughout the orbit. **Figure 13.11c** compares the velocity obtained for star 1 with the velocity obtained for star 2. Because the velocity is inversely proportional to the mass, that comparison gives the ratio of the masses of the two stars:

$$\frac{v_1}{v_2} = \frac{m_2}{m_1}$$

By observing the system's spectrum, we can find from that equation the relative masses of the two stars—star 2 is 2 times as massive as star 1—but we can't find the actual mass of either star from those observations alone.

The Masses of Binary Stars

In Chapter 4, we ignored the complexity of the motion of two objects around their common center of mass, because one object was so much more massive than the other. Now, however, that very complexity enables us to measure the masses of the two stars in a binary system. If we can measure the binary system's period and the average separation between the two stars, Newton's version of Kepler's third law gives us the total mass in the system: the sum of the two masses. Because the analysis in the previous subsection gives us the ratio of the two masses, we now have two relationships between two unknowns. We have all we need to determine the mass of each star separately. In other words, if we know that star 2 is 2 times as massive as star 1, and we know that star 1 and star 2 together are 3 times as massive as the Sun, we can calculate separate values for the masses of star 1 and star 2.

Depending on the type of system, there are several ways to measure the average separation and the orbital period. In a **visual binary** system, because the system is close enough to Earth, and the stars are far enough from each other, we can take pictures that show the two stars separately (**Figure 13.12**). Then, astronomers can directly measure the shapes and periods of the two stars' orbits just by watching them as they orbit each other. Those values can be used with Doppler measurements of the stars' radial (line-of-sight) velocities to solve for the ratio of the two masses.

In most binary systems, however, the two stars are so close together and so far away from us that we cannot see the stars separately. The identification of those stars as binary systems is indirect and comes from observing periodic variations in the *light* from the star or from observing periodic changes in the star's *spectrum*. If we view a binary system nearly edge-on, so that one star passes in front of the other, it is called an **eclipsing binary**. An observer will see a repeating dip in brightness as one star passes in front of (eclipses) the other (**Figure 13.13**). If the stars have different temperatures, a repeating pattern will occur, showing a smaller dip in brightness when the hotter star eclipses the cooler one, followed by a larger dip in brightness when the cooler star eclipses the hotter one. The pattern of those dips also gives an estimate of the two stars' relative sizes (radii). That procedure for identifying binary systems—similar to the transit method for finding exoplanets discussed in Chapter 7—works only when the system is viewed nearly edge-on. The Kepler space telescope has discovered thousands of eclipsing binaries in addition to finding new exoplanets.

In a **spectroscopic binary** system, the two stars' spectral lines show periodic changes as they are Doppler-shifted away from each other, first in one direction and then in the other (Figure 13.11). The orbit's period is determined from the time it takes for a set of spectral lines to go from approaching to receding and back again. The stars' orbital velocities and the period of the orbit give the orbit's size because distance equals velocity multiplied by time. Consequently, astronomers can estimate the combined masses of the two stars. To calculate the individual masses, an estimate of the orbit's tilt is needed. Thus, spectroscopic binary masses are more approximate than those in eclipsing binary systems.

A binary system can fall into more than one of those three categories, regardless of how it was originally discovered. If a spectroscopic binary system is also a visual or eclipsing binary, the orbit and masses of the stars can be completely

what if . . .

What if three stars were orbiting about their common center of mass, all in the same plane? Could you still use measured velocities to determine the orbital radii, and would Kepler's third law give you the system's mass?

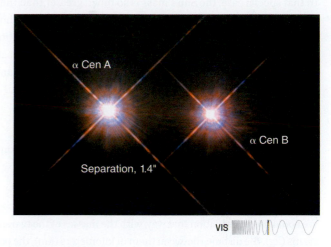

VIS

Figure 13.12 The two stars of this visual binary are resolved. These stars are two components of Alpha Centauri, the nearest star system to the Sun.

★ **WHAT AN ASTRONOMER SEES** When looking at images of stars, astronomers know that the brighter stars appear larger on the image. In this image, then, α Cen A is significantly brighter than α Cen B. Knowing that these two stars are part of a system of stars, an astronomer will further conclude that they are the same distance away. She will then know that α Cen A is not only brighter but also more luminous than α Cen B.

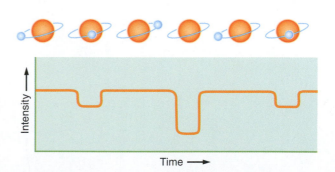

Figure 13.13 In an eclipsing binary system, the system is viewed nearly edge-on, so that the stars repeatedly pass behind each other, blocking some of the light. When the blue star passes in front of the larger, cooler star, less light is blocked than when the smaller, hotter blue star passes behind the red star. The shape of the dips in the light curve of an eclipsing binary can reveal information about the relative size and surface brightness of the two stars.

working it out 13.4

Measuring the Mass of an Eclipsing Binary Pair

In Working It Out 4.3, we used Newton's version of Kepler's third law to calculate the Sun's mass by observing the orbital period of one of its planets. In that special case, the Sun's mass is so much greater than the mass of the planet that the planet's mass is negligible. For two stars, however, both masses are significant, and we need to keep both in the equations. Newton showed that if two objects with masses m_1 and m_2 are in orbit about each other, the period of the orbit, P, is related to the average distance between the two masses, the semimajor axis A, by the equation

$$P^2 = \frac{4\pi^2 A^3}{G(m_1 + m_2)}$$

This equation can be rearranged into the following expression for the sum of the two objects' masses:

$$m_1 + m_2 = \frac{4\pi^2}{G} \times \frac{A^3}{P^2}$$

We could use the equation that way, with the masses of the two stars in kilograms (kg), the distance between them in kilometers (km), the period of their orbit in seconds (s), and the gravitational constant G in km³/(kg s²). However, astronomers often think about stellar masses in units of the Sun's mass. If we divide that equation by the "mass of the Sun" equation in Working It Out 4.3,

$$M_{Sun} = \frac{4\pi^2}{G} \times \frac{A^3}{P^2}$$

(where M_{Sun} = mass of the Sun, $A = 1$ AU, and $P = 1$ year), the constants cancel out and the equation simplifies to

$$\frac{m_1}{M_{Sun}} + \frac{m_2}{M_{Sun}} = \frac{A_{AU}^3}{P_{years}^2}$$

Therefore, if we know both m_1/m_2 from measuring velocities by Doppler shifts and $m_1 + m_2$ from the observed orbital properties, we can solve for the separate values of m_1 and m_2.

Suppose you are an astronomer studying a binary star system. After observing the star for several years, you accumulate the following information about the system:

1. The star is an eclipsing binary.
2. The period of the orbit is 2.63 years.
3. Star 1 has a Doppler velocity that varies between +20.4 and −20.4 km/s.
4. Star 2 has a Doppler velocity that varies between +6.8 and −6.8 km/s.
5. The stars are in circular orbits. You know that because the Doppler velocities about the star are symmetric—that is, for each star the approach and recession speeds are equal.

Those data are summarized in **Figure 13.14**. The star is an eclipsing binary, so the star's orbit is edge-on to your line of sight. The Doppler velocities tell you the total orbital velocity of each star, and you determine the size of the orbits by using the relationship

$$\text{Distance} = \text{Speed} \times \text{Time}$$

In one orbital period, star 1 travels around a circle—a distance of

$$d = (20.4 \text{ km/s}) \times (2.63 \text{ yr}) = 53.7 \text{ km} \times \text{yr/s}$$

solved (**Working It Out 13.4**). Historically, most stellar masses were measured for stars in eclipsing binary systems rather than for those in visual or spectroscopic binaries. New observational capabilities, however, have increased the number of known visual binaries by greatly improving the ability to directly see the stars in a binary. Accurate mass measurements have been obtained for several hundred binary stars, about half of which are eclipsing binaries. The range of stellar masses found in that way is not nearly as great as the range of stellar luminosities. The least massive stars have masses of about 0.08 M_{Sun}, whereas the most massive stars appear to have masses up to about 200 M_{Sun}.

CHECK YOUR UNDERSTANDING 13.3

Which of the following properties must be measured to determine the masses of stars in a typical binary system? (Choose all that apply.) (a) the period of the two stars' orbits; (b) the average separation between the two stars; (c) the two stars' radii; (d) the two stars' velocities.

Multiply by the number of seconds in a year:

$$d = 53.7 \frac{\text{km} \times \text{yr}}{\text{s}} \times \frac{3.16 \times 10^7 \text{s}}{\text{yr}} = 1.70 \times 10^9 \text{km}$$

That distance is the circumference of the star's orbit, or 2π times the radius of the star's orbit, A_1. Thus, star 1 is following an orbit with a radius of

$$A_1 = \frac{d}{2\pi} = \frac{1.70 \times 10^9 \text{km}}{2\pi} = 2.7 \times 10^8 \text{km}$$

To convert that result to astronomical units, use the relation $1 \text{AU} = 1.50 \times 10^8$ km:

$$A_1 = 2.7 \times 10^8 \text{km} \times \frac{1 \text{AU}}{1.50 \times 10^8 \text{km}} = 1.8 \text{AU}$$

A similar analysis of star 2 shows that its orbit has a radius of $A_2 = 0.6$ AU.

Next, apply Newton's version of Kepler's third law. Because the stars are always on opposite sides of the center of mass, the semimajor axis $A_{\text{AU}} = 1.8$ AU + 0.6 AU = 2.4 AU. Because you know A and the period P (measured as 2.63 years), you can calculate the total mass of the two stars:

$$\frac{m_1}{M_{\text{Sun}}} + \frac{m_2}{M_{\text{Sun}}} = \frac{(A_{\text{AU}})^3}{(P_{\text{years}})^2} = \frac{(2.4)^3}{(2.63)^2} = 2.0$$

Thus, the combined mass of the two stars is twice the mass of the Sun. To sort out the stars' individual masses, use the measured velocities and the fact that the mass and velocity are inversely proportional:

$$\frac{m_2}{m_1} = \frac{v_1}{v_2} = \frac{20.4 \text{km/s}}{6.8 \text{km/s}} = 3.0$$

Star 2 is 3 times as massive as star 1. In mathematical terms, $m_2 = 3 \times m_1$. Substituting into the equation

$$m_1 + m_2 = 2.0 M_{\text{Sun}}$$

gives

$$m_1 + 3m_1 = 2.0 M_{\text{Sun}}$$

or $4m_1 = 2.0 M_{\text{Sun}}$, so $m_1 = 0.5 M_{\text{Sun}}$. Because $m_2 = 3 \times m_1$, $m_2 = 1.5 M_{\text{Sun}}$.

Star 1 has a mass of $0.5 M_{\text{Sun}}$, and star 2 has a mass of $1.5 M_{\text{Sun}}$. You have just found the masses of two distant stars.

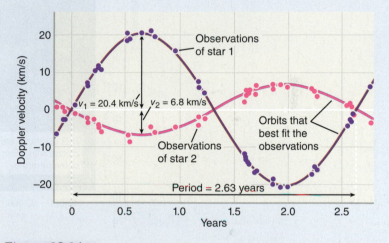

Figure 13.14 Doppler velocities of the stars in an eclipsing binary are used to measure the masses of the stars.

13.4 The Hertzsprung-Russell Diagram Is the Key to Understanding Stars

Measuring basic properties of stars is only the first step in understanding them; the next step is to look for patterns in their properties. In the early 20th century, Ejnar Hertzsprung (1873–1967) and Henry Norris Russell (1877–1957) independently studied the properties of stars. Both astronomers plotted stars' luminosities versus their surface temperatures—a diagram that came to be known as the *Hertzsprung-Russell diagram*, or simply the **H-R diagram**. We use H-R diagrams often to study stars. In this section, we provide a first look at that important diagram and how stars are organized within it.

The H-R Diagram

The H-R diagram (**Figure 13.15**) is a graph of luminosity versus temperature that can be used to study how stars change over time. The spectral type is plotted on

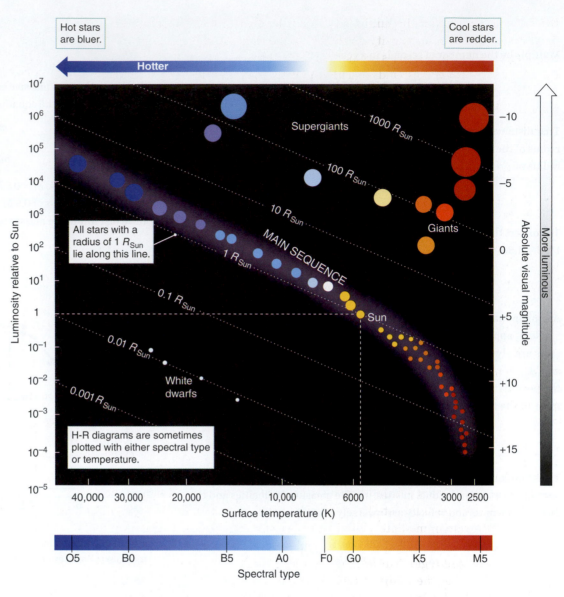

Figure 13.15 The Hertzsprung-Russell (H-R) diagram is used to plot the properties of stars. More luminous stars are at the top of the diagram. Hotter stars are on the left. Stars of the same radius (*R*) lie along the dotted lines moving from upper left to lower right. Absolute magnitudes are discussed in Working It Out 13.2 and Appendix 7.

the horizontal axis (the *x*-axis), along with the surface temperature plotted backward: temperature is higher on the left and lower on the right. Hot, blue stars are on the left side of the H-R diagram, whereas cool, red stars are on the right. Temperature is plotted logarithmically. For example, consider two points on the H-R diagram. One represents a star with a surface temperature of 40,000 K, and the other a star with a surface temperature of 20,000 K—a temperature change of a factor of 2. The interval along the axis between those points is the same size as the interval between points representing stars with surface temperatures of 10,000 and 5000 K—also a temperature change of a factor of 2. The horizontal axis is sometimes labeled with another characteristic that corresponds to temperature, such as the color.

Stars' luminosities are plotted along the vertical axis (the *y*-axis). More luminous stars are toward the top of the diagram; less luminous stars are toward

▶❙❙ **AstroTour:** H-R Diagram

▶▶ **Interactive Simulation:** H-R Diagram

the bottom. Sometimes the luminosity axis is labeled with the absolute visual magnitude instead of luminosity, as shown on the right-hand *y*-axis. As with the temperature axis, luminosities are plotted logarithmically. Here, each step along the left-hand *y*-axis corresponds to a multiplicative factor of 10 in the luminosity. To understand why the plotting is done that way, recall that the most luminous stars are 10 billion times more luminous than the least luminous stars, yet all those stars must fit on the same plot.

Each point on the H-R diagram is specified by a surface temperature and luminosity. Therefore, we can use the luminosity-temperature-radius relationship described earlier in the chapter to find a star's radius at that point as well. A star in the upper right corner of the H-R diagram is very cool, so each square meter of its surface radiates only a small amount of energy. That star is extremely luminous, too. It must be huge, then, to account for its high luminosity, despite the feeble radiation coming from each square meter of its surface. Conversely, a star in the lower left corner of the H-R diagram is very hot, which means that a large amount of energy is coming from each square meter of its surface. That star has a very low overall luminosity, however, so it must be very small. Moving up and to the right takes you to larger and larger stars; moving down and to the left takes you to smaller and smaller stars. All stars of the same radius lie along slanted lines across the H-R diagram. Astronomers can note a star's properties—its temperature, color, size, and luminosity—from a glance at its position on the H-R diagram. The discovery and study of those patterns led to an understanding of the astrophysics of stars (see the **Process of Science Figure**).

The Main Sequence

Figure 13.16 shows 4 million stars plotted on an H-R diagram. The data are based on observations of stars within 5000 light-years of the Sun, which makes them near enough for the *Gaia* satellite to obtain parallax measurements. A quick look at the diagram immediately shows what was first discovered in the original diagrams of Hertzsprung and Russell. About 90 percent of the stars in the sky lie in a well-defined region running across the H-R diagram from lower right to upper left, known as the **main sequence**. On the left end of the main sequence are the O stars: hotter, larger, and more luminous than the Sun. On the right end of the main sequence are the M stars: cooler, smaller, and fainter than the Sun. If you know where a star lies on the main sequence, you know its approximate luminosity, surface temperature, and size.

The H-R diagram offers a useful method for finding the distance to main-sequence stars. Astronomers can determine whether a star is on the main sequence by looking at the absorption lines in its spectrum. The spectral type, too, is determined from the spectral lines, and that spectral type indicates the star's temperature. Once that value on the *x*-axis is known, we can then read up to the main sequence and then across to the *y*-axis to find the star's luminosity. Recall that the luminosity, brightness, and distance are all connected, so we can find the star's distance by comparing its luminosity, obtained from the H-R diagram, with its apparent brightness. That method of determining distances to main-sequence stars from the spectra, luminosity, and brightness of stars is called **spectroscopic parallax**. Details of that method are discussed in Appendix 7. Despite the similarity between the names, the method is very different from the parallax method discussed in Section 13.1. Spectroscopic parallax is useful to much larger distances than the geometric method, although it is less precise.

what if . . .

What if Hertzsprung and Russell had made their diagram a plot of stellar radius versus temperature? How would the main sequence appear in such a diagram?

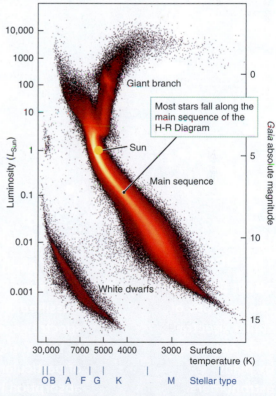

Figure 13.16 An H-R diagram for 4 million stars within 5000 light-years of the Sun plotted from data obtained by the *Gaia* satellite clearly shows the main sequence. Most of the stars lie along this band running from the lower right of the diagram toward the upper left. Credit: ESA/Gaia/DPAC, CC BY-SA 3.0 IGO, https://www.esa.int/ESA_Multimedia/Images/2018/04/Gaia_s_Hertzsprung-Russell_diagram. https://creativecommons.org/licenses/by-sa/3.0/igo/.

Science Is Collaborative

It took decades, and the contributions of dozens of people, all working toward a common goal, to understand the meaning behind stellar data.

Henry Norris Russell

Cecilia Payne-Gaposchkin

Ejnar Hertzsprung

Meghnad Saha

Annie Jump Cannon

1910s
The Graph

Hertzsprung and Russell independently developed what will later be called the H-R diagram. They did not understand why the *x*-axis, ordered O-B-A-F-G-K-M, revealed a nice band across the middle of the diagram. Russell hypothesized this result must be caused by a single stellar characteristic.

1920s
The Understanding

Meghnad Saha showed that the stellar characteristic in question was temperature. Cecilia Payne-Gaposchkin showed that stars were mostly composed of hydrogen and helium. Modern astrophysics was born; others went on to develop the under-standing of stellar atmospheres.

1800s
The Observations

About 500,000 photographs of stellar spectra were obtained by many astronomers at many telescopes.

1900s
The Classification

Annie Jump Cannon led a team that classified all the spectra according to the strengths of particular absorption lines at particular wavelengths.

Scientific discoveries sometimes seem to occur suddenly. Instead, new scientific knowledge usually results from the effort of many people working for many years to solve a problem.

From a combination of observations of binary star masses, parallax, luminosity measurements, and mathematical models, astronomers have determined that stars of different masses lie on different parts of the main sequence. Stellar mass increases smoothly from the lower right to the upper left along the main sequence (**Figure 13.17**). If a main-sequence star is *less* massive than the Sun, it is also smaller, cooler, redder, and less luminous than the Sun and is located to the lower right of the Sun on the main sequence. Conversely, if a main-sequence star is *more* massive than the Sun, it is also larger, hotter, bluer, and more luminous than the Sun and is located to the upper left of the Sun on the main sequence. The mass of a star determines where on the main sequence the star will lie.

Table 13.2 summarizes the properties of the spectral classes of main-sequence stars. *All* main-sequence stars with a mass of 1 M_{Sun} are G2 stars like the Sun and have the *same* surface temperature, size, and luminosity as the Sun. Similarly, if a main-sequence star is classified as B0, it has a surface temperature of about 30,000 K, a luminosity about 32,500 times that of the Sun, a mass of about 17.5 M_{Sun}, and a radius of about 6.7 R_{Sun}. If a different main-sequence star is classified as M5, then it has a surface temperature of 3170 K, a luminosity of about 0.008 L_{Sun}, a mass of about 0.21 M_{Sun}, and a radius of about 0.29 R_{Sun}.

The relationship between the mass and the luminosity of stars is very sensitive. Relatively small differences in the masses of stars result in large differences in their main-sequence luminosities. From determining the luminosities of binary stars with measured mass, a relationship between the mass and luminosity emerged. That **mass-luminosity relationship**, usually expressed as

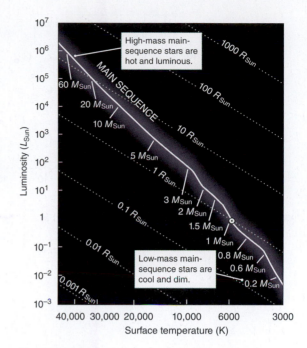

Figure 13.17 Mass determines the location of a star along the main sequence.

Table 13.2	Properties of Main-Sequence Stars			
Spectral Type	Temperature (K)	Mass (M_{Sun})	Radius (R_{Sun})	Luminosity (L_{Sun})
O5	42,000	60	13	500,000
B0	30,000	17.5	6.7	32,500
B5	15,200	5.9	3.2	480
A0	9800	2.9	2.0	39
A5	8200	2.0	1.8	12.3
F0	7300	1.6	1.4	5.2
F5	6650	1.4	1.2	2.6
G0	5940	1.05	1.06	1.25
G2 (Sun)	5780	1.00	1.00	1.0
G5	5560	0.92	0.93	0.8
K0	5150	0.79	0.93	0.55
K5	4410	0.67	0.80	0.32
M0	3840	0.51	0.63	0.08
M5	3170	0.21	0.29	0.008

unanswered questions

What is the *upper* limit for stellar mass? Both theory and observation have shown that the *lower* limit for stellar mass is approximately 0.08 M_{Sun} with a temperature of about 2000 K. However, neither theory nor observation has yielded a definitive value for the upper limit. Many astronomers believe that the upper limit lies somewhere around 150–200 M_{Sun}. The very first stars that formed in the universe might have been even larger.

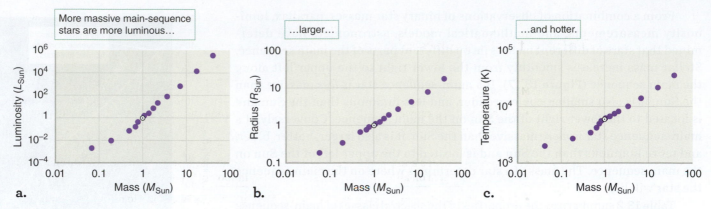

Figure 13.18 These graphs plot **a.** luminosity, **b.** radius, and **c.** temperature versus mass for stars along the main sequence. The mass of a main-sequence star determines its other properties.

$L \propto M^{3.5}$, is shown in **Figure 13.18a**. The exact exponent varies from about 3 to 4 for different ranges of stellar masses, but the method is useful for estimating masses of single stars. The mass also correlates to the size of a star (**Figure 13.18b**) and to its temperature (**Figure 13.18c**).

The mass and chemical composition of a main-sequence star determine how large it is, what its surface temperature is, how luminous it is, what its internal structure is, how long it will live, how it will evolve, and what its final fate will be. A star must have a balance between gravity trying to hold the star together and the energy released by nuclear reactions within the star trying to blow it apart. A star's mass determines the strength of its gravity, which in turn determines how much energy must be generated in its interior to prevent it from collapsing under its own weight. The mass of a star determines where that balance is struck.

Stars *Not* on the Main Sequence

Although 90 percent of stars are main-sequence stars, some stars are found in the upper right portion of the H-R diagram, well above the main sequence (see Figure 13.15). Those stars, which must be luminous, cool, and large, with radii hundreds or thousands of times the radius of the Sun, are called giants or supergiants. At the other extreme are stars found in the far lower left corner of the H-R diagram. Those stars are the tiny white dwarfs, comparable to the size of Earth. Their small surface areas explain why they have such low luminosities, despite having high temperatures.

Stars that lie off the main sequence on the H-R diagram can be identified by their luminosities (determined by their distance) or by slight differences in their spectral lines. The width of a star's spectral lines indicates the density and surface pressure of gas in the star's atmosphere. In general, denser stars have broader lines. Puffed-up stars above the main sequence have lower densities and lower surface pressure and narrower absorption lines than main-sequence stars.

When using the H-R diagram to estimate the distance to a star by the spectroscopic parallax method, astronomers must know whether the star is on, above, or below the main sequence to find the star's luminosity. The spectral

unanswered questions

Are planets with life likely to be orbiting around main-sequence stars of type M? Those low-luminosity stars are the most common type in the Milky Way, and the Kepler telescope has detected many planets orbiting such stars. However, the habitable zone of an M star is very close to the star, so that the planet may be strongly affected by streams of charged particles blowing off the star. Unless the planet has a strong protective magnetic field, that radiation may decrease or completely strip a planet of its atmosphere. Another complication is that a close-in planet is likely to be tidally locked to the star, so that one hemisphere of the planet receives light and the other hemisphere is permanently dark. That imbalance of light and heat might make the planet uninhabitable.

Table 13.3	Taking the Measure of Stars

Property	Methods
Luminosity	• For a star with a known distance, measure the brightness and then apply the inverse square law of radiation: $$\text{Luminosity} = 4\pi \times \text{Distance}^2 \times \text{Brightness}$$ • For a star *without* a known distance, take a spectrum of the star to determine its spectral and luminosity classes, plot them on an H-R diagram, and read the luminosity from the diagram.
Temperature	• Measure the star's color index by using blue and visual filters. Use Wien's law to relate the color to a temperature. • Take a spectrum of the star, and estimate the temperature from its spectral class by noting which spectral lines are present.
Distance	• For a relatively nearby star (within a few hundred parsecs), measure the star's parallax shift over the year. • For a more distant star, use the spectroscopic parallax method to find the luminosity from the H-R diagram. Then determine the distance from the luminosity and brightness.
Size	• For a few of the largest and closest stars, measure the size directly or by the length of the eclipse in eclipsing binary stars. • From the width of the star's spectral lines, estimate the luminosity class (supergiant, giant, or main sequence) for a star of given temperature. • For a star with known luminosity and temperature, use the Stefan-Boltzmann law to calculate the star's radius (the luminosity-temperature-radius relationship).
Mass	• Measure the motions of the stars in a binary system, use those to determine the stars' orbits, and then apply Newton's form of Kepler's third law. • For a non-binary star, use the mass-luminosity relationship to estimate the mass from the luminosity.
Composition	• Analyze the lines in the star's spectrum to measure chemical composition.

line widths of stars both on and off the main sequence indicate **luminosity class**, which tells us the star's relative *size* within each spectral class. Supergiant stars, the largest stars that we see, are luminosity class I, bright giants are class II, giants are class III, subgiants are class IV, main-sequence stars are class V, and white dwarfs are class WD. Luminosity classes I–IV lie above the main sequence, whereas class WD falls below and to the left of the main sequence (**Figure 13.19**). Thus, the complete spectral classification of a star includes both its spectral type (indicating temperature and color) and its luminosity class (indicating relative size).

The existence of the main sequence, together with the fact that the mass of a main-sequence star determines where on the sequence it will lie, creates a grand pattern that makes it possible to understand stars in fundamental ways. The existence of stars that do *not* follow that pattern raises yet more questions. In the coming chapters, we show that the main sequence tells us what stars are and how they work, and that stars off the main sequence reveal how stars form, evolve, and die. **Table 13.3** summarizes the techniques that astronomers use to determine some of the basic properties of stars. Of the properties listed in the table, only temperature, distance, and composition can be *measured*. Luminosity must be *inferred* from the H-R diagram or calculated from distance and brightness, and

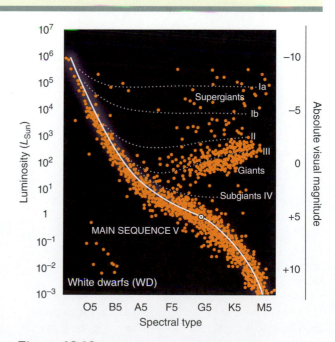

Figure 13.19 Stellar luminosity classes indicate the size (radius) of a star at each spectral type.

size and mass must be *calculated*. Other measurable properties include brightness, color, spectral type, and parallax shift.

CHECK YOUR UNDERSTANDING 13.4

Choose the two qualities that describe a star located in the lower right of the H-R diagram: (a) hot; (b) cool; (c) high luminosity; (d) low luminosity.

Origins: Habitable Zones

How might a star's basic property, such as its luminosity, color, mass, or surface temperature, affect the chance of a planet with life being in orbit around that star? The only known life is that on planet Earth, where liquid water was essential for life to form and evolve. Whether liquid water is an absolute requirement for life elsewhere is not known, but the presence of water is a good starting point for determining where to look. So astronomers look for planets at the right distance from their stars to have a planetary temperature that permits water to exist in a liquid state on their surfaces—a range of distances known as the **habitable zone**. On planets whose orbits are closer to their star than the habitable zone, water would exist only as a vapor—if at all. On planets that have orbits beyond the habitable zone, water would be permanently frozen as ice.

An important factor for estimating the temperature of a planet is the brightness of the sunlight that falls on that planet (see the Chapter 5 "Origins" section and Working It Out 5.4). That factor depends on the star's luminosity and the planet's distance from the star. In the Solar System, the habitable zone ranges from ~0.9 to ~1.4 AU, which includes Earth but just misses Venus and Mars. Main-sequence stars less luminous than the Sun are cooler and have narrower habitable zones, minimizing the chance that a planet will form within that slender zone. Main-sequence stars more massive than the Sun are hotter and have larger habitable zones. **Figure 13.20** illustrates those zones around Sun-like, hotter, and cooler stars.

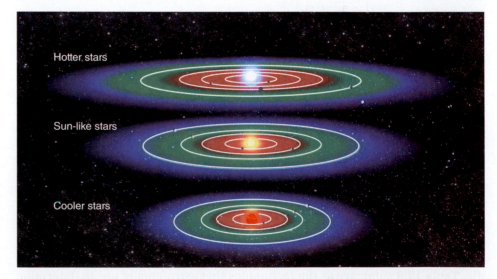

Figure 13.20 The distance and extent of a habitable zone (green) surrounding a star depends on the star's temperature. Regions too close to the star are too hot (red) and those too far away are too cold (blue) to permit the existence of liquid water. The orbits of Mercury, Venus, Earth, and Mars have been drawn around these stars for scale.

Astronomers are now finding planets in the habitable zones of their respective stars. Methods of planet detection, as discussed in Chapter 7, work best when the planet is close to its star. As of this writing, the Kepler Mission has identified and confirmed a few dozen planets in habitable zones by using the transit method and has found more candidate planets that need to be confirmed.

The distance from a star at a certain temperature is not the only consideration for whether a planet has water. The presence of a planetary atmosphere also is a factor. More massive planets can retain their atmospheres, which can trap heat and raise the planet's temperature, as we saw in Chapter 5 for Venus and Earth. Smaller planets have a lower gravitational pull and may not be able to retain an atmosphere. In addition, some habitable zones may be heated by planets, not stars. Some of the giant planets in the cold, outer part of our own Solar System have moons with liquid water (Chapters 11 and 24). The heat keeping the water liquid is due to the gravity of the nearby planet, not from the Sun.

Finally, *habitable* does not mean *inhabited*; it means only that the planet is at the right distance from its star that it could have liquid water. Identifying planets in their habitable zone is a first step to selecting which planets are most interesting for further study.

SUMMARY

Finding the distances to stars is a difficult but important task for astronomers. Parallax and spectroscopic parallax are two methods that astronomers use to determine distances to stars. Brightness and distance can be used to obtain the luminosity. Careful study of the light from a star, including its spectral lines, gives the temperature, size, and composition of the star. The study of binary systems gives the masses of stars of various spectral types, which we can extend to all stars of the same spectral type. The H-R diagram shows the relationship among the various physical properties of stars. The mass of a star is the major determining factor in its evolution. The habitable zone is the distance from a star in which a planet could have the right temperature for liquid water to exist on its surface. Stars of different luminosities and temperatures have habitable zones of different widths at different distances from the star.

(1) **Demonstrate how astronomers use parallax to determine the distances to stars, and explain how astronomers combine these distances with the brightness of nearby stars to determine how luminous the stars are.** The distance to a nearby star is measured by finding its parallax—by measuring how the star's apparent position changes in the sky over a year. The nearest star (other than the Sun) is about 4.2 light-years (1.3 parsecs) away. The brightness of a star in the sky can be measured directly, and brightness and distance can be used to obtain the star's luminosity—how much light the star emits.

(2) **Explain how astronomers obtain the temperatures, sizes, and composition of stars.** The color of a star depends on its temperature; blue stars are hotter, whereas red stars are cooler. The radius can be computed from the temperature and luminosity of the star. Small, cool stars greatly outnumber large, hot stars. Spectral lines carry a great deal of information about a star, including what chemical elements and molecules are present in the star.

(3) **Outline how astronomers estimate the masses of stars.** The motions of stars are observed in a binary system. Newton's universal law of gravitation and Kepler's laws connect the stars' motion to the forces they experience. Comparing the accelerations of the stars to the forces that cause them yields the masses of the stars in the system.

(4) **Categorize stars and organize that information on a Hertzsprung-Russell (H-R) diagram.** The H-R diagram shows the relationship among the various physical properties of stars. Temperature increases to the left, so that hotter stars lie on the left side of the diagram, whereas cooler stars lie on the right. Luminosity increases vertically, so that the most luminous stars lie near the top of the diagram. A star's luminosity class and temperature indicate its size. The mass and composition of a main-sequence star determine its luminosity, temperature, and size. Ninety percent of stars lie along the main sequence.

(5) **Determine the luminosity, temperature, and size of a star from its mass and composition.** The mass and composition of a main-sequence star determine its position on the H-R diagram. The main sequence on the H-R diagram is actually a sequence of masses. That position connects its other properties such as its luminosity, temperature, and size.

QUESTIONS AND PROBLEMS

TEST YOUR UNDERSTANDING

1. Star A and star B are nearly the same distance from Earth. If star A is half as bright as star B, which of the following statements must be true?
   a. Star B is farther away than star A.
   b. Star B is twice as luminous as star A.
   c. Star B is hotter than star A.
   d. Star B is larger than star A.

2. Star A and star B are two nearby stars. If star A is blue and star B is red, which of the following statements must be true?
   a. Star A is hotter than star B.
   b. Star A is cooler than star B.
   c. Star A is farther away than star B.
   d. Star A is more luminous than star B.

3. Star A and star B are two stars nearly the same distance from Earth. If star A is blue and star B is red, but they have equal brightness, which of the following statements is true?
   a. Star A is more luminous than star B.
   b. Star A is larger than star B.
   c. Star A is smaller than star B.
   d. Star A is less luminous than star B.

4. What does it most likely mean when a star has very weak hydrogen lines and is blue?
   a. The star is too hot for hydrogen lines to form.
   b. The star has no hydrogen.
   c. The star is too cold for hydrogen lines to form.
   d. The star is moving too fast to measure the lines.

5. Star A and star B are a binary system. If the Doppler shift of star A's absorption lines is 3 times the Doppler shift of star B's absorption lines, which of the following statements is true?
   a. Star A is 3 times as massive as star B.
   b. Star A is one-third as massive as star B.
   c. Star A is closer than star B.
   d. The binary pair is moving toward Earth, but star A is farther away.

6. Star A and star B are two red stars at nearly the same distance from Earth. If star A is many times brighter than star B, which of the following statements is true?
   a. Star A is a main-sequence star, whereas star B is a red giant.
   b. Star A is a red giant, whereas star B is a main-sequence star.
   c. Star A is hotter than star B.
   d. Star A is a white dwarf, whereas star B is a red giant.

7. Star A and star B are two blue stars at nearly the same distance from Earth. If star A is many times brighter than star B, which of the following statements is true?
   a. Star A is a main-sequence star, whereas star B is a red giant.
   b. Star A is a main-sequence star, whereas star B is a blue giant.
   c. Star A is a white dwarf, whereas star B is a blue giant.
   d. Star A is a blue giant, whereas star B is a white dwarf.

8. In which region of an H-R diagram would you find the main-sequence stars with the widest habitable zones?
   a. upper left
   b. upper right
   c. center
   d. lower left
   e. lower right

9. If star A is more massive than star B, and both are main-sequence stars, star A is _____ than star B. (Choose all that apply.)
   a. more luminous
   b. less luminous
   c. hotter
   d. colder
   e. larger
   f. smaller

10. A telescope on Mars could measure the distances to more stars than can be measured from Earth because
    a. the resolution of the telescope would be better.
    b. Mars has a thin atmosphere.
    c. it would be closer to the stars.
    d. the parallax "baseline" would be longer.

11. Star A and star B are two nearby stars. If star A has a parallax angle 4 times as large as star B's, which of the following statements is true?
    a. Star A is one-quarter as far away as star B.
    b. Star A is 4 times as far away as star B.
    c. Star A has moved through space one-quarter as far as star B.
    d. Star A has moved through space 4 times as far as star B.

12. If star A appears twice as bright as star B but is also twice as far away, star A is _____ as luminous as star B.
    a. 8 times   c. twice
    b. 4 times   d. half

13. In Table 13.1, the percentage of hydrogen in the Sun decreases when changing from percentage by number of atoms to percentage by mass, but the percentage of helium increases. Why?
    a. Hydrogen is more massive than helium.
    b. Helium is more massive than hydrogen.
    c. Hydrogen is located in a different part of the Sun.
    d. The mass of hydrogen is hard to measure.

14. Capella (in the constellation Auriga) is the sixth-brightest star in the sky. When viewed with a high-powered telescope, it turns out that Capella is actually two pairs of binary stars: the first pair are G-type giants, whereas the second pair are M-type main-sequence stars. What color does Capella appear to be?
    a. red
    b. yellow
    c. blue
    d. The color cannot be determined from the given information.

15. An eclipsing binary system has a primary eclipse (star A is eclipsed by star B) that is deeper (more light is removed from the light curve) than the secondary eclipse (star B is eclipsed by star A). What does that information tell you about stars A and B?
  a. Star A is hotter than star B.
  b. Star B is hotter than star A.
  c. Star B is larger than star A.
  d. Star B is moving faster than star A.

THINKING ABOUT THE CONCEPTS

16. The distances of nearby stars are determined by their parallaxes. Why are the distances of stars farther from Earth more uncertain?

17. To know certain properties of a star, you must first determine the star's distance. For other properties, knowing its distance is unnecessary. Explain why an astronomer does or does not need to know a star's distance to determine each of the following properties: size, mass, temperature, color, spectral type, and chemical composition.

18. Albireo, in the constellation Cygnus, is a visual binary system whose two components can easily be seen with even a small, amateur telescope. Viewers describe the brighter star as "golden" and the fainter one as "sapphire blue."
  a. What does that description tell you about the relative temperatures of the two stars?
  b. What does that description tell you about their respective sizes?

19. Very cool stars have temperatures around 2500 K and emit Planck spectra with peak wavelengths in the red part of the spectrum. Do those stars emit any blue light? Explain your answer.

20. The stars Betelgeuse and Rigel are both in the constellation Orion. Betelgeuse appears red, whereas Rigel is bluish white. To the eye, the two stars seem equally bright. If you can compare the temperature, luminosity, or size from just that information, do so. If not, explain why.

21. Explain why the stellar spectral types (O, B, A, F, G, K, M) are not in alphabetical order. What sequence of temperatures is defined by those spectral types?

22. Other than the Sun, the only stars whose mass astronomers can measure *directly* are those in eclipsing or visual binary systems. Why? How do astronomers estimate the masses of stars that are not in eclipsing or visual binary systems?

23. Once the mass of a certain spectral type of star located in a binary system has been determined, it can be assumed that all other stars of the same spectral type and luminosity class have the same mass. Why is that a reasonable assumption?

24. ★ **WHAT AN ASTRONOMER SEES** Suppose that Figure 13.12 showed a third star in the system of stars, with the third star being much smaller than the other two in the image. What could you conclude about that third component's brightness, distance, and luminosity? 👁

25. As discussed in the Process of Science Figure, scientific advances often require the participation of scientists from all over the world, working on the same problem over many decades, even centuries. Compare that mode of "collaboration" with collaborations in your courses (perhaps on final projects or papers). What mechanisms must be in place to allow scientists to collaborate across space and time in that way? 👁

26. How would our ability to measure stellar parallax change if we were on Mars? What if we were on Venus or Jupiter?

27. Figure 13.7 has an absorption line at about 410 nm that is weak for O stars and weak for G stars but very strong in A stars. That particular line is due to the transition from the second excited state of hydrogen up to the sixth excited state. Why is that line weak in O stars? Why is it weak in G stars? Why is it strongest in the middle of the range of spectral types? 👁

28. Which kinds of binary systems are detected edge-on? Which kinds are detected face-on?

29. In Figure 13.10, two stars orbit a common center of mass. 👁
  a. Explain why star 2 has a smaller orbit than star 1.
  b. Re-sketch that picture for the case in which star 1 has a very low mass—perhaps close to that of a planet.
  c. Re-sketch that picture for the case in which star 1 and star 2 have the same mass.

30. If our Sun were a blue main-sequence star, and Earth was still 1 AU from the Sun, would you expect Earth to be in the habitable zone? What about if our Sun were a red main-sequence star?

APPLYING THE CONCEPTS

31. Betelgeuse (in Orion) has a parallax of 0.00451 ± 0.00080 arcsec, as measured by the Hipparcos satellite. What is the distance to Betelgeuse, and what is the uncertainty in that measurement? 🟡—🟤—🟢
  a. Make a prediction: Working It Out 13.1 gives a few examples of the relationship between the parallax and the distance. Study those numbers for a moment, then compare the parallax of Betelgeuse to the numbers in Working It Out 13.1. Do you expect the distance to Betelgeuse to be more or less than 100 pc?
  b. Calculate: Find the distance to Betelgeuse, and then use the uncertainty of 0.00080 to find the range in distances that result from this uncertainty. (Hint: Adding 0.00080 to 0.00451 gives a slightly larger parallax, which corresponds to a slightly smaller distance and gives one end of the range.)
  c. Check your work: Compare your answer for the distance to Betelgeuse to your prediction, and check that the distance you calculate lies within the range of distances that you calculated from the uncertainty.

32. Wolf 359 has an apparent magnitude of 13.44, whereas Barnard's Star has an apparent magnitude of 9.53. Which star is brighter, and how many times brighter is it? ●–●–●

   a. Make a prediction: Roughly how many apparent magnitudes difference is there in the brightness of these two stars? Each magnitude corresponds to a factor of between 2 and 3 in brightness. What is the range of reasonable answers?

   b. Calculate: Follow Working It Out 13.2 to find the ratio of brightnesses from the difference in magnitudes.

   c. Check your work: Compare your answer to the range of reasonable answers that you found in part a. Does your answer lie in this range?

33. Suppose that you observe a star to have a luminosity 12,000 L_{Sun} and a temperature of 4500 K. What is the radius of this star in units of R_{Sun}? ●–●–●

   a. Make a prediction: Identify the approximate location of this star on the H-R diagram in Figure 13.17. Roughly what radius do you expect to calculate?

   b. Calculate: Follow Working It Out 13.3 to calculate the radius of this star.

   c. Check your work: Compare your answer to your prediction.

34. In an unusual twist of events, you discover a system that is almost exactly like the binary star system described in Working It Out 13.4! The only difference is that the speed of the second star (whose mass and velocity have a subscript "2") is $v_2 = 10.2$ km/s. What are the masses of the two stars in the system? ●–●–●

   a. Make a prediction: This star moves *faster* than the original star 2 from the system in Working It Out 13.4, but the size of the orbit is the same. Do you expect to find that star 1 in this system is more or less massive than the original star 1?

   b. Calculate: Follow the last few steps of Working It Out 13.4 to find the mass of the two stars in this binary system.

   c. Check your work: Compare your answer to your prediction.

35. Barnard's Star has the highest known motion across the sky for any star; it moves 10.3 arcseconds per year as it orbits the center of the Milky Way. This star is located 1.8 pc away. What is its parallax? Is this larger or smaller than the angle of its movement each year across the sky?

   a. Make a prediction: Working It Out 13.1 gives a few examples of the relationship between the parallax and the distance. Study those numbers for a moment, then compare the parallax of Barnard's Star to the numbers in Working It Out 13.1. Do you expect the parallax angle for this star to be "large" (something like 1), or "small" (something like 0.0001)?

   b. Calculate: Find the parallax of Barnard's Star. Is this larger or smaller than the 10.3 arcseconds that Barnard's Star moves every year?

   c. Check your work: Compare your answer to your prediction to check your work.

36. Suppose that Figure 13.1b included a third star, located 4 times as far away as star A. How much less than star A would the third star appear to move each year? How much less than star B? ★

37. Suppose you see an object jump from side to side by half a degree as you blink back and forth between your eyes. How much farther away is an object that moves only one-third of a degree?

38. Figure 13.5 is plotted logarithmically on both axes. The luminosities are in units of solar luminosities. ◉

   a. How much more luminous than the Sun is a star on the far right side of the plot?

   b. How much less luminous than the Sun is a star on the far left side of the plot?

39. Compared with the Sun, how luminous, large, and hot is a star that has 10 times the mass of the Sun? Use Figure 13.17 to answer that question. ◉

40. Sirius, the brightest star in the sky, has a parallax of 0.379 arcsec. What is its distance in parsecs? In light-years? How long does the light take to reach Earth?

41. Sirius is actually a binary pair of two A-type stars. The brighter star is called the "Dog Star" and the fainter is called the "Pup Star" because Sirius is in the constellation Canis Major (meaning "big dog"). The Dog Star appears about 6800 times brighter than the Pup Star, even though both stars are at the same distance from Earth. Compare the temperatures, luminosities, and sizes of those two stars.

42. Sirius and its companion orbit around a common center of mass with a period of 50 years. The mass of Sirius is 2 times the mass of the Sun.

   a. If the orbital velocity of the companion is 2.35 times greater than that of Sirius, what is the mass of the companion?

   b. What is the semimajor axis of the orbit?

43. The star Achernar has a Hipparcos parallax of 0.02339 arcsec and appears about as bright as Betelgeuse (Problem 31) in the sky. Which star is actually more luminous? Betelgeuse appears reddish, whereas Achernar appears bluish, so which star is hotter? Why is it hotter?

44. The Sun is about 16 trillion (1.6×10^{13}) times brighter than the faintest stars visible to the naked eye.

   a. How far away (in astronomical units) would an identical solar-type star be if it were just barely visible to the naked eye?

   b. What would be its distance in light-years?

45. Find the peak wavelength of blackbody emission for a star with a temperature of about 10,000 K. In what region of the spectrum does that wavelength fall? What color is that star?

EXPLORATION   H-R Diagram

digital.wwnorton.com/astro7

Visit the Digital Resources Page and on the Student Site open the "H-R Diagram" Interactive Simulation for Chapter 13. This simulation enables you to compare stars on the H-R diagram in two ways. You can compare an individual star (marked by a red *X*) to the Sun by varying its properties in the box in the left half of the window. Or you can compare groups of the nearest and brightest stars. Play around with the controls for a few minutes to familiarize yourself with the simulation.

Begin by exploring how changes to the properties of the individual star change its location on the H-R diagram. First, press the "Reset" button.

Decrease the temperature of the star by dragging the temperature slider to the left. Notice that the luminosity remains the same. Because the temperature has decreased, each square meter of star surface must be emitting less light. What other property of the star changes to keep the total luminosity of the star constant?

Predict what will happen when you move the temperature slider all the way to the right. Now do it. Did the star behave as you expected?

1 As you move to the left across the H-R diagram, what happens to the radius?

_____

_____

_____

2 What happens to the radius as you move to the right?

_____

_____

_____

Press "Reset" and experiment with the luminosity slider.

3 As you move up on the H-R diagram, what happens to the radius?

_____

_____

4 What happens to the radius as you move down?

_____

_____

_____

Press "Reset" again and predict how you would have to adjust the slider bars to move your star into the red giant portion of the H-R diagram (upper right). Adjust the slider bars until the star is in that area. Were you correct?

5 How would you adjust the slider bars to move the star into the white dwarf area of the H-R diagram (lower left)?

_____

_____

Press the "Reset" button and explore the right-hand side of the window. Add the nearest stars to the graph by clicking their radio button. Using what you have learned so far, compare the temperatures and luminosities of those stars with the Sun (marked by the yellow circle on the graph).

6 Are the nearest stars generally hotter or cooler than the Sun?

_____

7 Are the nearest stars generally more or less luminous than the Sun?

_____

Press the radio button for the brightest stars. That action will add the brightest stars in the sky to the plot. Compare those stars with the Sun.

8 Are the brightest stars generally hotter or cooler than the Sun?

_____

9 Are the brightest stars generally more or less luminous than the Sun?

_____

10 How do the temperatures and luminosities of the brightest stars in the sky compare with the temperatures and luminosities of the nearest stars? Does that information support the claim in the chapter that there are more low-luminosity stars than high-luminosity stars? Explain.

_____

_____

_____

_____

Our Star—The Sun

The Sun is more massive *and* significantly larger than everything else in the Solar System. You can measure the diameter of the Sun with a heavy sheet of paper or poster board, a long piece of string, a ruler, and a little help from a friend. On a sunny day, poke a tiny hole in the paper. Standing with her back to the Sun, your friend should hold up the paper so that the Sun shines through the hole. An image of the Sun will be projected onto a second piece of paper on the ground within the shadow of the paper. If the paper is close to the ground, the image will be small. If the paper is far from the ground, the image will be larger, but it will not be as bright. Move the paper until you can accurately measure the diameter (d) of the image on the ground with your ruler, in millimeters. Convert this number to meters. Use the long piece of string to measure how far (h) the paper is from the ground, by first marking the distance on the piece of string, and then using the ruler to measure the marked distance in meters. (You can use a tape measure, instead, if you have one.) To calculate the diameter of the Sun, all you need are these two measurements and the distance from Earth to the Sun (in meters), which is available in your textbook or on the Internet.

EXPERIMENT SETUP

On a sunny day, poke a tiny hole in the paper. Standing with her back to the Sun, your friend should hold up the paper so that the Sun shines through the hole. An image of the Sun will be projected onto the second piece of paper on the ground within the shadow of the paper.

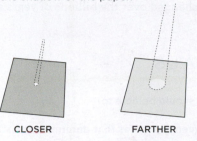

CLOSER **FARTHER**

If the paper is close to the ground, the image will be small. If the paper is far from the ground, the image will be larger, but it will not be as bright. Move the paper until you can accurately measure the diameter (d) of the image on the ground with your ruler, in millimeters. Convert this number to meters.

Use a long piece of string to measure how far (h) the paper is from the ground, by first marking the distance on the piece of string, and then using the ruler to measure the marked distance in meters.

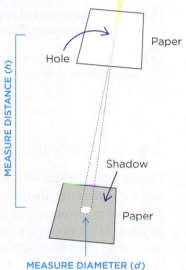

Sun

Hole

Paper

MEASURE DISTANCE (h)

Shadow

Paper

MEASURE DIAMETER (d)

SKETCH OF RESULTS

Diameter of image of Sun

$$\text{Diameter of Sun} = d \times \frac{\text{Distance to the Sun}}{h}$$

Distance from paper to ground

d: ___ millimeters; d: ___ meters

h: ___ meters

Distance from Earth to Sun: _____ meters

14

ecause the Sun is the only star close to Earth, much of our detailed knowledge about stars has come from studying the Sun. In Chapter 13, we looked at the physical properties of distant stars, including their mass, luminosity, size, temperature, and chemical composition. Here in Chapter 14, we ask fundamental questions about Earth's local star: Where does the Sun get its energy? How does energy move through the Sun? What is its atmosphere like? How has its luminosity changed over the billions of years since the Solar System formed?

LEARNING GOALS

By the end of this chapter, you should be able to:

① Describe the balance between the forces that determine the structure of the Sun.

② Diagram how mass is converted to energy in the Sun's core and estimate how long the Sun will take to use up its fuel.

③ Sketch a physical model of the Sun's interior, and list the ways that energy moves outward from the Sun's core toward its surface.

④ Relate observations of solar neutrinos and seismic vibrations on the surface of the Sun to astronomers' models of the Sun.

⑤ Describe the solar activity cycles of 11 and 22 years, and relate those cycles to the Sun's changing magnetic field and solar phenomena such as sunspots and flares.

14.1 The Sun Is Powered by Nuclear Fusion

Energy from the Sun is responsible for daylight, for Earth's weather and seasons, and for terrestrial life itself. At a luminosity of 3.85×10^{26} watts (W), the Sun produces more energy in a second than all the power plants on Earth could generate in a half-million years. In this section, we explain how the Sun releases this energy from deep within its core.

Hydrostatic Equilibrium

Like Earth's structure, the structure of the Sun is governed by several physical processes and relationships. Geologists learn about Earth's interior by using a combination of physics, detailed computer models, and experiments that test the predictions of those models. Similarly, astronomers use physics, chemistry, and the properties of matter and radiation to create a model of the Sun. One of the great successes of 20th century astronomy has been the construction of a theoretical model of the Sun that agrees with observations of the mass, composition, size, temperature, and luminosity of the real thing.

The structure of the Sun results from a balance between the outward force due to the pressure of radiation produced inside and the inward force of gravity. That balance is known as **hydrostatic equilibrium**. The outward pressure is a result of local energy that is finding its way to the Sun's surface from deep in its interior. To understand how hydrostatic equilibrium affects the Sun, we need to know how those forces are produced and how they continually change to balance each other.

The balance between the forces due to radiation pressure and gravity is illustrated in **Figure 14.1**. The Sun is a huge ball of hot gas. Deep in the Sun's interior, the outer layers press downward because of gravity, producing a large inward force. To maintain balance, the outward force due to pressure must be equally large. If the forces due to gravity exceeded the forces due to radiation pressure, the Sun would collapse. Conversely, if radiation pressure were greater than gravity, the Sun would blow itself apart. At every point within the Sun's interior, the radiation pressure must be just enough to hold up the weight of all the layers above that point. If the Sun were not in a stable hydrostatic equilibrium, forces within it would not be in balance, and the size of the Sun would change accordingly.

Hydrostatic equilibrium becomes an even more powerful concept when combined with the way gases behave. Deeper in the Sun's interior, the weight of the material above becomes greater, and hence the radiation pressure increases. In a gas, higher pressure means higher density, higher temperature, or both. **Figure 14.2** shows how conditions vary inside the Sun. The graphs in Figure 14.2b, which are based on calculations, show that as the radiation pressure increases toward the Sun's center, the density and temperature of the gas increase as well.

CHECK YOUR UNDERSTANDING 14.1a

Hydrostatic equilibrium in the Sun means that: (a) the Sun does not change; (b) the Sun absorbs and emits equal amounts of energy; (c) the outward force from radiation pressure balances the weight of overlying layers; (d) energy produced in the core per unit time equals energy emitted at the surface per unit time.

Answers to Check Your Understanding questions are in the back of the book.

Nuclear Fusion

A second fundamental balance within the Sun is the balance of energy (see Figure 14.1). Stars like the Sun are remarkably stable objects. To remain in balance, the Sun must produce just enough energy in its interior to replace the energy radiated away from its surface. That energy balance tells us how much energy must be produced in the Sun's interior and how that energy finds its way from the interior to the Sun's surface, where it is radiated away. Theoretical models of stellar evolution indicate that the Sun's luminosity is increasing, but very, very slowly. From geological records and these models, astronomers estimate that the Sun's luminosity 4.5 billion years ago was about 70 percent of its current luminosity.

One of the most basic questions facing the pioneers of stellar astrophysics was how the Sun and other stars get their energy. In the 19th century, physicists proposed that the Sun was slowly shrinking and that the core was heating up as a result of that gravitational contraction. Calculations soon showed, however, that gravitational contraction could power the Sun for only millions of years. Geological and biological evidence available at the time suggested that Earth was tens of millions or hundreds of millions of years old. In the early 20th century, radiometric dating suggested that Earth was more than a billion years old, and therefore gravitational contraction could not be the source of the Sun's energy. In the 1930s, using theoretical and laboratory physics, nuclear physicists concluded that the Sun's energy

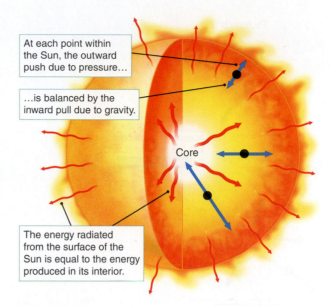

Figure 14.1 The structure of the Sun is determined by the balance between the forces due to pressure and gravity and the balance between the energy generated in its core and the energy radiated from its surface.

At each point within the Sun, the outward push due to pressure…

…is balanced by the inward pull due to gravity.

Core

The energy radiated from the surface of the Sun is equal to the energy produced in its interior.

Figure 14.2 **a.** This cutaway figure shows how the fraction of radius given in the *x*-axis of the graphs in b. is measured. The energy produced by the Sun is generated in the Sun's core. **b.** Pressure, density, and temperature all increase toward the center of the Sun.

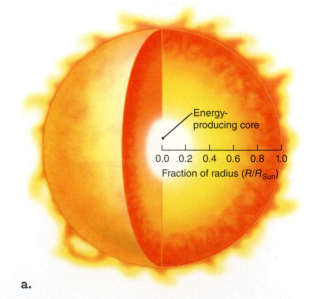

Energy-producing core

0.0   0.2   0.4   0.6   0.8   1.0
Fraction of radius (R/R_{Sun})

a.

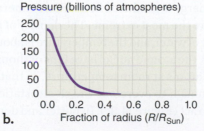

Pressure (billions of atmospheres)

250
200
150
100
50
0

0.0   0.2   0.4   0.6   0.8   1.0
Fraction of radius (R/R_{Sun})

b.

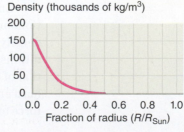

Density (thousands of kg/m³)

200
150
100
50
0

0.0   0.2   0.4   0.6   0.8   1.0
Fraction of radius (R/R_{Sun})

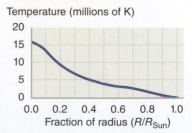

Temperature (millions of K)

20
15
10
5
0

0.0   0.2   0.4   0.6   0.8   1.0
Fraction of radius (R/R_{Sun})

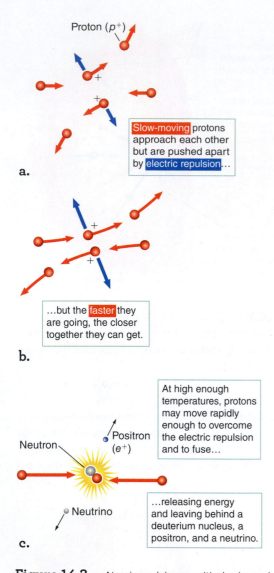

Proton (p^+)

Slow-moving protons approach each other but are pushed apart by electric repulsion...

a.

...but the faster they are going, the closer together they can get.

b.

At high enough temperatures, protons may move rapidly enough to overcome the electric repulsion and to fuse...

Neutron

Positron (e^+)

Neutrino

...releasing energy and leaving behind a deuterium nucleus, a positron, and a neutrino.

c.

Figure 14.3 **a.** Atomic nuclei are positively charged and electrostatically repel each other. **b.** The faster that two nuclei are moving toward each other, the closer they will get before veering away. **c.** At the temperatures and densities found in the centers of stars, nuclei can overcome that electrostatic repulsion, so fusion takes place.

comes from nuclear reactions at its core, capable of powering the star for billions of years.

The nucleus of a hydrogen atom consists of a single proton and, in most cases, no neutrons (Chapter 5). The nuclei of all other atoms are built from a mixture of protons and neutrons. Helium nuclei, for example, consist of two protons and usually two neutrons. Protons have a positive electric charge, whereas neutrons have no electric charge. Because like charges repel, and the closer they are the stronger the force, therefore all of the protons in an atomic nucleus continually repel each other with a tremendous force. The nuclei of atoms should fly apart because of electric repulsion. They don't, though, because they are held together by the **strong nuclear force**, which overcomes that repulsion. However, the strong nuclear force acts only over very short distances, on the order of 10^{-15} meters—about the size of the atomic nucleus, or about a hundred-thousandth the size of an atom.

Compared with the energy required to free an electron from an atom, the amount of energy required to tear a nucleus apart is enormous. Conversely, when an atomic nucleus (with a mass up to the nucleus of iron) is formed from component parts, energy is released. **Nuclear fusion**—the process of combining two less massive atomic nuclei into a single, more massive atomic nucleus—occurs when atomic nuclei are brought close enough together for the strong nuclear force to overcome the force of electric repulsion (**Figure 14.3**). Many kinds of nuclear fusion can occur in stars. In main-sequence stars like the Sun, the primary energy generation process is the fusion of hydrogen into helium. That process is sometimes called "hydrogen burning," even though it has nothing to do with fire or other chemical combustion. The fusion of hydrogen into helium takes several steps, but the net result is that four hydrogen nuclei become one helium nucleus plus energy.

The energy produced in nuclear reactions comes from converting mass to energy. The exchange rate between mass and energy is given by Einstein's equation, $E = mc^2$, in which E is energy, m is mass, and c^2 is the speed of light squared. For any nuclear reaction, we can determine the mass turned into energy by calculating the mass lost. To find that lost mass, we subtract the mass of the outputs from the mass of the inputs. In **hydrogen fusion**, the inputs are four hydrogen nuclei, and the output is a helium nucleus plus energy. The mass of four separate hydrogen nuclei is 1.007 times greater than the mass of a single helium nucleus (with two protons and two neutrons); so, when hydrogen fuses to make helium, 0.7 percent of the mass of the hydrogen is converted to energy.

Although each fusion reaction produces a small amount of energy, the Sun's total mass is very large, so a great deal of hydrogen is available to fuse. When the amount of energy produced by nuclear fusion is compared with the luminosity of the Sun, we see that those reactions can power the Sun for 10 billion years—a time frame longer than the 4.6-billion-year age of the Solar System (measured from radioactive dating of meteorites from space and the Moon). Details of that calculation are provided in **Working It Out 14.1**.

Energy is produced in the Sun's innermost region, the **core**, where the conditions are the most extreme. The density of matter in the core is about 150 times the density of water, and the temperature is about 15 million kelvin (K). Under those conditions, the atomic nuclei have tens of thousands of times more kinetic energy than that of atoms at room temperature, so that many of them can slam into each other hard enough to overcome the electrostatic repulsion, allowing the strong nuclear force to act (Figure 14.3c). In hotter and denser gases, these collisions happen more

working it out 14.1

The Source of the Sun's Energy

Like all stars, the Sun has a lifetime limited by the amount of fuel available. We can calculate how long the Sun will live by comparing the mass involved in nuclear fusion with the amount of mass available. Converting four hydrogen nuclei (protons) into a single helium nucleus results in a loss of mass. The mass of a single hydrogen nucleus is 1.6726×10^{-27} kilogram (kg). So, four hydrogen nuclei have a mass of 4 times that, or 6.6904×10^{-27} kg. The mass of a helium nucleus is 6.6447×10^{-27} kg, which is less than the mass of the four hydrogen nuclei. The amount of mass lost, m, is

$$m = 6.6904 \times 10^{-27}\ \text{kg} - 6.6447 \times 10^{-27}\ \text{kg} = 0.0457 \times 10^{-27}\ \text{kg}$$

We can write that result as 4.57×10^{-29} kg—a mass loss of about 0.7 percent. Converting 0.7 percent of the mass of the hydrogen into energy might not seem very efficient—until we compare it with other sources of energy and discover that it is millions of times more efficient than even the most efficient chemical reactions.

Using Einstein's equation $E = mc^2$, where c is the speed of light (3×10^8 m/s), along with the definition of a joule ($1\ \text{J} = 1\ \text{kg m}^2/\text{s}^2$), we can calculate the energy released by that mass-to-energy conversion:

$$E = mc^2 = (4.57 \times 10^{-29}\ \text{kg}) \times (3.00 \times 10^8\ \text{m/s})^2 = 4.11 \times 10^{-12}\ \text{J}$$

Each reaction that takes four hydrogen nuclei and turns them into a helium nucleus releases 4.11×10^{-12} J of energy, which doesn't seem like very much. Atoms are very small, however, so fusing a single kilogram of hydrogen into helium releases about 6.3×10^{14} J of energy—about the equivalent of the chemical energy released in burning 100,000 barrels of oil. To see how much the Sun must be fusing per second to produce its current luminosity, we divide the Sun's luminosity by that amount of energy per kilogram:

$$\frac{\text{Luminosity of Sun}}{\text{Energy per kilogram}} = \frac{3.9 \times 10^{26}\ \text{J/s}}{6.3 \times 10^{14}\ \text{J/kg}} = 6.2 \times 10^{11}\ \text{kg/s}$$

For the Sun to produce as much energy as it does, it must convert roughly 620 billion kg of hydrogen into helium every second (and about 4 billion kg of matter—0.7 percent—is converted to energy in the process). The Sun has been fusing hydrogen at that rate for at least the age of Earth and the Solar System—4.6 billion years. How much longer will the Sun last?

Astronomers estimate that only 10 percent of the Sun's total mass will ever be involved in fusion because the other 90 percent will never get hot enough or dense enough for the strong nuclear force to make fusion happen. Ten percent of the mass of the Sun is $(0.1) \times (2 \times 10^{30})$ kg, or 2×10^{29} kg. That is the amount of "fuel" the Sun has available. The Sun consumes hydrogen at a rate of 620 billion kg/s, so each year the Sun consumes about:

$$M_{\text{year}} = (6.2 \times 10^{11}\ \text{kg/s}) \times (3.16 \times 10^7\ \text{s/yr}) \approx 2 \times 10^{19}\ \text{kg/yr}$$

If we know how much fuel the Sun has (2×10^{29} kg), and we know how much the Sun fuses each year (2×10^{19} kg/yr), we can divide the amount by the rate to find the lifetime of the Sun:

$$\text{Lifetime} = \frac{M_{\text{fuel}}}{M_{\text{year}}} = \frac{2 \times 10^{29}\ \text{kg}}{2 \times 10^{19}\ \text{kg/yr}} = 10^{10}\ \text{yr}$$

When the Sun was formed, it had enough fuel to power it for about 10 billion years. The Sun is nearly halfway through its lifetime of hydrogen fusion.

often. For that reason, the rate of nuclear fusion reactions is extremely sensitive to the temperature and the density of the gas, which is why those energy-producing collisions are concentrated in the Sun's core. Half the energy produced by the Sun is generated within the inner part of the core: the inner 9 percent of the Sun's radius, or less than 0.1 percent of the Sun's volume.

The conversion of four hydrogen nuclei to one helium nucleus is the most significant source of energy in main-sequence stars. Hydrogen is the most abundant element in the universe, so it is the most abundant source of nuclear fuel at the beginning of a star's lifetime. Hydrogen also is the easiest atom to fuse. Hydrogen nuclei—protons—have an electric charge of +1. The electric barrier that must be overcome to fuse protons is the repulsion of one proton against another. To fuse two carbon nuclei, for example, requires overcoming the repulsion of the six protons in one carbon nucleus that are pushing against the six protons in the other carbon nucleus. As a result, the repulsion between two carbon nuclei is 36 times

▶ǁ **AstroTour:** The Solar Core

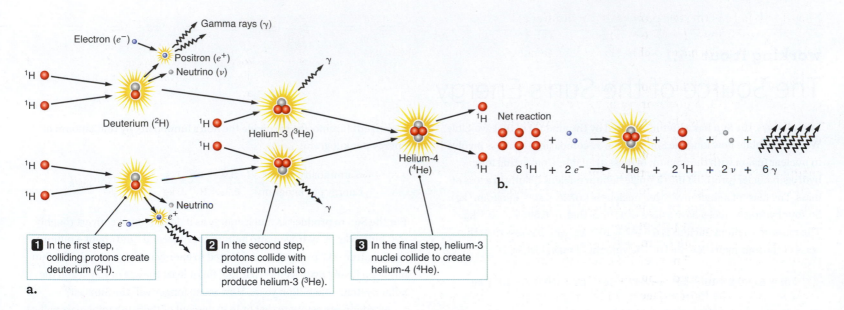

Figure 14.4 **a.** In the Sun, about 85 percent of the energy produced comes from the branch of the proton-proton chain illustrated here. **b.** The net reaction, in which four hydrogen atoms fuse to make a helium atom, is the primary source of energy for the Sun and all other main-sequence stars. The energy is released in the form of gamma rays, neutrinos, and kinetic energy.

▶▶ **Interactive Simulation:** Proton-Proton Chain

unanswered questions

Will nuclear fusion become a major source of energy production on Earth? Scientists have been working on controlled nuclear fusion for more than 60 years, since the first hydrogen bombs were developed. So far, however, replicating the conditions inside the Sun has posed too many difficulties. Nuclear fusion requires that we have hydrogen isotopes at very high temperature, density, and pressure, just as when a hydrogen bomb explodes. However, controlled nuclear fusion requires that we confine that material long enough to get more energy out than we put in. Several major experiments have attempted to fuse isotopes of hydrogen. So far, hydrogen fusion reactions have not been sustained long enough to be a commercially viable energy source; South Korea's KSTAR project set a record by sustaining the reaction for 20 seconds. An alternative approach is to fuse an isotope of helium, ³He, which has only three particles in the nucleus (two protons and one neutron). On Earth, ³He is found in very limited supply. It is in much greater abundance on the Moon, however, so some people propose setting up mining colonies on the Moon to extract ³He for use in fusion reactions on Earth or possibly even on the Moon.

stronger than that between two hydrogen nuclei. Therefore, hydrogen fusion occurs at a much lower temperature than any other type of nuclear fusion.

The Proton-Proton Chain

To test the theory that the Sun shines because of nuclear fusion, astronomers can analyze the predicted by-products of the nuclear reactions. In the Sun's core and in other low-mass stars, hydrogen fusion takes place in a series of nuclear reactions called the **proton-proton chain**, which has three branches. The most important branch, responsible for about 85 percent of the energy generated in the Sun, consists of the three steps illustrated in **Figure 14.4**. Each step produces particles and/or energy in the form of light. We begin by following the creation of the helium nucleus and then go back to find out what happens to the other products of the reaction.

Follow along in Figure 14.4a as we step through the proton-proton chain. The hydrogen nucleus consists of just one proton and is written as ¹H (where H is the element symbol for hydrogen and 1 is the *atomic mass number*—the total number of protons and neutrons in the nucleus). In the first step, two protons fuse. During that process, one proton is transformed into a neutron. To conserve spin and charge, two particles are emitted: a positively charged particle called a **positron** (e^+) and a neutral particle called a **neutrino** (ν). The positron is then annihilated when it encounters an electron, thereby producing energy in the form of gamma-ray photons (γ). The new atomic nucleus formed by the first step in the chain consists of a proton and a neutron. Recall from Chapter 5 that an *isotope* of an element has the same number of protons but a different number of neutrons. Thus, the new atomic nucleus, called **deuterium,** is still hydrogen because it has only one proton, but it is written as ²H because its atomic mass number is 2 (1 proton + 1 neutron = 2).

In the second step of the proton-proton chain, another proton slams into the deuterium nucleus, forming the nucleus of an isotope of helium, ³He, consisting of two protons and one neutron. The energy released in that step is carried away as a highly energetic gamma-ray photon. Those first two steps are shown twice in Figure 14.4a because two ³He nuclei are needed to produce a single ⁴He nucleus.

In the third and final step of the proton-proton chain, two ³He nuclei collide and fuse, producing an ordinary ⁴He nucleus (consisting of 2 protons and

2 neutrons) and ejecting two protons (^{1}H) in the process. The energy released in that step is the kinetic energy of the helium nucleus and two ejected protons. Overall, four hydrogen nuclei have combined to form one helium nucleus, as summarized in Figure 14.4b.

Now let's go back and look at what happens to the other products of the reaction. In step 1, a positron—a particle of antimatter—is produced. **Antimatter** particles have the same mass as a corresponding matter particle but have opposite values of other properties, such as charge. The positron (e^+) is the antimatter counterpart of an electron (e^-). When matter (electrons) and antimatter (positrons) meet, they annihilate each other, and their total mass is converted to energy in the form of gamma-ray photons (γ). That's what happens to the emitted positrons inside the Sun, and the emitted photons from the annihilation carry away part of the energy released when the two protons fuse. Those photons heat the surrounding gas. The gamma rays emitted in step 2 similarly heat the gas. The thermal energy produced in the core of the Sun typically takes one hundred thousand years or more to find its way to the Sun's surface, and so the light we see from the Sun indicates what the Sun's core was doing a very long time ago.

The neutrino emitted in step 1 has a very different fate. Neutrinos are particles that have no charge, have very little mass, and travel at nearly the speed of light. They interact so weakly with ordinary matter that the neutrino escapes from the Sun without further interactions with any other particles. The core of the Sun is buried beneath 700,000 kilometers of dense, hot matter, yet the Sun is transparent to neutrinos—essentially all of them travel into space as though the outer layers of the Sun did not exist. Because they travel at nearly the speed of light, neutrinos from the center of the Sun arrive at Earth after 8.3 minutes. Therefore, we can use them to probe what the Sun is doing today.

The dominant branch of the proton-proton chain can be written symbolically as follows:

Step 1: ^{1}H + ^{1}H → ^{2}H + e^+ + ν and then e^+ + e^- → γ + γ

Step 2: ^{2}H + ^{1}H → ^{3}He + γ

Step 3: ^{3}He + ^{3}He → ^{4}He + ^{1}H + ^{1}H

The rate of the proton-proton chain reaction depends on both temperature and density. At the temperature and pressure that exist within the Sun's core, the reaction rate is relatively slow, which is fortunate for life on Earth. If the hydrogen fused quickly, the Sun would have exhausted its supply long ago, and life might not have had time to evolve.

While the reactions in Figure 14.4 generate about 85 percent of the Sun's energy, variations of the proton-proton chain account for the other 15 percent. The most common variation happens in step 3, in which ^{3}He fuses with an existing ^{4}He to create beryllium (^{7}Be), which decays to lithium (^{7}Li) and energy, and then the ^{7}Li plus one ^{1}H become two ^{4}He. In a less common variation, the beryllium combines with hydrogen to become boron (^{8}B), which then decays to beryllium and then to two ^{4}He. In both variations, ultimately four hydrogen nuclei become one helium nucleus.

CHECK YOUR UNDERSTANDING 14.1b

When hydrogen fuses into helium, energy is released from: (a) gravitational collapse; (b) the conversion of mass to energy; (c) the increase in pressure; (d) the decrease in the gravitational field.

what if . . .

What if you had a neutrino detector in your classroom? How would you expect the signal of solar neutrinos to vary between night and day?

14.2 Energy Is Transferred from the Interior of the Sun

Although geologists cannot travel deep inside Earth to find out how it is structured, they can build a model of its interior by using data on how seismic waves travel during earthquakes. Similarly, astronomers can create a model of the Sun's interior by using their knowledge of the balance of forces and energy within the Sun and an understanding of how energy moves from one place to another. Those models can be tested by observations of waves traveling through the Sun and by studying neutrinos from the Sun.

Energy Transport

Some of the energy released by hydrogen fusion in the Sun's core escapes directly into space in the form of neutrinos. However, most of the energy heats the solar interior and then moves outward through the Sun to the surface, a process known as **energy transport**. Energy transport, a key determinant of the Sun's structure, can occur by *conduction*, *convection*, or *radiation*.

Conduction is important primarily in solids. When you pick up a hot object, for example, your fingers are heated by conduction. That happens because energetic thermal vibrations of atoms and molecules cause neighboring atoms and molecules to vibrate more rapidly as well. Conduction is typically ineffective in a gas because the atoms and molecules are too far apart to transmit vibrations to one another efficiently. Conduction does not play a key role in transporting energy from the Sun's core to its surface, but conduction is relevant when we discuss dying stars in Chapters 16 and 17.

In the Sun, energy is transported by convection and radiation through zones, as shown in **Figure 14.5**. The mechanism of energy transport from the center of the Sun outward depends on the decreasing temperature and density as the radius increases. First, energy moves outward through the inner layers of the Sun as radiation in the form of photons. Next, energy moves by convection in parcels of gas. Finally, energy radiates from the Sun's surface as light. We look at each process in turn.

Near the core, **radiation** transfers energy from hotter to cooler regions via photons (**Figure 14.6**), which carry the energy with them. Recall from your study of radiation in Chapter 5 that the hotter region contains more (and more energetic) photons than the cooler region. More photons move from the hotter, very crowded region to the cooler, less crowded region than in the reverse direction. A net transfer of photons and photon energy occurs from the hotter region to the cooler region, and radiation carries energy outward from the Sun's core.

The transfer of energy from one point to another by radiation also depends on how freely radiation can move from one point to another within a star. The degree to which matter blocks the flow of photons through it is called **opacity**. The opacity of a material depends on many things, including the density of the material, its composition, its temperature, and the wavelength of the photons moving through it.

Energy transfer by radiation is most efficient in regions with low opacity. The **radiative zone** (see Figure 14.5) is the region in the inner part of the Sun in which the opacity is relatively low, and radiation carries the energy produced in the core outward through the star. That radiative zone extends about 70 percent of the way out toward the surface of the Sun. Even though the region's opacity is low enough

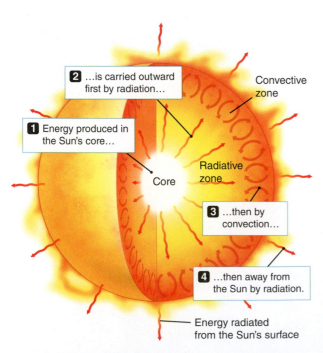

2 …is carried outward first by radiation…

1 Energy produced in the Sun's core…

Convective zone

Core

Radiative zone

3 …then by convection…

4 …then away from the Sun by radiation.

Energy radiated from the Sun's surface

Figure 14.5 The interior of the Sun is divided into zones on the basis of where energy is produced and how it is transported outward.

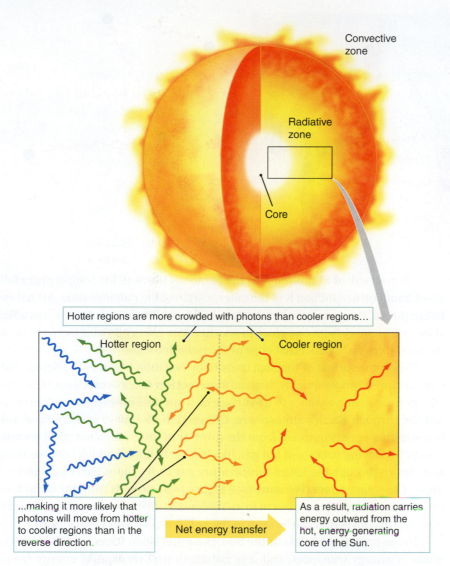

Figure 14.6 Higher-temperature regions deep within the Sun contain more radiation than do lower-temperature regions farther out. Although radiation flows in both directions, more radiation flows from the hotter regions to the cooler regions than from the cooler regions to the hotter regions. Therefore, radiation carries energy outward from the inner parts of the Sun. For simplicity, this illustration includes only a few common photons. Photons of wavelengths representing all colors are present in all regions, with more of all kinds in the hotter regions and fewer of all kinds in the cooler regions.

for radiation to dominate convection as an energy transport mechanism, photons still travel only a short distance within the region before being absorbed and then reemitted, or scattered by matter, much like a beach ball being batted about by a crowd of people (**Figure 14.7**). Each interaction sends the photon in an unpredictable direction—not necessarily toward the surface of the star. The distances between interactions are so short that, on average, the energy of a gamma-ray photon produced in the interior of the Sun takes about one hundred thousand or more years to find its way to the outer layers of the Sun. Opacity holds energy inside the Sun and lets it seep away only slowly. As it travels, the gamma-ray photon is gradually converted to lower-energy photons, emerging mostly as optical and infrared radiation from the surface.

Figure 14.7 **a.** When a crowd of people plays with a beach ball, the ball never travels very far before someone hits it, turning it in another direction. A ball often takes a long time to make its way from one edge of the crowd to the other. **b.** Similarly, a photon traveling through the Sun takes a long time to make its way out of the Sun.

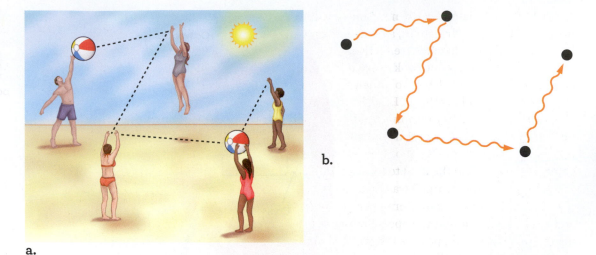

a.

b.

From a peak of 15 million K at the center of the Sun, the temperature falls to about 2 million to 3 million K at the outer margin of the radiative zone. At that cooler temperature, the opacity is higher than in the center, so radiation is less efficient at carrying energy from one place to another. The energy flowing outward through the Sun "piles up" against that edge of the radiative zone.

Nearer the surface of the Sun, transfer by radiation becomes inefficient and the temperature changes quickly. Instead, **convection** takes over. Convection carries energy from inside a planet to its surface or from the Sun-heated surface of Earth upward through Earth's atmosphere. Convection also plays an important role in transporting energy outward from the interior of the Sun. Convection transports energy by moving packets of hot gas, like hot-air balloons, which become buoyant and rise up through the lower-temperature gas above them, carrying energy with them. The solar **convective zone** (see Figure 14.5) extends from the outer boundary of the radiative zone outward to just below the visible surface of the Sun, where evidence of convection can be seen in the bubbling surface (**Figure 14.8**).

In the outermost layers of the Sun, radiation again takes over as the primary mode of energy transport, and it is radiation that transports energy from the Sun's outermost layers off into space.

Observing Neutrinos from the Core of the Sun

The model of energy production and energy transport in the Sun, as just presented, correctly matches observed global properties of the Sun such as its size, temperature, and luminosity. The nuclear fusion model of the Sun predicts exactly which nuclear reactions should be occurring in the Sun's core and at what rate. The nuclear reactions that make up the proton-proton chain produce a vast number of neutrinos. Because neutrinos barely interact with other ordinary matter, almost all the neutrinos produced in the heart of the Sun travel freely through the outer parts of the Sun and on into space as though the outer layers of the Sun were not there. Solar neutrinos produced in the core of the Sun, traveling at nearly the speed of light, take only 8.3 minutes to reach Earth—much quicker than the 100,000-year journey of photons leaving the core.

Neutrinos interact so weakly with matter that they are extremely difficult to observe. Nevertheless, an extremely large number of nuclear reactions take place in the Sun, so the Sun produces an enormous number of neutrinos. As you read this sentence, about 400 trillion solar neutrinos are passing through your body. That

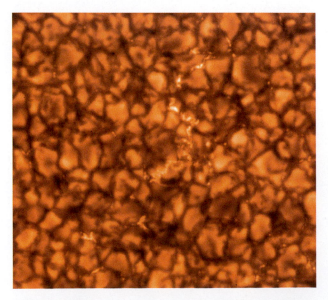

Figure 14.8 The top of the Sun's convective zone shows the bubbling of the surface caused by rising and falling packets of gas.

happens even at night because neutrinos easily pass through Earth, too. With that many neutrinos moving about, a neutrino detector does not have to detect a very large percentage of them to be effective.

A neutrino telescope looks very different from other telescopes. The first apparatus designed to detect solar neutrinos was built 1500 meters underground, within the Homestake Mine in Lead, South Dakota. Astronomers filled a tank with 100,000 gallons of a dry-cleaning fluid called perchloroethylene (C_2Cl_4). Astronomers predicted that, over 2 days, roughly 10^{22} solar neutrinos passed through the Homestake detector. Of those, on average only one neutrino interacted with a chlorine atom within the fluid to form a radioactive isotope of argon. Over time, a measurable amount of argon was produced.

The Homestake experiment operated from the late 1960s to the early 1990s and detected that argon isotope—evidence of neutrinos from the Sun, confirming that nuclear fusion powers the Sun. Astronomers noticed, however, that they were measuring only about one-third as many solar neutrinos as predicted by solar models. The difference between the predicted and measured number of solar neutrinos was called the **solar neutrino problem**.

One possible explanation of the solar neutrino problem was that the working model of the structure of the Sun was somehow wrong. That possibility seemed unlikely, however, because of the many other successful predictions of the solar model. A second possibility was that an understanding of the neutrino itself was incomplete. The neutrino was long thought to have zero mass, like photons, and to travel at the speed of light. But if neutrinos actually do have a tiny amount of mass, particle physics suggests that solar neutrinos should oscillate—alternate back and forth—among three kinds of neutrinos: the electron, muon, and tau neutrinos. According to that explanation, early neutrino experiments could detect only the electron neutrino and, consequently, observed only about a third of the expected number of neutrinos. Since then, many other neutrino detectors have been built, each using different reactions to detect neutrinos of different energies or different types. Experiments at high-energy physics labs, nuclear reactors, and neutrino telescopes around the world have shown that neutrinos do have a nonzero mass and do oscillate among neutrino types.

Solving the solar neutrino problem is a good example of how science works—how a better model of the neutrino showed that the solar neutrino problem was real and not merely an experimental mistake, and how a single set of anomalous observations was later confirmed by other, more sophisticated experiments. All that effort led to a better understanding of basic physics (see the **Process of Science Figure**).

Probing the Sun's Interior

Models of Earth's interior predict how density and temperature change from place to place. Those differences affect the seismic waves traveling through Earth, bending the paths that they travel. Geologists test models of Earth's interior by comparing measurements of seismic waves from earthquakes with model predictions of how seismic waves should travel through the planet.

Just as geologists use seismic waves from earthquakes to probe the interior of Earth, solar physicists use the surface oscillations of the Sun to probe the solar interior. The science that uses solar oscillations to study the Sun is called **helioseismology**. Detailed observations of the Doppler shifts caused by the motions of material from place to place across the Sun's surface show that the Sun vibrates

what if . . .

What if a typical photon produced in the Sun's center took 1000 rather than 100,000 years to escape from the Sun's surface? What could be different about this hypothetical Sun?

Learning from Failure

The first detection of solar neutrinos raised more questions than it answered.

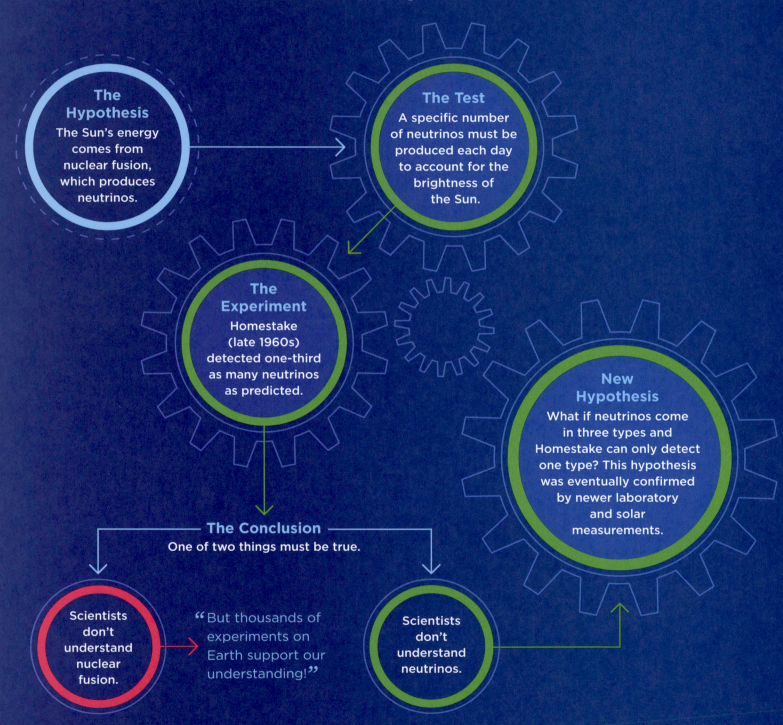

The Hypothesis
The Sun's energy comes from nuclear fusion, which produces neutrinos.

The Test
A specific number of neutrinos must be produced each day to account for the brightness of the Sun.

The Experiment
Homestake (late 1960s) detected one-third as many neutrinos as predicted.

New Hypothesis
What if neutrinos come in three types and Homestake can only detect one type? This hypothesis was eventually confirmed by newer laboratory and solar measurements.

The Conclusion
One of two things must be true.

Scientists don't understand nuclear fusion.

" But thousands of experiments on Earth support our understanding! "

Scientists don't understand neutrinos.

Part of the "scientific attitude" is to find failure exciting. When experiments do not turn out as expected, good scientists get excited because there is something new to understand!

or rings, something like a bell that has been struck. Unlike a well-tuned bell—which vibrates primarily at one frequency—the vibrations of the Sun are very complex. In the Sun, many frequencies of vibrations occur simultaneously, causing some parts of the Sun to bulge outward and some to draw inward. Those motions help researchers probe what lies below. **Figure 14.9** illustrates the motions of the different parts of the Sun, with red and blue areas moving in opposite directions. Some waves are amplified and some are suppressed, depending on how they overlap as they travel through the Sun. Astronomers study those waves by using the Doppler effect (see Chapter 5), which distinguishes between parts of the Sun that move toward the observer and those that move away.

To detect the disturbances of helioseismic waves on the surface of the Sun, astronomers must measure Doppler shifts of less than 0.1 m/s while detecting changes in brightness of only a few parts per million at any given location on the Sun. Tens of millions of wave motions are possible within the Sun. Some waves travel around the circumference of the Sun, yielding information about the density of the upper convection zone. Other waves travel through the interior of the Sun, revealing the Sun's density structure close to its core. Still others travel inward toward the center of the Sun, until they are bent by the changing solar density and return to the surface.

All those wave motions are going on at the same time, so sorting them out requires computer analysis of long, unbroken strings of solar observations from several sources. The Global Oscillation Network Group (GONG) is a network of six solar observation stations spread around the world that enables astronomers to observe the Sun's surface approximately 90 percent of the time.

To interpret helioseismology data, scientists compare the measurements of the strength, frequency, and wavelengths of the waves against predicted vibrations calculated from models of the solar interior. That technique serves as a powerful test of solar interior models and has led both to some surprises and to improvements in the models. For example, some scientists proposed that the solar neutrino problem might be solved if the models had overestimated the amount of helium in the Sun. That explanation was ruled out by analyzing the waves that penetrate to the Sun's core. Helioseismology showed that the value for opacity used in early solar models was too low. That realization led astronomers to recalculate the location of the bottom of the convective zone. Both theory and observation now put the base of the convective zone at 70 percent of the way out from the Sun's center, with an uncertainty in that number of less than 0.5 percent.

Working back and forth between observation and theory has enabled astronomers to probe the Sun's otherwise inaccessible interior. We now know that the energy is produced by nuclear fusion deep in the core and that it moves outward by radiation to a point about 70 percent of the Sun's radius. Then it travels outward by convection to the surface. We also know how the temperature, density, and pressure change with radius and how those factors change the opacity at different distances from the center. Even though researchers usually cannot sample directly or set up controlled experiments, collaboration between theory and observation is essential to observational sciences such as astronomy.

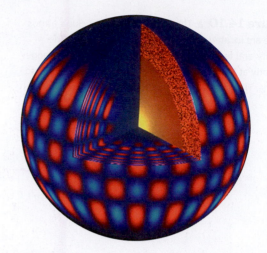

Figure 14.9 The interior of the Sun rings like a bell as helioseismic waves move through it. This figure shows one particular mode of the Sun's vibration, in false color. Red indicates regions where gas is traveling inward, whereas blue indicates regions where gas is traveling outward. Astronomers observe those motions via Doppler shifts.

CHECK YOUR UNDERSTANDING 14.2

How do neutrinos help us understand what is going on in the Sun's core? (a) Neutrinos from distant objects pass through the Sun, probing the interior. (b) Neutrinos from the Sun pass easily through Earth. (c) Neutrinos created in fusion reactions at the Sun's core easily escape.

Figure 14.10 a. The components of the Sun's atmosphere are located above the convective zone. **b.** The density and temperature of the Sun's atmosphere change abruptly at the boundary between the chromosphere and corona. The *y*-axes are logarithmic.

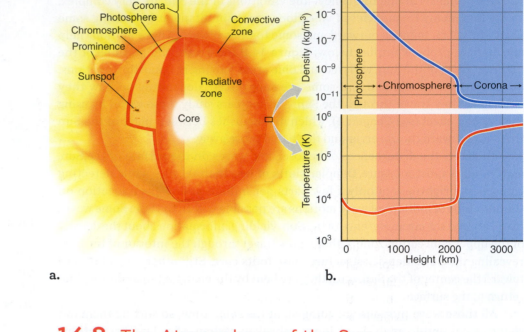

a.   b.

The Sun is "limb darkened." It is dimmer near its edge because near its edge we see the Sun at a steep angle and so do not see deeply into its atmosphere.

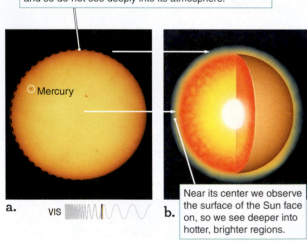

Near its center we observe the surface of the Sun face on, so we see deeper into hotter, brighter regions.

a.   VIS   b.

Figure 14.11 a. When viewed in visible light, the Sun appears to have a sharp outline, even though it has no true surface. The center of the Sun appears brighter, whereas the limb of the Sun is darker—an effect known as limb darkening. The small black dot indicated by the white circle is Mercury, transiting in front of the Sun. **b.** Looking at the center of the Sun allows us to see deeper into the Sun's interior than we do when looking at the edge of the Sun. Because higher temperature means more luminous radiation, the center of the Sun appears brighter than its limb.

14.3 The Atmosphere of the Sun

Beyond the convective zone lie the outer layers of the Sun, collectively known as the Sun's atmosphere. Those layers, shown in **Figure 14.10a**, include the *photosphere*, the *chromosphere*, and the *corona*. We can observe those layers of the Sun directly by using telescopes and satellites. Observations of the Sun's atmosphere are important because activity in the Sun's atmosphere has consequences for human infrastructure such as power grids and satellites in orbit around Earth.

Unlike Earth, the Sun has no solid surface. Its apparent surface is like a fog bank on Earth. After people walk into a fog bank and disappear from view, you would say they were inside the fog bank, even though they never passed through a definite boundary. The apparent surface of the Sun is similar. Light from the Sun's surface can escape directly into space, so we can see it. Light from below the Sun's surface cannot escape directly into space, so we cannot see it.

At the base of the atmosphere is the **photosphere**: the apparent surface of the Sun or its visible boundary. That is where features such as sunspots can be seen. Above the photosphere is the **chromosphere**, a region of strong emission lines. The top layer is the **corona**, which can be viewed during a solar eclipse as a halo around the Sun. In the Sun's atmosphere, the density of the gas drops very rapidly with increasing altitude. The graphs in **Figure 14.10b** show how density and temperature change across the Sun's atmosphere. The temperature increases sharply across the boundary between the chromosphere and the corona, whereas the density falls sharply across the same boundary. In this section, we explore each layer, beginning at the bottom with the photosphere.

The Photosphere

The **effective temperature** of a star is the temperature found from the spectrum of the light emitted from the photosphere. While underlying layers are much hotter, the effective temperature is an observable quantity that is useful for comparing stars with one another; when astronomers mention the temperature of a star, they are typically referring to the effective temperature. The effective temperature of the Sun is calculated from the Sun's luminosity and radius by using the Stefan-Boltzmann law (see Chapter 5). The photosphere has an effective temperature of 5780 K, ranging from 6600 K to 4500 K over a 500-km-thick zone. The Sun appears to have a well-defined surface and a sharp outline when viewed from Earth because 500 km does not look very thick when viewed from a distance of 150 million km.

In **Figure 14.11a**, the Sun appears fainter near its edges than near its center, an effect called **limb darkening**. That effect is an artifact of the structure of the Sun's photosphere. When looking near the edge of the Sun, you are looking through the photosphere at a steep angle. As a result, you do not see as deeply into the Sun as when you are looking directly down through the photosphere near the center of the Sun's disk. The light from the **limb** of the Sun comes from a shallower layer that is cooler and fainter, as shown in **Figure 14.11b**.

In the Sun's atmosphere, the gas density drops very rapidly with increasing altitude. All visible solar phenomena take place in the Sun's atmosphere. Most of the radiation from below the Sun's photosphere is absorbed by matter and reemitted at the photosphere as a blackbody spectrum.

As we examine the Sun's structure in more detail, however, that simple description of the spectra of stars turns out to be incomplete. Light from the solar photosphere must escape through the upper layers of the Sun's atmosphere, which affects the spectrum we observe. In Chapter 13, we discussed the presence of absorption lines in the spectra of stars. Now we can take a closer look at how those absorption lines form. As photospheric light travels upward, atoms in the solar atmosphere absorb the light at discrete wavelengths, forming absorption lines. Because the Sun appears so much brighter than any other star, its spectrum can be studied in far more detail, so specially designed telescopes and high-resolution spectrometers have been built specifically to study the Sun's light. The solar spectrum is shown in **Figure 14.12**. Absorption lines from more than 70 elements have been identified. Analysis of those lines forms the basis for much of astronomers' knowledge of the solar atmosphere, including the Sun's composition. That is also the starting point for understanding the atmospheres and spectra of other stars.

The Chromosphere and Corona

The temperature falls from 6600 K at the photosphere's bottom to 4400 K at its top. At that point, the trend reverses and the temperature slowly begins to climb, rising to about 6000 K at a height of 1500 km above the top of the photosphere (see Figure 14.10b). This region of increasing temperature is called the chromosphere (**Figure 14.13a**). Why the chromosphere's temperature increases from the bottom to the top is not well understood, but it may be caused by magnetic waves propagating through the region and depositing their energy at the top of the chromosphere.

The chromosphere was discovered in the 19th century during observations of total solar eclipses (**Figure 14.13b**). The chromosphere is seen most strongly at the solar limb as a source of emission lines, especially the Hα line (the "hydrogen alpha line"), which is produced when an electron falls from the third energy state of hydrogen to the second energy state. The deep red color of the Hα line is what gives the chromosphere its name ("chromosphere" means "the place where color

what if . . .

What if the temperature at the base of the corona were 10 million K rather than 1 million K? How would the light from this region be different in this hypothetical Sun?

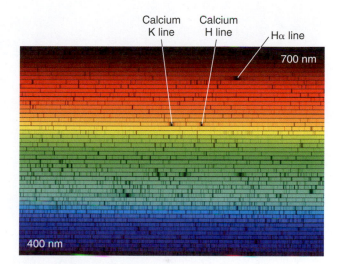

Figure 14.12 This high-resolution spectrum of the Sun stretches from 400 nanometers (nm) in the lower left corner to 700 nm in the upper right corner and shows black absorption lines.

★ **WHAT AN ASTRONOMER SEES** An astronomer will know that this spectrum was produced by passing the Sun's light through a prism-like device and then cutting and folding the single long spectrum (from blue to red) into rows so that it will fit in a single image taken by a camera. She will notice the particularly strong absorption lines, which show up as dark blotches. The one near the top in the red part of the spectrum is particularly noticeable, and an astronomer who looks at a lot of spectra will recognize this line by its color: It is the hydrogen alpha (Hα) line, marking the transition of electrons from the third down to the second energy state of hydrogen. She might also recognize the strong "calcium H" and "calcium K" lines in the orange part of the spectrum. These are recognizable from the combination of their color, their relative strength, and their nearness to each other. The Sun's spectrum is crowded with absorption lines, and an astronomer will immediately know that the outer layers of the Sun are cooler than the layers deep down, because atoms in those outer layers are absorbing energy as it makes its way out from the Sun. Credit: Nigel Sharp, NOAO/NSO/Kitt Peak FTS/AURA/NSF, https://noirlab .edu/public/images/noao-sun/. https://creativecommons.org/licenses/by/4.0.

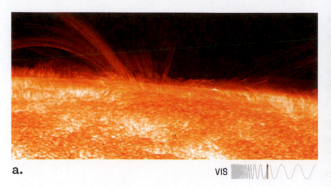

a. VIS

b. VIS

c. VIS

Figure 14.13 a. This spacecraft image of the Sun shows fine structure in the chromosphere extending outward from the photosphere. **b.** The chromosphere (seen in pink here) is visible during a total eclipse. **c.** This eclipse image shows the Sun's corona, consisting of million-kelvin gas that extends for millions of kilometers beyond the surface of the Sun.

comes from"). The element helium was discovered in 1868 from a spectrum of the chromosphere of the Sun nearly 30 years before it was found on Earth. Helium is named after *helios*, the Greek word for "Sun."

At the top of the chromosphere, across a transition region only about 100 km thick, the temperature suddenly soars, whereas the density abruptly drops (see Figure 14.10b). Above that transition lies the outermost region of the Sun's atmosphere, the corona, where temperatures reach 1 million to 2 million K. The corona is thought to be heated by magnetic fields and micro solar flares. NASA's *Parker Solar Probe* is currently in orbit around the Sun, investigating the flow of energy that heats and accelerates the solar corona and solar wind. As a result, scientists hope to be better able to predict changes in the solar wind known as "space weather," which affects satellites, spacecraft, and astronauts, even here in Earth orbit.

The Sun's corona has been known since ancient times: It is visible with the naked eye during total solar eclipses as an eerie glow stretching several solar radii beyond the Sun's surface (**Figure 14.13c**). Because it is so hot, the solar corona also is a strong source of X-rays. Those X-ray photons, which are invisible to the human eye, have so much energy that many electrons are stripped away from nuclei, leaving atoms in the corona highly ionized.

CHECK YOUR UNDERSTANDING 14.3

The Sun's surface appears sharp in visible light because: (a) the photosphere is cooler than the layers below it; (b) the photosphere is thinner than the other layers in the Sun; (c) the photosphere is less dense than the convection zone; (d) the Sun has a distinct surface.

14.4 The Atmosphere of the Sun Is Very Active

The Sun's atmosphere is a very turbulent place. The best-known features on the Sun's surface are relatively dark blemishes in the solar photosphere, called **sunspots**. Sunspots come and go over time, though they remain long enough for us to determine the rotation rate of the Sun. Those spots are associated with **active regions**: loops of material and explosions that fling particles far out into the Solar System. Long-term patterns have been observed in the variations of sunspots and active regions, revealing that the Sun's magnetic field is constantly changing.

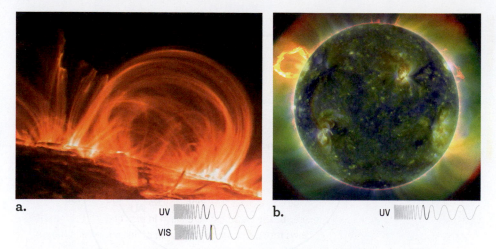

a. UV VIS

b. UV

Figure 14.14 **a.** This close-up image of the Sun shows the tangled structure of coronal loops. **b.** This image is a combination of several extreme ultraviolet images of the Sun from the Solar Dynamics Observatory (SDO). Coronal holes are dark in these images, which indicates that they are cooler and less dense than their surroundings. Several solar prominences are also visible in this image, as well as the wispy corona.

Solar Activity Is Caused by Magnetic Effects

The Sun's magnetic field (see Chapter 5) causes virtually all the structure seen in the Sun's atmosphere. High-resolution images of the Sun show *coronal loops* that make up much of the Sun's lower corona (**Figure 14.14a**). That texture is the result of magnetic structures called flux tubes. Magnetic fields are responsible for much of the corona's structure as well. The corona is far too hot to be held in by the Sun's gravity, but over most of the Sun's surface, coronal gas is confined by magnetic loops with both ends firmly anchored deep within the Sun. The magnetic field in the corona acts almost like a network of rubber bands that coronal gas is free to slide along but cannot cross. In contrast, about 20 percent of the Sun's surface is covered by an ever-shifting pattern of **coronal holes**, large regions where the magnetic field points outward, away from the Sun, and where coronal material is free to stream away into interplanetary space as the solar wind. In extreme ultraviolet images of the Sun, we see coronal holes as dark regions, which indicates that they are cooler and less dense than their surroundings (**Figure 14.14b**).

The relatively steady part of the solar wind consists of lower-speed flows with velocities of about 350 km/s and higher-speed flows with velocities up to about 700 km/s. The higher-speed flows originate in coronal holes. Depending on their speed, particles in the solar wind take 2–5 days to reach Earth. Often, 2–5 days after a coronal hole passes across the center of the face of the Sun, the speed and density of the solar wind reaching Earth increases. The solar wind drags the Sun's magnetic field along with it (**Figure 14.15**), so the magnetic field in the solar wind gets "wound up" by the Sun's rotation. Consequently, the magnetic field has a spiral structure resembling the stream of water from a rotating lawn sprinkler.

The effects of the solar wind are felt throughout the Solar System. The solar wind blows the tails of comets away from the Sun, shapes the magnetospheres of the planets, and supplies the energetic particles that power Earth's spectacular auroral displays. Using space probes, astronomers have observed the solar wind extending out to 100 astronomical units (AU) from the Sun. But the solar wind does not go on forever. The farther it gets from the Sun, the more it spreads out.

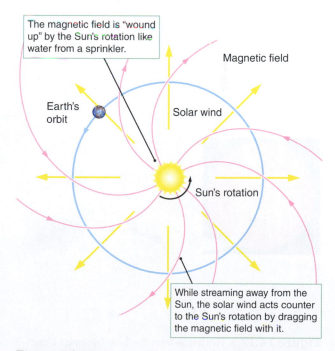

The magnetic field is "wound up" by the Sun's rotation like water from a sprinkler.

Magnetic field

Earth's orbit

Solar wind

Sun's rotation

While streaming away from the Sun, the solar wind acts counter to the Sun's rotation by dragging the magnetic field with it.

Figure 14.15 The solar wind (yellow arrows) streams away from active areas and coronal holes on the Sun. As the Sun rotates, the solar wind takes on a spiral structure, much like the spiral of water that streams away from a rotating lawn sprinkler.

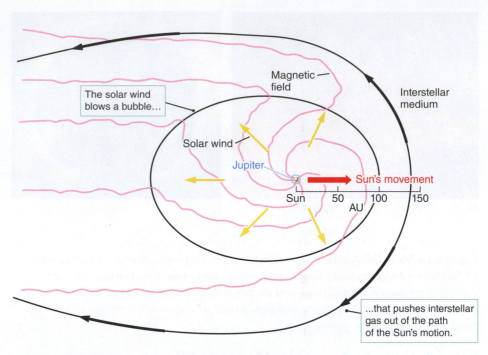

Figure 14.16 The solar wind (yellow arrows) streams away from the Sun for about 100 AU until it finally piles up against the pressure of the interstellar medium through which the Sun is traveling. The *Voyager 1* spacecraft crossed that boundary in 2012.

Just like radiation, the density of the solar wind follows an inverse square law. At a distance of about 100 AU from the Sun, the solar wind stops abruptly. Here it piles up against the pressure of the **interstellar medium**, the gas and dust that lie between stars in a galaxy. **Figure 14.16** shows the region of space over which the solar wind is measured. The *Voyager 1* and *Voyager 2* spacecraft have crossed the outer edge of that boundary and sent back the first direct measurements of true interstellar space. The *Interstellar Boundary Explorer* spacecraft, launched in 2008, is exploring the region, too.

Sunspots and Changes in the Sun

Sunspots have been noted since antiquity. Telescopic observations of sunspots date back almost 400 years, and records exist of naked-eye observations by Chinese, Greek, and medieval astronomers centuries before that. *But remember: Never look directly at the Sun!* Direct viewing through a commercial solar filter is safe, as is projecting the image through a telescope or binoculars onto a surface such as paper and looking only at the projection. Many websites have live images of the Sun viewed through ground and space telescopes. Sunspots are places where material is trapped at the surface of the Sun by magnetic-field lines. When that material cools, convection cannot carry it downward, so it makes a cooler (and therefore darker) spot on the surface of the Sun. **Figure 14.17** shows a large sunspot group. Sunspots appear dark, but only in contrast to the brighter surface of the Sun (**Working It Out 14.2**).

Early telescopic observations of sunspots made during the 17th century led to the discovery of the Sun's rotation, which has an average period of about 27 days as seen from Earth and 25 days relative to the stars. Because Earth orbits the Sun in the same direction that the Sun rotates, observers on Earth see a

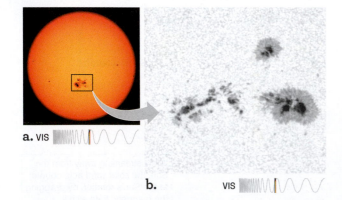

Figure 14.17 a. This image from the Solar Dynamics Observatory, taken in 2010, shows a large sunspot group. Sunspots are magnetically active regions cooler than the surrounding surface of the Sun. **b.** This high-resolution view shows the sunspots in this group.

working it out 14.2

Sunspots and Temperature

Sunspots are about 1500 K cooler than their surroundings. What does that lower temperature tell us about their luminosity? Think back to the Stefan-Boltzmann law in Chapter 5. The flux, $\mathcal{F}$, from a blackbody is proportional to the fourth power of the temperature, T:

$$\mathcal{F} = \sigma T^4$$

The constant of proportionality is the Stefan-Boltzmann constant, σ, which has a value of 5.67×10^{-8} W/(m² K⁴).

The flux is the amount of energy coming from a square meter of surface every second. How much less energy comes out of a sunspot than out of the rest of the Sun? Using 4500 and 6000 K for the temperatures of a typical sunspot and the surrounding photosphere, respectively, we can set up two equations:

$$\mathcal{F}_{spot} = \sigma T_{spot}^4 \quad \text{and} \quad \mathcal{F}_{surface} = \sigma T_{surface}^4$$

We could solve each of those separately and then divide the value of $\mathcal{F}_{spot}$ by $\mathcal{F}_{surface}$ to find out how much fainter the sunspot is, but solving for the ratio of the fluxes eliminates σ:

$$\frac{\mathcal{F}_{spot}}{\mathcal{F}_{surface}} = \frac{\sigma T_{spot}^4}{\sigma T_{surface}^4} = \frac{T_{spot}^4}{T_{surface}^4} = \left(\frac{T_{spot}}{T_{surface}}\right)^4$$

Plugging in our values for T_{spot} and $T_{surface}$ gives

$$\frac{\mathcal{F}_{spot}}{\mathcal{F}_{surface}} = \left(\frac{4500 \text{ K}}{6000 \text{ K}}\right)^4 = 0.32$$

Finally, multiplying both sides by $\mathcal{F}_{surface}$ gives

$$\mathcal{F}_{spot} = 0.32 \mathcal{F}_{surface}$$

So, the amount of energy coming from a square meter of sunspot every second is about one-third as much as the amount of energy coming from a square meter of surrounding surface every second. In other words, the sunspot is about one-third as bright as the surrounding photosphere. If you could cut out the sunspot and place it elsewhere in the sky, it would be brighter than the full Moon.

---

slightly longer rotation period. Observations of sunspots also show that the Sun's rotation period is shorter at its equator than at higher latitudes, which means that the equator rotates faster. This effect is called **differential rotation**. Differential rotation is possible only because the Sun is not a solid object.

Figure 14.18 shows the structure of a sunspot on the surface of the Sun. A sunspot consists of an inner dark core called the **umbra**, surrounded by a less dark region called the **penumbra**. These words are the same as those used when discussing brighter and darker areas within shadows (see Figure 2.24), but sunspots are not shadows. The words are simply used to refer to the brighter and darker regions of the sunspot. The penumbra shows an intricate radial pattern, reminiscent of a flower's petals. Sunspots are caused by magnetic fields thousands of times greater than the magnetic field at Earth's surface. They occur in pairs connected by loops in the magnetic field. Sunspots range in size from a few tens of kilometers across up to complex groups that may contain several dozen individual spots and span as much as 150,000 km. The largest sunspot groups can be seen without a telescope.

Although sunspots occasionally last 100 days or longer, half of all sunspots come and go in about 2 days, and 90 percent are gone within 11 days. The number and distribution of sunspots change in a pattern averaging 11 years called the **sunspot cycle**. **Figure 14.19a** shows data for several recent cycles. At the beginning of a cycle, sunspots appear at solar latitudes of about 30° north and south of the solar equator. Over the following years, sunspots are found closer to the equator as their number increases to a maximum and then declines. As the last few sunspots approach the equator, new sunspots again begin appearing at middle

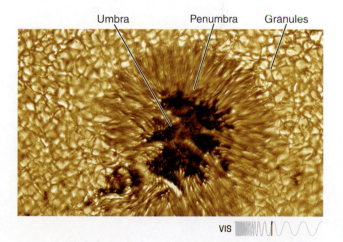

Umbra Penumbra Granules

VIS

Figure 14.18 This very high-resolution view of a sunspot shows the dark umbra surrounded by the lighter penumbra. The solar surface around the sunspot bubbles with separate cells of hot gas called *granules*. The smallest features are about 100 km across, whereas each granule is about the size of the state of Texas.
Credit: NOIRLab, https://noirlab.edu/public/images/noao9808a/, https://creativecommons.org/licenses/by/4.0/.

Figure 14.19 **a.** The number of sunspots varies, as shown in this graph of the past few solar cycles. **b.** The "solar butterfly" diagram shows the fraction of the Sun covered by sunspots at each latitude. The data are color coded to show the percentage of the strip at that latitude that is covered in sunspots at that time: black, 0 to less than 0.1 percent; red, 0.1–1.0 percent; yellow, greater than 1.0 percent. **c.** The Sun's magnetic poles flip every 11 years. Yellow indicates magnetic north, whereas blue indicates magnetic south.

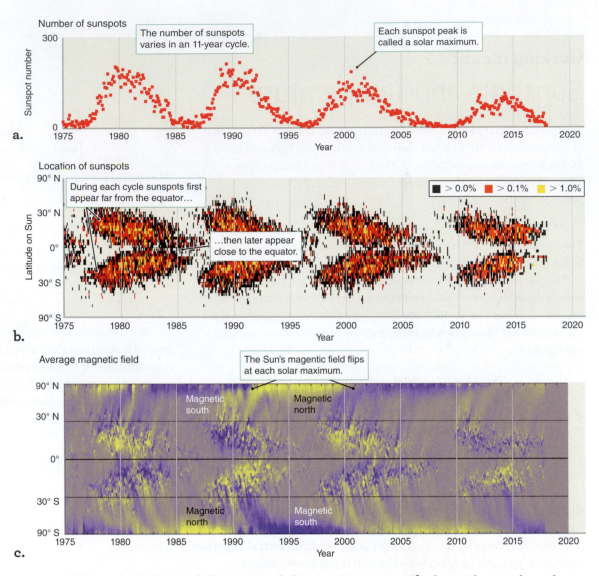

latitudes, and the next cycle begins. **Figure 14.19b** shows the number of sunspots at a given latitude plotted against time. The resulting diagram of opposing diagonal bands is called the sunspot "butterfly diagram."

In the early 20th century, solar astronomer George Ellery Hale (1868–1938) was the first to show that the 11-year sunspot cycle is actually half of a 22-year magnetic cycle during which the direction of the Sun's magnetic field reverses after each 11-year sunspot cycle. **Figure 14.19c** shows how the average strength of the magnetic field at every latitude has changed over more than 45 years. The direction of the Sun's magnetic field flips at the maximum of each sunspot cycle. Sunspots come in pairs, with one spot (the leading sunspot) in front of the other with respect to the Sun's rotation. In one 11-year sunspot cycle, the leading sunspot in each pair tends to be a north magnetic pole, whereas the trailing sunspot tends to be a south magnetic pole. In the next 11-year sunspot cycle, that polarity is reversed, so the leading sunspot in each pair is a south magnetic pole, whereas the trailing sunspot tends to be a north magnetic pole. The transition between those two magnetic polarities occurs near the peak of each sunspot cycle. Magnetic activity on the Sun affects the photosphere, chromosphere, and corona.

The graph in **Figure 14.20** shows 400 years of sunspot observations. The 11-year cycle is neither perfectly periodic nor especially reliable. The time between peaks in the number of sunspots actually varies between about 9.7 and 11.8 years. The

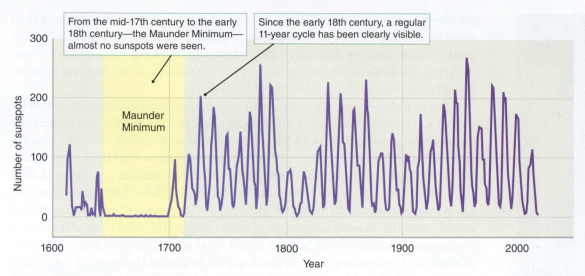

From the mid-17th century to the early 18th century—the Maunder Minimum—almost no sunspots were seen.

Since the early 18th century, a regular 11-year cycle has been clearly visible.

Figure 14.20 Sunspots have been observed for hundreds of years. In this plot, the 11-year cycle in the number of sunspots (half of the 22-year solar magnetic cycle) is clearly visible. Sunspot activity varies greatly. The period from the middle of the 17th century to the early 18th century, when almost no sunspots were seen, is called the *Maunder minimum*.

number of spots seen during a given cycle fluctuates as well, and in some periods sunspot activity has disappeared almost entirely. An extended lull in solar activity, called the **Maunder minimum**, lasted from 1645 to 1715. Typically, about six peaks of solar activity occur in 70 years, but virtually no sunspots were seen during the Maunder minimum, and auroral displays were less frequent than usual.

Sunspots are only one of several phenomena that follow the Sun's 22-year cycle of magnetic activity. The peaks of the cycle, called **solar maxima**, are times of intense activity. Sunspots are often accompanied by a brightening of the solar chromosphere that is seen most clearly in emission lines such as Hα. Those bright regions are known as solar active regions. The magnificent loops arching through the solar corona, shown in **Figure 14.21**, are solar **prominences**, magnetic flux tubes of

what if . . .

What if the Sun had only a very small magnetic field at the surface? How might that change the appearance of the Sun?

Approx. size of Earth

UV

Figure 14.21 Solar prominences are magnetically supported arches of hot gas that rise high above active regions on the Sun. Here, you can see a close-up view at the base of a large prominence. An image of Earth is included for scale (it is not actually that close to the Sun).

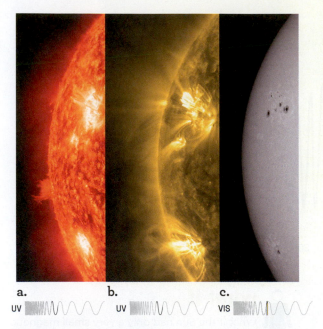

a. UV
b. UV
c. VIS

Figure 14.22 The Solar Dynamics Observatory (SDO) observed these active regions of the Sun that produced solar flares in August 2011. **a.** Activity near the surface at 60,000 K is visible in extreme ultraviolet light (along with a prominence rising up from the Sun's edge). **b.** Viewed at other ultraviolet wavelengths, many looping arcs and plasma heated to about 1 million K become visible. **c.** The dark spots in this image are the magnetically intense sunspots that are the sources of all the activity.

relatively cool (5000–10,000 K) but dense gas extending through the million-kelvin gas of the corona. Those prominences are anchored in the active regions. Although most prominences are relatively quiet, others can erupt out through the corona, towering a million kilometers or more over the surface of the Sun and ejecting material into the corona at velocities of 1000 km/s.

Solar flares—violent eruptions in which enormous amounts of magnetic energy are released over a few minutes to a few hours—are the most energetic form of solar activity. **Figure 14.22** shows solar flares erupting from two sunspot groups. The images in Figures 14.22a and b, taken in ultraviolet light, show material at very high temperatures. The spots in the visible-light image (Figure 14.22c) are at the base of the activity seen in Figures 14.22a and b. Solar flares can heat gas to temperatures of 20 million K, and they are the source of intense X-rays and gamma rays. Hot **plasma** (consisting of atoms stripped of some or all of their electrons) moves outward from flares at speeds that can reach 1500 km/s. Magnetic effects can then accelerate subatomic particles to almost the speed of light. Such events, called **coronal mass ejections (CMEs)** (**Figure 14.23**), send powerful bursts of energetic particles outward through the Solar System. CMEs occur about once per week during the minimum of the sunspot cycle and as often as several times per day near the maximum of the cycle.

Solar Activity Affects Earth

The amount of solar radiation received at the distance of Earth from the Sun has been measured to be 1361 watts per square meter (W/m²) on average. Satellite

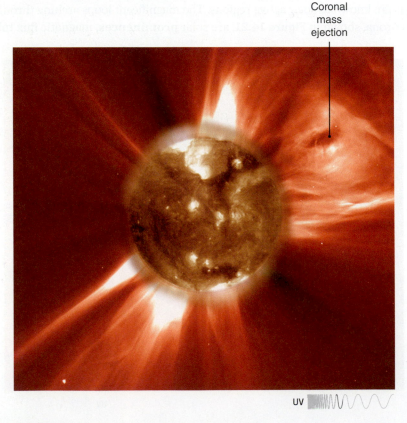

Coronal mass ejection

UV

Figure 14.23 This Solar and Heliospheric Observatory (SOHO) image shows a coronal mass ejection (upper right); a simultaneously recorded ultraviolet image of the solar disk is superimposed.

Reading Astronomy News

Carrington-Class CME Narrowly Misses Earth

Dr. Tony Phillips, Science@NASA

Last month (April 8–11, 2014), scientists, government officials, emergency planners, and others converged on Boulder, Colorado, for NOAA's Space Weather Workshop—an annual gathering to discuss the perils and probabilities of solar storms.

The current solar cycle is weaker than usual, so you might expect a correspondingly low-key meeting. On the contrary, the halls and meeting rooms were abuzz with excitement about an intense solar storm that narrowly missed Earth.

"If it had hit, we would still be picking up the pieces," says Daniel Baker of the University of Colorado, who presented a talk entitled *The Major Solar Eruptive Event in July 2012: Defining Extreme Space Weather Scenarios*.

The close shave happened almost two years ago. On July 23, 2012, a plasma cloud or "CME" rocketed away from the Sun as fast as 3000 km/s, more than 4 times faster than a typical eruption. The storm tore through the Earth's orbit, but fortunately Earth wasn't there. Instead it hit the *STEREO-A* spacecraft. Researchers have been analyzing the data ever since, and they have concluded that the storm was one of the strongest in recorded history. "It might have been stronger than the Carrington Event itself," says Baker.

The Carrington Event of September 1859 was a series of powerful CMEs that hit Earth head-on, sparking Northern Lights as far south as Tahiti. Intense geomagnetic storms caused global telegraph lines to spark, setting fire to some telegraph offices and disabling the "Victorian Internet." A similar storm today could have a catastrophic effect on modern power grids and telecommunication networks. According to a study by the National Academy of Sciences, the total economic impact could exceed $2 trillion or 20 times greater than the costs of a Hurricane Katrina. Multi-ton transformers fried by such a storm could take years to repair and impact national security.

A recent paper in *Nature Communications* authored by UC Berkeley space physicist Janet G. Luhmann and former postdoc Ying D. Liu describes what gave the July 2012 storm Carrington-like potency. For one thing, the CME was actually *two* CMEs separated by only 10 to 15 minutes. This double storm cloud traveled through a region of space that had been cleared out by another CME only four days earlier. As a result, the CMEs were not decelerated as much as usual by their transit through the interplanetary medium.

Had the eruption occurred just one week earlier, the blast site would have been facing Earth, rather than off to the side, so it was a relatively narrow escape.

When the Carrington Event enveloped Earth in the 19th century, technologies of the day were hardly sensitive to electromagnetic disturbances. Modern society, on the other hand, is deeply dependent on Sun-sensitive technologies such as GPS, satellite communications, and the Internet.

"The effect of such a storm on our modern technologies would be tremendous," says Luhmann.

During informal discussions at the workshop, Nat Gopalswamy of the Goddard Space Flight Center noted that "without NASA's *STEREO* probes, we might never have known the severity of the 2012 superstorm. This shows the value of having 'space weather buoys' located all around the Sun."

It also highlights the potency of the Sun even during so-called "quiet times." Many observers have noted that the current solar cycle is weak, perhaps the weakest in 100 years. Clearly, even a weak solar cycle can produce a very strong storm. Says Baker, "We need to be prepared."

QUESTIONS

1. What is a coronal mass ejection (CME)?

2. Why would a CME cause disruptions on Earth?

3. Explain how the 2012 storm missed Earth by 1 week.

4. More sensationalistic headlines for this story claimed that Earth almost "was sent back to the Dark Ages." What did they mean by that exaggeration?

5. Go to the NASA press release for this event (https://science.nasa.gov/science-news/science-at-nasa/2014/23jul_superstorm/), and click to watch the 4-minute "ScienceCast" video. What happened during the Carrington CME in 1859? Is that video effective at communicating the science information to the nonspecialist?

Source: https://science.nasa.gov/science-news/science-at-nasa/2014/02may_superstorm/.

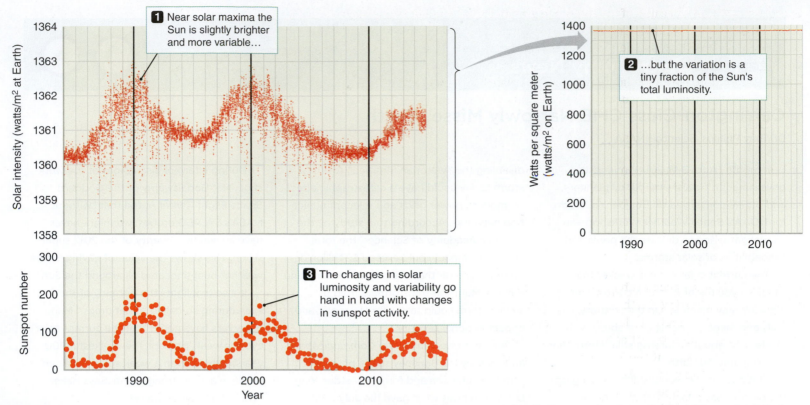

Figure 14.24 Measurements taken by satellites show that the amount of light from the Sun changes slightly.

measurements of the amount of radiation coming from the Sun (**Figure 14.24**) show that this value varies by as much as 0.2 percent over periods of a few weeks, as dark sunspots in the photosphere and bright spots in the chromosphere move across the disk. Overall, however, the increased radiation from active regions on the Sun more than makes up for the reduction in radiation from sunspots. On average, the Sun seems to be about 0.1 percent brighter during the peak of a solar cycle than it is at its minimum.

Solar activity affects Earth in many ways. Solar active regions are the source of most of the Sun's extreme ultraviolet and X-ray emissions—energetic radiation that heats Earth's upper atmosphere and, during periods of increased solar activity, causes Earth's upper atmosphere to expand. When that happens, the swollen upper atmosphere can significantly increase the atmospheric drag on spacecraft orbiting at relatively low altitudes, such as that of the Hubble Space Telescope, causing their orbits to decay. As a result, periodic boosts have been necessary to keep the Hubble Space Telescope in its orbit.

Earth's magnetosphere is the result of the interaction between Earth's magnetic field and the solar wind. Increases in the solar wind accompanying solar activity, especially CMEs directed at Earth, can disrupt Earth's magnetosphere. Spectacular auroras can accompany such events, as can magnetic storms that have disrupted electric power grids and caused blackouts across large regions. CMEs emitted in the direction of Earth also hinder radio communication and navigation, and they can damage sensitive satellite electronics, including communication satellites. In addition, energetic particles accelerated in solar flares pose one of the greatest dangers to human exploration of space.

Detailed observations from the ground and from space help astronomers understand the complex nature of the solar atmosphere. The Solar and Heliospheric Observatory (*SOHO*) spacecraft is a joint mission between NASA and the European

unanswered questions

Are large timescale variations in Earth's climate—ice ages—related to solar activity? Solar activity affects Earth's upper atmosphere and may affect weather patterns as well. Variations in the amount of radiation from the Sun might be responsible for past variations in Earth's climate. Current models indicate that observed variations in the Sun's luminosity could account for only about 0.1-K differences in Earth's average temperature—much less than the effects due to the ongoing buildup of carbon dioxide in Earth's atmosphere. Triggering the onset of an ice age may require a sustained drop in global temperatures of only 0.2–0.5 K, so astronomers are continuing to investigate a possible link between solar variability and long-timescale changes in Earth's climate.

Space Agency (ESA). Because of the combined gravity of Earth and the Sun, *SOHO* moves in lockstep with Earth at a location approximately 1,500,000 km from Earth that is almost directly in line between Earth and the Sun. *SOHO* carries 12 scientific instruments that monitor the Sun and measure the solar wind upstream of Earth. In addition, NASA's Solar Dynamics Observatory (SDO) studies the solar magnetic field to predict when major solar events will occur, instead of simply responding after they happen.

CHECK YOUR UNDERSTANDING 14.4

Sunspots appear dark because: (a) they have very low density; (b) magnetic fields absorb most of the light that falls on them; (c) they are regions of very high pressure; (d) they are cooler than their surroundings.

Origins: The Solar Wind and Life

Energetic particles accelerated in solar flares need to be considered when astronauts are orbiting Earth in a space station or traveling to the Moon or farther. Earth's magnetic field protects life on the surface from those energetic particles, which travel along the magnetic-field lines to Earth's poles, creating the auroras. But the Moon does not have that protection because its magnetic field is very weak. Astronauts on the lunar surface would be exposed to as much radiation as astronauts traveling in space. The strength of the solar wind varies with the solar cycle, as noted in Section 14.4, so the exposure danger varies as well.

The Solar System is surrounded by the **heliosphere (Figure 14.25)**, in which the solar wind blows against the interstellar medium and clears out an area like the inside of a bubble. As the Sun and Solar System move through the Milky Way Galaxy, passing in and out of interstellar clouds, that heliosphere protects the entire Solar System from galactic high-energy particles known as cosmic rays that originate primarily in high-energy explosions of massive dying stars. When the Sun is in its lower-activity state, the heliosphere is weaker, so more galactic cosmic rays enter the Solar System. In addition, the intensity of those cosmic rays depends on where the Sun and Solar System are located in their orbit about the center of the Milky Way Galaxy.

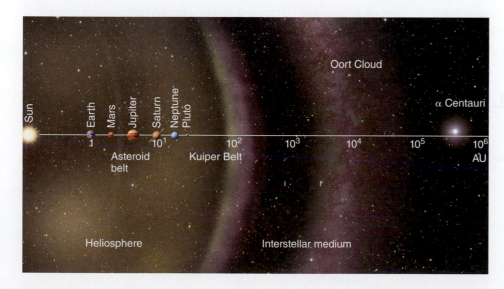

Figure 14.25 The heliosphere of the Sun is a bubble of charged particles covering the Solar System. The heliosphere is formed by the solar wind blowing against the interstellar medium. Both of the *Voyager* spacecraft are now past 100 AU. The scale, indicating how far objects are from the Sun, is logarithmic. Recall that a logarithmic graph has steps that are powers of ten: 10, 100, 1000, 10,000, etc.

Some scientists have theorized that at times when the Sun was quiet and the heliosphere was weaker than average, and the Solar System was passing through a particular part of the galaxy, the cosmic-ray flux in the Solar System—and on Earth—increased. That increased flux possibly led to a disruption in Earth's ozone layer and possibly contributed to a mass extinction in which many species died out on Earth.

Thus, in addition to heat and light on Earth, the extension of the Sun through the solar wind may have affected the evolution of life on Earth—and may affect the ability of humans to live and work in space.

SUMMARY

The forces due to pressure and gravity balance each other in hydrostatic equilibrium, maintaining the Sun's structure. Nuclear reactions converting hydrogen to helium are the source of the Sun's energy. Energy created in the Sun's core moves outward to the surface, first by radiation and then by convection. The solar wind may adversely affect astronauts located in space or on planets that lack a protective magnetic field, but it also has protected the Solar System from galactic high-energy cosmic-ray particles.

① Describe the balance between the forces that determine the structure of the Sun. The outward radiation pressure of the hot gas inside the Sun balances the inward pull of gravity at every point. That balance is dynamically maintained. An energy balance is also maintained, with the energy produced in the Sun's core balancing the energy lost from the surface.

② Diagram how mass is converted to energy in the Sun's core and estimate how long the Sun will take to use up its fuel. In the Sun's core, mass is converted to energy via the proton-proton chain. When four hydrogen atoms fuse to form one helium atom, some mass is lost. That mass is released as energy, nearly all of which leaves the Sun either as photons or as neutrinos. Neutrinos are elusive, almost massless particles that interact only very weakly with other matter. Observations of neutrinos confirm that nuclear fusion is the Sun's primary energy source. The innermost 10 percent of the Sun will participate in the fusion process, limiting the Sun's main-sequence lifetime to 10 billion years.

③ Sketch a physical model of the Sun's interior, and list the ways that energy moves outward from the Sun's core toward its surface. The Sun's interior is divided into zones defined by how energy is transported in that region. Energy moves outward through the Sun by radiation and by convection.

④ Relate observations of solar neutrinos and seismic vibrations on the surface of the Sun to astronomers' models of the Sun. The Sun has multiple layers, each with a characteristic pressure, density, and temperature. Neutrinos directly probe the interior of the Sun. That model of the interior of the Sun has been tested by helioseismology, in much the same way that the model of Earth's interior has been tested by seismology.

⑤ Describe the solar activity cycles of 11 and 22 years, and relate those cycles to the Sun's changing magnetic field and solar phenomena such as sunspots and flares. Activity on the Sun follows a cycle that peaks every 11 years but takes 22 full years for the magnetic field to reverse. Sunspots are photospheric regions cooler than their surroundings, and they reveal the cycles in solar activity. Material streaming away from the Sun's corona creates the solar wind, which moves outward through the Solar System until it meets the interstellar medium. Solar storms, including flares and other ejections of mass from the corona, produce auroras and can disrupt power grids and damage satellites.

QUESTIONS AND PROBLEMS

TEST YOUR UNDERSTANDING

1. The physical model of the Sun's interior has been confirmed by observations of
   a. neutrinos and seismic vibrations.
   b. sunspots and solar flares.
   c. neutrinos and positrons.
   d. sample returns from spacecraft.
   e. sunspots and seismic vibrations.

2. Place in order the following steps in the fusion of hydrogen into helium. If two or more steps happen simultaneously, use an equals sign (=).
   a. A positron is emitted.
   b. One gamma ray is emitted.
   c. Two hydrogen nuclei are emitted.
   d. Two ^{3}He collide and become ^{4}He.
   e. Two hydrogen nuclei collide and become ^{2}H.
   f. Two gamma rays are emitted.
   g. A neutrino is emitted.
   h. One deuterium nucleus and one hydrogen nucleus collide and become ^{3}He.

3. Sunspots, flares, prominences, and coronal mass ejections are all caused by
   a. magnetic activity on the Sun.
   b. electrical activity on the Sun.
   c. the interaction of the Sun's magnetic field and the interstellar medium.
   d. the interaction of the solar wind and Earth's magnetic field.
   e. the interaction of the solar wind and the Sun's magnetic field.

4. The structure of the Sun is determined by not only the balance between the forces due to _____ and gravity but also the balance between energy generation and energy _____.
   a. radiation pressure; production   c. ions; loss
   b. radiation pressure; loss   d. solar wind; production

5. In the proton-proton chain, four hydrogen nuclei are converted to a helium nucleus. That does not happen spontaneously on Earth because the process requires
   a. vast amounts of hydrogen.
   b. very high temperatures and densities.
   c. hydrostatic equilibrium.
   d. very strong magnetic fields.

6. The solar neutrino problem pointed to a fundamental gap in our knowledge of
   a. nuclear fusion.
   b. neutrinos.
   c. hydrostatic equilibrium.
   d. magnetic fields.

7. Sunspots change in number and location during the solar cycle. That phenomenon is connected to
   a. the rotation rate of the Sun.
   b. the temperature of the Sun.
   c. the magnetic field of the Sun.
   d. the tilt of the axis of the Sun.

8. Suppose an abnormally large amount of hydrogen suddenly fused in the core of the Sun. Which of the following would be observed first?
   a. The Sun would become brighter.
   b. The Sun would swell and become larger.
   c. The Sun would become bluer.
   d. The Sun would emit more neutrinos.

9. The solar corona has a temperature of 1 million to 2 million K, whereas the photosphere has a temperature of only about 6000 K. Why isn't the corona much, much brighter than the photosphere?
   a. The magnetic field traps the light.
   b. The corona emits only X-rays.
   c. The photosphere is closer to us.
   d. The corona has a much lower density.

10. The Sun rotates once every 25 days relative to the stars. The Sun rotates once every 27 days as seen from Earth. Why are those two numbers different?
    a. The stars are farther away.
    b. Earth is smaller.
    c. Earth moves in its orbit as the Sun rotates.
    d. The Sun moves relative to the stars.

11. Place the following regions of the Sun in order of increasing radius.
    a. corona   e. chromosphere
    b. core   f. photosphere
    c. radiative zone   g. a sunspot
    d. convective zone

12. Coronal mass ejections
    a. carry away 1 percent of the mass of the Sun each year.
    b. are caused by breaking magnetic fields.
    c. are always emitted in the direction of Earth.
    d. are unimportant to life on Earth.

13. As energy moves out from the Sun's core toward its surface, it first travels by _____, then by _____, and then by _____.
    a. radiation; conduction; radiation
    b. conduction; radiation; convection
    c. radiation; convection; radiation
    d. radiation; convection; conduction

14. Energy is produced primarily in the center of the Sun because
    a. the strong nuclear force is too weak elsewhere.
    b. that's where neutrinos are created.
    c. that's where most of the helium is.
    d. the temperature and density are high enough in the core.

15. The solar wind pushes on the magnetosphere of Earth, changing its shape, because
    a. the solar wind is so dense.
    b. the magnetosphere is so weak.
    c. the solar wind contains charged particles.
    d. the solar wind is so fast.

THINKING ABOUT THE CONCEPTS

16. Explain how hydrostatic equilibrium acts to keep the Sun at its constant size, temperature, and luminosity.

17. Two of the three atoms in a molecule of water (H_2O) are hydrogen. Why are Earth's oceans not fusing hydrogen into helium and setting Earth ablaze?

18. Why are neutrinos so difficult to detect?

19. Explain the proton-proton chain through which the Sun generates energy by converting hydrogen to helium.

20. On Earth, nuclear power plants use *fission* to generate electricity. In fission, a heavy element such as uranium is broken into many atoms, where the total mass of the fragments is less than that of the original atom. Explain why fission could not be powering the Sun today.

21. If an abnormally large amount of hydrogen suddenly fused in the core of the Sun, what would happen to the rest of the Sun? Would the Sun change as seen from Earth?

22. Study the Process of Science Figure. If the follow-up experiments did not detect the other types of neutrinos, what would have been the next step for scientists at that point? ★

23. What is the solar neutrino problem, and how was it solved?

24. How are orbiting satellites and telescopes affected by the Sun?

25. Describe the solar corona. Under what circumstances can it be seen without special instruments?

26. ★ **WHAT AN ASTRONOMER SEES**   In Figure 14.12, an astronomer would identify hydrogen and calcium lines immediately. Explain how these lines may have been identified the first time, and how experience might lead an astronomer to be able to instantly recognize them. ★

27. In the proton-proton chain, the mass of four protons is slightly greater than the mass of a helium nucleus. Explain what happens to that "lost" mass.

28. What have sunspots revealed about the Sun's rotation?

29. Why are different parts of the Sun best studied at different wavelengths? Which parts are best studied from space?

30. Why is studying the interaction of the solar wind with the interstellar medium important?

APPLYING THE CONCEPTS

31. The Sun shines by converting mass into energy according to $E = mc^2$. Show that if the Sun produces 3.85×10^{26} J of energy per second, it must convert 4.3 billion kg (4.3×10^9 kg) of mass per second into energy. ●—●—●
    a. Make a prediction: Is your final answer a rate or a mass? Therefore, what will the final units of your answer be?
    b. Calculate: Use the equation $E = mc^2$ to find the mass lost every second.
    c. Check your work: Verify that your units are correct and that you found the numerical answer stated in the problem statement.

32. If a sunspot has a temperature of 4800 K, while the surrounding photosphere has a temperature of 5780 K, what is the ratio of fluxes between the sunspot and the photosphere? ●—●—●
    a. Make a prediction: In Working It Out 14.2, you learned that the flux is proportional to T^4. Do you expect that the ratio of fluxes in this case will be larger or smaller than 1?
    b. Calculate: Follow Working It Out 14.2 to find the ratio of fluxes.
    c. Check your work: Verify that your answer is dimensionless (the units cancel out), and compare your answer to your prediction and to the answer in Working It Out 14.2.

33. Assume that the Sun has been producing energy at a constant rate over its lifetime of 4.6 billion years (1.4×10^{17} seconds). The current mass of the Sun is 2×10^{30} kg, and in Problem 31, you may have verified that the Sun loses mass at 4.3×10^{30} kg/s. What fraction of its current mass has been converted into energy over the lifetime of the Sun?
    a. Make a prediction: If the Sun has lost a large fraction of its mass over its lifetime, would that have had a large or small effect on planetary orbits? Therefore, do you expect this fraction of current mass to be large or small?
    b. Calculate: First, find the amount of mass lost by the Sun over its lifetime, and then express that as a fraction of the current mass.
    c. Check your work: Verify that your answer is dimensionless (all the units have canceled out), and compare your answer to your prediction.

34. The hydrogen bomb represents an effort to create a process similar to what takes place in the core of the Sun. The energy released by a 5-megaton hydrogen bomb is 2×10^{16} J. How much mass did Earth lose each time a 5-megaton hydrogen bomb was exploded?
    a. Make a prediction: The Sun loses about 4 billion kg of mass every second, to produce the amount of energy it emits. Do you expect your answer to be on that same scale (billions of kg) or much smaller?
    b. Calculate: Use $E = mc^2$ to find the mass lost in a 5-megaton hydrogen bomb explosion.
    c. Check your work: Verify that your answer has units of kg, and compare your answer to your prediction.

35. If a sunspot appears only 70 percent as bright as the surrounding photosphere, and the photosphere has a temperature of approximately 5780 K, what is the temperature of the sunspot?
    a. Make a prediction: What is a reasonable number for the temperature of a sunspot? A few kelvins? Hundreds of kelvins? Thousands

of kelvins? Millions of kelvins? Should the temperature that you calculate for a sunspot be larger or smaller than the temperature of the surrounding photosphere?
    b. Calculate: Follow Working It Out 14.2 in reverse to solve for the temperature of a sunspot from the ratio of brightness.
    c. Check your work: Verify that your answer has units of kelvins and compare your answer to your prediction.

36. In Figure 14.10, density and temperature are both graphed versus height. ◉
    a. Is the height axis linear or logarithmic? How do you know?
    b. Is the density axis linear or logarithmic? How do you know?
    c. Is the temperature axis linear or logarithmic? How do you know?

37. Use the data in Figures 14.19b and c to present an argument that sunspots occur in regions of strong magnetic field. ◉

38. Study Figure 14.17a and Figure 14.20. ◉
    a. Estimate the fraction of the Sun's surface covered by the large sunspot group in Figure 14.17a. (Remember that you are seeing only one hemisphere of the Sun.)
    b. From the graph in Figure 14.20, estimate the average number of sunspots that occurs at solar maximum.
    c. On average, what fraction of the Sun could be covered by sunspots at solar maximum? Is that a large fraction?
    d. Compare your conclusion with the graph of intensity in Figure 14.24. Does that graph make sense to you?

39. Assume that the Sun's mass is about 300,000 Earth masses and that its radius is about 100 times that of Earth. The density of Earth is about 5500 kg/m³.
    a. What is the average density of the Sun?
    b. How does that value compare with the density of Earth? With the density of water?

40. Suppose our Sun was an A5 main-sequence star, with twice the mass and 12 times the luminosity of the Sun, a G2 star. How long would that A5 star fuse hydrogen to helium? What would that mean for Earth?

41. Imagine that the source of energy inside the Sun changed abruptly.
    a. How long would it take before a neutrino telescope detected the event?
    b. When would a visible-light telescope see evidence of the change?

42. On average, how long do particles in the solar wind take to reach Earth from the Sun if they are traveling at an average speed of 400 km/s?

43. Verify the claim made at the start of this chapter that the Sun produces more energy per second than all the electric power plants on Earth could generate in a half-million years. Estimate or look up how many power plants are on the planet and how much energy an average power plant produces. Be sure to account for different kinds of power, such as coal, nuclear, and wind.

44. Let's examine the reason why the Sun cannot power itself by chemical reactions. Using Working It Out 14.1 and the fact that an average chemical reaction between two atoms releases 1.6×10^{-19} J, estimate how long the Sun could emit energy at its current luminosity. Compare that estimate with the known age of Earth.

45. The Sun could get energy from gravitational contraction for a time period of ($GM_{Sun}^2/R_{Sun}L_{Sun}$). How long would the Sun last at its current luminosity? (Pay careful attention to units!)

EXPLORATION The Proton-Proton Chain

digital.wwnorton.com/astro7

The proton-proton chain powers the Sun by fusing hydrogen into helium. That fusion process produces several types of particles as by-products, as well as energy. In this Exploration, we study the steps of the proton-proton chain in detail, with the intent of helping you keep them straight.

Visit the Digital Resources Page on the Student Site and open the "Proton-Proton Chain" Interactive Simulation in Chapter 14.

Watch the animation all the way through once.

Play the animation again, pausing after the first collision. Two hydrogen nuclei (both positively charged) have collided to produce a new nucleus with only one positive charge.

1 Which particle carried away the other positive charge?

_____

_____

2 What is a neutrino? Did the neutrino enter the reaction, or was the neutrino produced in the reaction?

_____

_____

_____

Compare the interaction on the top with the interaction on the bottom.

3 Did the same reaction occur in each instance?

_____

_____

Resume playing the animation, pausing it after the second collision.

4 What two types of nuclei entered the collision? What type of nucleus resulted?

_____

_____

5 Was charge conserved in that reaction, or did a particle have to carry charge away?

_____

_____

6 What is a gamma ray? Did the gamma ray enter the reaction, or was it produced by the reaction?

_____

_____

_____

Resume the animation again, and allow it to run to the end.

7 What nuclei enter the final collision? What nuclei are produced?

_____

_____

8 In chemistry, a catalyst is a species that facilitates a reaction but is not used up in the process. Do any nuclei act like catalysts in the proton-proton chain?

_____

_____

Make a table of inputs and outputs. Which particles in the final frame of the animation were inputs to the reaction? Which were outputs? Fill in your table with those inputs and outputs.

9 Which outputs are converted into energy that leaves the Sun as light?

_____

_____

10 Which outputs could become involved in another reaction immediately?

_____

_____

11 Which output is likely to stay in that form for a very long time?

_____

_____

Life

When astronomers look for worlds that might harbor life, they look for water, as you will learn here in Chapter 24. Life as we know it depends on water because water is a terrific solvent; it can carry chemicals, minerals, and nutrients around in an organism. It has this property because it is a polar molecule—the hydrogen end of the molecule is positively charged, while the oxygen end is negatively charged. Water, therefore, is attracted to lots of different molecules. It can be attracted, in fact, to anything that carries a charge. Rub a plastic object, like a pen, a ruler, or an inflated balloon, all over your dry hair, so that the plastic object becomes electrically charged. Turn on the water from a faucet so that a thin stream falls from the tap. Take a moment to predict what will happen when you bring the charged plastic object close to the water. Then slowly bring the plastic object close to the stream of water. Sketch the result.

EXPERIMENT SETUP

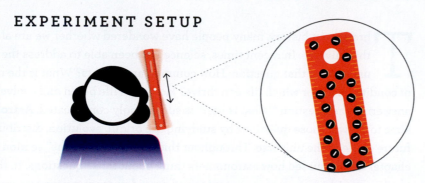

Rub a plastic object, like a pen, a ruler, or an inflated balloon, all over your hair, separating charges, so that the plastic object becomes electrically charged.

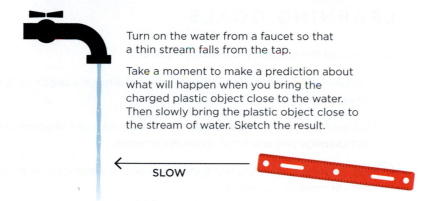

Turn on the water from a faucet so that a thin stream falls from the tap.

Take a moment to make a prediction about what will happen when you bring the charged plastic object close to the water. Then slowly bring the plastic object close to the stream of water. Sketch the result.

SLOW

PREDICTION

I predict that the stream of water will:
- ☐ bend away from the object.
- ☐ bend toward the object.
- ☐ fall straight down.
- ☐ yank the object out of my hand.

SKETCH OF RESULTS (in progress)

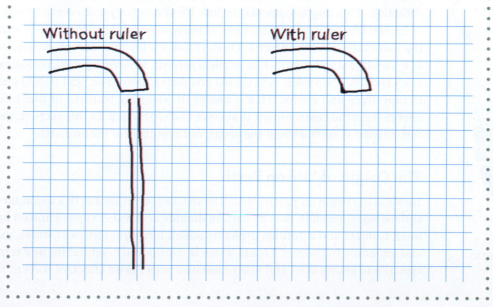

Without ruler

With ruler

24

Throughout history, many people have wondered whether we are alone in the universe. In recent times, science has been able to address the issues underlying that question: How common are planets? What is the range of conditions under which life can thrive? How does life begin and evolve? Even answering the question "What is life?" is surprisingly complicated. **Astrobiology** aims to answer those questions by studying the origin, evolution, distribution, and future of life in the universe. Throughout this book, the "Origins" section in each chapter has discussed how astronomers think about those questions. In this chapter, we expand on some of those topics and provide a more systematic overview of how scientists think about life in the universe and how they search for signs of it.

LEARNING GOALS

By the end of this chapter, you should be able to:

(1) Explain our current understanding of how and when life began on Earth and how it has evolved.

(2) Explain how life is a structure that has evolved through the action of the physical and chemical processes that shape the universe.

(3) List the locations in our Solar System and around other stars where astronomers think life might be possible.

(4) Describe some methods used to search for intelligent extraterrestrial life.

24.1 Life Evolves on Earth

What is **life**? Many scientists suggest that no single definition of life would encompass all the forms of life that may exist in the universe. To date, we have discovered only a single example of life: that found here on Earth. But even comparing varied organisms within that single example leads to complications in the definition of life. Viruses, for example, meet some criteria for life but not others. Life on Earth may be very different from life found in other places in the universe, and a complete definition of life may also one day include life-forms not yet discovered. From studies of known life, we conclude that like planets, stars, and galaxies, life is a structure that has evolved in the universe. On Earth, all life involves carbon-based chemistry and uses liquid water as its biochemical solvent, whereas specific biological molecules such as ribonucleic acid (RNA) and deoxyribonucleic acid (DNA) enable life to reproduce and evolve. This is one of the defining features of life: Life draws energy from the environment to survive and reproduce. In this section, we briefly review what is known about the origin and evolution of life on Earth.

The Origin of Life on Earth

How did life begin on Earth? Recall from Chapter 7 that Earth's secondary atmosphere was formed in part by carbon dioxide and water vapor emitted by volcanoes. Comets and asteroids probably added large quantities of water, methane, and ammonia to the mix. Liquid water is considered essential for any terrestrial-type life to get its start and evolve because water is an effective solvent that can

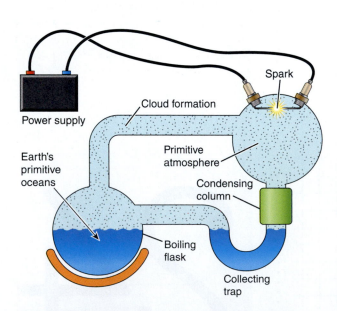

Figure 24.1 The Urey-Miller experiment was designed to simulate conditions in an early-Earth atmosphere.

move other atoms and molecules around, making them more accessible to cells. Early Earth had abundant sources of energy, such as lightning and ultraviolet solar radiation, that fragmented those simple molecules. Those fragments later reassembled into molecules of greater mass and complexity. Some of those were organic molecules—that is, molecules that contained hydrogen and carbon. Rain carried the heavier molecules out of the atmosphere into Earth's oceans, forming a primordial soup.

In 1952, chemists Harold Urey (1893–1981) and Stanley Miller (1930–2007) attempted to create something similar to those early-Earth conditions. Using equipment illustrated in **Figure 24.1**, they placed water in a sterilized laboratory jar to represent the ocean and then added methane, ammonia, and hydrogen as a primitive atmosphere; electric sparks simulated lightning as a source of energy. Within a week, the Urey-Miller experiment yielded molecules associated with life—namely, amino acids and components of nucleic acids. Proteins, the structural molecules of life, are made of 20 amino acids. Eleven of those acids were synthesized in the Urey-Miller experiment. Nucleic acids are the precursors of RNA and DNA.

Additional sealed samples from that old experiment were examined 50 years later. In those samples, hydrogen sulfide had been added to the "primitive atmosphere." When the samples were analyzed, 23 amino acids were found. That finding suggests that hydrogen sulfide, which would have come from volcanic plumes in the early Earth, was important. More recent experiments with carbon dioxide and nitrogen as the primitive atmosphere have produced results similar to those of Urey and Miller. A plausible atmospheric composition with an energy source can produce significant quantities of amino acids and other substances important to life.

From laboratory experiments such as those, scientists have developed various models to explain how life might have begun in an early-Earth environment. However, the details of how those precursor molecules evolved into the molecules of life are not yet clear. Some biologists think life began in the ocean depths, where volcanic vents provided the localized heat needed to create the highly organized molecules responsible for biochemistry (**Figure 24.2**). Other researchers think that life originated in tide pools, where lightning and ultraviolet radiation supplied the energy (**Figure 24.3**). Some researchers think life may have evolved in shallow lakes further inland. In either case, short strands of molecules that could replicate themselves may have formed first, later evolving into RNA, and finally into DNA, the huge molecule that serves as the biological "blueprint" for self-replicating organisms.

A few scientists have suggested that life on Earth may have been "seeded" from space in the form of microorganisms brought here by meteoroids or comets. This idea is called "panspermia." However, while quite long and complex carbon chains have been observed in nebulae, no scientific evidence exists to support this much more speculative hypothesis about *life* beginning in star-forming nebulae or in the outer reaches of the Solar System. Furthermore, seeding might explain how life came to Earth, but it does not satisfactorily answer the question: How did life begin?

When Life Began

If life did indeed get its start in Earth's oceans, when did it happen? Recall from Chapter 7 that Earth was bombarded by Solar System debris for several hundred

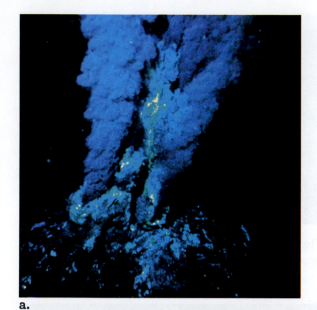

a.

b.

Figure 24.2 a. Life on Earth may have arisen near ocean hydrothermal vents such as this one. Similar environments might exist elsewhere in the Solar System. **b.** Living organisms around hydrothermal vents, such as the giant tube worms shown here, rely on hydrothermal rather than solar energy for their survival.

Figure 24.3 Life may have begun in tide pools, where lightning and ultraviolet light supply the energy for chemical processes.

million years after it formed roughly 4.6 billion years ago. Those conditions might have been too harsh for life to form and evolve on Earth. Once the bombardment abated and oceans formed, the opportunities for life to begin greatly improved. Terrestrial life seems to have quickly taken advantage of that more favorable environment. In the Nuvvuagittuq belt in Quebec, Canada, scientists recently found what they think are fossilized microorganisms that are 3.77 billion to 4.28 billion years old and seem similar to microorganisms from modern hydrothermal vents. Scientists debate whether carbonized material in Greenland rocks dating back 3.65 billion to 3.85 billion years provides indirect evidence of early life. Stronger and more direct evidence for early life appears in the form of fossilized **stromatolites** (masses of simple microorganisms) that date back about 3.5 billion years. Fossilized stromatolites have been found in western Australia and southern Africa, and living examples still exist today (**Figure 24.4**). An analysis of eleven 3.465-billion-year-old microfossils from western Australia found five species—two of which carried out primitive photosynthesis and the other three produced or consumed methane. That evidence suggests that the earliest life formed less than a billion years after the Solar System formed and within a few hundred million years of the end of the late heavy-bombardment period.

All life on Earth shares a similar genetic code that originated from a common ancestor. Close comparison of DNA of different species enables biologists to trace backward to the time when different types of life first appeared on Earth and to identify the species from which those life-forms evolved. The earliest organisms were **extremophiles**—life-forms that not just survive but actually thrive under extreme environmental conditions. Extremophiles include organisms such as thermophiles, which flourish in water temperatures as high as 120°C that occur near deep-ocean hydrothermal vents, such as the one shown in Figure 24.2a. Other extremophiles thrive in conditions of extraordinary cold, low oxygen, salinity, pressure, dryness, acidity, or alkalinity. Scientists today study extremophiles in boiling-hot sulfur springs in Yellowstone National Park, in salt crystals beneath the Atacama Desert in Chile, at the bottoms of glaciers, in ice fields in the Arctic, and in other extreme environments (**Figure 24.5**).

Among the early life-forms was an ancestral form of **cyanobacteria**, single-celled organisms otherwise known as **blue-green algae**. Those microorganisms form extensive sheets on the surface of bodies of water, as shown in **Figure 24.6a**.

Figure 24.4 These modern-day stromatolites are growing in colonies along an Australian shore.

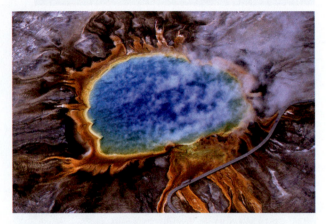

Figure 24.5 These thermophiles in the Grand Prismatic Spring in Yellowstone National Park live in temperatures of 70°C. The colors result from different amounts of chlorophyll.

a.

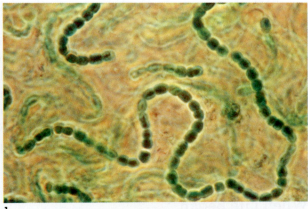

b.

Figure 24.6 a. Cyanobacteria today form sheets on lakes and other bodies of water. b. Under a microscope, the individual microorganisms are visible.

Under a microscope, it turns out that those sheets are colonies of individual microorganisms, as seen in **Figure 24.6b**. Cyanobacteria **photosynthesize**, using sunlight and carbon dioxide as food and generating oxygen as a waste product. Initially, the highly reactive oxygen that cyanobacteria produced was quickly removed from Earth's atmosphere by oxidation, or rusting, of surface minerals. Once most of the exposed minerals were oxidized and could no longer absorb oxygen, atmospheric levels of oxygen began to rise. Oxygenation of Earth's atmosphere and oceans began about 2 billion years ago, and the current level was reached only about 250 million years ago, as shown in **Figure 24.7**. Without cyanobacteria and other photosynthesizing organisms, Earth's atmosphere would be as oxygen-free as the atmospheres of Venus and Mars.

Biologists comparing DNA sequences find that terrestrial life is divided into two types: prokaryotes and eukaryotes. Prokaryotes, which include bacteria and archaea, are simple organisms that consist of free-floating DNA inside a cell wall; as shown in **Figure 24.8a**, they lack both cell structure and a nucleus. Eukaryotes (**Figure 24.8b**), which form the cells in animals, plants, and fungi, have a more complex form of DNA contained within the cell's membrane-enclosed nucleus. The first eukaryote fossils date from about 2 billion years ago, coincident with the rise of free oxygen in the oceans and atmosphere, although the first multicellular eukaryotes did not appear until a billion years later.

Increasing Complexity

Scientists have used DNA sequencing to establish what is known as the "phylogenetic tree of life," shown in **Figure 24.9**. That complex tree describes the evolutionary interconnectivity of all species of Bacteria, Archaea, and Eukarya and has revealed some interesting relationships. For example, Archaea were initially thought to be the same as Bacteria, but genetic studies show they diverged long ago, and the Archaea have genes and metabolic pathways more similar to those of Eukarya than to those of Bacteria. On the macroscopic scale, the phylogenetic tree places animals closest to fungi, which branched off the evolutionary tree after slime molds and plants.

Living creatures in Earth's oceans remained much the same—a mixture of single-celled and relatively primitive multicellular organisms—for more than 3 billion years after terrestrial life appeared. Between 540 million and 500 million years ago, the number and diversity of biological species increased spectacularly. Biologists call that event the **Cambrian explosion**. The trigger of that sudden surge in biodiversity remains unknown, but possibilities include rising oxygen levels, an increase in genetic complexity, major climate change, or some combination of those factors. The "Snowball Earth" hypothesis suggests that before the Cambrian explosion, Earth was in a period of extreme cold between about 750 million and 550 million years ago and was covered almost entirely by ice. During that period of extreme cold, predatory animals died out, making it easier for new species to adapt and thrive. Another possibility is that the marked increase in atmospheric oxygen (O_2) would have been accompanied by a corresponding increase in stratospheric ozone (O_3), which shields Earth's surface from deadly solar ultraviolet radiation. With that protective ozone layer in place, life was free to leave the oceans and move to land. *Tiktaalik*, a fish with limblike fins and ribs, was an animal in a

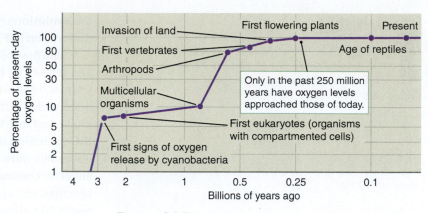

Figure 24.7 The amount of oxygen in Earth's atmosphere has built up as a result of photosynthesis by cyanobacteria and plant life on the planet.

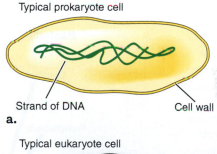

a.

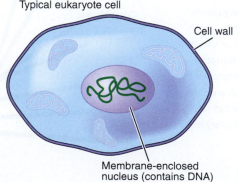

b.

Figure 24.8 **a.** A simple prokaryote cell contains little more than the cell's genetic material. **b.** A eukaryote cell contains several membrane-enclosed structures, including a nucleus that houses the cell's genetic material.

what if . . .

What if life formed in both the ocean depths and in tide pools? Would you expect all life on Earth to be as interconnected as is shown in Figure 24.9?

midevolutionary step of leaving the water for dry land, as shown in the artist's illustration in **Figure 24.10**.

The first plants appeared on land about 475 million years ago. Large forests and insects go back 360 million years. The age of dinosaurs began 230 million years ago and ended abruptly 65 million years ago, when an asteroid or comet collided with Earth (see the "Origins" feature in Chapter 8). The collision threw so much dust into the atmosphere that the sunlight was dimmed for months, causing the extinction of more than 70 percent of all existing plant and animal species. Mammals were the big winners in the aftermath. Primates evolved from the last ancestor common with other mammals about 70 million years ago. The great apes (gorillas, chimpanzees, bonobos, and orangutans) split off from the lesser apes about 20 million years ago (Figure 24.9, inset). DNA tests show that humans and chimpanzees share about 98 percent of their DNA, indicating that they evolved from a common ancestor about 6 million years ago. By comparison, all humans share 99.9 percent of their DNA. The earliest human ancestors appeared a few million years ago, and the first civilizations occurred a mere 10,000 years ago. Present-day industrial society, barely more than two centuries old, is but a moment in the history of life on Earth.

Humans are here today because of a series of events that occurred throughout the history of the universe. Some of those events are common in the universe, such as the formation of heavy elements in earlier generations of stars and the formation of planets. Other events in Earth's history may have been less likely to happen elsewhere, such as the formation of a planet with life-supporting conditions like Earth or the development of self-replicating molecules that led to Earth's earliest life. A few events stand out, such as major extinctions that allowed the evolution of mammalian life and, ultimately, human beings.

Figure 24.9 This simplified version of the phylogenetic tree has been constructed from analysis of the DNA strands of different life-forms. Humans are included in the "Animals" twig on the Eukarya branch. The primate branch, which includes humans, is shown in the inset on the bottom. By tracing those common ancestors through DNA and other means, scientists can reconstruct the evolutionary history of a species.

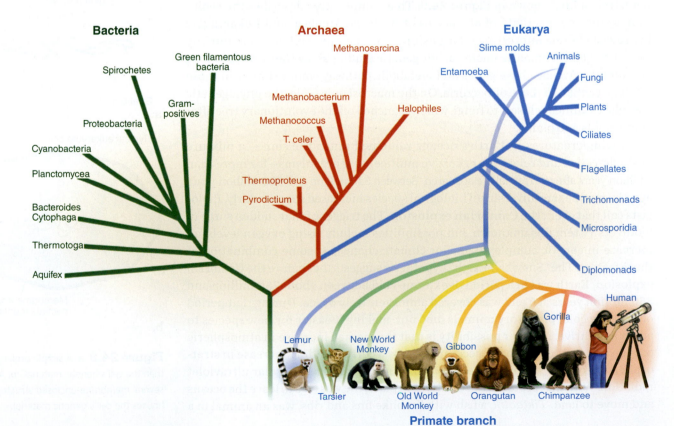

working it out 24.1

Exponential Growth

Self-replication is an example of exponential growth. The number of times a sample will double, n, for exponential growth is determined by the ratio of the original and final amounts:

$$\frac{P_F}{P_O} = 2^n$$

Assume a hypothetical self-replicating molecule makes one copy of itself each minute, and each copy in turn copies itself each minute. How many molecules will exist after an hour? Here, the number of generations is given by $n = 60$, because there are 60 minutes in an hour:

$$\frac{P_F}{P_O} = 2^{60} = 1.2 \times 10^{18}$$

Thus, a billion billion of those molecules will exist after 1 hour.

Now suppose a mutation occurs once every 50,000 times that a molecule reproduces itself, and one of every 200,000 mutations turns out to be beneficial. After 100 generations, how many molecules with those beneficial mutations might exist? That equation is similar to the previous equation, but here $n = 100$:

$$\frac{P_F}{P_O} = 2^{100} = 1.3 \times 10^{30}$$

The total number of molecules is 1.3×10^{30}. The number of mutations is that number divided by 50,000, or 2.6×10^{25} mutated molecules. The number of beneficially mutated molecules is that number divided by 200,000, or 1.3×10^{20} molecules. So, 100 million trillion (10^{20}) mutations will occur that, by chance, might improve the survivability of the original molecule. Because that number does not count earlier beneficial changes that themselves replicated, the total number of molecules with beneficial changes will be even larger!

Evolution as a Mechanism of Change

Imagine that just once during the first few hundred million years after Earth formed, a single molecule formed somewhere in Earth's oceans. That molecule had a very special property: chemical reactions between that molecule and other molecules in the surrounding water caused the molecule to make a copy of itself. The molecule became *self-replicating*. Chemical reactions then produced copies of each of those two molecules, making four molecules. Four molecules became eight, eight became 16, 16 became 32, and so on. By the time the original molecule had copied itself just 100 times, more than a million trillion trillion (10^{30}) of those molecules existed. That is about 100 million times more of those molecules than the number of stars in the observable universe. The molecules of DNA that make up the chromosomes in the nuclei of the cells of all advanced life today are direct descendants of those early self-replicating molecules that flourished in the oceans of the young Earth.

Over time, not all replications are exact. The likelihood that a copying variation will occur while a molecule is replicating increases significantly with the number of copies being made. For DNA, which contains the genetic code for an entire organism, a change in the genetic code is called a **mutation**. Sometimes a mutation has no effect. Other times, a mutation can prevent an organism from flourishing. In still other cases, a mutation can make an organism better suited to its environment. Organisms with those advantageous mutations will survive to reproduce successfully. Even if mutations are rare and only a small fraction turn out to be beneficial, after just 100 generations trillions of mutations occur that, by luck, might improve on the original (**Working It Out 24.1**). **Heredity**—the ability of one generation to pass on its genetic code to future generations—allow beneficial mutations to persist and be incorporated into a species' genetic code.

Figure 24.10 This illustration is an artist's reconstruction from a fossil of *Tiktaalik* found in the Canadian Arctic.

Figure 24.11 Fossils, such as this *Parasaurolophus* ("near crested lizard"), record the history of the evolution of life on Earth. This 10-meter-long plant-eating dinosaur lived in North America about 75 million years ago.

As the organisms of the early Earth continued to interact with their surroundings and make copies of themselves, mutations caused them to diversify into many species. Sometimes the resources they needed to reproduce became scarce. In the face of that scarcity, varieties that were more successful reproducers became more numerous. Competition for resources, predation by one species on another, and cooperation between organisms became important to the survival of different varieties. Some varieties were more successful and reproduced to become more numerous, whereas less successful varieties became less and less common. That process, in which better-adapted organisms reproduce and thrive, whereas less well-adapted organisms become extinct, is called **natural selection**.

Life has existed on Earth for about 4 billion years, which is a very long time—long enough for the combined effects of heredity and natural selection to shape the descendants of that early self-copying molecule into a huge variety of complex, competitive, successful structures. Geological processes on Earth have preserved a fossil record of the history of some of those structures (**Figure 24.11**). Among those descendants are human "structures" that can think about their own existence and unravel the mysteries of the stars.

CHECK YOUR UNDERSTANDING 24.1

Extremophiles are organisms that: (a) are extremely reactive; (b) are extremely rare; (c) have an extreme quality, such as mass or size; (d) live in extreme conditions.

Answers to Check Your Understanding questions are in the back of the book.

24.2 Life Involves Complex Chemical Processes

The evolution of life on Earth cannot be separated from the narrative of astronomy: it is one of many examples of the emergence of structure in an evolving universe (see the **Process of Science Figure**). The emergence of that structure then leads to the following question: Has life arisen elsewhere? Unlike the study of planets, stars, and galaxies, only one known case exists for the study of life—Earth—and scientists do not know how much of that example can be generalized to other places. To explore that question, we need to take a closer look at the processes that have led to life on Earth. In this section, we explore the chemical and physical properties of life on Earth.

The infant universe was composed basically of hydrogen and helium and very little else. After 9 billion years of stellar nucleosynthesis, all the heavier chemical elements essential to life were present and available in the molecular cloud that gave birth to the Solar System. Those heavier elements were formed by nuclear fusion in the cores of earlier generations of stars and were then dispersed into space. At times, that dispersal was passive. For example, low-mass stars such as the Sun, when they become puffed-up, dying red giants, may shed their extended atmospheres, sending some newly created carbon into space. Other dispersals were more violent. High-mass stars produce even heavier elements through nucleosynthesis in their cores—up to and including iron. But some of the trace elements essential to biology on Earth are even more massive than iron. They are produced within a matter of minutes during

All of Science Is Interconnected

More than most other subjects, astrobiology relies on concepts from a number of scientific fields—showing that all of science is interconnected.

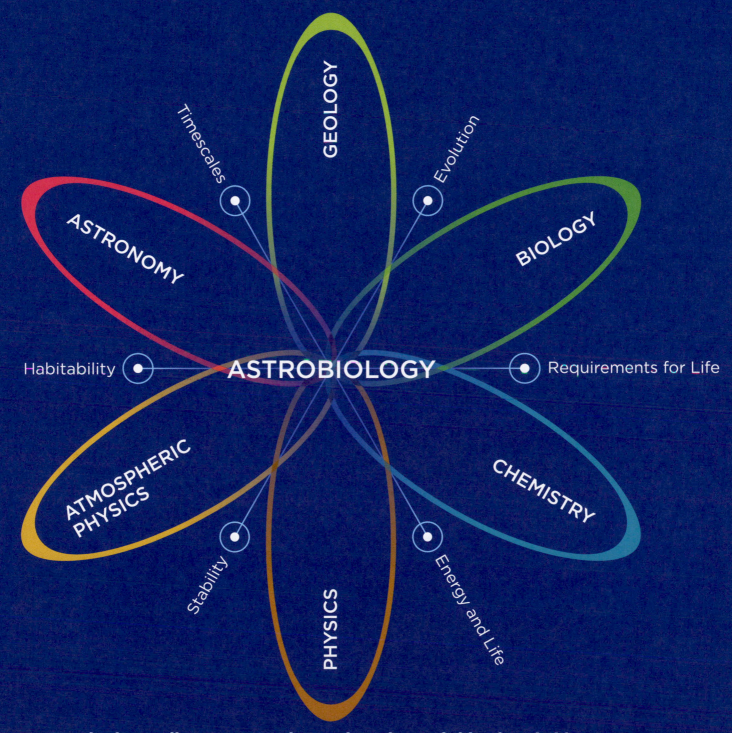

No science stands alone. All are connected. Interdisciplinary fields of study like astrobiology provide opportunities for new tests for theories of many fields.

what if . . .

What if we discovered self-replicating life on another planet, where the life is based on a molecule other than DNA? Would we expect life to be evolving on such a planet?

the violent supernova explosions that mark the death of high-mass stars and then are thrown into the chemical mix found in molecular clouds. Or they are produced in the collisions of neutron stars, the dense remains of prior supernova explosions (**Figure 24.12**).

All known living organisms on Earth are composed of a more or less common suite of complex chemicals. Approximately two-thirds of the atoms in the human body are hydrogen (H); about one-fourth are oxygen (O); a tenth are carbon (C); and a few hundredths are nitrogen (N). Carbon, nitrogen, and oxygen are the three most abundant products of stellar nucleosynthesis after helium (see the "astronomer's periodic table" in Figure 5.17). The several dozen remaining atomic elements in the human body make up only 0.2 percent of the total. All known living creatures are assemblages of molecules composed almost entirely of those four elements, sometimes called CHON (carbon, hydrogen, oxygen, nitrogen), along with small amounts of phosphorus and sulfur. Some of those molecules, such as RNA, DNA, and proteins, are enormous. A small piece of DNA, which is responsible for genetic codes, is illustrated in **Figure 24.13**. DNA is made up entirely of only five atomic elements—CHON and phosphorus—but the DNA in each cell of the human body is composed of combinations of *tens of billions* of atoms of those same five elements. Proteins, the huge molecules responsible for the structure and function of living organisms, are long chains of smaller molecules called amino acids. Terrestrial life uses 20 specific amino acids, which also consist of no more than five atomic elements—CHON plus sulfur.

Although life primarily uses the half-dozen atoms of CHON plus sulfur and phosphorus, some other elements, present in smaller amounts, are essential to the chemical processes that living organisms carry out. Those elements include sodium, chlorine, potassium, calcium, magnesium, iron, manganese, and iodine. Trace elements such as copper, zinc, selenium, and cobalt also play a crucial role in biochemistry but are needed in only tiny amounts.

Carbon, which can bond to four other atoms, forms the backbone of the DNA molecule shown in Figure 24.13. That is why carbon is so important to life on Earth. Forms of extraterrestrial life could exist that are also carbon-based but have chemistries different from that of life on Earth. For example, countless varieties of amino acids exist in addition to the 20 used by terrestrial life. Most other atoms are more limited than carbon in the number of bonds they can make, but silicon, like carbon, can bond to four other atoms, so many combinations are possible. As a potential life-enabling atom, silicon has both advantages and disadvantages in comparison with carbon. Silicon-based molecules remain stable at much higher temperatures than carbon-based molecules, perhaps enabling silicon-based life to thrive in high-temperature environments, such as on planets that orbit close to their parent star. But silicon is also a larger and more massive atom than carbon: It cannot form molecules as complex as those based on carbon. Any silicon-based life probably would be simpler than life-forms here on Earth, but it might exist in high-temperature niches somewhere within the universe. Although carbon's unique properties make it

Figure 24.12 This version of the periodic table shows the astronomical origin of the naturally occurring elements (up to uranium) in the Solar System.

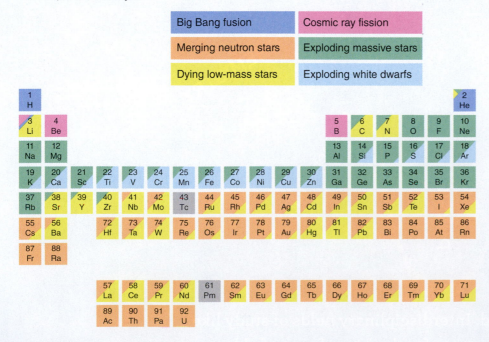

readily adaptable to the chemistry of life on Earth, other types of life might be found elsewhere.

CHECK YOUR UNDERSTANDING **24.2**

Carbon is a favorable base for life because: (a) it can bond to many other atoms in long chains; (b) it is nonreactive; (c) it forms weak bonds that can be readily reorganized as needed; (d) it is organic.

24.3 Where Do Astronomers Look for Life?

One approach to the scientific search for extraterrestrial life is to use robotic spacecraft to explore the planets and moons of the Solar System (see Chapter 6). Spacecraft have visited all the planets and some moons and sent back at least some information about the conditions on those worlds. Another approach is to use telescopes to detect planets outside our Solar System (see Chapter 7). Telescopes on the ground and in space have detected several thousand planets orbiting other stars. In this section, we survey the locations where life might be found, both in our own Solar System and around other stars.

Life within Our Solar System

Scientists start the search for evidence of extraterrestrial life here in our own Solar System. Early conjectures about life in our Solar System seem naïve, considering what we now know. Two centuries ago, the eminent astronomer Sir William Herschel, discoverer of Uranus, proclaimed, "We need not hesitate to admit that the Sun is richly stored with inhabitants." In 1877, astronomer Giovanni Schiaparelli (1835–1910) observed what appeared to be linear features on Mars and dubbed them *canali* ("channels" in Italian). Another observer of Mars, Percival Lowell (1855–1916), misinterpreted Schiaparelli's *canali* as "canals," suggesting that they were constructed by intelligent beings.

Because Mercury and the Moon lacked atmospheres, astronomers determined that those worlds were not conducive to life. The giant planets and their moons were thought to be too remote and too cold to sustain life. By the 1960s it was understood that the surface of Venus was too hot to permit life. At the same time, astronomers using ground-based telescopes had discovered that Mars has a thin atmosphere, water ice, and carbon dioxide ice. During the 1960s, the United States and the Soviet Union sent reconnaissance spacecraft to the Moon, Venus, and Mars, but the instruments on those spacecraft probed the physical and geological properties of those astronomical bodies instead of searching for life. Serious efforts to look for signs of life—past or present—require more advanced spacecraft with specialized instrumentation.

In the mid-1970s, two American *Viking* spacecraft were sent to Mars with detachable landers containing a suite of instruments designed to find evidence of a terrestrial type of life. When the *Viking* landers failed to find convincing evidence of life on Mars, hopes faded for finding life on any other body orbiting the Sun. Since then, however, further exploration of the Solar System has generated renewed optimism. A better understanding of the history of Mars indicates the planet's climate has changed. Mars was once wetter and warmer than it is today.

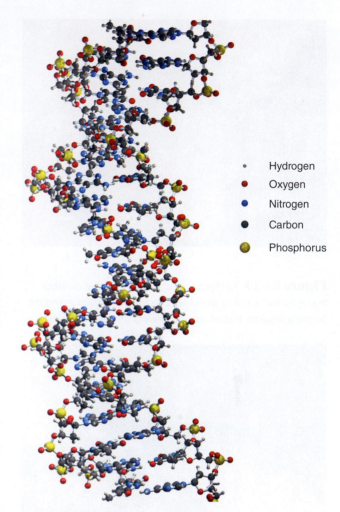

- Hydrogen
- Oxygen
- Nitrogen
- Carbon
- Phosphorus

Figure 24.13 DNA, the heritable molecule that forms the basis for life on Earth, contains the atoms of only five elements. Even so, those atoms are combined in billions of ways, giving rise to the diversity of life on Earth.

Figure 24.14 The Mars *Curiosity* rover detected evidence that Mars had a watery past. The rounded gravel surrounding the bedrock suggests that an ancient, flowing stream once existed.

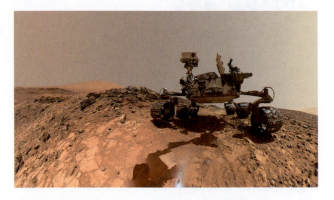

Figure 24.15 A "selfie" of the *Curiosity* Mars rover at the site where it drilled into some rocks and identified organic compounds.

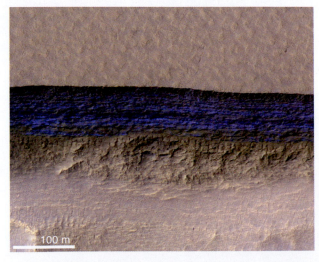

Figure 24.16 This image from the *MRO* shows 80 meters of exposed water ice on a steep slope. The color is exaggerated. Similar steep slopes with ice at other midlatitude locations also were seen.

In the 1990s, Mars missions began to map the planet's surface from the ground and from space. In 2008, NASA's *Phoenix* spacecraft landed at a far-northern latitude, inside the planet's arctic circle, where specialized instruments dug into and analyzed the Martian water-ice permafrost. *Phoenix* found that the Martian arctic soil has a chemistry similar to that of the Antarctic dry valleys on Earth, where life exists deep below the surface at the ice-soil boundary. Minerals that form in water, such as calcium carbonate, have been detected. That finding suggests that oceans existed in the past on Mars. However, *Phoenix* found no evidence of life.

The *Curiosity* rover (originally known as the *Mars Science Laboratory*) landed in Gale Crater on Mars in 2012. That rover studies the rocks and soil of Mars to acquire data for a better understanding of the history of the planet's climate and geology. Shortly after landing, *Curiosity* found evidence that a stream of liquid water had once flowed in the crater. The rover observed rounded, gravelly pebbles stuck together, which have been interpreted as coming from a stream that varied at times from ankle-deep to hip-deep, and moved at about 1 meter per second (**Figure 24.14**). The rover also found sedimentary rocks containing clay, which suggest the presence of a freshwater lake bed at one time. Later observations found that the surface soil contained up to 2 percent water by weight—or about 1 quart per cubic foot of Martian dirt. Those conditions are too dry to support Earth-like plant life, which permanently wilts in soil that is less than about 10 percent water by weight.

Curiosity also detected evidence of organic molecules in 3-billion-year-old rocks located in that crater (**Figure 24.15**). Those organic molecules could have come from space on asteroids and comets, from geological processes on Mars, or from ancient life. *Curiosity* also detected seasonal variation in the amount of methane gas within the crater; methane can come from water-rock chemistry or from life. Thus, a water lake inside Gale Crater could have been hospitable to life billions of years ago, but astrobiologists do not know whether any life existed.

The *Mars Atmosphere and Volatile EvolutioN* (*MAVEN*) mission, which arrived at Mars in September 2014, is studying the upper atmosphere to learn more about the escape of carbon dioxide, hydrogen, and nitrogen from the planet's atmosphere and how losing those gases affected surface pressure and the existence of liquid water. Recently the *Mars Reconnaissance Orbiter* (*MRO*) found water ice on eight steep, eroded slopes located at midlatitudes on Mars (**Figure 24.16**). *Mars Express* may have found evidence of a lake of liquid water under the south polar ice cap. Future experiments will look for liquid water—and fossil or living microorganisms—below the Martian surface. Eventually other missions will return samples of Martian rock and soil to Earth for more advanced analysis.

NASA's instrumented robotic spacecraft reached the outer Solar System starting in the 1970s, and many astrobiologists were surprised by the findings. Although the outer planets themselves did not appear to be habitats for life, some of their moons became objects of special interest. Jupiter's moon Europa is covered with a layer of water ice that appears to overlie a great ocean of briny liquid water (**Figure 24.17**). The water remains liquid because of high pressure and tidal heating by Jupiter. Impacts by comet nuclei may have added a mix of organic material, another essential ingredient for life. Once thought to be a frozen, inhospitable world, Europa is now a candidate for biological exploration. Recently, scientists using the Hubble Space Telescope observed water geysers taller than Mount Everest erupting from the icy surface of Europa. Ejected material from those geysers may make searching for life on Europa possible without drilling through the ice. NASA's *Europa Clipper* mission, scheduled for launch in the 2020s, will fly by Europa a few dozen times.

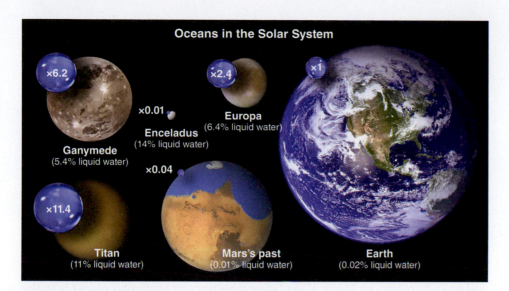

Figure 24.17 The total amount of liquid water (blue spheres) on Jupiter's moons and Saturn's moon Titan compared with the amount on Earth. All figures are drawn to scale and assume average ocean depths of 4 km (Earth), 100 km (Europa), and 200 km (Titan).

Jupiter's moons Ganymede and Calisto also have icy surfaces and possibly subsurface salty oceans.

Saturn's moon Titan has an atmosphere rich in organic chemicals, many of which are thought to be precursor molecules of a type that existed on Earth before life appeared here. A probe from the *Cassini* spacecraft in orbit around Saturn descended through Titan's atmosphere and found additional evidence for a variety of molecules that might be necessary for life, as well as liquid lakes of methane on the surface and probably a liquid-water ocean under the surface. The *Cassini* spacecraft also detected water-ice crystals spouting from cryovolcanoes (which erupt ice crystals instead of rocks) near the south pole of Saturn's tiny moon Enceladus. Liquid water must lie beneath its icy surface, so Enceladus is also a possible habitat of extremophile life—perhaps life similar to that found near hydrothermal vents deep within Earth's oceans.

The discovery of life on even one Solar System body beyond Earth would be exciting. If scientists discover that life arose independently *twice* in the same planetary system, that finding could suggest that the spontaneous appearance of life is not rare at all.

CHECK YOUR UNDERSTANDING 24.3a

Which of the following Solar System objects is a good candidate for future searches for life? (Choose all that apply.) (a) Mars; (b) Jupiter's moon Europa; (c) Saturn's moon Titan; (d) Uranus

Habitable Zones

Recall from Chapter 7 that thousands of confirmed and thousands of candidate exoplanets exist within the Milky Way Galaxy. To decide which planets to focus on for further study, astronomers are narrowing the possibilities by searching for planets with environments conducive to the formation and evolution of life, as we understand it, while eliminating clearly unsuitable planets. Astronomers consider issues such as each planet's orbit, inferred temperature, distance from its star, and location in the galaxy. One criterion astrobiologists look for is stability of planetary systems. As noted in Chapters 2, 3, and 4, astronomers think about the effects of a planet's rotation and orbit. Planets in stable systems have nearly circular orbits that preserve relatively uniform climatological environments. Planets in very elliptical

what if . . .

What if we find an exoplanet with a very small magnetic field and large amounts of greenhouse gases in its atmosphere? How would each of those factors increase or decrease the habitable zone distance for that exoplanet?

▶▶ **Interactive Simulation:** Habitable Zone

Figure 24.18 The habitable zone changes with the mass and temperature of a star. Habitable zones around hot, high-mass stars are larger and more distant than the zones around cooler, lower-mass stars.

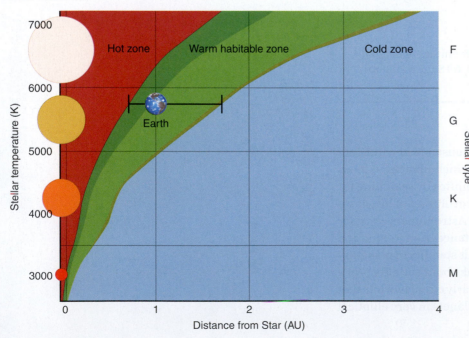

orbits or planets with a large axial tilt can experience more intense temperature swings that could be detrimental to the survival of life. A stable temperature that maintains the existence of water in a liquid state might be important. We know that liquid water was essential for life on Earth to form and evolve. We don't know whether liquid water is an absolute requirement for life elsewhere, but it's a reasonable starting assumption.

In the "Origins" sections of Chapters 7 and 13, we discussed the idea of the habitable zone, the location of a planet relative to its parent star that provides a range of temperatures in which liquid water can exist on the surface. On planets too close to their parent stars, water would exist only as a vapor—if at all. On planets too far from their stars, water would be permanently frozen as ice. Planet size is another consideration: Large gas giants retain most of their light gases during formation and have no solid surface. Small planets may be rocky or a mix of water, rock, and ice. Measuring the mass and radius enables scientists to estimate the density. Very small planets may not have enough surface gravity to retain their atmospheric gases and so end up like our Moon. Calculating whether any particular planet is in the habitable zone is complicated. Recall from Chapter 7 that even if a planet is located in the habitable zone, that means only that liquid water *could* exist on the surface: It does not mean that astronomers have confirmed the presence of liquid water or that the planet has inhabitants.

In our own Solar System, Venus, which orbits at 0.7 times Earth's distance from the Sun, has become an inferno because of its runaway greenhouse effect. Any liquid water that might once have existed on Venus has long since evaporated and been lost to space. Mars orbits about 1.5 times farther from the Sun than the orbit of Earth, and the water that we see on Mars today is nearly always frozen. But the orbit of Mars is more elliptical and variable than Earth's, giving Mars a greater variety of climate, including long-term cycles that might occasionally permit liquid water to exist. Most astrobiologists put the habitable zone of our Solar System at about 0.9–1.4 astronomical units (AU), which includes Earth but just misses Venus and Mars. However, that range ignores the possibility of liquid water under ice, as occurs on the moons of the outer Solar System.

Astronomers also must think about the type of star they are observing in their search for planets that could have liquid water. Stars less massive than the Sun and thus cooler will have narrower habitable zones. A planet in the habitable zone of a cool star is close in to its star. As a result, the planet is more likely to be tidally locked to the star so that no day/night cycle exists. Stars more massive than the Sun are hotter and will have a larger and more distant habitable zone (**Figure 24.18**). However, massive stars have shorter main-sequence lifetimes and might not last long enough for life to evolve in the first place. For example, a star of $3\,M_{Sun}$ has a lifetime of only a few hundred million years. Here on Earth, a billion years was long enough for bacterial life to form and cover the planet, but insufficient for anything more advanced to evolve. It took 3.5 billion years of evolution on Earth to reach the period of the Cambrian explosion. Even though evolution might happen at a different pace elsewhere, stellar lifetime is still a sufficiently strong consideration that astronomers focus their efforts on stars with longer lifetimes—specifically, stars of 0.6–1.4 M_{Sun}, which

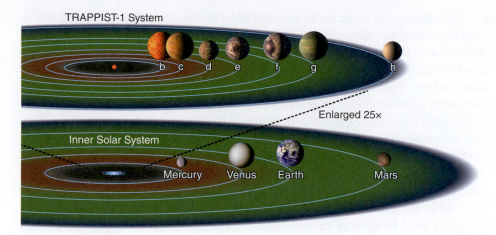

Figure 24.19 The Trappist-1 system. The Trappist planets are very close to their star: their distances are more comparable to those between Jupiter and its large moons than to those between the Sun and the terrestrial planets. (The star is not to scale.)

corresponds to spectral types F, G, K, and M. Most stars in our galaxy are low-mass M main-sequence stars, and many of those seem to have small, rocky planets. One example is the Trappist-1 system of seven planets in close orbits to their cool M dwarf star (**Figure 24.19**). Some of those planets could be tidally locked (see the Chapter 4 "Origins") to their star. Two or three of the planets are in the star's habitable zone.

Another factor to consider is a planet's atmosphere. A planet's ability to keep an atmosphere depends on the planet's mass and radius (and therefore its escape velocity) and its temperature. Very small planets may not have enough surface gravity to retain their atmospheric gases. In the inner Solar System, the Moon and Mercury were too small to keep any atmosphere. Mars lost most of its atmosphere, but the larger Earth and Venus kept thick atmospheres. Another important consideration is the greenhouse effect, which traps heat underneath an atmosphere and raises the temperature on a planetary surface. That has happened on Venus, Earth, and Mars, each of which has a higher surface temperature because of its atmosphere. The thickness and chemical content of the atmosphere affect the strength of the greenhouse effect, so that, for example, Venus is much hotter than its distance from the Sun would suggest because of its thick atmosphere of carbon dioxide (see Figure 5.25 and the Chapter 5 "Origins"). Some of those exoplanets might be more like Venus than like Earth if they have atmospheres filled with greenhouse gases. The total amount of atmosphere affects the atmospheric pressure at the surface, which, along with the temperature, determines whether water (or other molecules) can exist in a liquid state on the surface. The current thin atmosphere of Mars does not permit standing liquid water on its surface.

In the next few years, with new telescopes, many more observations will identify atmospheres on exoplanets. Water vapor, hydrogen, and carbon monoxide have been found on a few exoplanets already. In particular, discovering oxygen in the atmosphere of an exoplanetary atmosphere would be exciting, but not definitive. The oxygen in Earth's atmosphere makes it stand out from the rest of the planets and moons in the Solar System, and we know that most terrestrial oxygen was created by photosynthetic life. However, oxygen can also come from the breakup of water molecules, so the presence of oxygen alone doesn't necessarily mean life exists there. Aside from oxygen, astrobiologists are thinking about other possible **biosignatures**—chemical compounds that would show up in the spectra of an exoplanet atmosphere that would suggest life may be present (**Figure 24.20**).

Figure 24.20 Spectroscopy of rocky exoplanet atmospheres may pick up many chemicals important for life, including some that are biosignatures, thought to be produced overwhelmingly by life.

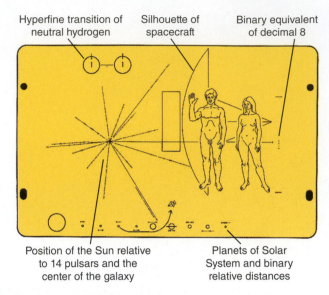

Hyperfine transition of neutral hydrogen

Silhouette of spacecraft

Binary equivalent of decimal 8

Position of the Sun relative to 14 pulsars and the center of the galaxy

Planets of Solar System and binary relative distances

Figure 24.21 This plaque is carried by the *Pioneer 11* probe, which launched in 1973 and will eventually leave the Solar System to travel in interstellar space.

Figure 24.22 This message was beamed toward the star cluster M13 in 1974. Reading right to left, this binary-encoded message contains the numbers 1–10, hydrogen and carbon atoms, some interesting molecules, DNA, a human figure and its size, the basics of the Solar System, and a depiction of the now-defunct Arecibo telescope.

Astrobiologists continue to develop ways to classify planets more clearly as they try to narrow the possibilities about where life might exist. One such classification uses the currently available data on an exoplanet to estimate how much it is like Earth. Factors include the radius, density, escape velocity, and surface temperature. That is an Earth-centric approach based on the experience of life on Earth. Another measure aims to be less Earth-centric and to broaden the options for habitability, but it depends on factors not yet measured or measurable for most exoplanets. That measure depends on whether the planet has a surface on which organisms can grow, as well as the right kind of chemistry, a source of energy, and the ability to hold a liquid solvent. Saturn's moon Titan or Jupiter's moon Europa might satisfy those conditions, and Mars might have done so in the past. Over the next decade, improvements in observations will probably lead to enough information that at least some exoplanets can be classified in that way. As observations of exoplanets become more complete, astrobiologists will undoubtedly develop new classification schemes that are more accurate and informative.

Astronomers also consider the *galactic habitable zone*—the idea that the Milky Way Galaxy may have some locations where planets might have a higher probability of hosting life. Stars situated too far from the galactic center may be without enough heavy elements—such as oxygen, silicon (silicates), iron, and nickel—in their protoplanetary disks to form rocky planets like Earth. Conversely, regions too close to the galactic center experience less star formation and therefore fewer opportunities to gather heavy elements into planetary environments. Stars too close to the galactic center may be affected by the high-energy radiation environment (X-rays and gamma rays from supermassive black holes or gamma-ray bursts), which can damage RNA and DNA. Stars that migrate within the galaxy and change their distance from the galactic center may move in and out of any galactic habitable zone.

CHECK YOUR UNDERSTANDING 24.3b

The habitable zone around a star depends most on the star's: (a) mass and age; (b) radius and distance; (c) age and radius; (d) color and distance; (e) luminosity and velocity.

24.4 Scientists Are Searching for Signs of Intelligent Life

Are we alone? Scientists approach that question from many directions. Biologists consider the origin and evolution of life and the definition of intelligence. Astronomers send messages and search for alien signals in the vast array of astronomical data. In this section, we describe the search for intelligent life and how scientists think about the probability of finding it.

Sending Messages

During the 1970s, messages were sent from Earth to space. The *Pioneer 11* spacecraft, which will probably spend eternity drifting through interstellar space, carries the plaque shown in **Figure 24.21**. It pictures humans and the location of Earth for any future interstellar traveler who might happen to find it and understand its content. Another message to the cosmos accompanied the two *Voyager* spacecraft on the "Golden Records"—identical phonograph records that contained greetings from planet Earth in 60 languages, samples of music, animal sounds, and a message from then-President Jimmy Carter. Some politicians were concerned that scientists were dangerously advertising our location in the galaxy, even though radio signals had already been broadcast into space for nearly 80 years. Some philosophers also worried that those messages contained anthropomorphic assumptions about aliens being enough like us to decode the messages. However, sending messages on spacecraft is an inefficient way to make contact with extraterrestrial life. The probability that an alien species will actually find any of those messages is very, very small.

In 1974, astronomers used the 300-meter-wide dish of the Arecibo radio telescope to beam a message in binary code (**Figure 24.22**) toward the star cluster M13, located 25,000 light-years away. That distance is far enough that by the time the message arrives, the core of M13 will have moved, and the radio signal will not actually arrive there. However, the intention of the experiment was to show that such a message could be sent, not to make contact, because 50,000 years would pass before any reply could come back. In 2008, a radio telescope in Ukraine sent a message to the exoplanet Gliese 581c. The message was composed of 501 digitized images and text messages selected by users on a social networking site and will arrive at Gliese 581c in 2029.

The Drake Equation

The first serious effort to quantify the probability of the existence of intelligent extraterrestrial life was made by astronomer Frank Drake in 1960. He developed the **Drake equation**, which estimates the likely number (N) of intelligent civilizations currently existing in the Milky Way Galaxy. The Drake equation is different from the other equations in this book because the values for many of the variables are uncertain. However, it is a useful way to categorize some of the factors that relate to the conditions that must be met for a civilization to exist. The equation is discussed further in **Working It Out 24.2.**

As illustrated in **Figure 24.23**, the conclusions we draw using the Drake equation depend a great deal on the assumptions we make and therefore on the numbers used in the equation. For the most pessimistic estimates, the Drake equation sets the number of technological civilizations in our galaxy at about 1, in which case we are the *only* technological civilization in the Milky Way at this time. Such a universe could still be full of intelligent life. With 100 billion galaxies in the observable universe, even those pessimistic assumptions mean that 100 billion technological civilizations could be out there somewhere. However, if the nearest neighbors are in another galaxy, they are *very* far away—millions of parsecs, on average.

At the other extreme—with the most optimistic numbers, which assume that intelligent life arises and survives everywhere it gets the chance—the Milky Way alone could have tens of millions of technological civilizations! If so, the nearest neighbors may be "only" 40 or 50 light-years away.

If humans did meet a technologically advanced civilization, what would it be like? The Drake equation suggests that neighbors nearby are highly unlikely, unless civilizations typically live for many thousands or even millions of years (see Working It Out 24.2). If so, any civilization we encountered would probably have been around for much longer than we have.

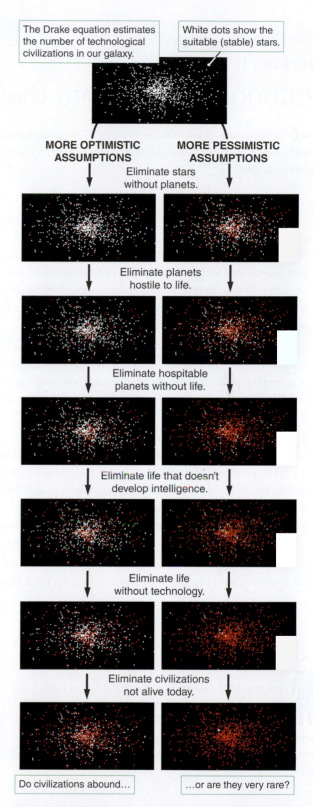

The Drake equation estimates the number of technological civilizations in our galaxy.

White dots show the suitable (stable) stars.

MORE OPTIMISTIC ASSUMPTIONS   **MORE PESSIMISTIC ASSUMPTIONS**

Eliminate stars without planets.

Eliminate planets hostile to life.

Eliminate hospitable planets without life.

Eliminate life that doesn't develop intelligence.

Eliminate life without technology.

Eliminate civilizations not alive today.

Do civilizations abound...   ...or are they very rare?

Figure 24.23 The two columns show estimates of the existence in the Milky Way Galaxy of intelligent, communicative civilizations as made using the Drake equation. White dots represent stars with possible civilizations. Notice how widely the estimates vary, given optimistic and pessimistic assumptions about the seven factors in the equation.

working it out 24.2

Putting Numbers into the Drake Equation

The Drake equation states that the number, N, of extraterrestrial civilizations in the Milky Way Galaxy that can communicate by electromagnetic radiation is given by

$$N = R^* \times f_p \times n_e \times f_l \times f_i \times f_c \times L$$

where the factors on the right-hand side of the equation are explained as follows:

1. R^* is the number of stars that form in the Milky Way Galaxy each year that are suitable for the development of intelligent life. Astronomers consider those to be F, G, K, or M spectral-type stars because their lifetimes are long enough. That is about five to seven stars per year, so $R^* = 5$ to 7.

2. f_p is the fraction of stars that form planetary systems. The discoveries of exoplanets over the past two decades have shown that planets form as a natural by-product of star formation and that many—perhaps most—stars have planets. For that factor, astronomers assume that f_p is between 0.5 and 1.

3. n_e is the number of planets and moons in each planetary system with an environment suitable for life. In the Solar System, that number is at least 1 (for Earth), but it could be more if Mars or an outer-planet moon or two has suitability for life. Only recently have stars with multiple planets been discovered, so astronomers are just starting to get data on that factor. Generally, they estimate that n_e is less than 3.

4. f_l is the fraction of suitable planets and moons on which life actually arises. Remember that just a single self-replicating molecule may be enough to start the process. Some biochemists think that life *will*

develop if the right chemical and environmental conditions are present, but others disagree. Values of f_l range from 100 percent (life always develops) to 1 percent (life is more rare). Astronomers use a range of 0.01–1 for f_l.

5. f_i is the fraction of those planets harboring life that eventually develop intelligent life. Intelligence is certainly the kind of survival trait that might often be strongly favored by natural selection. Yet on Earth, tool-building intelligence took about 4 billion years—nearly half the expected lifetime of the Sun—to evolve. The correct value for f_i might be close to 0.01 or it might be closer to 1. No one knows.

6. f_c is the fraction of intelligent life-forms that develop technologically advanced civilizations—that is, civilizations that send communications into space. With only one example of a technological civilization to work with, f_c also is unknown. Astronomers estimate f_c to be between 0.1 and 1.

7. L is the number of years that technologically advanced civilizations exist. That factor is certainly difficult to estimate because it depends on the long-term stability of those civilizations. On Earth, the longest-lived civilizations have existed for, at most, thousands of years. Those civilizations, however, were not at the level of technology that allows interstellar communication—thus far, the first "technologically advanced civilization" on Earth is less than 100 years old. We do not know whether life intelligent enough to manipulate its environment technologically can maintain its planetary resources for any extended length of time. Astronomers usually put a value of L between 1000 years and 1 million years in their estimates.

what if . . .

What if SETI succeeds in finding a signal from a distant civilization—one that clearly must consist of technologically advanced beings? What (if anything) should we say to them, and who should have the authority to say it?

Technologically Advanced Civilizations

During lunch with colleagues, the physicist Enrico Fermi (1901–1954), a firm believer in extraterrestrial life, is reported to have asked, "If the universe is teeming with aliens . . . where is everybody?" Fermi's question—first posed in 1950 and sometimes called the *Fermi paradox*—remains unanswered. If intelligent life-forms are common but interstellar travel is difficult or impossible, would the aliens send out messages—perhaps by electromagnetic waves instead? And if they did, why haven't astronomers detected their signals?

Drake used what was then astronomy's most powerful radio telescope to listen for signals from intelligent life around two nearby stars, but he found nothing unusual. His original project has grown over the years into a much more elaborate program called the Search for Extraterrestrial Intelligence, or **SETI**. Scientists from around the world have thought carefully about what strategies might be useful for finding life in the universe. Most of those endeavors use radio telescopes to listen for signals from space that bear an unambiguous signature of an intelligent source. Some have focused on significant parts of the spectrum, assuming that a civilization will broadcast on a channel that astronomers throughout the galaxy should find interesting—for example, the 21-centimeter (21-cm) line from hydrogen gas. More recent searches have used advances in technology to record as broad a range of radio signals from space as possible. Analysts search those databases for regular signals that might be intelligent in origin.

The SETI Institute's Allen Telescope Array (ATA) received much of its initial financing from Microsoft cofounder Paul Allen. The ATA consists of a "farm" of small, inexpensive radio dishes like those used to capture signals from orbiting television-broadcasting satellites. One key project of the ATA is to observe the planets discovered by the Kepler Mission. Each dish has a diameter of 6.1 meters, but all the telescopes working together have a total signal-receiving area greater than that of a 100-meter radio telescope. Just as your brain can sort out sounds coming from different directions, that array of radio telescopes can determine the direction a signal is coming from, allowing it to listen to many stars at the same time. Over several years' time, astronomers using the ATA are expected to survey as many as a million stars, hoping to find a civilization that has sent a signal toward Earth.

As stated earlier, finding even one nearby civilization in the Milky Way Galaxy—that is, a *second* technological civilization in Earth's small corner of the universe—will make scientists optimistic that the universe as a whole is teeming with intelligent life. The likelihood of SETI's success is difficult to predict, but its potential payoff is enormous. Few discoveries would have a more profound impact than the certain knowledge that we on Earth are not alone.

Science fiction is filled with tales of humans who leave Earth to "seek out new life and new civilizations." Unfortunately, those scenarios are not scientifically realistic. The distances to the stars and their planets are enormous: To explore a significant sample of stars would require extending the physical search over tens or hundreds of light-years. Special relativity limits how fast one can travel. The speed of light is the limit, and even at that rate reaching the *nearest* star would take more than 4 years. The relativistic effect of time dilation means that time would pass slower for astronauts traveling at very high speeds, and they would return to Earth younger than if they had stayed at home. For example, suppose astronauts visited a star 15 light-years distant. Even if they traveled at speeds close to the speed of light, by the time they returned to Earth, 30 years would have passed at home.

Some science fiction writers get around that problem by invoking "warp speed" or "hyperdrive," which enables travel faster than the speed of light, or by using wormholes as shortcuts across the galaxy—but absolutely no evidence exists that any of those options are possible. And most of those imaginative stories ignore the vast number of other complications that accompany human space travel. Humans are just beginning to learn how to live in space even for short periods. Much work remains to be done before we can realistically contemplate even voyages within the Solar System.

Some people claim that aliens have already visited Earth: Tabloid newspapers, books, and websites are filled with tales of sightings of unidentified flying objects (UFOs), government conspiracies and cover-ups, alleged alien abductions, and UFO religious cults. However, none of those reports meets the basic standards of science. They are not falsifiable—they lack verifiable evidence and repeatability—so we must conclude that no scientific evidence currently exists for any alien visitations.

unanswered questions

Will humans spread life into space? Some scientists have suggested that seeding from Earth may have already happened as Earth microorganisms scattered into space after giant impacts. Humans also may have unintentionally sent microorganisms to space aboard our spacecraft. More intentional methods of seeding include sending microorganisms from Earth to other planets or moons to try to jump-start evolution, "terraforming" Mars or a moon to change conditions on it to make it more habitable for humanity, or sending humans in spaceships to colonize the galaxy.

Reading Astronomy News

When Reporting News about Aliens, Caution Is Advised

Elizabeth Howell, airspacemag.com

Astronomers struggle with how to handle announcements (or nonannouncements) about extraterrestrial contact.

Speculation about extraterrestrials seems to be everywhere these days. Last week it was "Tabby's Star" (more officially known as KIC 8462852), whose mysterious dimming and brightening, according to the latest analysis, is likely due to dust blocking different wavelengths of light rather than "alien megastructures." Before that came reports of an interstellar asteroid—not a spacecraft—entering our solar system and a UFO monitoring program conducted by the Department of Defense.

The attention given to such stories has some scientists worried, especially as social media amplifies claims of alien contact over other, more prosaic explanations.

"Currently, most SETI-related news seems to be interfering with conventional scientific discoveries, stealing the limelight—without following basic rules of science," wrote Dutch exoplanet researcher Ignas Snellen of Leiden Observatory, on a Facebook exoplanets discussion group for professional astronomers.

Although he has "great respect for SETI scientists," Leiden wrote, "there is no place for alien civilizations in a scientific discussion on new astrophysical phenomena, in the same way as there is no place for divine intervention as a possible solution. One may view it as harmless fun, but I see parallels in athletes taking banned substances. It may lead to short-term fame and medals, but in the long run it harms the sport. Same for astronomy: we should be very careful not to be ridiculed. I really hope we can stop mentioning SETI for every unexplained phenomenon."

Such worries aren't new. Nearly 20 years ago researchers in the field came up with the Rio Scale to guide them in reporting the significance of any candidate SETI signal. Modeled after the Richter Scale for earthquakes, it assigns a value between 0—no significance—and 10—extraordinary significance—to any detected signal. The method was first proposed by Ivan Almar and Jill Tarter at the 51st International Astronautical Congress in Rio de Janeiro, in 2000. Although the International Academy of Astronautics' SETI Permanent Committee adopted the scale two years later, it hasn't seen widespread use.

Meanwhile, the news business has changed drastically since 2000. Today much of the public doesn't get its information from TV and newspapers, but from Facebook updates and Twitter posts, which move—and change—at a much faster pace. In recognition of the altered media landscape, a new paper submitted to the *International Journal of Astrobiology* proposes streamlining the Rio Scale and having scientists take a short quiz to answer a few questions about their discovery. Journalists and other news providers could use the answers in their stories.

"You can go through the steps and ask yourself, do I believe the instrumentation is working properly? Do I believe I am looking at it objectively? Do I have enough people who looked at the signal with different instruments? Has there been a lot of scientific discussion? Have there been alternative explanations? Is it a hoax?" says the paper's lead author, Duncan Forgan, a postdoctoral scholar at the University of St. Andrews' School of Physics and Astronomy in Scotland, whose work includes problems related to astrobiology and SETI.

Forgan says that although it can be difficult to convey uncertainty about a particular result to the public, the revised Rio Scale would make that job easier. Some scientific speculation can be "a bit wibbly-wobbly," he says, "and those wibbly-wobbly bits end up in the press." He agrees that spurious SETI claims can sometimes distract from more legitimate science.

In recognition of the sensitivity around alien signal detection, SETI has a voluntary list of protocols to follow when something interesting is found. The first principle urges researchers to "verify that the most plausible explanation for the evidence is the existence of extraterrestrial intelligence, rather than some other natural phenomenon or anthropogenic phenomenon, before making any public announcement."

Morris Jones, an Australian space observer with both scientific and journalistic training, says we should label "fringe SETI" claims for what they are. "The media is under pressure to deliver attention-grabbing news, but it's hard to expect them to judge fringe SETI as spurious when it comes from reputable institutions and qualified researchers. The best way to reduce these reports is to stop the production of questionable scientific papers in the first place."

QUESTIONS

1. Why are astrobiologists concerned about media coverage of SETI?

2. How do the suggestions for journalists fit with what you have learned about the process of science?

3. Think of an article you have seen recently on evidence for alien life. Was it overly speculative?

4. How do you think people on Earth would react to an actual discovery of alien life?

Source: https://www.airspacemag.com/daily-planet/when-reporting-news-about-aliens-caution-advised-180967777/.

CHECK YOUR UNDERSTANDING 24.4

The Drake equation enables astronomers to: (a) calculate precisely the number of alien civilizations; (b) organize their thoughts about probabilities for life; (c) locate the stars they should study to find life; (d) find new kinds of life.

Origins: The Fate of Life on Earth

Astronomers have used their understanding of physics and cosmology to look back through time and watch as structure formed throughout the universe and to look forward to the future of our Sun, our galaxy, and the ultimate fate of the universe. About 5 billion years from now, the Sun will end its long period of relative stability. The Sun will expand to become a red giant star, swelling to hundreds of times larger than it is now and thousands of times more luminous. The giant planets, orbiting outside the extended red giant atmosphere, will probably survive. But at least some of the planets of the inner Solar System will not. Just as an artificial satellite is slowed by drag in Earth's tenuous outer atmosphere and eventually falls to the ground, so, too, will a planet caught in the Sun's atmosphere be engulfed by the expanding Sun. If that happens to Earth, no trace of this planet will remain other than a slight increase in the amount of massive elements in the Sun's atmosphere.

Another possibility is that the red giant Sun will lose mass in a powerful wind, its gravitational pull on the planets will weaken, and the orbits of both the inner and outer planets will spiral outward. If Earth moves out far enough, it may survive as a seared cinder, orbiting the small, hot, white dwarf star that the Sun will become. Barely larger than Earth and with its nuclear fuel exhausted, the white dwarf Sun will slowly cool, eventually becoming a cold sphere of densely packed carbon, orbited by what remains of its planets. The ultimate outcome for Earth—consumed in the Sun or left behind as a cold, burned rock orbiting a long-dead white dwarf—is not yet known.

Alternatively, the threat to the future of Earth can come from within, as humanity itself changes conditions on Earth that are necessary for our habitability (for example, by releasing chemicals into the atmosphere that alter the ozone layer, or are greenhouse gases that alter the trapping of radiation).

In either case, however, Earth's status as a garden spot in the habitable zone will be at an end. If the Sun does not expand too far or Mars also migrates outward, Mars could become the habitable planet in the Solar System, at least for a while. As the dying Sun loses more and more of its atmosphere in a stellar wind, Earth's atoms might be expelled back into the reaches of interstellar space from which they came, perhaps to be recycled into new generations of stars, planets, and even life itself.

But even before the Sun's change into a red giant star, the Sun's luminosity will begin to rise. As solar luminosity increases, so will temperatures on all the planets. The inner edge of the Sun's habitable zone will slowly move out past the orbit of Earth. Eventually, Earth's temperatures will climb so high that all animal and plant life will perish. Even the extremophiles that inhabit the oceanic depths will die as the oceans themselves boil away. Models of the Sun's evolution are still not precise enough for astronomers to predict with certainty when that fatal event will occur, but the end of all terrestrial life may be 1 billion to 4 billion years away. In addition, the Milky Way Galaxy will collide with the Andromeda Galaxy in 4 billion to 5 billion years' time. Galaxies are mostly empty space, so the Sun is not likely to collide with another star, but one effect of the collision is that our Solar System may be gravitationally flung to a different part of our galaxy.

unanswered questions

Will humans themselves spread into space? At some point, humans must leave planet Earth if the species is to survive. But space is a dangerous place, and many of the problems that humans encounter in space have not yet been solved. Those problems are as diverse as purely physical issues, such as the loss of bone density in low gravity, and the societal problems that occur when a few people are confined together for long periods. We do not yet know whether humanity can overcome those problems and journey to other planets.

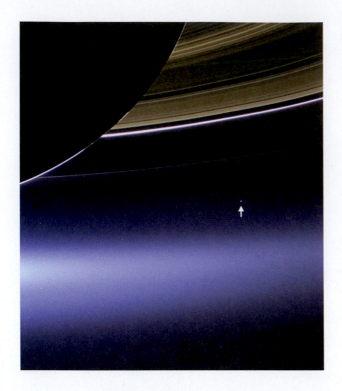

Figure 24.24 ★ **WHAT AN ASTRONOMER SEES** This image from the *Cassini* space-craft shows Earth as seen from Saturn. The pale dot in the lower right, indicated by the arrow, is Earth. For this last "What an Astronomer Sees" image, we give you the words of Carl Sagan, describing what he saw in a similar image of Earth from a distant perspective:

"We succeeded in taking that picture, and, if you look at it, you see a dot. That's here. That's home. That's us. On it, everyone you ever heard of, every human being who ever lived, lived out their lives.

The aggregate of all our joys and sufferings, thousands of confident religions, ideologies and economic doctrines, every hunter and forager, every hero and coward, every creator and destroyer of civilizations, every king and peasant, every young couple in love, every hopeful child, every mother and father, every inventor and explorer, every teacher of morals, every corrupt politician, every superstar, every supreme leader, every saint and sinner in the history of our species, lived there—on a mote of dust, suspended in a sunbeam.

The Earth is a very small stage in a vast cosmic arena. Think of the rivers of blood spilled by all those generals and emperors so that in glory and in triumph they could become the momentary masters of a fraction of a dot. Think of the endless cruelties visited by the inhabitants of one corner of the dot on scarcely distinguishable inhabitants of some other corner of the dot. How frequent their misunderstandings, how eager they are to kill one another, how fervent their hatreds. Our posturings, our imagined self-importance, the delusion that we have some privileged position in the universe, are challenged by this point of pale light ...

To my mind, there is perhaps no better demonstration of the folly of human conceits than this distant image of our tiny world. To me, it underscores our responsibility to deal more kindly and compassionately with one another and to preserve and cherish that pale blue dot, the only home we've ever known."

— *Carl Sagan, speech at Cornell University, October 13, 1994*

However, whether the descendants of today's humanity will even be around a billion years from now is far from certain. Some threats to life come from beyond Earth. For the rest of the Sun's life, the terrestrial planets, including Earth, will continue to be bombarded by asteroids and comets. Perhaps a hundred or more of those impacts will involve kilometer-sized objects, capable of causing the kind of devastation that eradicated the dinosaurs. Those impacts may create new surface scars but they will have little effect on the integrity of Earth itself. Earth's geological record is filled with such events, and each time they happen, life manages to recover and reorganize. But we do not know if human life or civilization could survive.

Humans might protect themselves from the fate of the dinosaurs, but in the long run humanity will either leave this world or die out. Planetary systems are common to other stars, and many other Earth-like planets may well exist throughout the Milky Way Galaxy. Colonizing other planets is currently the stuff of science fiction, but if the descendants of modern-day humans are ultimately to survive the death of the home planet, off-Earth colonization must at some point become science fact.

By studying astronomy, you have learned where you come from. You have learned about the self-correcting nature of science—how it continually adapts to new information to give us the ability to make better and better predictions about the behavior of the physical world. That predictive ability makes science an extremely powerful tool. No other species in the history of Earth has been able to understand its position, predict what will happen next, and therefore adapt its behavior to seek the best possible future for its members.

Figure 24.24 shows Earth as seen by the *Cassini* spacecraft, looking near Saturn's rings. That tiny dot, which is Earth, is the only place in the entire universe where we know that life exists. Compare the size of that dot with the size of the universe. Compare the history of life on Earth with the history of the universe. Compare Earth's future with the fate of the universe. Astronomy is humbling. We occupy a tiny part of space and time. Yet we are unique, as far as we know. Think for a moment about what that means to you. That may be the most important lesson the universe has to offer.

SUMMARY

Astrobiology seeks to answer the question, Are we alone? All the sciences have a part to play in answering that question, from astronomy to zoology. Theories of how life began and evolved on Earth help limit the number of targets astronomers study in their search for life elsewhere. Even before the Sun expands and evolves to a red giant star, its luminosity will increase enough to alter the location of its habitable zone. When that happens, Earth may no longer be the planet at the right location for maintaining liquid water.

(1) Explain our current understanding of how and when life began on Earth and how it has evolved. Life on Earth is a form of complex carbon-based chemistry, made possible by self-replicating molecules. Life probably formed in Earth's oceans and then evolved chemically from simple molecules into self-replicating organisms through a combination of mutation and heredity.

(2) Explain how life is a structure that has evolved through the action of the physical and chemical processes that shape the universe. All terrestrial life is composed primarily of only six elements: carbon, hydrogen, oxygen, nitrogen, sulfur, and phosphorus. However, life-forms very different from those on Earth, including those based on silicon chemistry, cannot be ruled out.

(3) List the locations in our Solar System and around other stars where astronomers think life might be possible. Within the Solar System, Mars and some moons of Jupiter and Saturn are the most promising candidates for life. A habitable zone around a star is a location within which the temperature of a planet will support liquid water on its surface. Thus, astronomers look for exoplanets that orbit in habitable zones surrounding solar-type stars. A habitable zone can be extended by special circumstances, such as tidal heating from a giant planet or atmospheres containing greenhouse gases.

(4) Describe some methods used to search for intelligent extraterrestrial life. The Drake equation includes factors that astronomers consider when thinking about the possibility of life in the universe. Astronomers use radio telescopes to search for signals from extraterrestrial life, particularly in astronomically important regions of the spectrum. In recent times, that effort has been expanded to a search through as broad a region of the radio spectrum as possible. No intelligent extraterrestrial life has yet been detected.

QUESTIONS AND PROBLEMS

TEST YOUR UNDERSTANDING

1. The study of life and the study of astronomy are connected because (select all that apply)
   a. life may be commonplace in the universe.
   b. studying other planets may help explain why life exists on Earth.
   c. explorations of extreme environments on Earth suggest where to look for life elsewhere.
   d. life is a structure that evolved through physical processes, and life on Earth may not be unique.
   e. life elsewhere is most likely to be found by astronomers.

2. Scientists look for water to indicate places where life might exist because
   a. water is a common molecule in interstellar space.
   b. life on Earth depends on it.
   c. no other molecules are solvents.
   d. the spectrum of water is very complex.

3. The Urey-Miller experiment produced _____ in a laboratory jar.
   a. life
   b. RNA and DNA
   c. amino acids
   d. proteins

4. A mutation is
   a. always a deadly change to DNA.
   b. always a beneficial change to DNA.
   c. a change to DNA that is sometimes beneficial and sometimes not.
   d. a change to DNA that cannot be inherited.

5. Scientists think that terrestrial life probably originated in Earth's oceans, rather than on land, because (select all that apply)
   a. all the chemical pieces were in the ocean.
   b. energy was available in the ocean.
   c. the earliest evidence for life on Earth is from fossils of ocean-dwelling organisms.
   d. the deepest parts of the ocean have hydrothermal vents.

6. Any system with the processes of heredity, mutation, and natural selection will, over time (choose all that apply),
   a. change.
   b. become larger.
   c. become more complex.
   d. develop intelligence.

7. The fact that no alien civilizations have yet been detected indicates that
   a. they are not there.
   b. they are rare.
   c. Earth is in a "blackout," and they are not talking to us.
   d. we don't know enough yet to draw any conclusions.

8. During the Cambrian explosion
   a. the dinosaurs were killed.
   b. all the carbon that is now here on Earth was produced.
   c. biodiversity increased significantly.
   d. a lot of carbon dioxide was released into the atmosphere.

9. The habitable zone is the place around a star where
   a. life has been found.
   b. atmospheres can contain oxygen.
   c. liquid water has been confirmed.
   d. liquid water can exist on the surface of a planet.

10. The difference between a prokaryote and a eukaryote is that prokaryotes
    a. have no DNA.
    b. have no cell wall.
    c. have no nucleus.
    d. do not exist today.

11. A thermophile is an organism that lives in extremely _____ environments.
    a. salty
    b. hot
    c. cold
    d. dry

12. The search for life elsewhere in the Solar System is carried out primarily by
    a. astronauts.
    b. robotic spacecraft.
    c. astronomers using optical telescopes.
    d. astronomers using radio telescopes.

13. Astronomers think that intelligent life is more likely to be found around stars of types F, G, K, and M because
    a. those stars are hot enough to have planets and moons with liquid water.
    b. those stars are cool enough to have planets and moons with liquid water.
    c. those stars live long enough for life to begin and evolve.
    d. those stars produce no ultraviolet radiation or X-rays.

14. Life first appeared on Earth
    a. billions of years ago.
    b. millions of years ago.
    c. hundreds of thousands of years ago.
    d. thousands of years ago.

15. In the phrase "theory of evolution," the word *theory* means that evolution
    a. is an idea that can't be tested scientifically.
    b. is an educated guess to explain natural phenomena.
    c. probably doesn't happen anymore.
    d. is a tested and corroborated scientific explanation of natural phenomena.

THINKING ABOUT THE CONCEPTS

16. How does the evolution of life on Earth depend on RNA and DNA?

17. How do scientists think that amino acids first formed on Earth?

18. Today, most organisms on Earth enjoy relatively moderate climates and temperatures. Compare that environment with some of the conditions in which early life developed.

19. The Process of Science Figure represents many areas of science that all inform the science of astrobiology. Choose any two of those, and explain how one depends on the other. ✪

20. How was Earth's carbon dioxide atmosphere changed into today's oxygen-rich atmosphere? How long did that transformation take?

21. What was the Cambrian explosion, and what might have caused it?

22. Why did plants and forests appear in large numbers before large animals did?

23. What is a habitable zone? What defines its boundaries?

24. Which general conditions were needed on Earth for life to arise?

25. Is evolution under way on Earth today? If so, how might humans continue to evolve?

26. Where did all the atoms in your body come from?

27. The two *Viking* spacecraft found no convincing evidence of life on Mars when they visited that planet in the late 1970s, nor did the *Phoenix* lander when it examined the martian soil in 2009. Do those results imply that life never existed on the planet? Why or why not?

28. Some scientists believe that humans may be the only advanced life in the galaxy today. If so, which factors in the Drake equation must be extremely small?

29. **★ WHAT AN ASTRONOMER SEES**   The image in Figure 24.24 shows Earth from a quite distant perspective. Take a moment to think back to the beginning of this course, and reflect on what you have learned. How has your concept of your place in the universe changed since then? How has your understanding of this image (and others like it) changed as you've learned about the universe beyond Earth? 👁

30. Why is life on Earth as we know it likely to end long before the Sun runs out of nuclear fuel?

APPLYING THE CONCEPTS

31. Suppose that an organism grows in number in the exponential fashion of Working It Out 24.1. How many times more of the organism will there be after 3 doubling times? ●—●—●
    a. Make a prediction: Do you expect this number to be more or less than 1? What approximate power of 10 do you expect: 10, 100, 1000, or more?
    b. Calculate: Use your calculator and follow Working It Out 24.1 to find the answer.
    c. Check your work: Compare your answer to your prediction, to the numbers in Working It Out 24.1, and to the process of multiplying $2 \times 2 \times 2$ in your calculator. This will help you verify that you know how to work your calculator correctly to raise 2 to a power.

32. Use the most optimistic numbers in the Drake equation for all the terms on the right, except L. What, then, does L have to be equal to in order for a million civilizations to exist in the Milky Way? ●—●—●
    a. Make a prediction: Most of the "optimistic" values for terms in the Drake equation are around 1. Do you expect L to be a small number, like a century, or a large number, like millions of years?
    b. Calculate: Determine what L must be, under these conditions, for there to be a million civilizations in the Milky Way.
    c. Check your work: Compare your number to your prediction, and work the problem the other way (plug in your value for L, and solve for N) in order to check your work.

33. Suppose that an organism replicates itself each second. If you start with a single specimen, what will the final population be after 10 seconds?

34. Suppose that an organism replicates itself each second. After how many seconds will the population increase by a factor of 1024?

35. The "rule of 70" states that you can approximate the doubling time for exponential growth by dividing 70 by the rate of increase. So, if a population increases by 7 percent per year, the doubling time is 10 years (70/7 = 10). Suppose Earth's human population continues to grow by 1 percent annually. What is the doubling time? How much time will pass before Earth has 4 times as many humans?

36. Several missions are searching for planets in the habitable zones of stars. Explain which factors in the Drake equation are affected by that search and how the final number N will be affected if these missions find that most stars have planets in their habitable zones.

37. The white blocks at the top of Figure 24.22 represent the numbers 1–10, in order from right to left. Each representation of a number uses a maximum of four rows, where the bottom row of the set of four is a placeholder, and the top three rows of the four represent the number. Explain the "rule" for the kind of counting shown here. (For example, how do three white blocks represent the number 7?) 👁

38. As discussed in Working It Out 24.1, the doubling time for exponential growth is given by $P_F/P_O = 2^n$. Assume a self-replicating molecule that makes one copy of itself each second. Make a graph of the number of molecules versus time for the first 60 seconds after the molecule begins replicating.

39. The doubling time for *Escherichia coli* is 20 minutes, and you start getting sick when just 10 bacteria enter your system. How many bacteria are in your body after 12 hours?

40. If the chance that a given molecule will mutate is 1 in 100,000, how many generations are needed before, on average, at least one mutation has occurred?

41. Study the Drake equation in Working It Out 24.2. Make your own most optimistic and most pessimistic assumptions for each variable in the equation. What values do you find for N?

42. Trace (or photocopy) the graph in Figure 5.25, and then add horizontal lines for the temperatures at which water freezes and boils. Which planets in the Solar System have measured temperatures that fall within those lines? Which planets have predicted temperatures (based on the equilibrium model) that fall within those lines? What does that tell you about assumptions about the habitable zone? 👁

43. Figure 5.25 shows that Mercury's measured range of temperatures overlaps the temperatures at which water is a liquid. Is Mercury in the habitable zone? Why or why not? 👁

44. Suppose astronomers announce the discovery of a new planet around a star with a mass equal to the Sun's. The planet has an orbital period of 87 days. Is that planet in the habitable zone for a Sun-like star?

45. As noted in Section 24.1, some scientists suspect the early Earth was "seeded" with primitive life stored in comets and meteoroids. Now that you know when and how our Solar System, galaxy, and universe formed, what time line is required for such seeding to be possible?

EXPLORATION Exploring the Habitable Zone

digital.wwnorton.com/astro7

The habitable zone depends on the properties of both the star and the planet's orbit. This can be a little bit tricky to calculate for main-sequence stars, because the temperature, luminosity, and radius of the star are all connected. Open the Habitable Zone simulation, and study the habitable zone diagram in the large window for a moment. Click on "Legend," and identify the inner and outer radius of the conservative habitable zone and the optimistic habitable zone.

1 Which is wider: the conservative habitable zone or the optimistic habitable zone?

_____

_____

2 What color is the scale bar that shows the radius of the outer edge of the optimistic habitable zone?

_____

_____

Set the "Star Properties" to be equal to those for the Sun, and set the "Planet Orbit Properties" to be equal to those for Earth. It is good review for you to try to remember those, and look them up if you cannot! Answer the following questions.

3 For a star like the Sun, what is the inner radius of the optimistic habitable zone? How does this compare to the radius of Earth's orbit?

_____

_____

4 For a star like the Sun, what is the outer radius of the optimistic habitable zone? How does this compare to the radius of Earth's orbit?

_____

_____

5 Use the slider to change the eccentricity of the planet's orbit. How eccentric can the orbit of a planet with a semimajor axis of 1 AU be and still have the planet always within the *optimistic* habitable zone?

_____

_____

6 How eccentric can the orbit of a planet with a semimajor axis of 1 AU be and still have the planet always within the *conservative* habitable zone?

_____

_____

7 Double the luminosity of the star, keeping the temperature the same. What happened to the habitable zone? Did it become narrower or wider than the habitable zone around the Sun? Did it move closer or farther from the star than the habitable zone of the Sun? Why?

_____

_____

8 Reset the luminosity to the luminosity of the Sun, and double the temperature. What happened to the habitable zone? Why?

_____

_____

9 Double-check that the luminosity of the star is equal to the luminosity of the Sun, and set the temperature to half the temperature of the Sun. What happened to the habitable zone? Why?

_____

_____

10 Now set the star's luminosity to half the luminosity of the Sun, and set the temperature to half the temperature of the Sun. What happened to the habitable zone? Why?

_____

_____

_____

Mathematics helps scientists understand the patterns they see and then communicate that understanding to others. Appendix 1 presents some tools that will be useful in our study of astronomy.

Powers of 10 and Scientific Notation

Astronomy is a science of both the very large and the very small. The mass of an electron, for example, is

0.0000000000000000000000000000009109 kilograms (kg)

whereas the distance to a galaxy far, far away might be about

100,000,000,000,000,000,000,000,000 meters

A quick glance at those two numbers shows why astronomers need a more convenient way to express numbers.

Our number system is based on powers of 10. Going to the left of the decimal place,

$$10 = 10 \times 1$$
$$100 = 10 \times 10 \times 1$$
$$1000 = 10 \times 10 \times 10 \times 1$$

and so on. Going to the right of the decimal place,

$$0.1 = \frac{1}{10} \times 1$$

$$0.01 = \frac{1}{10} \times \frac{1}{10} \times 1$$

$$0.001 = \frac{1}{10} \times \frac{1}{10} \times \frac{1}{10} \times 1$$

and so on. In other words, each place to the right or left of the decimal place in a number represents a power of 10. For example, 1 million can be written as

$$1 \text{ million} = 1,000,000 = 1 \times 10 \times 10 \times 10 \times 10 \times 10 \times 10$$

That is, 1 million is "1 multiplied by six factors of 10." **Scientific notation** combines those factors of 10 in convenient shorthand. Rather

than all being written out, the six factors of 10 are expressed using an exponent:

$$1 \text{ million} = 1 \times 10^6$$

which also means "1 multiplied by six factors of 10."

Moving to the right of the decimal place, each step *removes* a power of 10 from the number. One-millionth can be written as

$$1 \text{ millionth} = 1 \times \frac{1}{10} \times \frac{1}{10} \times \frac{1}{10} \times \frac{1}{10} \times \frac{1}{10} \times \frac{1}{10}$$

That divides by powers of 10, so that number can be written by use of a negative exponent

$$\frac{1}{10} = 10^{-1}$$

as

$$1 \text{ millionth} = 1 \times 10^{-1} \times 10^{-1} \times 10^{-1} \times 10^{-1} \times 10^{-1} \times 10^{-1}$$
$$= 1 \times 10^{-6}$$

Let's return to our earlier examples. The mass of an electron is 9.109×10^{-31} kg, and the distant galaxy is located 1×10^{26} meters away. Scientific notation is a much more convenient way of writing those values. Notice that *the exponent in scientific notation gives you a feel for the size of a number at a glance.* The exponent of 10 in the electron mass is −31, which quickly indicates that it is a very small number. The exponent of 10 in the distance to a remote galaxy, +26, quickly indicates that it is a very large number. The exponent is often called the *order of magnitude* of a number. When you see a number written in scientific notation while reading *21st Century Astronomy* (or elsewhere), just remember to look at the exponent to understand the size of the number.

Scientific notation also is convenient because it makes multiplying and dividing numbers easier. For example, 2 billion multiplied by eight-thousandths can be written as

$$2,000,000,000 \times 0.008$$

but writing those two numbers in scientific notation is:

$$(2 \times 10^9) \times (8 \times 10^{-3})$$

We can regroup those expressions in the following form:

$$(2 \times 8) \times (10^9 \times 10^{-3})$$

The first part of the problem is just $2 \times 8 = 16$. The more interesting part of the problem is the multiplication in the right-hand parentheses. The first number, 10^9, is just shorthand for $10 \times 10 \times 10 \ldots$ nine times. That is, it represents nine factors of 10. The second number stands for three factors of $1/10$—or removing three factors of 10 if you prefer to think of it that way. Altogether, that makes $9 - 3 = 6$ factors of 10. In other words,

$$10^9 \times 10^{-3} = 10^{(9-3)} = 10^6$$

Putting the problem together, we get

$$(2 \times 10^9) \times (8 \times 10^{-3}) = (2 \times 8) \times (10^9 \times 10^{-3})$$
$$= 16 \times 10^6$$

By convention, when a number is written in scientific notation, only one digit is placed to the left of the decimal point. Here, though, we have two. However, 16 is 1.6×10, so we can add that additional factor of 10 to the exponent at right, making the final answer

$$1.6 \times 10^7$$

Dividing is just the inverse of multiplication. Dividing by 10^3 means removing three factors of 10 from a number. Using the previous number,

$$(1.6 \times 10^7) \div (2 \times 10^3) = (1.6 \div 2) \times (10^7 \div 10^3)$$
$$= 0.8 \times 10^{(7-3)}$$
$$= 0.8 \times 10^4$$

This time we have only a zero to the left of the decimal point. To get the number into proper form, we can substitute 8×10^{-1} for 0.8, giving

$$0.8 \times 10^4 = (8 \times 10^{-1}) \times 10^4 = 8 \times 10^3$$

Adding and subtracting numbers in scientific notation is harder because all numbers must be written as values multiplied by the *same* power of 10 before they can be added or subtracted. Therefore, you will need to use a calculator that has scientific notation. Most scientific calculators have a button that says EXP or EE. Those abbreviations mean "times 10 to the ___." So for 4×10^{12}, you would type [4][EXP][1][2] or [4][EE][1][2] into your calculator. Usually, that number shows up in the window on your calculator either just as you see it written in this book or as a 4 with a smaller 12 all the way over in the right side of the window.

Significant Figures

In the previous example, we actually broke some rules in the interest of explaining how powers of 10 are treated in scientific notation. The rules we broke involve the *precision* of the numbers. When expressing quantities in science, it is extremely important to know not only the value of a number but also how precise that value is.

The most complete way to keep track of the precision of numbers is to actually write down the uncertainty in the number. For example, suppose you know that the distance to a store (call it d) is between 0.8 and 1.2 kilometers (km); you can then write

$$d = 1.0 \pm 0.2 \text{ km}$$

where the symbol "$\pm$" is pronounced "plus or minus." Here, d is between $1.0 - 0.2 = 0.8$ km and $1.0 + 0.2 = 1.2$ km. That is an unambiguous statement about the limitations on knowing the value of d, but carrying along the formal errors with every number written would be cumbersome at best. Instead, you keep track of the approximate precision of a number by using *significant figures*.

The convention for significant figures is this: Assume the written number has been rounded from a number that had one additional digit to the right of the decimal point. If a quantity d, which might represent the distance to the store, is "1.", then d is close to 1. It is probably not as small as "0.", and it is probably not as large as "2.". If instead it is written as

$$d = 1.0$$

then d is probably not 0.9 and is probably not 1.1. It is roughly 1.0 to the nearest tenth. The greater the number of significant figures, the more precisely the number is being specified. For example, 1.00000 is not the same number as 1.00. The first number, 1.00000, represents a value that is probably not as small as 0.99999 and probably not as large as 1.00001. The second number, 1.00, represents a value that is probably not as small as 0.99 nor as large as 1.01. The number 1.00000 is much more precise than the number 1.00.

In mathematical operations, significant figures are important. For example, consider $2.0 \times 1.6 = 3.2$. It does *not* equal 3.20000000000. *The product of two numbers cannot be known to any greater precision than the numbers themselves.* In general, when you multiply and divide, the answer should have the same number of significant figures as the less precise of the numbers being multiplied or divided. In other words, $2.0 \times 1.602583475 = 3.2$. Because all you know is that the first factor is probably closer to 2.0 than to 1.9 or 2.1, all you know about the product is that it is between about 3.0 and 3.4. It is 3.2. It is not 3.205166950 (*even if that is the answer on your calculator*). The rest of the digits to the right of 3.2 just do not mean anything.

When two numbers are added or subtracted, if one number has a significant figure with a particular place value but another number does not, their sum or difference cannot have a significant figure in that place value. For example,

$$1045.$$
$$+1.34567$$
$$\overline{1046.}$$

The answer is "1046." *not* "1046.34567." Again, the extra digits to the right of the decimal place have no meaning because "1045." is not known to that precision.

What is the precision of the number 1,000,000? As it is written, the answer is unclear. Are all those zeros really significant, or are they merely placeholders? If the number is written in scientific notation, however, you never have to wonder. Instead of 1,000,000, you write 1.0×10^6 for a number that is known to the nearest hundred thousand or so, or you write 1.00000×10^6 for a number that is known to the nearest 10.

So the earlier example would have been more correct if written as

$$(2.0 \times 10^9) \times (8.0 \times 10^{-3}) = 1.6 \times 10^7$$

Algebra

Mathematics has many branches. The branch that focuses on the relationships between quantities is called **algebra**. Basically, algebra begins by using symbols to represent quantities.

For example, you could write the distance you travel in a day as d. As it stands, d has no value. It might be 10,000 miles. It might be 30 feet. It does, however, have **units**—here, the units of distance. The average speed at which you travel is equal to the distance you travel divided by the time you take. By using the symbol v to represent your average speed and the symbol t to represent the time you take, instead of writing out "Your average speed is equal to the distance you travel divided by the time you take" you can write

$$v = \frac{d}{t}$$

The meaning of that algebraic expression is the same as the sentence quoted before it, but it is much more concise. As it stands, v, d, and t still have no specific values. No numbers are assigned to them yet. However, the expression indicates what the relationship between those numbers will be when you look at a specific example. For example, if you go 500 km ($d = 500$ km) in 10 hours ($t = 10$ hours), that expression tells you that your average speed is

$$v = \frac{d}{t} = \frac{500\,\text{km}}{10\,\text{h}} = 50\,\text{km/h}$$

Notice that the units in this expression act exactly like the numerical values. Dividing the two shows that the units of v are kilometers divided by hours, or km/h (pronounced "kilometers per hour").

We introduced algebra as shorthand for expressing relations between quantities, but it is far more powerful than that. Algebra provides rules for manipulating the symbols used to represent quantities. We begin with a bit of notation for *powers* and *roots*. Similar to powers of 10, raising a quantity to a power means multiplying the quantity by itself some number of times. For example, if S is a symbol for something (anything), S^2 (pronounced "S-squared" or "S to the second power") means $S \times S$, and S^3 (pronounced "S-cubed" or "S to the third power") means $S \times S \times S$. Suppose S represents the length of the side of a square. The area of the square is given by

$$\text{Area} = S \times S = S^2$$

If $S = 3$ meters (m), the area of the square is

$$S^2 = 3\,\text{m} \times 3\,\text{m} = 9\,\text{m}^2$$

(pronounced "9 square meters"). That is why raising a quantity to the second power is called *squaring* the quantity. We could have done the same thing for the sides of a cube and found that the volume of the cube is

$$\text{Volume} = S \times S \times S = S^3$$

If $S = 3$ meters, the volume of the cube is

$$S^3 = 3\,\text{m} \times 3\,\text{m} \times 3\,\text{m} = 27\,\text{m}^3$$

(pronounced "27 cubic meters"). That is why raising a quantity to the third power is called *cubing* the quantity.

Roots are the reverse of that process. The square root of a number is the value that, when squared, gives the original quantity. The square root of 4 is 2, which means that $2 \times 2 = 4$. The square root of 9 is 3, which means that $3 \times 3 = 9$. Similarly, the cube root of a quantity is the value that, when cubed, gives the original quantity. The cube root of 8 is 2, which means that $2 \times 2 \times 2 = 8$. Roots are written with the symbol $\sqrt{\ }$. For example, the square root of 9 is written as

$$\sqrt{9} = 3$$

and the cube root of 8 is written as

$$\sqrt[3]{8} = 2$$

If the volume of a cube is $V = S^3$, then

$$S = \sqrt[3]{V} = \sqrt[3]{S^3}$$

Roots can also be written as powers. Powers and roots behave like the exponents of 10 in our discussion of scientific notation. (The exponents used in scientific notation are just powers of 10.) For example, if a, n, and m are all algebraic quantities, then

$$a^n \times a^m = a^{n+m} \quad \text{and} \quad \frac{a^n}{a^m} = a^{n-m}$$

(The square root of a can also be written $a^{1/2}$ and the cube root of a can be written $a^{1/3}$.)

The rules of arithmetic can be applied to the symbolic quantities of algebra. As long as the rules of algebra are applied properly, the relationships among symbols arrived at through algebraic manipulation remain true for the physical quantities that those symbols represent.

Here we summarize a few algebraic rules and relationships. Here, a, b, c, m, n, r, x, and y are all algebraic quantities.

Associative rule:

$$a \times b \times c = (a \times b) \times c = a \times (b \times c)$$

Commutative rule:

$$a \times b = b \times a$$

Distributive rule:

$$a \times (b + c) = (a \times b) + (a \times c)$$

Cross-multiplication:

$$\text{If } \frac{a}{b} = \frac{c}{d}, \text{ then } ad = bc$$

Working with exponents:

$$\frac{1}{a^n} = a^{-n} \quad a^n a^m = a^{n+m}$$

$$\frac{a^n}{a^m} = a^{n-m} \quad (a^n)^m = a^{n \times m} \quad \left(\frac{a}{b}\right)^n = \frac{a^n}{b^n}$$

Equation of a line with slope m and y-intercept b:

$$y = mx + b$$

Equation of a circle with radius r centered at $x = 0$, $y = 0$:

$$x^2 + y^2 = r^2$$

Angles and Distances

The farther away something is, the smaller it appears. That is common sense and everyday experience. Because astronomers cannot walk up to the object they are studying and measure it with a meter-stick, their knowledge about the sizes of things usually depends on relating the size of an object, its distance, and the angle it covers in the sky.

One way to measure angles is to use a unit called the **radian**. As shown in **Figure A1.1a**, the size of an angle in radians is the length of the arc subtending the angle, divided by the radius of the circle. In the figure, the angle $x = S/r$ radians.

Because the circumference of a circle is 2π multiplied by the radius ($C = 2\pi r$), a complete circle has an angular measure of $(2\pi r)/r = 2\pi$ radians. In more conventional angular measure, a complete circle is $360°$, so

$$360° = 2\pi \text{ radians}$$

or

$$1 \text{ radian} = \frac{360°}{2\pi} = 57.2958°$$

Often, seconds of arc (**arcseconds**) are used to measure angles for stars and galaxies. A degree is divided into 60 minutes of arc (**arcminutes**), each of which is divided into 60 seconds of arc—so 3600 seconds of arc are in a degree. Therefore,

$$3600 \frac{\text{arcseconds}}{\text{degree}} \times 57.2958 \frac{\text{degree}}{\text{radian}} = 206{,}265 \frac{\text{arcseconds}}{\text{radian}}$$

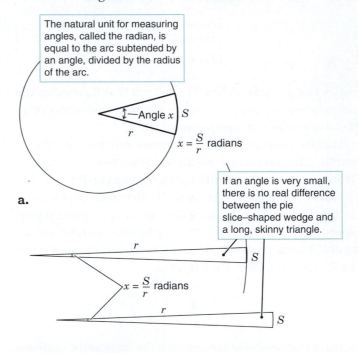

The natural unit for measuring angles, called the radian, is equal to the arc subtended by an angle, divided by the radius of the arc.

—Angle x S

r

$x = \dfrac{S}{r}$ radians

If an angle is very small, there is no real difference between the pie slice–shaped wedge and a long, skinny triangle.

r

S

$x = \dfrac{S}{r}$ radians

r

r

S

a.

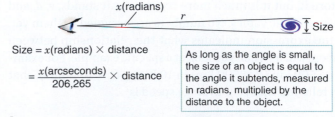

x(radians)

r

Size

Size $= x$(radians) $\times$ distance

$= \dfrac{x(\text{arcseconds})}{206{,}265} \times$ distance

As long as the angle is small, the size of an object is equal to the angle it subtends, measured in radians, multiplied by the distance to the object.

b.

Figure A1.1 Measuring angles.

If the angle is small enough (which it usually is in astronomy), very little difference exists between the pie slice just described and a long skinny triangle with a short side of length S, as **Figure A1.1b** illustrates. So, if you know the distance d to an object and you can measure the angular size x of the object, the size of the object is given by

$$S = x \text{ (in radians)} \times d = \frac{x \text{ (in degrees)}}{57.2958 \text{ degrees/radian}} \times d$$

$$S = \frac{x \text{ (in arcseconds)}}{206{,}265 \text{ arcseconds/radian}} \times d$$

That is how astronomers relate an object's angular size, distance, and physical size.

Circles and Spheres

To round out those mathematical tools, here are a few useful formulas for circles and spheres. The circle or sphere in each case has a radius r.

$$\text{Circumference}_{\text{circle}} = 2\pi r$$
$$\text{Area}_{\text{circle}} = \pi r^2$$
$$\text{Surface area}_{\text{sphere}} = 4\pi r^2$$
$$\text{Volume}_{\text{sphere}} = \frac{4}{3}\pi r^3$$

Working with Proportionalities

Most of the mathematics in *21st Century Astronomy* involves **proportionalities**—statements about how one physical quantity changes when another quantity changes. We began a discussion of proportionality in Working It Out 1.1; here, we offer a few examples of working with proportionalities.

To use a statement of proportionality to compare two objects, begin by turning the proportionality into a ratio. For example, the price of a bag of apples is **proportional** to the weight of the bag:

$$\text{Price} \propto \text{Weight}$$

Here, the symbol $\propto$ is pronounced "is proportional to." That means the ratio of the prices of two bags of apples is equal to the ratio of the weights of the two bags:

$$\text{Price} \propto \text{Weight means}$$

$$\frac{\text{Price of A}}{\text{Price of B}} = \frac{\text{Weight of A}}{\text{Weight of B}}$$

Let's work a specific example. Suppose bag A weighs 2 pounds and bag B weighs 1 pound. That means bag A will cost twice as much as bag B. We can turn that proportionality into the following equation:

$$\frac{\text{Price of A}}{\text{Price of B}} = \frac{\text{Weight of A}}{\text{Weight of B}} = \frac{2 \text{ lb}}{1 \text{ lb}} = 2$$

In other words, the price of bag A is 2 times the price of bag B. The price per pound is an example of a **constant of proportionality**.

Now let's work another, more complicated example. In Chapter 13, we discuss how the luminosity, brightness, and distance of stars are related. The luminosity of a star—the total energy that the star radiates each second—is proportional to the star's brightness multiplied by the square of its distance:

$$\text{Luminosity} \propto \text{Brightness} \times \text{Distance}^2$$

We can turn that proportionality into a ratio for two stars, A and B:

$$\frac{\text{Luminosity of A}}{\text{Luminosity of B}} = \frac{\text{Brightness of A}}{\text{Brightness of B}} \times \left(\frac{\text{Distance of A}}{\text{Distance of B}}\right)^2$$

If we use the symbols L, b, and d to represent luminosity, brightness, and distance, respectively, that equation becomes

$$\frac{L_A}{L_B} = \frac{b_A}{b_B} \times \left(\frac{d_A}{d_B}\right)^2$$

As an example, suppose that star A appears twice as bright in the sky as star B, but star A is located 10 times as far away as star B. How do the luminosities of the two stars compare? We know that

$$\text{Luminosity} \propto \text{Brightness} \times \text{Distance}^2$$

we write

$$\frac{\text{Luminosity of A}}{\text{Luminosity of B}} = \frac{\text{Brightness of A}}{\text{Brightness of B}} \times \left(\frac{\text{Distance of A}}{\text{Distance of B}}\right)^2$$

$$\frac{\text{Luminosity of A}}{\text{Luminosity of B}} = \frac{2}{1} \times \left(\frac{10}{1}\right)^2 = 200$$

In other words, star A is 200 times as luminous as star B.

Two quantities may also be inversely proportional, such that making one of them smaller makes the other larger. For example, when you are driving to another town, if you drive twice as fast, getting there takes half the time. The travel time is inversely proportional to the travel speed. We write that relationship as

$$\text{Time} \propto \frac{1}{\text{Speed}}$$

Proportionalities are used to compare one object with another. Constants of proportionality are used to calculate actual values. In *21st Century Astronomy*, the proportionality is usually what is important.

APPENDIX 2

PHYSICAL CONSTANTS AND UNITS

Fundamental Physical Constants

Constant	Symbol	Value
Speed of light in a vacuum	c	2.99792×10^8 m/s
Universal gravitational constant	G	6.6743×10^{-11} m³/(kg s²) 6.6743×10^{-20} km³/(kg s²)
Planck's constant	h	6.62607×10^{-34} J-s
Boltzmann constant	k	1.38065×10^{-23} J/K
Stefan-Boltzmann constant	σ	5.67037×10^{-8} W/(m² K⁴)
Mass of electron	m_e	9.10938×10^{-31} kg
Mass of proton	m_p	1.67262×10^{-27} kg
Mass of neutron	m_n	1.67493×10^{-27} kg
Electric charge of electron or proton	e	1.60218×10^{-19} C

Source: Data from the Particle Data Group (http://pdg.lbl.gov).

Unit Prefixes

Prefix*	Name	Factor†
n	nano-	10^{-9}
μ	micro-	10^{-6}
m	milli-	10^{-3}
k	kilo-	10^{3}
M	mega-	10^{6}
G	giga-	10^{9}
T	tera-	10^{12}

*When appended to a unit, these prefixes change the size of the unit by the factor (†) given. For example, 1 km is 10^3 meters.

Units and Values

Quantity	Fundamental Unit	Values
Length	meters (m)	radius of Sun (R_{Sun}) = 6.957×10^8 m astronomical unit (AU) = 1.49598×10^{11} m 1 AU = 149,598,000 km light-year (ly) = 9.4607×10^{15} m 1 ly = 6.324×10^4 AU 1 parsec (pc) = 3.262 ly = 3.0857×10^{16} m 1 m = 3.281 feet
Volume	cubic meters (m^3)	$1\ m^3$ = 1000 liters = 264.2 gallons
Mass	kilograms (kg)	1 kg = 1000 grams mass of Earth (M_{Earth}) = 5.9724×10^{24} kg mass of Sun (M_{Sun}) = 1.9885×10^{30} kg
Time	seconds (s)	1 hour (h) = 60 minutes (min) = 3600 s solar day (noon to noon) = 86,400 s sidereal day (Earth rotation period) = 86,164.2 s tropical year (equinox to equinox) = 365.24219 days = 3.15569×10^7 s sidereal year (Earth orbital period) = 365.25636 days = 3.15581×10^7 s
Speed	meters/second (m/s)	1 m/s = 2.237 miles/h 1 km/s = 1000 m/s = 3600 km/h c = 2.99792×10^8 m/s = 299,792 km/s
Acceleration	meters/second^2 (m/s^2)	g = gravitational acceleration on Earth = 9.81 m/s^2
Energy	joules (J)	1 J = 1 kg m^2/s^2 1 megaton = 4.18×10^{15} J
Power	watts (W)	1 W = 1 J/s solar luminosity (L_{Sun}) = 3.828×10^{26} W
Force	newtons (N)	1 N = 1 kg m/s^2 1 pound (lb) = 4.448 N 1 N = 0.22481 lb
Pressure	newtons/meter^2 (N/m^2)	atmospheric pressure at sea level = 1.013×10^5 N/m^2 = 1.013 bar
Temperature	kelvins (K)	absolute zero = 0 K = −273.15°C = −459.67°F

Source: Data from the Particle Data Group (http://pdg.lbl.gov).

APPENDIX 3

PERIODIC TABLE OF THE ELEMENTS

Legend:
- 1 — Atomic number
- H — Symbol
- Hydrogen — Name
- 1.00794 — Average atomic mass

- Metals
- Metalloids
- Nonmetals

1 / 1A																	18 / 8A
1 H Hydrogen 1.00794	**2** / 2A											**13** / 3A	**14** / 4A	**15** / 5A	**16** / 6A	**17** / 7A	**2** He Helium 4.002602
3 Li Lithium 6.941	**4** Be Beryllium 9.012182											**5** B Boron 10.811	**6** C Carbon 12.0107	**7** N Nitrogen 14.0067	**8** O Oxygen 15.9994	**9** F Fluorine 18.9984032	**10** Ne Neon 20.1797
11 Na Sodium 22.98976928	**12** Mg Magnesium 24.3050	**3** / 3B	**4** / 4B	**5** / 5B	**6** / 6B	**7** / 7B	**8** / 8B	**9** / 8B	**10** / 8B	**11** / 1B	**12** / 2B	**13** Al Aluminum 26.9815386	**14** Si Silicon 28.0855	**15** P Phosphorus 30.973762	**16** S Sulfur 32.065	**17** Cl Chlorine 35.453	**18** Ar Argon 39.948
19 K Potassium 39.0983	**20** Ca Calcium 40.078	**21** Sc Scandium 44.955912	**22** Ti Titanium 47.867	**23** V Vanadium 50.9415	**24** Cr Chromium 51.9961	**25** Mn Manganese 54.938045	**26** Fe Iron 55.845	**27** Co Cobalt 58.933195	**28** Ni Nickel 58.6934	**29** Cu Copper 63.546	**30** Zn Zinc 65.38	**31** Ga Gallium 69.723	**32** Ge Germanium 72.64	**33** As Arsenic 74.92160	**34** Se Selenium 78.96	**35** Br Bromine 79.904	**36** Kr Krypton 83.798
37 Rb Rubidium 85.4678	**38** Sr Strontium 87.62	**39** Y Yttrium 88.90585	**40** Zr Zirconium 91.224	**41** Nb Niobium 92.90638	**42** Mo Molybdenum 95.96	**43** Tc Technetium [98]	**44** Ru Ruthenium 101.07	**45** Rh Rhodium 102.90550	**46** Pd Palladium 106.42	**47** Ag Silver 107.8682	**48** Cd Cadmium 112.411	**49** In Indium 114.818	**50** Sn Tin 118.710	**51** Sb Antimony 121.760	**52** Te Tellurium 127.60	**53** I Iodine 126.90447	**54** Xe Xenon 131.293
55 Cs Cesium 132.9054519	**56** Ba Barium 137.327	**57** La Lanthanum 138.90547	**72** Hf Hafnium 178.49	**73** Ta Tantalum 180.94788	**74** W Tungsten 183.84	**75** Re Rhenium 186.207	**76** Os Osmium 190.23	**77** Ir Iridium 192.217	**78** Pt Platinum 195.084	**79** Au Gold 196.966569	**80** Hg Mercury 200.59	**81** Tl Thallium 204.3833	**82** Pb Lead 207.2	**83** Bi Bismuth 208.98040	**84** Po Polonium [209]	**85** At Astatine [210]	**86** Rn Radon [222]
87 Fr Francium [223]	**88** Ra Radium [226]	**89** Ac Actinium [227]	**104** Rf Rutherfordium [261]	**105** Db Dubnium [262]	**106** Sg Seaborgium [266]	**107** Bh Bohrium [264]	**108** Hs Hassium [277]	**109** Mt Meitnerium [268]	**110** Ds Darmstadtium [271]	**111** Rg Roentgenium [272]	**112** Cn Copernicium [285]	**113** Uut Ununtrium [284]	**114** Fl Flerovium [289]	**115** Uup Ununpentium [288]	**116** Lv Livermorium [292]	**117** Uus Ununseptium [294]	**118** Uuo Ununoctium [294]

6 Lanthanides

58 Ce Cerium 140.116	**59** Pr Praseodymium 140.90765	**60** Nd Neodymium 144.242	**61** Pm Promethium [145]	**62** Sm Samarium 150.36	**63** Eu Europium 151.964	**64** Gd Gadolinium 157.25	**65** Tb Terbium 158.92535	**66** Dy Dysprosium 162.500	**67** Ho Holmium 164.93032	**68** Er Erbium 167.259	**69** Tm Thulium 168.93421	**70** Yb Ytterbium 173.05	**71** Lu Lutetium 174.967

7 Actinides

90 Th Thorium 232.03806	**91** Pa Protactinium 231.03588	**92** U Uranium 238.02891	**93** Np Neptunium [237]	**94** Pu Plutonium [244]	**95** Am Americium [243]	**96** Cm Curium [247]	**97** Bk Berkelium [247]	**98** Cf Californium [251]	**99** Es Einsteinium [252]	**100** Fm Fermium [257]	**101** Md Mendelevium [258]	**102** No Nobelium [259]	**103** Lr Lawrencium [262]

We have used the U.S. system as well as the system recommended by the International Union of Pure and Applied Chemistry (IUPAC) to label the groups in this periodic table. The system used in the United States includes a letter and a number (1A, 2A, 3B, 4B, etc.), which is close to the system developed by Mendeleev. The IUPAC system uses numbers 1–18 and has been recommended by the American Chemical Society (ACS). While we show both numbering systems here, we use the IUPAC system exclusively in the book. Elements with atomic numbers higher than 112 have been reported but not yet fully authenticated.

PROPERTIES OF PLANETS, DWARF PLANETS, AND MOONS

Physical Data for Planets and Dwarf Planets

Planet	EQUATORIAL RADIUS		MASS		Average Density (relative to water*)	Rotation Period (days)	Tilt of Rotation Axis (degrees, relative to orbit)	Equatorial Surface Gravity (relative to Earth†)	Escape Velocity (km/s)	Average Surface Temperature (K)§
	(km)	(R/R_{Earth})	(kg)	(M/M_{Earth})						
Mercury	2440	0.383	3.30×10^{23}	0.055	5.427	58.79	0.03	0.378	4.3	340 (100, 700)
Venus	6052	0.950	4.87×10^{24}	0.815	5.243	243.02‡	177.4	0.905	10.36	737
Earth	6378	1.000	5.97×10^{24}	1.000	5.514	1.000	23.44	1.000	11.19	288 (185, 331)
Mars	3396	0.532	6.42×10^{23}	0.107	3.933	1.027	25.19	0.379	5.03	210 (120, 293)
Ceres	473	0.074	9.39×10^{20}	0.0002	2.09	0.378	4	0.26	0.5	168
Jupiter	71,492	11.209	1.90×10^{27}	317.8	1.326	0.4135	3.13	2.530	59.5	165
Saturn	60,268	9.449	5.68×10^{26}	95.16	0.687	0.444	26.73	1.065	35.5	134
Uranus	25,559	4.007	8.68×10^{25}	14.54	1.271	0.7183‡	97.77	0.905	21.3	76
Neptune	24,764	3.883	1.02×10^{26}	17.15	1.638	0.6713	28.32	1.14	23.5	72
Pluto	1188	0.186	1.30×10^{22}	0.0022	1.854	6.387‡	122.53	0.063	1.21	31
Haumea	~800	0.13	4.0×10^{21}	0.0007	~2	0.163	?	0.025	0.71	<50
Makemake	715	0.11	3.1×10^{21}	0.0005	~2	0.95	3	~ 0.5	~ 0.9	~ 30
Eris	1163	0.182	1.65×10^{22}	0.0028	2.4	1.08	?	0.084	1.38	42

*The density of water is 1000 kg/m³.
†The surface gravity of Earth is 9.81 m/s².
‡Venus, Uranus, and Pluto rotate opposite to the directions of their orbits. Their north poles are south of their orbital planes.
§Values in parentheses give extremes of recorded temperatures.

Orbital Data for Planets and Dwarf Planets

Planet	MEAN DISTANCE FROM SUN (A*)		Orbital Period (P) (sidereal years)	Eccentricity	Inclination (degrees, relative to ecliptic)	Average Speed (km/s)
	(10⁶ km)	(AU)				
Mercury	57.9	0.387	0.241	0.2056	7.0	47.36
Venus	108.2	0.723	0.615	0.0067	3.395	35.02
Earth	149.6	1.000	1.000	0.0167	0.000	29.78
Mars	227.9	1.524	1.881	0.0935	1.851	24.07
Ceres	414.0	2.768	4.603	0.076	10.59	17.90
Jupiter	778.57	5.204	11.862	0.0489	1.30	13.06
Saturn	1433.53	9.582	29.457	0.0565	2.485	9.68
Uranus	2872.46	19.20	84.011	0.0457	0.772	6.80
Neptune	4495.06	30.048	164.79	0.0113	1.769	5.43
Pluto	5869.656	39.237	247.94	0.2444	17.16	4.67
Haumea	6450.1	43.116	284.12	0.196	28.21	4.53
Makemake	6796.2	45.43	309.1	0.161	28.98	4.42
Eris	10,152	67.86	558	0.436	44.0	3.43

*A is the semimajor axis of the planet's elliptical orbit.

Properties of Selected Moons*

Planet	Moon	ORBITAL PROPERTIES		PHYSICAL PROPERTIES		Relative Density (g/cm³) (water = 1.00)
		P (days)	A (10³ km)	R (km)	M (10²⁰ kg)	
Earth (1 moon)	Moon	27.32	384.4	1738.1	734.6	3.34
Mars (2 moons)	Phobos	0.319	9.38	11.27	0.0001	1.88
	Deimos	1.26	23.46	6.2	0.00002	1.48
Jupiter (79 known moons)	Metis	0.29	128	21.5	0.0004	0.86
	Amalthea	0.50	181.4	83.5	0.021	0.86
	Io	1.77	421.7	1822	893	3.53
	Europa	3.55	670.9	1561	480	3.01
	Ganymede	7.15	1070	2634	1482	1.94
	Callisto	16.69	1883	2410	1076	1.83
	Himalia	248.3	11,389	70	0.042	1.6
	Pasiphae	722†	23,209	19	0.0030	2.6
	Callirrhoe	787†	24,583	4.8	0.00001	2.6
Saturn (82 known moons)	Pan	0.58	133.58	14.1	0.00005	0.42
	Prometheus	0.61	139.38	43.1	0.0016	0.48
	Pandora	0.63	141.70	40.7	0.0014	0.49
	Mimas	0.94	185.54	198	0.37	1.15
	Enceladus	1.37	237.95	252	1.08	1.6
	Tethys	1.89	294.62	531	6.18	0.98
	Dione	2.74	377.40	561	11.0	1.48
	Rhea	4.52	527.12	764	23.1	1.24
	Titan	15.95	1222	2575	1346	1.88
	Hyperion	21.28	1481	135	0.056	0.54
	Iapetus	79.33	3561	735	18.1	1.09
	Phoebe	550†	12,960	107	0.08	1.64
	Paaliaq	687	15,200	13	0.0007	2.3
Uranus (27 known moons)	Cordelia	0.34	49.8	20	0.0004	1.3
	Miranda	1.41	129.4	236	0.64	1.20
	Ariel	2.52	191.0	579	12.5	1.59
	Umbriel	4.14	266.0	585	12.8	1.39
	Titania	8.71	435.9	788	34.0	1.71
	Oberon	13.46	583.5	761	30.8	1.63
	Setebos	2225	17,418	24	0.0008	1.3

(continued)

Properties of Selected Moons*

(continued)

Planet	Moon	ORBITAL PROPERTIES		PHYSICAL PROPERTIES		Relative Density (g/cm³) (water = 1.00)
		P (days)	*A* (10³ km)	*R* (km)	*M* (10²⁰ kg)	
Neptune (14 known moons)	Naiad	0.29	48.2	30	0.002	1.3
	Larissa	0.55	73.5	97	0.04	1.2
	Proteus	1.12	117.6	210	0.44	1.3
	Triton	5.88†	354.8	1353	214	2.06
	Nereid	360.11	5513.9	170	0.3	1.5
Pluto (5 moons)	Charon	6.39	19.60	606	15.9	1.70
Haumea (2 moons)	Namaka	18.3	25.66	85	0.018	~1
	Hi'iaka	49	49.88	160	0.179	~1
Eris	Dysnomia	15.8	37.3	350	1.43	0.8

*Innermost, outermost, largest, and/or a few other moons for each planet.
†Irregular moon (has retrograde orbit).

SPACE MISSIONS

Selected Recent and Current Solar System Missions

Spacecraft	Sponsoring Nation(s)*	Destination	Launch Year	Type	Status (mid-2021)
Voyager 1 and *2*	USA	Jupiter, Saturn, Uranus (2), Neptune (2)	1977	Flyby	Actively exploring outer edge of Solar System
Galileo	USA	Jupiter	1989	Orbiter/probe	Ended 2003
Ulysses	USA, Europe	Sun	1990	Solar polar orbiter	Ended 2008
SOHO	USA, Europe	Sun	1995	Orbiter	Active
Mars Global Surveyor	USA	Mars	1996	Orbiter	Ended 2006
Cassini-Huygens	USA, Europe, Italy	Saturn, Titan	1997	Saturn orbiter, Titan probe/lander	Ended 2017
Stardust	USA	Comets	1999	Sample return/flyby	Ended 2011
Mars Odyssey	USA	Mars	2001	Orbiter	Active
Mars Exploration Rover	USA	Mars	2003	Two landers	Ended 2019
Hayabusa	Japan	Asteroid	2003	Sample return	Ended 2010
Mars Express	Europe	Mars	2003	Orbiter	Active
Messenger	USA	Mercury (2011)	2004	Orbiter	Ended 2015
Rosetta	Europe	Comet 67P/Churyumov–Gerasimenko (2014)		Orbiter and lander	Ended 2016
Venus Express	Europe	Venus	2005	Orbiter	Ended 2014
Mars Reconnaissance Orbiter (MRO)	USA	Mars	2005	Orbiter	Active
Deep Impact/EPOXI	USA	Comet Hartley (2010)	2005	Impactor/flyby	Ended 2010
STEREO	USA	Sun	2006	Two orbiters	Active
New Horizons	USA	Pluto (2015)	2006	Flyby	Active
Chang'e 1	China	Moon	2007	Orbiter	Ended 2009
Kayuga	Japan	Moon	2007	Orbiter	Ended 2009
Themis	USA	Moon, solar wind	2007	Multiple Orbiters	Active
Dawn	USA	Vesta (2011), Ceres (2015)	2007	Orbiter	Ended 2018
Chandrayaan	India	Moon	2008	Orbiter/impactor	Ended 2009
Lunar Reconnaissance Orbiter (LRO)	USA	Moon	2009	Orbiter	Active
Lunar Crater Observation and Sensing Satellite (LCROSS)	USA	Moon	2009	Impactor	Ended 2009

(continued)

Selected Recent and Current Solar System Missions

(continued)

Spacecraft	Sponsoring Nation(s)*	Destination	Launch Year	Type	Status (mid-2021)
Chang'e 2	China	Moon	2010	Orbiter	Ended 2011
Akatsuki	Japan	Venus (2015)	2010	Orbiter	Active
Juno	USA	Jupiter (2016)	2011	Orbiter	Active
Gravity Recovery and Interior Laboratory (*GRAIL*)	USA	Moon	2011	Two orbiters	Ended 2012
Mars Science Laboratory (*Curiosity* rover)	USA	Mars	2011	Lander	Active
Mars Atmosphere and Volatile EvolutioN (*MAVEN*) mission	USA	Mars	2013	Orbiter	Active
Chang'e 3	China	Moon	2013	Lander	On lunar surface
Mangalyaan	India	Mars	2013	Orbiter	Active
Hayabasu 2	Japan	Asteroid 162173 Ryugu	2014	Orbiter, scheduled sample return	Delivered sample, on extended mission
DSCOVR	USA	Sun	2015	Orbiter	Active
Osiris-REx	USA	Asteroid 101955 Bennu	2016	Orbiter, scheduled sample return	Returning
ExoMars Trace Gas Orbiter (*TGO*)	Europe, Russia	Mars (2016)	2016	Orbiter (lander crashed)	Active
InSight	USA	Mars	2018	Lander	Active
BepiColombo	Europe, Japan	Mercury	2018	Orbiter	En route
Parker Solar Probe	USA	Sun	2018	Orbiter	Active
Solar Orbiter	Europe	Sun	2020	Orbiter	En route
Emirates Mars Mission	United Arab Emirates	Mars	2020	Orbiter	Active
Tianwen-1	China	Mars	2020	Orbiter	Active
Zhurong	China	Mars	2020	Rover	Active
Perseverance	USA	Mars	2020	Rover	Active
Ingenuity	USA	Mars	2020	Helicopter	Active

*Countries are represented by the following agencies: China = CNSA (China National Space Administration); Europe = ESA (European Space Agency); India = ISRO (Indian Space Research Organisation); Italy = Italian Space Agency; Japan = JAXA (Japan Aerospace Exploration Agency); Russia = ROSCOSMOS (State Corporation for Space Activities); USA = NASA (National Aeronautics and Space Administration).

BRIGHTEST STARS

The 25 Brightest Stars in the Sky

Name	Common Name	Distance (ly)	Spectral Type	Relative Visual Luminosity* (Sun = 1.000)	Apparent Visual Magnitude	Absolute Visual Magnitude
Sun	Sun	1.58×10^{-5}	G2V	1.000	−26.74	4.83
Alpha Canis Majoris	Sirius	8.60	A1V	22.9	−1.46	1.43
Alpha Carinae	Canopus	309	F0II	14,900	−0.72	−5.60
Alpha^1 Centauri	Rigil Kentaurus A	4.36	G2V	1.51	−0.01	4.38
Alpha^2 Centauri	Rigil Kentaurus B	4.36	K1V	0.44	1.33	5.71
Alpha Bootis	Arcturus	37	K1.5III	113	−0.04	−0.30
Alpha Lyrae	Vega	25	A0Va	49.2	0.03	0.60
Alpha Aurigae	Capella	43	G5IIIe+G0III	137	0.08	−0.51
Beta Orionis	Rigel	860	B8Iab	54,000	0.12	−7.0
Alpha Canis Minoris	Procyon	11.5	F5IV-V	7.73	0.34	2.61
Alpha Eridani	Achernar	140	B3Vpe	1030	0.46	−2.70
Beta Centauri	Hadar	392	B1III	7180	0.61	−4.81
Alpha Orionis	Betelgeuse	570	M2Iab	13,600	0.7	−5.5
Alpha Aquilae	Altair	16.7	A7V	11.1	0.77	2.22
Alpha Crucis	Acrux	325	B0.5IV+B1V	3100	1.3	−3.9
Alpha Tauri	Aldebaran	67	K5III	163	0.85	−0.70
Alpha Scorpii	Antares	550	M1.5Ib	16,300	0.96	−5.7
Alpha Virginis	Spica	250	B1IV+B4V	1920	1.04	−3.38
Beta Geminorum	Pollux	34	K0III	32.2	1.14	1.06
Alpha Piscis	Fomalhaut	25	A3V	17.4	1.16	1.73
Beta Crucis	Mimosa	280	B0.5III	1980	1.25	−3.41
Alpha Cygni	Deneb	1425	A2Ia	58,600	1.25	−7.09
Alpha Leonis	Regulus	79	B7V	146	1.35	−0.58
Epsilon Canis Majoris	Adhara	405	B2II	3400	1.50	−4.0
Alpha Gemini	Castor	51	A1V+A5Vm	49	1.58	0.61
Gamma Crucis	Gacrux	88	M3.5III	138	1.63	−0.52

Sources: Data from Jim Kaler's *STARS* page (http://stars.astro.illinois.edu/sow/bright.html); SIMBAD Astronomical Database (http://simbad.u-strasbg.fr/simbad).
*Luminosity in this table refers only to radiation in "visual" light.

APPENDIX 7

OBSERVING THE SKY

This appendix gives you enough information to make sense of a star chart or list of astronomical objects and find a few objects in the sky.

Celestial Coordinates

In Chapter 2, we discuss the **celestial sphere**—the imaginary sphere with Earth at its center on which celestial objects appear to lie. Several coordinate systems are used to specify the positions of objects on the celestial sphere. The simplest of these is the *altitude-azimuth coordinate system*. The altitude-azimuth coordinate system is based on the "map" direction to an object (the object's azimuth, with north = 0°, east = 90°, south = 180°, and west = 270°) combined with how high the object is above the horizon (the object's altitude, with the horizon at 0° and the zenith at 90°). For example, an object 10° above the eastern horizon has an altitude of 10° and an azimuth of 90°. An object 45° above the horizon in the southwest is at altitude 45°, azimuth 225°.

The altitude-azimuth coordinate system is the simplest way to tell someone where in the sky to look at the moment, but it is not a good coordinate system for cataloging the positions of objects. The altitude and azimuth of an object are different for each observer, depending on the observer's position on Earth, and they are constantly changing as Earth rotates on its axis. To specify the direction to an object in a way that is the same for everyone requires a coordinate system that is fixed relative to the celestial sphere. The most common such coordinates are called *celestial coordinates*.

Celestial coordinates are illustrated in **Figure A7.1**. Celestial coordinates are much like the traditional system of latitude and longitude used on the surface of Earth. On Earth, latitude specifies how far you are from Earth's equator, as discussed in Chapter 2. If you are on Earth's equator, your latitude is 0°. If you are at Earth's North Pole, your latitude is 90° north. If you are at Earth's South Pole, your latitude is 90° south.

The latitude-like coordinate on the celestial sphere is called **declination**, often signified with the lowercase Greek letter δ (delta). The celestial equator has $\delta = 0°$. The north celestial pole has $\delta = +90°$. The south celestial pole has $\delta = -90°$. (See Chapter 2 if you need to refresh your memory about the celestial equator or celestial poles.) Declination is usually expressed in degrees, minutes of arc, and seconds of arc. For example, Sirius, the brightest star in the sky,

has $\delta = -16°42'58''$ meaning that it is located not quite 17° south of the celestial equator.

On Earth, east–west position is specified by longitude. Lines of constant longitude run north–south from one pole to the other. Unlike latitude, for which the equator provides a natural place to call "zero," longitude has no natural starting point, so one was invented. By arbitrary convention, the Royal Observatory in Greenwich, England, is defined to lie at a longitude of 0°. On the celestial sphere, the longitude-like coordinate is called **right ascension**, often signified with the lowercase Greek letter α (alpha). Unlike longitude, right ascension *does* have a natural starting point on the celestial sphere: the vernal equinox, or the point at which the ecliptic crosses the celestial equator with the Sun moving from the southern sky into the northern sky. The (Northern Hemisphere) vernal equinox defines the line of right ascension at which $\delta = 0°$. The (Northern Hemisphere) autumnal equinox, located on the opposite side of the sky, is at $\delta = 180°$.

Normally, right ascension is measured in units of time rather than degrees. Earth takes 24 hours (of sidereal time) to rotate on its axis, so the celestial sphere is divided into 24 hours of right ascension, with each hour of right ascension corresponding to 15°. Hours of right ascension are then subdivided into minutes and seconds of time. Right ascension increases going to the east. The right ascension of Sirius, for example, is $\alpha = 06^h45^m08.9^s$, meaning that Sirius is about 101° (that is, 06^h45^m) east of the vernal equinox. Time is a natural unit for measuring right ascension because time naturally tracks the motion of objects due to Earth's rotation on its axis. If stars on the meridian at a certain time have $\alpha = 06^h$, then an hour later the stars on the meridian will have $\alpha = 07^h$, and an hour after that they will have $\alpha = 08^h$. The *local sidereal time*, or *star time*, at your location right now is equal to the right ascension of the stars on your meridian at the moment. Because of Earth's motion around the Sun, a sidereal day is about 4 minutes shorter than a solar day, so local sidereal time constantly gains on solar time. At midnight on September 22, the local sidereal time is 0^h. By midnight on December 21, local sidereal time has advanced to 06^h. On March 20, local sidereal time at midnight is 12^h. And at midnight on June 20, local sidereal time is 18^h.

Putting all that together, right ascension and declination provide a convenient way to specify the location of any object on the celestial sphere. Sirius is located at $\alpha = 06^h45^m08.9^s$, $\delta = -16°42'58''$ which means that at midnight on December 21 (local sidereal

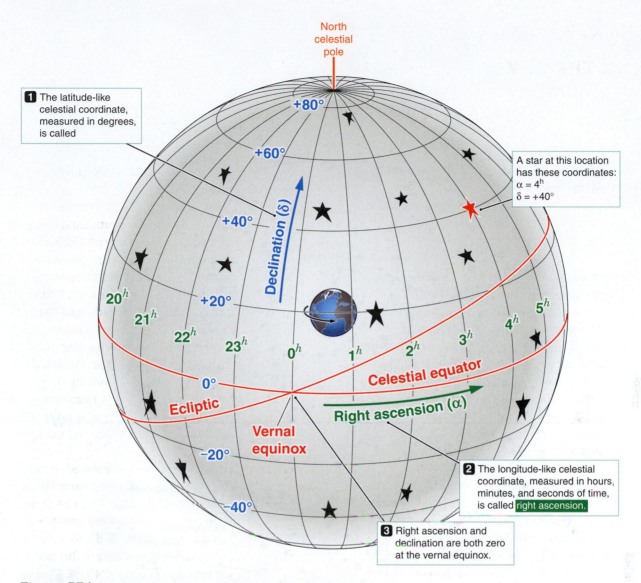

1 The latitude-like celestial coordinate, measured in degrees, is called

A star at this location has these coordinates:
$\alpha = 4^h$
$\delta = +40°$

North celestial pole

Declination (δ)

Celestial equator

Right ascension (α)

Ecliptic

Vernal equinox

2 The longitude-like celestial coordinate, measured in hours, minutes, and seconds of time, is called right ascension.

3 Right ascension and declination are both zero at the vernal equinox.

Figure A7.1 Celestial coordinates.

time = 06^h), you will find Sirius about 45^m east of the meridian, not quite 17° south of the celestial equator.

Just one final caveat remains. As we discussed in Chapter 2, the directions of the celestial equator, celestial poles, and vernal equinox are constantly changing as Earth's axis wobbles like the axis of a spinning top. In Chapter 2, we called that 26,000-year wobble the **precession of the equinoxes**, meaning that the location of the equinoxes is slowly advancing along the ecliptic. So when we specify the celestial coordinates of an object, we need to specify the date at which the positions of the vernal equinox and celestial poles were measured. By convention, coordinates are usually referred to with the position of the vernal equinox on January 1, 2000. A complete, formal specification of the coordinates of Sirius would then be $\alpha(2000) = 06^h45^m08.9^s$, $\delta(2000) = -16°42'58''$, where the "2000" in parentheses refers to the equinox of the coordinates.

Constellations and Names

Although it is certainly possible to specify any location on the surface of Earth exactly by giving its latitude and longitude, using a more descriptive address is usually more convenient. We might say, for example, that one of the coauthors of this book works near latitude 37° north, longitude 122° west; but it would probably mean a lot more to you if we said that George Blumenthal works in Santa Cruz, California.

Just as the surface of Earth is divided into nations and states with shared boundaries, the celestial sphere is divided into 88 **constellations**, the names of which are often used to refer to objects within their boundaries (see the star charts in **Figure A7.2**). The brightest stars within the boundaries of a constellation are named using a Greek letter combined with the name of the constellation.

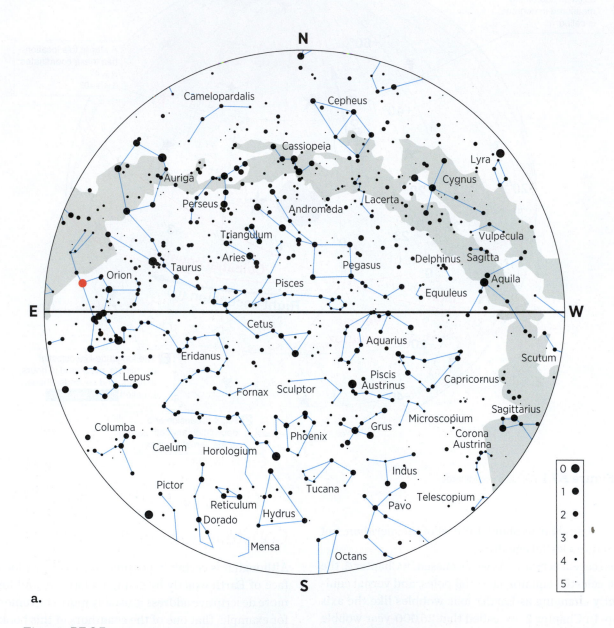

Figure A7.2A The night sky on the northern autumnal equinox, approximately September 21 each year. The celestial equator runs across the middle of the map. A red dot indicates a star that is noticeably red in the sky. The shaded area is the Milky Way.

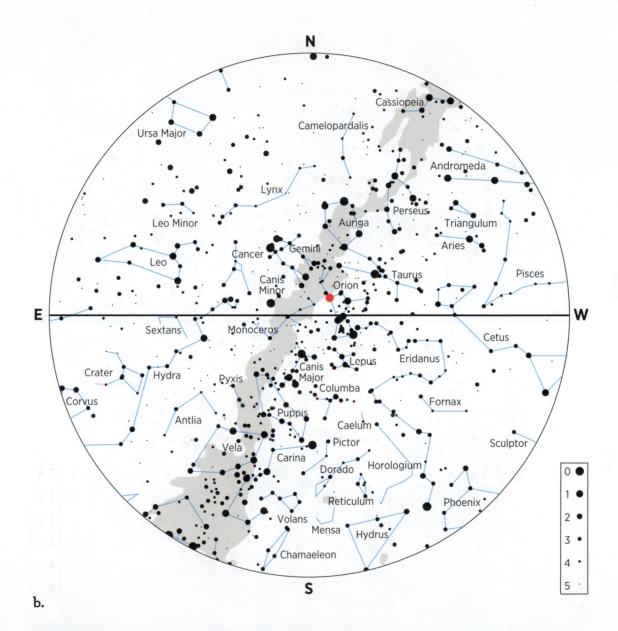

Figure A7.2B The night sky on the northern winter solstice, approximately December 21 each year. The celestial equator runs across the middle of the map. A red dot indicates a star that is noticeably red in the sky. The shaded area is the Milky Way.

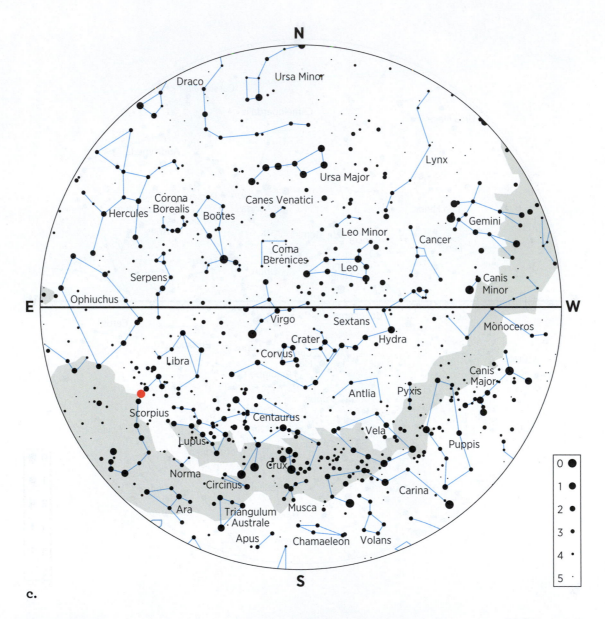

c.

Figure A7.2C The night sky on the northern vernal equinox, approximately March 21 each year. The celestial equator runs across the middle of the map. A red dot indicates a star that is noticeably red in the sky. The shaded area is the Milky Way.

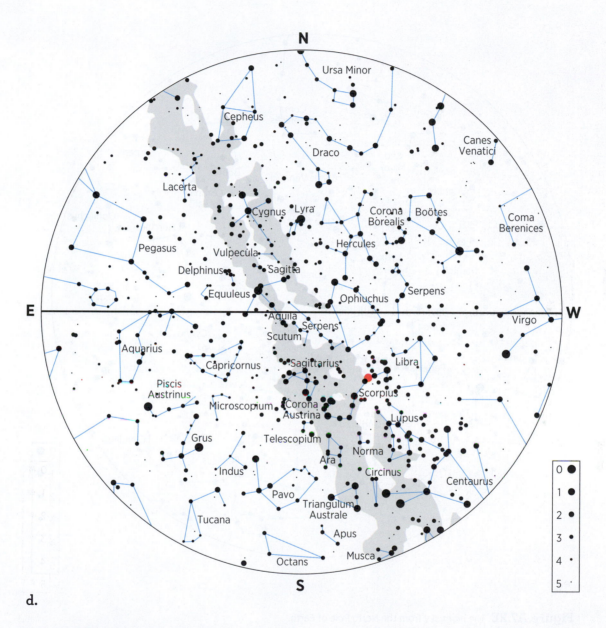

Figure A7.2D The night sky on the northern summer solstice, approximately June 21 each year. The celestial equator runs across the middle of the map. A red dot indicates a star that is noticeably red in the sky. The shaded area is the Milky Way.

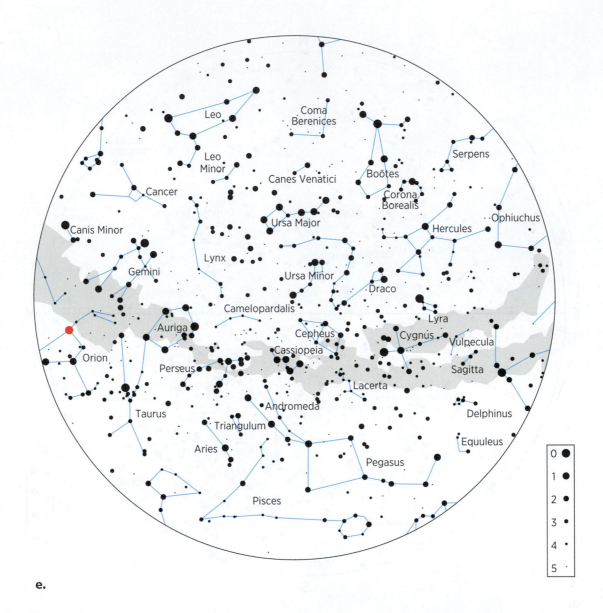

e.

Figure A7.2E The night sky from the North Pole of Earth.

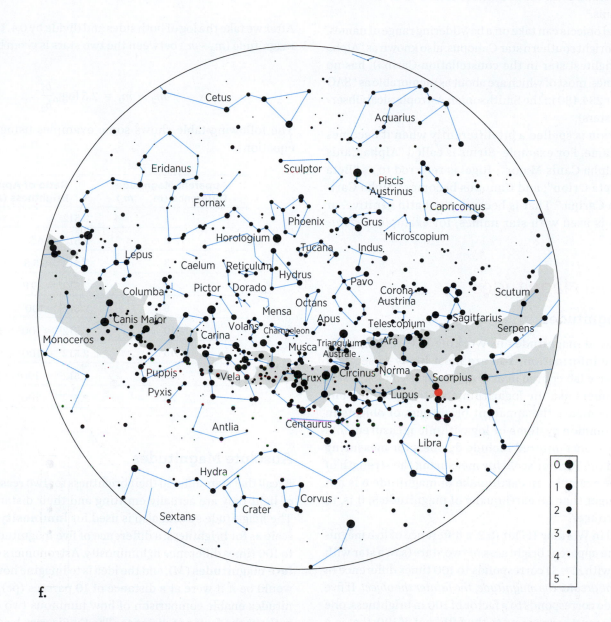

Figure A7.2F The night sky from the South Pole of Earth.

For example, the star Sirius is the brightest star in the constellation Canis Major (literally, the "great dog"), so it is called "Alpha Canis Majoris." The bright red star in the northeastern corner of the constellation Orion is called "Alpha Orionis," also known as Betelgeuse. Rigel, the bright blue star in the southwest corner of Orion, is also called "Beta Orionis."

Astronomical objects can take on a bewildering range of names. For example, the bright southern star Canopus, also known as "Alpha Carinae" (the brightest star in the constellation Carina), has no fewer than 34 names, most of which are about as memorable as "SAO 234480" (number 234,480 in the Smithsonian Astrophysical Observatory catalog of stars).

A constellation is spelled a bit differently when it becomes part of a star's name. For example, Sirius is called "Alpha Canis Majoris," not "Alpha Canis Major"; Rigel is referred to as "Beta Orionis," not "Beta Orion"; and Canopus becomes "Alpha Carinae," not "Alpha Carina." That is because the Latin genitive, or possessive, case is used with star names; for example, *Orionis* means "of Orion."

Astronomical Magnitudes

Apparent Magnitudes

We first introduced magnitudes in Working It Out 13.2; here, we provide more information. You are most likely to see that system if you take a lab course in astronomy or if you use a star catalog. Astronomers use the logarithmic system of **apparent magnitudes** to compare the apparent brightness of objects in the sky. Other common systems of logarithmic measurements that you may have encountered include decibels for measuring sound levels and the Richter scale for measuring the strength of earthquakes. For example, an earthquake of magnitude 6 is not just a little stronger than an earthquake of magnitude 5; it is, in fact, 10 times stronger.

As discussed in Working It Out 13.2, a difference of five magnitudes between the apparent brightness of two stars (say, a star with $m = 6$ and a star with $m = 1$) corresponds to 100 times difference in brightness, and *the greater the magnitude, the fainter the object.* If five steps in magnitude corresponds to a factor of 100 in brightness, one step in magnitude must correspond to the fifth root of 100; that is, a factor of $100^{1/5} =$ approximately 2.512 in brightness ($100^{1/5} \times 100^{1/5} \times 100^{1/5} \times 100^{1/5} \times 100^{1/5} = 100$).

If star 1 has a brightness of b_1 and star 2 has a brightness of b_2, the ratio of the brightness of the stars is given by

$$\frac{b_1}{b_2} = (2.512)^{m_2 - m_1} = 100^{\frac{(m_2 - m_1)}{5}}$$

We can put that into the more common base 10 by noting that $100 = 10^2$, so the expression becomes

$$\frac{b_1}{b_2} = 10^{2 \times \frac{(m_2 - m_1)}{5}} = 10^{0.4(m_2 - m_1)}$$

After we take the log of both sides and divide by 0.4, the difference in magnitude ($m_2 - m_1$) between the two stars is given by

$$m_2 - m_1 = 2.5 \log_{10} \frac{b_1}{b_2}$$

The following table shows some examples using the preceding equations.

Apparent Magnitude Difference ($m_2 - m_1$)	Ratio of Apparent Brightness (b_1/b_2)
1	2.512
2	$2.512^2 = 6.3$
3	$2.512^3 = 15.8$
4	$2.512^4 = 39.8$
5	$2.512^5 = 100$
10	$2.512^{10} = 100^2 = 10,000$
15	$2.512^{15} = 100^3 = 1,000,000$
20	$2.512^{20} = 100^4 = 10^8$
25	$2.512^{25} = 100^5 = 10^{10}$

Absolute Magnitudes

Recall that stars differ in their brightness for two reasons: the amount of light they are actually emitting and their distance from Earth. The magnitude system also is used for **luminosity**, with the same scale as for brightness: a difference of five magnitudes corresponds to 100 times difference in luminosity. Astronomers call those **absolute magnitudes** (M), and the idea is to imagine how bright the star would be if it were at a distance of 10 parsecs (pc). Absolute magnitudes enable comparison of how luminous two stars really are, without the factor of distance. The Sun is very bright because it is so close (apparent visual magnitude = –27), but if the Sun were at a distance of 10 pc, its magnitude would be only about 5.[1] Thus, the absolute magnitude of the Sun is $M = 5$. Recall that the luminosity of a star is usually expressed by comparing it with the luminosity of the Sun. As with apparent magnitudes, higher magnitude numbers correspond to lower luminosity. Thus, a star 1/100 as luminous as the Sun will be 5 absolute magnitudes fainter, or $M = 10$. A star 10,000

[1]The apparent and absolute magnitudes of the Sun are –26.74 and +4.83, respectively. We use +5 for the Sun's absolute magnitude as an approximation.

times more luminous than the Sun will be 10 absolute magnitudes brighter, or $M = -5$.

Absolute magnitudes and luminosities follow the same equations as those we provided already, using L instead of b and M instead of m:

$$\frac{L_1}{L_2} = 10^{2 \times \frac{M_{(2)} - M_{(1)}}{5}} = 10^{0.4(M_{(2)} - M_{(1)})}$$

and

$$M_{(2)} - M_{(1)} = 2.5 \log_{10} \frac{L_1}{L_2}$$

Most often, astronomers think about the luminosity of a star in comparison with the luminosity of the Sun. Here $L_1 = L_{star}$ and $L_2 = L_{Sun}$. The following table compares luminosity (where $L_{Sun} = 1$) with absolute magnitude of a star.

L_{Star}/L_{Sun}	M
1,000,000	-10
10,000	-5
100	0
1	5
1/100	10
1/10,000	15

Distance Modulus

The difference between the apparent magnitude and the absolute magnitude depends on the star's distance. By definition, a star at a distance of exactly 10 pc will have an apparent magnitude equal to its absolute magnitude. Astronomers can always measure the brightness of a star and thus its apparent magnitude and can estimate the luminosity of a star and thus its absolute magnitude by using the Hertzsprung-Russell (H-R) diagram. That is how the distances to most stars are found.

Using the preceding equations and the definition of absolute magnitude, we can get to the following relatively simple expression:

$$m - M = 5 \log_{10} d - 5$$

where distance d is in parsecs.

We can rewrite that equation to solve for distance as follows:

$$d = 10^{\frac{(m - M_{abs} + 5)}{5}}$$

The following table shows how the difference between an object's apparent and absolute magnitudes leads to its distance in parsecs.

$m - M$	Distance (pc)
-3	2.5
-2	4.0
-1	6.3
0	10
1	16
2	25
3	40
4	63
5	100
10	1000
15	10,000
20	100,000

Although the system of astronomical magnitudes is convenient in many ways—which is why astronomers continue to use it—it can also be confusing to new students. Just remember three things and you will probably get by:

1. The greater the magnitude, the fainter the object.
2. One magnitude *smaller* means about two and a half times *brighter*.
3. The brightest stars in the sky have magnitudes of less than 1, and the faintest stars visible to the naked eye on a dark night have magnitudes of about 6.

A final note: Astronomers sometimes use "colors" based on the ratio of the brightness of a star as seen in two parts of the spectrum. The "b_B/b_V color," for example, is the ratio of the brightness of a star seen through a blue filter, divided by the brightness of a star seen through a yellow-green (visual) filter. Normally, astronomers instead discuss the "$B - V$ color" of a star, which is equal to the difference between a star's blue magnitude and its visual magnitude. We can use the previous expression for a magnitude difference to write

$$B - V \text{ color} = m_B - m_V = -2.5 \log_{10} (b_B/b_V)$$

Thus, a star with a b_B/b_V color of 1.0 has a $B - V$ color of 0.0, and a star with a b_B/b_V color of 1.4 has a $B - V$ color of -0.37. Notice that, as with magnitudes, $B - V$ colors are "backward": the bluer a star, the greater its b_B/b_V color but the less its $B - V$ color.

UNIFORM CIRCULAR MOTION AND CIRCULAR ORBITS

Uniform Circular Motion

In Chapter 4 (see Section 4.2 and Figure 4.7), we discuss the motion of an object moving in a circle at a constant speed. That motion, called **uniform circular motion**, is the result of the fact that centripetal force always acts toward the center of the circle. The key question when thinking about uniform circular motion is, How hard does something have to pull to keep the object moving in a circle? Part of the answer to that question is obvious: the more massive an object is, the harder it will be to keep it moving on its circular path. According to Newton's second law of motion, $F = ma$, or here, the centripetal force equals the mass multiplied by the centripetal acceleration. The larger the mass, the greater the force required to keep it moving in its circle.

The centripetal force needed to keep an object moving in constant circular motion also depends on two other quantities: the speed of the object and the size of the circle. The faster an object is moving, the more rapidly it has to change direction to stay on a circle of a given size. The second quantity that influences the needed acceleration is the radius of the circle. The smaller the circle, the greater the pull needed to keep it on track. You can understand that relationship by looking at the motion. A small circle requires a continuous "hard" turn, whereas a larger circle requires a more gentle change in direction. It takes more force to keep an object moving faster in a smaller circle than it does to keep the same object moving more slowly in a larger circle. (To get a better feel for how that works, think about the difference between riding in a car taking a tight curve at high speed and a car moving slowly around a gentle curve.)

To arrive at the circular velocity and other results discussed in Chapter 4, those intuitive ideas about uniform circular motion are turned into a quantitative expression of exactly how much centripetal acceleration is needed to keep an object moving in a circle with radius r at speed v. **Figure A8.1** shows a ball moving around a circle of radius r at a constant speed v at two times. The centripetal acceleration keeping the ball on the circle is a. Remember that the acceleration is always directed toward the center of the circle, whereas the velocity of the ball is always perpendicular to the acceleration. The ball's velocity and its acceleration are always at right angles to each other. As the object moves around the circle,

the direction of motion and the direction of the acceleration change together in lockstep.

Figure A8.1 contains two triangles. Triangle 1 shows the velocity (speed and direction) at both times. The arrow labeled "Δv" connecting the heads of the two velocity arrows shows how much the velocity changed between time 1 (t_1) and time 2 (t_2). That change is the effect of the centripetal acceleration. If you imagine that points 1 and 2 are very close together—so close that the direction of the centripetal acceleration does not change by much between the two—the centripetal acceleration equals the change in the velocity divided by the time between the two, $\Delta t = t_2 - t_1$. So, $\Delta v = a\Delta t$.

Triangle 2 shows something similar. Here, the arrow labeled "Δr" indicates the change in the position of the ball between time 1 and time 2. Again, if you imagine that the time between the two points is very short, Δr is equal to the velocity multiplied by the time, or $\Delta r = v\Delta t$.

The line between the center of the circle and the ball is always perpendicular to the velocity of the ball. So if the direction of the ball's velocity changes by an angle θ, the direction of the line between the ball and the center of the circle must also change by the same angle α. In other words, triangles 1 and 2 are "similar triangles." They have

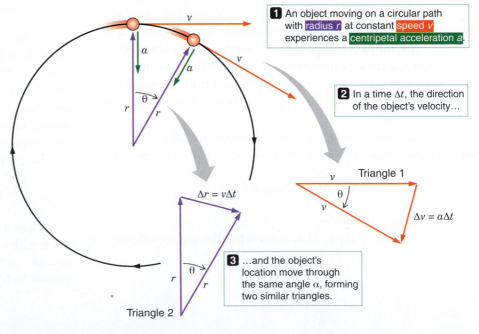

1 An object moving on a circular path with radius r at constant speed v experiences a centripetal acceleration a.

2 In a time Δt, the direction of the object's velocity…

Triangle 1

$\Delta v = a\Delta t$

$\Delta r = v\Delta t$

3 …and the object's location move through the same angle α, forming two similar triangles.

Triangle 2

Figure A8.1 Similar triangles are used to find the centripetal force needed to keep an object moving at a constant speed on a circular path.

the same *shape*. If the triangles are the same shape, the ratio of two sides of triangle 1 must equal the ratio of the two corresponding sides of triangle 2. Then,

$$\frac{a\Delta t}{v} = \frac{v\Delta t}{r}$$

If we divide by Δt on both sides of the equation and then cross-multiply, we obtain

$$ar = v^2$$

which, after we divide both sides of the equation by r, becomes

$$a_{\text{centripetal}} = \frac{v^2}{r}$$

The subscript "centripetal" is added to a to signify that this is the centripetal acceleration needed to keep the object moving in a circle of radius r at speed v. The centripetal force required to keep an object of mass m moving on such a circle is then

$$F_{\text{centripetal}} = ma_{\text{centripetal}} = \frac{mv^2}{r}$$

Circular Orbits

For an object moving in a circular orbit, no string is available to hold the ball on its circular path. Instead, that force is provided by **gravity**.

Think about an object with mass m in orbit about a much larger object with mass M. The orbit is circular, and the distance between the two objects is given by r. The force needed to keep the smaller object moving at speed v in a circle with radius r is given by the previous expression for $F_{\text{centripetal}}$. The force actually provided by gravity (see Chapter 4) is

$$F_{\text{grav}} = G\frac{Mm}{r^2}$$

If gravity is responsible for holding the mass in its circular motion, it should be true that $F_{\text{grav}} = F_{\text{centripetal}}$. That is, if mass m is moving in a circle under the force of gravity, the force provided by gravity must equal the centripetal force needed to explain that circular motion. Setting the two expressions for $F_{\text{centripetal}}$ and F_{grav} equal to each other gives

$$\frac{mv^2}{r} = G\frac{Mm}{r^2}$$

All that remains is a bit of algebra. Dividing by m on both sides of the equation and multiplying both sides by r gives

$$v^2 = G\frac{M}{r}$$

Taking the square root of both sides then yields the desired result:

$$v_{\text{circ}} = \sqrt{\frac{GM}{r}}$$

That is the **circular velocity** we presented in Chapter 4. It is the velocity at which an object in a circular orbit must be moving. If the object were not moving at that velocity, gravity would not be providing the needed centripetal force, and the object would not move in a circle.

A

aberration of starlight A star's apparent displacement in position due to the finite speed of light and Earth's orbital motion around the Sun. (Ch. 18)

absolute magnitude A measure of the intrinsic brightness, or luminosity, of a celestial object, generally a star. Specifically, the apparent magnitude an object would have if it were located at a standard distance of 10 parsecs (pc). Compare *apparent magnitude*. (Ch. 13)

absolute zero The temperature at which thermal motions cease. The lowest possible temperature. Zero on the Kelvin temperature scale. (Ch. 5)

absorption The process by which an atom captures energy from a passing photon. Compare *emission*. (Ch. 5)

absorption line A minimum in the intensity of a spectrum that is due to the absorption of electromagnetic radiation at a specific wavelength determined by the energy levels of an atom or molecule. Compare *emission line*. (Ch. 5)

acceleration (*a*) The rate at which the speed and/or direction of an object's motion is changing. (Ch. 3)

accretion disk A flat, rotating disk of gas and dust surrounding an object, such as a young stellar object, a forming planet, a collapsed star in a binary system, or a black hole. (Ch. 7)

achondrite A stony meteorite that does not contain chondrules. Compare *chondrite*. (Ch. 12)

active comet A comet nucleus that approaches close enough to the Sun to show signs of activity, such as the production of a coma and tail. (Ch. 12)

active galactic nucleus (AGN) A highly luminous, compact galactic nucleus whose luminosity may exceed that of the rest of the galaxy. (Ch. 19)

active region An area of the Sun's chromosphere anchoring bursts of intense magnetic activity. (Ch. 14)

adaptive optics Electro-optical systems that largely compensate for image distortion caused by Earth's atmosphere. (Ch. 6)

AGB See *asymptotic giant branch*. (Ch. 16)

AGN See *active galactic nucleus*. (Ch. 19)

albedo The fraction of electromagnetic radiation striking a surface that is reflected by that surface. (Ch. 5)

algebra A branch of mathematics in which letters represent numeric variables. (Appendix 1)

alpha particle A ^{4}He nucleus, consisting of two protons and two neutrons. Alpha particles are given off in the type of radioactive decay referred to as *alpha decay*. (Ch. 16)

altitude The location of an object above the horizon, measured by the angle formed between an imaginary line from an observer to the object and a second line from the observer to the point on the horizon directly below the object. (Ch. 2)

Amors A group of asteroids whose orbits cross the orbit of Mars but not the orbit of Earth. Compare *Apollos* and *Atens*. (Ch. 12)

amplitude In a wave, the maximum deviation from its undisturbed or relaxed position. For example, in a water wave the amplitude is the vertical distance from the wave's crest to the undisturbed water level. (Ch. 5)

angular momentum In a rotating or revolving system, a conserved property whose value depends on the velocity and distribution of the system's mass. (Ch. 7)

angular resolution The ability of an imaging device such as a telescope (or the eye) to separate two objects that appear close together. (Ch. 6)

annular solar eclipse The type of solar eclipse that occurs when the apparent diameter of the Moon is less than that of the Sun, leaving a visible ring of light ("annulus") surrounding the dark disk of the Moon. Compare *partial solar eclipse* and *total solar eclipse*. (Ch. 2)

Antarctic Circle The circle on Earth with latitude 66.5° south, marking the northern limit where at least one day per year is in 24-hour daylight. Compare *Arctic Circle*. (Ch. 2)

anthropic principle The idea that this universe (or this bubble in the universe) must have physical properties that allow intelligent life to develop. (Ch. 22)

anthropogenic climate change The release of greenhouse gases into the atmosphere from human activities, such as the burning of fossil fuels. (Ch. 9)

anticyclonic motion The rotation of a weather system resulting from the Coriolis effect as air moves outward from a region of high atmospheric pressure. Compare *cyclonic motion*. (Ch. 9)

antimatter Matter made up of antiparticles. (Ch. 14, 22)

antiparticle An elementary particle of antimatter identical in mass but opposite in charge and all other properties to its corresponding ordinary matter particle. (Ch. 22)

aperture The clear diameter of a telescope's objective lens or primary mirror. (Ch. 6)

aphelion (pl. aphelia) The point in a solar orbit that is farthest from the Sun. Compare *perihelion*. (Ch. 12)

Apollos A group of asteroids whose orbits cross the orbits of both Earth and Mars. Compare *Amors* and *Atens*. (Ch. 12)

apparent daily motion As seen from Earth's surface, the path along which each object seems to move across the sky. (Ch. 2)

apparent magnitude A measure of the apparent brightness of a celestial object, generally a star. Compare *absolute magnitude*. (Ch. 13)

arcminute (arcmin) A minute of arc ('), a unit used to measure angles. An arcminute is 1/60 of a degree of arc. (Ch. 6, 13)

arcsecond (arcsec) A second of arc ("), a unit used to measure very small angles. An arcsecond is 1/60 of an arcminute, or 1/3,600 of a degree of arc. (Ch. 6, 13)

Arctic Circle The circle on Earth with latitude 66.5° north, marking the southern limit where at least one day per year is in 24-hour daylight. Compare *Antarctic Circle*. (Ch. 2)

asteroid Also called *minor planet*. A primitive rocky or metallic body (planetesimal) that has survived planetary accretion. Asteroids are parent bodies of meteoroids. (Ch. 12)

asteroid belt The region between the orbits of Mars and Jupiter that contains most of the asteroids in our Solar System. (Ch. 12)

astrobiology An interdisciplinary science combining astronomy, biology, chemistry, geology, and physics to study life in the cosmos. (Ch. 1, 24)

astrology The belief that the positions and aspects of stars and planets influence human affairs and characteristics, as well as terrestrial events. (Ch. 1)

astronomical seeing A measurement of the degree to which Earth's atmosphere degrades the resolution of a telescope's view of astronomical objects. (Ch. 6)

astronomical unit (AU) The average distance from the Sun to Earth: approximately 150 million kilometers (km). (Ch. 3)

astronomy The scientific study of planets, stars, galaxies, and the universe as a whole. (Ch. 1)

astrophysics The application of physical laws to the understanding of planets, stars, galaxies, and the universe as a whole.

asymptotic giant branch (AGB) The path on the H-R diagram that goes from the horizontal branch toward higher luminosities and lower temperatures, asymptotically approaching and then rising above the red giant branch. (Ch. 16)

Atens A group of asteroids whose orbits cross the orbit of Earth but not the orbit of Mars. Compare *Amors* and *Apollos*. (Ch. 12)

atmosphere The gravitationally bound outer gaseous envelope surrounding a planet, moon, or star. (Ch. 9)

atmospheric greenhouse effect A warming of planetary surfaces produced by atmospheric gases that transmit optical solar radiation but partially trap infrared radiation. Compare *greenhouse effect*. (Ch. 9)

atmospheric probe An instrumented package designed to provide on-site measurements of the chemical and/or physical properties of a planetary atmosphere. (Ch. 6)

atmospheric window A region of the electromagnetic spectrum in which radiation can penetrate a planet's atmosphere. (Ch. 6)

atom The smallest unit of a chemical element that retains the properties of that element. Each atom is composed of a nucleus (neutrons and protons) surrounded by a cloud of electrons.

AU See *astronomical unit*. (Ch. 3)

aurora Emission in the upper atmosphere of a planet from atoms that have been excited by collisions with energetic particles from the planet's magnetosphere. (Ch. 9)

autumnal equinox 1. One of two points where the Sun crosses the celestial equator. 2. The day on which the Sun appears at that location, marking the first day of autumn (about September 22 in the Northern Hemisphere and March 20 in the Southern Hemisphere). Compare *vernal equinox*. See also *summer solstice* and *winter solstice*. (Ch. 2)

axion A hypothetical elementary particle first proposed to explain certain properties of the neutron and now considered a candidate for cold dark matter. (Ch. 23)

B

backlighting Illumination from behind a subject as seen by an observer. Fine material such as human hair and dust in planetary rings stands out best when viewed under backlighting conditions. (Ch. 11)

bar A unit of pressure. One bar is equivalent to 10^5 newtons per square meter—approximately equal to Earth's atmospheric pressure at sea level. (Ch. 9)

barred spiral (BS) galaxy A spiral galaxy with a bulge having an elongated, barlike shape. Compare *elliptical galaxy*, *irregular galaxy*, *S0 galaxy*, and *spiral galaxy*. (Ch. 19)

basalt Gray to black volcanic rock, rich in iron and magnesium. (Ch. 8)

beta decay 1. The decay of a neutron into a proton by emission of an electron (beta ray) and an antineutrino. 2. The decay of a proton into a neutron by emission of a positron and a neutrino.

Big Bang The event that occurred 13.8 billion years ago that marks the beginning of time and the universe. (Ch. 1, 21)

Big Bang nucleosynthesis The formation of low-mass nuclei (H, He, Li, Be) during the first few minutes after the Big Bang. (Ch. 21)

Big Crunch A hypothetical cosmic future in which the expansion of the universe reverses and the universe collapses onto itself. (Ch. 22)

Big Rip A hypothetical cosmic future in which all matter in the universe, from stars to subatomic particles, is progressively torn apart by expansion of the universe. (Ch. 22)

binary star A system in which two stars are in gravitationally bound orbits about their common center of mass. (Ch. 13)

binding energy The minimum energy required to separate an atomic nucleus into its component protons and neutrons. (Ch. 17)

biosignature Chemical compounds detectable in the spectra of an exoplanet's atmosphere that might indicate the presence of life. (Ch. 24)

biosphere The global sum of all living organisms on Earth (or any planet or moon). Compare *hydrosphere* and *lithosphere*.

bipolar outflow Material streaming away in opposite directions from both sides of the accretion disk of a young star. (Ch. 15)

black hole An object so dense that its escape velocity exceeds the speed of light; a *singularity* in spacetime. (Ch. 18)

blackbody An object that absorbs and can reemit all electromagnetic energy it receives. (Ch. 5)

blackbody spectrum See *Planck spectrum*. (Ch. 5)

blue-green algae See *cyanobacteria*. (Ch. 24)

blueshift The Doppler shift toward shorter (bluer) wavelengths of light from an approaching object. Compare *redshift*. (Ch. 5)

Bohr model A model of the atom, proposed by Niels Bohr in 1913, in which a small positively charged nucleus is surrounded by orbiting electrons, similar to a miniature solar system. (Ch. 5)

bound orbit An orbit in which an object is gravitationally bound to the body it is orbiting. A bound orbit's velocity is less than the escape velocity. Compare *unbound orbit*. (Ch. 4)

bow shock The boundary at which the speed of the solar wind abruptly drops from supersonic to subsonic in its approach to a planet's magnetosphere; the boundary between the region dominated by the solar wind and the region dominated by a planet's magnetosphere.

brightness The apparent intensity of light from a luminous object. Brightness depends on both the *luminosity* of a source and its distance. Units at the detector: watts per square meter (W/m^2).

brown dwarf A "failed star" without enough mass to fuse hydrogen in its core. An object whose mass is intermediate between that of the least massive stars and that of supermassive planets. (Ch. 7, 15, 16)

bulge The central region of a spiral galaxy that is similar in appearance to a small elliptical galaxy. (Ch. 19)

C

C See *Celsius*.

C-type asteroid An asteroid made of material that has remained mostly unmodified since the formation of the Solar System; the most primitive type of asteroid. Compare *M-type asteroid* and *S-type asteroid*. (Ch. 12)

caldera The summit crater of a volcano.

Cambrian explosion The spectacular rise in the number and diversity of biological species that occurred between 540 million and 500 million years ago. (Ch. 24)

carbon-nitrogen-oxygen (CNO) cycle One way in which hydrogen is converted to helium (hydrogen fusion in the interiors of main-sequence stars. Compare *proton-proton chain*. (Ch. 17)

carbon star A cool red giant or asymptotic giant branch star that has an excess of carbon in its atmosphere.

carbonaceous chondrite A primitive stony meteorite that contains chondrules and is rich in carbon and volatile materials. (Ch. 12)

Cassini Division The largest gap in Saturn's rings, discovered by Jean-Dominique Cassini in 1675. (Ch. 11)

catalyst An atomic and molecular structure that permits or encourages chemical and nuclear reactions but does not change its own chemical or nuclear properties. (Ch. 9)

CCD See *charge-coupled device*. (Ch. 6)

celestial equator The imaginary great circle that is the projection of Earth's equator onto the celestial sphere. (Ch. 2)

celestial sphere An imaginary sphere with celestial objects on its inner surface and Earth at its center. The celestial sphere has no physical existence but is a convenient tool for picturing the directions in which celestial objects are seen from Earth's surface. (Ch. 2)

Celsius (C) Also called *centigrade scale*. The arbitrary temperature scale, formulated by Anders Celsius (1701–1744), that defines 0°C as the freezing point of water and 100°C as the boiling point of water at sea level. Unit: degrees Celsius (°C). Compare *Fahrenheit* and *Kelvin temperature scale*.

center of mass 1. The weighted average location of all the mass in a system of objects. The point in any isolated system that moves according to Newton's first law of motion. 2. In a binary star system, the point between the two stars that is the focus of both their elliptical orbits. (Ch. 4, 13)

centigrade scale See *Celsius*.

centripetal force A force directed toward the center of curvature of an object's curved path. (Ch. 4)

Cepheid variable An evolved high-mass star with an atmosphere that is pulsating, leading to variability in the star's luminosity and color. (Ch. 17)

Chandrasekhar limit The upper limit on the mass of an object supported by electron degeneracy pressure; approximately 1.4 solar masses (M_{Sun}). (Ch. 16)

chaotic Behavior in complex systems in which a small change in the initial state of a system can lead to a large change in the final state of the system. (Ch. 7)

charge-coupled device (CCD) A common type of solid-state detector of electromagnetic radiation that transforms the intensity of light directly into electric signals. (Ch. 6)

charge destruction Electrons are absorbed by protons in atomic nuclei, forming neutrons and releasing neutrinos. (Ch. 17)

chondrite A stony meteorite that contains chondrules. Compare *achondrite*. (Ch. 12)

chondrule A small, crystallized, spherical inclusion of rapidly cooled molten droplets found inside some meteorites. (Ch. 12)

chromatic aberration A detrimental property of a lens in which rays of different wavelengths are brought to different focal distances from the lens. (Ch. 6)

chromosphere The region of the Sun's atmosphere located between the *photosphere* and the *corona*. (Ch. 14)

circular velocity The orbital velocity needed to keep an object moving in a circular orbit. (Ch. 4)

circumpolar Describing the part of the sky, near either celestial pole, that can always be seen above the horizon from a specific location on Earth. (Ch. 2)

climate The state of an atmosphere averaged over an extended time. Compare *weather*. (Ch. 9)

closed universe A finite universe with a curved spatial structure such that the sum of the angles of a triangle always exceeds 180 degrees. Compare *flat universe* and *open universe*. (Ch. 22)

CMB See *cosmic microwave background radiation*. (Ch. 6, 21)

CNO cycle See *carbon-nitrogen-oxygen cycle*. (Ch. 17)

cold dark matter Particles of dark matter that move slowly enough to be gravitationally bound even in the smallest galaxies. Compare *hot dark matter*. (Ch. 23)

color index The color of a celestial object, generally a star, based on the ratio of its brightness in blue light to its brightness in "visual" (yellow-green) light. The difference between an object's blue (B) magnitude and visual (V) magnitude, $B − V$. (Ch. 13, Appendix 7)

coma (pl. **comae**) The nearly spherical cloud of gas and dust surrounding the nucleus of an active comet. (Ch. 12)

comet A complex object consisting of a small, solid, icy nucleus; an atmospheric halo; and a tail of gas and dust. (Ch. 7)

comet nucleus A primitive planetesimal composed of ices and refractory materials that has survived planetary accretion. The "heart" of a comet, containing nearly the entire mass of the comet. A "dirty snowball." (Ch. 7)

comparative planetology The study of planets by comparing their chemical and physical properties. (Ch. 8)

composite volcano A large, cone-shaped volcano formed by viscous, pasty lava flows alternating with pyroclastic (explosively generated) rock deposits. Compare *shield volcano*. (Ch. 8)

compound lens A lens made up of two or more elements with different refractive indices, the purpose of which is to minimize chromatic aberration. (Ch. 6)

concave mirror A telescope mirror with a surface that curves inward toward the incoming light. (Ch. 6)

conduction The transfer of energy in which the thermal energy of particles is transferred to adjacent particles by collisions or other interactions. Conduction is the most important way that thermal energy is transported in solid matter. Compare *convection*. (Ch. 14)

conservation law A physical law stating that the amount of a particular physical quantity (such as energy or angular momentum) of an isolated system does not change.

conservation of angular momentum The physical law stating that the amount of angular momentum of an isolated system does not change. (Ch. 7)

conservation of energy The physical law stating that the amount of energy of an isolated, closed system does not change. (Ch. 7)

constant of proportionality The number by which one quantity is multiplied to get another number.

constellation An imaginary image formed by patterns of stars; any of 88 defined areas on the celestial sphere that astronomers use to locate celestial objects. (Ch. 2)

continental drift The slow motion (centimeters per year) of Earth's continents relative to one another and to Earth's mantle. See also *plate tectonics*. (Ch. 8)

continuous radiation Electromagnetic radiation with intensity that varies smoothly over a wide range of wavelengths.

continuous spectrum A spectrum containing all wavelengths, without specific spectral lines. (Ch. 5)

convection The transport of thermal energy from the lower (hotter) to the higher (cooler) layers of a fluid by motions within the fluid driven by variations in buoyancy. Compare *conduction*. (Ch. 8, 14)

convective zone A region within a star where energy is transported outward by convection. Compare *radiative zone*. (Ch. 14)

core 1. The innermost region of a planetary interior. Compare *crust* and *mantle*. 2. The innermost part of a star. (Ch. 8, 14)

core accretion–gas capture A process for forming giant planets, in which large amounts of surrounding hydrogen and helium gas are gravitationally captured onto a massive rocky core. (Ch. 7)

Coriolis effect The apparent displacement of objects in a direction perpendicular to their true motion as viewed from a rotating frame of reference. On a rotating planet, that effect arises from different latitudes' rotating at different speeds. (Ch. 2)

corona The hot, outermost part of the Sun's atmosphere. Compare *chromosphere* and *photosphere*. (Ch. 14)

coronal hole A low-density region in the solar corona containing "open" magnetic-field lines along which coronal material is free to stream into interplanetary space. (Ch. 14)

coronal mass ejection (CME) An eruption on the Sun that ejects hot gas and energetic particles at much higher speeds than are typical in the solar wind. (Ch. 14)

cosmic microwave background radiation (CMB) Also called simply *cosmic background radiation*. Isotropic microwave radiation from every direction in the sky having a 2.73-kelvin (K) blackbody spectrum. The CMB is residual radiation from the Big Bang. (Ch. 6, 21)

cosmic ray A very fast-moving particle (usually protons or another atomic nucleus) that originated in outer space; cosmic rays fill the disk of the Milky Way. (Ch. 18, 20)

cosmological constant A constant, introduced into general relativity by Einstein, that characterizes an extra, repulsive force in the universe due to the vacuum of space itself. (Ch. 22)

cosmological principle The (testable) assumption that the same physical laws that apply here and now also apply everywhere and at all times and that the universe has no special locations or directions. (Ch. 1)

cosmological redshift The redshift that results from the expansion of the universe rather than from the motions of galaxies or gravity. Compare *gravitational redshift*.

cosmology The study of the large-scale structure and evolution of the universe as a whole. (Ch. 21)

crescent Any phase of the Moon, Mercury, or Venus in which the object appears less than half illuminated by the Sun. Compare *gibbous*.

Cretaceous-Paleogene (K-Pg) boundary The boundary between the Cretaceous and Paleogene periods in Earth's history. That boundary corresponds to the time of the impact of an asteroid or comet and the extinction of the dinosaurs. (Ch. 8)

critical density The value of mass density of the universe that, ignoring any cosmological constant, can just barely halt expansion of the universe. (Ch. 22)

crust The relatively thin, outermost, hard layer of a planet, which is chemically distinct from the interior. Compare *core* and *mantle*. (Ch. 8)

cryovolcanism Low-temperature volcanism in which the magmas are composed of molten ices rather than rocky material. (Ch. 11)

cyanobacteria Also called *blue-green algae*. Single-celled organisms that created oxygen in Earth's atmosphere by photosynthesizing carbon dioxide and releasing oxygen as a waste product. (Ch. 24)

cyclonic motion The rotation of a weather system resulting from the Coriolis effect as air moves toward a region of low atmospheric pressure. Compare *anticyclonic motion*. (Ch. 9)

D

Dark Ages The epoch in the history of the universe during which no visible "light" came from astronomical objects. (Ch. 23)

dark energy A form of energy that permeates all space (including the vacuum), producing a repulsive force that accelerates the expansion of the universe. (Ch. 22)

dark matter Matter in galaxies that does not emit or absorb electromagnetic radiation. Dark matter is thought to constitute most of the mass in the universe. Compare *luminous matter*. (Ch. 19)

dark matter halo The centrally condensed, greatly extended dark matter component of a galaxy that accounts for up to 95 percent of the galaxy's mass. (Ch. 19)

daughter product An element resulting from radioactive decay of a more massive *parent element*. (Ch. 8)

decay 1. The process of a radioactive nucleus changing into its daughter product. 2. The process of an atom or molecule dropping from a higher energy state to a lower energy state. 3. The process of a satellite's orbit losing energy. (Ch. 5)

declination A measure, analogous to *latitude*, that tells you the angular distance of a celestial body north or south of the celestial equator (from 0° to ±90°). Compare *right ascension*. (Ch. 2)

degeneracy pressure Pressure exerted by closely packed electrons in the collapsing core of a star. Such pressure pushes outward against gravity and keeps the star from collapsing further. (Ch. 16)

density The measure of an object's mass per unit of volume. Possible units include kilograms per cubic meter (kg/m³). (Ch. 4, 7)

deuterium An isotope of hydrogren; the nucleus contains one proton and one neutron. (Ch. 14)

differential rotation Rotation of different parts of a system at different rates. (Ch. 14)

differentiation The process by which materials of higher density sink toward the center of a molten or fluid planetary interior. (Ch. 8)

diffraction The spreading of a wave after it passes through an opening or beyond the edge of an object. (Ch. 6)

diffraction grating An optical component with many narrow parallel lines that separate the wavelengths of light to produce a spectrum. (Ch. 6)

diffraction limit The limit of a telescope's angular resolution caused by diffraction. (Ch. 6)

diffuse ring A sparsely populated planetary ring spread out both horizontally and vertically.

dispersion The separation of rays of light into their component wavelengths. (Ch. 6)

distance ladder A sequence of techniques for measuring cosmic distances: each method is calibrated using the results from other methods that have been applied to closer objects. (Ch. 19)

Doppler effect The change in wavelength of sound or light as a result of the relative motion of the source toward or away from the observer. (Ch. 5)

Doppler shift The amount by which the Doppler effect shifts the wavelength of light. (Ch. 5)

Drake equation A prescription for estimating the number of intelligent civilizations existing in the Milky Way Galaxy. (Ch. 24)

dust devil A small tornado-like column of air containing dust or sand. (Ch. 9)

dust tail A type of comet tail consisting of dust particles pushed away from the comet's head by radiation pressure from the Sun. Compare *ion tail*. (Ch. 12)

dwarf galaxy A small galaxy with a luminosity ranging from 1 million to 1 billion solar luminosities (L_{Sun}). Compare *giant galaxy*. (Ch. 19)

dwarf planet A body with characteristics similar to those of a planet except that it has not cleared smaller bodies from the neighboring regions around its orbit. Compare *planet* (definition 2). (Ch. 12)

dynamic equilibrium A state in which a system is constantly changing but its configuration remains the same because one source of change is exactly balanced by another source of change. Compare *static equilibrium*.

dynamo theory A theory postulating that Earth's magnetic field (and those of other planets) is generated from a rotating and electrically conducting liquid core. (Ch. 8)

E

eccentricity (*e*) The ratio of the distance between the two foci of an ellipse to the length of its major axis, which measures how noncircular the ellipse is. (Ch. 3)

eclipse 1. The total or partial obscuration of one celestial body by another. 2. The total or partial obscuration of light from one celestial body as it passes through the shadow of another celestial body. (Ch. 2)

eclipse season Any time during the year when the Moon's line of nodes points towards the Sun and eclipses can occur. (Ch. 2)

eclipsing binary A binary system in which the orbital plane is oriented such that the two stars appear to pass in front of each other as seen from Earth. Compare *spectroscopic binary* and *visual binary*. (Ch. 13)

ecliptic 1. The apparent annual path of the Sun against the background of stars. 2. The projection of Earth's orbital plane onto the celestial sphere. (Ch. 2)

ecliptic plane The plane of Earth's orbit around the Sun. The ecliptic is the projection of that plane onto the celestial sphere. (Ch. 2)

effective temperature The temperature at which a blackbody, such as a star, appears to radiate. (Ch. 14)

Einstein ring Light bent by gravitational lensing into a ring. (Ch. 18)

ejecta 1. Material thrown outward by the impact of an asteroid or comet on a planetary surface, leaving a crater behind. 2. Material thrown outward by a stellar explosion. (Ch. 8)

electric field A field that can exert a force on a charged object, whether at rest or moving. Compare *magnetic field*. (Ch. 5)

electric force The force exerted on electrically charged particles such as protons and electrons, arising from their electric charges. Compare *magnetic force*. See also *electromagnetic force*. (Ch. 5)

electromagnetic force The force, including both electric and magnetic forces, that acts on electrically charged particles. One of four fundamental forces of nature, along with the *strong nuclear force*, *weak nuclear force*, and *gravity* (definition 1). The force is mediated by the exchange of photons. (Ch. 22)

electromagnetic radiation A traveling disturbance in the electric and magnetic fields caused by accelerating electric charges. In quantum mechanics, a stream of photons. Light.

electromagnetic spectrum The spectrum made up of all possible frequencies or wavelengths of electromagnetic radiation, ranging from gamma rays through radio waves and including the portion our eyes can use. (Ch. 5)

electromagnetic wave A wave consisting of oscillations in the electric-field strength and the magnetic-field strength. (Ch. 5)

electron (e⁻) A subatomic particle having a negative electric charge of 1.6×10^{-19} coulomb (C), a rest mass of 9.1×10^{-31} kilogram (kg), and a rest energy of 8×10^{-14} joule (J). The antiparticle of the *positron*. Compare *proton* and *neutron*. (Ch. 5)

electron-degenerate Describing matter, compressed to the point at which electron density reaches the limit imposed by the rules of quantum mechanics. (Ch. 16)

electroweak theory The quantum theory that combines descriptions of both the electromagnetic force and the weak nuclear force. (Ch. 22)

element One of 92 naturally occurring substances (such as hydrogen, oxygen, and uranium) and more than 20 human-made ones (such as plutonium). Each element is chemically defined by the specific number of protons in the nuclei of its atoms. (Ch. 5)

elementary particle One of the basic building blocks of nature that is not known to have substructure, such as the *electron* or the *quark*. (Ch. 23)

ellipse A conic section produced by the intersection of a plane with a cone when the plane is passed through the cone at an angle to the axis other than 0° or 90°. The shape that results when you attach the two ends of a piece of string to a piece of paper, stretch the string tight with the tip of a pencil, and then draw around those two points while keeping the string taut. (Ch. 3)

elliptical galaxy A galaxy of Hubble type "E" class, with a circular to elliptical outline on the sky, and containing almost no disk and a population of old stars. Compare *barred spiral galaxy, irregular galaxy, S0 galaxy*, and *spiral galaxy*. (Ch. 19)

emission The production of a photon when an atom decays to a lower energy state. Compare *absorption*. (Ch. 5)

emission line A peak in the intensity of a spectrum that is due to the emission of electromagnetic radiation at a specific wavelength determined by the energy levels of an atom or molecule. Compare *absorption line*. (Ch. 5)

empirical science Descriptive scientific investigation based primarily on observations and experimental data rather than on theoretical inference. (Ch. 3)

energy The conserved quantity that gives objects and systems the ability to do work. Possible units include joules (J). (Ch. 5)

energy transport The transfer of energy from one location to another. In stars, energy transport is carried out mainly by radiation or convection. (Ch. 14)

entropy A measure of the disorder of a system related to the number of ways a system can be rearranged without its appearance being affected.

equator The imaginary great circle on the surface of a body midway between its poles that divides the body into northern and southern hemispheres. The equatorial plane passes through the center of the body and is perpendicular to its rotation axis. Compare *meridian*. (Ch. 2)

equilibrium The state of an object in which physical processes balance each other so that its properties or conditions remain constant.

equinox Literally, "equal night." 1. One of two positions on the ecliptic where it intersects the celestial equator. 2. Either of the two times of year (the *autumnal equinox* and *vernal equinox*) when the Sun is at one of these two positions. At this time, night and day are of the same length everywhere on Earth. Compare *solstice*.

equivalence principle The principle stating that no difference exists between a frame of reference freely floating through space and one freely falling within a gravitational field. (Ch. 18)

erosion The degradation of a planet's surface topography by the mechanical action of wind, water, or living organisms. (Ch. 8)

escape velocity The minimum velocity needed for an object to achieve a parabolic trajectory and thus permanently leave the gravitational grasp of another mass. (Ch. 4)

eternal inflation The idea that a universe might inflate forever. In such a universe, quantum effects could randomly cause regions to slow their expansion, eventually stop inflating, and experience an explosion resembling the Big Bang. (Ch. 22)

event Something that happens at a particular location in spacetime. (Ch. 18)

event horizon The effective "surface" of a black hole. Nothing inside that surface—not even light—can escape from a black hole. (Ch. 18)

evolutionary track The path that a star follows across the H-R diagram as it evolves through its lifetime. (Ch. 15)

excited state Any energy level of a system or part of a system, such as an atom, molecule, or particle, that is higher than its ground state. Compare *ground state*. (Ch. 5)

exoplanet A planet orbiting a star other than the Sun. (Ch. 7)

exosphere A very thin atmosphere or layer of atmosphere, where the molecules are bound by gravity to the moon or planet but their density is too low to behave like a gas of colliding particles. (Ch. 9)

extrasolar planet See *exoplanet*. (Ch. 7)

extremophile A life-form that thrives under extreme environmental conditions. (Ch. 24)

eyepiece A lens that is closest to the eye in a telescope. Changing the eyepiece will change the magnification of the image in the telescope. (Ch. 6)

F

F See *Fahrenheit*.

Fahrenheit (F) The arbitrary temperature scale, formulated by Daniel Gabriel Fahrenheit (1686–1736), that defines 32°F as the melting point of water and 212°F as the boiling point of water at sea level. Unit: degrees Fahrenheit (°F). Compare *Celsius* and *Kelvin temperature scale*.

falsified A hypothesis shown to be false. (Ch. 1)

fault A fracture in the crust of a planet or moon along which blocks of material can slide. (Ch. 8)

filter An instrument element that transmits a limited wavelength range of electromagnetic radiation. For the optical range, such elements are typically made of different kinds of glass and take on the hue of the light they transmit. (Ch. 13)

first quarter Moon The phase of the Moon in which only the western half of the Moon, as viewed from Earth, is illuminated by the Sun. It occurs about a week after a new Moon. Compare *third quarter Moon*. See also *full Moon* and *new Moon*. (Ch. 2)

fissure A fracture in the planetary lithosphere from which magma emerges. (Ch. 8)

flat universe An infinite universe whose spatial structure obeys Euclidean geometry, such that the sum of the angles of a triangle always equals 180 degrees. Compare *closed universe* and *open universe*. (Ch. 22)

flatness problem The surprising result that the sum of Ω_m plus Ω_Λ is extremely close to 1 in the present-day universe; equivalent to saying that it is surprising the universe is so close to being exactly flat.

flux (*f*) The total amount of energy passing through each square meter of a surface each second. Unit: watts per square meter (W/m²). (Ch. 5)

flux tube A strong magnetic field contained within a tubelike structure. Flux tubes are found in the solar atmosphere and connecting the space between Jupiter and its moon Io. (Ch. 10)

flyby A spacecraft that first approaches and then continues flying past a planet or moon. Flybys can visit multiple objects, but they remain near their targets only briefly. Compare *orbiter*. (Ch. 6)

focal length The optical distance between a telescope's objective lens or primary mirror and the plane (called the focal plane) on which the light from a distant object is focused. (Ch. 6)

focal plane The plane, perpendicular to the optical axis of a lens or mirror, on which an image is formed. (Ch. 6)

focus (pl. foci) 1. One of two points that define an ellipse. 2. A point in the focal plane of a telescope. (Ch. 3)

force (*F*) A push or a pull on an object. (Ch. 3)

frame of reference A coordinate system within which an observer measures positions and motions. (Ch. 2)

free fall The motion of an object when the only force acting on it is gravity. (Ch. 4)

frequency (*f*) The number of times per second that a periodic process occurs. Unit: hertz (Hz), or cycles per second (1/s). (Ch. 5)

full Moon The phase of the Moon in which the near side of the Moon, as viewed from Earth, is fully illuminated by the Sun. It occurs about two weeks after a new Moon. See also *first quarter Moon* and *third quarter Moon*. (Ch. 2)

G

galaxy A gravitationally bound system that consists of stars and star clusters, gas, dust, and dark matter; typically greater than 1,000 light-years across and recognizable as a discrete, single object. (Ch. 19)

galaxy cluster A large, gravitationally bound collection of galaxies containing hundreds to thousands of members; typically 3–5 megaparsecs (Mpc) across. Compare *galaxy group* and *supercluster*. (Ch. 23)

galaxy group A small, gravitationally bound collection of galaxies containing from several to a hundred members; typically 1–2 megaparsecs (Mpc) across. Compare *galaxy cluster* and *supercluster*. (Ch. 20)

gamma ray Also called *gamma radiation*. Electromagnetic radiation with higher frequency, higher photon energy, and shorter wavelength than all other types of electromagnetic radiation. (Ch. 5)

gamma-ray burst (GRB) A brief, intense burst of gamma rays from a distant energetic explosion. (Ch. 18)

gas giant A giant planet formed mostly of hydrogen and helium. In the Solar System, Jupiter and Saturn are the gas giants. Compare *ice giant*. (Ch. 10)

general relativistic time dilation The verified prediction that time passes more slowly in a gravitational field than in the absence of a gravitational field. Compare *time dilation*. (Ch. 18)

general relativity See *general theory of relativity*. (Ch. 18)

general theory of relativity Sometimes referred to as simply *general relativity*. Einstein's theory explaining gravity as the distortion of spacetime by massive objects, such that particles travel on the shortest path between two events in spacetime. This theory deals with all types of motion. Compare *special theory of relativity*. (Ch. 18)

geocentric model A historical cosmological model with Earth at its center, and all the other objects in the universe in orbit around Earth. Compare *heliocentric model*. (Ch. 3)

geodesic The path an object will follow through spacetime in the absence of external forces. (Ch. 18)

giant molecular cloud An interstellar cloud composed primarily of molecular gas and dust and having hundreds of thousands of solar masses. (Ch. 15)

giant planet Also called *Jovian planet*. One of the largest planets in the Solar System (Saturn, Jupiter, Uranus, or Neptune), typically 10 times the size and many times the mass of any *terrestrial planet* and lacking a solid surface. (Ch. 7)

gibbous Any phase of the Moon, Mercury, or Venus in which the object appears more than half illuminated by the Sun. Compare *crescent*.

global circulation The overall, planetwide circulation pattern of a planet's atmosphere. (Ch. 9)

globular cluster A spherically symmetric, highly condensed group of stars, containing tens of thousands to a million members. Compare *open cluster*. (Ch. 17)

gluon The particle that carries (or, equivalently, mediates) interactions due to the strong nuclear force. (Ch. 22)

grand unified theory (GUT) A unified quantum theory that combines the strong nuclear, weak nuclear, and electromagnetic forces but does not include gravity. (Ch. 22)

granite Rock that is cooled from magma and is relatively rich in silicon and oxygen. (Ch. 8)

grating An optical surface containing many narrow, closely and equally spaced parallel grooves or slits that spectrally disperse reflected or transmitted light. (Ch. 6)

gravitational lens A massive object that gravitationally focuses the light of a more distant object to produce multiple brighter, magnified, possibly distorted images.

gravitational lensing The bending of light by gravity. (Ch. 7, 18)

gravitational potential energy The stored energy in an object that is due solely to its position within a gravitational field. (Ch. 7)

gravitational redshift The shifting to longer wavelengths of radiation from an object deep within a gravitational well. Compare *cosmological redshift*. (Ch. 18)

gravitational wave A wave in the faric of spacetime emitted by accelerating masses. (Ch. 6, 18)

gravity 1. The mutually attractive force between massive objects. One of four fundamental forces of nature, along with the *electromagnetic force*, the *strong nuclear force*, and the *weak nuclear force*. 2. An effect arising from the bending of spacetime by massive objects. (Ch. 4)

GRB See *gamma-ray burst*. (Ch. 18)

Great Red Spot The giant, oval, brick red anticyclone seen in Jupiter's southern hemisphere. (Ch. 10)

greenhouse effect The solar heating of air in an enclosed space, such as a closed building or car, resulting primarily from the inability of the hot air to escape. Compare *atmospheric greenhouse effect*. (Ch. 9)

greenhouse gas One of a group of atmospheric gases such as carbon dioxide that are transparent to visible radiation but absorb infrared radiation. (Ch. 9)

greenhouse molecule A molecule such as water vapor or carbon dioxide that transmits visible radiation but absorbs infrared radiation.

Gregorian calendar The modern calendar. A modification of the Julian calendar decreed by Pope Gregory XIII in 1582. By then, the less accurate Julian calendar had developed an error of 10 days over the 13 centuries since its inception. (Ch. 2)

ground state The lowest possible energy state for a system or part of a system, such as an atom, molecule, or particle. Compare *excited state*. (Ch. 5)

GUT See *grand unified theory*. (Ch. 22)

H

H II region A region of interstellar gas that has been ionized by ultraviolet radiation from nearby hot, massive stars. (Ch. 15)

H-R diagram The Hertzsprung-Russell diagram, a plot of the luminosities versus the surface temperatures of stars. The evolving properties of stars are plotted as tracks across the H-R diagram. (Ch. 13)

habitable zone The distance from its star at which a planet must be located to have a temperature suitable for water to exist in a liquid state. (Ch. 7, 13)

Hadley circulation A simplified, and therefore uncommon, atmospheric global circulation that carries thermal energy directly from the equator to the polar regions of a planet. (Ch. 9)

half-life The time that half a sample of a particular radioactive parent element takes to decay into a daughter product. (Ch. 8)

halo The spherically symmetric, low-density distribution of stars and dark matter that defines the outermost regions of a galaxy.

harmonic law See *Kepler's third law*. (Ch. 3)

Hawking radiation Radiation from a black hole. (Ch. 18)

Hayashi track The path that a protostar follows on the H-R diagram as it contracts toward the main sequence. (Ch. 15)

head The part of a comet that includes both the nucleus and the inner part of the coma. (Ch. 12)

heavy element Also called *massive element*. Any element more massive than helium. (Ch. 10, 13)

heliocentric model A model of the Solar System, with the Sun at its center, and the planets, including Earth, in orbit around the Sun. Compare *geocentric model*. (Ch. 3)

helioseismology The use of solar oscillations to study the interior of the Sun. (Ch. 14)

heliosphere A region surrounding the Solar System in which the solar wind blows against the interstellar medium and clears out an area like the inside of a bubble. The heliosphere protects the Solar System from cosmic rays. (Ch. 14)

helium capture A helium nucleus is captured by another nucleus during nucleosynthesis. (Ch. 17)

helium flash The runaway explosive fusion of helium in the degenerate helium core of a red giant star. (Ch. 16)

Herbig-Haro (HH) object A glowing, rapidly moving knot of gas and dust that is excited by bipolar outflows in very young stars. (Ch. 15)

heredity The process by which one generation passes on its characteristics to future generations. (Ch. 24)

hertz (Hz) A unit of frequency equivalent to cycles per second. (Ch. 5)

Hertzsprung-Russell diagram See *H-R diagram*. (Ch. 13)

HH object See *Herbig-Haro object*. (Ch. 15)

hierarchical clustering The "bottom up" process of forming large-scale structure. Small-scale structure first produces groups of galaxies, which in turn form clusters, which then form superclusters. (Ch. 23)

high-mass star A star with a main-sequence mass of greater than about 8 solar masses (M_{Sun}). Compare *low-mass star* and *medium-mass star*. (Ch. 16)

homogeneous In cosmology, describing a universe in which observers in any location would observe the same properties. Compare *isotropic*. (Ch. 21)

horizon The boundary that separates the sky from the ground. (Ch. 2)

horizon problem The puzzling observation that the cosmic background radiation is so uniform in all directions, even though widely separated regions should have been "over the horizon" from each other in the early universe. (Ch. 22)

horizontal branch A region on the H-R diagram defined by stars fusing helium to carbon in a stable core. (Ch. 16)

hot dark matter Particles of dark matter that move so fast that gravity cannot confine them to the volume occupied by a galaxy's normal luminous matter. Compare *cold dark matter*. (Ch. 23)

hot Jupiter A large, Jupiter-type extrasolar planet located very close to its parent star. (Ch. 7)

hot spot A place where hot plumes of mantle material rise near the surface of a planet. (Ch. 8)

Hubble constant (H_0) The constant of proportionality relating the recession velocities of galaxies to their distances. Compare *Hubble time*. (Ch. 19)

Hubble flow The motion of galaxies as a result of the expanding universe. (Ch. 21)

Hubble's law The law stating that the speed at which a galaxy is moving away from Earth is proportional to the distance of that galaxy. (Ch. 19)

Hubble time An estimate of the age of the universe from the inverse of the *Hubble constant*, $1/H_0$. (Ch. 21)

hurricane A large tropical cyclonic system circulating counterclockwise in the Northern Hemisphere and clockwise in the Southern Hemisphere. Hurricanes can extend outward from their center to more than 600 kilometers (km) and generate winds in excess of 300 kilometers per hour (km/h). (Ch. 9)

hydrogen fusion The release of energy from the nuclear fusion of four hydrogen atoms into a single helium atom. (Ch. 14)

hydrogen shell fusion The fusion of hydrogen in a shell surrounding a stellar core that may be either degenerate or fusing more massive elements. (Ch. 16)

hydrosphere The portion of Earth that is largely liquid water. Compare *biosphere* and *lithosphere*. (Ch. 8)

hydrostatic equilibrium The condition in which the weight bearing down at a particular point within an object is balanced by the pressure within the object. (Ch. 8, 14)

hypernova (pl. **hypernovae**) A very energetic supernova from a very high-mass star.

hypothesis A well-considered idea, based on scientific principles and knowledge, that leads to testable predictions. Compare *theory*. (Ch. 1)

Hz See *hertz*. (Ch. 5)

I

ice The solid form of a volatile material; sometimes the *volatile material* itself, in any form. (Ch. 7)

ice giant A giant planet formed mostly of the liquid form of volatile substances (ices). In the Solar System, Uranus and Neptune are the ice giants. Compare *gas giant*. (Ch. 10)

impact crater The scar of the impact left on a solid planetary or moon surface by collision with another object. Compare *secondary crater*. (Ch. 8)

impact cratering The process in which solid planetary objects collide with each other, leaving distinctive scars. (Ch. 8)

index of refraction (n) The ratio of the speed of light in a vacuum (c) to the speed of light in an optical medium (v). (Ch. 6)

inert gas A gaseous element that combines with other elements only under conditions of extreme temperature and pressure. Examples include helium, neon, and argon.

inertia The tendency for objects to retain their state of motion. (Ch. 3)

inertial frame of reference 1. A frame of reference moving in a straight line at constant speed, that is, not accelerating. 2. In general relativity, a frame of reference falling freely in a gravitational field. (Ch. 3)

inertial reference frame See *inertial frame of reference*. (Ch. 18)

inferior planet A Solar System planet that orbits closer to the Sun than Earth does. Compare *superior planet*. (Ch. 3)

inflation An extremely brief phase of ultra-rapid expansion of the very early universe. After inflation, the standard Big Bang models of expansion apply. (Ch. 22)

infrared (IR) radiation Electromagnetic radiation with frequencies, photon energies, and wavelengths between those of visible light and microwaves. (Ch. 5)

instability strip A region of the H-R diagram containing stars that pulsate with a periodic variation in luminosity. (Ch. 17)

integration time The time interval during which photons are collected and added up in a detecting device. (Ch. 6)

intensity Of light, the amount of radiant energy emitted per second per unit area. Units for electromagnetic radiation: watts per square meter (W/m²).

intercloud gas A low-density region of the interstellar medium that fills the space between interstellar clouds. (Ch. 15)

interfere Typically pertaining to light, two sets of waves mutually interact to either amplify or reduce each other. See also *interference*. (Ch. 6)

interference The interaction of two sets of waves producing high and low intensity, depending on whether their amplitudes reinforce (*constructive interference*) or cancel (*destructive interference*). See also *interfere*.

interferometer Linked optical or radio telescopes whose overall separation determines the angular resolution of the system. (Ch. 6)

interferometric array An interferometer made up of several telescopes arranged in an array. (Ch. 6)

interstellar cloud A discrete, high-density region of the interstellar medium made up mostly of atomic or molecular hydrogen and dust. (Ch. 15)

interstellar dust Small particles or grains (0.01–10 microns [μm] in diameter) of matter, primarily carbon and silicates, distributed throughout interstellar space. (Ch. 15)

interstellar extinction The dimming of visible and ultraviolet light by interstellar dust. (Ch. 15)

interstellar gas The tenuous gas, far less dense than air, composing 99 percent of the matter in the interstellar medium. (Ch. 15)

interstellar medium The gas and dust that fill the space between the stars within a galaxy. (Ch. 14, 15)

inverse square law The rule stating that a quantity or effect diminishes with the square of the distance from the source. (Ch. 4)

ion An atom or molecule that has lost or gained one or more electrons. (Ch. 15)

ionize see *ionization*. (Ch. 9, 15)

ionization The process by which electrons are stripped free from an atom or molecule, resulting in free electrons and a positively charged atom or molecule. (Ch. 9, 15)

ionosphere A layer high in Earth's atmosphere in which most atoms are ionized by solar radiation. (Ch. 9)

ion tail A type of comet tail consisting of ionized gas. Particles in the ion tail are pushed directly away from the comet's head in the antisolar direction at high speeds by the solar wind. Compare *dust tail*. (Ch. 12)

IR Infrared. See *infrared radiation*. (Ch. 5)

iron meteorite A metallic meteorite composed mostly of iron-nickel alloys. Compare *stony-iron meteorite* and *stony meteorite*. (Ch. 12)

irregular galaxy A galaxy without regular or symmetric appearance. Compare *barred spiral galaxy*, *elliptical galaxy*, *S0 galaxy*, and *spiral galaxy*. (Ch. 19)

irregular moon A moon that has been captured by a planet instead of having formed along with that planet. Some irregular moons revolve in a direction opposite to the rotation of the planet, and many are in distant, unstable orbits. Compare *regular moon*. (Ch. 11)

isotope A forms of the same chemical element that has the same number of protons but a different number of neutrons. (Ch. 5, 8)

isotropic In cosmology, having the same appearance to an observer in all directions. Compare *homogeneous*. (Ch. 21)

J

J See *joule.*

jansky (Jy) The basic unit of flux density. Unit: watts per square meter per hertz (W/m²/Hz). (Ch. 6)

jet 1. A stream of gas and dust ejected from a comet nucleus by solar heating. 2. A stream of material that moves away from a protostar or active galactic nucleus at hundreds of kilometers per second. (Ch. 15)

joule (J) A unit of energy or work. 1 J = 1 newton meter.

Jovian planet See *giant planet.*

Jy See *jansky.* (Ch. 6)

K

K See *kelvin.*

K-Pg boundary See *Cretaceous-Paleogene boundary.* (Ch. 8)

KBO See *Kuiper Belt object.* (Ch. 12)

kelvin (K) The basic unit of the Kelvin scale of temperature. (Ch. 5)

Kelvin temperature scale The temperature scale, formulated by William Thomson, better known as Lord Kelvin (1824–1907), that uses Celsius-sized degrees but defines 0 K as absolute zero instead of as the melting point of water. Unit: kelvins (K). Compare *Celsius* and *Fahrenheit.* (Ch. 5)

Kepler's first law A rule of planetary motion that Johannes Kepler inferred, stating that planets move in elliptical orbits with the Sun at one focus. (Ch. 3)

Kepler's laws The three rules of planetary motion that Johannes Kepler inferred from data collected by Tycho Brahe. (Ch. 3)

Kepler's second law Also called *law of equal areas.* A rule of planetary motion that Johannes Kepler inferred, stating that a line drawn from the Sun to a planet sweeps out equal areas in equal times as the planet orbits the Sun. (Ch. 3)

Kepler's third law Also called *harmonic law.* A rule of planetary motion that Johannes Kepler inferred, describing the relationship between the period of a planet's orbit and its distance from the Sun. The law states that the square of the period of a planet's orbit, measured in years, is equal to the cube of the semimajor axis of the planet's orbit, measured in astronomical units: $(P_{years})^2 = (A_{AU})^3$. (Ch. 3)

kiloparsec A unit of distance equal to 1,000 parsecs, or 3,260 light-years.

kinetic energy (E_K) The energy of an object resulting from its motions. $E_K = \frac{1}{2}mv^2$ Possible units include joules (J). (Ch. 5, 7)

Kirkwood gap A gap in the main asteroid belt related to orbital resonances with Jupiter. (Ch. 12)

Kuiper Belt A disk-shaped population of comet nuclei extending from Neptune's orbit to perhaps several thousand astronomical units (AU) from the Sun. The highly populated innermost part of the Kuiper Belt has an outer edge approximately 50 AU from the Sun. (Ch. 12)

Kuiper Belt object (KBO) Also called *trans-Neptunian object.* An icy planetesimal (comet nucleus) that orbits within the Kuiper Belt beyond the orbit of Neptune. (Ch. 12)

L

Lambda-CDM The standard model of the Big Bang universe in which most of the energy density of the universe is dark energy (similar to Einstein's cosmological constant), and most of the mass in the universe is cold dark matter. (Ch. 23)

lander An instrumented spacecraft designed to land on a planet or moon. Compare *rover.* (Ch. 6)

large-scale structure Observable aggregates on the largest scales in the universe, including galaxy groups, clusters, and superclusters. (Ch. 23)

latitude The angular distance north (+) or south (–) from the equatorial plane of a nearly spherical body. Compare *longitude.* (Ch. 2)

lava Molten rock flowing out of a volcano during an eruption; also the rock that solidifies and cools from that liquid. (Ch. 8)

law of equal areas See *Kepler's second law.* (Ch. 3)

law of gravitation See *universal law of gravitation.* (Ch. 4)

leap year A year that contains 366 days. Leap years occur every 4 years when the year is divisible by 4, correcting for the accumulated excess time in a normal year, which is approximately 365 1/4 days long. (Ch. 2)

length contraction The relativistic compression of moving objects in the direction of their motion. (Ch. 18)

Leonids A November meteor shower associated with the dust debris left by comet Tempel-Tuttle. (Ch. 12)

life A biochemical process in which living organisms can reproduce, evolve, and sustain themselves by drawing energy from their environment. All terrestrial life involves carbon-based chemistry, assisted by the self-replicating molecules ribonucleic acid (RNA) and deoxyribonucleic acid (DNA). (Ch. 24)

light All electromagnetic radiation, which composes the entire electromagnetic spectrum.

light-year (ly) The distance that light travels in 1 year—about 9.5 trillion kilometers (km). (Ch. 1)

limb The outer edge of the visible disk of a planet, moon, or the Sun. (Ch. 14)

limb darkening The darker appearance caused by increased atmospheric absorption near the limb of a planet or star. (Ch. 14)

limestone A common sedimentary rock composed of calcium carbonate. (Ch. 9)

line of nodes 1. A line defined by the intersection of two orbital planes. 2. The line defined by the intersection of Earth's equatorial plane and the plane of the ecliptic. (Ch. 2)

lithosphere The solid, brittle part of Earth (or any planet or moon), including the crust and the upper part of the mantle. Compare *biosphere* and *hydrosphere.* (Ch. 8)

lithospheric plate A separate piece of Earth's lithosphere that can move independently. See also *continental drift* and *plate tectonics.* (Ch. 8)

Local Group The group of galaxies that includes the Milky Way and Andromeda galaxies. (Ch. 1, 20)

long-period comet A comet with an orbital period of greater than 200 years. Compare *short-period comet.* (Ch. 12)

longitude The angular distance east (+) or west (−) from the prime meridian at Greenwich, England. Compare *latitude.* (Ch. 2)

longitudinal wave A wave that oscillates parallel to the direction of the wave's propagation. Compare *transverse wave.* (Ch. 8)

look-back time The amount of time that the light from an astronomical object has taken to reach Earth. (Ch. 21)

low-mass star A star with a main-sequence mass of less than about 3 solar masses (M_{Sun}). Compare *high-mass star* and *medium-mass star.* (Ch. 16)

luminosity The total amount of light emitted by an object. Unit: watts (W). Compare *brightness.* (Ch. 5)

luminosity class A spectral classification based on stellar size, ranging from supergiants at the large end to white dwarfs at the small end. (Ch. 13)

luminosity-temperature-radius relationship A relationship among those three properties of stars indicating that if any two are known, the third can be calculated. (Ch. 13)

luminous matter Also called *normal matter.* Matter in galaxies—including stars, gas, and dust—that emits electromagnetic radiation. Compare *dark matter.* (Ch. 19)

lunar eclipse An eclipse that occurs when the Moon is partially or entirely in Earth's shadow. Compare *solar eclipse.* (Ch. 2)

lunar tide A tide on Earth caused by the differential gravitational pull of the Moon. Compare *solar tide.* (Ch. 4)

lunisolar calendar Calendar created by the Babylonians in which a month began with the first sighting of the lunar crescent, and a 13th month was added when needed to catch up to the solar year. (Ch. 2)

ly See *light-year.* (Ch. 1)

M

μm See *micron.* (Ch. 5)

M-type asteroid An asteroid made of material that was once part of the metallic core of a larger, differentiated body that has since broken into pieces; made mostly of iron and nickel. Compare *C-type asteroid* and *S-type asteroid*. (Ch. 12)

MaCHO Short for *massive compact halo object*. MaCHOs include brown dwarfs, white dwarfs, and black holes and are candidates for dark matter. Compare *WIMP*. (Ch. 19)

magma Molten rock, often containing dissolved gases and solid minerals.

magnetic field A field that can exert a force on a moving electric charge. Compare *electric field*. (Ch. 5)

magnetic force The force exerted on electrically charged particles such as protons and electrons, arising from their motion. Compare *electric force*. See also *electromagnetic force*. (Ch. 5)

magnetosphere The region surrounding a planet that is filled with relatively intense magnetic fields and plasmas. (Ch. 8)

magnitude A system used by astronomers to describe the brightness or luminosity of stars. The brighter the star, the lower its magnitude. (Ch. 13)

major axis The long axis of an ellipse. (Ch. 3)

main asteroid belt See *asteroid belt*. (Ch. 12)

main sequence The strip on the H-R diagram where most stars are found. Main-sequence stars are fusing hydrogen to helium in their cores. (Ch. 13)

main-sequence lifetime The amount of time a star spends on the main sequence, fusing hydrogen into helium in its core. (Ch. 16)

main-sequence turnoff The location on the main sequence of an H-R diagram made from a population of stars of the same age (such as a star cluster) where stars are just evolving off the main sequence. That location is determined by the age of the population of stars. (Ch. 17)

mantle The solid portion of a rocky planet that lies between the *crust* and the *core*. (Ch. 8)

mare (pl. **maria**) A dark region on the Moon composed of basaltic lava flows. (Ch. 8)

mass 1. Inertial mass: the property of matter that determines its resistance to changes in motion. Compare *weight*. 2. Gravitational mass: the property of matter defined by its attractive force on other objects. According to general relativity, the two are equivalent. (Ch. 3, 4)

mass-luminosity relationship An empirical relationship between the luminosity (L) and mass (M) of main-sequence stars; for example, $L \propto M^{3.5}$. (Ch. 13, 16)

mass transfer The transfer of mass from one member of a binary star system to its companion. Mass transfer occurs when one of the stars evolves to the point that it overfills its Roche lobe, so that its outer layers are pulled toward its binary companion. (Ch. 16)

massive element Also called *heavy element*. Any element more massive than helium. (Ch. 13)

matter 1. Objects made of particles that have mass, such as protons, neutrons, and electrons. 2. Anything that occupies space and has mass. (Ch. 5)

Maunder Minimum The period from 1645 to 1715 during which very few sunspots were observed. (Ch. 14)

medium The substance that a wave, such as light, travels through—for example, air or glass. Compare *vacuum*. (Ch. 5)

medium-mass star A star with a main-sequence mass between 3 and 8 solar masses (M_{Sun}). Compare *high-mass star* and *low-mass star*. (Ch. 16)

megaparsec (**Mpc**) A unit of distance equal to 1 million parsecs, or 3.26 million light-years. (Ch. 19)

meridian The imaginary arc in the sky running from the horizon at due north through the zenith to the horizon at due south. The meridian divides the observer's sky into eastern and western hemispheres. Compare *equator*. (Ch. 2)

mesosphere The layer of Earth's atmosphere immediately above the stratosphere, extending from an altitude of 50 kilometers (km) to about 90 km. Compare *troposphere*, *stratosphere*, and *thermosphere*. (Ch. 9)

meteor A meteoroid that enters and burns up in a planetary atmosphere, often leaving an incandescent trail. Compare *meteorite* and *meteoroid*. (Ch. 8)

meteor shower A larger-than-normal display of meteors, occurring when Earth passes through the orbit of a disintegrating comet, sweeping up its debris. Compare *sporadic meteor*. (Ch. 12)

meteorite A piece of rock or other fragment of material (a meteoroid) that survives to reach a planet's surface. Compare *meteor* and *meteoroid*. (Ch. 7, 8)

meteoroid A small cometary or asteroidal fragment, ranging in size from 100 microns (μm) to 100 meters. When entering a planetary atmosphere, the meteoroid creates a *meteor*. Compare *meteor* and *meteorite*; also *planetesimal* and *zodiacal dust*. (Ch. 8)

micrometer (μm) See *micron*. (Ch. 5)

micron (μm) One-millionth (10^{-6}) of a meter; a unit of length used for the wavelength of infrared light. (Ch. 5)

microwave radiation Electromagnetic radiation with frequencies, photon energies, and wavelengths between those of infrared radiation and radio waves. (Ch. 5)

Milky Way Galaxy The galaxy in which the Sun and Solar System reside. (Ch. 1)

minor axis The short axis of an ellipse, perpendicular to the major axis. (Ch. 3)

minor planet See *asteroid*.

model A simplified mathematical or conceptual representation of a physical system used to carry out calculations or predictions. (Ch. 1)

molecular cloud An interstellar cloud composed primarily of molecular hydrogen. (Ch. 15)

molecular-cloud core A dense clump within a molecular cloud that forms as the cloud collapses and fragments. Protostars form from molecular-cloud cores. (Ch. 15)

molecule Generally, the smallest particle of a substance that retains its chemical properties and is composed of two or more atoms. (Ch. 5)

momentum The product of the mass and velocity of a particle. Possible units include kilograms times meters per second (kg m/s). (Ch. 5)

moon A less massive satellite orbiting a more massive object. Moons are found around planets, dwarf planets, asteroids, and Kuiper Belt objects. The term is usually capitalized when referring to Earth's Moon. (Ch. 7)

Mpc See *megaparsec*. (Ch. 19)

multiverse A collection of parallel universes that together make up all that exists. (Ch. 22)

mutation In biology, an imperfect reproduction of self-replicating material. (Ch. 24)

N

N See *newton*. (Ch. 3)

nadir The point on the celestial sphere located directly below an observer, opposite the *zenith*. (Ch. 2)

nanometer (**nm**) One-billionth (10^{-9}) of a meter; a unit of length used for the wavelength of visible light. (Ch. 5)

natural selection The process by which forms of structure, ranging from molecules to whole organisms, that are best adapted to their environment become more common than less well-adapted forms. (Ch. 24)

NCP See *north celestial pole*. (Ch. 2)

neap tide An especially weak tide that occurs around the time of the first or third quarter Moon, when the gravitational forces of the Moon and the Sun on Earth are at right angles to each other. Compare *spring tide*. (Ch. 4)

near-Earth asteroid An asteroid whose orbit brings it close to the orbit of Earth. See also *near-Earth object*. (Ch. 12)

near-Earth object (**NEO**) An asteroid, comet, or large meteoroid whose orbit intersects Earth's orbit. (Ch. 12)

nebula (pl. **nebulae**) A cloud of interstellar gas and dust, either illuminated by stars (for bright nebulae) or seen in silhouette against a brighter background (for dark nebulae). (Ch. 7)

nebular hypothesis The first plausible theory of the formation of the Solar System, proposed by Immanuel Kant in 1755, which stated that the Solar System formed from the collapse of an interstellar cloud of rotating gas. (Ch. 7)

NEO See *near-Earth object*. (Ch. 12)

neutrino A very low-mass, electrically neutral particle emitted during beta decay. Neutrinos interact with matter only very feebly and so can penetrate great quantities of matter. (Ch. 6, 14)

neutron capture The process in which an atomic nucleus forms a heavier nucleus after colliding with a neutron. (Ch. 17)

neutrino cooling The process in which thermal energy is carried out of the center of a star by neutrinos rather than by electromagnetic radiation or convection. (Ch. 17)

neutron A subatomic particle having no net electric charge and a rest mass and rest energy nearly equal to that of the proton. Compare *electron* and *proton*. (Ch. 5)

neutron star The neutron-degenerate stellar core left behind by a Type II supernova. (Ch. 17)

new Moon The phase of the Moon in which the Moon is between Earth and the Sun; from Earth, we see only the side of the Moon not being illuminated by the Sun. Compare *full Moon*. See also *first quarter Moon* and *third quarter Moon*. (Ch. 2)

newton (N) The force required to accelerate a 1-kilogram (kg) mass at a rate of 1 meter per second per second (m/s^2). Unit: kilograms multiplied by meters per second squared ($kg\,m/s^2$). (Ch. 3)

Newton's first law of motion The law, formulated by Isaac Newton, stating that an object will remain at rest or will continue moving along a straight line at a constant speed until an unbalanced force acts on it. (Ch. 3)

Newton's laws The three physical laws of motion that Isaac Newton formulated.

Newton's second law of motion The law that Isaac Newton formulated, stating that if an unbalanced force acts on a body, the body will accelerate in proportion to the unbalanced force and in inverse proportion to the object's mass: $a = F/m$. The acceleration will be in the direction of the unbalanced force. (Ch. 3)

Newton's third law of motion The law that Isaac Newton formulated, stating that for every force an equal force in the opposite direction exists. (Ch. 3)

nm See *nanometer*. (Ch. 5)

normal matter See *luminous matter*. (Ch. 19)

north celestial pole (NCP) The northward projection of Earth's rotation axis onto the celestial sphere. Compare *south celestial pole*. (Ch. 2)

North Pole The location in the Northern Hemisphere where Earth's rotation axis intersects Earth's surface. Compare *South Pole*. (Ch. 2)

nova (pl. **novae**) A stellar explosion that results from runaway nuclear fusion in a layer of material on the surface of a white dwarf in a binary system. (Ch. 16)

nuclear burning See nuclear fusion.

nuclear fusion The combination of two less massive atomic nuclei into a single more massive atomic nucleus. (Ch. 14)

nucleosynthesis The formation of more massive atomic nuclei from less massive nuclei, either in the Big Bang (Big Bang nucleosynthesis) or in the interiors of stars (stellar nucleosynthesis). (Ch. 17)

nucleus (pl. **nuclei**) 1. The dense, central part of an atom. 2. The central core of a galaxy, comet, or other diffuse object. (Ch. 5)

 O

objective lens The primary optical element in a telescope or camera that produces an image of an object. (Ch. 6)

oblate Pertaining to a sphere, flattened or squashed along the polar axis. (Ch. 10)

obliquity The inclination of a celestial body's equator to its orbital plane.

observable universe The part of the universe from which light has had time to reach us since shortly after the Big Bang. (Ch. 21)

observational uncertainty The fact that real measurements are never perfect; all observations are uncertain by some amount.

Occam's razor The principle that the simplest hypothesis is the most likely, named after William of Occam (circa 1285–1349), the medieval English cleric to whom the idea is attributed. (Ch. 1)

Oort Cloud A spherical distribution of comet nuclei stretching from beyond the Kuiper Belt to more than 50,000 astronomical units (AU) from the Sun. (Ch. 12)

opacity A measure of how effectively a material blocks the radiation going through it. (Ch. 14)

open cluster A loosely bound group of a few dozen to a few thousand stars that formed together in the disk of a spiral galaxy. Compare *globular cluster*. (Ch. 17)

open universe An infinite universe with a negatively curved spatial structure (much like the surface of a saddle) such that the sum of the angles of a triangle is always less than 180 degrees. Compare *closed universe* and *flat universe*. (Ch. 22)

orbit The path taken by one object moving around another object under the influence of their mutual gravitational or electric attraction. (Ch. 4)

orbital resonance A situation in which the orbital periods of two objects are related by a ratio of small integers. (Ch. 11)

orbiter A spacecraft placed in orbit around a planet or moon. Compare *flyby*. (Ch. 6)

organic Containing the element carbon. (Ch. 7)

P

P wave See *primary wave*. (Ch. 8)

pair production The creation of a particle-antiparticle pair from a source of electromagnetic energy. (Ch. 22)

paleoclimatology The study of changes in Earth's climate throughout its history. (Ch. 9)

parallax Also called *parallactic angle*. The displacement in the apparent position of a nearby star caused by the changing location of Earth in its orbit. (Ch. 13)

parent element A radioactive element that decays to form more-stable *daughter products*. (Ch. 8)

parsec (pc) Short for *parallax second*. The distance to a star with a parallax of 1 arcsecond (arcsec) using a base of 1 astronomical unit (AU). One parsec is approximately 3.26 light-years. (Ch. 13)

partial lunar eclipse An eclipse that occurs when the Moon is partially in Earth's shadow. (Ch. 2)

partial solar eclipse The type of eclipse that occurs when Earth passes through the penumbra of the Moon's shadow, so that the Moon blocks only a portion of the Sun's disk. Compare *annular solar eclipse* and *total solar eclipse*. (Ch. 2)

pc See *parsec*. (Ch. 13)

peculiar velocity The motion of a galaxy relative to the overall expansion of the universe. (Ch. 23)

penumbra (pl. **penumbrae**) 1. The outer part of a shadow, where the source of light is only partially blocked. Compare *umbra* (definition 1). 2. The region surrounding the umbra of a sunspot. The penumbra is cooler and darker than the surrounding surface of the Sun but is not as cool or dark as the umbra. Compare *umbra* (definition 2). (Ch. 2, 14)

penumbral lunar eclipse A lunar eclipse in which the Moon passes through the penumbra of Earth's shadow. Compare *total lunar eclipse*. (Ch. 2)

perihelion (pl. **perihelia**) The point in a solar orbit that is closest to the Sun. Compare *aphelion*. (Ch. 12)

period The time that a regularly repetitive process takes to complete one cycle. (Ch. 17)

period-luminosity relationship The relationship between the period of variability of a pulsating variable star, such as a Cepheid or RR Lyrae variable, and the luminosity of the star. Longer-period pulsating variable stars are more luminous than shorter-period ones. (Ch. 17)

Perseids A prominent August meteor shower associated with the dust debris left by comet Swift-Tuttle. (Ch. 12)

phase One of the various appearances of the sunlit surface of the Moon or a planet caused by the change in viewing location of Earth relative to both the Sun and the object. Examples include crescent phase and gibbous phase. (Ch. 2)

photino A hypothetical subatomic particle related to the photon. One of the candidates for cold dark matter. (Ch. 23)

photochemical Resulting from light acting on chemical systems.

photodisintegration An atomic nucleus absorbs an energetic gamma ray and emits some particles. (Ch. 17)

photodissociation The breaking apart of molecules into smaller fragments or individual atoms by the action of photons. Compare *recombination* (definition 1). (Ch. 11)

photoelectric effect The emission of electrons from a substance illuminated by electromagnetic radiation greater than a certain critical frequency.

photon Also called *quantum of light*. A discrete unit or particle of electromagnetic radiation. The energy of a photon is equal to Planck's constant (h) multiplied by the frequency (f) of its electromagnetic radiation: $E_{photon} = h \times f$. The photon is the carrier of the electromagnetic force. (Ch. 5)

photosphere The apparent surface of the Sun as seen in visible light. Compare *chromosphere* and *corona*. (Ch. 14)

photosynthesis The process by which green plants use sunlight to create food from water and carbon dioxide. (Ch. 9)

photosynthesize The process by which plants and algae absorb sunlight, water, and carbon dioxide, and release oxygen. (Ch. 24)

physical law A broad statement that predicts a particular aspect of how the physical universe behaves and that is supported by many empirical tests. See also *theory*.

pixel The smallest picture element in a digital image array. (Ch. 6)

Planck era The early time, just after the Big Bang, for which the universe as a whole must be described with quantum mechanics. (Ch. 22)

Planck spectrum Also called *blackbody spectrum*. The spectrum of electromagnetic energy emitted by a blackbody per unit area per second, which is determined only by the temperature of the object. (Ch. 5)

Planck's constant (h) The constant of proportionality between the energy and the frequency of a photon. This constant defines how much energy a single photon of a given frequency or wavelength has. Value: 6.63×10^{-34} joule-second.

planet 1. A large body that orbits the Sun or other star that shines only by light reflected from the Sun or star. 2. In the Solar System, a body that orbits the Sun, has enough mass for self-gravity to overcome rigid body forces so that it assumes a spherical shape, and has cleared smaller bodies from the neighborhood around its orbit. Compare *dwarf planet*. (Ch. 7)

planet migration The theory that a planet can move to a location away from where it formed, through gravitational interactions with other bodies or loss of orbital energy from interaction with gas in the protoplanetary disk. (Ch. 7)

planetary nebula The expanding shell of material ejected by a dying asymptotic giant branch star. A planetary nebula glows from fluorescence caused by intense ultraviolet light coming from the hot, stellar remnant at its center. (Ch. 16)

planetary system A system of planets and other smaller objects in orbit around a star. (Ch. 7)

planetesimal A primitive body of rock and ice, 100 meters or more in diameter, that combines with others to form a planet. Compare *meteoroid* and *zodiacal dust*. (Ch. 7)

plasma A gas composed largely of charged particles but that also may include some neutral atoms. (Ch. 10, 14)

plate tectonics The geological theory concerning the motions of lithospheric plates, which in turn serves as the theoretical basis for *continental drift*. (Ch. 8)

positron A positively charged subatomic particle; the antiparticle of the *electron*. (Ch. 14)

power Energy per unit time. Possible units include watts (W) and joules per second (J/s). (Ch. 15)

precession of the equinoxes The slow change in orientation between the ecliptic plane and the celestial equator caused by the wobbling of Earth's axis. (Ch. 2)

pressure Force per unit area. Possible units include newtons per square meter (N/m^2) and bars.

primary atmosphere An atmosphere, composed mostly of hydrogen and helium, that forms at the same time as its host planet. Compare *secondary atmosphere*. (Ch. 7)

primary mirror The principal optical mirror in a reflecting telescope. The primary mirror determines the telescope's light-gathering power and resolution. Compare *secondary mirror*. (Ch. 6)

primary wave Also called *P wave*. A longitudinal seismic wave, in which the oscillations involve compression and decompression parallel to the direction of travel. Compare *secondary wave*. (Ch. 8)

principle A general idea or sense about the universe that guides us in constructing new scientific theories. Principles can be testable theories. (Ch. 1)

prograde motion 1. Rotational or orbital motion of a moon that is in the same direction as the planet it orbits. 2. The counterclockwise orbital motion of Solar System objects as seen from above Earth's orbital plane. Compare *retrograde motion*. (Ch. 3)

prominence An archlike projection above the solar photosphere often associated with a sunspot. (Ch. 14)

proportional See *proportionality*. (Appendix 1)

proportionality A relationship between two things whose ratio is a constant. (Appendix 1)

proton (p or p^+) A subatomic particle having a positive electric charge of 1.6×10^{-19} coulomb (C), a rest mass of 1.67×10^{-27} kilogram (kg), and a rest energy of 1.5×10^{-10} joule (J). Compare *electron* and *neutron*. (Ch. 5)

proton-proton chain One way in which hydrogen fusion can take place. That path is the most important for hydrogen fusion in low-mass stars such as the Sun. Compare *carbon-nitrogen-oxygen cycle*. (Ch. 14)

protoplanetary disk The remains of the accretion disk around a young star from which a planetary system may form. Sometimes called *circumstellar disk*. (Ch. 7)

protostar A young stellar object that derives its luminosity from converting gravitational energy to thermal energy rather than from nuclear reactions in its core. (Ch. 7, 15)

pulsar A rapidly rotating neutron star that beams radiation into space in two searchlight-like beams. To a distant observer, the star appears to flash on and off. (Ch. 17)

pulsating variable star A variable star that undergoes periodic radial pulsations. (Ch. 17)

Q

QCD See *quantum chromodynamics*. (Ch. 22)

QED See *quantum electrodynamics*. (Ch. 22)

quantized Existing as discrete, irreducible units.

quantum chromodynamics (QCD) The quantum theory describing the strong nuclear force and its mediation by gluons. Compare *quantum electrodynamics*. (Ch. 22)

quantum efficiency The likelihood that a particular photon falling on a detector will actually produce a response in the detector. (Ch. 6)

quantum electrodynamics (QED) The quantum theory describing the electromagnetic force and its mediation by photons. Compare *quantum chromodynamics*. (Ch. 22)

quantum mechanics The branch of physics that deals with the quantized and probabilistic behavior of atoms and subatomic particles. (Ch. 5)

quantum of light See *photon*. (Ch. 5)

quark The building block of protons and neutrons. (Ch. 22)

quasar Short for *quasi-stellar radio source*. The most luminous of the active galactic nuclei, seen only at great distances from the Milky Way. (Ch. 19)

R

radial velocity (v_r) The component of velocity directed toward or away from the observer. (Ch. 5)

radian The angle at the center of a circle subtended by an arc equal to the length of the circle's radius; 2π radians equals 360°, and 1 radian equals approximately 57.3°. (Appendix 1)

radiant The point in the sky from which the meteors in a meteor shower appear to come. (Ch. 12)

radiation Waves or particles of energy traveling through space or a medium. (Ch. 14)

radiation belt A toroidal ring of high-energy particles surrounding a planet. (Ch. 9)

radiative transfer The transport of energy from one location to another by electromagnetic radiation.

radiative zone A region within a star where energy is transported outward by radiation. Compare *convective zone*. (Ch. 14)

radio galaxy A type of elliptical galaxy that has an active galactic nucleus at its center and very strong emission (10^{35} to 10^{38} watts [W]) in the radio part of the electromagnetic spectrum. Compare *Seyfert galaxy*. (Ch. 19)

radio telescope An instrument for detecting and measuring radio frequency emissions from celestial sources. (Ch. 6)

radio wave Electromagnetic radiation in the extreme long-wavelength region of the spectrum, beyond the region of microwaves. (Ch. 5)

radioisotope A radioactive element. (Ch. 8)

radiometric dating Use of the radioactive decay of elements to measure the ages of materials such as minerals. (Ch. 8)

ratio The relationship in quantity or size between two or more things.

ray 1. A beam of electromagnetic radiation. 2. A bright streak emanating from a young impact crater.

recombination 1. The combining of ions and electrons to form neutral atoms. Compare *photodissociation*. 2. An event early in the evolution of the universe in which hydrogen and helium nuclei combined with electrons to form neutral atoms. The removal of electrons caused the universe to become transparent to electromagnetic radiation. (Ch. 11, 21)

red giant A low-mass star that has evolved beyond the main sequence and is now fusing hydrogen in a shell surrounding a degenerate helium core. (Ch. 16)

red giant branch A region on the H-R diagram defined by low-mass stars evolving from the main sequence toward the horizontal branch. (Ch. 16)

reddening The effect by which stars and other objects, when viewed through interstellar dust, appear redder than they actually are. Reddening occurs because blue light is more strongly absorbed and scattered than red light. (Ch. 15)

redshift The Doppler shift toward longer (redder) wavelengths of light from an approaching object. Compare *blueshift*. (Ch. 5)

reflecting telescope A telescope that uses mirrors to collect and focus incoming electromagnetic radiation to form an image in their focal planes. The size of a reflecting telescope is defined by the diameter of the primary mirror. Compare *refracting telescope*. (Ch. 6)

reflection The redirection of a beam of light that strikes, but does not cross, the surface between two media having different refractive indices. If the surface is flat and smooth, the angle of incidence equals the angle of reflection. Compare *refraction*. (Ch. 6)

refracting telescope A telescope that uses objective lenses to collect and focus incoming electromagnetic radiation to form an image. Compare *reflecting telescope*. (Ch. 6)

refraction The redirection or bending of a beam of light when it crosses the boundary between two media

having different refractive indices. Compare *reflection*. (Ch. 6)

refractory material Material that remains solid at high temperatures. Compare *volatile material*. (Ch. 7)

regular moon A moon that formed together with the planet it orbits. Compare *irregular moon*. (Ch. 11)

reionization A period after the Dark Ages during which objects formed that radiated enough energy to ionize neutral hydrogen, at redshift $6 < z < 20$. (Ch. 23)

relative humidity The amount of water vapor held by a volume of air at a given temperature compared (stated as a percentage) to the total amount of water that could be held by the same volume of air at the same temperature. (Ch. 9)

relative motion The difference in motion between two individual frames of reference. (Ch. 2)

relativistic Describing systems that travel at nearly the speed of light or are located near very strong gravitational fields. (Ch. 18)

relativistic beaming The effect created when material moving at nearly the speed of light beams the radiation it emits in the direction of its motion. (Ch. 19)

relativistic speed A speed high enough that special relativity, rather than Newtonian physics, is needed to describe the motion. Speeds greater than about 10% the speed of light are relativistic. (Ch. 18)

remote sensing The use of images, spectra, radar, or other techniques to measure the properties of an object from a distance.

resolution The ability of a telescope to separate two point sources of light. Resolution is determined by the telescope's aperture and the wavelength of light it receives. (Ch. 6)

rest wavelength The wavelength of light that is seen coming from an object at rest with respect to the observer. (Ch. 5)

retrograde motion 1. Rotation or orbital motion of a moon that is in the opposite direction to the rotation of the planet it orbits. 2. The clockwise orbital motion of Solar System objects as seen from above Earth's orbital plane. Compare *prograde motion*. 3. Apparent retrograde motion is a motion of the planets with respect to the "fixed stars," in which the planets appear to move westward for a time before resuming their normal eastward motion. (Ch. 3)

revolve Motion of one object in orbit around another. (Ch. 2)

right ascension A measure, analogous to *longitude*, that tells you the angular distance of a celestial body eastward along the celestial equator from the vernal equinox. Compare *declination*. (Ch. 2)

ring An aggregation of small particles orbiting a planet or star. The rings of the four giant planets of the Solar System are composed variously of silicates, organic materials, and ices. (Ch. 11)

ring arc A discontinuous, higher-density region within an otherwise continuous, narrow ring. (Ch. 11)

ringlet A narrowly confined concentration of ring particles. (Ch. 11)

Roche limit The distance at which a planet's tidal forces exceed the self-gravity of a smaller object—such as a moon, asteroid, or comet—causing the object to break apart. (Ch. 4)

Roche lobes The hourglass-shaped or figure eight–shaped volume of space surrounding two stars, which constrains material that is gravitationally bound by one or the other. (Ch. 16)

rotation curve A plot showing how the orbital velocity of stars and gas in a galaxy changes with radial distance from the galaxy's center. (Ch. 19)

rover A remotely controlled instrumented vehicle designed to move and explore the surface of a terrestrial planet or moon. Compare *lander*. (Ch. 6)

RR Lyrae variable A variable giant star whose regularly timed pulsations are good predictors of its luminosity. RR Lyrae variables are used in measuring the distance to globular clusters. (Ch. 17)

S

S-type asteroid An asteroid made of material that was once part of the outer layer of a larger, differentiated body that has since broken into pieces. Compare *C-type asteroid* and *M-type asteroid*. (Ch. 12)

S wave See *secondary wave*. (Ch. 8)

S0 galaxy A galaxy with a bulge and a disk-like spiral, but smooth in appearance like ellipticals. Compare *barred spiral galaxy*, *elliptical galaxy*, *irregular galaxy*, and *spiral galaxy*. (Ch. 19)

satellite An object in orbit around a more massive body, for example, a moon of any planet or an artificial satellite. (Ch. 4)

scale factor (R_v) A dimensionless number proportional to the distance between two points in space. The scale factor increases as the universe expands. (Ch. 21)

scattering The random change in the direction of travel of photons, caused by their interactions with molecules or dust particles. (Ch. 9)

Schwarzschild radius The distance from the center of a nonrotating, spherical black hole at which the escape velocity equals the speed of light. (Ch. 18)

scientific method The formal procedure—including hypothesis, prediction, and experiment or observation—used to test (that is, to attempt to falsify) the validity of scientific hypotheses and theories. (Ch. 1)

scientific notation The standard expression of numbers with one digit (which can be zero) to the left of the decimal point and multiplied by 10 to the exponent required to give the number its correct value. Example: $2.99 \times 10^8 = 299,000,000$. (1, Appendix 1)

SCP See *south celestial pole*. (Ch. 2)

second law of thermodynamics The law stating that the entropy or disorder of an isolated system always increases as the system evolves.

secondary atmosphere An atmosphere that forms—as a result of volcanism, comet impacts, or another process—sometime after its host planet has formed. Compare *primary atmosphere*. (Ch. 7)

secondary crater A crater formed from ejecta thrown from an *impact crater*. (Ch. 8)

secondary mirror A small mirror placed on the optical axis of a reflecting telescope that returns the beam back through a small hole in the *primary mirror*, thereby shortening the mechanical length of the telescope. (Ch. 6)

secondary wave Also called *S wave*. A transverse seismic wave, which involves the sideways motion of material. Compare *primary wave*. (Ch. 8)

sediment Solid material such as rock or sand that has been affected by erosion and weather, and transported by water, wind, or ice. (Ch. 8)

seismic wave A vibration due to an earthquake, a large explosion, or an impact on the surface that travels through a planet's interior. (Ch. 8)

seismometer An instrument that measures the amplitude and frequency of seismic waves. (Ch. 8)

self-gravity The gravitational attraction among all parts of the same object. (Ch. 4)

semimajor axis Half of the longer axis of an ellipse. (Ch. 3)

SETI The Search for Extraterrestrial Intelligence project, which uses advanced technology combined with radio telescopes to search for evidence of intelligent life elsewhere in the universe. (Ch. 24)

Seyfert galaxy A type of spiral galaxy with an active galactic nucleus at its center; first discovered in 1943 by Carl Seyfert. Compare *radio galaxy*. (Ch. 19)

shepherd moon A moon that orbits close to rings and gravitationally confines the orbits of the ring particles. (Ch. 11)

shield volcano A volcano formed by very fluid lava flowing from a single source and spreading out from that source. Compare *composite volcano*. (Ch. 8)

short gamma-ray burst Bursts of gamma rays lasting a few milliseconds to a few minutes, thought to originate in neutron star mergers. (Ch. 17)

short-period comet A comet with an orbital period of less than 200 years. Compare *long-period comet*. (Ch. 12)

sidereal day Earth's period of rotation with respect to the stars—about 23 hours 56 minutes—which is the time Earth takes to make one rotation and face the same star on the meridian. It differs from the *solar day* because of Earth's motion around the Sun. (Ch. 2)

sidereal period An object's orbital or rotational period measured with respect to the stars. Compare *synodic period*. (Ch. 2)

silicate One of the family of minerals composed of silicon and oxygen in combination with other elements. (Ch. 7)

singularity The point at which a mathematical expression or equation becomes meaningless, such as a fraction whose denominator approaches zero See also *black hole*. (Ch. 18)

solar abundance The relative amount of an element detected in the atmosphere of the Sun, expressed as the ratio of the number of atoms of that element to the number of hydrogen atoms. (Ch. 5)

solar day The day in common use—24 hours, which is Earth's period of rotation that brings the Sun back to the same local meridian where the rotation started. Compare *sidereal day*. (Ch. 2)

solar eclipse An eclipse that occurs when the Moon partially or entirely blocks the Sun. Compare *lunar eclipse*. (Ch. 2)

solar flare An explosion on the Sun's surface associated with a complex sunspot group and a strong magnetic field. (Ch. 14)

solar maximum (pl. **maxima**) The time, occurring about every 11 years, when the Sun is at its peak activity, meaning that sunspot activity and related phenomena (such as prominences, flares, and coronal mass ejections) are at their peak. (Ch. 14)

solar neutrino problem The historical observation that only about a third as many neutrinos as predicted by theory seemed to be coming from the Sun. (Ch. 14)

Solar System The gravitationally bound system made up of the Sun, planets, dwarf planets, moons, asteroids, comets, and Kuiper Belt objects, along with their associated gas and dust. (Ch. 1)

solar tide A tide on Earth caused by the differential gravitational pull of the Sun. Compare *lunar tide*. (Ch. 4)

solar wind The stream of charged particles emitted by the Sun and that flows at high speeds through interplanetary space. (Ch. 9)

solstice Literally, "Sun standing still." 1. One of the two most northerly and southerly points on the ecliptic. 2. Either of the two times of year (the *summer solstice* and *winter solstice*) when the Sun is at one of these two positions. Compare *equinox*.

south celestial pole (SCP) The southward projection of Earth's rotation axis onto the celestial sphere. Compare *north celestial pole*. (Ch. 2)

South Pole The location in the Southern Hemisphere where Earth's rotation axis intersects Earth's surface. Compare *North Pole*. (Ch. 2)

spacetime A concept that combines space and time into a four-dimensional continuum with three spatial dimensions plus one dimension of time. (Ch. 18)

special relativity See *special theory of relativity*. (Ch. 18)

special theory of relativity Sometimes referred to as simply *special relativity*. Einstein's theory explaining how the fact that the speed of light is a constant affects nonaccelerating frames of reference. Compare *general theory of relativity*. (Ch. 18)

spectral type A classification system for stars that is based on the presence and relative strength of absorption lines in their spectra. Spectral type is related to the surface temperature of a star. (Ch. 13)

spectrograph Also called *spectrometer*. A device that spreads out the light from an object into its component wavelengths. (Ch. 6)

spectrometer See *spectrograph*. (Ch. 6)

spectroscopic binary A binary star system whose existence and properties are revealed to astronomers only by the Doppler shift of its spectral lines. Most spectroscopic binaries are close pairs. Compare *eclipsing binary* and *visual binary*. (Ch. 13)

spectroscopic parallax Use of the spectroscopically determined luminosity and the observed brightness of a star to determine the star's distance. (Ch. 13)

spectroscopy The study of an object's electromagnetic radiation in terms of its component wavelengths. (Ch. 6)

spectrum (pl. **spectra**) Waves sorted by wavelength. See also *electromagnetic spectrum*. (Ch. 5)

speed The rate of change of an object's position with time, without regard to the direction of movement. Possible units include meters per second (m/s) and kilometers per hour (km/h). Compare *velocity*. (Ch. 3)

spherically symmetric Describing an object whose properties depend only on distance from the object's center, so that the object has the same form viewed from any direction. (Ch. 4)

spin-orbit resonance A relationship between the orbital and rotation periods of an object such that the ratio of their periods can be expressed with simple integers.

spiral density wave A stable, spiral-shaped change in the local gravity of a galactic disk that can be produced by periodic gravitational kicks from neighboring galaxies or from nonspherical bulges and bars in spiral galaxies. (Ch. 20)

spiral galaxy A galaxy of Hubble type "S" class, with a discernible disk in which large spiral patterns exist. Compare *barred spiral galaxy*, *elliptical galaxy*, *irregular galaxy*, and *S0 galaxy*. (Ch. 19)

spoke One of several narrow radial features seen occasionally in Saturn's B Ring. Spokes appear dark in backscattered light and bright in forward, scattering light, indicating that they are composed of tiny particles. Their origin is not well understood. (Ch. 11)

spreading center A zone from which two tectonic plates diverge. (Ch. 8)

spring tide An especially strong tide that occurs around the time of a new or full Moon, when lunar tides and solar tides reinforce each other. Compare *neap tide*. (Ch. 4)

sputtering Collision of fast-moving ions with a surface or an atmosphere that causes the loss of material from the surface or atmosphere. (Ch. 9)

stable equilibrium An equilibrium state in which the system returns to its former condition after a small disturbance. Compare *unstable equilibrium*.

standard candle An object whose luminosity either is known or can be predicted in a distance-independent way, so its brightness can be used to determine its distance via the inverse square law of radiation. (Ch. 19)

standard model The theory of particle physics that combines electroweak theory with quantum chromodynamics to describe the structure of known forms of matter. (Ch. 22)

star A luminous ball of gas held together by gravity. A normal star is powered by nuclear reactions in its interior.

star cluster A group of stars that all formed at the same time and in the same general location. (Ch. 15)

static equilibrium A state in which the forces within a system are all in balance so that the system does not change. Compare *dynamic equilibrium*.

Stefan-Boltzmann constant (σ) The constant of proportionality that relates the flux emitted by an object to the fourth power of its absolute temperature. Value: 5.67×10^{-8} W/(m^2 K^4) (W = watts, m = meters, K = kelvins). (Ch. 5)

Stefan-Boltzmann law The law, formulated by Josef Stefan and Ludwig Boltzmann, stating that the amount of electromagnetic energy emitted from the surface of a body (flux), summed over the energies of all photons of all wavelengths emitted, is proportional to the fourth power of the temperature of the body: $\mathcal{F} = \sigma T^4$. (Ch. 5)

stellar evolution The stages of development that a star goes through over its life cycle. Astronomers observe many stars at the same time to build a picture of how stars change. (Ch. 16)

stellar-mass loss The loss of mass from the outermost parts of a star's atmosphere during the star's evolution. (Ch. 16)

stellar occultation An event in which a planet or other Solar System body moves between the observer and a star, eclipsing the light emitted by that star. (Ch. 10)

stellar population A group of stars with similar ages, chemical compositions, and dynamic properties. (Ch. 17)

stereoscopic vision The way an animal's brain combines the different information from its two eyes to perceive the distances to objects around it. (Ch. 13)

stony-iron meteorite A meteorite consisting of a mixture of silicate minerals and iron-nickel alloys. Compare *iron meteorite* and *stony meteorite*. (Ch. 12)

stony meteorite A meteorite composed primarily of silicate minerals, similar to those found on Earth. Compare *iron meteorite* and *stony-iron meteorite*. (Ch. 12)

stratosphere The atmospheric layer immediately above the troposphere. On Earth, it extends upward to an altitude of 50 kilometers (km). Compare *troposphere*, *mesosphere*, and *thermosphere*. (Ch. 9)

string theory See *superstring theory*. (Ch. 22)

stromatolite A structure created by living or fossilized cyanobacteria. (Ch. 24)

strong nuclear force The attractive short-range force between protons and neutrons that holds atomic nuclei together. One of the four fundamental forces of nature, along with the *electromagnetic force*, the *weak nuclear force*, and *gravity* (definition 1). The force is mediated by the exchange of gluons. (Ch. 14, 22)

subduction zone A region where two tectonic plates converge, with one plate sliding under the other and being drawn downward into the interior. (Ch. 8)

subgiant A giant star that is smaller and lower in luminosity than normal giant stars of the same spectral type. Subgiants evolve to become giants. (Ch. 16)

subgiant branch A region of the H-R diagram defined by stars that have left the main sequence but have not yet reached the red giant branch. (Ch. 16)

sublimation The process by which a solid becomes a gas without first becoming a liquid. (Ch. 12)

summer solstice 1. One of two points where the Sun is at its greatest distance from the celestial equator. 2. The day on which the Sun appears at that location, marking the first day of summer (about June 20 in the Northern Hemisphere and December 21 in the Southern Hemisphere). Compare *winter solstice*. See also *autumnal equinox* and *vernal equinox*. (Ch. 2)

Sun The star at the center of the Solar System. (Ch. 1)

sungrazer A comet whose perihelion is within a few solar diameters of the surface of the Sun. (Ch. 12)

sunspot A cooler, transitory region on the solar surface produced when loops of magnetic flux break through the surface of the Sun. (Ch. 14)

sunspot cycle The approximate 11-year cycle during which sunspot activity increases and then decreases. This is one-half of a full 22-year cycle, in which the magnetic polarity of the Sun first reverses and then returns to its original configuration. (Ch. 14)

supercluster A large conglomeration of galaxy clusters and galaxy groups, typically more than 100–300 megaparsecs (Mpc) in size and containing tens of thousands to hundreds of thousands of galaxies. Compare *galaxy cluster* and *galaxy group*. (Ch. 1, 23)

super-Earth An extrasolar planet with about 2–10 times the mass of Earth. (Ch. 7)

superior planet A Solar System planet that orbits farther from the Sun than Earth does. Compare *inferior planet*. (Ch. 3)

supermassive black hole A black hole of 1,000 solar masses (M_{Sun}) or more that resides in the center of a galaxy and whose gravity powers active galactic nuclei. (Ch. 19)

supernova (pl. **supernovae**) A stellar explosion resulting in the release of tremendous amounts of energy, including the high-speed ejection of matter into the interstellar medium. See also *Type Ia supernova* and *Type II supernova*.

supernova remnant The material ejected from the outer layers of a star after a supernova explosion. (Ch. 16)

supersonic Moving within a medium at a speed faster than the speed of sound in that medium. Compare *subsonic*.

superstring theory The theory that conceives of particles as strings in 10 dimensions of space and time; the current contender for a theory of everything. (Ch. 22)

surface brightness The amount of electromagnetic radiation emitted or reflected per unit area. (Ch. 6, 19)

surface wave A seismic wave that travels on the surface of a planet or moon. (Ch. 8)

surface gravity The acceleration (downward) due to gravity measured at the surface of a planet or star. (Ch. 9)

symmetry In theoretical physics, the properties of physical laws that remain constant when certain things change, such as the symmetry between matter and antimatter even though their charges may be different.

synchronous rotation The case that occurs when a body's rotation period equals its orbital period around another body. A special type of spin-orbit resonance. (Ch. 2)

synchrotron radiation Radiation from electrons moving at close to the speed of light as they spiral in a strong magnetic field, so named because that kind of radiation was first identified on Earth in particle accelerators called synchrotrons. (Ch. 10)

synodic period An object's orbital or rotational period measured with respect to the Sun. Compare *sidereal period*. (Ch. 2)

T

T Tauri star A young stellar object that has dispersed enough of the material surrounding it to be seen in visible light. (Ch. 15)

tail A stream of gas and dust swept away from the coma of a comet by the solar wind and by radiation pressure from the Sun. See also *ion tail* and *dust tail*. (Ch. 12)

tectonism Deformation of the lithosphere of a planet. (Ch. 8)

telescope The basic tool of astronomers. Working over the entire range from gamma rays to radio waves, astronomical telescopes collect and concentrate electromagnetic radiation from celestial objects. (Ch. 6)

temperature A measure of the average kinetic energy of the atoms or molecules in a gas, solid, or liquid. (Ch. 5)

terrestrial planet An Earth-like planet, made of rock and metal and having a solid surface. In the Solar System, the terrestrial planets are Mercury, Venus, Earth, and Mars. Compare *giant planet*. (Ch. 7)

theoretical model A description of the properties of a particular object or system in terms of known physical laws or theories; often, a computer calculation of predicted properties based on such a description.

theory A well-developed idea or group of ideas that are tied solidly to known physical laws and make testable predictions about the world. A very well-tested theory may be called a *physical law*, or simply a fact. Compare *hypothesis*. (Ch. 1)

theory of everything (TOE) A theory that unifies all four fundamental forces of nature: strong nuclear, weak nuclear, electromagnetic, and gravitational forces. (Ch. 22)

thermal conduction See *conduction*. (Ch. 14)

thermal energy The energy that resides in the random motion of atoms, molecules, and particles, by which we measure their temperature. (Ch. 5, 7)

thermal equilibrium The state in which the rate of thermal-energy emission by an object is equal to the rate of thermal-energy absorption. (Ch. 5)

thermal motion The random motion of atoms, molecules, and particles that gives rise to thermal radiation. (Ch. 5)

thermal radiation Electromagnetic radiation resulting from the random motion of the charged particles in every substance. (Ch. 5)

thermosphere The layer of Earth's atmosphere at altitudes greater than 90 kilometers (km), above the mesosphere. Near its top, at an altitude of 600 km, the temperature can reach 1000 K. Compare *troposphere*, *stratosphere*, and *mesosphere*. (Ch. 9)

thick disk Extends above and below the thin disk of a spiral galaxy, and contains the older population of disk stars, distinguishable by lower abundances of massive elements. (Ch. 20)

thin disk Central layer of the disk of a spiral galaxy. The youngest stars are concentrated in the thin disk because that is where the molecular clouds and gas are most concentrated. (Ch. 20)

third quarter Moon The phase of the Moon in which only the eastern half of the Moon, as viewed from Earth, is illuminated by the Sun. It occurs about one week after the full Moon. Compare *first quarter Moon*. See also *full Moon* and *new Moon*. (Ch. 2)

tidal bulge Distortion of a body resulting from tidal stresses. (Ch. 4)

tidal force A force caused by the change in the strength of gravity across an object. (Ch. 4)

tidal locking Synchronous rotation of an object caused by internal friction as the object rotates through its tidal bulge. (Ch. 4)

tide On Earth, the rise and fall of the oceans as Earth rotates through a tidal bulge caused by the Moon and the Sun. See also *lunar tide*, *neap tide*, *solar tide*, and *spring tide*. (Ch. 4)

time dilation The relativistic "stretching" of time. Compare *general relativistic time dilation*. (Ch. 18)

TOE See *theory of everything*. (Ch. 22)

topography The study of the natural physical features of the surface of a planet or moon. (Ch. 8)

tornado A violent rotating column of air, typically 75 meters across with 200-kilometer-per-hour (km/h) winds. Some tornadoes can be more than 3 km across, and winds up to 500 km/h have been observed. (Ch. 9)

torus (pl. **tori**) A three-dimensional, doughnut-shaped ring. (Ch. 10, 19)

total lunar eclipse A lunar eclipse in which the Moon passes through the umbra of Earth's shadow. Compare *penumbral lunar eclipse*. (Ch. 2)

total solar eclipse The type of eclipse that occurs when Earth passes through the umbra of the Moon's shadow, so that the Moon blocks the disk of the Sun. Compare *annular solar eclipse* and *partial solar eclipse*. (Ch. 2)

transit method A method of detecting extrasolar planets by measuring the decrease in light from a star as its orbiting planet passes in front of the star as viewed from Earth. (Ch. 7)

trans-Neptunian object See *Kuiper Belt object*.

transverse wave A wave that oscillates perpendicular to the direction of the wave's propagation. Compare *longitudinal wave*. (Ch. 8)

triple-alpha process The nuclear fusion reaction that combines three ^{4}He nuclei (alpha particles) into a single nucleus of carbon. (Ch. 16)

Trojans A group of asteroids orbiting in the L_4 and L_5 Lagrangian points of Jupiter's orbit. (Ch. 12)

tropical year The time between two crossings of the vernal equinox. Because of the precession of the equinoxes, a tropical year is slightly shorter than the time Earth takes to orbit once around the Sun. Compare *year*. (Ch. 2)

Tropics The region on Earth between latitudes 23.5° south and 23.5° north, where the Sun appears directly overhead twice during the year. (Ch. 2)

tropopause The top of a planet's troposphere. (Ch. 9)

troposphere The convection-dominated layer of a planet's atmosphere. On Earth, the atmospheric region closest to the ground within which most weather phenomena take place. Compare *stratosphere*, *mesosphere*, and *thermosphere*. (Ch. 9)

turbulence The random motion of blobs of gas within a larger cloud of gas.

Type Ia supernova A supernova explosion with a calibrated peak luminosity that occurs as a result of runaway carbon fusion in a white dwarf star that accretes mass from a companion and approaches the Chandrasekhar mass limit of 1.4 M_{Sun}. (Ch. 16)

Type II supernova A supernova explosion in which the degenerate core of an evolved massive star suddenly collapses and rebounds. (Ch. 17)

U

ultrafaint dwarf galaxy A dim dwarf galaxy with only 1,000–100,000 times the Sun's luminosity. Ultrafaint dwarf galaxies differ from globular clusters in that they are composed of large amounts of dark matter. (Ch. 23)

ultraviolet (UV) radiation Electromagnetic radiation having frequencies and photon energies greater than those of visible light but less than those of X-rays and having wavelengths shorter than those of visible light but longer than those of X-rays. (Ch. 5)

umbra (pl. **umbrae**) 1. The darkest part of a shadow, where the source of light is blocked. Compare *penumbra* (definition 1). 2. The darkest, innermost part of a sunspot. Compare *penumbra* (definition 2). (Ch. 2, 14)

unbound orbit An orbit in which an object is no longer gravitationally bound to the body it was orbiting. An unbound orbit's velocity is greater than the escape velocity. Compare *bound orbit*. (Ch. 4)

uncertainty principle The physical limitation that the product of the position and the momentum of a particle cannot be smaller than a well-defined value, Planck's constant (h). (Ch. 22)

unified model of AGN A model in which many types of activity in the nuclei of galaxies are all explained by accretion of matter around a supermassive black hole. (Ch. 19)

uniform circular motion Motion in a circular path at a constant speed. (Ch. 4)

unit A fundamental quantity of measurement. The meter is an example of a metric unit; the foot is an example of an English unit. (Appendix 1)

universal gravitational constant (G) The constant of proportionality in the universal law of gravitation. Value: 6.67×10^{-11} meters cubed per kilogram second squared [m³/kg s² = N m²/kg²]. (Ch. 4)

universal law of gravitation The law, formulated by Isaac Newton, stating that the gravitational force between any two objects is proportional to the product of their masses and inversely proportional to the square of the distance between them: $F_{grav} = G \times \dfrac{m_1 \times m_2}{r^2}$. (Ch. 4)

universe 1. All of space and everything contained therein. 2. Our own universe in a collection of parallel universes that together make up all that exists.

unstable equilibrium An equilibrium state in which a small disturbance will cause a system to move away from equilibrium. Compare *stable equilibrium*.

UV Ultraviolet. See *ultraviolet radiation*. (Ch. 5)

V

vacuum A region of space devoid of matter. In quantum mechanics and general relativity, however, even a perfect vacuum has physical properties. (Ch. 5, 22)

variable star A star with varying luminosity. Many periodic variables are found within the instability strip on the H-R diagram. (Ch. 17)

velocity (*v*) The rate and direction of change of an object's position with time. Possible units include meters per second (m/s) and kilometers per hour (km/h). Compare *speed*. (Ch. 3)

vernal equinox 1. One of two points where the Sun crosses the celestial equator. 2. The day on which the Sun appears at that location, marking the first day of spring (about March 20 in the Northern Hemisphere and September 22 in the Southern Hemisphere). Compare *autumnal equinox*. See also *summer solstice* and *winter solstice*. (Ch. 2)

very low-mass star A star with mass between 0.08 and 0.5 M_{Sun}. (Ch. 16)

virtual particle A particle that, according to quantum mechanics, comes into existence only momentarily. According to theory, fundamental forces are mediated by the exchange of virtual particles. (Ch. 18)

visual binary A binary system in which the two stars can be seen individually from Earth. Compare *eclipsing binary* and *spectroscopic binary*. (Ch. 13)

void A region in space containing little or no matter. Examples include regions in cosmological space that are largely empty of galaxies. (Ch. 23)

volatile material Generally called *ice* in its solid form. Material that remains gaseous at moderate temperature. Compare *refractory material*. (Ch. 7)

volcanism A form of geological activity on a planet or moon in which molten rock (magma) erupts onto the surface. (Ch. 8)

W

W See *watt*.

waning Describing the changing phases of the Moon as it becomes less fully illuminated between full Moon and new Moon as seen from Earth. Compare *waxing*. (Ch. 2)

waning crescent Moon The phases of the Moon between third quarter and new Moon. (Ch. 2)

waning gibbous Moon The phases of the moon between full Moon and third quarter. (Ch. 2)

water cycle The flow of water on, above, and through Earth's surface. (Ch. 9)

watt (W) A measure of *power*. Unit: joules per second (J/s).

wave A disturbance moving along a surface or passing through a space or a medium. (Ch. 5)

wavefront The imaginary surface of an electromagnetic wave, either plane or spherical, oriented perpendicular to the direction of travel.

wavelength (λ) The distance on a wave between two adjacent points having identical characteristics. The distance a wave travels in one period. Possible units include meters (m). (Ch. 5)

waxing Describing the changing phases of the Moon as it becomes more fully illuminated between new Moon and full Moon as seen from Earth. Compare *waning*. (Ch. 2)

waxing crescent Moon The phases of the Moon between new and first quarter. (Ch. 2)

waxing gibbous Moon The phases of the Moon between first quarter and full. (Ch. 2)

weak nuclear force The force underlying some forms of radioactivity and certain interactions between subatomic particles. It is responsible for radioactive beta decay and for the initial proton-proton interactions that lead to nuclear fusion in the Sun and other stars. One of the four fundamental forces of nature, along with the *electromagnetic force*, the *strong nuclear force*, and *gravity* (definition 1). The force is mediated by the exchange of *W* and *Z* particles. (Ch. 22)

weather The state of an atmosphere at any given time and place. Compare *climate*. (Ch. 9)

weight The gravitational force acting on an object, that is, the force equal to the mass of an object multiplied by the local acceleration due to gravity. In general relativity, the force equal to the mass of an object multiplied by the acceleration of the frame of reference in which the object is observed. Compare *mass*. (Ch. 4)

white dwarf The stellar remnant left at the end of the evolution of a low-mass star. A typical white dwarf has a mass of 0.6 solar mass (M_{Sun}) and a size about equal to that of Earth; it is made of non-fusing, electron-degenerate carbon. (Ch. 16)

Wien's law A law, named for Wilhelm Wien, stating that location of the peak wavelength in the electromagnetic spectrum of an object is inversely proportional to the temperature of the object. (Ch. 5)

WIMP Short for *weakly interacting massive particle*. A hypothetical massive particle that interacts through the weak nuclear force and gravity but not with electromagnetic radiation. WIMPs are candidates for dark matter. Compare *MaCHO*. (Ch. 19)

winter solstice 1. One of two points where the Sun is at its greatest distance from the celestial equator. 2. The day on which the Sun appears at that location, marking the first day of winter (about December 21 in the Northern Hemisphere and June 20 in the Southern Hemisphere). Compare *summer solstice*. See also *autumnal equinox* and *vernal equinox*. (Ch. 2)

X

X-ray Electromagnetic radiation having frequencies and photon energies greater than those of ultraviolet (UV) light but less than those of gamma rays and having wavelengths shorter than those of UV light but longer than those of gamma rays. (Ch. 5)

X-ray binary A binary system in which mass from an evolving star spills over onto a collapsed companion, such as a neutron star or black hole. The material falling in is heated to such high temperatures that it glows brightly in X-rays. (Ch. 17)

Y

year The time Earth takes to make one revolution around the Sun. A solar year is measured from equinox to equinox. A sidereal year, Earth's true orbital period, is measured relative to the stars. Compare *tropical year*. (Ch. 2)

Z

zenith The point on the celestial sphere located directly overhead of an observer. Compare *nadir*. (Ch. 2)

zero-age main sequence The strip on the H-R diagram plotting where stars of all masses in a cluster begin their lives.

zodiac The 12 constellations lying along the plane of the ecliptic. (Ch. 2)

zodiacal dust Particles of cometary and asteroidal debris smaller than 100 microns (μm) that orbit the inner Solar System close to the plane of the ecliptic. Compare *meteoroid* and *planetesimal*. (Ch. 12)

zodiacal light A band of light in the night sky caused by sunlight reflected by zodiacal dust. (Ch. 12)

zonal wind The east–west component of a wind. (Ch. 9)

These pages contain selected answers for all 24 chapters in the complete volume of *21st Century Astronomy,* Seventh Edition. The Solar System Edition does not include Chapters 15–23. The Stars and Galaxies Edition does not include Chapters 8–12.

Chapter 1

CHECK YOUR UNDERSTANDING

1.1. d, a, e, c, f, b

1.2. b

1.3. c

WHAT IF . . .

p. 4. If our Solar System were located in the Andromeda Galaxy our cosmic address would be the same except for replacing Milky Way with Andromeda.

p. 9. A theory that is not likely to be falsifiable within many human lifetimes can still be a scientific theory if our knowledge and technology need those lifetimes to falsify it.

p. 11. If there were a region in the universe where the laws of physics were different than those on Earth, we would call into question the cosmological principle and begin a search for other regions that might also be different.

p. 17. If Earth formed a mere 4 billion years after the Big Bang, it would probably not have the abundances of elements that are formed in stars that it has today.

TEST YOUR UNDERSTANDING

3. b

5. d

9. d

12. c

15. b, d, a, c, e

THINKING ABOUT THE CONCEPTS

16. Figure 1.1 places each object in reference to its location: Earth in the Solar System, Solar System in the Milky Way, Milky Way in the Local Group, Local Group in the Virgo Supercluster, and our supercluster in Laniakea. Figure 1.3 directly relates difference sizes of these objects to an equivalent time—the longer the time, the larger the object. The sizes of the distance steps in Figure 1.1 are greater than those of Figure 1.3. Figure 1.1 is simply showing us our cosmic address.

20. 2.5 million years

23. A scientific theory has a more defined, stringent definition than the everyday use of "theory." A scientific theory puts together a well-developed idea that must agree with known physical laws and accounts for all of the relevant data leading to its development. It must make predictions that are

testable and must stand up to those tests. Everyday usage is more fluent, meaning "theory" can be an idea or applied to something tossed aside as unimportant.

30. Our composition is the same as what remains after the death of a star. The current theory of how stars are born, live, and die—and observations of stellar remnants—support this statement.

APPLYING THE CONCEPTS

32. (a) Hours are units of time. (b) The units would be km^2/h, which are not units of time. (c) Division of speed by distance gives $\frac{km/h}{km} = \frac{1}{h}$, which is not units of time. (d) Dividing the distance, km, by the speed, km/h, will give units of time. (e) $\frac{100\ km}{30\ km/h} = 3.3\ h.$

45. (a) 9.42×10^8 km. (b) 1.1×10^5 km/hr. (c) 2.6×10^6 km/dy.

Chapter 2

CHECK YOUR UNDERSTANDING

2.1a. b

2.1b. d

2.2. d

2.3. full moon; third quarter

2.4. longer

2.5. b

WHAT IF . . .

p. 30. If Earth did not rotate with respect to the stars (it is not in a synchronous orbit), then a day would last a year. Any given star would take half a year to rise and set. The seasons would be more extreme.

p. 37. With a slightly larger tilt of its axis, the seasons on Mars would have larger variations. This is because the tilt toward and away from the Sun is greater.

p. 43. If the Moon were a cube, it would still go through phases, but the shape of the Moon would depend on the orientation of the cube.

p. 45. The dates of the "wandering" holidays would not change.

p. 51. There would be more eclipses. With the two moons being 120° apart in their orbit, it would take about 10 days after the first eclipse for the second moon to pass through the node. The eclipses would occur in the same eclipse season.

TEST YOUR UNDERSTANDING

3. b

6. e

10. a

14. b

15. a

THINKING ABOUT THE CONCEPTS

18. From Figure 2.13 we can see that the constellation Gemini is in the sky during the daytime.

25. (a) Earth would always remain in the same position in the sky. (b) The phases of Earth would be opposite the current phase of the Moon as seen from Earth.

29. We can determine that Earth is curved, no matter what the location or time of the observations; that is, it is curved "all the way around," making it a sphere.

APPLYING THE CONCEPTS

37. 50° for both the altitude of Polaris and the latitude

42. (a) 66.5° N. (b) summer solstice.

Chapter 3

CHECK YOUR UNDERSTANDING

3.1a. While the stars always remained in the same location, the planets moved against those stars.

3.1b. c

3.2. c, b, d, a

3.3. d

3.4. b

WHAT IF . . .

p. 62. Because the astronauts would be orbiting Earth faster than the Moon, they would observe retrograde motion of the Moon.

p. 68. The Northern Hemisphere winters would be much warmer and the Southern Hemisphere summers much hotter. When Earth was at its farthest, the Northern Hemisphere summers would be much colder and Southern Hemisphere winters cooler for a much longer period of time.

p. 71. Given $A = 2$ AU, $P = 4$ years, $P^2 = A^3$; $4^2 \neq 2^3$.

p. 73. One might hypothesize that the relationship is different for moons orbiting planets versus planets orbiting the Sun. Kepler applied his third law to the planets.

TEST YOUR UNDERSTANDING

3. a

8. d

12. c

15. c

THINKING ABOUT THE CONCEPTS

16. Mars is larger at that time because it is closer to Earth. The changing size of Venus is due to its changing distance from Earth.

22. The Moon's orbital eccentricity of 0.05 is greater than Earth's, which is 0.017. Since Earth's orbit is very close to a circle, whether there is a total

eclipse or an annular eclipse of the Sun will depend on the Moon's distance from Earth.

27. We wear seat belts to protect us from acceleration. Due to our inertia, we resist a change in our state and if not held by a seat belt, we would react to the force being applied.

APPLYING THE CONCEPTS

33. (a) Because it is closer to the Sun, Venus's sidereal orbital period will be slightly shorter than Earth's. (b) $P = 0.62$ yr. (c) The answer agrees with the prediction. (d) $P = 1.9$ yr.

37. It is approximately three times larger at the thin crescent versus the gibbous phase. The ratio of distances gives $\dfrac{1.72\,\text{AU}}{0.28\,\text{AU}} = 6$.

Chapter 4

CHECK YOUR UNDERSTANDING

4.1a. d
4.1b. c, a
4.2. c, a, b
4.3. d, b, a, c
4.4. a

WHAT IF . . .

p. 86. If the gravitational force did not depend on an object's mass, then we could assume the force would be the same for all objects. More massive objects would still have more inertia, would resist acceleration, and thus would fall more slowly.

p. 89. Assuming you did not reach terminal velocity, you would free-fall like an astronaut in orbit. Astronauts train for weightless conditions in deep pools.

p. 100. As the Moon's orbital radius shrank to half, the tidal forces would increase by a factor of eight. If it shrank to one-third, the tidal forces would increase by a factor of 27. Life on Earth's surface would need to move to higher ground and ocean life would have to deal with very strong currents during the extremes of the changing tides.

p. 102. If the Moon were orbiting as close as two Earth radii, then the probability that the two worlds would be tidally locked (like Pluto and Charon) would greatly increase.

TEST YOUR UNDERSTANDING

3. c
8. a
12. d
14. d

THINKING ABOUT THE CONCEPTS

20. It would be unbound.
25. Because Earth rotates from west to east faster than the Moon orbits, the high tide gets dragged away from the location of the Moon.
30. The tidal tail shows the trajectory of the galaxy on the right and indicates that that galaxy has

entered the Roche limit of the other galaxy and the two will eventually collide.

APPLYING THE CONCEPTS

38. $\dfrac{F_{\text{ground}}}{F_{\text{space}}} = 1.11$

39. 7698 m/s, or a little over 17,000 mph

Chapter 5

CHECK YOUR UNDERSTANDING

5.1a. e, c, b, d, a
5.1b. c
5.2. Each element has a unique pattern of spectral lines.
5.3. d
5.4. d, b, a, c, e
5.5. c

WHAT IF . . .

p. 112. Assuming the medium was the same all through space, the speed of light would not change.

p. 114. If we could see at infrared wavelengths, we would be able to tell the temperatures of objects and see them in the dark. Our lives would not change if we had X-ray vision because very few objects on Earth emit that radiation and our atmosphere blocks X-rays from space.

p. 126. The Sun is almost all hydrogen and helium. There would not be enough carbon, nitrogen, or oxygen for life to evolve.

p. 127. If an ambulance exceeded the speed of sound, we would see it go by first and then hear it.

p. 132. If a planet emitted less radiation as it heated up, and if its star kept heating it, then eventually the planet would reach very high temperatures and never reach equilibrium.

p. 135. The length of time Earth would take to reach equilibrium would depend on its radius and other factors such as ocean and mantle depth.

TEST YOUR UNDERSTANDING

2. b
5. b
11. c
13. d

THINKING ABOUT THE CONCEPTS

16. Ultraviolet radiation runs from about 50 to 350 nm. The radio part of the spectrum is very broad, running from around 10 cm to over 1 km in wavelength. When combining multiple parts of the spectrum, astronomers gain a great deal more information about an object.
21. Astronomers find the same elements across the entire universe—locally as well as in the farthest objects observed.
28. The brightnesses of stars do not reveal their distances because stars vary greatly in their luminosities.

APPLYING THE CONCEPTS

36. 10,000 km/s. Away from you.
38. 0.00005 times or 0.005 percent as bright

Chapter 6

CHECK YOUR UNDERSTANDING

6.1a. a, c
6.1b. a
6.2. c
6.3. d
6.4. c
6.5. a, b, d

WHAT IF . . .

p. 146. One possibility is that our eyes would capture more light.

p. 156. With a quantum efficiency of 100% and an integration time of 1 ms, we would be able to see in the dark and distinguish separate events that happen far faster than we can distinguish now.

p. 159. With almost no atmosphere, astronomers on this planet would see celestial objects far more sharply and would not have to worry about atmospheric blurring.

p. 165. A primary goal would be to find out if there is life on any planets in those planetary systems.

p. 166. We might conclude that mergers of black holes and neutron stars were more rare than expected.

TEST YOUR UNDERSTANDING

1. c
7. a
10. b
13. d
14. b

THINKING ABOUT THE CONCEPTS

19. Information gathered from observations at all wavelengths tells us so much more about the objects we observe.
23. An adaptive optic system will observe a laser-generated star and use sophisticated computer programs to control a deformable mirror to keep the image steady. Figure 6.14 reveals the fine details that become apparent with the use of adaptive optics.
29. Neutrino detectors are located deep inside mines. Neutrinos are neutral and interact very weakly with matter, and the only way to detect one is to watch the result of it impacting another atom.

APPLYING THE CONCEPTS

36. (a) 14 arcsec. (b) This limit is 5 to 10 times smaller than the stated resolution. (c) The diffraction limit gives the theoretical limit. The lens of an eye and the placement of the rods and cones on the retina are far from the perfection needed for such high resolution.
43. (a) 2.1×10^{10} km. (b) 19 hr. (c) 5.2×10^{-4}, or 0.05 percent.

Chapter 7

CHECK YOUR UNDERSTANDING

7.1. a, b, d

7.2. a

7.3. b

7.4. a

7.5. c

WHAT IF . . .

p. 178. A planetary system with planets orbiting in oblique (slanted to each other), separate planes could still support the nebular hypothesis. We would expect there to be multiple collisions between planets during formation. Perhaps collisions would be the cause for the different orbital planes. For example, Pluto's orbit.

p. 186. The dwarf planets far from the Sun have no atmospheres and their compositions are mostly solid ices. So far, none has a mass as great as Earth. With the paucity of planet-forming material far from its central star, the planet probably formed close to its star and was gravitationally flung out by interactions with one or more massive planets.

p. 194. One would conclude that solar systems are quite rare; fewer than 10 million of them in our galaxy. We would not yet conclude that there are no other planetary systems in the Milky Way. More stars may have planets, but the planets would be of very low mass and/or very far from their central stars; in other words, not similar to those that we have discovered with current technology.

TEST YOUR UNDERSTANDING

2. c

7. a

8. a

12. c

15. d

THINKING ABOUT THE CONCEPTS

16. What an Astronomer Sees, Figure 7.8.

19. A protoplanetary disk is the spinning disk of gas and dust out of which planets form around a central star. The inner part will be hotter because it is closer to the central star and because the inner material gained energy when it moved in from the outer regions.

28. Stars are very bright and far away, and the very faint reflected light from an exoplanet would come from a position extremely close to its star. Thus, it is difficult to mask out the light of the star and still see close enough to it to see the reflected light of an exoplanet.

30. The Kepler satellite used the planetary transit method and the monitoring of hundreds of millions of stars in the constellation of Cygnus to search for exoplanets. By "Earth-like" we mean planets that are terrestrial and have roughly the same mass as Earth, have thick protective atmospheres, and orbit in the habitable zones of their stars.

APPLYING THE CONCEPTS

36. 30 m/s × 2.23694 mph/1 m/s = 67 mph. Earth's orbital speed is 1000 times that.

38.
$$\frac{L_{Venus}}{L_{Earth}} = \frac{m_{Venus}}{m_{Earth}} \times \left(\frac{R_{Venus}}{R_{Earth}}\right)^2 \times \frac{P_{Earth}}{P_{Venus}}$$
$$= 0.815 \times 0.950^2 \times \frac{1}{243} = 0.003$$

42. $\triangle\lambda = 500 \text{ nm} \frac{0.09 \text{ m/s}}{3 \times 10^8 \text{ m/s}} = 1.5 \times 10^{-7} \text{ nm}$

Chapter 8

CHECK YOUR UNDERSTANDING

8.1. b

8.2. d

8.3a. d

8.3b. c

8.4. c

8.5. a, c, d

8.6. a, c, d, e

WHAT IF . . .

p. 207. We would expect equal or greater surface cratering.

p. 212. Yes, we would adjust for the beginning ratio.

p. 216. We would want the poles to migrate over time and not leave Earth unprotected from charged particles from space.

p. 219. There would be just one extremely large Hawai'ian island.

p. 225. Observing a volcanic eruption on an Earth-like planet might lead to our concluding that the planet has a hot interior and is still cooling.

p. 228. If large quantities of liquid water on Mars obeyed the water cycle, we might observe the erosion of Olympus Mons and features in Valles Marineris.

TEST YOUR UNDERSTANDING

1. b

5. d

10. c

14. c

THINKING ABOUT THE CONCEPTS

22. We can detect a shadow zone for secondary earthquake waves.

25. Mars has the largest volcanoes with Olympus Mons standing 27 km high and others that reach over 14 km. Mauna Kea on Earth reaches over 9 km high from its base.

27. Mercury and the Moon have many more craters of all sizes than Earth, Venus, and Mars.

30. Given that the region surrounding the large crater has very few craters, we can state that the surface is relatively young.

APPLYING THE CONCEPTS

36. (a) Up until about 3 billion years ago, the cratering rate fell off abruptly. Since then, the rate has fallen gradually. (b) It is hard to quantify without a y-axis scale, but today's rate is very close to zero. (c) The drop in the cratering rate indicates that almost all of the leftover material from the formation of the Solar System has been incorporated into the larger worlds.

45. 1.2×10^8, or 120 million years

Chapter 9

CHECK YOUR UNDERSTANDING

9.1. a, b, d

9.2. c

9.3a. c, a, e, b, d

9.3b. a

9.4. d, c, b. Mercury's variations are due to rotation, not revolution.

9.5. a, b, c

WHAT IF . . .

p. 243. Sublimation of CO_2 and H_2O from polar caps and impacts from asteroids and comets would help restore the atmosphere.

p. 247. The secondary atmospheres would be composed of outgassing from volcanoes and any volatiles from asteroid impacts.

p. 249. We would look for oxygen, O_2, because life is the primary producer of this molecule and it is short-lived in an atmosphere.

p. 253. Without the protection of Earth's magnetic field, astronauts would be bombarded with charged particles.

p. 258. With Mars's rotation period being similar to Earth's, we would expect the zonal winds to be more like Earth's.

p. 263. A few changes might include rising sea levels and more frequent and violent storms. Some areas on Earth will be so hot as to be inhabitable.

TEST YOUR UNDERSTANDING

3. c

6. b

7. b

11. c

14. d

THINKING ABOUT THE CONCEPTS

18. The secondary atmospheres were formed by cometary impacts and outgassing from volcanoes.

25. Venus is surrounded by a thick opaque cloud of gas.

26. The three terrestrial planets with atmospheres all have surface temperatures that are much higher than can be explained through the balance of energy received by the Sun and given off by the heated surface.

APPLYING THE CONCEPTS

37. From Figure 9.7, we see that we would have to go at least 40 km above Earth's surface to experience the same atmospheric pressure as the surface of Mars.

45. (a) 3×10^{15} kg. (b) 0.01. (c) 4×10^{38} molecules/yr. (d) Because it is the number of molecules that matters, not the percentage.

Chapter 10

CHECK YOUR UNDERSTANDING

10.1. a
10.2. b
10.3. b
10.4. d
10.5. a, d, b, c

WHAT IF . . .

p. 280. As with seasons on Uranus, Earth's poles would have 3 months of more direct Sun and 3 months of darkness. Weather during each season would be more extreme.

p. 284. Calculations show that Jupiter has enough gravity to hold on to its atmosphere of hydrogen and helium should its orbit shrink to that of Venus. However, it might get "puffed out" and become larger.

p. 290. The smaller shrinking rate would produce less thermal energy than is produced now. The shrinking rate would increase if Jupiter had more mass.

p. 294. Searching for magnetospheres would be a good way to detect other large planets due to the large sizes of these magnetic fields and the radiation they emit.

p. 296. Earth might have been either ejected from the Solar System or sent into a highly elliptical orbit. Either way, life would probably not develop.

TEST YOUR UNDERSTANDING

3. b
6. b
9. d
15. d

THINKING ABOUT THE CONCEPTS

21. Giant planets rotate very rapidly, and rapidly rotating spheres tend to expand along their equator and contract along their poles. This would make these planets oblate, or "squished."

22. Colorful regions on Jupiter are those for which we see deeper into its atmosphere and effectively see "smog"; that is, chemical pollutants that have caused the gases and ices to take on reddish colors. Methane gas, which is abundant in the atmospheres of Uranus and Neptune, strongly absorbs the longer red wavelengths of visible light but reflects the shorter blue and green wavelengths.

28. Jupiter moved from right to left in the image, reflecting its orbital prograde motion of west to east.

APPLYING THE CONCEPTS

36. Figure 10.1a shows the relative sizes of the planets and Figure 10.1b shows the relative angular diameters when viewed from Earth.

42. (a) 1320 Earths. (b) $p = \dfrac{M}{V} = \dfrac{318 \text{ Earths}}{1{,}320 \text{ Earths}} = 0.24.$

Chapter 11

CHECK YOUR UNDERSTANDING

11.1. a, b, c
11.2a. b, c, d
11.2b. a, d, b, c
11.3. a
11.4. b

WHAT IF . . .

p. 306. Possible answer: The sixth moon orbits either inside all other moons or it orbits outside them.

p. 314. Answers will vary. Possible moons are those thought to have salty oceans beneath their icy crusts.

p. 320. The resulting ring would fall along the Moon's orbital path. The ring could persist for a length of time as the Moon orbits outside Earth's Roche limit. We would see light reflected from the volcanic dust particles emitted.

TEST YOUR UNDERSTANDING

2. b
7. c
11. a
15. c

THINKING ABOUT THE CONCEPTS

19. Ultraviolet photons from the Sun photodissociate the methane molecules, producing organic molecules such as ethane.

24. Planetary ring material is likely produced by the breakup of moons, asteroids, or comets that fall inside a planet's Roche limit. It is also produced as nearby moons shed material through volcanic processes or impact events.

25. Gaps in Saturn's rings are created by small moons that clear out a lane of material as they orbit within a ring. Gaps are also created by orbital resonances of shepherd moons.

APPLYING THE CONCEPTS

37. (a) Both scales are linear. (b) Figure 11.2b covers 225 times as much area.

45. (a) A more massive planet will have a higher escape velocity. (b) The escape velocity is less. (c) Find the planet's radius.

Chapter 12

CHECK YOUR UNDERSTANDING

12.1. because they orbit the Sun and not another planet
12.2. c
12.3a. c
12.3b. c
12.4. a, b, c, e
12.5. d

WHAT IF . . .

p. 336. round moons, additional Kuiper Belt objects

p. 344. lots of dust, tenuous surface conditions, low gravity

p. 349. There would be two tails. Both comets would orbit together.

p. 355. Stony meteorites are hard to distinguish from regular rock. You could use a magnet to test for stony irons and irons.

p. 358. Instigate a slow, steady change in the asteroid's orbit.

TEST YOUR UNDERSTANDING

4. a
9. b
12. b
15. d

THINKING ABOUT THE CONCEPTS

19. An old surface would have had time to become heavily cratered.

22. Asteroids that cross Earth's orbit are potentially hazardous.

27. Comet tails have extremely low densities.

APPLYING THE CONCEPTS

36. 340 km/hr.

43. (a) 29 times. (b) 9×10^{12} kg. (c) 0.04 or 4 percent.

Chapter 13

CHECK YOUR UNDERSTANDING

13.1. b
13.2. d
13.3. a, b, d
13.4. b, d

WHAT IF . . .

p. 370. We would expect stars at very different distances, implying a range of luminosities.

p. 376. The stars have different compositions, but similar temperature.

p. 381. You could still get velocities, and Kepler's laws apply.

p. 385. as a linear relationship resembling Figure 13.18b, since higher masses mean higher temperatures

TEST YOUR UNDERSTANDING

1. b
5. b
8. a
10. d
15. a

THINKING ABOUT THE CONCEPTS

21. The original spectral sequence was based on the strengths of the hydrogen lines in the spectra and was developed before we understood atoms. When that knowledge came, we realized that O and B stars had weak hydrogen lines due to ionization, and they were moved ahead of A stars. The temperature sequence runs from the very hottest stars to the very coolest.

24. That third star would be Proxima Centauri. Since it is at the same distance as A and B, it would be dimmer with a much lower luminosity.

26. On Mars, we will be able to measure the distances to about 1.5 times as far. On Jupiter, we would get accurate parallaxes to about 5 times farther away. If we were on Venus, we would measure accurate parallaxes only for stars closer than about 70 percent of what we currently measure.

APPLYING THE CONCEPTS

36. Star B has ½ the parallax as star A. The third star would have ¼ the parallax. The third star has ½ the parallax as star B.

38. (a) Almost 10^6 (1 million) times as luminous.
    (b) About 10^{-4} (0.0001) times as luminous.

Chapter 14

CHECK YOUR UNDERSTANDING

14.1a. c
14.1b. b
14.2. c
14.3. b
14.4. d

WHAT IF . . .

p. 403. There would be no variation between night and day.

p. 407. Either the opacity or the method of energy transport or both could be different.

p. 411. The photon energy would be much higher.

p. 417. There would be fewer sunspots.

TEST YOUR UNDERSTANDING

1. a
7. c
11. b, c, d, f, g, e, a
15. c

THINKING ABOUT THE CONCEPTS

20. Fission requires heavy elements to exist in the core, such as uranium with atomic number 92. The Sun is made up almost exclusively of light elements; mainly H and He.

24. The winds from solar flares contain huge amounts of charged particles that can disrupt or destroy electronics on satellites and orbiting telescopes. The periods of increased solar activity cause Earth's upper atmosphere to expand slightly, increasing the drag on orbiting satellites and causing their orbits to decay. Unless the satellites are boosted back up again, this effect can cause them to crash to the ground.

28. Data from sunspots give us the Sun's rotation period, the direction of rotation, and the fact that it experiences differential rotation.

APPLYING THE CONCEPTS

37. Figure 14.19 shows (b) the location and number of sunspots versus time and (c) the strength and position of magnetic fields versus time. If we are to argue that spots result from strong magnetic fields, then they should appear in roughly the same regions over time. Inspection of the figures shows that the butterfly shape seen in panel (b) is reproduced in panel (c), confirming that the two effects are correlated.

42. 3.75×10^5 s ~ 4.3 dy

Chapter 15

CHECK YOUR UNDERSTANDING

15.1. a, d
15.2. c
15.3. a
15.4. b

WHAT IF . . .

p. 433. The dust in the cloud would be "blown" away by the high-energy photons.

p. 439. O stars that form fast and first tend to clear out material from the star-forming region and thus prevent some stars from forming.

p. 445. Jupiter would be much brighter as a brown dwarf, especially at infrared wavelengths.

p. 448. That it was still hidden by lots of dust.

TEST YOUR UNDERSTANDING

4. c
8. c
11. a
12. c
15. a

THINKING ABOUT THE CONCEPTS

17. The regions where stars are forming are usually shrouded by a lot of dust that blocks visible light. Infrared wavelengths, however, are not blocked so astronomers can observe details otherwise hidden.

22. Low-density regions have very few collisions that lead to the cooling of the gas, unlike high-density regions, where collisions are frequent.

26. The core of the star eventually starts fusing hydrogen to helium, which provides the pressure needed to halt the gravitational collapse.

27. Any regions that contain a lot of molecular hydrogen, H_2, would be seen at radio wavelengths. Visible wavelengths may show the dust disk surrounding the star, blocking the star's light, and maybe regions where the jets are creating shock waves in the interstellar medium.

APPLYING THE CONCEPTS

36. Reddening by the interstellar dust will make stars appear redder than they are by reducing the amount of blue light observed.

41. $\frac{v_e}{v_p} = 43$

Chapter 16

CHECK YOUR UNDERSTANDING

16.1. a
16.2. b
16.3. a, c
16.4. a
16.5. d

WHAT IF . . .

p. 461. The stars would live a lot longer on the main sequence.

p. 462. The orbit of the Moon would not change, but the acceleration due to gravity on Earth's surface would be 100^2 or 10,000 times greater.

p. 467. The core would keep contracting until the temperature and density were high enough to start helium-to-carbon fusion.

p. 473. The inner planets might be engulfed. Mass loss from the Sun might mean the size of planetary orbits would increase.

p. 475. The planets survived the evolution of their star and are probably all rocky.

p. 476. Each star would evolve according to its mass.

p. 478. It would depend on how much mass it accreted from its companion and what its current mass is. With a companion having 0.8 times Sun's mass, there would probably be no worry about it going supernova.

TEST YOUR UNDERSTANDING

2. a, e, d, c, f, b, g
8. d
10. c
14. a
15. c

THINKING ABOUT THE CONCEPTS

20. For having the same brightness, the derived distances to Type Ia supernovae would increase if it were found that they were more luminous than previously thought.

24. The horizontal branch star has a smaller radius.

27. It is cooling while staying the same size.

30. Low-mass stars spend enough time on the main sequence to increase the possibility that complex life has formed.

APPLYING THE CONCEPTS

41. (a) 1.5. (b) 45.

45. 112,000 m = 112 km

Chapter 17

CHECK YOUR UNDERSTANDING

17.1. b

17.2. b

17.3. d

17.4. a

WHAT IF . . .

p. 492. The white dwarf must have been more massive originally as it evolved faster. The currently variable star must have gained mass from its companion.

p. 497. neutrinos and then the visible light

p. 506. The gas may be totally dispersed. The neutron star would be rotating more slowly.

p. 510. There were many more generations of stars that formed and died that enriched the interstellar medium from which this star formed.

TEST YOUR UNDERSTANDING

2. e, a, f, b, c, d

4. d

7. e

12. a

14. b

THINKING ABOUT THE CONCEPTS

16. The bumps most noticeably change in brightness. It doesn't appear that the ring changed size.

18. High-mass stars begin fusing helium to carbon in their core when the temperatures and pressures get high enough. There is no helium flash.

21. We just need to measure the period of the variability to get the luminosity of the star through the observed period-luminosity relation.

28. There is a relationship between the mass of a main-sequence star and how long it stays on the main sequence.

APPLYING THE CONCEPTS

38. The bottom image of Figure 17.4 is not to scale.
$$\frac{0.01\,R_{Sun}}{1,000\,R_{Sun}} = 0.00001$$

42. (a) 3.8×10^{23} kg/min. (b) 5.2 (Moon masses).

Chapter 18

CHECK YOUR UNDERSTANDING

18.1. d

18.2. b

18.3. d

18.4. c

WHAT IF . . .

p. 520. Converting the speed of light to meters per second: 200 km/hr = 0.056 km/s = 56 m/s. The length of a soccer field is ~110 m. It would take ~2 s for a goalie to see what was happening at the other goal.

p. 524. Using $E = mc^2$ and solving for m: $m = E/c^2$, leading to the conclusion that a change in energy of the containers results in a change in its mass, but one that is extremely small.

p. 528. The ball would remain in orbit and would thus be accelerated by Earth's gravity. If the space station is an inertial frame, then so is the ball.

p. 538. Detection could occur through the black hole's gravitational influence on other stars or through gravitational lensing of starlight.

TEST YOUR UNDERSTANDING

1. c, a, b

6. d

8. b

11. c

15. c, d

THINKING ABOUT THE CONCEPTS

16. There is no difference between the shifts in wavelengths due to gravity or radial motion.

19. Twin A could never return before twin B was born, because time continues to pass along normally for twin B.

23. The astronaut is moving with the ship and is in the same reference frame, thus he will see no length contraction. An outside viewer who is *not* moving at that high speed would observe length contraction.

27. The color of the star would change to that of longer and longer wavelengths of light, say from yellow to orange to red to infrared and to radio.

APPLYING THE CONCEPTS

40. We might not get the *exact* number, but γ does not deviate significantly from 1 until the speed is about 0.33 to 0.5c. The Lorentz factor goes to infinity as the speed of an object approaches the speed of light.

44. (a) Working It Out 18.1 tells us for that spaceship, $v = 0.99c$, $\gamma = 7.09$. It will take 25.25 years (25/0.99)

to reach the alien civilization from both Earth's and the aliens' viewpoints. A spaceship traveling at 0.99c would take 25 ly/7.09 = 3.53 years to get there. (b) A return will happen 50.5 years later for the people on Earth.

Chapter 19

CHECK YOUR UNDERSTANDING

19.1a. d

19.1b. c

19.2. c

19.3. a

19.4. c

WHAT IF . . .

p. 555. If two giant spiral galaxies merged perpendicularly, then a giant elliptical galaxy might form with stars orbiting in all directions. If they merged edge on, and the disks were rotating in the same direction, then maybe most stars would eventually make up an even larger giant spiral.

p. 556. The distances to galaxies would all be greater.

p. 563. WIMPS would be streaming from all directions due to the random motions of halo objects. Since WIMPS would be detected by their mass, sometimes most of the particles would be coming at 250 km/s and 6 months later most of the particles would be coming at 190 km/s.

p. 568. Even before the supermassive black holes powering the AGNs merged, we would expect to observe gravitational waves.

TEST YOUR UNDERSTANDING

2. c

5. a

10. d

11. c

15. b

THINKING ABOUT THE CONCEPTS

16. Between 2 spirals: 7. Between 2 ellipticals: 0. Between an elliptical and a spiral: 1

19. Observing different kinds of standard candles gives one more confidence in the results.

25. Quasars appear to be associated with mergers and interactions between galaxies. Because they are observed at great distance, and thus back to an earlier time, these mergers must have been much more prevalent in the past when the universe was younger and the distances between galaxies smaller.

29. A nearby rung is needed to calibrate a more distant one; there must be overlap so that we can reliably use the more distant rungs.

APPLYING THE CONCEPTS

41. Because the object varies within an 83-minute period, it can be no larger than 83 light-minutes

across. Light travel time for 1 AU is 8.3 light-minutes, implying the object is 10 AU across.

45. (a) $E = \triangle mc^2 = (5.97 \times 10^{24}\,\text{kg}) \times 0.1 \times (3 \times 10^8\,\text{m/s})^2$

$= 5.4 \times 10^{40}$ J. (b) $\dfrac{5.4 \times 10^{40}\,\text{J}}{3.85 \times 10^{26}\,\text{J}} = 1.4 \times 10^{14}$

Chapter 20

CHECK YOUR UNDERSTANDING

20.1. The Milky Way has lots of star formation and dust. Observations over a wide range of wavelengths show spiral structure and the motion of stars orbiting the center on a disk.

20.2. The bulge, halo, and thick disk contain mostly old stars whereas the thin disk contains young stars.

20.3. c

20.4. The gravitational force between the Andromeda and Milky Way galaxies is strong enough to overcome other motions.

WHAT IF . . .

p. 583. There would not be a lot of dust and gas in these arms as no new star formation is occurring.

p. 585. The star would be low mass (a red dwarf) and crossing the disk in its halo orbit. We would expect this star to be a high-velocity star.

p. 593. If the Large Magellanic Cloud were to merge with our galaxy and came close enough to the center to merge with the black hole there, the black hole may then become an active galactic nucleus.

p. 597. With a much more crowded cluster of galaxies, the Milky Way might come close enough to merge with one or more of the other galaxies. Stars with planets would be affected, and if the mergers occurred early enough, and Earth were ejected from solar system, life may not have had time enough for humans to appear.

TEST YOUR UNDERSTANDING

2. a
7. b
8. b
13. a, b, c

THINKING ABOUT THE CONCEPTS

16. We observe many spiral galaxies and note that there are central bright areas surrounded by relatively flat disks. These disks contain dust and hot, bright stars. Since we observe many spiral galaxies with a range of viewing angles, we can assume that the Milky Way would be similar if, for example, it were viewed face on. If we lived in an elliptical galaxy, we would not see any dust or formation of massive stars. The stars we do see would be distributed spherically around us.

21. Halo stars are distinguished from disk stars by their low abundances of elements heavier than helium and their relatively fast orbits that trace back to the halo.

23. Lots and lots of dust between Earth and the center of the galaxy is opaque at visible wavelengths but transparent at X-ray, IR, and radio wavelengths.

26. The Large and Small Magellanic clouds are dwarf satellite galaxies that have interacted many times with the Milky Way, exchanging gas and dust and possibly stars.

APPLYING THE CONCEPTS

40. (a) 480 Myr. (b) 27 orbits.

45. Ratio: $\dfrac{158\,\text{kpc}^3}{296\,\text{kpc}^3} = 0.53$.

Chapter 21

CHECK YOUR UNDERSTANDING

21.1. b, a
21.2. d
21.3a. c
21.3b. b
21.4. a

WHAT IF . . .

p. 607. We would still not live in a special place because anyone anywhere else would see the same patterns that we do.

p. 612. Still, nothing special would be implied because the same relationship would be observed from a distant galaxy looking back at the Milky Way.

p. 615. If the universe were contracting, the light from other galaxies would be blueshifted.

p. 618. A dark cloud having a temperature of 1.5K would have a very high density.

TEST YOUR UNDERSTANDING

3. c
5. d
10. c
11. d
13. d

THINKING ABOUT THE CONCEPTS

16. The features in each view are in the same areas at the same relative scales. What has improved is the resolution, something with which astronomers are very familiar.

18. The galaxies are fixed in space, and it is that space that is expanding.

22. The expansion of the universe applies to the spaces among galaxy clusters. Gravity is in control for the Sun (as it is for all stars) and the Milky Way (as it is for all galaxies).

29. As it expanded, the universe became too cool to fuse any elements heavier than hydrogen to helium.

APPLYING THE CONCEPTS

38. (a) theory. (b) observations. (c) Within the ranges given, the theory and observations match.

41. (a) 4100 Mpc. (b) The age of the universe times the scale factor gives an age of about 2 billion years. Table 23.1 gives an age between 900 million ($z = 6$) and 1.2 billion ($z = 5$) years. (c) $R_U = 0.15$ The universe has expanded by about 7 times.

Chapter 22

CHECK YOUR UNDERSTANDING

22.1. c
22.2. c
22.3. The flatness and horizon problems.
22.4. d
22.5. a

WHAT IF . . .

p. 631. With $\Omega > 1$, the universe is closed and astronomers far in the future would observe the universe contracting and the light from galaxies blueshifted.

p. 632. All the galaxies would be much farther apart.

p. 641. The four forces would still break, but the ripples in the cosmic microwave background would not be smoothed out. There would be no flatness or horizon problem.

p. 645. Yes. Since an interaction of dark matter particles and antiparticles today would be observable at some distance away.

TEST YOUR UNDERSTANDING

2. c
5. a, f, j, b, e, g, d, i, h, c
10. b
14. b
15. b

THINKING ABOUT THE CONCEPTS

18. Dark energy is vacuum energy—that is, a repulsive force that results from the energy present in totally empty space.

19. Gravity is in control for the Milky Way (and all galaxies and galaxy clusters), the Solar System (and all planetary systems), and for all planets.

23. The early universe was *hot* and *dense*, which are situations that are replicated within the realm of high-energy particle physics. At these high temperatures and densities, all sorts of exotic forms of sub-atomic matter can exist that may have had a substantial influence on how the early universe evolved.

28. Grand unified theories (GUTs) attempt to unite the electroweak and strong nuclear forces. Theories of everything (TOEs) attempt to unify suitable GUTs with gravity.

APPLYING THE CONCEPTS

34. Each gamma ray has an energy: $E = 8.2 \times 10^{-14}$ J, and 2 gamma rays are emitted.

40. (a) The time axis is approximately logarithmic.
    (b) Density has dropped by 125 orders of magnitude. (c) Temperature has dropped by about 31 orders of magnitude.

Chapter 23

CHECK YOUR UNDERSTANDING

23.1. c, b, d, a
23.2. b
23.3. c
23.4. b

WHAT IF . . .

p. 659. With dark matter filling the voids, the total mass of the universe would increase, thereby increasing the density, perhaps exceeding the critical level.

p. 666. Without the gravity from dark matter, it would take longer for structure to form. More time would be needed for intelligent life to appear.

p. 668. We should look at "blank" regions of the sky at infrared wavelengths.

p. 678. Infrared wavelengths would be used to observe galaxies that are far away, at higher redshifts, and thus younger than those seen at visible wavelengths. We would need to pause the simulation to match the age of the universe corresponding to the redshifts.

TEST YOUR UNDERSTANDING

3. b
6. a
8. d
9. a
14. d

THINKING ABOUT THE CONCEPTS

17. If we could see the early universe when galaxies were first forming, it would be smaller and more active. There was a large amount of hot gas but little dust. Smaller, irregularly shaped protogalaxies dominated the visible universe. Galaxies were merging, making irregular galaxies a common sight.

23. Inhomogeneities in the density of normal matter were never strong enough to cause the structure seen today. There had to be another source of gravity.

27. As we look back in time, we should see more spiral galaxies than elliptical galaxies. As we look back even farther in distance and time, we should see galaxies that have not yet formed spiral arms, and galaxies should have a lot of massive stars in them because they have just formed and are not yet old enough to have exploded. This is what is observed.

30. The strong and weak nuclear forces are too short ranged to work over cosmic distances. Electromagnetic forces work only for charged particles, and most matter in the universe is neutral. This leaves only gravity.

APPLYING THE CONCEPTS

34. $z = 6.6$
43. Early universe is on the right, at a redshift of 10. The star formation rate now is about 0.01 M_{Sun}/yr Mpc^3 whereas the star formation rate at its peak, some 2.5 billion years after the Big Bang, was approximately 0.2 M_{Sun}/yr Mpc^3. This is 20 times higher than today's rate.

Chapter 24

CHECK YOUR UNDERSTANDING

24.1. d
24.2. a
24.3a. a, b, c
24.3b. a
24.4. b

WHAT IF . . .

p. 689. The environments of the ocean depths and tide pools are very different and thus we would expect that perhaps two different strands of life would evolve.

p. 694. Yes. Even a small change in a basic molecule of self-replicating life may lead to evolution if that change is beneficial and can be passed on.

p. 699. Even a weak magnetic field would offer some protection against cosmic rays, and the greenhouse gases would keep the planet warmer if it is at the outer edge of the star's habitable zone, thus increasing the size of the zone.

p. 702. These are good questions posed by Carl Sagan with many, many answers.

TEST YOUR UNDERSTANDING

4. c
6. a, c
9. d
13. c
15. d

THINKING ABOUT THE CONCEPTS

20. Cyanobacteria slowly broke CO_2 molecules apart through photosynthesis to form oxygen. This required about 2.25 billion years to reach the oxygen levels we have today.

26. The hydrogen and most of the helium in our bodies are from the Big Bang. All the other atoms are from the winds produced by stars as they ascended the giant branches and supernovae.

30. Main-sequence stars slowly become more luminous over time, pushing the inner edge of the habitable zone out past Earth. In a billion years or so, the increase in the Sun's luminosity will cause the oceans to evaporate and scorch all of Earth.

APPLYING THE CONCEPTS

39. 6.87×10^{11} or 687 billion bacteria.
43. Mercury is not in the habitable zone. Its range of temperatures is due to its lack of an atmosphere and the differences between its "day" and its "night."

These pages contain credits for all 24 chapters in the complete volume of *21st Century Astronomy*, Seventh Edition. The Solar System Edition does not include Chapters 15–23. The Stars and Galaxies Edition does not include Chapters 8–12.

Photos

FRONT MATTER

Page ix: Left_Coast_Photographer/Istockphoto/Getty Images Plus; **p. xi**: All Canada Photos/Alamy Stock Photo; **p. xiv**: SDO/AIA; **p. xvi**: NG Images/Alamy Stock Photo; **p. xxi**: Anne DeMarinis Design for W. W. Norton & Company; **p. xxii**: NASA, ESA, and the Hubble 20th Anniversary Team (STScI); **p. xxiv both**: W. W. Norton & Company; **p. xxxiv author Palen**: Courtesy Stacy Palen; **p. xxxiv author Blumenthal**: Courtesy George Blumenthal.

CHAPTER 1

Pages 2–3 composite: Anne DeMarinis Design for W. W. Norton & Company; **p. 3 trees**: Andrew Bret Wallis/Getty Images; **p. 6 moon**: NASA; **p. 6 Saturn**: NASA, ESA, J. Clarke (Boston University, USA), and Z. Levay (STScI), https://esahubble.org/products/calendars/cal200604/, https://creativecommons.org/licenses/by/4.0/; **p. 7 sea growth**: Chase Studio/Science Source; **p. 7 Earth**: NASA; **p. 13**: GL Archive/Alamy Stock Photo; **p. 15 tree**: Don Hammond/Design Pics/Science Source; **p. 17**: ESA/Hubble & NASA, https://esahubble.org/images/potw1208a/, https://creativecommons.org/licenses/by/4.0/.

CHAPTER 2

Pages 22–23 composite: Anne DeMarinis Design for W. W. Norton & Company; **p. 22**: Deanna Truesdale/EyeEm/Getty Images; **p. 23 moon day 14**: Anne DeMarinis Design for W. W. Norton & Co.; **p. 23 moon day 5**: Anthony Qualkinbush/EyeEm/Getty Images; **p. 24 top**: robertharding/Alamy Stock Photo; **p. 24 bottom**: Daniela Constantinescu/Shutterstock; **p. 28 left**: Pekka Parviainen/Science Source; **p. 28 right**: D. Nunuk/Science Source; **p. 39**: Dr. Juerg Alean/Science Source; **p. 44**: akg images/Rabatti – Domingie; **p. 47 top**: © Laura Kay; **p. 47 center**: HEINZ-PETER BADER/REUTERS/Newscom; **p. 47 bottom**: Laura Kay and Lisa Rand; **p. 51 left**: knickohr/Getty Images; **p. 51 right**: Courtesy © Wang Letian.

CHAPTER 3

Pages 58–59 composite: Anne DeMarinis Design for W. W. Norton & Company; **p. 58 lightbulb**: Jakub Gojda/Alamy Stock Photo; **p. 58 paper balls**: Anne DeMarinis Design for W. W. Norton & Co.; **p. 60**: Image Select/Art Resource, NY; **p. 61 top**: © Tunc Tezel; **p. 61 bottom**: GL Archive/Alamy Stock Photo; **p. 62**: Bettmann/Getty Images; **p. 66**: Art Collection 2/Alamy Stock Photo; **p. 67 top**: Science History Images/Alamy Stock Photo; **p. 67 bottom**: North Wind Picture Archives/Alamy Stock Photo; **p. 69 Copernicus**: Science History Images/Alamy Stock Photo; **p. 69 Tycho**: Pictorial Press Ltd/Alamy Stock Photo; **p. 69 Kepler**: GL Archive/Alamy Stock Photo; **p. 69 Newton**: Heinz-Dieter Falkenstein/age fotostock/Superstock; **p. 71**: FineArt/Alamy Stock Photo; **p. 72 top**: GRANGER; **p. 72 bottom**: David Hajnal/Shutterstock; **p. 73**: Lebrecht Music & Arts/Alamy Stock Photo.

CHAPTER 4

Pages 82–83 composite: Anne DeMarinis for W. W. Norton & Company; **p. 92 Earth**: NASA Johnson Space Center; **p. 92 moon**: NASA/JPL/USGS; **p. 92 lunar subsatellite**: JSC/NASA; **p. 92 David Scott**: JSC/NASA; **p. 92 Galileo**: ITAR-TASS News Agency/Alamy Stock Photo; **p. 99 left**: Eric Carr/Alamy Stock Photo; **p. 99 right**: Sandi Cullifer/Shutterstock; **p. 103**: NASA, Holland Ford (JHU), the ACS Science Team and ESA, https://esahubble.org/images/heic0206b/, https://creativecommons.org/licenses/by/4.0/.

CHAPTER 5

Pages 108–109 composite: Anne DeMarinis for W. W. Norton & Company; **p. 108 pencils**: Julladit Portfolio/Shutterstock; **p. 108 single slit diffraction**: Ted Kinsman/Science Source; **p. 109 tape**: fStop Images GmbH/Alamy Stock Photo; **p. 116 Romer**: Paul Fearn/Alamy Stock Photo; **p. 116 Bradley**: Heinz-Dieter Falkenstein/age fotostock/Superstock; **p. 116 Maxwell**: Iberfoto/Superstock; **p. 116 Einstein**: Underwood Photo Archives/Superstock; **p. 124**: ESO, https://www.eso.org/public/images/eso1413a/, https://creativecommons.org/licenses/by/4.0/; **p. 137 Mercury**: NASA/Johns Hopkins University Applied Physics Laboratory/Carnegie Institution of Washington; **p. 137 Venus**: NASA/JPL-Caltech/ASU; **p. 137 Earth**: NASA; **p. 137 Mars**: NASA/Hubble Heritage Team (STScI/AURA); **p. 137 Jupiter, Saturn, Uranus, Neptune & Pluto**: NASA/NSSDC/GSFC.

CHAPTER 6

Pages 142–143 composite: Anne DeMarinis for W. W. Norton & Company; **p. 142 water and glass**: Richard Sharrocks/Getty Images; **p. 145**: ASU Physics Instructional Resource Team. © 2009 Arizona State Board of Regents. Used with permission; **p. 146**: Richard Dreiser, Yerkes Observatory; **p. 148 light**: ASU Physics Instructional Resource Team. © 2009 Arizona State Board of Regents. Used with permission; **p. 148 telescope**: Jim Sugar/Getty Images; **p. 150 Keck reflectors**: Enrico Sacchetti/Science Source; **p. 150 angular resolution, both**: The Space Telescope Science Institute; **p. 152 both**: Nick Law (Caltech) and Craig Mackay (Univ. of Cambridge); **p. 153 top**: ESO/L. Calçada, https://www.eso.org/public/images/eso1440e/?lang=no, https://creativecommons.org/licenses/by/4.0/; **p. 153 bottom**: NASA, Earth Observatory; **p. 154**: Courtesy, Victoria Girgis/Lowell Observatory; **p. 155 top**: Hulton-Deutsch Collection/CORBIS/Corbis via Getty Images; **p. 155 bottom**: Jean-Charles Cuillandre (CFHT); **p. 157 Fermi**: NASA E/PO, Sonoma State University, Aurore Simonnet; **p. 157 Chandra**: NASA (Illustration: NASA/CXC/NGST); **p. 157 HST**: NASA; **p. 157 Keck**: Left_Coast_Photographer/Istockphoto/Getty Images Plus; **p. 157 Spitzer**: NASA/JPL-Caltech/R. Hurt (IPAC); **p. 157 JCMT**: POLARBEAR Consortium, UC Berkeley; **p. 157 EVLA**: NRAO/AUI/NSF; **p. 157 Green Bank**: ANDREW CABALLERO-REYNOLDS/AFP via Getty Images; **p. 157 FAST**: © Liu Xu/Xinhua/Alamy Live News/Alamy Stock photo; **p. 157 Green Bank**: ANDREW CABALLERO-REYNOLDS/AFP via Getty Images; **p. 157 FAST**: Liu Xu/Xinhua/Alamy Live News/Alamy Stock photo; **p. 158 top**: Photo by Dave Finley, courtesy National Radio Astronomy Observatory and Associated Universities, Inc.; **p. 158 bottom**: Markus Thomenius/Alamy Stock Photo; **p. 159 top**: NASA; **p. 159 bottom**: NASA/Tom Tschida; **p. 161**: NASA; **p. 164 top**: NASA/JPL-Caltech/MSSS; **p. 164 bottom**: NASA/GSF and NASA's Scientific Visualization Studio; **p. 165 photograph**: Maximilien Brice, © 2005–2021 CERN, http://cds.cern.ch/record/910381, https://creativecommons.org/licenses/by/4.0/; **p. 166**: Yuya Makino, IceCube/National Science Foundation; **p. 167 Weber**: Volker Steger/Science Source; **p. 167 LIGO**: Courtesy Caltech/MIT/LIGO Laboratory; **p. 167 LISA**: NASA; **p. 168 all**: Patrik Jonsson, Greg Novak & Joel Primack, UC Santa Cruz, 2008.

CHAPTER 7

Pages 174–175 composite: Anne DeMarinis for W. W. Norton & Company; **p. 177 left**: NASA/Hubble/STScI; **p. 177 right**: ALMA (NRAO/ESO/NAOJ); C. Brogan, B. Saxton (NRAO/AUI/NSF)/Science Source; **p. 178**: NASA; **p. 180**: John Schults/REUTERS/Newscom; **p. 183**: NASA, ESA, and the Hubble 20th Anniversary Team (STScI); **p. 188**: NASA/Johns Hopkins University Applied Physics Laboratory/Carnegie Institution of Washington; **p. 193 top**: ESO/A.-M. Lagrange et al., https://www.eso.org/public/images/eso0842b/, https://creativecommons.org/licenses/by/4.0/; **p. 193 bottom**: NASA, ESA and P. Kalas (University of California, Berkeley and Seti Institute); **p. 197**: NASA.

CHAPTER 8

Pages 202–203 composite: Anne DeMarinis for W. W. Norton & Co.; **pp. 202–203 moon, background, and inset**: NASA/GSFC/Arizona State University; **p. 205**: NASA/GFSC/Arizona State University; **p. 206 sequence**: RGB Ventures/SuperStock/Alamy Stock Photo; **p. 207 top**: Kit Leong/Shutterstock; **p. 207 bottom**: NASA/Goddard/MIT/Brown; **p. 208**: European Space Agency/Science Source; **p. 209**: NASA's Scientific Visualization Studio; **p. 212 both**: NASA; **p. 216**: Spencer Grant/Science Source; **p. 218**: Francois Gohier/Science Source; **p. 222 top four**: NASA; **p. 222 bottom left**: NASA/USGS; **p. 222 bottom right**: European Space Agency/DLR/FU Berlin/G. Neukum/Science Source; **p. 223**: NASA/JPL; **p. 224 Mauna Kea**: WorldStock/Shutterstock; **p. 224 Mt. Fuji**: Jukurae/Shutterstock; **p. 224 rock**: NASA/JSC; **p. 225 top**: NASA/JSC/Arizona State University; **p. 225 bottom**: NASA/Johns Hopkins University Applied Physics Laboratory/Carnegie Institution of Washington; **p. 226 top**: NASA/JPL/Malin Space Science Systems; **p. 226 center and bottom**: NASA/JPL-Caltech/ESA; **p. 227**: NASA/JPL-Caltech/Univ of Arizona; **p. 228 top**: NASA/JPL/University of Arizona; **p. 228 bottom**: NASA/JPL-Caltech/MSSS; **p. 229 top**: NASA/JPL-Caltech/University of Arizona/Texas A&M University; **p. 229 bottom**: NASA/JPL-Caltech/UA/USGS; **p. 230 both**: NASA; **p. 231**: NASA; **p. 233**: Joe Tucciarone/Science Source.

CHAPTER 9

Pages 238–239 composite: Anne DeMarinis Design for W. W. Norton & Company; **p. 245**: NASA's Goddard Space Flight Center; **p. 249**: Seth White; **p. 253 left**: NASA; **p. 253 right**: All Canada Photos/Alamy Stock Photo; **p. 256**: NASA/JPL-Caltech; **p. 258 top**: Russian Academy of Sciences/Ted Stryk; **p. 258 bottom**: NASA/JPL; **p. 259 top**: NASA/JPL/USGS; **p. 259 bottom**: NASA, James Bell (Cornell Univ.), Michael Wolff (Space Science Inst.), and The Hubble Heritage Team (STScI/AURA); **p. 260**: NASA/JPL-Caltech/University of Arizona; **p. 267 Venus**: NASA/JPL-Caltech; **p. 267 Earth**: NASA, Johnson Space Center; **p. 267 Mars**: Phil James (Univ. Toledo), Todd Clancy (Space Science Inst., Boulder, CO), Steve Lee (Univ. Colorado), and NASA/ESA, https://esahubble.org/images/opo9715c/, https://creativecommons.org/licenses/by/4.0/.

CHAPTER 10

Pages 272–273 composite: Anne DeMarinis Design for W. W. Norton & Company; **pp. 272–273 background spread**: Jordana Meilleur/Alamy Stock Photo; **p. 276 Jupiter**: NASA, ESA, STScI, A. Simon (Goddard Space Flight Center), M.H. Wong (University of California, Berkeley), and the OPAL team; **p. 276 Saturn**: NASA/JPL/Space Science Institute; **p. 276 Uranus**: NASA/JPL-Caltech; **p. 276 Neptune**: NASA/JPL; **p. 277 Jupiter**: NASA, ESA, STScI, A. Simon (Goddard Space Flight Center), M.H. Wong (University of California, Berkeley), and the OPAL team; **p. 277 Saturn**: NASA/JPL/Space Science Institute; **p. 277 Uranus**: NASA/JPL-Caltech; **p. 277 Neptune**: NASA/JPL; **p. 278 Uranus**: NASA/JPL-Caltech; **p. 278 Neptune**: NASA/JPL; **p. 279 top**: NASA/JPL/Space Science Institute; **p. 279 bottom**: NASA/JPL; **p. 280 top**: NASA, ESA, and the Hubble Heritage Team (STScI/AURA); **p. 280 bottom**: NASA/JPL-Caltech/SwRI/MSSS/Gerald Eichstädt/© Seán Doran; **p. 281**: Enhanced image by Jason Major based on images provided courtesy of NASA/JPL-Caltech/SwRI/MSSS; **p. 282 top**: NASA/JPL-Caltech/SwRI/ASI/INAF/JIRAM; **p. 282 bottom left**: NASA/JPL-Caltech/Space Science Institute; **p. 282 bottom right**: NASA/JPL-Caltech/SSI; **p. 283 Uranus**: Lawrence Sromovsky, University of Wisconsin-Madison/W.W. Keck Observatory; **p. 283 Neptune**: NASA, ESA, and M.H. Wong and J. Tollefson (UC Berkeley); **p. 283 Neptune spot**: NASA/JPL; **p. 287**: NASA/JPL; **p. 288**: NASA, ESA, L. Sromovsky and P. Fry (University of Wisconsin), H. Hammel (Space Science Institute), and K. Rages (SETI Institute); **p. 293 water over rocks**: Sigur/Shutterstock; **p. 293 Jupiter**: NASA/ESA, https://esahubble.org/images/heic1613a/, https://creativecommons.org/licenses/by/4.0/; **p. 293 inset top**: John Clarke (University of Michigan), and NASA/ESA, https://esahubble.org/images/opo9804b/, https://

creativecommons.org/licenses/by/4.0/; **p. 293 inset bottom**: NASA, JPL-Caltech, SwRI, ASI, INAF, JIRAM; **p. 293 Saturn**: NASA/Hubble/Z. Levay and J. Clarke; **p. 294**: Nick Schneider, Image obtained at Catalina Observatory with an instrument developed by John Trauger; **p. 295**: NASA's Goddard Space Flight Center/S. Wiessinger; **p. 301 both**: NASA/JPL.

CHAPTER 11

Pages 302–303 composite: Anne DeMarinis design for W. W. Norton & Company; **pp. 302–303 blackboard background**: Manbetta/ Shutterstock; **p. 302 socks**: oatautta/Shutterstock; **p. 302 string**: andersphoto/Shutterstock; **p. 305**: NASA; **p. 306 left**: ESA/DLR/ FU Berlin (G. Neukum), https://www.esa.int/ESA_Multimedia /Images/2004/11/Phobos_in_colour_close-up2, https:// creativecommons.org/licenses/by/4.0/; **p. 306 right**: NASA/ JPL-Caltech/University of Arizona; **p. 308 top**: NASA/JPL/ University of Arizona; **p. 308 bottom all three**: NASA/JPL; **p. 309 top**: NASA/JPL/University of Arizona; **p. 309 center**: NASA/JPL; **p. 309 bottom**: NASA/JPL-Caltech; **p. 310 top**: Cassini Imaging Team, SSI, JPL, ESA, NASA; **p. 310 bottom**: NASA/JPL-Caltech; **p. 312 top left**: NASA/JPL/Space Science Institute; **p. 312 top right**: NASA/JPL/University of Arizona/University of Idaho; **p. 312 bottom**: NASA/JPL-Caltech/ASI; **p. 313**: © ESA/NASA/ JPL/University of Arizona, https://nssdc.gsfc.nasa.gov/planetary /titan_images.html, https://creativecommons.org/licenses /by/4.0/; **p. 314 top**: NASA/JPL/USGS; **p. 314 bottom**: Paul M. Schenk, Lunar and Planetary Institute, Houston, TX; **p. 315**: NASA/JPL/Space Science Institute; **p. 316 top**: NASA/JPL/ Ted Stryk; **p. 316 bottom**: NASA/JPL/Space Science Institute; **p. 318**: NASA and The Hubble Heritage Team (STScI/AURA) Acknowledgment: R.G. French (Wellesley College), J. Cuzzi (NASA/Ames), L. Dones (SwRI), and J. Lissauer (NASA/Ames); **p. 320**: NASA/JPL/Space Science Institute; **p. 321 left**: NASA/ JPL/Space Science Institute; **p. 321 right**: NASA and The Hubble Heritage Team (STScI/AURA)Acknowledgment: R.G. French (Wellesley College), J. Cuzzi (NASA/Ames), L. Dones (SwRI), and J. Lissauer (NASA/Ames); **p. 322 top left**: Stephanie Swartz/ Dreamstime; **p. 322 top right**: NASA/JPL/Space Science Institute; **p. 322 bottom**: NASA/JPL-Caltech; **p. 323 top and center**: NASA/JPL/Space Science Institute; **p. 323 bottom**: NASA/ JPL/Cornell University; **p. 324 left**: NASA/JPL-Caltech; **p. 324 center and right**: NASA/JPL/Space Science Institute; **p. 325 top**: NASA/JPL/Space Science Institute; **p. 325 bottom**: NASA/ JPL; **p. 326**: NASA; **p. 331 all**: NASA/JPL.

CHAPTER 12

Pages 332–333 composite: Anne DeMarinis Design for W. W. Norton & Company; **pp. 332–333 background**: Daniel Pludowski/ EyeEm/Getty Images; **p. 334**: NASA; **p. 335 all**: NASA/ JPL-Caltech/UCLA/MPS/DLR/IDA; **p. 336 all**: NASA/Johns Hopkins University Applied Physics Laboratory/Southwest Research Institute; **p. 337 all**: JPL/ NASA/Johns Hopkins University Applied Physics Laboratory/Southwest Research Institute; **p. 339 both**: NASA; **p. 343 top**: NASA/NSSDCA/ Galileo Mission; **p. 343 bottom and inset**: NASA/JPL; **p. 344 both**: NASA/JPL-Caltech/UCLA/MPS/DLR/IDA; **p. 349 top**: Robert McNaught/Science Source; **p. 349 bottom**: Courtesy of Terry Acomb; **p. 350 top**: NASA/JPL-Caltech/University of Maryland/Cornell; **p. 350 bottom both**: NASA/JPL/Caltech/ UMD; **p. 351**: NASA/JPL-Caltech/UMD; **p. 354 top**: Allexxandar/ Shutterstock; **p. 354 bottom**: Kenneth Keifer/Shutterstock; **p. 355 chondrite**: Mirko Graul/Shutterstock; **p. 355 achondrite**: The Natural History Museum/Alamy Stock Photo; **p. 355 iron meteorite**: Matteo Chinellato/Shutterstock; **p. 355 stony-iron meteorite**: Scott Camazine/Alamy Stock Photo; **p. 355**: Mars meteorite: NASA/JPL-Caltech/LANL/CNES/IRAP/LPGNantes/ CNRS/IAS/MSSS; **p. 356**: ESO/Y. Beletsky, https://www.eso .org/public/hungary/images/zodiacal_beletsky_potw/, https:// creativecommons.org/licenses/by/4.0/; **p. 357 top**: R. Evans, J. Trauger, H. Hammel and the HST Comet Science Team and NASA; **p. 357 bottom left**: RIA NOVOSTI/Science Source; **p. 357 bottom right**: mmedp/Deposit Photos; **p. 363 all**: Larry Marschall, Gettysburg College and Christy Metzger, Towson University.

CHAPTER 13

Pages 364–365 composite: Anne DeMarinis Design for W. W. Norton and Company; **pp. 364–365 background**: Orchidpoet/ Getty Images; **p. 381**: ESA/NASA; **p. 385**: ESA/Gaia/DPAC, CC BY-SA 3.0 IGO, https://creativecommons.org/licenses/by-sa/3.0 /igo/, https://www.esa.int/ESA_Multimedia/Images/2018/04 /Gaia_s_Hertzsprung-Russell_diagram; **p. 386 Cannon**: GL Archive/Alamy Stock Photo; **p. 386 Payne-Gaposchkin**: Science History Images/Alamy Stock Photo; **p. 386 Saha**: Paul Fearn/ Alamy Stock Photo; **p. 386 Russell**: GL Archive/Alamy Stock Photo; **p. 386 Hertzsprung**: INTERFOTO /Alamy Stock Photo; **p. 390**: Art by Dana Berry, NASA Ames Research Center.

CHAPTER 14

Pages 396–397 composite: Anne DeMarinis Design for W. W. Norton & Company; **pp. 396–397 board and tape**: Anne DeMarinis Design for W. W. Norton & Company; **p. 406**: Hinode JAXA/NASA/PPARC; **p. 410**: Makarov Konstantin/Shutterstock; **p. 411**: Nigel Sharp, NOAO/NSO/Kitt Peak FTS/AURA/NSF, https:// noirlab.edu/public/images/noao-sun/, https://creativecommons .org/licenses/by/4.0/; **p. 412 left**: Hinode JAXA/NASA; **p. 412 center**: John Chumack/Science Source; **p. 412 right**: Westend61 GmbH/Alamy Stock Photo; **p. 413 left**: NASA/LMSAL; **p. 413 right**: SDO/AIA; **p. 414 both**: NASA/SDO; **p. 415**: NOIRLab. https://noirlab .edu/public/images/noao9808a/. https://creativecommons .org/licenses/by/4.0/; **p. 417**: NASA/SDO; **p. 418 top row all**: Image montage courtesy of NASA, SDO, and the AIA/HMI Science Teams; **p. 418 bottom**: SOHO Consortium, ESA, NASA; **p. 421**: NASA/JPL.

CHAPTER 15

Pages 426–427 composite: Anne DeMarinis Design for W. W. Norton & Company; **pp. 426–427 background**: CharlineXia Ontario Canada Collection/Alamy Stock Photo; **p. 429 top**: zorazhuang/Getty Images; **p. 429 bottom**: NASA/JPL- Caltech/ IRAS/2MASS/COBE; **p. 430**: Nik Taylor Sport/Alamy Stock Photo; **p. 432 top left**: NASA/JPL-Caltech/UCLA; **p. 432 top right**: IRAS/COBE (NASA); **p. 432 bottom**: Jeremy Sanders, Hermann Brunner, Andrea Merloni and the eSASS team (MPE); Eugene Churazov, Marat Gilfanov (on behalf of IKI); **p. 433**: Dr. Douglas Finkbeiner; **p. 434 top left**: Jean-Charles Cuillandre (CFHT), Hawaiian Starlight, CFHT; **p. 434 top center**: Courtesy of Emmanuel Mallart www.astrophoto-mallart.com & www .axisinstruments.com; **p. 434 top right**: Stefan Seip – photomeeting .de; **p. 434 30 Doradus Nebula and inset**: NASA, ESA, D. Lennon and E. Sabbi (ESA/STScI), J. Anderson, S.E. de Mink, R. van der Marel, T. Sohn, and N. Walborn (STScI), N. Bastian (Excellence Cluster, Munich), L. Bedin (INAF, Padua), E. Bressert (ESO), P. Crowther (University of Sheffield), A. de Koter (University of Amsterdam), C. Evans (UKATC/STFC, Edinburgh), A. Herrero (IAC, Tenerife), N. Langer (AifA, Bonn), I. Platais (JHU), and H. Sana (University of Amsterdam), Upper Right Image Credit: NASA, ESA, R. O'Connell (University of Virginia), and the WFC3 Science Oversight Committee, https://esahubble. org/images/opo1235d/, https://creativecommons.org/licenses /by/4.0/; **p. 435**: © ESO, 2016, https://www.aanda.org/articles/aa /full_html/2016/10/aa29178-16/F3.html, https://creativecommons .org/licenses/by/4.0/; **p. 437 left**: ESO, https://www.eso.org/public /images/eso0102a/, https://creativecommons.org/licenses/by/4.0/; **p. 437 right**: ESO, https://www.eso.org/public/images/eso0102b/, https://creativecommons.org/licenses/by/4.0/; **p. 438**: ESA/NASA/ Planck Collaboration; **p. 441 left**: Far-infrared: ESA/Herschel/ PACS/SPIRE/Hill, Motte, HOBYS Key Programme Consortium; X-ray: ESA/XMM-Newton/EPIC/XMM-Newton-SOC/Boulanger; **p. 441 center and right**: NASA, ESA/Hubble and the Hubble Heritage Team; **p. 445**: NASA/JPL-Caltech; **p. 447 inset**: NASA/Hubble/ STScI; **p. 449**: ESO/M. McCaughrean, https://www.eso.org/public /images/potw1541a/, https://creativecommons.org/licenses/by/4.0/; **p. 450**: Science Photo Library/Alamy Stock Photo; **p. 451**: NASA/ DLR/USRA/DSI/FORCAST team/NASA/Caltech – JPL.

CHAPTER 16

Pages 456–457 composite and pencils: Anne DeMarinis Design for W. W. Norton & Company; **pp. 456–457 Antares Region**: knickohr/iStock/Getty Images Plus; **p. 471 Helix Nebula**: NASA/ JPL-Caltech; **p. 471 Butterfly Nebula**: NASA/ESA/Hubble; **p. 471 Ant Nebula**: NASA/Space Telescope Science Institute; **p. 471 Cat's Eye Nebula**: NASA, ESA, HEIC, and The Hubble Heritage Team (STScI/AURA); **p. 479 SN1006**: NASA, ESA, Zolt Levay (STScI); **p. 479 Tycho Supernova**: X-ray: NASA/ CXC/Rutgers/K.Eriksen et al.; Optical: DSS; **p. 479 Kepler Supernova**: X-ray: NASA/CXC/NCSU/M.Burkey et al; Infrared: NASA/JPL-Caltech.

CHAPTER 17

Pages 486–487 composite and pencils: Anne DeMarinis Design for W. W. Norton & Company; **p. 493**: NASA/CXC/GSFC/M. Corcoran et al., Optical: NASA/STScI; **p. 497**: NASA/CXC/SAO; **p. 498 left and center**: © Australian Astronomical Optics, Macquarie University, photographs by David Malin; **p. 498 right**: Image: NASA, ESA, and A. Angelich (NRAO/AUI/NSF), Hubble data: NASA, ESA, and R. Kirshner (Harvard-Smithsonian Center for Astrophysics and Gordon and Betty Moore Foundation), Chandra data: NASA/CXC/Penn State/K. Frank et al., ALMA data: ALMA (ESO/NAOJ/NRAO) and R. Indebetouw (NRAO/AUI/NSF); **p. 499**: NASA, ESA, and R. Kirshner (Harvard-Smithsonian Center for Astrophysics and Gordon and Betty Moore Foundation), and P. Challis (Harvard-Smithsonian Center for Astrophysics); **p. 503 Jocelyn Bell**: Science and Society/Superstock; **p. 503 Antony Hewish**: PA Images/ Alamy Stock Photo; **p. 503 Franco Pacini**: ANGELO PALMA/ A3/CONTRASTO/Redux; **p. 503 Thomas Gold**: SPL/Science Source; **p. 503 both aliens**: Science Photo Library/Superstock; **p. 506**: Frank Rossoto/ Stocktrek/Getty Images; **p. 507 left**: ESA/ Hubble & NASA, https://esahubble.org/images/potw1140a/, https://creativecommons.org/licenses/by/4.0/; **p. 507 right**: NASA/ESA/STScI; **p. 508 top left**: Michael Miller/Stocktrek/ Getty Images; **p. 508 top right**: ESO, https://www.eso.org /public/images/m55-3point6-m_copy/, https://creativecommons .org/licenses/by/4.0/.

CHAPTER 18

Pages 516–517 composite, background, scissor and string: Anne DeMarinis Design for W. W. Norton & Company; **pp. 516–517 paper**: 32 pixels/Shutterstock; **p. 516 basketball**: Volodymyr Melnyk/Alamy Stock Photo; **p. 531 Einstein**: Underwood Photo Archives/Superstock; **p. 531 Newton**: Heinz-Dieter Falkenstein/ age fotostock/Superstock; **p. 532 top right**: NASA/ESA/STScI/ UCLA; **p. 532 bottom**: ESA/Hubble & NASA, https://apod.nasa .gov/apod/ap111221.html, https://creativecommons.org/licenses /by/4.0/; **p. 534**: NASA and the Night Sky Network; **p. 536 and p. 537**: NASA/CXC/M. Weiss; **p. 538 top**: The Event Horizon Telescope Collaboration; **p. 538 bottom**: LIGO-Virgo Collaboration/ Frank Elavsky, Aaron Geller/Northwestern; **p. 540 left**: NASA/CXC/M. Weiss; **p. 540 right**: ESO/A. Roquette, https://www.eso.org/public /images/eso0917a/, https://creativecommons.org/licenses/by/4.0/.

CHAPTER 19

Pages 546–547 composite: Anne DeMarinis Design for W. W. Norton & Company; **p. 546**: Hadafee/Shutterstock; **p. 549**: NASA, ESA, and the Hubble Heritage Team (STScI/AURA), R. Gendler; **p. 550 UGC 12591**: ESA/Hubble & NASA; **p. 550 Sombrero galaxy**: Hubble Heritage Team (AURA/STScI /NASA); **p. 550 NGC 253**: ESO, https://www.eso.org/public/hungary /images/eso0902c/, https://creativecommons.org/licenses /by/4.0/; **p. 550 NGC 5949**: ESA/Hubble & NASA; **p. 550 NGC 4911**: NASA, ESA, and the Hubble Heritage Team (STScI/AURA); **p. 550 NGC 6814**: ESA/Hubble & NASA, Acknowledgement: Judy Schmidt; **p. 550 elliptical galaxies all**: NASA/JPL-Caltech; **p. 550 spiral galaxies all**: NASA/JPL-Caltech; **p. 551 S0 galaxy**: ESA/ Hubble & NASA Acknowledgement: Judy Schmidt (Geckzilla, http://geckzilla.com/), https://esahubble.org/images/potw1620a/, https://creativecommons.org/licenses/by/4.0/; **p. 551 irregular galaxies**: ESA/Hubble & NASA, A. Adamo et al., https://esa -hubble.org/images/heic2101a/, https://creativecommons.org /licenses/by/4.0/; **p. 552 elliptical galaxy**: NOIRLab/NSF/AURA, https://noirlab.edu/public/images/noao-m59/, https://creative -commons.org/licenses/by/4.0/; **p. 553 spiral galaxy**: NOIR-Lab/NSF/AURA, https://noirlab.edu/public/images/noao-m109/, https://creativecommons.org/licenses/by/4.0/; **p. 553 Sombrero Galaxy**: NASA/Hubble Heritage Team; **p. 554 large spiral**: NASA, ESA, K. Kuntz (JHU), F. Bresolin (University of Hawaii), J. Trauger (Jet Propulsion Lab), J. Mould (NOAO), Y.-H. Chu (University of Illinois, Urbana) and STScI; **p. 554 small spiral**: ESO, https://www.eso.org/public/hungary/images/eso9949a/, https:// creativecommons.org/licenses/by/4.0/; **p. 554 dwarf elliptical galaxy**: NOIRLab/NSF/AURA, https://noirlab.edu/public

This index contains page references for all 24 chapters in the complete volume of 21st Century Astronomy, Seventh Edition. The Solar System Edition does not include Chapters 15–23 (pp. 426–683). The Stars and Galaxies Edition does not include Chapters 8–12 (pp. 202–363).